AF520619

Quorum Sensing vs Quorum Quenching: A Battle with No End in Sight

Vipin Chandra Kalia
Editor

Quorum Sensing vs Quorum Quenching: A Battle with No End in Sight

Editor
Vipin Chandra Kalia
Microbial Biotechnology and Genomics
CSIR-Institute of Genomics and Integrative Biology
Delhi, India

ISBN 978-81-322-1981-1 ISBN 978-81-322-1982-8 (eBook)
DOI 10.1007/978-81-322-1982-8
Springer New Delhi Heidelberg New York Dordrecht London

Library of Congress Control Number: 2014947883

Printed on acid-free paper

Springer is part of Springer Science+Business Media (www.springer.com)

Preface

Human beings encounter scenarios where they develop severe symptoms of a disease. The discovery of antibiotics was thought to relieve human beings of their miseries caused by microbial infections. However, microbes have been evolving rapidly to become resistant to antibiotics. It has been realized that bacteria have a unique system of multiplying silently through the phenomenon termed Quorum Sensing (QS). QS operates through signal molecules, which enable bacteria to sense their population density. At high cell density, bacteria activate their arsenal of virulence factors. QS mediated biofilms formed by pathogenic bacteria allow them to with stand high doses of antibiotics. This provoked scientist to look for novel alternatives to antibiotics. It led to the discovery of QS inhibitors, both natural and synthetic. Recent studies have indicated that QSIs may meet the same fate as antibiotics. Although a lot of scientific literature is being published rapidly since the last few years, it is limited largely to scientific research journals. A compilation of these important findings is not available to students at graduate and post-graduate levels. The best thing about this book is that the various chapters have been written by experts in the respective areas. As it demands tremendous and dedicated effort, we are extremely thankful to all the authors for their prompt responses and their contributions. I was inspired by my parents (Mr. R.B. Kalia and Mrs. Kanta Kalia), wife (Amita), Sunita (Sister), Ravi and Satyendra (Brothers), children (Daksh and Bhrigu), teachers, and Rup and Hemant (friends) to write this book. Throughout the preparation of this work, I was supported by Mr. Prasun Kumar, my Ph.D. student.

Delhi, India   Vipin Chandra Kalia

Contents

About the Editor

Dr. Vipin Chandra Kalia is presently working as Chief Scientist, Microbial Biotechnology and Genomics, CSIR-Institute of Genomics and Integrative Biology, Delhi. He is a Professor, AcSIR who obtained his M.Sc. and Ph.D. in Genetics, from Indian Agricultural Research Institute, New Delhi. He has been elected as: (1) Fellow of the Association of Microbiologists of India (FAMI), and (2) Fellow of the National Academy of Sciences (FNASc). His main areas of research are microbial biodiversity, bioenergy, biopolymers, genomics, microbial evolution, quorum sensing, quorum quenching, drug discovery and antimicrobials. He has published 65 papers in scientific journals such as (1) Nature Biotechnology, (2) Biotechnology Advances, (3) Trends in Biotechnology, (4) Critical Reviews in Microbiology, (5) Bioresource Technology, (6) PLoS ONE, and (vii) BMC Genomics. His works have been cited 1750 times with an *h* index of 23 and an i10 index of 36. He is presently the editor in chief of the Indian Journal of Microbiology and editor of (1) Journal of Microbiology & Biotechnology (Korea), (2). Appl. Biochem. & Biotechnology (USA), (3) International Scholarly Res. Network Renewable Energy, (4) Dataset Papers in Microbiology, and (5) PLoS ONE. He is a life member of the following scientific societies: (1) Society of Biological Chemists of India (2) Society for Plant Biochemistry and Biotechnology, India; (3) Association of Microbiologists of India; (4) Indian Science Congress Association; (5) BioEnergy Society of India, and (6) the Biotech Research Society of India (BRSI). He is also a member of the American Society for Microbiology. He can be contacted at: vckalia@igib.res.in; vc_kalia@yahoo.co.in

Microbes: The Most Friendly Beings?

Vipin C. Kalia

Introduction

Microbes, plants, animals, and human beings are intricately related to each other. In quite a few scenarios, these organisms live in close association with each other, for example, in rhizosphere, phyllosphere, etc. (Lindow and Brandl 2003). Epiphytes harbor microbes which prove helpful to the host (Hempel et al. 2008). Endophytic fungi prevent diseases and provide drought tolerance to *Poaceae* (grasses), and mycorrhizal fungi form symbiosis with vascular plants and protect them from diseases and enable them to sequester phosphate from the soil (Khan 2006; Rodriguez et al. 2009). Among all the symbiotic relationships, nitrogen fixation by rhizobia-legume interaction can be listed as the most beneficial outcomes (Drevon et al. 1987, 1988; Zhuang et al. 2013; Gao et al. 2014).

Human beings are conscientiously accomplished to exploit others for their personal benefits. Associations of microbes and human beings are not viewed in the positive light. Most microbes with an ability to infect human beings are perceived to lead to an unhealthy situation (http://ca-biomed.org/csbr/pdf/connect.pdf). However, each organism has been bestowed with a mechanism to protect itself from attack (Peterson et al. 2014). All organisms have distinct genetic makeup that they are able to maintain their identity and integrity with high precision. Incidentally, recent developments in microbial sciences have revealed that microbes stay on and inside our body (Walter et al. 2011). It has been realized that we harbor ten times more bacterial cells than our own (10 trillion) (http://gotsomescienceonyou.com/2013/05/08/there-isnt-that-much-you-in-you/). Skin microbiota is dominated by members of Bacteroidetes, Corynebacteria, Propionibacterium, Proteobacteria, Staphylococcus, etc. A second major group of microbes (around 160 species) is present in the human gut (Qin et al. 2010). Intensive analysis of gut microbes has provided very interesting insights for human health. In fact, our well-being is dependent upon the composition of this community (Gerritsen et al. 2011). It may not be inappropriate to state that their meticulous functioning is important for our health and welfare. We feel happy and are able to enjoy life while bacteria inside the gut are working silently. Our physical fitness and healthy appearance can be assigned to Firmicutes and Bacteroidetes (Wexler 2007). When these two groups compose between 44 and 48 % of the total gut microbiota, our metabolic activities allow us to have a lean appearance. However, if these two dominant groups reach a threshold level of around, 82–86 % of the total population, our body loses track of its growth. Our body starts expanding outward like a half balloon and we become obese. Firmicutes grow rapidly if

V.C. Kalia (✉)
Microbial Biotechnology and Genomics, CSIR-Institute of Genomics and Integrative Biology, Mall Road, Delhi 110007, India
e-mail: vckalia@igib.res.in; vc_kalia@yahoo.co.in

V.C. Kalia (ed.), *Quorum Sensing vs Quorum Quenching: A Battle with No End in Sight*,
DOI 10.1007/978-81-322-1982-8_1, © Springer India 2015

the fat intake is high, whereas Bacteroidetes multiply on a diet rich in carbohydrates (David et al. 2014). Actinobacterial, Fusobacterial, and Verrucomicrobial members assist the gut bacteria to function normally. While bacterial activities are going on in full swing, we do not take any notice of their presence. In addition to these, two relatively gentle set of bacteria, our body is also inhabited by a third group. It is composed of the pathogenic bacteria, which infect different body parts and our body becomes diseased. This pathogenic association is caused by two kinds of bacteria. They "love" our body and stay there for short or long duration depending upon their nature. The first group infects the body, multiplies fast, and exits hurriedly. These rapidly flushing out bacteria cause diseases like diarrhea, cholera, septicemia, etc. These gastroenteric infections are caused by enterotoxigenic *Escherichia coli*, *Klebsiella*, *Pseudomonas aeruginosa*, *Salmonella typhi*, *Serratia*, *Shigella*, *Vibrio cholerae*, Bacteroides, and *Proteus* (http://www2.hawaii.edu/ johnb/micro/m130/eadings/Septic Shock.htm). The worst part is their high resistance to antibiotics (White et al. 2002). The second group is composed of those bacteria which seem to have permanently "migrated" to the human body. They settle in different body parts and feel "comfortable" as there is no crunch of nutrients. Bacteria like *Mycobacterium tuberculosis, P. aeruginosa, and Burkholderia cepacia* cause infections and multiply in an uninterrupted manner by constantly evading the human immune system (Costerton et al. 1999; Mahenthiralingam et al. 2002; Purohit et al. 2007; Bhushan et al. 2013). Hence, we take notice of the presence of bacteria in and on our body when they cause diseases which we can no longer afford to avoid.

The Rise and Fall of Antibiotics

In the last few centuries, human beings have encountered severe bacterial infections. In the absence of any effective mechanism to get rid of these pathogenic organisms, the mortality rates used to be skyrocketing (Schmid-Hempel 2009). Health departments worldwide are constantly on their toes and making efforts to control the growth of these organisms and achieve a major reduction in human mortality and morbidity rate. During these epidemics, there were instances of entire human communities getting wiped out. It was realized that toxin-producing organisms such as *Staphylococcus aureus, Clostridium* spp., *Yersinia enterocolitica, Streptococcus mutans, S. pneumoniae,* etc., were the major culprits. A great discovery came to the rescue of patients infected with chronic and acute diseases. Sir Alexander Fleming was honored with the Nobel Prize for his contribution – the discovery of the wonderful antibiotic, penicillin. It totally shifted the way people could envision the future – a world free from diseases. All the same, this notion was soon shattered as bacteria were quick to respond. Their arsenal of genetic reservoir came to their deliverance. By undergoing genetic mutations, bacteria became resistant to penicillin. *Staphylococcus* resistant to penicillin emerged in 1940, which was soon followed by *Shigella* showing resistance to tetracycline, and evolution of erythromycin-resistant *Streptococcus,* etc. This trend of evolution of bacterial resistance to the newly discovered antibiotics was seen in the last few decades (Davies and Davies 2010). Now, we are in an era where the prevalence of multiple drug resistant (MDR) and extremely drug resistant (XDR) bacteria seems to be the norm of the day (Bhardwaj et al. 2013). It reckons a very depressing picture for scientists and pharmaceutical companies. Industrial houses are ordained to think before investing in R&D for novel antibiotics. It appears that bacterial genetic makeup is quite subsistent after having been exposed to antibiotics produced by other organisms. Pharmaceutical companies find investing in scouting drugs for chronic diseases more lucrative than in short-term diseases, which can be treated with a few doses of an antibiotic (Kalia et al. 2007).

Biofilm: The Bacterial Shield

It has been realized that the biofilm-forming bacteria are responsible for 80 % of the infectious diseases. This biofilm formation happens only under specific conditions and allows bacteria to survive in the presence of antibiotic doses up to 1,000 times more than those required to kill them than their free-living counterparts. Biofilm is a unique structure which is formed only when certain genes get expressed by a phenomenon called quorum sensing (QS) (Kalia and Purohit 2011). Bacteria grow and remain "silent" while their cell density is low. It enables them to evade the human immune system. Nevertheless, once above a threshold population density, they are able to sense it. This requisite cell density is paramount to survive under stress conditions, and they kick-start expressing genes which are responsible for biofilm formation, production of virulence factors, toxins, etc. This process of gene expression at high cell density is mediated by signal molecules produced by genes responsible for synthase enzymes. These signal molecules are sensed by the bacteria and form a complex with intended receptor molecules, which in turn transcribes genes for pathogenic behavior of the bacteria. The system was initially detected in *Vibrio fischeri*, which produced light and helped to eliminate the shadow of the host (squid) organism, cast by the moonlight and enables it to save itself from praying fishes (Nyholm et al. 2000). It was realized that if bacteria are not expressing bioluminescence, they are involved in phenomena such as antibiotic production, biofilm formation, nitrogen fixation, motility, sporulation, toxins, virulence, etc. (Nakayama et al. 2006; Kolodkin-Gal et al. 2007, Kolodkin-Gal and Engelberg-Kulka 2008; Rinaudi and Giordano 2010; Kalia and Purohit 2011, Kalia et al. 2011; Li and Tian 2012; Bogino et al. 2013). QS-regulated biofilm-forming bacteria – *Aeromonas, Vibrio,* and *Yersinia* spp. – are a major cause of diseases in fish (Bhargava et al. 2012; Chu et al. 2013). It affects the economies of Departments of Fishery and Aquacultures. Another adversely affected industrial sector that is involved in the reclamation of wastewater and producing drinking water. Biofilms formed on the membranes used for filtering water lead to inadequate recovery and add to economic losses (Bereschenko et al. 2010). The growth of bacteria on archaeological buildings and monuments leads to their rapid deterioration (Scheerer et al. 2009).

A New Era of Antibacterials

In view of the fact that we are losing a battle against bacteria, scientists have embarked upon a search as to how to circumvent this problem. It has been envisaged that the QSS can be interrupted at one or more stages, which consequently may enable us to inhibit their pathogenicity (Kalia and Purohit 2011, Kalia 2013). This strategy of employing quorum-sensing inhibitors (QSIs) can prove effective in designing novel antibacterials (Romero et al. 2012; LaSarre and Federle 2013). The presence of bioactive molecules with properties to act as QSIs have been detected in prokaryotes, animal, and plants (Huma et al. 2011; Kalia and Purohit 2011, Kalia et al. 2011; Desouky et al. 2013; Kalia 2013; Kumar et al. 2013; Nakayama et al. 2013). Chemists, with their specialty in synthesizing analogues of natural QSIs, have contributed significantly in improving the efficiency of these molecules (Mai et al. 2011; Klein et al. 2012; Kalia et al. 2013; Lu et al. 2012; Sahner et al. 2013). Only those molecules which exclusively affect QS without affecting bacterial growth are categorized as QSI. Thus, unlike antibiotics, which affect bacterial growth and eventually kill them, QSIs do not put much selective pressure (Kalia and Purohit 2011, Kalia 2013). This has become a strong basis for scientists to jump to the conclusion that bacteria will not evolve resistance to QSIs. It is no surprise that bacteria persistently tailor their genetic potential to blunt any attack on their survival. For bacteria the encounter with "novel" antibiotics is a situation of déjà vu. Questions are conjuring around the fate of QSIs. There are evidences which suggest that bacteria may evolve mechanisms to evade QSIs? (Defoirdt et al. 2013; Kalia et al. 2014). So the great query is: Are we doomed to drop dead at the hands of bacteria and lose the battle of life? Will we be the mute witnesses to human beings dying at an annual pace of millions of deaths by these pathogens (http://www.civilwar.org/education/pdfs/civil-was-curriculum-medicine.pdf)? How or will we be able to survive the onslaughts of this never-ending battle?

Acknowledgment The author wishes to thank the Director of CSIR-Institute of Genomics and Integrative Biology (IGIB), CSIR-INDEPTH (BSC0111), and the government of India for providing the necessary funds and facilities.

References

Bereschenko LA, Stams AJ, Euverink GJ, van Loosdrecht MC (2010) Biofilm formation on reverse osmosis membranes is initiated and dominated by *Sphingomonas* spp. Appl Environ Microbiol 76:2623–2632. doi:10.1128/AEM.01998-09

Bhardwaj AK, Vinothkumar K, Rajpara N (2013) Bacterial quorum sensing inhibitors: attractive alternatives for control of infectious pathogens showing multiple drug resistance. Recent Pat Antiinfect Drug Discov 8:68–83. doi:10.2174/1574891X11308010012

Bhargava N, Sharma P, Capalash N (2012) N-acyl homoserine lactone mediated interspecies interactions between *A. baumannii* and *P. aeruginosa*. Biofouling 28:813–822. doi:10.1080/08927014.2012.714372

Bhushan A, Joshi J, Shankar P, Kushwah J, Raju SC, Purohit HJ, Kalia VC (2013) Development of genomic tools for the identification of certain *Pseudomonas* up to species level. Ind J Microbiol 53:253–263. doi:10.1007/s12088-013-0412-1

Bogino P, Oliva MM, Sorroche FG, Giordano W (2013) The role of bacterial biofilms and surface components in plant-bacterial associations. Int J Mol Sci 14:15838–15859. doi:10.3390/ijms140815838

Chu WH, Liu YW, Jiang Y, Zhu W, Zhuang XY (2013) Production of N-acyl homoserine lactones and virulence factors of waterborne *Aeromonas hydrophila*. Indian J Microbiol 53:264–268. doi:10.1007/s12088-013-0381-4

Costerton JW, Stewart PS, Greenberg EP (1999) Bacterial biofilms: a common cause of persistent infections. Science 284:1318–1322. doi:10.1126/science.284.5418.1318

David LA, Maurice CF, Carmody RN, Gootenberg DB, Button JE, Wolfe BE, Ling AV, Devlin AS, Varma Y, Fischbach MA, Biddinger SB, Dutton RJ, Turnbaugh PJ (2014) Diet rapidly and reproducibly alters the human gut microbiome. Nature 505:559–563. doi:10.1038/nature12820

Davies J, Davies D (2010) Origins and evolution of antibiotic resistance. Microbiol Mol Biol Rev 74:417–433. doi:10.1128/MMBR.00016-10

Defoirdt T, Brackman G, Coenye T (2013) Quorum sensing inhibitors: how strong is the evidence? Trends Microbiol 21:619–624. doi:10.1016/j.tim.2013.09.006

Desouky SE, Nishiguchi K, Zendo T, Igarashi Y, Williams P, Sonomoto K, Nakayama J (2013) High-throughput screening of inhibitors targeting Agr/Fsr quorum sensing in *Staphylococcus aureus* and *Enterococcus faecalis*. Biosci Biotechnol Biochem 77:923–927. doi:10.1271/bbb.120769

Drevon JJ, Kalia VC, Heckmann MO, Salsac L (1987) Influence of the *Bradyrhizobium japonicum* hydrogenase on the growth of *Glycine* and *Vigna* species. Appl Environ Microbiol 53:610–612

Drevon JJ, Kalia VC, Heckmann MO, Pedelahore P (1988) In situ open-flow assay of acetylene reduction activity by soybean root nodules: influence of acetylene and oxygen. Plant Physiol Biochem 26: 73–78

Gao J, Ma A, Zhuang X, Zhuang G (2014) An *N*-acyl homoserine lactone synthase in the ammonia-oxidizing bacterium *Nitrosospira multiformis*. Appl Environ Microbiol 80:951–958. doi:10.1128/AEM.03361-13

Gerritsen J, Smidt H, Rijkers GT, de Vos WM (2011) Intestinal microbiota in human health and disease: the impact of probiotics. Genes Nutr 6:209–240. doi:10.1007/s12263-011-0229-7

Hempel M, Blume M, Blindow I, Gross EM (2008) Epiphytic bacterial community composition on two common submerged macrophytes in brackish water and freshwater. BMC Microbiol 8:58. doi: 10.1186/1471-2180-8-58. http://ca-biomed.org/csbr/pdf/connect.pdf, http://www.civilwar.org/education/pdfs/civil-was-curriculum-medicine.pdf

Huma N, Shankar P, Kushwah J, Bhushan A, Joshi J, Mukherjee T, Raju SC, Purohit HJ, Kalia VC (2011) Diversity and polymorphism in AHL-lactonase gene (*aiiA*) of *Bacillus*. J Microbiol Biotechnol 21:1001–1011. doi:10.4014/jmb.1105.05056

Kalia VC (2013) Quorum sensing inhibitors: an overview. Biotechnol Adv 31:224–245. doi:10.1016/j.biotechadv.2012.10.004

Kalia VC, Purohit HJ (2011) Quenching the quorum sensing system: potential antibacterial drug targets. Crit Rev Microbiol 37:121–140. doi:10.3109/1040841X.2010.532479

Kalia VC, Rani A, Lal S, Cheema S, Raut CP (2007) Combing databases reveals potential antibiotic producers. Expert Opin Drug Discov 2:211–224. doi:10.1517/17460441.2.2.211

Kalia VC, Raju SC, Purohit HJ (2011) Genomic analysis reveals versatile organisms for quorum quenching enzymes: acyl-homoserine lactone-acylase and -lactonase. Open Microbiol J 5:1–13. doi:10.2174/1874285801105010001

Kalia D, Merey G, Nakayama S, Zheng Y, Zhou J, Luo Y, Guo M, Roembke BT, Sintim HO (2013) Nucleotide, c-di-GMP, c-di-AMP, cGMP, cAMP, (p)ppGpp signaling in bacteria and implications in pathogenesis. Chem Soc Rev 42:305–341. doi:10.1039/c2cs35206k

Kalia VC, Wood TK, Kumar P (2014) Evolution of resistance to quorum-sensing inhibitors. Microb Ecol 68:13–23. doi:10.1007/s00248-013-0316-y

Khan AG (2006) Mycorrhizoremediation – an enhanced form of phytoremediation. J Zhejiang Univ Sci B 7:503–514. doi:10.1631/jzus.2006.B0503

Klein T, Henn C, de Jong JC, Zimmer C, Kirsch B, Maurer CK, Pistorius D, Muller R, Steinbach A, Hartmann RW (2012) Identification of small-molecule antagonists of the *Pseudomonas aeruginosa* transcriptional regulator PqsR: biophysically guided hit discovery and optimization. ACS Chem Biol 7:1496–1501. doi:10.1021/Cb300208g

Kolodkin-Gal I, Engelberg-Kulka H (2008) The extracellular death factor: physiological and genetic factors influencing its production and response in *Escherichia coli*. J Bacteriol 190:3169–3175. doi:10.1128/JB.01918-07

Kolodkin-Gal I, Hazan R, Gaathon A, Carmeli S, Engelberg-Kulka H (2007) A linear pentapeptide is a quorum-sensing factor required for mazEF-mediated cell death in *Escherichia coli*. Science 318:652–655. doi:10.1126/science.1147248

Kumar P, Patel SKS, Lee JK, Kalia VC (2013) Extending the limits of *Bacillus* for novel biotechnological applications. Biotechnol Adv 31:1543–1561. doi:10.1016/j.biotechadv.2013.08.007

LaSarre B, Federle MJ (2013) Exploiting quorum sensing to confuse bacterial pathogens. Microbiol Mol Biol Rev 77:73–111. doi:10.1128/MMBR.00046-12

Li YH, Tian XL (2012) Quorum sensing and bacterial social interactions in biofilms. Sensors 12:2519–2538. doi:10.3390/s120302519

Lindow SE, Brandl MT (2003) Microbiology of the phyllosphere. Appl Environ Microbiol 69:1875–1883. doi:10.1128/AEM.69.4.1875-1883.2003

Lu CB, Kirsch B, Zimmer C, de Jong JC, Henn C, Maurer CK, Musken M, Haussler S, Steinbach A, Hartmann RW (2012) Discovery of antagonists of PqsR, a key player in 2-Alkyl-4-quinolone-dependent quorum sensing in *Pseudomonas aeruginosa*. Chem Biol 19:381–390. doi:10.1016/j.chembiol.2012.01.015

Mahenthiralingam E, Baldwin A, Vandamme P (2002) *Burkholderia cepacia* complex infection in patients with cystic fibrosis. J Med Microbiol 51: 533–538

Mai J, Tian XL, Gallant JW, Merkley N, Biswas Z, Syvitski R, Douglas SE, Junqi Ling JQ, Li YH (2011) A novel target-specific, salt-resistant antimicrobial peptide against the cariogenic pathogen *Streptococcus mutans*. Antimicrob Agents Chemother 55:5205–5213. doi:10.1128/AAC.05175-11

Nakayama J, Chen S, Oyama N, Nishiguchi K, Azab EA, Tanaka E, Kariyama R, Sonomoto K (2006) Revised model for *Enterococcus faecalis fsr* quorum-sensing system: the small open reading frame *fsrD* encodes the gelatinase biosynthesis-activating pheromone propeptide corresponding to staphylococcal AgrD. J Bacteriol 188:8321–8326. doi:10.1128/JB.00865-06

Nakayama J, Yokohata R, Sato M, Suzuki T, Matsufuji T, Nishiguchi K, Kawai T, Yamanaka Y, Nagata K, Tanokura M, Sonomoto K (2013) Development of a peptide antagonist against *fsr* quorum sensing of *Enterococcus faecalis*. ACS Chem Biol 8:804–811. doi:10.1021/cb300717f

Nyholm SV, Stabb EV, Ruby EG, McFall-Ngai MJ (2000) Establishment of an animal–bacterial association: recruiting symbiotic vibrios from the environment. Proc Natl Acad Sci U S A 97:10231–10235. doi:10.1073/pnas.97.18.10231

Peterson RM, Huang T, Rudolf JD, Smanski MJ, Shen B (2014). Mechanisms of self-resistance in the platensimycin- and platencin-producing *Streptomyces platensis* MA7327 and MA7339 strains. Chem Biol 21:389–397. doi:10.1016/j.chembiol.2014.01.005, doi:10.1016/j.chembiol.2014.01.005#_blank

Purohit HJ, Cheema S, Lal S, Raut CP, Kalia VC (2007) In search of drug targets for *Mycobacterium tuberculosis*. Infect Disord Drug Targets 7:245–250. doi:10.2174/187152607782110068

Qin J, Li R, Raes J, Arumugam M, Burgdorf KS, Manichanh C, Nielsen T, Pons N, Levenez F, Yamada T, Mende DR, Li J, Xu J, Li S, Li D, Cao J, Wang B, Liang H, Zheng H, Xie Y, Tap J, Lepage P, Bertalan M, Batto JM, Hansen T, Le Paslier D, Linneberg A, Nielsen HB, Pelletier E, Renault P, Sicheritz-Ponten T, Turner K, Zhu H, Yu C, Li S, Jian M, Zhou Y, Li Y, Zhang X, Li S, Qin N, Yang H, Wang J, Brunak S, Doré J, Guarner F, Kristiansen K, Pedersen O, Parkhill J, Weissenbach J, MetaHIT Consortium, Bork P, Ehrlich SD, Wang J (2010) A human gut microbial gene catalogue established by metagenomic sequencing. Nature 464:59–65. doi:10.1038/nature08821

Rinaudi L, Giordano W (2010) An integrated view of biofilm formation in Rhizobia. FEMS Microbiol Lett 304:1–11. doi:10.1111/j.1574-6968.2009.01840.x

Rodriguez RJ, White JF Jr, Arnold AE, Redman RS (2009) Fungal endophytes: diversity and functional roles. New Phytol 182:314–330. doi:10.1111/j.1469-8137.2009.02773.x

Romero M, Acuna L, Otero A (2012) Patents on quorum quenching: interfering with bacterial communication as a strategy to fight infections. Recent Pat Biotechnol 6:2–12. doi:10.2174/187220812799789208

Sahner JH, Brengel C, Storz MP, Groh M, Plaza A, Muller R, Hartmann RW (2013) Combining *in silico* and biophysical methods for the development of *Pseudomonas aeruginosa* quorum sensing inhibitors: an alternative approach for structure-based drug design. J Med Chem 56:8656–8664. doi:10.1021/Jm401102e

Scheerer S, Ortega-Morales O, Gaylarde C (2009) Microbial deterioration of stone monuments an updated overview. Adv Appl Microbiol 66:97–139. doi:10.1016/S0065-2164(08)00805-8

Schmid-Hempel P (2009) Immune defence parasite evasion strategies and their relevance for 'macroscopic phenomena' such as virulence. Philos Trans R Soc Lond B Biol Sci 364:85–98. doi:10.1098/rstb.2008.0157

Walter J, Britton RA, Roos S (2011) Host Host-microbial symbiosis in the vertebrate gastrointestinal tract and the *Lactobacillus reuteri* paradigm. Proc Natl Acad Sci U S A 108:4645–4652. doi:10.1073/pnas.1000099107

Wexler HM (2007) *Bacteroides*: the good, the bad, and the nitty-gritty. Clin Microbiol Rev 20:593–621. doi:10.1128/CMR.00008-07

White DG, Zhao S, Simjee S, Wagner DD, McDermott, PF (2002) Antimicrobial resistance of foodborne pathogens. Microbes Infect 4:405–412. doi:10.1016/S1286-4579(02)01554-X, doi:10.1016/S1286-4579(02)01554-X#doilink

Zhuang X, Gao J, Ma A, Fu S, Zhuang G (2013) Bioactive molecules in soil ecosystems: masters of the underground. Int J Mol Med Sci 14:8841–8868. doi:10.3390/ijms14058841

Part I

Quorum Sensing Mediated Processes

Evolution of MDRs

Ashima Kushwaha Bhardwaj and Kittappa Vinothkumar

Introduction

With the evolution of multidrug-resistant bacteria at threatening rates, the mankind has witnessed the glorious rise and fall of antibiotics. Antibiotics, the miracle drugs and the magic bullets that appeared to have marked the end of the infectious diseases, are fast losing their charm and effectiveness in human medicine. The swift and untimely demise of these wonder molecules has been attributed chiefly to the resistance mounted by bacteria against them. The phenomenon of antibiotic resistance is inevitable and was something that was cautioned in the Noble Prize lecture by Sir Alexander Fleming in 1945. Dr. Joshua Lederberg very accurately fathomed the seriousness of these resistant bacteria whom he considered much more dangerous a threat as compared to Ebola and West Nile virus. Resistance to any molecule or drug intended to kill a target organism is a very natural phenomenon for the survival of that organism; a cancer cell being subjected to chemotherapeutic treatment, a fungal cell subjected to anti-fungals and, similarly, anti-parasitic and antibacterial compounds are all likely to face resistance from their target cells. Thus, all the popular drugs including antimalarials, anti-tuberculosis, anti-parasitic, antivirals, anti-fungals and antibacterial drugs are facing the risk of becoming obsolete. Consequently, the human race faces the risk of an apocalypse in the hands of these invincible bugs that no drug is able to kill. This chapter describes various genetic and some of the non-genetic factors such as environmental, social and political factors that have led to the evolution of a phenomenon called multidrug resistance (MDR). The threat of antibiotic resistance now spans a wide range of infectious agents including Gram-positive and Gram-negative bacteria, all the infectious diseases and all the geographical locations on this planet.

There has been an evolution of a myriad of resistant bacteria such as methicillin-resistant *Staphylococcus aureus* (MRSA), vancomycin-resistant *Enterococci* (VRE), vancomycin-resistant *Staphylococcus aureus* (VRSA), extremely drug-resistant tuberculosis (XDR), totally drug-resistant tuberculosis (TDR), New Delhi metallo-β-lactamases (NDM)-carrying superbugs, extended spectrum β-lactamases (ESBLs)-carrying bugs and carbapenem-resistant *Klebsiella pneumoniae* (CRKP) to name a few. Having thrived in hospital settings at operation theatres and intensive care units or in community settings, these superbugs have wreaked havoc and led to the number of deaths spiralling high. This exhaustive list also deserves the mention of major threats posed by *Pseudomonas aeruginosa* and *Acinetobacter baumannii* in

A.K. Bhardwaj (✉) • K. Vinothkumar
Department of Human Health and Disease, School of Biological Sciences and Biotechnology, Indian Institute of Advanced Research, Koba Institutional Area, Gandhinagar 382007, Gujarat, India
e-mail: ashima.bhardwaj@gmail.com

V.C. Kalia (ed.), *Quorum Sensing vs Quorum Quenching: A Battle with No End in Sight*,
DOI 10.1007/978-81-322-1982-8_2, © Springer India 2015

nosocomial infections. In addition to this, *Vibrio cholerae* actually stands at the top of the superbug list (Davies and Davies 2010). The world has witnessed seven pandemics due to this pathogen which has shown continuous evolution in terms of virulence factors and antibiotic resistance traits.

Causes of MDR

MDR is a complex phenomenon that arises due to interplay of large number of factors that work in synergy to manifest this problem. Some of the causes of MDR are described below.

Genetic Factors

There is a plethora of genes responsible for the evolution and dissemination of MDR. These genes could be chromosome borne or carried by molecular vehicles such as plasmids, viruses, integrons and transposons, some of which will be described in sections "General mechanisms involved in the evolution of antibiotic resistance" and "Role of MGEs in evolution of MDR".

Social and Political Factors

The underuse, overuse as well as misuse of antibiotics can lead to serious consequences in treatment of infectious diseases. Inappropriate prescriptions due to callousness of the medical doctors like prescribing drugs with improper dosage or prescribing wrong drugs could be one of the reasons. Using antibiotics to treat viral infections could lead to the development of resistance in bacteria residing in the human body. Poor compliance to the drug courses, sub-inhibitory concentrations or premature abrogation of antibiotic usage by humans serve as crucial factors for the development of resistant bugs. Profligate use of antibiotics in human medicine, aquaculture, agriculture and poultry adds to the reservoir of resistance-conferring genes from drug-resistant bacteria in environment, also termed as resistome. Through the extensive and persistent use of antibiotics, the selective pressures continue to be exerted on the bacterial communities. In this scenario, the probability of infection of any individual with a drug-resistant bacterium is much higher as compared to infection with a drug-susceptible bacterium. Therefore, the problem that started with a single patient or a small group of people assumes the proportion of a public health problem. Lack of government policies for proper disease surveillance, antibiotic usage as well as containment of the infectious diseases often leads to the spread of MDR. Pharmaceutical industries are losing interest in the antibiotic development due to the lack of government policies that could give incentives to the pharmaceutical sector for the research and development of new antibiotics. Inability to detect the newer and more subtle antibiotic resistance phenotypes with the available laboratory diagnostic techniques may lead to longer survival and circulation of MDR pathogens in human populations often hindering successful treatment regimens. For example, pneumococcal resistance to β-lactams and staphylococcal resistance to vancomycin are the difficult phenotypes to detect.

Environmental Factors

Natural disasters or calamities like earthquakes, floods, tsunamis and famines and political situations like civil wars or unrest where the medical facilities are heavily impaired can also lead to the high case fatality rates due to the thriving MDR bacteria. Many drugs such as antibiotics, antidepressants, chemotherapeutics and their residues often escape purification by water treatment plants and, therefore, contaminate drinking water supplies. Considerable amounts of these antibiotics are released into the biosphere by hospitals, research laboratories, pharmaceutical industries and domestic use. It is not surprising that the microbial world in soil, water and food has to resort to myriad resistance determinants to avert the catastrophe due to these contaminants. The Environmental Protection Agency (EPA) and Food and Drug

Administration (FDA) have not yet formulated any rules and regulations on this aspect of drug contamination in drinking water. This has serious consequences not only for humans but also for the aquatic ecosystems. At many places such as the United States and India, drugs have been dumped by pharmaceutical companies in the rivers (http://www.purewaterfreedom.com/osc/pharma_water_contamination.php). In addition to the spurt in the appearance and dissemination of drug resistance genes, their toxicity to all the organisms in water or land or air is probably unfathomable.

Consequences of MDR

The dissemination of MDR traits in bacteria has different consequences for mankind as well as the pathogens.

For the human hosts that fall prey to the super bugs, it leads to:

- Treatment failure
- Prolonged stays in the hospitals escalating the health-care budgets
- Reduction in manpower that has both social as well as economical consequences

For the bacterial populations, this MDR translates into:

- Increased virulence of the bacterium. For example, the studies on *A. baumannii* have revealed that genomic islands in this organism also harbour virulence determinants in addition to the antibiotic resistance determinants (Barbe et al. 2004). Similarly, the community-acquired MRSA has equipped itself with a wide range of genes that endow the bacteria with pathogenicity genes as well as antibiotic resistance genes (DeLeo and Chambers 2009).
- More efficient transmission of bacterium.
- Dissemination of the resistance genes to all other pathogens in their vicinity leading to amplification of the resistance genes in the nature.
- Transfer of resistance genes to the commensal organisms residing in the host affecting the microflora often leading to some other outcomes in the host health.

General Mechanisms Involved in the Evolution of Antibiotic Resistance

The evolutionary history of resistant bacteria predates the introduction of the antibiotic era. It is understandable that the antibiotic producers were actually the reservoirs of drug resistance genes. These antibiotic resistance genes were part of the paraphernalia involved in the production of antibiotics by the bacteria where they provided protection to the producers.

As antibiotics target the vital processes of a bacterial cell, they create a do-or-die situation for a bacterium. Hence, it is indispensable for the bugs to resist the action of antibiotics at any cost by devising various tactics. The molecular mechanisms of resistance exerted by bacteria to overcome drugs have been well studied, and they employ any one or a combination of the following strategies (Alekshun and Levy 2007).

Chromosomal Mutations at the Target Sites of Antibiotics

Mutations at the antibiotic target sites are the main mode of resistance to most of the antibiotics. Mutations occurring as a result of replication errors reduce the affinity of the antibiotics to their targets resulting in the resistant phenotype. For example, quinolone and fluoroquinolone resistance occurs through the mutations at the DNA gyrase and topoisomerase IV genes. Similarly, mutations in the gene encoding dihydropteroate synthase decrease the enzyme affinity to the sulphonamides. In *Mycobacterium tuberculosis*, resistance to the common drugs such as rifampin, streptomycin, ethambutol and fluoroquinolones used to treat the pathogen arises due to mutations in the genes that are involved in metabolic pathways or in housekeeping. Additional mutations in the already mutated genes result in increasing the minimum inhibitory concentration (MIC) of the antibiotic for the pathogen or extending the spectrum of resistance such as the development of extended spectrum β-lactamases (ESBLs) in the pneumococcus (Medeiros 1997).

Increased Efflux and Reduced Influx of Antibiotics in the Bacterial Cell

Efflux pumps play a major role in conferring resistance to antibiotics by efficiently recognising and throwing them out of the cells. Efflux pumps in bacteria can be classified into five different families, namely, the resistance nodulation cell division (RND), major facilitator super family (MFS), small multidrug resistance (SMR), ATP-binding cassette (ABC) and multidrug and toxic compound extrusion (MATE) families (Bhardwaj and Mohanty 2012). Among these five pumps, ABC pumps utilise ATP as their energy source, whereas others are driven by the proton-motive force (PMF). Generally efflux pumps are known to extrude out a wide range of substances including antibiotics, and therefore, this is a non-specific mechanism of resistance. However, few pumps are shown to have high specificity towards particular drugs. For example, TetA and NorM are found to be more specific towards tetracycline and norfloxacin, respectively, whereas AcrB, VcmA and MdfA have multiple substrate specificities. Efflux pumps confer only low level resistance to the bacteria towards drugs but their over-expression or cooperativity with other mechanisms could result in moderate to high-level resistance (Bhardwaj and Mohanty 2012).

Porins present in the cell membrane of bacteria are the passages which facilitate the entry and exit of antibiotics and other small organic molecules. Decrease in the expression of porins results in reduced uptake of antibiotics. For example, mutations that caused reduced expression of OprD porins contributed to imipenem resistance (Alekshun and Levy 2007).

Enzymatic Drug Modification or Degradation

This mechanism of resistance involves enzymes that either degrade or chemically modify the antibiotics so that they cannot exert their action. β-lactamases are the well-known examples for the enzymes that degrade β-lactam antibiotics. Few of them behave as extended spectrum β-lactamases (ESBLs) and as carbapenemases and show wider spectrum of resistance to newer generation β-lactam antibiotics (Alekshun and Levy 2007). There are a large number of aminoglycoside-modifying enzymes which chemically modify (acetylate or adenylate or phosphorylate) the aminoglycosides. Similarly, chloramphenicol is inhibited by chloramphenicol acetyltransferases and tetracycline by a flavin-dependent monooxygenase TetX (Alekshun and Levy 2007).

Protection and Alteration of Drug Target

Resistance to fluoroquinolones is mediated by a large number of pentapeptide repeat proteins, quinolone resistance (Qnr) proteins, which protect target DNA gyrase and topoisomerase IV from the antibiotic action. As these proteins mimic the structure of DNA, they occupy the DNA-binding portion of the topoisomerases and prevent the antibiotics from exerting their effect on these protected topoisomerase targets. The altered penicillin-binding protein (PBP) of methicillin-resistant *S. aureus*, PBP2a, confers resistance to most of the β-lactams by contributing the transpeptidase activity when exposed to methicillin (Fig. 1a).

Other Mechanisms

Sometimes resistance to different antibiotics can be conferred by a single determinant. For example, aminoglycoside acetyl transferase (aac (6′)-Ib) generally acetylates aminoglycosides like amikacin, kanamycin and tobramycin. But its mutant form aac (6′)-Ib-cr is known to acetylate quinolones like ciprofloxacin also (Robicsek et al. 2006). Therefore, a single protein renders resistance to aminoglycosides as well as quinolone class of antibiotics (Fig. 1b).

The tandem duplication of the resistance-conferring gene results in overexpression which

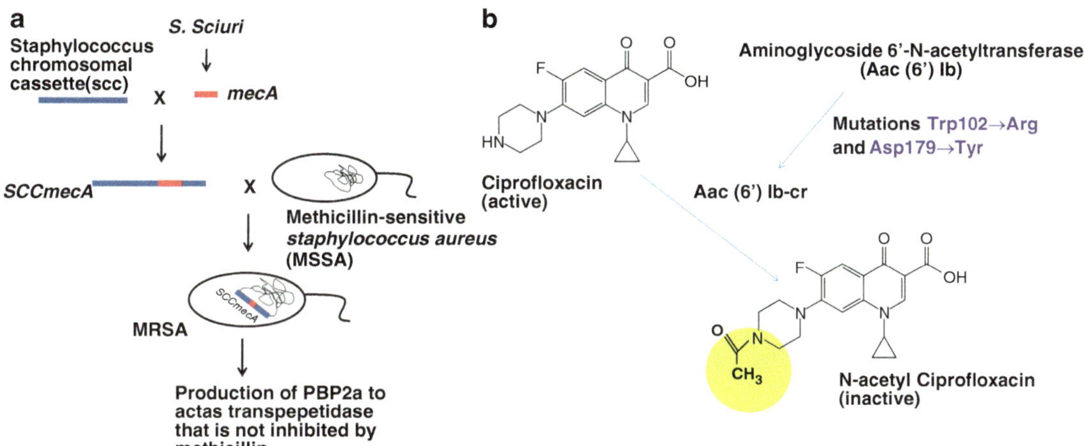

Fig. 1 Evolution of methicillin-resistant *Staphylococcus aureus* (MRSA) and Aac (6′) Ib-cr, an enzyme showing promiscuous drug resistance. (**a**). The evolution of MRSA by the successful acquisition and expression of *mecA* from *Staphylococcus sciuri*. The evolved *S. aureus* expresses *mecA*-derived PBP2a that acts as an alternate transpeptidase that is not inhibited by methicillin; (**b**). By acquiring mutations at the active sites, the modifying enzyme Aac (6′)-Ib evolves as Aac (6′)-Ib-cr with the additional ability to modify ciprofloxacin

eventually helps the bacteria to exhibit a high-level resistance to the antibiotics. In one instance, the overexpression of tandem duplicated genes of AcrAB drug efflux pumps in *E. coli* led to an MDR phenotype (Alekshun and Levy 2007).

Processes That Drive Evolution of MDR

There are chiefly two processes through which the mechanisms of resistance described in section "General mechanisms involved in the evolution of antibiotic resistance" lead to the evolution and persistence of MDR. These processes described below are either the horizontal gene transfers or the pressures due to environment. SOS responses mounted in a bacterium due to antibiotic exposure or HGT are also related to these processes and therefore deserve a special mention in this section (Fig. 2).

Horizontal Gene Transfer (HGT)

The process of HGT enables bacteria to exchange genetic material within themselves without the requirement of cell division. Different kinds of mobile genetic elements (MGEs) are transferred between bacteria through this process leading to the adaptation and evolution of bacteria/bacterial communities in tune with the changing environments. HGT is mediated by the processes of transformation, transduction or conjugation, and different types of MGEs could move through these processes of HGT (Fig. 2). These MGEs as agents of evolution will be described in section "Role of MGEs in evolution of MDR".

Selective Pressure due to Environment

Environment plays a vital role in the selection and spread of antibiotic resistance among bacterial communities that would be discussed at many places in this chapter with examples. These selective pressures lead to induction of mutations in the drug target genes conferring the mutant bacteria, a resistant phenotype (Fig. 2). The transmission dynamics of MDR is hugely responsive to the environmental factors at the hospitals or the communities.

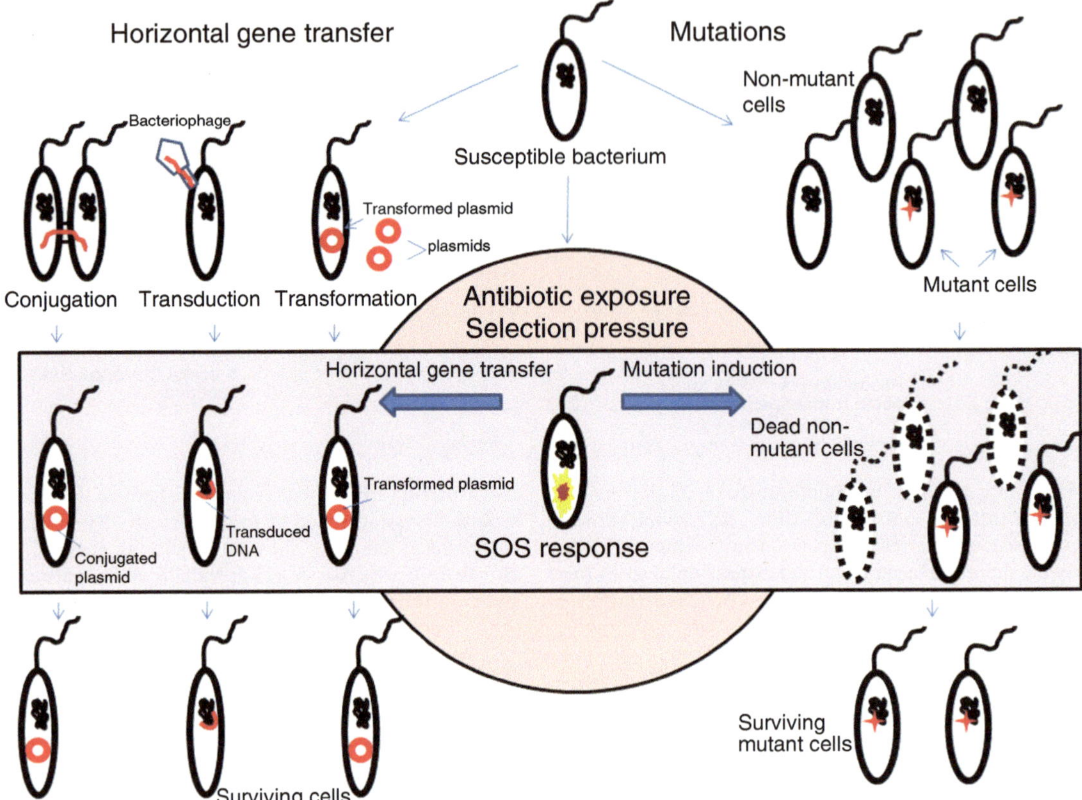

Fig. 2 The process of evolution of MDR bugs. The susceptible bacteria attain some mutations at their antibiotic target sites prior to the exposure of antibiotics. These mutants are selected under antibiotic pressure when they emerge, persist and disseminate as resistant bugs. The acquisition of resistance genes by the bacteria through horizontal gene transfer (HGT) also helps bacteria to resist antibiotics under selection pressure. Antibiotic exposure elicits SOS responses which facilitate both mutations and HGT processes. Therefore, evolution of MDR is an interplay of three processes: HGT, selection pressure and SOS responses

SOS Responses in Bacterium on HGT/Antibiotic Exposure

Any type of HGT through conjugation, transformation and transduction or any type of antibiotic challenge induces SOS response (Fig. 2) through events mediated by single-stranded DNA, RecA protein and LexA repressor (Baharoglu et al. 2013). On antibiotic exposure/HGT, RecA gets activated which leads to autoproteolysis of LexA repressor that keeps the SOS regulon in the repressed state under normal conditions. LexA inactivation thus leads to the expression of a diverse array of genes that were repressed by it. Integrases associated with integrons and integrating conjugative elements (ICEs) are examples of the genes that are induced during SOS due to LexA inactivation (Baharoglu et al. 2013). This leads to the escape of integrons and ICEs from the bacterial cell under crisis. Similarly, the regulation of expression of *qnrB2* (a quinolone resistance determinant) through SOS response is induced by ciprofloxacin in LexA-/RecA-dependent manner. Even sub-inhibitory concentration of ciprofloxacin was found to cleave LexA repressor so that it was prevented from binding on the LexA binding site present in the promoter region of *qnrB2* gene (Da Re et al. 2009).

Therefore, under ciprofloxacin pressure, the bacterial cell expressed resistance gene for this antibiotic through SOS-mediated pathway.

Role of MGEs in Evolution of MDR

MDR evolves through a large spectrum of genetic elements that could either reside on the chromosomes of a bacterium or reside on the pieces of DNA that are mobile. The latter types are called mobile genetic elements (MGEs) and include a diverse class of genetic elements such as integrons, bacteriophages, integrating conjugative elements (ICE) and conjugative plasmids (Fig. 3). These MGEs play an important role in reshaping and in the evolution of the bacterial genomes enabling bacteria to thrive in a variety of ecological niches. In the subsequent sections, the MGEs that have resulted in fast acquisition and dissemination of MDR genes have been described.

Integrons

These MGEs are capable of capturing the gene cassettes by site-specific recombination, integrating them and expressing them using a common promoter (Stokes and Hall 1989; Recchia and Hall 1995). Integrons therefore convert the acquired open reading frames (ORFs) into their functional form. Integrons consist of an integrase gene (*intI*), a recombination site (*attI*) and a promoter Pc. There are numerous classes of integrons known that are classified based on the sequences of their integrase genes. Class 1 integrons have been studied most extensively, and these integrons have been characterised vis-à-vis their role in dispersal of the MDR genes in clinical isolates of Gram-negative bacteria. Integrons have been an important part of bacterial evolution, are widespread among all the bacteria and have a wider role to play in bacterial physiology and adaptation than simply antibiotic resistance (Rapa and Labbate 2013). The structure of a class

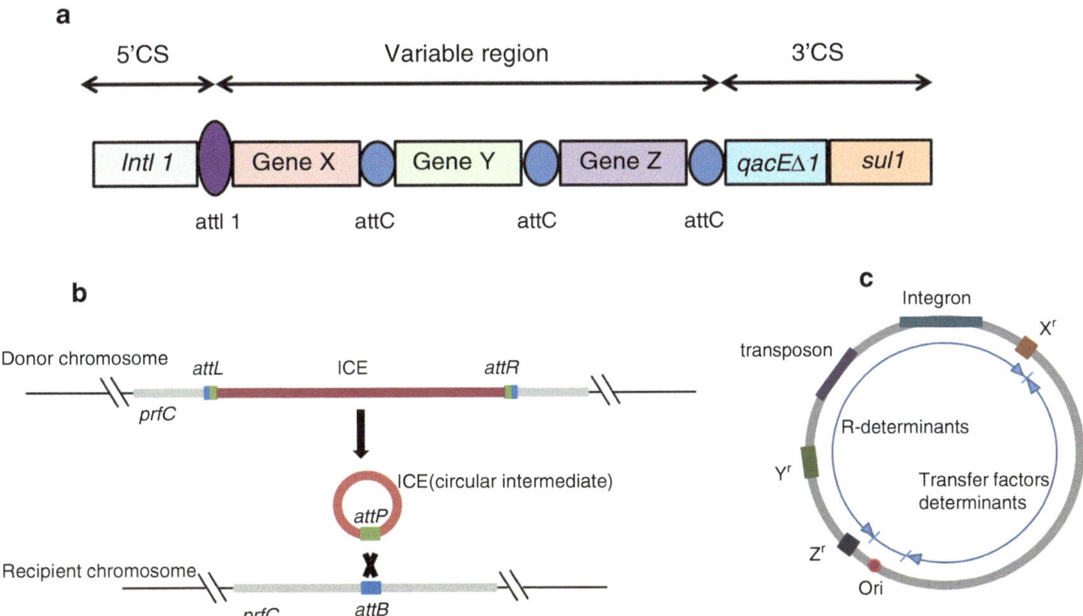

Fig. 3 MGEs that facilitate evolution of MDR. (**a**). Structure of a class 1 integron having conserved segments at its 5′ and 3′ ends (5′CS and 3′CS, respectively) and variable region consisting of extraneous gene cassettes that encode various functions including antibiotic resistance. *intI 1* encodes an integrase and *attI 1,* and *attC* sites are formed with the insertion of the extraneous gene cassettes. (**b**). Insertion and excision of integrating conjugative elements (ICEs) in the *prfC* region of a bacterial genome. The recombination process is mediated by a circular intermediate. (**c**). A conjugative R plasmid that may carry resistance genes, integrons and transposons along with transfer factor determinants

1 integron consists of the conserved segments at the 5′ and 3′ ends of the integron (5′ CS and 3′CS). These conserved segments encompass a variable region which varies with the number and nature of the gene cassettes captured by the integrons (Fig. 3a). The 5′ CS consists of *intI* gene, the *attI* site and the promoter, whereas the 3′CS consists of two genes that encode resistance for ethidium bromide and sulphonamides. The extraneous gene cassettes captured by integrons are usually promoterless, and they recombine with the *attI* site through a recombination site called *attC* or 59-base element (Fig. 3a). Each gene cassette captured in an integron is thus bound by the *attI* site on its 5′ end and an *attC* site on its 3′ end. The *attC* sites share a common set of characteristics that enable them to be identified despite the diversity of their sequences and sizes (Hall et al. 1991). They are characterised by a palindrome of variable length and sequence between the RYYYAAC (R = Purines; Y = Pyrimidine) inverse core site and the GTTRRRY core site. The size of these recombination sites vary in length from 57 to 141 bp.

Integrons have been reported from a wide variety of bacteria such as *V. cholerae*, *V. fluvialis*, *V. parahaemolyticus*, *P. aeruginosa* and *K. pneumoniae*. They have been shown to harbour a diverse array of genes including antibiotic resistance genes. Resistance genes for chloramphenicol (*catB*, *cmlA*), trimethoprim (*dfrA*, *dfrB*), β-lactam antibiotics (*bla*, *oxa*), aminoglycosides (*aac* and *aad*) and many ORFs of unknown functions have been observed in integrons. Two types of integrons are known to exist: chromosomal integrons (CIs) or superintegrons (SIs) that are sedentary in nature and the mobile integrons (MIs) that are associated with mobile DNA elements and involved in the spread of antibiotic resistance genes (Rowe-Magnus et al. 2002). CIs are located on the chromosomes of a large number of bacteria. MIs usually contain less than 20, while SIs/CIs contain more than twenty-gene cassettes. The nature of gene cassettes harboured by MIs and SIs also varies. While MIs usually contain antibiotic resistance gene cassettes, majority of cassettes associated with SIs are of unknown functions.

Integrating Conjugative Elements (ICEs)

ICEs are a type of conjugative transposons that integrate and replicate with the chromosomal DNA of the host bacterium (Burrus and Waldor 2004). ICEs are not capable of autonomous replication, and therefore, they have to depend on the host cell machinery for its survival. They excise themselves from the host chromosome, form a circular intermediate and then get transferred to the recipient cell during conjugation (Fig. 3b). OriT, a *cis*-acting site, is required on ICEs for their translocation to the recipient through the mating bridge formed during conjugation. ICE known as SXT element was first reported from Madras, India, in 1992, in *V. cholerae* O139 strains where they imparted resistance to drugs like trimethoprim, sulphamethoxazole, streptomycin and chloramphenicol (Waldor et al. 1996). Since then, these elements have been reported from a large number of bacteria such as *V. cholerae*, *Providencia alcalifaciens* and *P. rettgeri* at many places as important vehicles for spreading of antibiotic resistance. The integration in the host genome is mediated by an integrase, and ICEs also encode other functions required for their maintenance. These functions include conjugative transfer of these elements, their excision and integration and regulation of the events related to ICE transfer and maintenance. ICEs harbour a wide array of genes for diverse functions such as antibiotic resistance, heavy metal resistance and complex degradation pathways for toxic compounds. Two different ICE elements can also recombine to produce a tandem array of ICE elements called hybrid ICEs. One such hybrid is an SXT/R391 family of ICEs which is the largest family of ICEs detected in clinical as well as environmental strains of many bacteria. Through the process of recombination mediated mainly by RecA protein, these hybrids are known to promote their own diversity resulting in the formation of novel mosaics with new combinations of antibiotic resistance genes (Garriss et al. 2009).

Plasmids

Plasmids are autonomously replicating extrachromosomal DNA molecules that are transferred from donor to the recipient bacterium through conjugation. Resistance plasmids also known as R plasmids, harbouring genes conferring antibiotic resistance, have been well known for their role in the transfer of resistance traits from a drug-resistant bacterium to a drug-sensitive bacterium (Fig. 3c). Plasmids appear to have a major contribution in the spread of drug resistance, and several pathogens have been reported that harbour plasmids with multiple resistance traits. Bacteria carry either conjugative plasmids that are large in size or non-conjugative small-sized plasmids. Non-conjugative plasmids can be mobilised with the help of other conjugative plasmids present in the same cell or by the process of transformation. In some cases, these plasmids may carry integrons or transposons on them facilitating the dissemination of antibiotic resistance gene cassettes in different species of bacteria (Fig. 3c). In enteric pathogens *V. cholerae* and *Shigella dysenteriae*, the multidrug resistance plasmids have been responsible for MDR thus complicating the treatment of diarrhoeal diseases (Ries et al. 1994; Sack et al. 2001). In another pathogen *V. fluvialis*, plasmids have been shown to confer resistance to a large number of drugs (Rajpara et al. 2009; Singh et al. 2012).

Role of Environmental Factors in Evolution of MDR

The presence of an antibiotic in the environment accentuates the appearance of bacteria resistant to this antibiotic. Often, there is a direct correlation between the antibiotic consumption and the appearance of strains resistant to that antibiotic. Antibiotics promote evolution of MDR by the random genetic drift or by induction of large mutational events selecting for the survival of resistant bacteria (Baquero et al. 1998). Random genetic drift occurs during the crisis situations where the random variations acquired by the bacteria may improve the chances of bacterial survival. Apart from antibiotic usage, other environmental factors such as epidemiological features, other drugs being used at the time of study, host immunity and pollutants present in an environment also induce selective pressures for the development of MDR (Baquero et al. 1998). The resistant strain may have higher possibility of surviving in an immunocompromised host as compared to an immunocompetent host. Similarly, the presence of some other non-antibiotic drugs could alter the expression of porins or efflux pumps of bacteria eventually affecting the antibiotic concentrations inside the bacterial cell. This will lead to a change/evolution in the resistance phenotype of this bacterium. For example, drugs such as salicylate lead to the increase in efflux pump expression. Cumulative effect of all these environmental factors allows the survival of bacteria which have acquired the mutations to face the antibiotic pressure. These factors also promote the proliferation and dissemination of such novel bacterial mutants. Profligate use of antibiotics in all the spheres of life including human health, veterinary medicine, food industry and aquaculture actually seems to have provided selective pressures for the evolution of MDR in frightening proportions as we witness it today. Each new generation of antibiotics has spawned new generations of bacterial proteins/mechanism to thwart the effect of antibiotics. As described in Sect. "Extended spectrum β-lactamase (ESBL)-producing bacteria", when new β-lactams such as cefotaxime were produced to face the challenges imposed by early β-lactam-resistant bacteria, the β-lactamases acquired some additional mutations to inactivate these newer drugs. Higher mutated variants like TEM-10 of the β-lactamase TEM were evolved to provide higher resistance. Especially interesting is the scenario in intensive care units where multiple antibiotics are used at varying concentrations for different patients and different pathogens. This leads to selection of a large number of MDR bacteria due to the vast availability of resistome (the population of resistance genes in nature) with the potential to get incorporated into the genome of any bacterial cell and to express the trait.

Case Studies on the Evolution of MDR Bugs

As described in the earlier sections, the evolution of MDR bugs is mediated by the processes such as HGT, selective pressure and SOS response, and these processes are induced by several genetic and environmental factors. The antibiotic era has witnessed the evolution of several resistant bugs, and the following examples can explain the evolution of MDR bugs as a result of interplay of the above-mentioned factors.

Methicillin-Resistant *Staphylococcus aureus* (MRSA)

The rise of MRSA could explain the extraordinary ability of bacteria to evolve as a menace for public health. *S. aureus* is an omnipresent bacterium mostly found in the human nostrils and skin. They often cause respiratory diseases (e.g. nosocomial pneumonia) and skin diseases (e.g. impetigo) in humans. During early 1940s, the infections caused by this bacterium were treated using penicillin as they were extremely sensitive to these wonder drugs at that time. But soon the upsurge of penicillinase-mediated penicillin-resistant strains of *S. aureus* led to the arrival of an alternative drug, methicillin, a semisynthetic penicillin. Methicillin with a power to resist the penicillinase action came to therapeutic use in 1959. But within a short time span, the first case of MRSA was reported. The spectacular mechanism of resistance exhibited by MRSA to fight methicillin was a new penicillin-binding protein, PBP2a. PBPs, the targets of penicillin, methicillin and other β-lactams, are transpeptidases which are responsible for the cross-linking of the cell wall of bacteria. But the new variant of PBP, PBP2a, has low affinity for methicillin and other β-lactams and could substitute the role of native PBPs for cell wall formation (Fig. 1a). PBP2a was encoded by *mecA* gene which is a distinctive feature of MRSA and hence the methicillin resistance. The *mecA* gene was found in the chromosome of MRSA but associated with a large mobile genetic element called staphylococcal chromosome cassette [SCC] (Pantosti and Venditti 2009). The *mecA* gene seemed to have originated from *Staphylococcus sciuri* and then got incorporated into SCC to become SCC*mec* (Fig. 1a). The successful acquisition and expression of SCC*mec* in *S. aureus* gave rise to the strain of MRSA (de Lencastre et al. 2007). The first MRSA clone appeared in the 1960s, spread widely in the hospitals and clinical settings for about 17 years and new clonal types with different SCC*mec* elements were reported subsequently. The epidemic hospital-acquired MRSA (HA-MRSA) clones reported so far mainly fall into three types (type I, II and III) based on the multilocus sequence typing (MLTS) method. Subsequent to the acquisition of SCC*mec*, MRSA further evolved to resist other classes of antibiotics such as aminoglycosides, tetracycline, sulphonamides and quinolones as a result of selective pressure on exposure to antibiotics and acquisition of various resistance genes through HGT. In the 1990s the enigmatic emergence of community-acquired MRSA (CA-MRSA) has been reported with different epidemiological and molecular profile than that of HA-MRSA. Initially CA-MRSA clones carried the single trait of *mecA* mainly in two SCC*mec* element types (type IV and V) and were susceptible to non-β-lactam antibiotics. But few typical CA-MRSA have been reported now to evolve as multidrug-resistant strains (e.g. USA 300 and ST80 clone) (Pantosti and Venditti 2009; de Lencastre et al. 2007).

Vancomycin-Resistant *Staphylococcus aureus* (VRSA) and Vancomycin-Resistant Enterococci (VRE)

Vancomycin served as a possible alternate therapy for the infections caused by the MRSA. Vancomycin inhibits cell wall synthesis by blocking the transglycosylation and transpeptidation reactions as it binds to the C-terminal peptide of D-Ala-D-Ala of pentapeptide precursor for the formation of bacterial peptidoglycan. The *van* gene mediating vancomycin resistance was first observed in enterococci only. These genes are of seven types (*vanA*, *vanB*, *vanC*, *vanD*, *vanE*, *vanG*, *vanL*) which are known to synthesise a new target (peptidoglycan precursor) which replaces the normal D-Ala-D-Ala precursor, and hence, the antibiotic cannot find its target. The *vanA*-, *vanB*- and *vanD*-type genes produce the D-Ala-D-*Lac* target, whereas *vanC*, *vanE*, *vanG* and *vanL* gene types synthesise the D-Ala-D-Ser target. The acquisition of plasmid-borne *vanA* gene through conjugation from *Enterococcus* to *S. aureus* resulted in the development of VRSA. The evolution of *S. aureus* which were already resistant to multiple drugs into VRSA further complicated the treatment of infections caused by such bacteria (Perichon and Courvalin 2009).

Extended Spectrum β-Lactamase (ESBL)-Producing Bacteria

The emergence of β-lactamases served as a common mechanism of resistance for β-lactam antibiotics in Gram-negative bacteria. In the 1970s to 1980s, β-lactamases such as TEM-1, TEM-2 and SHV-1 that hydrolysed penicillin, ampicillin and early generation cephalosporins were detected. TEM-1 and TEM-2 were predominant in *E. coli* and SHV-1 was prevalent in *K. pneumoniae* (Chong et al. 2011). During the early 1980s, the emergence of modified β-lactamases carrying amino acid mutations in TEM-1, TEM-2 and SHV-1 enzymes was detected. As they were able to hydrolyse the third-generation cephalosporins such as cefotaxime, ceftriaxone, ceftazidime, cefuroxime and cefepime, apart from penicillin and ampicillin, they were termed as ESBLs. The TEM and SHV ESBLs were genetically evolved by amino acid substitutions from their non-ESBL progenitors TEM-1, TEM-2 and SHV-1, whereas another ESBL called CTX-M evolved independently of this lineage. Some other ESBLs different from TEM, SHV and CTX-M are OXA, BEL-1, BES-1, GES/IBC, SFO-1, TLA-1, TLA-2, PER and VEB enzyme families. ESBLs soon became pervasive and were reported all across the globe within two decades. So far more than 300 ESBLs have been described (Lynch et al. 2013). The increased incidents of dissemination of ESBL genes among bacteria through various MGEs which carry other antibiotic resistance genes have reduced the therapeutic options and caused an emerging threat to public health.

Quinolone-Resistant Bacteria

The increased drug resistance among bacteria towards various natural and semisynthetic antibiotics led to the introduction of synthetic drugs like quinolones and fluoroquinolones due to their broad spectrum of activity and possibilities of the absence of resistance mechanisms in bacteria to these synthetic drugs. Quinolones inhibit nucleic acid synthesis in bacteria by targeting DNA gyrase and topoisomerase IV enzymes which are involved in the essential activities of bacterial cell such as replication, transcription, recombination and repair. The mutations at the quinolone resistance determining regions (QRDRs) of subunits of DNA gyrase (GyrA, GyrB) and topoisomerase IV (ParC, ParE) enzymes cause their reduced affinity towards quinolones which lead to quinolone resistance phenotypes. Mutations that occur at the target sites of the antibiotics prior to the exposure of antibiotics help in selection and subsequent evolution of resistant bacteria. The accumulation of multiple mutations in the drug target facilitates the development of high-level resistance to quinolones in bacteria. The quinolone resistance can also be mediated by efflux pumps encoded by chromosome-borne genes such as *norM*, *norA*, *vcmA* and *bmrA* (Bhardwaj and Mohanty 2012). When quinolones were introduced, it was imprudently predicted that there would not be any quinolone resistance genes as these antibiotics were not naturally produced by any bacteria (Hernandez et al. 2011). Hence, the dissemination of resistance among bacterial communities through HGT was not expected. But the emergence of factors like target-protecting quinolone resistance proteins (Qnr), drug-modifying enzyme and plasmid-borne efflux pump genes falsified the latter belief.

Qnr proteins are pentapeptide repeat proteins which are believed to be evolved from the proteins like McbG that protect topoisomerases from the naturally occurring toxins like microcin B17, a topoisomerase poison. These pentapeptide repeat proteins occupy the DNA-binding region of the enzyme and protect them from the drug action. Though the Qnr proteins are of chromosomal origin, they are more often found in plasmids through which they tend to disseminate among different bacterial species. Similarly, as shown in Fig. 1b, a variant aminoglycoside acetyl transferase (AAC(6′)-Ib-cr) borne on plasmid was found to inactivate (by acetylation) ciprofloxacin and norfloxacin apart from aminoglycosides due to two amino acid changes (Trp102Arg and Asp179Tyr) in the active site of the enzyme (Robicsek et al. 2006). Two plasmid-mediated quinolone transporters (OqxAB and QepA) have been described to effectively efflux out quinolone antibiotics. All the above-mentioned resistance mechanisms may work alone or in synergy to combat the quinolone drugs. Apart from the mutation in the target sites, other genetic factors of quinolone resistance such as *qnr* genes, *aac(6′)Ib-cr* gene and *oqxAB* and *qepA* genes are often harboured by plasmids and cause plasmid-mediated quinolone resistance (PMQR). Though higher-level resistance through PMQR has not been reported, they could help the isolates to attain clinical breakpoint of resistance in combination with other mechanisms. Hence, the great plasticity of the bacterial systems allows them to educe their armaments to battle against these drugs.

Conclusions

Evolution of MDR is a very natural process, and from the discussions above, it can be concluded that the resistance determinants for antibiotics were always present in nature even before the miracle drugs were introduced for human use. These determinants just made their public appearances with the escalating use of antibiotics in many applications. With the selective pressures rising due to antibiotic use, the resistance genes kept appearing in their new avatars. Evolution of antibiotic resistance mechanisms paralleled evolution of antibiotics. Therefore, new strategies need to be used for keeping one-step ahead of the MDR pathogens. The advent of technologies based on microbial genomics, proteomics, combinatorial chemistry and high-throughput screening could lead to success stories in the development of new anti-infectives inspite of the large funds required for them. There are many innovative strategies being used to curb the problem of MDR (Breithaupt 1999; Tegos and Hamblin 2013). In pathogens *Neisseria gonorrhoeae* and *N. meningitides*, lipooligosaccharides on the bacterial surface have been shown to be crucial for their virulence-associated functions such as colonisation and immune evasion. The glycosyltransferases and hydrolases involved in the synthesis of these lipooligosaccharides have been used as targets for synthesis of small inhibitors as antibiotics. Companies such as TerraGen Diversity Inc. (Canada), ChromaXome Corp.(California) and GLYCODesign (Toronto) have been actively involved in the innovations pertaining to antibiotic research and development. Strategies of quorum-sensing inhibition and efflux pump inhibition also provide attractive alternatives to solve the problems of MDR (Bhardwaj et al. 2013; Kalia 2013; Bhardwaj and Mohanty 2012) though the possibility of evolution of resistance to these new molecules cannot be ruled out (Bhardwaj et al. 2013; Kalia et al. 2014). The governments should realise the seriousness of impending disaster due to MDR bacteria and urgency of the situation (Finch and Hunter 2006). Accordingly, new policies should be made to deal with this problem. In India, a national policy has been made for containment of antimicrobial resistance by the Directorate General of Health Services (2011), and a task force was constituted to work on various aspects related to antimicrobial resistance and its monitoring.

Opinion

From the concepts and facts presented in this chapter and reiterated throughout this book, it would be amply clear that the problem of multidrug resistance is real and threatening and in all possibilities here to stay. Prudent choices need to be made to keep this problem to its lowest or least dangerous level where solutions for this problem are easier to make. As Dr. Stuart Levy pointed out in one of his writings, the mankind should understand how this equation of drug resistance should be balanced. Improving laboratory techniques for diagnosis of drug resistance profiles of prevailing pathogens or environmental organisms, proper surveillance and molecular epidemiology studies, striking a balance between the antibiotic use and abuse and utilisation of novel anti-virulent and anti-infective strategies are some of the approaches that should be utilised in synergy to keep the magic of antibiotics alive. Obviously, this herculean task can only be realised with the concerted efforts by people from different spheres of life including clinicians, policymakers, academicians, researchers, clinical microbiologists, citizens and pharmaceutical companies.

Acknowledgements The laboratory is supported by the grants from the Department of Biotechnology (DBT), Ministry of Science and Technology, Government of India (No. BT/PR/11634/INF/22/104/2008), Gujarat State Biotechnology Mission (GSBTM), Department of Science and Technology, Government of Gujarat (No. GSBTM/MD/PROJECTS/SSA/1535/2013–14) and Indian Council of Medical Research, New Delhi, India (No. AMR/49/11-ECDI). K. Vinothkumar is a JRF in the above-mentioned GSBTM grant. The authors are grateful to Dr. Amit Ghosh, Dr. T. Ramamurthy and Dr. S. K. Niyogi, National Institute of Cholera and Enteric Diseases (NICED), Kolkata, India, for their support and advice. The authors thankfully acknowledge the Puri Foundation for Education in India for providing infrastructure facilities and Ms. Neha Rajpara, Mr. Priyabrata Mohanty, Mr. Braj M. R. N. S. Kutar and Ms. Aneri Shah for their invaluable help, advice and support.

References

Alekshun MN, Levy SB (2007) Molecular mechanisms of antibacterial multidrug resistance. Cell 128:1037–1050

Baharoglu Z, Garriss G, Mazel D (2013) Multiple pathways of genome plasticity leading to development of antibiotic resistance. Antibiotics 2:288–315

Baquero F, Negri MC, Morosini MI, Blazquez J (1998) Antibiotic-selective environments. Clin Infect Dis 27(Suppl 1):S5–S11

Barbe V, Vallenet D, Fonknechten N, Kreimeyer A, Oztas S, Labarre L et al (2004) Unique features revealed by the genome sequence of *Acinetobacter* sp. ADP1, a versatile and naturally transformation competent bacterium. Nucleic Acids Res 32: 5766–5779

Bhardwaj AK, Mohanty P (2012) Bacterial efflux pumps involved in multidrug resistance and their inhibitors: rejuvinating the antimicrobial chemotherapy. Recent Pat Antiinfect Drug Discov 7:73–89

Bhardwaj AK, Vinothkumar K, Rajpara N (2013) Bacterial quorum sensing inhibitors: attractive alternatives for control of infectious pathogens showing multiple drug resistance. Recent Pat Antiinfect Drug Discov 8:68–83

Breithaupt H (1999) The new antibiotics. Nat Biotechnol 17:1165–1169

Burrus V, Waldor MK (2004) Shaping bacterial genomes with integrative and conjugative elements. Res Microbiol 155:376–386

Chong Y, Ito Y, Kamimura T (2011) Genetic evolution and clinical impact in extended-spectrum beta-lactamase-producing *Escherichia coli* and *Klebsiella pneumoniae*. Infect Genet Evol 11:1499–1504

Da Re S, Garnier F, Guerin E, Campoy S, Denis F, Ploy MC (2009) The SOS response promotes *qnrB* quinolone-resistance determinant expression. EMBO Rep 10:929–933

Davies J, Davies D (2010) Origins and evolution of antibiotic resistance. Microbiol Mol Biol Rev 74: 417–433

de Lencastre H, Oliveira D, Tomasz A (2007) Antibiotic resistant *Staphylococcus aureus*: a paradigm of adaptive power. Curr Opin Microbiol 10:428–435

DeLeo FR, Chambers HF (2009) Reemergence of antibiotic-resistant *Staphylococcus aureus* in the genomics era. J Clin Invest 119:2464–2474

Directorate General of Health Services (2011) National policy for containment of antimicrobial resistance, India. http://nicd.nic.in/ab_policy.pdf

Finch R, Hunter PA (2006) Antibiotic resistance-action to promote new technologies: report of an EU Intergovernmental Conference held in Birmingham, UK, 12–13 December 2005. J Antimicrob Chemother 58(Suppl 1):i3–i22

Garriss G, Waldor MK, Burrus V (2009) Mobile antibiotic resistance encoding elements promote their own diversity. PLoS Genet 5:e1000775

Hall RM, Brookes DE, Stokes HW (1991) Site-specific insertion of genes into integrons: role of the 59-base element and determination of the recombination cross-over point. Mol Microbiol 5:1941–1959

Hernandez A, Sanchez MB, Martinez JL (2011) Quinolone resistance: much more than predicted. Front Microbiol 2:22

Kalia VC (2013) Quorum sensing inhibitors: an overview. Biotechnol Adv 31:224–245

Kalia VC, Wood TK, Kumar P (2014) Evolution of resistance to quorum sensing inhibitors. Microb Ecol 68(1):13–23

Lynch JP 3rd, Clark NM, Zhanel GG (2013) Evolution of antimicrobial resistance among Enterobacteriaceae (focus on extended spectrum beta-lactamases and carbapenemases). Expert Opin Pharmacother 14: 199–210

Medeiros AA (1997) Evolution and dissemination of beta-lactamases accelerated by generations of beta-lactam antibiotics. Clin Infect Dis 24(Suppl 1): S19–S45

Pantosti A, Venditti M (2009) What is MRSA? Eur Respir J 34:1190–1196

Perichon B, Courvalin P (2009) VanA-type vancomycin-resistant *Staphylococcus aureus*. Antimicrob Agents Chemother 53:4580–4587

Rajpara N, Patel A, Tiwari N, Bahuguna J, Antony A, Choudhury I et al (2009) Mechanism of drug resistance in a clinical isolate of *Vibrio fluvialis*: involvement of multiple plasmids and integrons. Int J Antimicrob Agents 34:220–225

Rapa RA, Labbate M (2013) The function of integron-associated gene cassettes in species: the tip of the iceberg. Front Microbiol 4:385

Recchia GD, Hall RM (1995) Gene cassettes: a new class of mobile element. Microbiology 141(Pt 12): 3015–3027

Ries AA, Wells JG, Olivola D, Ntakibirora M, Nyandwi S, Ntibakivayo M et al (1994) Epidemic *Shigella dysenteriae* type 1 in Burundi: panresistance and implications for prevention. J Infect Dis 169:1035–1041

Robicsek A, Strahilevitz J, Jacoby GA, Macielag M, Abbanat D, Park CH et al (2006) Fluoroquinolone-modifying enzyme: a new adaptation of a common aminoglycoside acetyltransferase. Nat Med 12:83–88

Rowe-Magnus AD, Davies J, Mazel D (2002) Impact of integrons and transposons on the evolution of resistance and virulence. Curr Top Microbiol Immunol 264:167–188

Sack DA, Lyke C, Mc Laughlin C, Suwanvanichkij V (2001) Antimicrobial resistance in shigellosis, cholera and campylobacteriosis. World Health Organization, Geneva, Switzerland, pp 1–20

Singh R, Rajpara N, Tak J, Patel A, Mohanty P, Vinothkumar K et al (2012) Clinical isolates of *Vibrio fluvialis* from Kolkata, India, obtained during 2006: plasmids, the qnr gene and a mutation in gyrase A as mechanisms of multidrug resistance. J Med Microbiol 61: 369–374

Stokes HW, Hall RM (1989) A novel family of potentially mobile DNA elements encoding site-specific gene-integration functions: integrons. Mol Microbiol 3:1669–1683

Tegos GP, Hamblin MR (2013) Disruptive innovations: new anti-infectives in the age of resistance. Curr Opin Pharmacol 13:673–677

Waldor MK, Tschape H, Mekalanos JJ (1996) A new type of conjugative transposon encodes resistance to sulfamethoxazole, trimethoprim, and streptomycin in Vibrio cholerae O139. J Bacteriol 178:4157–4165

Biofilms: Maintenance, Development, and Disassembly of Bacterial Communities Are Determined by QS Cascades

Hadas Ganin, Eliane Hadas Yardeni, and Ilana Kolodkin-Gal

Introduction

Unicellular organisms use a variety of mechanisms to coordinate activity within communities, called biofilms, and across species to accomplish complex multicellular processes (Aguilar et al. 2007; Kolter and Greenberg 2006; Miller and Bassler 2001; Stoodley et al. 2002). Informed by chemical communication, motile cells of the myxobacteria and filamentous cells of the streptomycetes organize themselves into conspicuous multicellular structures that carry out specialized tasks in spore formation and dispersal. Furthermore, most bacteria have evolved elaborate mechanisms for adhering to solid surfaces and thereby establishing complex communities referred to as biofilms. Biofilms can be viewed as differentiated communities in which an extracellular matrix holds the cells together in the multicellular community. Bacterial biofilms are of a high significance in agricultural (Chen et al. 2013), environmental (Cha et al. 2012; Sanchez 2011), and clinical (Bryers 2008; Costerton et al. 1999) settings. In many instances they provide beneficial effects to other organisms. Such is the case for biofilms of *Bacillus subtilis* that form on the surface of plant roots, thereby preventing the growth of fungal pathogens (Nagorska et al. 2007). However, in other situations, bacterial biofilms can have deadly effects; in a clinical context, biofilms in human hosts are inherently resistant to antimicrobial agents (Costerton et al. 1999) and are thus the cause of many persistent and chronic bacterial infections.

For decades, it has been mysterious how bacteria in these biofilm communities communicate with each other to coordinate their behavior. This chapter sheds new light on cell-to-cell signaling during the development of a bacterial biofilms in the most prominent models of Gram-negative and Gram-positive bacteria.

Quorum sensing is an efficient type of cell-to-cell communication between bacteria. This process is concentration dependent and regulated by small chemical signals produced by bacteria. These small molecules are termed autoinducers, and when bacterial concentration is high enough, their concentration raises up to a threshold concentration in which different genes are being transcribes and expressed by the group of bacteria (Miller and Bassler 2001).

Below we are going to discuss QS cascades determining the fates of bacterial biofilms belonging to five fascinating examples. Three belong to the family of Gram-negative bacteria, generally communicating via small diffusible organic molecules such as homoserine lactones: (1) the deadly opportunistic Gram-negative pathogen, *Pseudomonas aeruginosa*; (2a) the Gram-negative pathogen *Vibrio cholerae*, and his immediate classic QS model, (2b)

H. Ganin • E.H. Yardeni • I. Kolodkin-Gal (✉)
Department of Molecular Genetics, Weizmann Institute of Science, Rehovot, Israel
e-mail: Ilana.Kolodkin-Gal@weizmann.ac.il

V.C. Kalia (ed.), *Quorum Sensing vs Quorum Quenching: A Battle with No End in Sight*,
DOI 10.1007/978-81-322-1982-8_3, © Springer India 2015

the Gram-negative *Vibrio fischeri*. Two are Gram-positives, generally communicating via peptide autoinducers: the agriculturally relevant bacterium (3) *Bacillus subtilis* and the Gram-positive pathogen (4) *Staphylococcus aureus*.

Chemical Communication Cascades Regulate Biofilm Development in *Pseudomonas aeruginosa*

Pseudomonas aeruginosa is a Gram-negative, opportunistic human pathogen and is one of the most common bacteria found in nosocomial and life-threatening infections of immunocompromised people (Hentzer et al. 2003b). Patients with cystic fibrosis (CF), burn victims, and patients with implanted medical devices (Sadikot et al. 2005) are especially sensitive to get infected by this bacterium. The threat of *P. aeruginosa* relies in its ability to produce diverse virulence factors such as elastase, alkaline protease, exotoxin A, rhamnolipids, pyocyanin, and biofilm formation which will be further discussed here in this chapter. *P. aeruginosa* uses two main QS systems to control its pathogenicity: the Las and Rhl systems in addition to other regulators which will be discussed later (Pesci et al. 1997) (Schuster et al. 2003a).

The *P. aeruginosa* LasI–LasR/RhlI–RhlR Systems

The lasR–lasI system consists of the *lasI* gene which produces the signal molecule N-(3-oxododecanoyl)-L-homoserine lactone (3-oxo-C12-HSL) that is detected by a transcriptional regulatory protein LasR, and the rhlR–rhlI system (also called vsmR–vsmI) in a similar way produces and recognizes the signaling molecule N-butyryl-L-homoserine lactone (C4-HSL) that is detected by the transcriptional regulator RhlR (Latifi et al. 1996; Pesci et al. 1997). The *las* and *rhl* systems are organized in hierarchy. The LasI–LasR system controls the expression of *lasI* for autoregulation and also activates the RhlI–RhlR system by activating the production of RhlR (Koch et al. 2005; Ochsner et al. 1994). The role of these quorum-sensing systems in *P. aeruginosa* was described, for the first time, by Davies and colleagues in 1998 (Davies et al. 1998). Biofilm formation of wild-type (WT) *P. aeruginosa* PAO1 and a *lasI–rhlI* double mutant that makes neither of the quorum-sensing signals was analyzed. Though both the mutant and the wild type adhered to the surface and made a biofilm, the mutant biofilm was thinner and about 20 % of the WT thickness. The WT formed characteristic microcolonies composed of groups of cells separated by water channels, whereas the mutant appeared to grow rather as continuous sheets on the glass surface. Thus, although *lasI–rhlI* were not involved in the initial attachment and growth stages of biofilm formation, the *P. aeruginosa* quorum-sensing systems jointly participated in the subsequent biofilm differentiation process. The flat biofilms of the quorum-sensing mutant were susceptible to treatment with the surfactant sodium dodecyl sulfate, while the structured biofilms were resistant. The authors concluded that *las*-regulated functions were required for biofilm formation.

Subsequent research indicated that quorum sensing's role in *P. aeruginosa* biofilm formation was not always as dramatic (Kirisits and Parsek 2006). Yet, Purevdorj and colleagues showed that although under high flow conditions, both wild-type and quorum-sensing mutant strains formed structured biofilms, the biofilms differed in their microscopic appearance (Purevdorj et al. 2002). Independently, it was demonstrated that AHL signal analogues called furanones, known to inhibit *P. aeruginosa* quorum sensing, impaired biofilm development when added to the growth medium (Hentzer et al. 2002; Hentzer et al. 2003a). In addition, it was shown that a *lasR–rhlR* double mutant strain produced biofilms which are more susceptible to the clinically relevant antibiotic tobramycin than the isogenic wild-type strain (Bjarnsholt et al. 2005).

Notably, the *las*- and *rhl*-based quorum-sensing systems regulate many different functions, and their control of these functions can change depending on environmental conditions. Supporting this notion, several different quorum-sensing-regulated functions have been shown to impact biofilm formation at different stages; the quorum-sensing-regulated surfactant rhamnolipid is necessary for maintaining the open spaces between cell aggregates in structured biofilms (Davey et al. 2003). In addition, rhamnolipid production may aid in the formation of mature mushroom structures (Lequette and Greenberg 2005). Another quorum-sensing-regulated factor shown to contribute to biofilm formation is the siderophore pyoverdine. Siderophores are small, high-affinity iron chelators secreted by bacteria. In *Pseudomonas* pyoverdine is critical for acquiring iron, and mutants unable to make pyoverdine formed flat biofilms, while an isogenic wild-type strain formed structured biofilms (Banin et al. 2005). *P. aeruginosa* also uses quorum sensing to regulate the production of two sugar-binding lectins, LecA and LecB, which are secreted from the cell. These lectins are expressed in biofilms and both *lecA* and *lecB* mutant strains formed aberrant biofilms (Diggle et al. 2006; Tielker et al. 2005).

Since so many different quorum-sensing-regulated functions affect biofilm development, it is especially clear that in *Pseudomonas* both critical behaviors are intervened.

Several open questions remain to be answered to explain the discrepancies between different studies. One is whether quorum-sensing response may not be induced or active in biofilms grown under these conditions. It was shown by De Kievitand and colleagues that the expression of *lasI* and *rhlI* is highest near the attachment surface and decreases toward the periphery of the biofilm (De Kievit et al. 2001). This spatiotemporal distribution of quorum-sensing autoinducers producing community members suggests that there may be conditions, such as high liquid flow, where signal concentrations may not reach an inducing level in the biofilm.

The PQS System

The PQS system which is comprises from the *Pseudomonas* quinolone signal (PQS); 2-heptyl-3-hydroxy-4-quinolone, the synthase PqsH and the response regulator PqsR (also called MvfR). The PQS structure is very similar to the Pyo compounds, which had been identified as antibiotics in 1945 (Dietrich et al. 2006; Hays et al. 1945) and shown to belong to the family of 4-quinolones in 1952 (Wells 1952; Wells et al. 1952).

The three QS systems in *P. aeruginosa* are arranged in a temporal manner, with AHLs and PQS being released in the early and late exponential phase, respectively (Lepine et al. 2003). The expression of the PQS requires LasR, and the PQS in turn induces transcription of *rhlI*. These data indicate that the PQS is an additional link between the Las and Rhl circuits (Miller and Bassler 2001). Thus, the PQS initiates the Rhl cascade by allowing the production of the RhlI-derived autoinducer only after the establishment of the LasI–LasR signaling cascade (Pesci et al. 1999a).

The Las, Rhl, and Pqs systems mutually regulate the production of virulence factors such as elastase, alkaline protease, exotoxin A, rhamnolipids, pyocyanin biofilm formation, and others (Smith and Iglewski 2003). Specifically, PqsR/MvfR positively regulates the production of a number of virulence factors as well as the expression of PA2274, a putative monooxygenase, and the *mexGHI–opmD* operon that encodes proton-driven efflux pumps of the resistance-nodulation-cell division (RND) transporter super family (Deziel et al. 2005; Dietrich et al. 2006). It is thought that this response is mediated through PqsE and the PQS (Deziel et al. 2005). *P. aeruginosa* releases a 4-quinolone signal molecule into the extracellular milieu, as culture supernatants were found to contain approximately 6 μM (Pesci et al. 1999b).

Importantly, the PQS has recently been isolated from the lungs of CF patients infected with *P. aeruginosa* (Collier et al. 2002;

Guina et al. 2003), and the presence of the molecule in vivo may be a factor in allowing *P. aeruginosa* to develop or maintain a chronic state (Collier et al. 2002), involving biofilm formation. In support of this hypothesis, it was shown that PQS concentrations of 60 μM and above significantly enhanced surface coverage and biofilm formation of PAO1 on stainless steel coupons. The PQS expression was shown to be involved in DNA release in *P. aeruginosa* biofilms, and the expression of the *pqsA* reporter occurred specifically in the microcolonies in the early phase of biofilm formation (Allesen-Holm et al. 2006; Yang et al. 2007). A later study has suggested that the *pqsA* gene is expressed specifically in the stalk-forming subpopulation, suggesting strongly that a subpopulation of quorum-sensing producers dramatically affects the development of the biofilm structure as a whole (Yang et al. 2009).

Phenazines

P. aeruginosa releases phenazines which are a group of small heterocyclic, redox-active compounds that are toxic to both prokaryotes and eukaryotes (Mavrodi et al. 2006; Mazzola et al. 1992; Price-Whelan et al. 2006).

Phenazines are important virulence factors (Lau et al. 2004) that serve as antibiotics toward microbial competitors (Baron and Rowe 1981) and damage mammalian cells (Britigan et al. 1992). Phenazines can benefit *P. aeruginosa* by serving as signaling molecules (Dietrich et al. 2006), regulating persister cell formation (Moker et al.), influencing colony morphology (Dietrich et al. 2008; Kempes et al. 2014), and promoting iron acquisition and biofilm development (Glasser et al. 2014; Wang et al. 2011). Like quinolones, phenazines are excreted from cells at specific points following exponential growth (Diggle et al. 2003). The phenazine pyocyanin (PYO) is a terminal signaling molecule in the *P. aeruginosa* QS network (Dietrich et al. 2006). D. K. Newman and coworkers reported that micromolar concentrations of phenazines can support anaerobic survival by transferring electrons to an extracellular oxidant (Wang et al. 2010). This is a critical observation for developing biofilms. Indeed a phenazine-null mutant makes an especially rugose morphology. Using a variety of approaches, it was demonstrated that the rugose morphotype increases colony surface area and access to oxygen for resident bacteria when phenazines and other electron acceptors are absent (Dietrich et al. 2008). Consistent with this, the production of phenazines or medium amendment with the alternate electron acceptor nitrate promotes colony smoothness. Furthermore, in the phenazine-null mutant, an increase in the cellular NADH/NAD$^+$ coincides with the onset of wrinkling and a decrease occurs as wrinkles develop (Dietrich et al. 2013). This pioneering work was followed by several independent works that suggested that indeed colony wrinkling is an adaptation that supports redox balancing in response to electron acceptor limitation (Kolodkin-Gal et al. 2013; Okegbe et al. 2014). Furthermore, it provided a novel missing link between QS and redox balancing.

QscR

QscR is an orphan LuxR homolog that does not have a partner LuxI homolog, although QscR can bind the AI 3-oxo-C12-HSL (Lequette et al. 2006; Lintz et al. 2011; Oinuma and Greenberg 2011).

QscR forms mixed dimers with LasR and RhlR and inactives them (Ledgham et al. 2003). QscR has broader signal specificity than LasR and can respond to some non-self-signals. Synthetic 3OHC10 activated QscR much more strongly than 3OC12 did (Ha et al. 2012). The acyl side chain of C2 is ten-carbon as with that of C10-HSL, although it is a non-self-produced signal. The bacterial species *P. fluorescens, Burkholderia vietnamiensis,* and *Roseobacter gallaeciensis* are capable of producing signals such as C10, C12, and 3OHC10, which can preferentially activate QscR to LasR (Wagner-Dobler et al. 2005). The difference between QscR and LasR is the broader signal specificity, suggesting that QscR might respond to extrinsic signals by autoactivating its own expression. In this situation, *P. aeruginosa* can preferentially activate QscR. The QscR regulon may be turned on independently of the LasR system in the presence of its preferred signal. The earlier and stronger activation of QscR may antagonize the conventional QS signaling pathway led by LasR and RhlR. The mechanism of C2 inhibition of PAO1 biofilm formation is through repression of the Las and Rhl systems by QscR (Weng et al. 2014). QscR activation was also able to block the antibiotic tolerance of biofilms and, when combined with antibiotics, abolish biofilm formation completely. This indicates a possible QSCr-based treatment strategy for *P. aeruginosa* biofilms.

Parallel QS Cascades Initiate Dispersal in *Vibrio* Biofilms

Vibrio species are natural inhabitants of aquatic environments and form symbiotic or pathogenic relationships with eukaryotic hosts. Recent studies reveal that the ability of *Vibrio* to form biofilms depends on specific structural genes (flagella, pili, and exopolysaccharide biosynthesis) and regulatory processes (two-component regulators, quorum sensing, and c-di-GMP signaling (Ng and Bassler 2009b; Tischler and Camilli 2004)). While many *Vibrio* species are free living, a small group can form pathogenic or symbiotic interactions with eukaryotic hosts. These *Vibrios* change modes between growth within their hosts and prolonged survival in aquatic habitats (Yildiz and Visick 2009). Adaptation of *Vibrio* species to changes in the aquatic ecosystem and changes of their hosts is critical to their survival and colonization success. One key factor for environmental survival is the ability to form biofilms.

Vibrio cholerae

V. cholerae is a Gram-negative bacterium, which usually inhabits natural aquatic environments and is best known as the causative agent of the human disease cholera, and its QS mechanisms have been investigated extensively in recent years. This pathogen triggers the disease cholera in humans, which is characterized by acute enteric infection and severe diarrhea (Yildiz and Visick 2009), and it is a major cause of death in developing countries. Factors that are important for *V. cholerae* virulence are the cholera enterotoxin (CT) (enterotoxin is a protein toxin released by a microorganism in the intestine), the intestinal colonization factor known as the toxin coregulated pilus (TCP) and the regulatory protein ToxR, which regulates their expression. The expression of CT and TCP in vivo is affected by environmental signals such as optimum temperature, sunlight, and osmotic conditions (Faruque et al. 1998; Lee et al. 1999). *V. cholera* uses cell-to-cell communication to control pathogenicity and biofilm formation (Miller et al. 2002) (Zhu et al. 2002). Several structural components play a cardinal role in pathogenicity and in biofilm formation. For example, the type IV pilus, MSHA, which is responsible for mannose-sensitive hemagglutination by *V. cholerae* El Tor, has been implicated in biofilm formation on nonnutritive, abiotic surfaces (Watnick et al. 1999), as well as in host colonization. Also, the colony morphology of the wrinkled form of *V. cholerae* El Tor, which forms thicker biofilms than non-wrinkled El Tor, results from an exopolysaccharide (EPS) encoded by the *vps* locus (Yildiz and Schoolnik 1999). Importantly, mutants lacking vps clusters exhibited a defect in intestinal colonization. The specific quorum-sensing systems most critical for biofilm regulations are AI-2 and CAI-1.

Bassler and coworkers elucidated the structure of *V. cholerae's* autoinducer, (*S*)-3-hydroxytridecan-4-one (CAI-1), and demonstrated its control of virulence factor production (Higgins et al. 2007); another important study, by Kelly et al. showed that (*S*)-3-aminotridecan-4-one (amino-CAI-1) is the precursor of CAI-1 produced by the synthase CqsA (Kelly et al. 2009). Recently, Wei et al. proposed that 3-aminotridec-2-en-4-one (Ea-CAI-1) is the precursor of CAI-1 (Perez et al. 2012; Wei et al. 2011).

CAI-1 is produced by the enzyme CqsA and sensed by the receptor CqsS, and AI-2 is synthesized by the enzyme LuxS and its receptor is the LuxPQ complex (Higgins et al. 2007). Besides CAI-1, *V. cholerae* also uses the autoinducer AI-2 to control virulence and biofilm formation. The CqsA/CqsS system is found in several *Vibrio* species (Henke and Bassler 2004; Miller et al. 2002), which suggests that it functions as an intragenus signal and used for communication between *Vibrios* (Ng et al. 2011).

CAI-1 and AI-2 operate synergistically to control gene regulation in *V. cholerae*, although CAI-1 was shown to be the dominant signal by Higgins et al. (Higgins et al. 2007).

At low cell density, when AI concentrations are below the detection limit, CqsS acts as a kinase and phosphorylates the response regulator LuxO. As a result *V. cholerae* expresses virulence factors and forms biofilms (Hammer and Bassler 2003; Miller et al. 2002; Zhu and Mekalanos 2003).

This pattern of gene expression enables host colonization and contributes to persistence in the environment. At high cell density, when AI concentration is sufficient, CAI-1 binds CqsS, which converts from kinase to phosphatase, leading to dephosphorylation and inactivation of LuxO and thus suppression of both the expression of virulence factors and the formation of biofilms, through activation of the negative regulator HapR and repression of the positive regulator VpsT. Bassler and coworkers suggest that these events allow *V. cholerae* to exit from the host, reenter the environment in large numbers, and initiate a new cycle of infection (Higgins et al. 2007). We note that in Gram-negative bacteria, CAI-1 is probably the strongest known trigger of dispersal, the last stage in a biofilm life cycle (Oppenheimer-Shaanan et al. 2013), allowing planktonic cells to leave behind the differentiated colony and colonize a new environment.

Vibrio fischeri

The best described QS system is the Lux system of *Vibrio fischeri*, a bioluminescent Gram-negative bacterium. The *V. fischeri* QS mechanism consists of a synthase (LuxI) that produces the autoinducer signal, an acyl-homoserine lactone (AHL), 3-oxo-hexanoyl-homoserine lactone (3-oxo-C6-HSL) (Eberhard et al. 1981), and a transcriptional activator(LuxR) that recognizes the signal, leading to activation of genes in the lux operon (Engebrecht et al. 1983). Sensor kinases LuxQ (in association with the periplasmic protein LuxP), LuxN, and CqsS (not depicted), the histidine phosphotransferase LuxU and the response regulator LuxO (Ng and Bassler 2009a) are involved in transmitting the signal. Under low cell densities (low AI concentrations), the Sensor Kinases exhibit net kinase activity and serve as phosphoryl donors to LuxU, which serves as a phosphoryl donor to the response regulator LuxO. Biofilm formation plays a key role in host colonization by *V. fischeri* (Nyholm et al. 2000; Yip et al. 2006). *Vibrio fischeri* is known to promote biofilm formation through the symbiotic polysaccharide (*syp*) locus. The *syp* locus is a set of 18 genes thought to be involved in the regulation, production, and transport of a polysaccharide involved in biofilm formation (Shibata et al. 2012). It was recently demonstrated that *syp* is regulated by QS pathway and more specifically by LuxU. The loss of LuxQ and LuxU resulted in a delay in wrinkled colony formation. In the aforementioned study, LuxU played a more critical role than LuxO in controlling biofilm formation suggesting that LuxU may function independently of LuxO (Ray and Visick 2012).

Bacillus subtilis, Biofilm as a Multicellular Organism: Differentiation and Paracrine Signaling Orchestrated by QS Cascades

Bacterial communities and bacterial biofilms, a higher order community, thrive in their natural habitats as a result of their ability to respond accordingly to environmental changes. Some microorganisms are capable of differentiating into subpopulations of phenotypically distinct but genetically identical cells (Aguilar et al. 2007). These subpopulations of cells produce or respond to different signals and serve distinct functions within the community. A classic example is the soil bacterium *Bacillus subtilis* that responds to different environmental cues by differentiating into subpopulations of specialized cell types, which coexist within a biofilm (Lopez et al. 2009b). Each subpopulation must have the ability to sense one particular extracellular signal and integrate it with the rest, discarding irrelevant signals. For this purpose, *B. subtilis* possesses at least three different master regulators, Spo0A, DegU, and ComA, that coordinate the activation and regulation of the developmental programs that result in distinct cell fates within the biofilm (Vlamakis et al. 2013). The initiation of the production of the extracellular matrix, which is essential for biofilm formation, is carried out by a subpopulation of specialized cells in *B. subtilis*, activating a protein named SinI, derepressing SinR, the master regulator of the matrix genes (Chai et al. 2008; Kearns et al. 2005). All of the cells are encased within the matrix in a mature biofilm. Thus, the matrix serves as a "public good," benefiting the community as a whole (Vlamakis et al. 2008). Matrix-producing cells specialize to secrete the main components of the matrix through dedicated machinery: the extracellular polysaccharide (EPS) (Branda et al. 2004) and the structural matrix-associated proteins TasA (Branda et al. 2006), forming amyloid fibers (Romero et al. 2010), and BslA, forming a hydrophobic layer. The expression of both EPS and TasA is simultaneously triggered upon induction of *sinI*. sinI induction commences at low levels of Spo0A ~ P (Branda et al. 2001; Chai et al. 2008; Fujita et al. 2005). Low levels of spo0A ~ P in the cell are reached by the action of four membrane-bound and cytoplasmic sensor histidine kinases: KinA, KinB, KinC, and KinD (McLoon et al. 2011; Vlamakis et al. 2013). KinA and KinB synergistically sense the redox state of the biofilm cells. KinB is activated via a redox switch involving interaction of its second transmembrane segment with one or more cytochromes under conditions of reduced electron transport. In parallel KinA is activated by a decrease in the nicotinamide adenine dinucleotide (NAD^+)/NADH ratio via binding of NAD(+) to the kinase in a PAS domain A-dependent manner (Kolodkin-Gal et al. 2013). KinD is a canonical membrane kinase with two transmembrane segments connected by a 211-amino acid extracellular sensing domain that is presumably involved in signal recognition and binding to a specific extracellular signal. KinD was suggested to specifically respond to small secreted molecules produced by plants (Beauregard et al. 2013), as well as nonspecific signals, such as osmotic pressure (Rubinstein et al. 2012). The membrane kinase KinC harbors two transmembrane segments with no extracellular sensor domain. Instead, KinC has a PAS–PAC sensor domain in the cytoplasmic region of the kinase. PAS–PAC sensor domain of KinC somehow senses the leakage of cytoplasmic potassium cations (Lopez et al. 2009a). Diverse small molecules that are able to form pores in the membrane of the bacterium can induce this potassium leakage. This triggers the phosphorylation of Spo0A ~ P, which leads to matrix production. Because of the nature of the stimulus, the various small molecules identified that induce matrix production via KinC differ vastly in their molecular structure that share the ability to cause potassium leakage by making pores in the membrane of *B. subtilis* (Lopez et al. 2009a). The most important small molecule described to trigger matrix production via KinC is the self-generated lipopeptide, surfactin (Arima et al. 1968; Lopez et al. 2009b). Surfactin production is via a nonribosomal peptide synthetase machinery termed *SrfAA–AB–AC–AD* (henceforth, Srf). In this process, multidomain enzymes coordinately catalyze several of the reactions needed to synthesize surfactin (Kluge et al. 1988).

Once produced, surfactin causes the leakage of potassium with the formation of pores in the membrane (Sheppard et al. 1991), and that is sensed as a trigger to the subpopulation of matrix producers to differentiate. Surfactin thus serves as an autoinducer signal. Surfactin is recognized by its specific mode of action, generating pores in the membrane, promoting membrane leakage rather than by its chemistry, and offering

Fig. 1 QS signals and biofilm formation by *Pseudomonas aeruginosa*

an efficient strategy to allow sensing of quite a large repertoire of signals. This mechanism allows *B. subtilis* to respond not only to self-produced molecules but also to natural products secreted by other soil-dwelling organisms. This mechanism may also suggest that ECM secretion may have developed primarily as a defense strategy versus neighboring enemies.

The production of the quorum-sensing molecule surfactin is tightly regulated by another quorum-sensing pathway, mediated by the bacterial pheromone ComX, and the subsequent phosphorylation of the response regulator ComA (Roggiani and Dubnau 1993). ComA ~ P activates the expression of the operon responsible for surfactin production (Magnuson et al. 1994). Only after ComX is sensed and surfactin is produced can surfactin go on to trigger matrix production via activation of KinC, suggesting the need for several sequential cascades for proper biofilm development (Lopez et al. 2009d).

The activation of ComA is at some extent controlled by bimodal regulation because only a subpopulation of cells senses ComX and becomes surfactin producers (Lopez et al. 2009d). The subpopulation of surfactin producers is different from the matrix producers, responding to surfactin. Therefore, surfactin acts as a unidirectional signal in which one population produces the molecule and another population responds to it by producing an extracellular matrix. This mechanism adds sophistication to the concept of "quorum sensing," where all cells are physiologically similar and thus able to produce the signal and response (Bassler and Losick 2006; Miller and Bassler 2001). In the case of surfactin, the signaling can be referred to as paracrine signaling because there is a producing cell that is distinct from the nearby cell that can sense the signal.

The "paracrine signaling" system of *B. subtilis* can be compared with other autocrine quorum-sensing signaling systems described previously in bacteria. For example, as we discussed, *Pseudomonas aeruginosa* possesses two interrelated acyl-homoserine lactone quorum-sensing signaling systems. These systems, the LasR–LasI system and the RhlR–RhlI system, are global regulators of the expression of a large number of genes involved in diverse developmental processes. The analysis of quorum-induced genes suggests that the gene expression is sequential and time dependent (certain genes are activated early in growth, most genes are activated during the transition, and some genes are activated at the stationary phase) (Schuster et al. 2003b) (Fig. 1).

QS Signaling Promotes Cannibalism, Tightly Linked with Biofilm Formation

Bacillus subtilis responds to nutrient depletion by sporulating, a developmental process that results in the formation of two distinct cell types (McKenney et al. 2013). However, sporulation is a time- and energy-consuming process. *B. subtilis* delays the commitment to initiate sporulation under nutrient-limited conditions by forming a subpopulation of cells termed cannibals, obtaining nutrients by lysing their surrounding sensitive cells. Cannibal cells secrete two peptide toxins, Skf and Sdp, while at the same time expressing the immunity machinery to resist the action of these toxins. The toxins kill their sensitive siblings in a process termed cannibalism because the dead cells can be used as food to temporarily overcome the nutritional limitation and delay the onset of sporulation (Ellermeier et al. 2006; Engelberg-Kulka et al. 2006; Gonzalez-Pastor et al. 2003). The expression of the Skf and Sdp toxins is positively regulated by the transcriptional regulator Spo0A, when the cell has low levels of Spo0A ~ P (Fujita et al. 2005). Spo0A ~ P directly induces the expression of the operon responsible for Skf production (*skfA–H*) and indirectly, by repressing the repressor AbrB, induces the expression of the operon responsible for Sdp production (*sdpABC*). Because both cannibalism and matrix production are triggered by low levels of Spo0A ~ P, the expression of cannibalism and matrix production have been reported to occur in the same subpopulation of cells. This subpopulation specializes to produce the extracellular matrix required for biofilm formation at the same time that the cannibalism toxins and the immunity to the action of the toxins are expressed. Additionally, as described for matrix production, the differentiation of cannibal cells is also triggered by the quorum-sensing signal surfactin (Lopez et al. 2009c). Because only cells that have achieved high-enough levels of Spo0A ~ P express the cannibalism genes, any cells that do not have Spo0A ~ P will be sensitive to the toxins and lyse. The nutrients released by these lysed cells are used to promote the growth of the matrix producers/cannibals, because those are the only cells immune to the action of the cannibalism toxins. In this way, the representation of matrix producers within the community increases, allowing them to thrive at the expense of the rest of the cell types (sporulation is delayed and the other cell types are killed). Cannibalism in *B. subtilis* may also play a role in regulating the differentiation of matrix producers/cannibals. The subpopulation of cells that produce surfactin (the molecule responsible for the differentiation of matrix producers/cannibals) will also benefit from cannibalism. Because a fraction of surfactin producers will ultimately differentiate into competent cells, they might take up the DNA released when cells are killed by the cannibalism toxins (Lopez et al. 2009c, d).

QS Signaling Through AI-2 Controls Biofilm Formation

As mentioned previously, LuxS-/AI-2-dependent QSS has been proposed to act as a universal lexicon that mediates intra- and interspecific bacterial behavior. *B. subtilis luxS* produces active AI-2 able to mediate the interspecific activation of light production in *Vibrio harveyi*. It was demonstrated that in *B. subtilis*, *luxS* expression was negatively regulated by the master regulatory proteins of biofilm development, SinR and Spo0A. *B. subtilis* cells, from the undomesticated natto strain, required the LuxS-dependent QSS to form robust and differentiated biofilms and also to swarm on solid surfaces. Furthermore, LuxS activity was required for the formation of complex colonies. AI-2 production and spore morphogenesis were spatially regulated at different sites of the developing architectonically complex colony (Lombardia et al. 2006). Though the research of this QS cascade in *Bacillus* is still scarce, it is highly feasible that AI-2 behaves as a morphogen that coordinates the social behavior and biofilm development of *B. subtilis*.

Bacillus subtilis Biofilms as a Model for Multicellularity Regulated by QS Cascades

Differentiation of distinct cells types in *B. subtilis* is necessary for the proper development of the bacterial community. This differentiation is regulated, at least partially, by sensing several extracellular signals. Most of these signals are produced by *B. subtilis* itself. Secretion and sensing of these extracellular signals might regulate the timing of development in concordance with the surroundings.

Staphylococcus aureus: A Cross Talk Between the AIP Pheromone and the *Agr* Regulon

Biofilm formation plays a critical role in many device-related infections, infective endocarditis, urinary tract infections, and acute septic arthritis through pathogens such as *Staphylococcus aureus. S. aureus* forms complex and highly heterogeneous communities in the presence of glucose to enhance as a result of acidification of the media caused by increased excretion of metabolites. Other supplements such as serum and high salt were also reported to induce biofilm formation. Once a biofilm forms, strikingly, over 60 % of the total cells become phenotypic variants, making heterogeneity at the molecular level in the staphylococci, perhaps the highest of all Gram-positive bacteria.

Biofilms are known to be heterogeneous structures, with channels running throughout that facilitate the transport of nutrients and water. The cells are held together by extracellular matrix that includes the following:

(a) polysaccharides similar or identical to staphylococcal polysaccharide intercellular adhesion polysaccharides (PIA) (Mack et al. 1992). PIA-related polymers are produced by *Staphylococcus epidermidis* and *S. aureus* (PNAG) (Foreman et al. 2013);
(b) proteins, mostly surface adhesins, such as Bap and SasG (Roche et al. 2003); and
(c) extracellular DNA (eDNA) are the primary matrix components.

The reason for these assignments is straightforward; enzymes that degrade each of these materials, such as polysaccharide hydrolases, proteases, and DNases, can disassemble staphylococcal biofilms.

Staphylococci regulate biofilm formation and dispersal using the *agr* quorum-sensing system. The *agr* system responds to the extracellular concentration of an autoinducing peptide (AIP) signal (Boles and Horswill 2008), which is a cyclic thiolactone-containing peptide of varying amino acid composition depending on the strain (McDowell et al. 2001). Once the local AIP concentration reaches a critical level, usually in the low nanomolar range, AIP binds to the membrane-bound receptor domain of the AgrC histidine kinase, activating the AgrCA two-component system. This activation alters global gene expression and leads to increased detachment cells from a mature biofilm and returns them to a planktonic state, completing the biofilm life cycle (Boles and Horswill 2008, 2011). *Agr* activation can result in increased levels of staphylococcal proteases that cut cell surface proteins and disrupt cell–cell interactions within the biofilm, and proteases can also be added exogenously to cause biofilm dispersal. Matrix-degrading materials, such as dispersin B, can lead to biofilm disassembly by weakening the structural integrity of the biofilm matrix (Boles and Horswill 2011; Tegmark et al. 1998; Tsompanidou et al. 2011).

The *agr* system controls staphylococcal biofilm formation in several scenarios. *S. aureus* biofilm formation in some in vivo models was pronounced in *agr* mutants accumulated suggesting that the quorum-sensing mechanism was inhibitory toward biofilm formation. Importantly, pockets of *agr*-activated *S. aureus* cells within an established biofilm were observed to detach under in vitro flow conditions, while the *agr* inactive cells remained in the biofilm (Dai et al. 2012). Exogenous AIP addition was found to activate the *agr* system throughout a mature biofilm, leading to complete disassembly and conversion of biofilm-associated cells back to a planktonic phenotype. Thus, QS via the AIP system triggers a biofilm disassembly mechanism (Boles and Horswill 2008). Notably, across the *staphylococci*, activation of the *agr* system by QS is known to induce the expression of phenol-soluble modulins (PSMs). PSMs are surfactant-like molecules and have an important role in the structuring of *staphylococci* biofilms, a property achieved by their shared physicochemical properties. PSM expression can also result in biofilm dispersal (i.e., the detachment of cells or cellular clusters from biofilms), which is a key mechanism leading to the systemic dissemination of infections involving biofilms (Periasamy et al. 2012). Thus, the AIP-mediated QS system may have a dual role in the structuring of biofilms, as well in their dispersal.

Conclusion

The plasticity of transiting between unicellular and multicellular lifestyle renders bacterial cells similar to many other types of living cells, which are capable of unicellular existence, yet generally reside within multicellular communities. Biofilms offer their member cells several benefits: they protect their residents from environmental insults and assaults, improve their attachment to many different hosts, and allow efficient access to oxygen and nutrients (Chen et al. 2012; Costerton et al. 1987; Dietrich et al. 2013; Kolodkin-Gal et al. 2013). Importantly, the formation, maintenance, and disassembly of structured multicellular communities critically depend upon the chemical communication between cells. Those chemical autoinducers are various: They can be HSLs determining the maturation of *Pseudomonas* biofilms, or CAI-1 signal triggering *Vibrio* to disperse, ComX pheromone initiating complex development of *Bacillus* biofilms, or the AI-P autoinducers controlling the stability of *staphylococcal* communities. But, whatever the signal is, it seems that bacterial multicellular communities critically depend on a chemical cross talk between resident bacteria.

References

Aguilar C, Vlamakis H, Losick R, Kolter R (2007) Thinking about *Bacillus subtilis* as a multicellular organism. Curr Opin Microbiol 10:638–643

Allesen-Holm M, Barken KB, Yang L, Klausen M, Webb JS, Kjelleberg S, Molin S, Givskov M, Tolker-Nielsen T (2006) A characterization of DNA release in Pseudomonas aeruginosa cultures and biofilms. Mol Microbiol 59:1114–1128

Arima K, Kakinuma A, Tamura G (1968) Surfactin a crystalline peptidelipid surfactant produced by Bacillus Subtilis – isolation characterization and its inhibition of fibrin clot formation. Biochem Biophys Res Commun 31:488–494

Banin E, Vasil ML, Greenberg EP (2005) Iron and Pseudomonas aeruginosa biofilm formation. Proc Natl Acad Sci U S A 102:11076–11081

Baron SS, Rowe JJ (1981) Antibiotic action of pyocyanin. Antimicrob Agents Chemother 20:814–820

Bassler BL, Losick R (2006) Bacterially speaking. Cell 125:237–246

Beauregard PB, Chai Y, Vlamakis H, Losick R, Kolter R (2013) Bacillus subtilis biofilm induction by plant polysaccharides. Proc Natl Acad Sci U S A 110:E1621–1630

Bjarnsholt T, Jensen PO, Burmolle M, Hentzer M, Haagensen JAJ, Hougen HP, Calum H, Madsen KG, Moser C, Molin S et al (2005) Pseudomonas aeruginosa tolerance to tobramycin, hydrogen peroxide and polymorphonuclear leukocytes is quorum-sensing dependent. Microbiology 151:373–383

Boles BR, Horswill AR (2008) Agr-mediated dispersal of Staphylococcus aureus biofilms. PLoS Pathog 4:e1000052

Boles BR, Horswill AR (2011) Staphylococcal biofilm disassembly. Trends Microbiol 19:449–455

Branda SS, Gonzalez-Pastor JE, Ben-Yehuda S, Losick R, Kolter R (2001) Fruiting body formation by Bacillus subtilis. Proc Natl Acad Sci U S A 98:11621–11626

Branda SS, Gonzalez-Pastor JE, Dervyn E, Ehrlich SD, Losick R, Kolter R (2004) Genes involved in formation of structured multicellular communities by Bacillus subtilis. J Bacteriol 186:3970–3979

Branda SS, Chu F, Kearns DB, Losick R, Kolter R (2006) A major protein component of the Bacillus subtilis biofilm matrix. Mol Microbiol 59:1229–1238

Britigan BE, Roeder TL, Rasmussen GT, Shasby DM, McCormick ML, Cox CD (1992) Interaction of the Pseudomonas aeruginosa secretory products pyocyanin and pyochelin generates hydroxyl radical and causes synergistic damage to endothelial cells. Implications for Pseudomonas-associated tissue injury. J Clin Invest 90:2187–2196

Bryers JD (2008) Medical biofilms. Biotechnol Bioeng 100:1–18

Cha M, Hong S, Kang MY, Lee JW, Jang J (2012) Gas-phase removal of biofilms from various surfaces using carbon dioxide aerosols. Biofouling 28:681–686

Chai Y, Chu F, Kolter R, Losick R (2008) Bistability and biofilm formation in Bacillus subtilis. Mol Microbiol 67:254–263

Chen Y, Cao S, Chai Y, Clardy J, Kolter R, Guo JH, Losick R (2012) A Bacillus subtilis sensor kinase involved in triggering biofilm formation on the roots of tomato plants. Mol Microbiol 85:418–430

Chen Y, Yan F, Chai Y, Liu H, Kolter R, Losick R, Guo JH (2013) Biocontrol of tomato wilt disease by Bacillus subtilis isolates from natural environments depends on conserved genes mediating biofilm formation. Environ Microbiol 15:848–864

Collier DN, Anderson L, McKnight SL, Noah TL, Knowles M, Boucher R, Schwab U, Gilligan P, Pesci EC (2002) A bacterial cell to cell signal in the lungs of cystic fibrosis patients. Fems Microbiol Lett 215:41–46

Costerton JW, Cheng KJ, Geesey GG, Ladd TI, Nickel JC, Dasgupta M, Marrie TJ (1987) Bacterial biofilms in nature and disease. Annu Rev Microbiol 41:435–464

Costerton JW, Stewart PS, Greenberg EP (1999) Bacterial biofilms: a common cause of persistent infections. Science 284:1318–1322

Dai L, Yang L, Parsons C, Findlay VJ, Molin S, Qin ZQ (2012) Staphylococcus epidermidis recovered from indwelling catheters exhibit enhanced biofilm dispersal and "self-renewal" through downregulation of agr. BMC Microbiol 12:102

Davey ME, Caiazza NC, O'Toole GA (2003) Rhamnolipid surfactant production affects biofilm architecture in Pseudomonas aeruginosa PAO1. J Bacteriol 185:1027–1036

Davies DG, Parsek MR, Pearson JP, Iglewski BH, Costerton JW, Greenberg EP (1998) The involvement of cell-to-cell signals in the development of a bacterial biofilm. Science 280:295–298

De Kievit TR, Gillis R, Marx S, Brown C, Iglewski BH (2001) Quorum-sensing genes in Pseudomonas aeruginosa biofilms: their role and expression patterns. Appl Environ Microbiol 67:1865–1873

Deziel E, Gopalan S, Tampakaki AP, Lepine F, Padfield KE, Saucier M, Xiao G, Rahme LG (2005) The contribution of MvfR to Pseudomonas aeruginosa pathogenesis and quorum sensing circuitry regulation: multiple quorum sensing-regulated genes are modulated without affecting lasRI, rhlRI or the production of N-acyl-L-homoserine lactones. Mol Microbiol 55: 998–1014

Dietrich LE, Price-Whelan A, Petersen A, Whiteley M, Newman DK (2006) The phenazine pyocyanin is a terminal signalling factor in the quorum sensing network of Pseudomonas aeruginosa. Mol Microbiol 61:1308–1321

Dietrich LE, Teal TK, Price-Whelan A, Newman DK (2008) Redox-active antibiotics control gene expression and community behavior in divergent bacteria. Science 321:1203–1206

Dietrich LE, Okegbe C, Price-Whelan A, Sakhtah H, Hunter RC, Newman DK (2013) Bacterial community morphogenesis is intimately linked to the intracellular redox state. J Bacteriol 195:1371–1380

Diggle SP, Winzer K, Chhabra SR, Worrall KE, Camara M, Williams P (2003) The Pseudomonas aeruginosa quinolone signal molecule overcomes the cell density-dependency of the quorum sensing hierarchy, regulates rhl-dependent genes at the onset of stationary phase and can be produced in the absence of LasR. Mol Microbiol 50:29–43

Diggle SP, Stacey RE, Dodd C, Camara M, Williams P, Winzer K (2006) The galactophilic lectin, LecA, contributes to biofilm development in Pseudomonas aeruginosa. Environ Microbiol 8:1095–1104

Eberhard A, Burlingame AL, Eberhard C, Kenyon GL, Nealson KH, Oppenheimer NJ (1981) Structural identification of autoinducer of Photobacterium fischeri luciferase. Biochemistry 20:2444–2449

Ellermeier CD, Hobbs EC, Gonzalez-Pastor JE, Losick R (2006) A three-protein signaling pathway governing immunity to a bacterial cannibalism toxin. Cell 124:549–559

Engebrecht J, Nealson K, Silverman M (1983) Bacterial bioluminescence: isolation and genetic analysis of functions from Vibrio fischeri. Cell 32:773–781

Engelberg-Kulka H, Amitai S, Kolodkin-Gal I, Hazan R (2006) Bacterial programmed cell death and multicellular behavior in bacteria. PLoS Genet 2:e135

Faruque SM, Albert MJ, Mekalanos JJ (1998) Epidemiology, genetics, and ecology of toxigenic Vibrio cholerae. Microbiol Mol Biol Rev 62:1301

Foreman A, Jervis-Bardy J, Boase SJ, Tan L, Wormald PJ (2013) Noninvasive Staphylococcus aureus biofilm determination in chronic rhinosinusitis by detecting the exopolysaccharide matrix component poly-N-acetylglucosamine. Int Forum Allergy Rhinol 3:83–88

Fujita M, Gonzalez-Pastor JE, Losick R (2005) High- and low-threshold genes in the Spo0A regulon of Bacillus subtilis. J Bacteriol 187:1357–1368

Glasser NR, Kern SE, Newman DK (2014) Phenazine redox cycling enhances anaerobic survival in Pseudomonas aeruginosa by facilitating generation of ATP and a proton-motive force. Mol Microbiol 92:399–412

Gonzalez-Pastor JE, Hobbs EC, Losick R (2003) Cannibalism by sporulating bacteria. Science 301:510–513

Guina T, Purvine SO, Yi EC, Eng J, Goodlett DR, Aebersold R, Miller SI (2003) Quantitative proteomic analysis indicates increased synthesis of a quinolone by Pseudomonas aeruginosa isolates from cystic fibrosis airways. Proc Natl Acad Sci U S A 100:2771–2776

Ha C, Park SJ, Im SJ, Park SJ, Lee JH (2012) Interspecies signaling through QscR, a quorum receptor of Pseudomonas aeruginosa. Mol Cells 33:53–59

Hammer BK, Bassler BL (2003) Quorum sensing controls biofilm formation in Vibrio cholerae. Mol Microbiol 50:101–104

Hays EE, Wells IC, Katzman PA, Cain CK, Jacobs FA, Thayer SA, Doisy EA, Gaby WL, Roberts EC, Muir RD et al (1945) Antibiotic substances produced by Pseudomonas Aeruginosa. J Biol Chem 159:725–750

Henke JM, Bassler BL (2004) Three parallel quorum-sensing systems regulate gene expression in Vibrio harveyi. J Bacteriol 186:6902–6914

Hentzer M, Riedel K, Rasmussen TB, Heydorn A, Andersen JB, Parsek MR, Rice SA, Eberl L, Molin S, Hoiby N et al (2002) Inhibition of quorum sensing in Pseudomonas aeruginosa biofilm bacteria by a halogenated furanone compound. Microbiology 148:87–102

Hentzer M, Wu H, Andersen JB, Riedel K, Rasmussen TB, Bagge N, Kumar N, Schembri MA, Song Z, Kristoffersen P et al (2003a) Attenuation of Pseudomonas aeruginosa virulence by quorum sensing inhibitors. EMBO J 22:3803–3815

Hentzer M, Wu H, Andersen JB, Riedel K, Rasmussen TB, Bagge N, Kumar N, Schembri MA, Song Z, Kristoffersen P et al (2003b) Attenuation of *Pseudomonas aeruginosa* virulence by quorum sensing inhibitors. EMBO J 22:3803–3815

Higgins DA, Pomianek ME, Kraml CM, Taylor RK, Semmelhack MF, Bassler BL (2007) The major Vibrio cholerae autoinducer and its role in virulence factor production. Nature 450:883

Kearns DB, Chu F, Branda SS, Kolter R, Losick R (2005) A master regulator for biofilm formation by Bacillus subtilis. Mol Microbiol 55:739–749

Kelly RC, Bolitho ME, Higgins DA, Lu W, Ng WL, Jeffrey PD, Rabinowitz JD, Semmelhack MF, Hughson FM, Bassler BL (2009) The Vibrio cholerae quorum-sensing autoinducer CAI-1: analysis of the biosynthetic enzyme CqsA. Nat Chem Biol 5:891–895

Kempes CP, Okegbe C, Mears-Clarke Z, Follows MJ, Dietrich LE (2014) Morphological optimization for access to dual oxidants in biofilms. Proc Natl Acad Sci U S A 111:208–213

Kirisits MJ, Parsek MR (2006) Does Pseudomonas aeruginosa use intercellular signalling to build biofilm communities? Cell Microbiol 8:1841–1849

Kluge B, Vater J, Salnikow J, Eckart K (1988) Studies on the biosynthesis of surfactin, a lipopeptide antibiotic from Bacillus-subtilis Atcc-21332. FEBS Lett 231:107–110

Koch B, Liljefors T, Persson T, Nielsen J, Kjelleberg S, Givskov M (2005) The LuxR receptor: the sites of interaction with quorum-sensing signals and inhibitors. Microbiology 151:3589–3602

Kolodkin-Gal I, Elsholz AK, Muth C, Girguis PR, Kolter R, Losick R (2013) Respiration control of multicellularity in Bacillus subtilis by a complex of the cytochrome chain with a membrane-embedded histidine kinase. Genes Dev 27:887–899

Kolter R, Greenberg EP (2006) Microbial sciences: the superficial life of microbes. Nature 441:300–302

Latifi A, Foglino M, Tanaka K, Williams P, Lazdunski A (1996) A hierarchical quorum-sensing cascade in Pseudomonas aeruginosa links the transcriptional activators LasR and RhIR (VsmR) to expression of the stationary-phase sigma factor RpoS. Mol Microbiol 21:1137–1146

Lau GW, Hassett DJ, Ran H, Kong F (2004) The role of pyocyanin in Pseudomonas aeruginosa infection. Trends Mol Med 10:599–606

Ledgham F, Ventre I, Soscia C, Foglino M, Sturgis JN, Lazdunski A (2003) Interactions of the quorum sensing regulator QscR: interaction with itself and the other regulators of Pseudomonas aeruginosa LasR and RhlR. Mol Microbiol 48:199–210

Lee SH, Hava DL, Waldor MK, Camilli A (1999) Regulation and temporal expression patterns of Vibrio cholerae virulence genes during infection. Cell 99:625–634

Lepine F, Deziel E, Milot S, Rahme LG (2003) A stable isotope dilution assay for the quantification of the Pseudomonas quinolone signal in Pseudomonas aeruginosa cultures. Biochim Biophys Acta 1622:36–41

Lequette Y, Greenberg EP (2005) Timing and localization of rhamnolipid synthesis gene expression in Pseudomonas aeruginosa biofilms. J Bacteriol 187:37–44

Lequette Y, Lee JH, Ledgham F, Lazdunski A, Greenberg EP (2006) A distinct QscR regulon in the Pseudomonas aeruginosa quorum-sensing circuit. J Bacteriol 188:3365–3370

Lintz MJ, Oinuma K, Wysoczynski CL, Greenberg EP, Churchill ME (2011) Crystal structure of QscR, a Pseudomonas aeruginosa quorum sensing signal receptor. Proc Natl Acad Sci U S A 108:15763–15768

Lombardia E, Rovetto AJ, Arabolaza AL, Grau RR (2006) A LuxS-dependent cell-to-cell language regulates social behavior and development in Bacillus subtilis. J Bacteriol 188:4442–4452

Lopez D, Fischbach MA, Chu F, Losick R, Kolter R (2009a) Structurally diverse natural products that cause potassium leakage trigger multicellularity in Bacillus subtilis. Proc Natl Acad Sci U S A 106:280–285

Lopez D, Vlamakis H, Kolter R (2009b) Generation of multiple cell types in Bacillus subtilis. FEMS Microbiol Rev 33:152–163

Lopez D, Vlamakis H, Losick R, Kolter R (2009c) Cannibalism enhances biofilm development in Bacillus subtilis. Mol Microbiol 74:609–618

Lopez D, Vlamakis H, Losick R, Kolter R (2009d) Paracrine signaling in a bacterium. Genes Dev 23:1631–1638

Mack D, Siemssen N, Laufs R (1992) Parallel induction by glucose of adherence and a polysaccharide antigen specific for plastic-adherent Staphylococcus epidermidis: evidence for functional relation to intercellular adhesion. Infect Immun 60:2048–2057

Magnuson R, Solomon J, Grossman AD (1994) Biochemical and genetic-characterization of a competence pheromone from Bacillus-subtilis. Cell 77:207–216

Mavrodi DV, Blankenfeldt W, Thomashow LS (2006) Phenazine compounds in fluorescent Pseudomonas spp. biosynthesis and regulation. Annu Rev Phytopathol 44:417–445

Mazzola M, Cook RJ, Thomashow LS, Weller DM, Pierson LS 3rd (1992) Contribution of phenazine antibiotic biosynthesis to the ecological competence of fluorescent pseudomonads in soil habitats. Appl Environ Microbiol 58:2616–2624

McDowell P, Affas Z, Reynolds C, Holden MTG, Wood SJ, Saint S, Cockayne A, Hill PJ, Dodd CER, Bycroft BW et al (2001) Structure, activity and evolution of the group I thiolactone peptide quorum-sensing system of Staphylococcus aureus. Mol Microbiol 41:503–512

McKenney PT, Driks A, Eichenberger P (2013) The Bacillus subtilis endospore: assembly and functions of the multilayered coat. Nat Rev Microbiol 11:33–44

McLoon AL, Kolodkin-Gal I, Rubinstein SM, Kolter R, Losick R (2011) Spatial regulation of histidine kinases governing biofilm formation in Bacillus subtilis. J Bacteriol 193:679–685

Miller MB, Bassler BL (2001) Quorum sensing in bacteria. Annu Rev Microbiol 55:165–199

Miller MB, Skorupski K, Lenz DH, Taylor RK, Bassler BL (2002) Parallel quorum sensing systems converge to regulate virulence in Vibrio cholerae. Cell 110:303–314

Moker N, Dean CR, Tao J (1946–1955) Pseudomonas aeruginosa increases formation of multidrug-tolerant persister cells in response to quorum-sensing signaling molecules. J Bacteriol(192)

Nagorska K, Bikowski M, Obuchowski M (2007) Multicellular behaviour and production of a wide variety of toxic substances support usage of Bacillus subtilis as a powerful biocontrol agent. Acta Biochim Pol 54: 495–508

Ng WL, Bassler BL (2009a) Bacterial quorum-sensing network architectures. Annu Rev Genet 43:197–222

Ng WL, Bassler BL (2009b) Bacterial quorum-sensing network architectures. Annu Rev Genet 43:197–222

Ng WL, Perez LJ, Wei Y, Kraml C, Semmelhack MF, Bassler BL (2011) Signal production and detection specificity in Vibrio CqsA/CqsS quorum-sensing systems. Mol Microbiol 79:1407–1417

Nyholm SV, Stabb EV, Ruby EG, McFall-Ngai MJ (2000) Establishment of an animal-bacterial association: recruiting symbiotic vibrios from the environment. Proc Natl Acad Sci U S A 97:10231–10235

Ochsner UA, Koch AK, Fiechter A, Reiser J (1994) Isolation and characterization of a regulatory gene affecting rhamnolipid biosurfactant synthesis in Pseudomonas aeruginosa. J Bacteriol 176:2044–2054

Oinuma K, Greenberg EP (2011) Acyl-homoserine lactone binding to and stability of the orphan Pseudomonas aeruginosa quorum-sensing signal receptor QscR. J Bacteriol 193:421–428

Okegbe C, Price-Whelan A, Dietrich LE (2014) Redox-driven regulation of microbial community morphogenesis. Curr Opin Microbiol 18C:39–45

Oppenheimer-Shaanan Y, Steinberg N, Kolodkin-Gal I (2013) Small molecules are natural triggers for the disassembly of biofilms. Trends Microbiol 21(11): 594–601

Perez LJ, Ng WL, Marano P, Brook K, Bassler BL, Semmelhack MF, Perez LJ, Ng WL, Marano P, Brook K, Bassler BL, Semmelhack MF (2012) Role of the CAI-1 fatty acid tail in the Vibrio cholerae quorum sensing response. J Med Chem 55(22):9669–81

Periasamy S, Joo HS, Duong AC, Bach TH, Tan VY, Chatterjee SS, Cheung GY, Otto M (2012) How Staphylococcus aureus biofilms develop their characteristic structure. Proc Natl Acad Sci U S A 109:1281–1286

Pesci EC, Pearson JP, Seed PC, Iglewski BH (1997) Regulation of las and rhl quorum sensing in Pseudomonas aeruginosa. J Bacteriol 179:3127–3132

Pesci EC, Milbank JB, Pearson JP, McKnight S, Kende AS, Greenberg EP, Iglewski BH (1999a) Quinolone signaling in the cell-to-cell communication system of Pseudomonas aeruginosa. Proc Natl Acad Sci U S A 96:11229–11234

Pesci EC, Milbank JBJ, Pearson JP, McKnight S, Kende AS, Greenberg EP, Iglewski BH (1999b) Quinolone signaling in the cell-to-cell communication system of Pseudomonas aeruginosa. Proc Natl Acad Sci U S A 96:11229–11234

Price-Whelan A, Dietrich LE, Newman DK (2006) Rethinking 'secondary' metabolism: physiological roles for phenazine antibiotics. Nat Chem Biol 2:71–78

Purevdorj B, Costerton JW, Stoodley P (2002) Influence of hydrodynamics and cell signaling on the structure and behavior of Pseudomonas aeruginosa biofilms. Appl Environ Microbiol 68:4457–4464

Ray VA, Visick KL (2012) LuxU connects quorum sensing to biofilm formation in Vibrio fischeri. Mol Microbiol 86:954–970

Roche FM, Meehan M, Foster TJ (2003) The Staphylococcus aureus surface protein SasG and its homologues promote bacterial adherence to human desquamated nasal epithelial cells. Microbiology 149:2759–2767

Roggiani M, Dubnau D (1993) Coma, a phosphorylated response regulator protein of Bacillus-subtilis, binds to the promoter region of Srfa. J Bacteriol 175: 3182–3187

Romero D, Aguilar C, Losick R, Kolter R (2010) Amyloid fibers provide structural integrity to Bacillus subtilis biofilms. Proc Natl Acad Sci U S A 107:2230–2234

Rubinstein SM, Kolodkin-Gal I, McLoon A, Chai L, Kolter R, Losick R, Weitz DA (2012) Osmotic pressure can regulate matrix gene expression in Bacillus subtilis. Mol Microbiol 86:426–436

Sadikot RT, Blackwell TS, Christman JW, Prince AS (2005) Pathogen-host interactions in Pseudomonas aeruginosa pneumonia. Am J Respir Crit Care Med 171:1209–1223

Sanchez C (2011) Microbial ecology: bacteria reinforce plant defences. Nat Rev Microbiol 9:483

Schuster M, Lostroh CP, Ogi T, Greenberg EP (2003a) Identification, timing, and signal specificity of Pseudomonas aeruginosa quorum-controlled genes: a transcriptome analysis. J Bacteriol 185:2066–2079

Schuster M, Lostroh CP, Ogi T, Greenberg EP (2003b) Identification, timing, and signal specificity of Pseudomonas aeruginosa quorum-controlled genes: a transcriptome analysis. J Bacteriol 185:2066–2079

Sheppard JD, Jumarie C, Cooper DG, Laprade R (1991) Ionic channels induced by surfactin in planar lipid bilayer-membranes. Biochim Biophys Acta 1064:13–23

Shibata S, Yip ES, Quirke KP, Ondrey JM, Visick KL (2012) Roles of the structural symbiosis polysaccharide (syp) genes in host colonization, biofilm formation, and polysaccharide biosynthesis in Vibrio fischeri. J Bacteriol 194:6736–6747

Smith RS, Iglewski BH (2003) P. aeruginosa quorum-sensing systems and virulence. Curr Opin Microbiol 6:56–60

Stoodley P, Sauer K, Davies DG, Costerton JW (2002) Biofilms as complex differentiated communities. Annu Rev Microbiol 56:187–209

Tegmark K, Morfeldt E, Arvidson S (1998) Regulation of agr-dependent virulence genes in Staphylococcus aureus by RNAIII from coagulase-negative staphylococci. J Bacteriol 180:3181–3186

Tielker D, Hacker S, Loris R, Strathmann M, Wingender J, Wilhelm S, Rosenau F, Jaeger KE (2005) Pseudomonas aeruginosa lectin LecB is located in the outer membrane and is involved in biofilm formation. Microbiology 151:1313–1323

Tischler AD, Camilli A (2004) Cyclic diguanylate (c-di-GMP) regulates Vibrio cholerae biofilm formation. Mol Microbiol 53:857–869

Tsompanidou E, Sibbald MJ, Chlebowicz MA, Dreisbach A, Back JW, van Dijl JM, Buist G, Denham EL (2011) Requirement of the agr locus for colony spreading of Staphylococcus aureus. J Bacteriol 193:1267–1272

Vlamakis H, Aguilar C, Losick R, Kolter R (2008) Control of cell fate by the formation of an architecturally complex bacterial community. Genes Dev 22:945–953

Vlamakis H, Chai Y, Beauregard P, Losick R, Kolter R (2013) Sticking together: building a biofilm the Bacillus subtilis way. Nat Rev Microbiol 11: 157–168

Wagner-Dobler I, Thiel V, Eberl L, Allgaier M, Bodor A, Meyer S, Ebner S, Hennig A, Pukall R, Schulz S (2005) Discovery of complex mixtures of novel long-chain quorum sensing signals in free-living and host-associated marine alphaproteobacteria. Chembiochem 6:2195–2206

Wang Y, Kern SE, Newman DK (2010) Endogenous phenazine antibiotics promote anaerobic survival of Pseudomonas aeruginosa via extracellular electron transfer. J Bacteriol 192:365–369

Wang Y, Wilks JC, Danhorn T, Ramos I, Croal L, Newman DK (2011) Phenazine-1-carboxylic acid promotes bacterial biofilm development via ferrous iron acquisition. J Bacteriol 193:3606–3617

Watnick PI, Fullner KJ, Kolter R (1999) A role for the mannose-sensitive hemagglutinin in biofilm formation by Vibrio cholerae El Tor. J Bacteriol 181: 3606–3609

Wei Y, Perez LJ, Ng WL, Semmelhack MF, Bassler BL (2011) Mechanism of Vibrio cholerae autoinducer-1 biosynthesis. ACS Chem Biol 6:356–365

Wells IC (1952) Antibiotic substances produced by Pseudomonas aeruginosa; syntheses of Pyo Ib, Pyo Ic, and Pyo III. J Biol Chem 196:331–340

Wells IC, Elliott WH, Thayer SA, Doisy EA (1952) Ozonization of some antibiotic substances produced by Pseudomonas aeruginosa. J Biol Chem 196:321–330

Weng LX, Yang YX, Zhang YQ, Wang LH (2014) A new synthetic ligand that activates QscR and blocks antibiotic-tolerant biofilm formation in Pseudomonas aeruginosa. Appl Microbiol Biotechnol 98: 2565–2572

Yang L, Barken KB, Skindersoe ME, Christensen AB, Givskov M, Tolker-Nielsen T (2007) Effects of iron on DNA release and biofilm development by Pseudomonas aeruginosa. Microbiology 153:1318–1328

Yang L, Nilsson M, Gjermansen M, Givskov M, Tolker-Nielsen T (2009) Pyoverdine and PQS mediated subpopulation interactions involved in Pseudomonas aeruginosa biofilm formation. Mol Microbiol 74:1380–1392

Yildiz FH, Schoolnik GK (1999) Vibrio cholerae O1 El Tor: identification of a gene cluster required for the rugose colony type, exopolysaccharide production, chlorine resistance, and biofilm formation. Proc Natl Acad Sci U S A 96:4028–4033

Yildiz FH, Visick KL (2009) Vibrio biofilms: so much the same yet so different. Trends Microbiol 17:109–118

Yip ES, Geszvain K, DeLoney-Marino CR, Visick KL (2006) The symbiosis regulator rscS controls the syp gene locus, biofilm formation and symbiotic aggregation by Vibrio fischeri. Mol Microbiol 62:1586–1600

Zhu J, Mekalanos JJ (2003) Quorum sensing-dependent biofilms enhance colonization in Vibrio cholerae. Dev Cell 5:647–656

Zhu J, Miller MB, Vance RE, Dziejman M, Bassler BL, Mekalanos JJ (2002) Quorum-sensing regulators control virulence gene expression in Vibrio cholerae. Proc Natl Acad Sci U S A 99:3129–3134

Quorum Sensing in Pathogenesis and Virulence

Pragasam Viswanathan, Suneeva S.C., and Prasanth Rathinam

Introduction

The bacterial species have co-evolved with the human race and account for more than 3,000 distinct species of which most are beneficial to the humans. They are a few micrometres in length and outnumber the eukaryotic cells by a ratio of 10,000:1. They have evolved their own strategies for countering the defences of the human host. Pathogenicity of microorganisms is through the expression of the virulence factors, which could be attributed to their genetic or biochemical or structural features, which are responsible for the organism to cause an infection in the host (Oelschlaeger et al. 2002). The dynamic relationship between the host and pathogen is an outcome of the virulence of the pathogen and the resistance or susceptibility of the host to the invading pathogen (Casadevall and Pirofski 2000).

Colonization of the Host Surfaces

Penetration of microbes through the skin unaided by surgery, catheters or other events that breach the integrity of skin proves to be an impossible task with the exception of the pathogen that causes Lyme disease, *Borrelia burgdorferi*, which enters the skin through the wound inflicted by the tick bite (Kung et al. 2013). Mucous lines the body of the host in most parts and are referred to as skin-associated lymphoid tissue (SALT), gastrointestinal-associated lymphoid tissue (GALT) and mucosa-associated lymphoid tissue (MALT). Mucin and a meshwork of proteins that protect the mucosal cells act as a lubricant (intestinal and vaginal regions) and serve to trap pathogens from gaining access to the mucous cells present in between the SALT, GALT or MALT in the host cells (Godaly et al. 2000). Bacterial pathogens that lack surface proteins or carbohydrates bind to the mucin components and penetrate this layer without being trapped. In addition, the fact that mucin is highly viscous in nature and replaced every few days acts as a continuous defence mechanism of the body, making the propagation of pathogens in this layer difficult (Lievin-Le Moal and Servin 2006).

Communication Amongst Bacteria

The adherent bacteria rapidly multiply to produce intracellular bacterial communities within the host cells and communication or quorum sensing (QS) within the bacteria in this communal living established through the production, detection and response to extracellular signalling molecules called autoinducers (AI) (Platt and Fuqua 2010). These AI bring about genetic reprogramming in the pathogenic bacteria in a cell density-dependent manner (Ng and Bassler 2009).

P. Viswanathan (✉) • S. S.C. • P. Rathinam
Renal Research Laboratory, Centre for Bio Medical Research, School of Bio Sciences and Technology, VIT University, Vellore 632 014, Tamil Nadu, India
e-mail: pragasam.v@vit.ac.in

V.C. Kalia (ed.), *Quorum Sensing vs Quorum Quenching: A Battle with No End in Sight*,
DOI 10.1007/978-81-322-1982-8_4, © Springer India 2015

Once the threshold concentrations are attained, the release of AI works on a feedback mechanism, and the activity of the target sensor kinase is modulated to induct the repressors of the transcriptional regulators (Carnes et al 2010). QS regulates alteration of gene expression amongst groups of bacteria acting in synchrony and between the individual cells their bioluminescence, sporulation, competence, virulence, symbiosis, biofilm production, motility, production and resistance to antibiotics and transfer of genetic material (Fuqua et al 2001) and environmental factors such as diffusion and confinement (Platt and Fuqua 2010).

QS, amongst the groups of bacteria, works on three basic principles although they use different types of systems; firstly all bacteria produce AI, which diffuse away at low cell density and remain undetected but, at high local concentration, enable detection and response. Secondly, the receptors for these AI are cytoplasmic or membrane bound in nature, and finally, their detection is necessary to bring about their coordinated gene expression and repression (Novick et al. 1995; Seed et al. 1995).

In gram-negative bacteria, QS is by the amphiphilic N-acyl homoserine lactones (AHLs) (Pearson et al. 1999) that are synthesized by Luxl-type AL synthases from acyl-acyl carrier proteins and S-adenosylmethionine and differ from each other by the length of their acyl side chains that contain C3 substitutions. The AHLs move across the bacterial membranes through efflux pumps or by passive diffusion (Pearson et al. 1999) to bind to their LuxR transcriptional regulators and modulate their activity (Williams 2007).

The autoinducing peptides (AIPs) are responsible for QS in gram-positive bacteria. When the concentration of AIPs is high, they bind to their membrane-bound two-component histidine kinase (HK) receptors; HK receptor autophosphorylation activates the cytoplasmic regulator, which in turn leads to the transcription of the genes associated with QS. Upon release from the HK receptor, the AIPs are recycled back in to the cell cytoplasm where they modulate the activity of the transcription factors (Rutherford and Bassler 2012).

Virulence Factors of the Pathogens

As we all know that bacterial cells have three distinct regions in their structure: the flagella and the pili (proteins attached to the cell surface); the capsule, cell wall and the plasma membrane that constitute the envelope of the cell; and the cell genome, ribosomes and the inclusions which form the cytoplasmic content of the bacterial cell. Virulence factors (Fig. 1, Table 1) are determined by the exact molecular composition of the plasma membrane, cell wall and LPS and the role of flagella, fimbriae and capsule. They serve as permeability membranes to allow the passage of nutrients and prevent the passage of harmful substances, adhesins to attach to specific surfaces or tissues, protection against phagocytosis and secretion of various enzymes on the cell surface and sensing proteins that respond to temperature, osmolarity, salinity, nutrients, etc. (Johnson and Russo 2002).

Surface Components

Flagella are filamentous structures that aid in the motility of the pathogen and for adherence to the host cell. Motility is through the rotator action of the motor apparatus powered by the proton motive force and ATP hydrolysis in the plasma membrane. This allows the cell to swim across the fluid environment in response to the chemotaxis (Antunez-Lamas et al. 2009). The *fimbriae* and *pili* are short hairlike structures that are stiffer and smaller than the flagella and found all over the surface of the pathogen. They are involved in the adherence, colonization and resistance of the phagocytic attack by the leukocytes. The cell wall of the pathogen protects the protoplasm from osmolysis and is composed of peptidoglycan. On the periphery of the cell wall is the endotoxin, lipopolysaccharide. The bacterial membranes consist of saturated or monounsaturated fatty acids, which are branched with repeating isoprenoid subunits that attach to the glycerol through the ether linkage, which serves as an adaptation to exist in host environment. The capsule (also known as glycocalyx or slimy

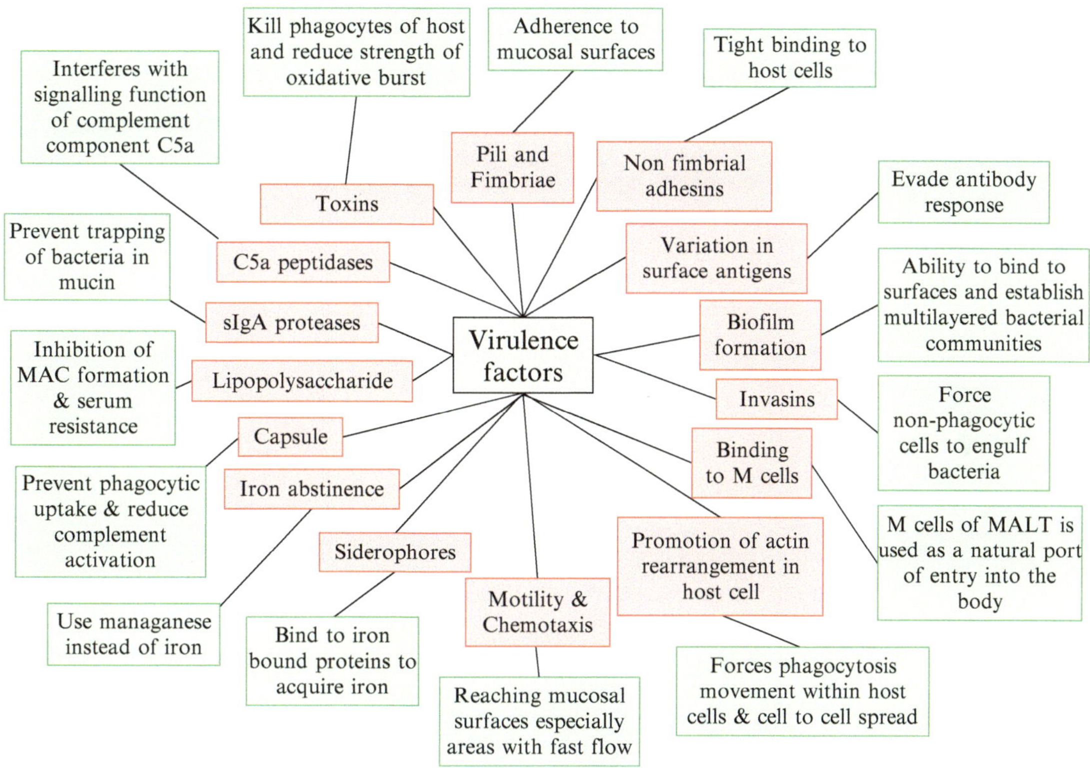

Fig. 1 Various virulence factors utilized by pathogens during infection

layer) mediates the attachment of the bacteria to surfaces and protects it from phagocytosis by the antimicrobial agents.

Cytoplasmic Constituents

The chromosome and the plasmids contribute to the genetic material of the pathogen. The protoplasm of the pathogen also contains the ribosomes, involved in the protein synthesis and the inclusions that serve as the energy reserves for the bacteria. Bacteria undergo genetic recombination, through conjugation, transduction and transformation, where their genetic material is exchanged, although sexual reproduction is not evident. Horizontal gene transfer across genera serves as the single responsible factor for the evolution of new species. Genes for drug resistance are incorporated into the extrachromosomal genetic material. There are several factors that contribute for the evolution and adaptability of the microbes such as their multiplication rate, number of cells within their area, mutation, species and gene transfer mechanisms in the host niche (Juhas et al. 2009).

Pathogens have developed mechanisms to evade peptides found at the base of the tongue to the mucosal layer in the urinary tract; these host peptides intercalate into bacterial membranes and create pores that allow essential molecules from the protoplasm of the bacterium to leach out. The presence of lipopolysaccharide layer in the gram-negative bacteria provides protection against the action of the antimicrobial peptide molecules. In addition, the presence of peptidases and granules in the cytoplasm of the bacteria that degrade the proteins provides a defence mechanism to the pathogen. It is interesting that certain bacteria like *Helicobacter pylori* have developed a slow yet effective mechanism to digest mucin and survive in this region before

Table 1 Common pathogens with their prominent virulence mechanisms that have been identified for the establishment of an infection in the human host

Organism	Virulence factors
Bacillus anthracis	1. Unique protein capsule (polymer of γ-D-glutamic acid: antiphagocytic)
	2. Nonmotile
	3. Virulence depends on acquiring 2 plasmids; one carries gene for protein
	4. Capsule, other carries gene for exotoxin
Bacteroides fragilis	1. Does not contain lipid A
Bordetella pertussis	1. Pertussis toxin
	2. Extracytoplasmic toxin
	3. FHA: binding to ciliated epithelial cells
	4. Tracheal cytotoxin: kills ciliated epithelial cells
Borrelia burgdorferi	1. Antigenic variation of outer membrane Vmp lipoproteins
Borrelia recurrentis	
Borrelia hermsii	
Campylobacter jejuni	1. Flagella (H antigen)
	2. Invasiveness
Chlamydia trachomatis	1. Resistant to lysozyme
	2. Prevents phagosome-lysosome fusion
	3. Nonmotile
	4. No pili
	5. No exotoxins
Clostridium perfringens	1. Nonmotile
Clostridium tetani	1. Flagella (H-Ag (+))
Clostridium botulinum	
Clostridium difficile	
Enterococci	1. Extracellular dextran helps bind to heart valve
Escherichia coli	1. Fimbriae (pili): colonization factor
	2. Siderophore
	3. Adhesins
	4. Capsule (K antigen)
	5. Flagella (H antigen)
Haemophilus ducreyi	1. No capsule
Haemophilus influenzae	1. Capsule
	2. Attachment pili (FHA)
	3. IgA_1 protease
Klebsiella pneumoniae	1. Capsule
	2. Nonmotile
Legionella pneumophila	1. Capsule
	2. Flagella and multiple fimbriae
	3. Haemolysin
Mycoplasma pneumoniae	1. Produces hydrogen peroxide
	2. Ab's against RBCs (cold agglutinins), brain, lung and liver cells produced during infection
Neisseria gonorrhoeae	1. Pili
	(a) Adherence
	(b) Antigenic variation
	(c) Antiphagocytic: binds bacteria tightly to host cell
	2. IgA_1 protease

(continued)

Table 1 (continued)

Organism	Virulence factors
	3. Outer membrane proteins:
	(a) Protein I: porin
	(b) Protein II
Neisseria meningitides	1. Capsule:
	2. IgA_1 protease
	3. Siderophores
	4. Pili (adherence)
Pseudomonas aeruginosa	1. Polar flagellum (H antigen)
	2. Haemolysin
	3. Collagenase
	4. Elastase
	5. Fibrinolysin
	6. Phospholipase C
	7. DNAase
	8. Antiphagocytic capsule
Salmonella species	1. Flagella (H antigen)
	2. Capsule (Vi antigen)
	3. Siderophore
Shigella dysenteriae	1. Nonmotile (no H antigen)
	2. Invades submucosa not lamina propria
Shigella dysenteriae	1. Nonmotile (no H antigen)
	2. Invades submucosa not lamina propria
Staphylococcus aureus	1. Protective
	(a) Microcapsule
	(b) Protein A: binds IgG
	(c) Coagulase
	(d) Haemolysins
	(e) Leukocidins
	(f) Penicillinase
	2. Tissue destroying
	(a) Hyaluronidase
	(b) Staphylokinase
	(c) Lipase
Staphylococcus epidermidis	1. Protective
Staphylococcus saprophyticus	2. Polysaccharide capsule
Streptococcus pneumoniae	1. Capsule
Streptococcus pyogenes	1. M protein (adherence factor, antiphagocytic, antigenic)
	2. Lipoteichoic acid (adherence factor)
	3. Steptokinase
	4. Hyaluronidase
	5. DNAase
	6. Anti-C5a peptidase
Treponema pallidum	1. Motility
Vibrio cholera	1. Flagellum (H antigen)
	2. Mucinase
	3. Fimbriae
	4. Noninvasive

causing an infection through the production of extracellular enzymes and toxins (Yoshiyama and Nakazawa 2000).

Adherence

Adherence of the pathogen through the pili, fimbriae and afimbrial adhesins to the host surface is the most important dogma in the pathogenesis of infections (Mulvey et al. 2000). Bacteria adhere and stay for long periods in the mouth, small intestine, bladder and mucosal surfaces and rapidly multiply, although these surfaces are constantly in a fluid environment, which helps in the removal of the non-adherent and weakly adhering bacteria from the host. Therefore, adhesion to the surface becomes an essential feature of virulence to cause an infection (Mulvey 2002).

Pili and fimbriae are proteinaceous, rod-shaped structures composed of repeated subunits of pilin (20 kDa) that link into a helical array to form a long cylindrical structure. The fimbriae are shorter, thinner structures, while pili are generally thicker in nature and seen as long outgrowths from the bacterial surface. Pili establish contact with the host cell surface without getting close to the electrostatic repulsion that prevents attachment, due to the negativity of both the surfaces. Also, it has been identified that the shorter fimbriae and the surface proteins are required for the irreversible attachment between the host cells and the bacteria (Tomme et al. 1995). The highly specific receptors for the pili on the host surface are the carbohydrate moieties of the glycoproteins or glycolipids. Bacteria propagating in the host constantly lose and reform pili due to their fragile nature. Pilus replacement also proves to be an effective evasion against the host immune response as the host produces antibodies against the bound pili, due to which these pili are lost and replaced (Giltner et al. 2012).

The adhesion of the bacteria to the host cell through the pili brings an altered expression in the virulence genes due to the physical changes in the conformation of the pilus tip proteins and the host surface receptors, which results in a conformational change in the shaft subunit proteins, which reflect along the bacterial surface. In gram-positive bacteria, such as *Streptococcus pyogenes*, a non-fibrillar adhesion through M protein-mediated attachment to fibronectin occurs. The afimbrial adhesins are proteins that mediate lighter binding of the bacteria to the host cell. Bacterial surface proteins allow attachment and invasion of host cells. Although only a few bacterial surface proteins have been identified with regard to their role during attachment, these proteins have been reported to recognize proteins on the host cell surface in contrast to the carbohydrate moieties (Kuehn 1997).

Another adhesion that paves way for infections in the host is the exopolysaccharide-mediated attachment between the bacteria in the host milieu that leads to the formation of biofilm or intracellular bacterial communities within the host cells. The metabolism within the biofilm shifts closely to a dormant state even though there is a constant supply of nutrients through channels that reach most of the microbial community that are deep seated. Biofilms in lungs (*Legionella pneumophila*), dental caries or periodontal disease (plague) or plastic tubing (catheters, stents) or implants (heart valves or hip joint replacements) have become primary source of concern due to the refractory potential of the biofilm to antibiotics and disinfectants, the protection it offers the pathogens from phagocytosis and the added risk of sepsis, make it difficult to eliminate a biofilm once established (Anderson et al. 2003).

sIgA Proteases

The mucin surrounding the host cells contains the sIgA molecules that simultaneously bind to the bacterial antigens via their antigen binding sites and interact with the mucin via the Fc portions. The pathogens avoid sIgA-mediated trapping through the production of sIgA proteases that cleave human IgA in the hinge region, which separates the IgA part that binds to the bacteria from the mucin (Pastorello et al. 2013).

Iron Acquisition Mechanisms

Bacterial pathogens being no respecters of human paradigms defy the statement that iron is essential for their growth. Some bacteria have evolved mechanisms to by-pass the problem of iron acquisition from the host by not using iron at all for their growth. The concentration of free iron in the host is generally low as it exists in chelated forms as lactoferrin, transferrin ferritin and haemoglobin. To survive in such conditions, bacteria possess iron acquisition mechanisms known as siderophores that chelate to iron with high affinity (Skaar 2010). These siderophores are low-molecular-weight compounds that exist in two forms, catechols and hydroxamates, that are recognized by the siderophore receptors on the bacterial surface in a bound state to the host iron. Once the complexes are internalized, into the bacterial cell, they are cleaved for iron utilization. The receptors for siderophores are not pathogen specific, and bacteria have been reported to utilize siderophores produced by other species for free iron loading.

Recently it has been identified that pathogenic strains have more than one iron-sequestering system, a mutant-deficient system where the pathogenic bacteria rely on iron abstinence mechanisms, where they require no iron at all. Few bacteria have been reported to contain iron-sulphur centres associated with their respiratory mechanism that are indispensable for the bacteria in their anaerobic lifestyles as in the case of *B. burgdorferi,* the causative of Lyme disease (Ouyang et al. 2009), where the bacteria survive on manganese, to replace iron as a cofactor in indispensable enzymes (Whittaker 2012).

Invasion and Intracellular Homing

Bacteria enter host epithelia that are not naturally phagocytic by adhering to the host cell and making alterations in the host cell cytoskeleton. This causes polymerization and depolymerization of the actin in the host cell to form pseudopods that result in the engulfing of the bacteria. This pseudopod formation by the bacteria is also enhanced by the bacterial surface proteins, invasins (Bohdanowicz et al. 2010). Bacteria also invade non-phagocytic cells by vesicle formation as mode of transiting to reach underlying host tissues. Bacteria that force ingestion by the host cell continue to interact with the host actin by moving along the pre-existing fibres that form the cytoskeleton of the host cell as in the case of *Listeria* sp. (Martinez et al. 2000).

Bacteria also induce their uptake by phagocytes and are present as either membrane-bound vesicles in the phagolysosomes where they are killed, or they prevent the fusion of phagosome and lysosome, or they disrupt the membrane vesicle they are bound in by crating pores or disrupting the membrane.

Evading Host Innate Immune Response

A loose, unstructured network composed of polysaccharides or polypeptides or protein-carbohydrate mixtures forms the capsule of the bacteria. The capsule serves to protect the bacterium from the host immune response and prevent the formation of C3 convertase by failing to bind to serum protein B due to high affinity towards serum protein H and due to their high content of sialic acid (Almagro-Moreno and Boyd 2009a, b). The failure to form C3 convertase offers protection to the encapsulated pathogens as only the capsule is digested and this prevents the pathogen from being engulfed by the phagocytes. Subversion of the host response also occurs through the presence of capsules that resemble host polysaccharides such as sialic acid (e.g. *Neisseria meningitidis*) and sialic acid (e.g. *Streptococcus pyogenes*) (Groisman 1994).

The pathogens also circumvent the host response by resistance to host nitric oxide, an important antibacterial mechanism of the host, through the flavohaemoglobin of the respiratory chain as in *E. coli*. Flavohaemoglobin utilizes NADPH, FAD and oxygen to convert nitric oxide to NO^{3-} (Bower et al. 2009).

Lipopolysaccharide serves not only as the target for complement system but also for the attachment of C3b and C5b and the nucleation site for the formation of MAB. This is inhibited by the binding of the sialic acid to LPS O antigen, which inhibits the formation of C3 convertase, and also the changes in the length of the LPS O antigen prevent MAC formation. Bacteria that are not killed by the MAC become serum resistant and cause systemic infections. The other strategies prevent migration of the phagocytes to where the bacteria are residing within the host cell, through the bacterial enzymes found in gram-positive cocci (Brodsky and Medzhitov 2009). Toxic proteins that kill phagocytes and inhibit their migration or reduce the strength of oxidative burst and production of LPS-like molecule that does not elicit the strong host response as in the case of *H. pylori* are the other mechanisms (Terao et al. 2008).

Prevention of Phagocytosis

Bacteria survive inside the leukocytes, macrophages and monocytes through fusion with the lysosome, thereby avoiding phagocytosis (Duclos and Desjardins 2000). These pathogens are the most dangerous as they are immune to all the defence strategies of the host, making their eradication difficult (Smith 1998). Acidification of the membrane-bound vesicle before phagolysosome formation and releasing of toxins into the phagolysosome, production of catalase and superoxide dismutase that detoxify the reactive oxygen species and production of cell surface polysaccharides that reduce the effect of oxidative burst and cell walls that are refractory to destruction by the lysosomal proteases and lysosomes are some of the other mechanisms of virulence that pathogens have evolved to subvert phagocytosis (Lu and Zhou 2012).

Toxins

All the methods of bacterial virulence discussed above are invasive methods through which bacteria are responsible in the pathogenesis of an infection. Toxins produced by the bacteria on the other hand are non-invasive and have a limited capacity to disseminate infection in the host. They act on the host at a distance from the site of infection and are target specific (Schmitt et al. 1999). Heat-stable enterotoxins are a group of toxins that are proteinaceous in nature and expressed by pathogenic *E. coli* and *V. cholera* (Galen et al. 1992). They accumulate in the intestine and elicit their response in the host.

Pore-forming toxins such as colicins, and diphtheria toxins, are the other group of toxins that form pores on the surface of the host cells and alter the membrane permeability, thereby leading to ion imbalance. They are soluble units, which form stable multimeric complexes on the target membranes and translocation across lipid membranes (Schmitt et al. 1999).

The other class of toxins are the superantigens that are secreted by *Staphylococcus aureus, Streptococcus* sp., *Mycoplasmas* and *Yersinia* sp. that bind to the major histocompatibility complex II and stimulate the peptide-independent MHC II/T cell receptor interaction and immune activation. They are responsible for toxic shock syndrome and food poisoning (Schmitt et al. 1999).

Quorum Sensing and Regulation of Virulence Factors

Clinically relevant pathogens utilize QS to regulate the collective production of VFs. In gram-positive bacteria, the AIPs are oligopeptide precursors and are diverse in sequence and structure (Thoendel et al. 2011). These AIP require transporters that not only process the pro-AIPs but also help in their translocation. The post-translationally modified AIPs are generally 5–17 amino acids long and are linear or cyclic in nature (Bouillaut et al. 2008). These AIPs are detected by the HK sensor kinases, which are autophosphorylated due to the transfer of the phosphoryl group from the histidine to a conserved aspartate, which thereby controls the QS target genes (Peterson et al. 2000). Expression of these genes is activated by the phosphorylated response regulator, resulting in an autoinducing feedforward loop that coordinates the QS response. This QS mechanism controls the virulence factor production in gram-positive pathogens such as *Streptococcus pneumonia* (competence), *Bacillus subtilis* (competence and sporulation), *S. aureus, Listeria monocytogenes, Enterococcus faecalis* and *Clostridium perfringens* (Kleerebezem et al. 1997; Autret et al. 2003; Podbielski and Kreikemeyer 2004; Ohtani et al. 2009; Riedel et al. 2009).

S. aureus is a commensal of the human skin and is responsible for minor skin infections that lead to pneumonia, bacteraemia and sepsis (Massey et al. 2006). The array of VFs such as adhesion molecules and toxins that affect the immune system are responsible for its pathogenesis, and two-component QS system encoded by *agr* locus regulates the expression of these genes. The P2 promoter drives the expression of the RNAII transcript, which encodes the four-component QS system (Novick et al. 1995), while the *agr*D encodes for the pro-AIP, which is processed into the final cyclic AIP (7–9 peptide long) by the transmembrane protein AgrB (Thoendel and Horswill 2010). When the AIP accumulates, it binds to the membrane-bound AgrC, which autophosphorylates and transfers the phosphate group to AgrA, which subsequently autoinduces the *agr* operon (Lina et al. 1998) and activates the P3 promoter, which encodes the RNAIII promoter (Novick et al. 1993). RNAIII activates the α-toxin production and simultaneously represses the expression of *rot,* fibronectin binding proteins A and B, protein A, coagulase and other surface proteins (Dunman et al. 2001). Repression of *rot* leads to the activation of genes that encode for the production of additional toxins, lipases, proteases, enterotoxins, superantigens and ureases (Geisinger et al. 2006). As a result of this, the surface virulence factors and biofilm formation are downregulated at high cell density of the pathogen, to facilitate its dispersal (Yarwood et al. 2001; Boles and Horswill 2008). This behaviour of *S. aureus* is analogous to the strategy of *V. cholera.*

In more than 100 gram-negative bacterial species such as *Vibrio fischeri, P. aeruginosa, Serratia marcescens, Brucella melitensis, Chromobacterium violaceum,* etc., the QS circuits control the VFs. The LuxI/LuxR-type QS systems are present. The LuxI homolog is an AI synthase that catalyses the reaction between SAM and the ACP to produce the diffusible AHL (Ng and Bassler 2009). At high concentrations, AHL binds to cytoplasmic LuxR transcriptions factors and activates luxI expression leading to the feedforward autoinduction loop (Zhu and Winans 1999, 2001). The AHLs produced by different bacteria vary in their side chains at the C3 position and are highly specific in their interactions (Chen et al. 2011).

In *P. aeruginosa,* a ubiquitous gram-negative pathogen that is responsible for both acute and chronic infections (Mena and Gerba 2009) in immunocompromised hosts, cystic fibrosis patients (Zemanick et al. 2011) and patients with severe burns, tracheal intubation and mechanical ventilation, where up to 10 % of its genome is under the influence of QS, that is collectively responsible for the production of VFs that contribute its pathogenicity. Three QS circuits (two LuxI/LuxR-type QS and one *Pseudomonas* quinolone signal (PQS)) are present in *P. aeruginosa.* The LuxI circuit is responsible for the transcription of the target genes that encode for the VFs elastase, proteases and exotoxin A through the production of the homolog 3-oxo-C12-homoserine lactone (Schuster et al. 2003, 2004). The second AI butanoyl homoserine lactone activates genes responsible for the production of pyocyanin and siderophores along with elastase and proteases (Schuster and Greenberg 2007). The other VFs that are encoded by the QS system are lectin and rhamnolipid production, swarming motility and toxins. The third non-LuxI/LuxR QS system in *P. aeruginosa* is used to control the VF gene expression (Deziel et al. 2004).

Virulence Mechanisms of Opportunists

Opportunists are members of the commensal flora of the humans that cause infections in people whose defence system is compromised by acquiring virulence genes through horizontal gene transfer mechanisms. The common opportunists are *Staphylococcus epidermidis, Enterococcus faecalis, Streptococci, Bacteroides fragilis* and *Burkholderia cepacia*. Some of the virulence mechanisms that are acquired in this class of pathogens are capsule formation, presence in high numbers at the site of infection in the host, ability to escape from the immune system of the host and the resistance to antibiotics (Brown et al. 2012).

Conclusion

The virulence factors present in pathogenic bacteria aid in the dissemination of infections within the host. A clear understanding of the underlying mechanisms of virulence helps identify ways in which the pathogens are responsible for the pathological changes they induce in the host and the signalling pathways that are up- or downregulated during QS to bring in this effect. With an alarming increase in resistance to antibiotics, knowledge on how virulence factors and QS interact with the host receptors and proteins helps in designing new diagnostic kits that can identify the presence of infections at a quicker pace and develop new drug targets that help combat this menace. In addition, an idea on the structural and genetic characterization of the virulence factors sheds light to design possible drug targets that can combat infections and act as inhibitors of infection through the QS mechanism in individuals genetically prone to infections. A complete analysis also provides insight into the evolutionary interplay involved in the generation of new species and the interactions of the pathogens with the commensal species in the niche they adapt to cause potential infections.

References

Almagro-Moreno S, Boyd EF (2009a) Insights into the evolution of sialic acid catabolism among bacteria. BMC Evol Biol 9:118

Almagro-Moreno S, Boyd EF (2009b) Sialic acid catabolism confers a competitive advantage to pathogenic in the *Vibrio cholerae* mouse intestine. Infect Immun 77:3807–3816

Anderson GG, Palermo JJ, Schilling JD, Roth R, Heuser J, Hultgren SJ (2003) Intracellular bacterial biofilm-like pods in urinary tract infections. Science 301(5629):105–107

Antunez-Lamas M et al (2009) Role of motility and chemotaxis in the pathogenesis of Dickeya dadantii 3937 (ex Erwinia chrysanthemi 3937). Microbiology 155:434–442

Autret N, Raynaud C, Dubail I, Berche P, Charbit A (2003) Identification of the agr locus of Listeria monocytogenes: role in bacterial virulence. Infect Immun 71:4463–4471

Bohdanowicz M, Cosio G, Backer JM, Grinstein S (2010) Class I and class III phosphoinositide 3-Kinases are required for actin polymerization that propels phagosomes. J Cell Biol 191(5):999–1012

Boles BR, Horswill AR (2008) Agr-mediated dispersal of *Staphylococcus aureus* biofilms. PLoS Pathog 4:e1000052

Bouillaut L, Perchat S, Arold S, Zorrilla S, Slamti L, Henry C, Gohar M, Declerck N, Lereclus D (2008) Molecular basis for group-specific activation of the virulence regulator PlcR by PapR heptapeptides. Nucleic Acids Res 36:3791–3801

Bower JM, Gordon-Raagas HB, Mulvey MA (2009) Conditioning of uropathogenic *Escherichia coli* for enhanced colonization of host. Infect Immun 77: 2104–2112

Brodsky IE, Medzhitov R (2009) Targeting of immune signalling networks by bacterial pathogens. Nat Cell Biol 11:521–526

Brown SP et al (2012) Evolution of virulence in opportunistic pathogens: generalism, plasticity, and control. Trends Microbiol 20:336–342

Carnes EC, Lopez DM, Donegan NP, Cheung A, Gresham H et al (2010) Confinement-induced quorum sensing of individual *Staphylococcus aureus* bacteria. Nat Chem Biol 6:41–45

Casadevall A, Pirofski LA (2000) Host-pathogen interactions: basic concepts of microbial commensalism, colonization, infection, and disease. Infect Immun 68:6511–6518

Chen G, Swem LR, Swem DL, Stauff DL, O'Loughlin CT, Jeffrey PD et al (2011) A strategy for antagonizing quorum sensing. Mol Cell 42:199–209

Deziel E, Lepine F, Milot S, He J, Mindrinos MN, Tompkins RG, Rahme LG (2004) Analysis of *Pseudomonas aeruginosa* 4-hydroxy-2-alkylquinolines (HAQs) reveals a role for 4-hydroxy-2-heptylquinoline in cell-to-cell communication. Proc Natl Acad Sci USA 101:1339–1344

Duclos S, Desjardins M (2000) Subversion of a young phagosome: the survival strategies of intracellular pathogens. Cell Microbiol 2:365–377

Dunman PM, Murphy E, Haney S, Palacios D, Tucker-Kellogg G, Wu S, Brown EL, Zagursky RJ, Shlaes D, Projan SJ (2001) Transcription profiling-based identification of *Staphylococcus aureus* genes regulated by the agr and/or sarA loci. J Bacteriol 183: 7341–7353

Fuqua C, Parsek MR, Greenberg EP (2001) Regulation of gene expression by cell-to-cell communication: acyl-homoserine lactone quorum sensing. Annu Rev Genet 35:439–468

Galen JE et al (1992) Role of *Vibrio cholerae* neuraminidase in the function of cholera toxin. Infect Immun 60(2):406–415

Geisinger E, Adhikari RP, Jin R, Ross HF, Novick RP (2006) Inhibition of rot translation by RNAIII, a key feature of agr function. Mol Microbiol 61:1038–1048

Giltner CL, Nguyen Y, Burrows LL (2012) Type IV pilin proteins: versatile molecular modules. Microbiol Mol Biol Rev 76(4):740–772

Godaly G, Bergsten G, Frendéus B et al (2000) Innate defences and resistance to gram-negative mucosal infection. Adv Exp Med Biol 485:9–24

Groisman EA (1994) How bacteria resist killing by host defence peptides. Trends Microbiol 2:444–449

Johnson JR, Russo TA (2002) Extra-intestinal pathogenic Escherichia coli: "the other bad *E. coli*". J Lab Clin Med 139:155–162

Juhas M, van der Meer JR, Gaillard M, Harding RM, Hood DW, Crook DW (2009) Genomic islands: tools of bacterial horizontal gene transfer and evolution. FEMS Microbiol Rev 33(2):376–393

Kleerebezem M, Quadri LE, Kuipers OP, de Vos WM (1997) Quorum sensing by peptide pheromones and two-component signal-transduction systems in Gram-positive bacteria. Mol Microbiol 24:895–904

Kuehn MJ (1997) Establishing communications *via* gram-negative bacterial pili. Trends Microbiol 5:130–135

Kung F, Anguita J, Pal U (2013) *Borrelia burgdorferi* and tick proteins supporting pathogen persistence in the vector. Future Microbiol 8:41–56

Lievin-Le Moal V, Servin AL (2006) The front line of enteric host defence against unwelcome intrusion of harmful micro-organisms: mucins, antimicrobial peptides, and microbiota. Clin Microbiol Rev 19: 315–337

Lina G, Jarraud S, Ji G, Greenland T, Pedraza A, Etienne J, Novick RP, Vandenesch F (1998) Transmembrane topology and histidine protein kinase activity of AgrC, the agr signal receptor in *Staphylococcus aureus*. Mol Microbiol 28:655–662

Lu N, Zhou Z (2012) Membrane trafficking and phagosome maturation during the clearance of apoptotic cells. Int Rev Cell Mol Biol 293:269–309

Martinez JJ, Mulvey MA, Schilling JD, Pinkner JS, Hultgren SJ (2000) Type 1 pilus mediated bacterial invasion of bladder epithelial cells. EMBO J 19(12):2803–2812

Massey RC, Horsburgh MJ, Lina G, Hook M, Recker M (2006) The evolution and maintenance of virulence in *Staphylococcus aureus*: a role for host-to-host transmission? Nat Rev Microbiol 4:953–958

Mena KD, Gerba CP (2009) Risk assessment of *Pseudomonas aeruginosa* in water. Rev Environ Contam Toxicol 201:71–115

Mulvey MA (2002) Adhesion and entry of uropathogenic *Escherichia coli*. Cell Microbiol 4(5):257–271

Mulvey MA, Schilling JD, Martinez JJ, Hultgren SJ (2000) Bad bugs and beleaguered bladders: interplay between uropathogenic *Escherichia coli* and innate host defences. Proc Natl Acad Sci U S A 97(16): 8829–8835

Ng WL, Bassler BL (2009) Bacterial quorum-sensing network architectures. Annu Rev Genet 43:197–222

Novick RP, Ross HF, Projan SJ, Kornblum J, Kreiswirth B, Moghazeh S (1993) Synthesis of staphylococcal virulence factors is controlled by a regulatory RNA molecule. EMBO J 12:3967–3975

Novick RP, Projan SJ, Kornblum J, Ross HF, Ji G, Kreiswirth B, Vandenesch F, Moghazeh S (1995) The agr P2 operon: an autocatalytic sensory transduction system in *Staphylococcus aureus*. Mol Gen Genet 248:446–458

Oelschlaeger TA, Dobrindt U, Hacker J (2002) Pathogenicity islands of uropathogenic *E. coli* and the evolution of virulence. Int J Antimicrob Agents 19(6):517–521

Ohtani K, Yuan Y, Hassan S, Wang R, Wang Y, Shimizu T (2009) Virulence gene regulation by the agr system in *Clostridium perfringens*. J Bacteriol 191:3919–3927

Ouyang Z, He M, Oman T, Yang XF, Norgard MV (2009) A manganese transporter, BB0219 (BmtA), is required for virulence by the Lyme disease spirochete, *Borrelia burgdorferi*. Proc Natl Acad Sci U S A 106: 3449–3454

Pastorello I, Paccani SR, Rosini R, Mettera R, Navarro F et al (2013) EsiB, a novel pathogenic *Escherichia coli* secretory immunoglobulin A-binding protein impairing neutrophil activation. MBio 4(4):e00206–e00213

Pearson JP, Van Delden C, Iglewski BH (1999) Active efflux and diffusion are involved in transport of *Pseudomonas aeruginosa* cell-to-cell signals. J Bacteriol 181:1203–1210

Peterson S, Cline RT, Tettelin H, Sharov V, Morrison DA (2000) Gene expression analysis of the *Streptococcus pneumoniae* competence regulons by use of DNA microarrays. J Bacteriol 182:6192–6202

Platt TG, Fuqua C (2010) What's in a name? The semantics of quorum sensing. Trends Microbiol 18: 383–387

Podbielski A, Kreikemeyer B (2004) Cell density-dependent regulation: basic principles and effects on the virulence of Gram-positive cocci. Int J Infect Dis 8:81–95

Riedel CU, Monk IR, Casey PG, Waidmann MS, Gahan CG, Hill C (2009) AgrD-dependent quorum sensing affects biofilm formation, invasion, virulence and global gene expression profiles in *Listeria monocytogenes*. Mol Microbiol 71:1177–1189

Rutherford ST, Bassler BL (2012) Bacterial quorum sensing: its role in virulence and possibilities for its control. Cold Spring Harb Perspect Med 2:a012427

Schmitt CK, Meysick KC, O'Brien AD (1999) Bacterial toxins: friends or foes? Emerg Infect Dis 5(2): 224–235

Schuster M, Greenberg EP (2007) Early activation of quorum sensing in *Pseudomonas aeruginosa* reveals the architecture of a complex regulon. BMC Genomics 8:287

Schuster M, Lostroh CP, Ogi T, Greenberg EP (2003) Identification, timing, and signal specificity of *Pseudomonas aeruginosa* quorum-controlled genes: a transcriptome analysis. J Bacteriol 185:2066–2079

Schuster M, Urbanowski ML, Greenberg EP (2004) Promoter specificity in *Pseudomonas aeruginosa* quorum sensing revealed by DNA binding of purified LasR. Proc Natl Acad Sci USA 101:15833–15839

Seed PC, Passador L, Iglewski BH (1995) Activation of the *Pseudomonas aeruginosa* lasI gene by LasR and the *Pseudomonas* autoinducer PAI: an autoinduction regulatory hierarchy. J Bacteriol 177:654–659

Skaar EP (2010) The battle for iron between bacterial pathogens and their vertebrate hosts. PLoS Pathog 6:e1000949

Smith H (1998) What happens to bacterial pathogens in vivo? Trends Microbiol 6:239–241

Terao Y, Mori Y, Yamaguchi M, Shimizu Y, Ooe K, Hamada S, Kawabata S (2008) Group A streptococcal cysteine protease degrades C3 (C3b) and contributes to evasion of innate immunity. J Biol Chem 283: 6253–6260

Thoendel M, Horswill AR (2010) Biosynthesis of peptide signals in gram-positive bacteria. Adv Appl Microbiol 71:91–112

Thoendel M, Kavanaugh JS, Flack CE, Horswill AR (2011) Peptide signaling in the staphylococci. Chem Rev 111:117–151

Tomme P, Warren RA, Miller RC et al (1995) Cellulose binding domains: classification and properties. In: Saddler JN, Penner MH (eds) Enzymatic degradation of insoluble carbohydrates. American Chemical Society, Washington, DC, pp 42–161

Whittaker JW (2012) Non-heme manganese catalase – the other catalase. Arch Biochem Biophys 525(2): 111–120

Williams P (2007) Quorum sensing, communication and cross-kingdom signalling in the bacterial world. Microbiology 153:3923–3938

Yarwood JM, McCormick JK, Schlievert PM (2001) Identification of a novel two-component regulatory system that acts in global regulation of virulence factors of *Staphylococcus aureus*. J Bacteriol 183: 1113–1123

Yoshiyama H, Nakazawa T (2000) Unique mechanism of *Helicobacter pylori* for colonizing the gastric mucus. Microbes Infect 2:55–60

Zemanick ET, Sagel SD, Harris JK (2011) The airway microbiome in cystic fibrosis and implications for treatment. Curr Opin Pediatr 23:319–324

Zhu J, Winans SC (1999) Autoinducer binding by the quorum- sensing regulator TraR increases affinity for target promoters in vitro and decreases TraR turnover rates in whole cells. Proc Natl Acad Sci USA 96: 4832–4837

Zhu J, Winans SC (2001) The quorum-sensing transcriptional regulator TraR requires its cognate signalling ligand for protein folding, protease resistance, and dimerization. Proc Natl Acad Sci USA 98: 1507–1512

Quorum Sensing in Nitrogen Fixation

Jie Gao, Anzhou Ma, Xuliang Zhuang, and Guoqiang Zhuang

Introduction

Soil microbial activity in the rhizosphere has the effects on soil biodiversity and quality and the health of plants, thus affecting the stability and productivity of above-ground plant parts. Amazingly complex interactions exist within the unseen underground environment, including root-root, root-insect, and root-microbe interactions, which can have both positive and negative outcome (Bais et al. 2006). Over the past decade, studies of the rhizosphere have revealed that when roots, microorganisms, and soil fauna physically contact one another, bioactive molecular exchanges often mediate these interactions as intercellular signal, which prepare the partners for successful interactions. These signal molecules are derived from multiple types of biosynthesis and provide cell-signaling networks to control the individual physiological process.

Quorum sensing (QS) is an important communication system used during symbiosis, defense, forming biofilm, and other interactions between plants and microorganisms, and it appears that this cross-talk system in microorganisms provides information in complex unseen networks. By studying QS between plants and microbes, scientists can better understand how communication occurs. The autoinducing signals, mediated by bacteria, cover the *N*-acyl homoserine lactones (AHLs) as well as other molecules, including quinolone, *p*-coumarate, and 3OH palmitic acid methyl ester (3OH PAME) in several gram-negative bacteria and oligopeptides in gram-positive bacteria (Waters and Bassler 2005). Moreover, furanosyl borate diester produced by *Vibrio harveyi* allows bacteria to communicate between species. One of the best-studied examples of quorum sensing is AHL-mediated cell-cell communication. The key stages involved in QS include secretion of signal molecular, accumulation of signals, recognition process, autoinduction synthesis process, transcriptional activation by the operon of QS, and QS signal withering. These processes, involving a number of signal transduction cascades, regulate enzymatic synthesis, transport protein synthesis, transduction by membrane, transcriptional activation, gene expression, and signal degradation.

While we have been able to describe the QS system in the individual microorganism for some time now, understanding the positive and negative effects on different microbial communities is more important. Signaling molecules involved in QS are strongly associated with stimulating biological activities and triggering a range of signal transduction cascades during root-microbe

Note: Part of this article was published in *International Journal of Molecular Sciences* (ISSN 1422–0067), doi:10.3390/ijms14058841.

J. Gao (✉) • A. Ma • X. Zhuang • G. Zhuang
Key Laboratory of Environmental Biotechnology, Research Center for Eco-Environmental Sciences, Chinese Academy of Sciences, Shuangqing Road 18, Beijing 100085, People's Republic of China
e-mail: giji0812@126.com

V.C. Kalia (ed.), *Quorum Sensing vs Quorum Quenching: A Battle with No End in Sight*,
DOI 10.1007/978-81-322-1982-8_5, © Springer India 2015

interaction processes (Zhuang et al. 2013). Plant-growth promoting and biocontrol bacteria, such as certain *Pseudomonas* biocontrol strains, are affected by QS systems. And the regulation of QS and the positive interaction effect can improve overall plant growth and health (Alavi et al. 2013). Biosynthesis of antibiotics and other antifungal compounds, such as phenazines, pyrrolnitrin, 2,4-diacetylphloroglucinol, hydrogen cyanide, and pyoluteorin, is related to the *phzIR* QS system of *Pseudomonas aureofaciens* in the wheat rhizosphere (Wood et al. 1997). Quorum sensing was also found to modulate expression of virulence genes in *Pseudomonas aeruginosa*, a plant pathogen that infects the roots of *Arabidopsis* and sweet basil. And it is possible that higher plants may also synthesize and secrete compounds that mimic the activity of bacterial AHL signaling compounds. The AHL signal-mimic activities detected in pea (*Pisum sativum*) exudates might play important roles in stimulating AHL-regulated behaviors in certain bacterial strains while inhibiting these behaviors in others (Teplitski et al. 2000). Also, the R-type proteins of some plant pathogens are stimulated by a compound present in rice, which does not make AHLs and does not possess a LuxI-family AHL synthase themselves (Ferluga and Venturi 2009). Furthermore, the response of plants to AHLs depends on various external factors, such as AHL type and concentration. Plants or parts of the plant will react differently to treatment with AHLs. These sophisticated information feedback systems show the importance of QS in cross-kingdom signaling in the rhizosphere.

Biological nitrogen fixation is mainly driven by soil bacteria, called rhizobia, that form symbiotic associations with roots. Nitrogen fixation by rhizobia is one of the best-studied examples of root-microbe interactions (Jackson et al. 2008). Nitrogen availability is important to nitrogen cycling and plant productivity within the ecosystem. The formation of nitrogen-fixing nodules in legumes involves complex molecular interactions and recognition (Oldroyd and Downie 2008). We will discuss the process of this important symbiosis controlled by QS in the followed section.

QS in Symbiotic Nitrogen-Fixation Process

The symbiotic relationship formed between the nitrogen-fixing rhizobia and their legume host is complex signaling communication process. Many bioactive molecules are involved in the regulatory network, such as root exudates (flavonoids), exopolysaccharides (EPSs), Nod factors, and AHLs. A review by Gonzalez and Marketon (Gonzalez and Marketon 2003) reported that quorum sensing is involved in the signal exchange process and perhaps plays a major role in preparing and coordinating the behavior of nitrogen-fixing rhizobia during the establishment of the symbiosis. During the course of rhizobial nodulation, the bacteria undergo chemotaxis toward the plant roots, leading to an increase in cell density, and the subsequent phenomena, including nodulation, symbiosome development, exopolysaccharide production, and nitrogen fixation, all involve the QS process (Wisniewski-Dyé and Downie 2002; Sanchez-Contreras et al. 2007) (Fig. 1).

QS System in Rhizobia

Sinorhizobium sp. and *Rhizobium* sp., able to fix atmospheric nitrogen, are two main types of soil bacteria that are capable of nodulation with alfalfa plants, including *Medicago*, *Melilotus* spp., *Vicia*, and *Pisum*. These strains possess multiple quorum-sensing systems that can synthesis of AHLs with different chains (Braeken et al. 2008). These AHL-based QS systems usually include a *luxI*-homolog AHL synthase gene and one gene of *luxR* family detecting the presence of AHLs and regulating transcription accordingly.

For example, in *Rhizobium leguminosarum* bv. Viciae, four different AHL-based QS systems (*tra*, *rai*, *rhi* and *cin*), synthesis of seven AHLs, have been identified, and *cinI*-*cinR*-determined quorum-sensing regulation is at the top of a hierarchy that induces the other QS regulatory systems (Frederix et al. 2011). Mutations in *cinR* and *cinI* abolish the production of *N*-(3-hydroxy-7-*cis*-tetradecenoyl)-l-homoserine

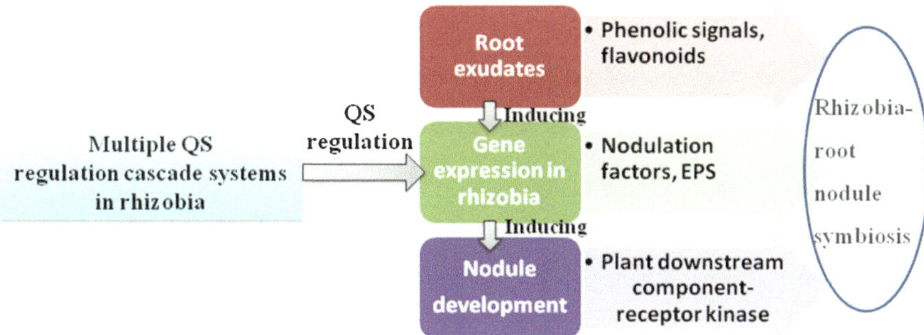

Fig. 1 A series of complex signal transmission of rhizobia-root nodule symbiosis. Plant roots secrete chemicals (flavonoids) that attract the correct plant symbiont. These exudates induce nodule bacteria to recognize the plant and produce Nod signaling molecules. The Nod factors elicit root hair curling and trigger the root cortical cell divisions, and lead to the development of plant nodules. And then plant receptor kinases are involved in the perception of Nod factors to generate suitable rhizobia-root nodule symbiosis. The invasion produced by rhizobia requires the action of chemotaxis, an increase in cell density, nodulation, symbiosome development, exopolysaccharide production, symbiotic plasmid transfer, and nitrogen fixation, all involve the QS process

lactone (3OH-$C_{14:1}$-HSL) and also reduce the production of several other AHLs produced by *raiI*, *traI*-like, or *rhiI*, whereas mutations in *cinI* have little effect on growth or nodulation of the host plant (Lithgow et al. 2000). The AHL synthetase CinI produces 3-OH-$C_{14:1}$-HSL, and this signal can induce a strong growth-inhibitory response of several strains of *R. leguminosarum* (Lithgow et al. 2000). A gene (*cinS*), coexpressed with *cinI*, acts downstream of a *luxR*-type regulator (*expR*) for *raiR* induction. It appears that the link between the *cin* QS system and *rai* QS system is regulated by CinS and CinS and ExpR act to increase the glycanase (PlyB) levels that influence the surface polysaccharide (Edwards et al. 2009). The *raiI* and *raiR* genes seem no observed effect on nodulation process by producing 3-OH-C_8-HSL. The other QS element TraR is responsible for transfer of the symbiotic plasmid pRL1JI. pRL1JI from *R. leguminosarum* is the first identified plasmids in the rhizobia including genes required for nodulation and nitrogen fixation. It has the *traI-trbBCDEJKLFGHI* plasmid transfer operon, and two regulatory BisR and TraR take charge of conjugation for transferring between different rhizobial strains. The *rhiR* and *rhiI* genes found in *R. leguminosarum* bv. Viciae induce the *rhiABC* genes by RhiI-made C_6, C_7, and C_8-HSLs. And the operon plays a role in nodulation efficiency.

Sinorhizobium meliloti is also a gram-negative soil bacterium that is capable of nodulation with leguminous plants. In *Sinorhizobium meliloti* strain Rm1021, the Sin QS system depends upon two genes, *sinR* and *expR*, which can respond to AHLs made by signal synthetase SinI. The AHL signals, from C_{12}- to C_{18}-HSL, can regulate motility, swarming, nodule formation, and other important processes in symbiosis with host plant. The QS system in some typical rhizobia has demonstrated in Table 1.

QS for Exopolysaccharide Production and Motility

AHL-mediated QS systems have been extensively characterized in different rhizobia and various aspects of the legume symbiosis. Exopolysaccharide (EPS) is indispensable for the invasion of leguminous plants (infection, attachment, and biofilm formation), which is regulated by QS.

Rhizobia strains are capable of synthesizing the distinct EPS succinoglycan and second exopolysaccharide (EPS II), which are both involved in symbiosis. The precise physiological function of EPS, as produced by nodule bacteria, has been investigated in an exopolysaccharide-deficient mutant. As a result of this mutation,

Table 1 Quorum-sensing system in some types of rhizobia

Rhizobial strain	QS element	AHL signal	QS regulation	References
Rhizobium leguminosarum bv. Viciae	*cinR/cinI/cinS*	3-OH-$C_{14:1}$-HSL	Growth inhibition; regulation of *plyB* encoding an extracellular glycanase	Edwards et al. (2009)
	rhiR/rhiI	C_6-HSL, C_7-HSL, C_8-HSL	Nodulation efficiency	Rodelas et al. (1999)
	traR/traI	3-oxo-C_8-HSL, C_8-HSL	Plasmid transfer	Danino et al. (2003)
	raiR/raiI	3-OH-C_8-HSL	Unknown	Wisniewski-Dye et al. (2002)
Rhizobium etli CNPAF512	*cinR/cinI*	3-OH-(slc)-HSL (slc, saturated long chain)	Nitrogen fixation, symbiosome development, growth inhibition, swarming	Daniels et al. (2002)
	raiR/raiI	C_8-HSL, 3OH-C_8-HSL	Restriction of nodule number, growth inhibition	Rosemeyer et al. (1998)
Sinorhizobium meliloti Rm1021	*sinR/sinI*	C_{12}-HSL, 3-oxo-C_{14}-HSL, $C_{16:1}$-HSL, 3-oxo-$C_{16:1}$-HSL, C_{18}-HSL	EPSII production, swarming	Gurich and González (2009)
	expR	$C_{16:1}$-HSL	EPSII production, swarming, motility	Marketon et al. (2003) and Hoang et al. (2004)

these bacteria failed to invade legumes and establish symbiosis due to a defect in initiating the production of infection threads (Jones et al. 2008). In *S. meliloti*, the ExpR/Sin quorum-sensing system is involved in the regulation of genes responsible for succinoglycan biosynthesis. At least five different AHLs (which are produced by SinI) can induce expression of several genes involved in the biosynthesis of EPS and EPS-II, which play important roles during symbiosis (Glenn et al. 2007). The regulatory protein MucR also plays a key role in the control of EPS and EPS II production by binding to promoter regions in both exopolysaccharide biosynthesis gene clusters (Downie 2010). Exopolysaccharides are secreted in a range of sizes, represented by two major fractions: high-molecular-weight and low-molecular-weight fractions. Moreover, the low-molecular-weight fraction represents symbiotically active forms of EPS, and it has been suggested that they may act as signaling molecules during this process (Pellock et al. 2000). During nitrogen-fixing symbiosis, lectins respond to polysaccharides produced by nodule bacteria; lectins are carbohydrate-binding proteins that promote the aggregation of rhizobia on the surface of legume root hairs, and they bind polysaccharides of the rhizobia with specific sugar-binding sites (Díaz et al. 1989). According to some reports, lectins are likely necessary within the nodule primordium to sustain cortical mitotic activity and increase the concentrations of Nod factors, which are necessary for the nodulation process (Mathesius and Watt 2011).

The ExpR/Sin quorum-sensing system in *S. meliloti* also regulates motility and chemotaxis. In order to make a successful symbiosis with its legume host, the microorganism needs to activate the chemotactic movement to the host. Many genes encode activators of the motility status, and the ExpR/Sin QS system controls the VisN/VisR regulators, which appear to regulate the expression of proteins used for flagella assembling and mobility and available energy for movement. One recent research has reported a series of genes expression was dependent on the level of AHLs in *S. meliloti* symbiosis with legume, and the QS-based expression activity was controlled in a type of temporal expression program (Charoenpanich et al. 2013). The result also showed that activation of the expression of genes responsible for exopolysaccharide production prior to mediating genes required for the repression of flagella production (Charoenpanich et al. 2013).

Nod Factors and Plant Receptor Kinases

Nod factors are lipochitooligosaccharide-based signaling molecules secreted by rhizobia that initiate nodule development in legumes. Several rhizobia genes, for example, the nodulation (*nod*) genes, are essential for successful interaction with the host, similarly to the genes involved in exopolysaccharide synthesis. Once the plant recognizes the nodulation factors, transcriptional and developmental changes occur in the root, such as cortical cell divisions, which allow bacterial invasion and nodule formation. Some scientists have tested whether quorum sensing is involved in Nod factor production in *Mesorhizobium tianshanense* by root hair deformation assays. The QS mutants have showed no difference to induce root hair deformation with the wild-strain (Zheng et al. 2006). While it had been shown that MucR could play a key role in the control of EPS production, the production of Nod factor could be increased by the transcriptional activator MucR (Mueller and González 2011). Quorum-sensing regulatory systems could modulate a series of MucR functions to coordinate some symbiosis behavior of *S. meliloti*, like the biosynthesis of EPS II. However, in *S. meliloti*, quorum sensing appeared not to be involved in MucR regulatory effects on Nod factor production.

Due to the specificity of nodulation in hosts and the low concentrations of Nod factors, receptors for the bacterial signals are necessary. Different types of plant receptor kinases are involved in the perception of Nod factors, such as the lysine motif (LysM)-type kinase gene *NFR5* from *Lotus japonicus*, which encodes a transmembrane serine/threonine receptor-like kinase and is required for the earliest detectable plant responses to Nod factor (Madsen et al. 2003), as well as the downstream component *SymRK*, which encodes a leucine-rich-repeat receptor kinase involved in nodulation symbiosis (Gherbi et al. 2008). Studies have shown that individual bacterial strains can enable nodulation in a range of hosts and can synthesize a mixture of several different Nod factor molecules (Mathesius and Watt 2011). Furthermore, plants also choose specific rhizobium species through recognition of Nod factors (Radutoiu et al. 2007). It has been hypothesized that receptor kinases in legumes co-evolved with the structure of Nod factors to generate suitable rhizobia-root nodule symbiosis by selective perception (Radutoiu et al. 2007).

Symbiotic Plasmid Transfer

In the soil rhizobia, a large proportion of beneficial genes required for establishment of nitrogen-fixing associations with their host plants are located on one or more extrachromosomal plasmids, known as symbiotic plasmids (pSyms). Conjugal transfer of rhizobial plasmids has been shown to play adaptive roles in environmental conditions and in the evolution of these soil bacteria (Nogales et al. 2013). In general, conjugation is mediated by potential genes for DNA transfer and replication: Mpf (mating pair formation) system and Dtr (DNA transfer and replication, containing transfer origins *oriT* sequence) components. At least two different regulatory mechanisms regulating the conjugal transfer of rhizobial plasmids have been identified: the quorum-sensing-regulated system and the RctA-regulated system. The QS-regulated conjugation systems show that LuxR-type regulators are involved in the transfer of the symbiotic plasmids in response to bacterial cell density. Ti plasmid and pAoF64/95 plasmid from *Agrobacterium tumefaciens* strains, pRL1JI from *R. leguminosarum* strains, and p42a from *Rhizobium etli* CFN42 are regulated by QS system through inducing expression of plasmid transfer operons. RctA-type plasmids, like pSymA from *S. meliloti*, contain the *rctA* gene to prevent the plasmid transfer. Recently, the third and fourth groups of conjugal transfer regulation systems were proposed by some plasmids (pRL10JI, pRL11JI, pRL12JI, pRleVF39d, pRleVF39e, pRleVF39f) of *R. leguminosarum* bv Viciae based on the phylogenetic analyses of the rhizobial relaxase sequences (Ding et al. 2013).

Signals from Host Plants

Interestingly, the activation of the bacterial nodulation process is induced by root exudates. The first step toward establishing a successful symbiosis is to attract the correct plant symbiont (Mathesius and Watt 2011). Flavonoids released from legume roots affect root nodulation by inducing chemo-attraction of rhizobia toward the root, enhancing the growth rate of bacterial cells and inducing transcription of rhizobial *nod* genes (Phillips and Tsai 1992). This suggests that molecular communication within the rhizosphere is complex and interactive flavonoids secreted by the legume root cause nodule bacteria to recognize the plant and produce Nod signaling molecules, which in turn trigger a number of processes within the root, including division of root cortical cells and nodule morphogenesis. Quorum sensing helps rhizobia to adjust themselves to the different type and concentration of phenolic signals and regulate their gene expression in response to changes in their population density for successful symbiosis with their host plants. However, some eukaryotes, including a variety of higher plants, can secrete bioactive molecules to interfere with bacterial QS system, so-called AHL mimics. These compounds in plant seed and root exudates are able to inhibit or activate some AHL report biosensors. A recent work indicated AHL-mimic molecules from seed and root extracts of bean and rice and bean did not have the typical lactone ring of the AHL molecules, and these signals enhanced biofilm formation of *Sinorhizobium fredii* SMH12, which suggested the AHL-mimic molecules played important roles in the process of interactions between legume roots and bacterium (Pérez-Montaño et al. 2013).

Other Nitrogen-Fixation and Nodulation Form

Researchers have also analyzed the effects of AHLs on the growth and nitrogen metabolism of the cyanobacterium *Anabaena* sp. PCC7120. However, the AHL signals tested strongly inhibited nitrogen fixation (Romero et al. 2011). The genomic analysis of sugarcane nitrogen-fixing endophyte *Gluconacetobacter diazotrophicus* Pal5 has demonstrated that three *luxI-luxR*-type quorum-sensing genes, present in genome, might play roles in nodulation and nitrogen fixation (Bertalan et al. 2009).

In addition, nitrogen fixation by actinomycete-nodulated plants is a major source of biological fixation of atmospheric nitrogen. The actinomycete genus *Frankia* contains nitrogen-fixing symbionts of many species of actinorhizal plants belonging to eight dicotyledonous families, which is in contrast to the rhizobium-legume symbiosis in which the host plants mainly belong to the leguminous plant family (Benson and Silvester 1993). Recently, scientists have sequenced the genomes of several *Frankia* strains and uncovered no evidence of the dissemination of nodulating ability by symbiotic genes (i.e., *nod* genes) in *Frankia* (Normand et al. 2007), which suggests that novel interaction mechanisms may be used during actinorhizal symbioses (Popovici et al. 2010).

QS in the Next Transformation of Nitrogen

Rhizobia plays a vital role in the nitrogen cycling and the chemolithoautotrophic ammonia-oxidizing bacteria (AOB) is also involved in affecting the process of nitrogen conversion. *Nitrosospira multiformis*, oxidizing ammonia to obtain energy for growth, is an agriculturally important soil bacteria that is commonly used as model organism for soil AOB. The ecological importance of *N. multiformis* and other ammonia-oxidizing bacteria is affecting the biological oxidation of inorganic nitrogen compounds in the environment. During the ammonia oxidation process in water or soil, biofilm formed by AOB can greatly affect the nitrification efficiency and the ecological behavior of nitrifying bacteria. In recent study, we have shown a LuxI/R-type quorum-sensing signal synthase and regulator in this AOB strain (Gao et al. 2014). With research of QS in *N. multiformis*, it has offered the possibility of exploring whether QS effects nitrification rates and other performance in the AOB strain and the interaction with rhizobial communities in soil.

Conclusion

Quorum sensing is a well-known system to control cell-to-cell communication, used by many species microorganisms. The QS typical circuit was first discovered in the gram-negative marine bacteria *Vibrio fischeri*; however, numerous nitrogen-fixing rhizobia have been described that QS controls a variety of processes in the development of symbiosis with legume plants. These plant-associated bacteria, harboring LuxI-LuxR-type QS system, are keen on the production of signal molecules to communicate, by acting as multicellular organisms. AHL is one type of symbiotic association-dependent signals which is commonly found in bacterial density-dependent behaviors. A hierarchical QS cascade controls motility and swarming, exopolysaccharide production, colonization, plasmid transfer, root nodulation efficiencies, and nitrogen fixation in these plant symbionts. These results indicate that bacterial density-dependent behaviors play multiple roles to show different regulatory capabilities. There remain lots of phenomenon seemed to linked with quorum sensing, and the increasing range of signal compounds and the responses of both bacterium and plants are benefiting for further study of the QS effects on soil nitrogen cycling (Lindemann et al. 2011). To integrate the complete contribution of quorum sensing in soil nitrogen cycling, in a sense, it is also important to identify what kind of QS function before or after the stage of symbiosis.

Opinion

The signal transduction mechanism is accurate and effective in the soil-plant-microorganism ecological system. A lot of work have been expended to elucidate the complex communication systems used by plants and microorganisms and to identify the various bioactive compounds generated by different species. In the nitrogen fixation process, a successful symbiotic interaction of the indigenous rhizobia and host plant relies upon the AHL-regulated action of chemotaxis, exopolysaccharide production, symbiotic plasmid transfer, and so on. Because many genes in bacterium are induced by QS, understanding the specific biological chemicals between microbes and plants is a critical first step to achieve sustainable production of crops, due to some limiting nutrients in soil. By taking advantage of the plants response of QS, it is possible for us to create transgenic plants able to secrete QS signals into soil for attracting rhizobia. However, various genes associated with virulence factor of pathogenic bacterium are regulated by QS system. According to the biocontrol point, it seems quorum quenching (QQ) is a better mean of keeping plant health. In fact, soil-plant-microorganism interaction network mediated by QS has proven to be far more sophisticated in a true soil environment. The bio-communities use "sophisticated algorithms" to generate signal maps and layout of the location where the responders can develop statistically based plans. Understanding how QS affects the rhizosphere interactions and the responses triggered by this system should be the priority before it is used.

References

Alavi P, Müller H, Cardinale M, Zachow C, Sánchez MB, Martínez JL, Berg G (2013) The DSF quorum sensing system controls the positive influence of *Stenotrophomonas maltophilia* on plants. PLoS One 8(7):e67103

Bais HP, Weir TL, Perry LG, Gilroy S, Vivanco JM (2006) The role of root exudates in rhizosphere interactions with plants and other organisms. Annu Rev Plant Biol 57:233–266

Benson DR, Silvester W (1993) Biology of *Frankia* strains, actinomycete symbionts of actinorhizal plants. Microbiol Rev 57(2):293–319

Bertalan M, Albano R, de Pádua V, Rouws L, Rojas C, Hemerly A, Teixeira K, Schwab S, Araujo J, Oliveira A (2009) Complete genome sequence of the sugarcane nitrogen-fixing endophyte *Gluconacetobacter diazotrophicus* Pal5. BMC Genomics 10:450

Braeken K, Daniels R, Ndayizeye M, Vanderleyden J, Michiels J (2008) Quorum sensing in bacteria-plant interactions. In: Nautiyal CS, Dion P (eds) Molecular mechanisms of plant and microbe coexistence, vol 15. Springer, Berlin/Heidelberg, pp 265–289

Charoenpanich P, Meyer S, Becker A, McIntosh M (2013) Temporal expression program of quorum sensing-based transcription regulation in *Sinorhizobium meliloti*. J Bacteriol 195(14):3224–3236

Daniels R, De Vos DE, Desair J, Raedschelders G, Luyten E, Rosemeyer V, Verreth C, Schoeters E, Vanderleyden J, Michiels J (2002) The *cin* quorum sensing locus of *Rhizobium etli* CNPAF512 affects growth and symbiotic nitrogen fixation. J Biol Chem 277(1):462–468

Danino VE, Wilkinson A, Edwards A, Downie JA (2003) Recipient-induced transfer of the symbiotic plasmid pRL1JI in *Rhizobium leguminosarum* bv. viciae is regulated by a quorum-sensing relay. Mol Microbiol 50(2):511–525

Díaz CL, Melchers LS, Hooykaas PJJ, Lugtenberg BJJ, Kijne JW (1989) Root lectin as a determinant of host-plant specificity in the *Rhizobium*-legume symbiosis. Nature 338:579–581

Ding H, Yip CB, Hynes MF (2013) Genetic characterization of a novel rhizobial plasmid conjugation system in *Rhizobium leguminosarum* bv. viciae Strain VF39SM. J Bacteriol 195(2):328–339

Downie JA (2010) The roles of extracellular proteins, polysaccharides and signals in the interactions of rhizobia with legume roots. FEMS Microbiol Rev 34(2):150–170

Edwards A, Frederix M, Wisniewski-Dyé F, Jones J, Zorreguieta A, Downie JA (2009) The *cin* and *rai* quorum-sensing regulatory systems in *Rhizobium leguminosarum* are coordinated by ExpR and CinS, a small regulatory protein coexpressed with CinI. J Bacteriol 191(9):3059–3067

Ferluga S, Venturi V (2009) OryR is a LuxR-family protein involved in interkingdom signaling between pathogenic *Xanthomonas oryzae* pv. oryzae and rice. J Bacteriol 191(3):890–897

Frederix M, Edwards A, McAnulla C, Downie JA (2011) Co-ordination of quorum-sensing regulation in *Rhizobium leguminosarum* by induction of an anti-repressor. Mol Microbiol 81(4):994–1007

Gao J, Ma A, Zhuang X, Zhuang G (2014) An *N*-acyl homoserine lactone synthase in the ammonia-oxidizing bacterium *Nitrosospira multiformis*. Appl Environ Microbiol 80(3):951–958

Gherbi H, Markmann K, Svistoonoff S, Estevan J, Autran D, Giczey G, Auguy F, Péret B, Laplaze L, Franche

C (2008) SymRK defines a common genetic basis for plant root endosymbioses with arbuscular mycorrhiza fungi, rhizobia, and *Frankia* bacteria. Proc Natl Acad Sci USA 105(12):4928

Glenn SA, Gurich N, Feeney MA, Gonzalez JE (2007) The ExpR/Sin quorum-sensing system controls succinoglycan production in *Sinorhizobium meliloti*. J Bacteriol 189(19):7077–7088

Gonzalez JE, Marketon MM (2003) Quorum sensing in nitrogen-fixing rhizobia. Microbiol Mol Biol Rev 67(4):574–592

Gurich N, González JE (2009) Role of quorum sensing in *Sinorhizobium meliloti*-alfalfa symbiosis. J Bacteriol 191(13):4372–4382

Hoang HH, Becker A, González JE (2004) The LuxR homolog ExpR, in combination with the Sin quorum sensing system, plays a central role in *Sinorhizobium meliloti* gene expression. J Bacteriol 186(16): 5460–5472

Jackson LE, Burger M, Cavagnaro TR (2008) Roots, nitrogen transformations, and ecosystem services. Annu Rev Plant Biol 59(1):341–363

Jones KM, Sharopova N, Lohar DP, Zhang JQ, VandenBosch KA, Walker GC (2008) Differential response of the plant *Medicago truncatula* to its symbiont *Sinorhizobium meliloti* or an exopolysaccharide-deficient mutant. Proc Natl Acad Sci USA 105(2):704

Lindemann A, Pessi G, Schaefer AL, Mattmann ME, Christensen QH, Kessler A, Hennecke H, Blackwell HE, Greenberg EP, Harwood CS (2011) Isovaleryl-homoserine lactone, an unusual branched-chain quorum-sensing signal from the soybean symbiont *Bradyrhizobium japonicum*. Proc Natl Acad Sci USA 108(40):16765–16770

Lithgow JK, Wilkinson A, Hardman A, Rodelas B, Wisniewski-Dyé F, Williams P, Downie JA (2000) The regulatory locus *cinRI* in *Rhizobium leguminosarum* controls a network of quorum-sensing loci. Mol Microbiol 37(1):81–97

Madsen EB, Madsen LH, Radutoiu S, Olbryt M, Rakwalska M, Szczyglowski K, Sato S, Kaneko T, Tabata S, Sandal N (2003) A receptor kinase gene of the LysM type is involved in legume perception of rhizobial signals. Nature 425(6958):637–640

Marketon MM, Glenn SA, Eberhard A, González JE (2003) Quorum sensing controls exopolysaccharide production in *Sinorhizobium meliloti*. J Bacteriol 185(1):325–331

Mathesius U, Watt M (2011) Rhizosphere signals for plant-microbe interactions: implications for field-grown plants. In: Lüttge UE, Beyschlag W, Büdel B, Francis D (eds) Progress in botany, vol 72. Springer, Berlin/Heidelberg, pp 125–161

Mueller K, González JE (2011) Complex regulation of symbiotic functions is coordinated by MucR and quorum sensing in *Sinorhizobium meliloti*. J Bacteriol 193(2):485–496

Nogales J, Blanca-Ordóñez H, Olivares J, Sanjuán J (2013) Conjugal transfer of the *Sinorhizobium meliloti* 1021 symbiotic plasmid is governed through the concerted action of one-and two-component signal transduction regulators. Environ Microbiol 15(3): 811–821

Normand P, Lapierre P, Tisa LS, Gogarten JP, Alloisio N, Bagnarol E, Bassi CA, Berry AM, Bickhart DM, Choisne N (2007) Genome characteristics of facultatively symbiotic *Frankia* sp. strains reflect host range and host plant biogeography. Genome Res 17(1): 7–15

Oldroyd GED, Downie JA (2008) Coordinating nodule morphogenesis with rhizobial infection in legumes. Annu Rev Plant Biol 59:519–546

Pellock BJ, Cheng HP, Walker GC (2000) Alfalfa root nodule invasion efficiency is dependent on *Sinorhizobium meliloti* polysaccharides. J Bacteriol 182(15):4310

Pérez-Montaño F, Jiménez-Guerrero I, Contreras Sánchez-Matamoros R, López-Baena FJ, Ollero FJ, Rodríguez-Carvajal MA, Bellogín RA, Espuny MR (2013) Rice and bean AHL-mimic quorum-sensing signals specifically interfere with the capacity to form biofilms by plant-associated bacteria. Res Microbiol 164(7):749–760

Phillips DA, Tsai SM (1992) Flavonoids as plant signals to rhizosphere microbes. Mycorrhiza 1(2):55–58

Popovici J, Comte G, Bagnarol É, Alloisio N, Fournier P, Bellvert F, Bertrand C, Fernandez MP (2010) Differential effects of rare specific flavonoids on compatible and incompatible strains in the *Myrica gale-Frankia* actinorhizal symbiosis. Appl Environ Microbiol 76(8):2451–2460

Radutoiu S, Madsen LH, Madsen EB, Jurkiewicz A, Fukai E, Quistgaard EMH, Albrektsen AS, James EK, Thirup S, Stougaard J (2007) LysM domains mediate lipochitin-oligosaccharide recognition and *Nfr* genes extend the symbiotic host range. EMBO J 26(17):3923–3935

Rodelas B, Lithgow JK, Wisniewski-Dye F, Hardman A, Wilkinson A, Economou A, Williams P, Downie JA (1999) Analysis of quorum-sensing-dependent control of rhizosphere-expressed (*rhi*) genes in *Rhizobium leguminosarum* bv. viciae. J Bacteriol 181(12): 3816–3823

Romero M, Muro-Pastor AM, Otero A (2011) Quorum sensing *N*-acylhomoserine lactone signals affect nitrogen fixation in the cyanobacterium *Anabaena* sp. PCC7120. FEMS Microbiol Lett 315(2): 101–108

Rosemeyer V, Michiels J, Verreth C, Vanderleyden J (1998) *luxI*-and *luxR*-homologous genes of *Rhizobium etli* CNPAF512 contribute to synthesis of autoinducer molecules and nodulation of *Phaseolus vulgaris*. J Bacteriol 180(4):815–821

Sanchez-Contreras M, Bauer WD, Gao M, Robinson JB, Downie JA (2007) Quorum-sensing regulation in rhizobia and its role in symbiotic interactions with legumes. Philos Trans R Soc Lond B Biol Sci 362(1483): 1149–1163

Teplitski M, Robinson JB, Bauer WD (2000) Plants secrete substances that mimic bacterial *N*-acyl homoserine lactone signal activities and affect population density-dependent behaviors in associated bacteria. Mol Plant Microbe Interact 13(6):637–648

Waters CM, Bassler BL (2005) Quorum sensing: cell-to-cell communication in bacteria. Annu Rev Cell Dev Biol 21:319–346

Wisniewski-Dyé F, Downie JA (2002) Quorum-sensing in Rhizobium. Antonie Van Leeuwenhoek 81(1–4): 397–407

Wisniewski-Dye F, Jones J, Chhabra S, Downie J (2002) *raiIR* genes are part of a quorum-sensing network controlled by *cinI* and *cinR* in *Rhizobium leguminosarum*. J Bacteriol 184(6):1597–1606

Wood DW, Gong F, Daykin MM, Williams P, Pierson L (1997) *N*-acyl-homoserine lactone-mediated regulation of phenazine gene expression by *Pseudomonas aureofaciens* 30-84 in the wheat rhizosphere. J Bacteriol 179(24):7663–7670

Zheng H, Zhong Z, Lai X, Chen W-X, Li S, Zhu J (2006) A LuxR/LuxI-type quorum-sensing system in a plant bacterium, *Mesorhizobium tianshanense*, controls symbiotic nodulation. J Bacteriol 188(5): 1943–1949

Zhuang X, Gao J, Ma A, Fu S, Zhuang G (2013) Bioactive molecules in soil ecosystems: masters of the underground. Int J Mol Sci 14(5):8841–8868

Quorum Sensing in Competence and Sporulation

Navneet Rai, Rewa Rai, and K.V. Venkatesh

Introduction

In several Gram-positive bacteria, competence and sporulation are few of several physiological processes controlled by quorum sensing (QS). Competence is a phenomenon wherein a bacterium acquires extracellular DNA for its maintenance. Cells committed for competence halt DNA replication and do not form stable RNA. Only a fraction of cells (10–20 %), in a population, develop competence, at a particular window of growth phase, and in response upregulate expression of genes involved in the uptake and processing of extracellular DNA (Maamar and Dubnau 2005). Sporulation, second QS-controlled phenotype, occurs under extreme stress and nutritional scarcity. Prolonged nutrient deprivation compels the cell to enter the process of sporulation, the outcome of which is the production of a metabolically dormant endospore that resumes growth once the conditions become favorable again. Endospores can sustain high temperature, pressure, and anhydrous conditions. Spore formation is a complex and tightly regulated phenomenon, where several hundred genes are directly and indirectly involved (Narula et al. 2012). Some examples of well-studied Gram-positive bacteria, where QS regulates competence and sporulation, are *Streptococcus species*, *Bacillus subtilis*, and sporulation in *B. subtilis* (Lazazzera et al. 1997).

Autoinducer peptides (AIPs) are the QS molecules, which regulate competence and sporulation. AIPs are posttranslationally modified oligopeptides, produced and recognized by a diverse range of Gram-positive bacteria. Some AIPs freely diffuse across the cell membrane, and some are actively transported. These signaling molecules are recognized by a membrane-bound histidine kinase receptor. Reception signaling initiates a phosphorylation cascade that alters the activity of a DNA binding transcriptional regulatory protein termed as the response regulator. Each Gram-positive bacterium responds to a specific type of signaling peptide by using a specific type of membrane-bound receptor (Waters and Bassler 2005).

N. Rai (✉)
Genome Center, University of California Davis, Davis, CA, USA
e-mail: nrai.iitb@gmail.com

R. Rai
Department of Chemistry, Indian Institute of Technology Delhi, Hauz Khas, New Delhi, India

K.V. Venkatesh
Department of Chemical Engineering, Indian Institute of Technology Bombay, Powai, Mumbai, India

Competence in *Streptococcus pneumoniae*

S. pneumoniae infection causes pneumonia and sepsis. During the last couple of decades, this bacterium has acquired resistance against several antibiotics. One of the many reasons responsible for the development of antibiotic resistance is the horizontal transfer of antibiotic resistance genes

V.C. Kalia (ed.), *Quorum Sensing vs Quorum Quenching: A Battle with No End in Sight*,
DOI 10.1007/978-81-322-1982-8_6, © Springer India 2015

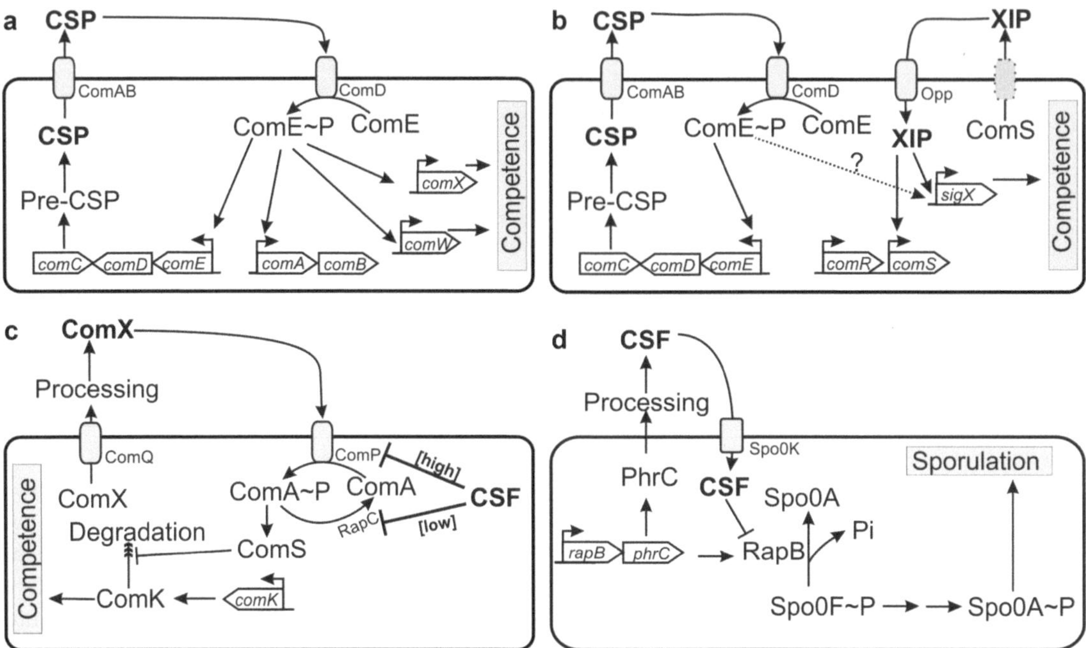

Fig. 1 Quorum-sensing-regulated competence and sporulation pathways. (**a**) CSP-mediated competence development in *S. pneumoniae*. (**b**) CSP- and XIP-mediated competence development in *S. mutans*. (**c**) ComX- and CSF-mediated competence development in *B. subtilis*. (**d**) CSF-mediated sporulation in *B. subtilis*

among bacteria. Bacteria susceptible to antibiotic get transformed into antibiotic resistance form during their competent phase by acquiring foreign antibiotic resistance gene(s). *S. pneumoniae* develops competence during mid-log phase of growth. QS signaling molecule, competence using competence and sporulating peptide (CSP), drives the competence development in *S. pneumoniae* (Fig. 1a) (Weng et al. 2013). CSP is the product of *comC* gene, which once synthesized, is modified, and transported out by the membrane-bound transporter, ComAB. Extracellular concentration of CSP increases as the cell density increases. Once cell density reaches a threshold value (10^7 cells/ml, corresponds to 1–10 ng/ml CSP), CSP binds to a membrane-bound histidine kinase, ComD. CSP bound ComD gets autophosphorylated and subsequently transfers the phosphate group to a response regulator protein, ComE. Phosphorylated ComE activates early and late competence genes, directly and indirectly (Johnsborg and Havarstein 2009). It has been observed that expression of early competence genes is maximum during 6–7 min and late genes during 9–10 min, after activation of this pathway (Peterson et al. 2000). ComE ~ P activates early competence genes, including *comAB* and *comCDE*. ComE is the product of *comE* of *comCDE* operon, and hence, by activating *comCDE*, ComE positively regulates its own expression, creating a sensitive autocatalytic loop. ComE ~ P activates late competence genes indirectly by upregulating transcriptions of *comX* and *comW*, which in turn upregulate late competence genes involved in the binding, uptake, and recombination of the extracellular DNA (Luo et al. 2004).

Competence in *Streptococcus mutans*

S. mutans forms biofilms on dental surfaces, preventing the removal of bacteria, consequently, leading to the decay of teeth. In *S. mutans*, QS regulates competence using two pathways, ComRS and ComCDE (Fig. 1b). ComRS pathway was discovered recently and is the major QS signaling system involved in the development of competence in *S. mutans*. In ComRS regulatory network, *comS* encodes an immature AIP, ComS, which is transported and processed by the unknown mechanism and converted in to mature hydrophobic AIP, XIP (*sigX inducing peptide*). XIP is internalized via Opp/Ami class of ABC transporter. Once above the threshold concentration inside the cell, XIP binds with an Rgg family transcription factor, ComR, and forms a complex. XIP-ComR complex activates *comS*, early, and late competence genes by upregulating the transcription of *sigX* (also called *comX*). Concentration of SigX is maximum during transition from late log phase to stationary phase, indicating that cells are more competent during this phase. Second competence regulatory machinery in *S. mutans*, ComCDE, is similar to the ComCDE competence development machinery of *S. pneumoniae*. But in *S. mutans*, it is not clear, how does ComE regulate the expression of late competence genes. It has been hypothesized that at high cell density, ComE upregulates the expression of *sigX*, directly or indirectly, which in turn, activates competence genes (Mashburn-Warren et al. 2010).

Competence and Sporulation in *Bacillus subtilis*

B. subtilis is a soil bacterium, uses peptide quorum-sensing signal molecules to choose one of two physiological stages, competence and sporulation, depending on environmental conditions and growth phase. *B. subtilis* develops competence during transition between logarithmic and stationary phases of growth. Two QS peptides, ComX and CSF (competence and sporulation factor), facilitate competence development in *B. subtilis*. ComX is derived from pre-ComX, a 55 amino acids long peptide. Mature ComX is 10 amino acids long peptide with an isoprenyl modification at a tryptophan residue (Okada et al. 2005). ComQ is required for the posttranslational modification of ComX and its secretion into the extracellular medium (Bacon Schneider et al. 2002). Extracellular ComX is detected by the membrane-bound histidine sensor kinase, ComP. Once above threshold concentration, ComX binds to ComP and stimulates its autophosphorylation, which later transfers phosphate to the DNA-binding response regulator ComA. ComA ~ P induces the expression of *comS*. ComS plays a significant role in maintaining the high concentration of the competence master regulator protein, ComK, by preventing its targeted degradation in the cell (Fig. 1c). High level of ComK produced in the cell, regulates the late competence operons, leading to the synthesis of proteins required for the DNA uptake, processing, and integration into the genome. Late competence operons include *comG, comE, comF*, and *comC*. The DNA uptake system of *B. subtilis* does not show any specificity for the foreign DNA sequences or the origin of the foreign DNA and is internalized with same specificity, as a single-stranded DNA, which is further recombined with the host genome. A nuclease located in the cell membrane, encoded by *nucA*, cleaves the DNA resulting in a fragment of about 20 kb. The complementary strand is degraded, with the release of nucleotides into the medium and the single-stranded DNA is taken up into the cell. After internalization of the single-stranded DNA, it is associated with bacterial recombination proteins, such as RecA and AddAB, which integrates the single-stranded DNA into the host genome (Hamoen et al. 2003).

Second QS molecule CSF is encoded by *phrC*. CSF is a pentapeptide derived from the C-terminal end of a 40 amino acid long secreted polypeptide, PhrC. CSF is released in the extracellular medium by the simple diffusion. With the increase in cell density the extracellular level of CSF increases. CSF is transported back into the cell by the Opp peptide transporter and counteracts the action of RapC protein, subsequently, leading to the synthesis of ComS by ComA ~ P (Fig. 1d). Presence of ComS stops targeted degradation of ComK, hence activates competence. CSF inhibits RapC phosphatase when present at a relatively lower concentration of 1–10 nM. But, at a relatively higher concentration of >20 nM, CSF binds to and inhibits the phosphorylation of ComP, hence, decreasing the level of active ComA ~ P in the cell, which subsequently leads to an inhibition of ComS synthesis, resulting in the inhibition of competence development. Also, at such higher concentration, CSF inhibits the activity of an aspartyl-phosphate phosphatase, RapB. RapB dephosphorylates sporulation response regulator protein, Spo0F ~ P. Inhibition of RapB results in the accumulation of high levels of Spo0F ~ P which further drives the synthesis of high levels of Spo0A ~ P in the cell. The presence of high level of Spo0A ~ P favors the establishment of sporulation in the cell. Hence, at higher concentrations, CSF compels the cell to progress toward the sporulation pathway by two paths, inhibition of continued synthesis of ComK that is switching off the synthesis of competence machinery and activation of components of the sporulation phosphorelay (Lopez and Kolter 2010).

References

Bacon Schneider K, Palmer TM, Grossman AD (2002) Characterization of comQ and comX, two genes required for production of ComX pheromone in *Bacillus subtilis*. J Bacteriol 184(2):410–419

Hamoen LW, Venema G, Kuipers OP (2003) Controlling competence in *Bacillus subtilis*: shared use of regulators. Microbiology 149(Pt 1):9–17

Johnsborg O, Havarstein LS (2009) Regulation of natural genetic transformation and acquisition of transforming DNA in *Streptococcus pneumoniae*. FEMS Microbiol Rev 33(3):627–642

Lazazzera BA, Solomon JM, Grossman AD (1997) An exported peptide functions intracellularly to contribute to cell density signaling in *B. subtilis*. Cell 89(6): 917–925

Lopez D, Kolter R (2010) Extracellular signals that define distinct and coexisting cell fates in *Bacillus subtilis*. FEMS Microbiol Rev 34(2):134–149

Luo P, Li H, Morrison DA (2004) Identification of ComW as a new component in the regulation of genetic transformation in *Streptococcus pneumoniae*. Mol Microbiol 54(1):172–183

Maamar H, Dubnau D (2005) Bistability in the *Bacillus subtilis* K-state (competence) system requires a positive feedback loop. Mol Microbiol 56(3): 615–624

Mashburn-Warren L, Morrison DA, Federle MJ (2010) A novel double-tryptophan peptide pheromone controls competence in *Streptococcus* spp. via an Rgg regulator. Mol Microbiol 78(3):589–606

Narula J, Devi SN, Fujita M, Igoshin OA (2012) Ultrasensitivity of the *Bacillus subtilis* sporulation decision. Proc Natl Acad Sci U S A 109(50): E3513–E3522

Okada M, Sato I, Cho SJ, Iwata H, Nishio T, Dubnau D, Sakagami Y (2005) Structure of the *Bacillus subtilis* quorum-sensing peptide pheromone ComX. Nat Chem Biol 1(1):23–24

Peterson S, Cline RT, Tettelin H, Sharov V, Morrison DA (2000) Gene expression analysis of the *Streptococcus pneumoniae* competence regulons by use of DNA microarrays. J Bacteriol 182(21):6192–6202

Waters CM, Bassler BL (2005) Quorum sensing: cell-to-cell communication in bacteria. Annu Rev Cell Dev Biol 21:319–346

Weng L, Piotrowski A, Morrison DA (2013) Exit from competence for genetic transformation in *Streptococcus pneumoniae* is regulated at multiple levels. PLoS One 8(5):e64197

How Important Is the Absolute Configuration to Bacteria Quorum Sensing and Quorum Quenching?

Francisca Diana da Silva Araújo, Armando Mateus Pomini, and Anita Jocelyne Marsaioli

Bacteria communicate with each other by producing and detecting small diffusible molecules in a process called quorum sensing (QS). This mechanism allows bacteria to coordinate their activities in response to their populations. For instance, QS regulates diverse phenotypes, such as virulence factors bioluminescence, biofilm formation, and antibiotic production, in various bacteria that survived in plants and mammals, only after reaching critical populations (Waters and Bassler 2005).

Some of the chiral signaling molecules shown in Fig. 1 belong to the acyl-homoserine lactone group (AHSLs) (**1-3**) used by Gram-negative bacteria (Eberhard et al. 1981; Fuqua and Winans 1994; Schaefer et al. 2008; Thiel et al. 2009; Pomini et al. 2009; Lindemann et al. 2011; Kai et al. 2012), cyclic peptides and linear peptides used by Gram-positive bacteria (Ji et al. 1995; Ansaldi et al. 2002), autoinducer-2 (AI-2, furanosyl borate diester) (**4**) used by Gram-positive and Gram-negative bacteria (Chen et al. 2002; Miller et al. 2002; Miller et al. 2004), and CAI-1 (**5,** (*S*)-3-hydroxytridecan-4-one) found in *Vibrio* species (Higgins et al. 2007). Each bacterium species has its own QS mechanism; however, they share a general stepwise process beginning with the signal production followed by signal diffusion, signal accumulation, and signal detection.

Signal detection involves the interaction of signal molecules with biological receptors, usually proteins and or DNA, causing phenotype expression (virulence, color, etc.) (Waters and Bassler 2005). However, how stereogenic elements of signal molecules modulate communication has been approached in a few publications and will be discussed from now on.

In higher organisms, chirality is used as an additional parameter to communicate and some populations of distinguishing each other by the enantiomeric ratio of some volatile compounds. A specific example is provided by the bark beetle, which (*Ips pini*) appears to be using enantiomeric ratios of (+)- and (−)-ipsdienol to distinguish populations over the geographic range of *I. pini*; the western population (British Columbia) averaged nearly 100 % (−)-ipsdienol, while the eastern population (New York) produced about 65 % (+)-ipsdienol. Thus, the response can be modulated by the subtle change of enantiomeric ratio in higher organisms, and it can be concluded that the same should occur in bacteria, even though such phenomenon has never been revealed in microorganisms (Miller et al. 1989).

Observing major classes of bacterial signal molecules, few are devoid of stereogenic elements; thus, this asset should be important in communication. Focusing on AHSL, one can observe that there are at least two chiral centers, one at position 5 of the homoserine lactone ring and another at the fatty acid moiety (Fig. 1).

F.D. da Silva Araújo • A.J. Marsaioli (✉)
Chemistry Institute, State University of Campinas, UNICAMP, POB 6154, 13084-971 Campinas, SP, Brazil
e-mail: anita@iqm.unicamp.br

A.M. Pomini
Department of Chemistry, Center of Exact Sciences, State University of Maringá, 87020900 Maringá, PR, Brazil

V.C. Kalia (ed.), *Quorum Sensing vs Quorum Quenching: A Battle with No End in Sight*,
DOI 10.1007/978-81-322-1982-8_7, © Springer India 2015

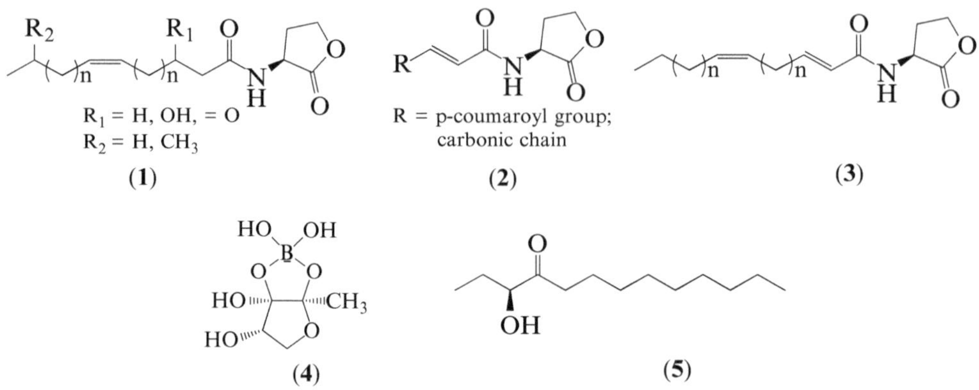

Fig. 1 Some structures of chiral quorum sensing signaling molecules in bacteria

In the first work describing the chemical characterization of a signaling substance from a Gram-negative bacterium, the racemic synthetic compound (±)-*N*-(3-oxo-hexanoyl)-HSL was less active than the (*S*) natural product in triggering *Vibrio fischeri* bioluminescence. At that time, the smaller induction capacity was attributed to the presence of the less active enantiomer, in the racemic mixture (Eberhard et al. 1981). *Pectobacterium carotovorum* also responded differently to AHSL enantiomers; the synthetic (*R*)-*N*-(3-oxohexanoyl)-HSL was 90 % less active than the natural enantiomer (*S*) in the regulation of carbapenem antibiotic biosynthesis (Chhabra et al. 1993). Despite the importance of absolute configuration of biological activity and the large number of Gram-negative bacteria studied, only in a few cases, the absolute configuration of AHSLs was directly accessed from the natural products (Bainton et al. 1992; Lithgow et al. 2000).

The absolute configuration determination at the lactone moiety of natural AHSLs has been achieved applying different techniques including comparison of optical and circular dichroism of natural products with authentic standards. However, gas chromatography-flame ionization detection with a chiral column is certainly a technique of choice when dealing with small amount of natural AHSLs (Pomini et al. 2006; Pomini and Marsaioli 2008; Araújo et al. 2012).

It should be mentioned that up to now all signaling AHS possess (*S*) stereochemistry (Bainton et al. 1992; Lithgow et al. 2000), with the exception of a small amount of (*R*)-*N*-heptanoyl-homoserine lactone detected in the cultivation media of *Pantoea* sp. (Pomini and Marsaioli 2008). The biosynthetic origin of these metabolites was fully investigated in the marine luminescent bacterium *V. fischeri*. The acyl side chain came from acyl-ACP (acyl carrier protein), while the homoserine lactone moiety came from SAM (*S*-adenosyl-methionine) (Schaeffer et al. 1996). The conserved (*S*) configuration seems to derive from the (*S*) configuration of the amino acid portion of SAM. However, these findings do not exclude the characterization of the absolute configurations of AHSLs from other producers. Based on the smaller activity of synthetic (*R*) enantiomers in some quorum-sensing systems (Chhabra et al. 1993) and on the hypothesis that isomerization is one of the possible quorum quenching (disruption of quorum sensing) mechanisms (Roche et al. 2004) the search of (*R*)-AHSL producing bacterium or a microorganism capable of isomerizing these metabolites is of importance to science in general.

Rare unsaturated AHSLs (*S*)-*N*-(7*Z*)-tetradecenoyl (**6**), (*S*)-*N*-(2*E*,7*Z*)-tetradecadienyl-HSL (**7**) and a new natural product (*S*)-*N*-(2*E*)-dodecenoyl-HSL (**8**) were detected in the extracts of *Methylobacterium mesophilicum* isolated from orange trees infected with the citrus variegated chlorosis (CVC) disease. The double-bond position at C-7 was obtained by microderivatization with dimethyl disulfide (DMDS) (Buser et al. 1983), and the double bond in C-2 position in compounds **7** and **8** did

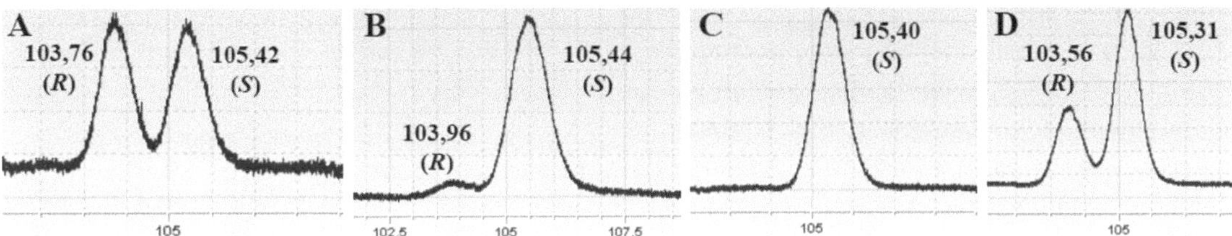

Fig. 2 Chromatograms (GC-FID) with a chiral column (Chrompack CP, chirasil-dex) for (**a**) synthetic product (±)-*N*-tetradecanoyl-HSL, (**b**) synthetic product (*S*)-*N*-tetradecanoyl-HSL, (**c**) fraction containing hydrogenated AHSL **6** and **7** of the *M. mesophilicum* extract, and (**d**) co-injection of A and C (1/1, 1 mg/each). (This figure is part of the experimental section of Prof. Armando Mateus Pomini, PhD thesis [Portuguese])

not react with DMDS due to its conjugation with the carbonyl group. The double-bond geometries were determined by the ^{1}H NMR spectrum as *trans* at C-2 and *cis* configuration at C-7 position. The absolute configuration at the lactone moiety of the natural unsaturated AHSLs produced by *M. mesophilicum* was analyzed by chiral GC-FID using saturated homologues obtained from a microscale catalytic hydrogenation reaction using Pd/C, and all natural AHSLs showed *S* absolute configuration (Pomini et al. 2009) (Fig. 2).

The acyl chain of AHSLs (**1**, Fig. 1) may carry other asymmetric centers as the 3-hydroxy-substituted which have been identified in species as in *Vibrio scophthalmi* (Garcia-Aljaro et al. 2008), *Vibrio harveyi* (Cao and Meighen 1993), *Aeromonas culicola* (Kai et al. 2012), *Rhizobium leguminosarum* (Schripsema et al. 1996), and *Acinetobacter baumannii* (Stacy et al. 2012), among others. The (*R*) stereochemistry at 3-OH position predominated in most cases, and this stereoisomer was the more efficient autoinducer than the corresponding (*S*)-stereoisomer. In *A. baumannii*, for example, the (*R*) and (*S*) diastereoisomers of *N*-(3′-hydroxydodecanoyl)-homoserine lactone (3-OH-C12-HSL) were tested using *A. baumannii* reporter strain (ΔabaI) lacking AbaI synthase and reports AbaR activity via β-galactosidase production, and the results showed that (*R*)-3-OH-C12-HSL was ~40-fold more active than (*S*)-3-OH-C12-HSL (Stacy et al. 2012).

The predominance of (*R*)-stereochemistry at AHSL 3-OH position can be explained in the light of the AHSL biosynthesis, which involves the coupling of S-adenosyl-L-methionine and acyl-acyl carrier protein (acyl-ACP) by LuxI-type synthases. In the biosynthesis of the substrate acyl-ACP, the enzyme FabG reduces stereoselectivity of β-ketoacyl-ACP to (*R*)-3-hydroxyacyl-ACP, which can be preferentially selected for enzyme AbaI, as substrates, to generate the *R*-stereoisomer of 3-hydroxy-substituted AHSLs (Stacy et al. 2012).

The antimicrobial activities of enantiomeric AHSLs against three Gram-positive bacteria (*Bacillus cereus, B. subtilis*, and *Staphylococcus aureus*) are less dependent on the absolute configuration, and *S* and *R* enantiomers have equivalent antimicrobial activity; however, the racemic mixture was less active than the pure enantiomers. The *N*-heptanoylhomoserine lactone was considerably less active than the 3-oxo derivatives (Pomini and Marsaioli 2008).

AI-2 QS pheromone (**4**, Fig. 1) is also chiral and produced by a wide variety of bacteria. It is part of the AI-2 pool, a group of equilibrium-connected compounds derived from (*S*)-4,5-dihydroxy-2,3-pentanedione (DPD), and different bacterial species recognize different signals within this AI-2 pool. Bioassays developed with *Vibrio harveyi* showed that LuxP receptor protein binds the 2,3-borate diester, demonstrating the necessity of boron for induction of bioluminescence (Meijler et al. 2004). However, in *Salmonella typhimurium*, the LsrB receptor protein recognizes a chemically distinct form of the AI-2 signal with opposite configuration at C-2, (2*R*,4*S*)-2-methyl-2,3,3,4-tetrahydroxytetrahydrofuran (Miller et al. 2004).

Investigation of AI-2 has employed indirect methods of detection based on fluorescence of receptor proteins modified with fluorophores or by bioluminescence of sensor strain as *V. harveyi* mutant (BB170). This mutant contains the sensor protein (LuxP) for AI-2 and the LuxN receptor for type 1 autoinducers (AHLs) is absent. A direct method for identification, quantification, and determination of the AI-2 absolute configuration of DPD was developed by Thiel et al. (2009) employing chiral gas chromatography and quinoxaline derivatives. This method is not influenced by the presence or absence of borate and showed similar sensitivity to or even better than currently employed biological sensor systems.

One could not finish this chapter without mentioning CAI-1 (**5**, Fig. 1) which was discovered in *V. cholerae*, which uses synergistically two quorum-sensing systems, CAI-1 and AI-2, to assess its population density. CAI-1 is produced by several *Vibrio* species, probably functioning as an intragenus signal. The structure was determined as *S* stereoisomer of 3-hydroxytridecan-4-one, using chiral chromatographic methods associated with synthetic standards. Bioassays with synthetic CAI-1 and homologues for the ability to stimulate bioluminescence production in the *V. cholerae* CAI-1 reporter strain confirmed the QS role, concluding that molecules with longer acyl chains are more active than those with shorter acyl chains and the *S* stereoisomer at the C3 position is more active than corresponding *R* stereoisomer (Higgins et al. 2007). Additional studies showed that the membrane-bound receptor CqsS in this bacterium tolerated several structural changes in the fatty acid tail portion of CAI-1 (Perez et al. 2012). Higgins et al. (2007) also demonstrated that CAI-1 acts repressing production of the canonical virulence factor TCP (toxin co-regulated pilus), suggesting that CAI-1 could be used as a therapy to prevent cholera infection.

It is easy to conclude that absolute configuration is an important asset in QS, however, not fully explored and we hope that the next decades will reveal more about this issue.

References

Ansaldi M, Marolt D, Stebe T, Mandic-Mulec I, Dubnau D (2002) Specific activation of the *Bacillus* quorum-sensing systems by isoprenylated pheromone variants. Mol Microbiol 44:1561–1573. doi:10.1046/j.1365-2958.2002.02977.x

Araújo FDS, Esper LMR, Kuaye AY, Sircili MP, Marsaioli AJ (2012) *N*-acyl-homoserine lactones from *Enterobacter sakazakii* (*Cronobacter* spp.) and their degradation by *Bacillus cereus* enzymes. J Agric Food Chem 60:585–592. doi:10.1021/jf203846f

Bainton NJ, Stead P, Chhabra SR, Bycroft BW, Salmond GP, Stewart GS, Williams PN (1992) *N*-(3-Oxohexanoyl)-L-homoserine lactone regulates carbapenem antibiotic production in *Erwinia carotovora*. Biochem J 288:997–1004

Buser HR, Arn H, Guerin P, Rauscher S (1983) Determination of double bond position in mono-unsaturated acetates by mass spectrometry of dimethyl disulfide adducts. Anal Chem 55:818–822. doi:10.1021/ac00257a003

Cao J, Meighen E (1993) Biosynthesis and stereochemistry of the autoinducer controlling luminescence in *Vibrio harveyi*. J Bacteriol 175:3856–3862. doi:10.1002/cmdc.200900180

Chen X, Schauder S, Potier N, Van Dorsselaer A, Pelczer I, Bassler BL, Hughson FM (2002) Structural identification of a bacterial quorum-sensing signal containing boron. Nature 415:545–549. doi:10.1038/415545a

Chhabra SR, Stead P, Bainton NJ, Salmond GPC, Stewart GSAB, Williams PJ (1993) Autoregulation of carbapenem biosynthesis in *Erwinia carotovora* by analogues of *N*-(3-oxohexanoyl)-L-homoserine lactone. J Antibiot 46:441–454

Eberhard A, Burlingame AL, Eberhard C, Kenyon GL, Nealson KH, Oppenheimer NJ (1981) Structural identification of autoinducer of *Photobacterium fischeri* luciferase. Biochemistry 20:2444–2449. doi:10.1021/bi00512a013

Fuqua WC, Winans SC (1994) A LuxR-LuxI type regulatory system activates Agrobacterium Ti plasmid conjugal transfer in the presence of a plant tumor metabolite. J Bacteriol 176:2796–2806

Garcia-Aljaro C, Eberl L, Riedel K, Blanch AR (2008) Detection of quorum-sensing-related molecules in *Vibrio scophthalmi*. BMC Microbiol 8:138. doi:10.1186/1471-2180-8-138

Higgins DA, Pomianek ME, Kraml CM, Taylor RK, Semmelhack MF, Bassler BL (2007) The major *Vibrio cholerae* autoinducer and its role in virulence factor production. Nature 450:883–886. doi:10.1038/nature06284

Ji G, Beavis RC, Novick RP (1995) Cell density control of staphylococcal virulence mediated by an octapeptide pheromone. Proc Natl Acad Sci U S A 92:12055–12059. doi:10.1073/pnas.92.26.12055

Kai K, Kasamatsu K, Hayashi H (2012) (*Z*)-*N*-(4-Decenoyl)homoserine lactone, a new quorum-sensing molecule, produced by endobacteria associated with *Mortierella alpina* A-178. Tetrahedron Lett 53:5441–5444. doi:10.1016/j.tetlet.2012.07.133

Lindemann A, Pessi G, Schaefer AL, Mattmann ME, Christensen QH, Kessler A, Hennecke H, Blackwell HE, Greenberg EP, Harwood CS (2011) Isovaleryl-homoserine lactone, an unusual branched-chain quorum-sensing signal from the soybean symbiont

Bradyrhizobium japonicum. Proc Natl Acad Sci U S A 108:16765–16770. doi:10.1073/pnas.1114125108

Lithgow JK, Wilkinson A, Hardman A, Rodelas B, Wisniewski-Dyé F, Williams P, Downie JA (2000) The regulatory locus *cinRI* in *Rhizobium leguminosarum* controls a network of quorum-sensing loci. Mol Microbiol 37:81–97. doi:10.1046/j.1365-2958.2000.01960.x

Meijler MM, Hom LG, Kaufmann GF, McKenzie KM, Sun C, Moss JA, Matsushita M, Janda KD (2004) Synthesis and biological validation of a ubiquitous quorum-sensing molecule. Angew Chem Int Ed Engl 43:2106–2108. doi:10.1002/anie.200353150

Miller DR, Borden JH, Slessor KNJ (1989) Inter- and intrapopulation variation of the pheromone, ipsdienol produced by male pine engravers, Ips pini (Say) (Coleoptera: Scolytidae). J Chem Ecol 15:233–247. doi:10.1007/BF02027785

Miller MB, Skorupski K, Lenz DH, Taylor RK, Bassler BL (2002) Parallel quorum sensing systems converge to regulate virulence in *Vibrio cholerae*. Cell 110:303–314. doi:10.1016/S0092-8674(02)00829-2

Miller ST, Xavier KB, Campagna SR, Taga ME, Semmelhack MF, Bassler BL, Hughson FM (2004) *Salmonella typhimurium* recognizes a chemically distinct form of the bacterial quorum-sensing signal AI-2. Mol Cell 15:677–687. doi:10.1016/j.molcel.2004.07.020

Perez LJ, Ng WL, Marano P, Brook K, Bassler BL, Semmelhack MF (2012) Role of the CAI-1 fatty acid tail in the *Vibrio cholerae* quorum sensing response. J Med Chem 55:9669–9681. doi:10.1021/jm300908t

Pomini AM, Marsaioli AJ (2008) Absolute configuration and antimicrobial activity of acylhomoserine lactones. J Nat Prod 71:1032–1036. doi:10.1021/np800127b

Pomini AM, Araújo WL, Marsaioli AJ (2006) Structural elucidation and biological activity of acylhomoserine lactones from the phytopathogen *Pantoea ananatis* Serrano 1928. J Chem Ecol 32:1769–1778. doi:10.1007/s10886-006-9108-x

Pomini AM, Gai C, Cruz PR, Araujo WL, Marsaioli AJJ (2009) Long-chain acyl-homoserine lactones from *Methylobacterium mesophilicum*: synthesis and absolute configuration. J Nat Prod 72:2125–2129. doi:10.1021/np900043j

Roche DM, Byers JT, Smith DS, Glansdorp FG, Spring DR, Welch M (2004) Communications blackout? Do N-acylhomoserine-lactone-degrading enzymes have any role in quorum sensing? Microbiology 150:2023–2028. doi:10.1099/mic.0.26977-0

Schaefer AL, Greenberg EP, Oliver CM, Oda Y, Huang JJ, Bittan-Banin G, Peres CM, Schmidt S, Juhaszova K, Sufrin JR, Harwood CS (2008) A new class of homoserine lactone quorum-sensing signals. Nature 454:595–599. doi:10.1038/nature07088

Schaeffer AL, Val DL, Hamzelka BL, Cronan JE, Greenberg EP (1996) Generation of cell-to-cell signals in quorum sensing: acyl homoserine lactone synthase activity of a purified *Vibrio fischeri* LuxI protein. Proc Natl Acad Sci U S A 93:9505–9509. doi:10.1073/pnas.93.18.9505

Schripsema J, deRudder K, vanVliet T, Lankhorst P, deVroom E, Kijne J, vanBrussel A (1996) Bacteriocin small of *Rhizobium leguminosarum* belongs to the class of *N*-acyl-L-homoserine lactone molecules, known as autoinducers and as quorum sensing co-transcription factors. J Bacteriol 178:366–371

Stacy D, Welsh MA, Rather PN, Blackwell HE (2012) Attenuation of quorum sensing in the pathogen *Acinetobacter baumannii* using non-native *N*-Acyl homoserine lactones. ACS Chem Biol 7:1719–1728. doi:10.1021/cb300351x

Thiel V, Kunze B, Verma P, Wagner-Döbler I, Schulz S (2009) New structural variants of homoserine lactones in bacteria. ChemBioChem 10:1861–1868. doi:10.1002/cbic.200900126

Waters CM, Bassler BL (2005) Quorum sensing: cell-to-cell communication in bacteria. Annu Rev Cell Dev Biol 21:319–346. doi:10.1146/annurev.cellbio.21.012704.131001

Part II

Quorum Sensing Systems in Microbes

Quorum-Sensing Systems in *Pseudomonas*

Jamuna Bai Aswathanarayan and V. Ravishankar Rai

Introduction

Quorum sensing (QS) or cell-to-cell communication is a mechanism used by bacteria to control a broad range of activities in bacteria. The modulation of gene expression by quorum sensing causes phenotypic changes in bacteria leading to their better adjustment to environmental conditions and stress during growth (Turovskiy et al. 2007). Quorum sensing involves the production, secretion, and response to small diffusible signaling molecules also known as autoinducers. Bacteria produce signaling molecules at a basal level during the stationary phase of their growth, and with the increase in cell density, the concentration of the signaling molecule in the environmental medium increases; and on reaching a threshold level, it induces phenotypic effects by regulating quorum-sensing-dependent target gene expression (Czajkowski and Jafra 2009). Quorum sensing is involved mainly in the regulation of virulence, development of genetic competence, transfer of conjugative plasmids, sporulation, biofilm formation, antimicrobial peptide synthesis, and symbiosis (Bai and Rai 2011).

There are two groups of signal molecules involved in bacterial quorum sensing. One is the peptide derivatives typically used by Gram-positive bacteria, while the fatty acid derivatives are utilized by the Gram-negative bacteria. Most bacteria utilize two general mechanisms for detecting and responding to quorum-sensing signals and in modulating the target gene expression. In the acyl-homoserine lactone (AHL)-dependent quorum-sensing systems, the quorum-sensing signal is detected by a cytosolic transcription factor, whereas the quorum-sensing signal autoinducing peptide (AIP) is detected by a membrane-associated two-component response regulatory system (Dong et al. 2005).

In AHL-mediated quorum sensing, AHL synthase (I-protein) encoded by LuxI homologue synthesizes AHL molecules using S-adenosylmethionine (SAM) and acyl chains derived from the common fatty acid biosynthesis pathway. The short-chain AHL signal passively diffuses across bacterial membranes and accumulates in the environment, and the long-chain AHL signals require active transportation mechanisms for their efflux. The bacteria produce signaling molecules at a basal level during the stationary phase of their growth. With an increase in bacterial population, the concentration of AHL signal reaches a threshold level, resulting in signal accumulation and recognition by the cognate receptors. The signal reception involves R protein which belongs to the LuxR family of transcriptional regulators and acts as a receptor for the AHLs synthesized by the LuxI proteins. The R-AHL complex is a dimer and

J.B. Aswathanarayan • V. Ravishankar Rai (✉)
Department of Studies in Microbiology,
University of Mysore, Mysore 570 006, Karnataka, India
e-mail: raivittal@gmail.com

V.C. Kalia (ed.), *Quorum Sensing vs Quorum Quenching: A Battle with No End in Sight*,
DOI 10.1007/978-81-322-1982-8_8, © Springer India 2015

binds to conserved palindromic sequences of the quorum-controlled promoters, including the promoter of the luxI-type gene, and boosts AHL production (autoinduction) and expression of other genes in the quorum-sensing regulon. Thus, the R-AHL complex is involved in autoinduction and control of quorum-sensing regulons. The AHL degradation enzyme and the cognate regulatory transcription factor(s) are involved in signal decay (Schuster et al. 2004).

The LuxI-LuxR system was first discovered in *Vibrio fischeri* during the investigation of the phenomenon of bioluminescence. Now, the LuxI/LuxR system has become the model system upon that the other quorum-sensing systems have been based. Homologous LuxI/LuxR systems have been identified in many Gram-negative bacteria, each capable of producing specific AHLs. In the opportunistic pathogens, such as *P. aeruginosa* and *Serratia marcescens*, these signaling mechanisms control the expression of the virulence factors. *Pseudomonas aeruginosa* contains two systems homologous to LuxI/LuxR. LasI/LasR has been shown to control biofilm formation and the production of extracellular enzymes, as well as transcription of another quorum-sensing system, RhlI/RhlR, adding an additional level of control through AHL signaling (de Kievit and Iglewski 2000).

Pseudomonads are ubiquitous Gram-negative bacteria capable of surviving in several environmental niches. The genus comprises of important plant pathogens (e.g., *P. syringae*) and human opportunistic pathogens (*P. aeruginosa*). Some of them are able to colonize plant-related niches, such as the rhizosphere (e.g., *P. aeruginosa*, *P. fluorescens*, *P. putida*, *P. aureofaciens*, and *P. chlororaphis*), where they can act as biocontrol agents through the production of traits that directly influence plant disease resistance and growth (Venturi 2006). Therefore, it is important to study the quorum-sensing systems in the genus *Pseudomonas* as it comprises of opportunistic pathogens, plant pathogens, biocontrol agents, and industrially relevant organisms. The quorum-sensing systems, the signaling molecules, and the QS regulated phenotypes in the genus *Pseudomonas* have been summarized in Table 1.

Quorum Sensing in *Pseudomonas aeruginosa*

P. aeruginosa is an opportunistic human pathogen responsible for microbial keratitis, burn wound, and pulmonary infections in cystic fibrosis and immunocompromised patients. It is a highly environmentally adaptable pathogen having a large dynamic genome of which 10 % codes for regulatory elements including a complex multi-signal QS system. Quorum sensing plays a key role in regulating a majority of genes related to various physiological processes, virulence factor production, motility, biofilm formation, and the expression of antibiotic efflux pumps, while the QS signal molecules are involved in the host–pathogen interactions (Williams and Camara 2009).

In *P. aeruginosa*, the QS circuit is integrated by two complete sets of LuxI-LuxR systems, LasI-LasR and RhlI-RhlR, and one incomplete system, QscR which has no cognate "I" protein (Lee et al. 2006). LasR responds to *N*-3-oxododecanoyl homoserine lactone (3-oxo-C12-HSL) synthesized by LasI, while RhlR responds to *N*-butyryl homoserine lactone (C4-HSL) produced by RhlI, and QscR responds to 3-oxo-C12-HSL, sharing it with LasR (Liu et al. 2009). Through this QS circuit, the LasI-LasR system controls expression of many extracellular virulence factors, the RhlI-RhlR system regulates production of rhamnolipid and secondary metabolites, and QscR induces PA1897 and its own gene, *qscR* (Choi et al. 2011). These QS systems are well regulated in a hierarchical cascade in which LasR activates expression of the RhlI-RhlR system and LasI-producing 3-oxo-C12-HSL activates QscR, which autoactivates its own expression (Ha et al. 2012). In certain cases, the LasR and RhlR regulate some genes in a mutually antagonistic manner, and QscR represses many genes induced by LasI-LasR and RhlI-RhlR systems (Choi et al. 2011; Park et al. 2013). Thus, *P. aeruginosa* has two acyl-HSL synthases and three receptors. The LasI synthase produces 3-oxo-C12-HSL, for which there are two receptors, LasR and QscR. The

Table 1 An overview of quorum-sensing systems in *Pseudomonas* species

Pseudomonas species	Quorum-sensing systems	Quorum-sensing signals	Quorum-sensing-regulated phenotypes	References
Pseudomonas aeruginosa	LasI/R RhlI/R QscR PQS	3-oxo-C12-HSL C4-HSL PQS, HHQ	Biofilm formation, virulence factors, elastase, lipase, exotoxin A, lectins, alkaline protease, hydrogen cyanide, swarming, twitching, pyocyanin, pyoverdine, rhamnolipids, and others	Winzer et al. (2000) Chugani et al. (2001) Wade et al. (2005) Lee et al. (2006) Williams and Camara (2009) Dubern and Diggle (2008) Liu et al. (2009)
Pseudomonas fluorescens	MpuI/R HdtS	3-OH-C14:1-HSL, C-10-HSL C6-HSL, C8-HSL, C4-HSL 3-OH-C6-HSL 3-OH-C8-HSL Diketopiperazines	Mupirocin and metalloprotease biosynthesis	Laue et al. (2000) Shaw et al. (1997) Cha et al. (1998) Cui (2004) Liu et al. (2007)
Pseudomonas putida	PpuI/R	3-oxo-C12-HSL	Biofilm development	Bertani and Venturi (2004)
Pseudomonas syringae	AhlI/R	3-oxo-C6-HSL	Cell aggregation and epiphytic fitness	Chatterjee et al. (2003), Quinones et al. (2004)
Pseudomonas aureofaciens	PhzI/R CsaI/R	C6-HSL	Phenazine antibiotics synthesis, cell surface components, and rhizosphere colonization	Zhang and Pierson (2001)
Pseudomonas chlororaphis	PhZI/R	C6-HSL	Phenazine-1-carboxamide biosynthesis	Chin et al. (2005)
Pseudomonas mediterranea	PmeI/R	C6-HSL	Virulence factor and lipodepsipeptides production	Licciardello et al. (2012)
Pseudomonas corrugata	PcoI/R	C6-HSL, 3-oxo-C6-HSL, and C8-HSL	Virulence factor and lipodepsipeptides production	Licciardello et al. (2009)

RhlI synthase produces C4-HSL, for which the receptor is RhlR. Integrated into the acyl-HSL quorum-sensing circuits is a third signal, 2-heptyl-3-hydroxy-4-quinolone, known as the *Pseudomonas* quinolone signal (PQS). The structures of the major signaling molecules of *P. aeruginosa* have been represented in Fig. 1. Transcriptome analyses have shown that quinolone signaling directly or indirectly controls the expression of at least 90 genes. The acyl-HSL and PQS signaling systems influence each other; the *las* system activates synthesis of PQS, which in turn activates *rhlI* expression. In addition, LasR, RhlR, and QscR influence expression of genes that can potentially alter intracellular levels of the PQS biosynthesis precursor anthranilate. Together these quorum-sensing systems regulate hundreds of *P. aeruginosa* genes. Different elements of the *P. aeruginosa* quorum-sensing circuit also influence each other at multiple levels; for example, LasR-3-oxo-C12-HSL activates *rhlR* and *rhlI* transcription, and QscR influences expression of a subset of *las-* and *rhl*-controlled genes. In fact, the regulons of LasR, RhlR, and QscR are partially overlapping (Chugani and Greenberg 2010).

The primary system is the Las system, which encodes the proteins LasI and LasR. The LasI protein catalyzes the production of the AHL molecule *N*-3-oxododecanoyl-L-homoserine lactone (3-oxo-C12-HSL). The 3-oxo-C12-HSL molecule docks with the DNA-binding transcription regulator LasR, which allows LasR to bind to the promoters of QS-regulated genes to control virulence factor such as *lasB* (elastase), *lasA* (staphylolysin), *aprA* (alkaline protease),

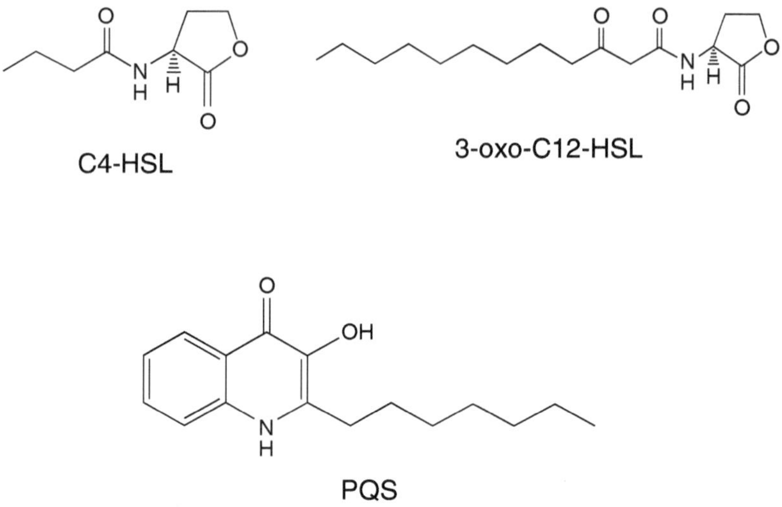

Fig. 1 Structures of the *P. aeruginosa* major quorum-sensing signal molecules C4-HSL, 3-oxo-C12-HSL, and PQS

toxA (exotoxin A), *hcnABC* (hydrogen cyanide synthase), and *lasI*. The Las circuit induces a positive feedback loop to produce more AHL and also induces a secondary QS circuit, the Rhl system. The Rhl system consists of RhlI, which synthesizes *N*-butyryl-L-homoserine lactone (C4-HSL), and the receptor RhlR. As with the Las system, C4-HSL accumulates to a sufficient concentration and binds to RhlR. The Rhl system induces expression of *rhlAB* (rhamnolipid synthesis genes), *rhlI*, *lasB*, *rpoS* (the stationary-phase sigma factor), *lecA* (type 1 lectin), *lecB* (type II lectin), *hcnABC*, and genes involved in pyocyanin production (Winzer et al. 2000).

The additional *P. aeruginosa* gene that codes for a homologue of LasR and RhlR, QscR, is an orphan quorum-sensing signal receptor. QscR mutants are hypervirulent, and a number of genes controlled by the other AHL-based QS systems are repressed by QscR. There are several possible mechanisms for QscR repression of LasR- or RhlR-activated genes. The QscR protein forms homomultimers and also heteromultimers with LasR and RhlR. The heteromultimer formation could interfere with the activity of LasR and RhlR. QscR might also bind to the AHLs and compete with LasR and RhlR for these signals. QscR can also function by direct binding as a homomultimer to specific promoters and function in an acyl-HSL-independent manner, or it could utilize the signal produced by LasI or RhlI. It has been shown that QscR does not bind to LasR-dependent promoters. QscR can repress the activation of selected LasR- and RhlR-dependent quorum-sensing responsive genes. This could be the result of competition for signal, competition for binding sites on the regulatory DNA, or heterodimer formation (Chugani et al. 2001). Thus, it is clear that the *qscR* gene codes for an orphan AHL transcription factor. Unlike *lasR* and *rhlR*, which are linked to *lasI* and *rhlI*, genes that code for the production of acyl-HSL signals, there is no *I* gene linked to *qscR*. However, the DNA-binding activity of QscR is dependent on the presence of a long-chain acyl-HSL. The 3-oxo-C12-HSL produced by LasI and to which LasR responds is an effective ligand for QscR. Like LasR, QscR requires 3-oxo-C12-HSL to fold into an active conformation, but unlike LasR signal binding to QscR is not irreversible. However, binding of purified QscR to DNA is dependent on added acyl-HSL. That two 3-oxo-C12-HSL-responsive transcription factors differ fundamentally in their ability to exist in the absence of the signal. This would allow for a very rapid response of the QscR regulon to sudden decreases in environmental levels of 3-oxo-C12-HSL where the LasR regulon may respond more slowly. Like other transcriptional activators in the LuxR family that have been studied, QscR requires the presence of an acyl-HSL in the culture growth medium for folding in an active state. However, unlike LasR, which also responds to 3-oxo-C12-HSL, purified QscR requires exogenous addition of 3-oxo-C12-HSL for binding to target DNA. QscR has a broader signal specificity than does LasR, and QscR may even respond to 3-oxo-C10-HSL, C10, and C12 better than it does to 3-Oxo-C12 HSL. This suggests that QscR might respond to signals produced by other bacteria that coexist with *P. aeruginosa*. This also shows the possibility that *qscR* and the genes surrounding it may be relatively recent acquisitions in the *P. aeruginosa* genome (Lequette et al. 2006). The structure of the transcription factor, QscR, bound to N-3-oxododecanoyl-homoserine lactone has been elucidated at a resolution of 2.55 Å. The ligand-bound QscR is a dimer with a unique symmetric "cross-subunit" arrangement containing multiple dimerization interfaces involving both domains of each subunit. The QscR dimer appears poised to bind DNA. QscR recognizes 3-oxo-C12-HSL in almost exactly the same way as LasR. However, QscR shows greater promiscuity in its response to AHLs than does LasR in *P. aeruginosa*. Although the binding pocket surface areas, packing densities, and pocket volumes are nearly identical in QscR and LasR, different interactions involving the 3-oxo position of the acyl chain may be responsible for the more relaxed specificity of QscR relative to LasR (Lintz et al. 2011).

The *Pseudomonas* quinolone signal (PQS) is a third *P. aeruginosa* QS signal that is dependent on the balanced production of 3-oxo-C12-HSL and C4-HSL. The PQS molecule (2-heptyl-3-hydroxy-4-quinolone) plays a significant role in the transcription of Rhl-dependent *P. aeruginosa* virulence genes encoding the production of pyocyanin and rhamnolipid. PQS production is intimately linked to the QS hierarchy, with its production and bioactivity requiring both the *las* and *rhl* QS systems. Additionally, PQS was recently shown to be solubilized by rhamnolipids, the production of which is controlled by the Rhl system, which may be important for the activity of PQS as an extracellular signal. LasR has been shown to regulate PQS production, and the provision of exogenous PQS induces expression of *lasB* (coding for elastase), *rhlI*, and *rhlR* implying that PQS activity constitutes a regulatory link between the *las* and *rhl* quorum-sensing systems. It is possible that PQS upregulates the *rhl* quorum-sensing system in late stationary-phase cultures. The structural genes required for PQS have been identified (*pqsABCDH*) along with a transcriptional regulator (*pqsR*) and a response effector (*pqsE*). The transcription of *pqsH* is regulated by the *las* QS system, linking QS and PQS regulation. Mutations in the PQS genes result in a loss of PQS synthesis and a corresponding loss of pyocyanin production. A mutation in the *pqsE* gene also results in a loss of pyocyanin even though PQS synthesis remains intact. This suggests that *pqsE* is not required for PQS biosynthesis and may have a role in the cellular response to PQS (Gallagher et al. 2002). PQS is synthesized via a "head-to-head" condensation of anthranilate and β-keto dodecanoate and requires the products of the *pqsA*, *pqsB*, *pqsC*, and *pqsD* genes, which also generate over 50 other 2-alkyl-4-quinolones (AHQs) including 2-heptyl-4(1*H*)-quinolone (HHQ). The function of the last gene in the *pqs* operon (*pqsE*) is not known, but while *pqsE* mutants produce parental levels of AHQs, they do not exhibit any PQS-associated phenotypes; consequently, PqsE is considered to facilitate the response to PQS. The immediate precursor of PQS is HHQ, and its conversion to PQS depends on the action of PqsH, a putative monooxygenase that is LasR regulated so linking the AHL and AHQ regulatory systems. Expression of the *pqsABCDE* operon and hence AHQ production are controlled by the LysR-type regulator PqsR (MvfR), which binds directly to the *pqsA* promoter. As PqsR binding is enhanced in the presence of PQS, it implies that PQS acts as a PqsR coinducer. The *pqsR* gene is itself positively regulated by *lasR* and negatively regulated by *rhlR*, establishing a further link between AHL-dependent quorum sensing and AHQ biosynthesis and hence AHQ signaling. Among the many different AHQs produced by *P. aeruginosa*, two of the major compounds are PQS and its precursor, HHQ, although similar concentrations of 2-nonyl-4-quinolone (HNQ), 2-nonenyl-4-quinolone, and 2-heptyl-4-quinolone-*N*-oxide (HQNO) have been

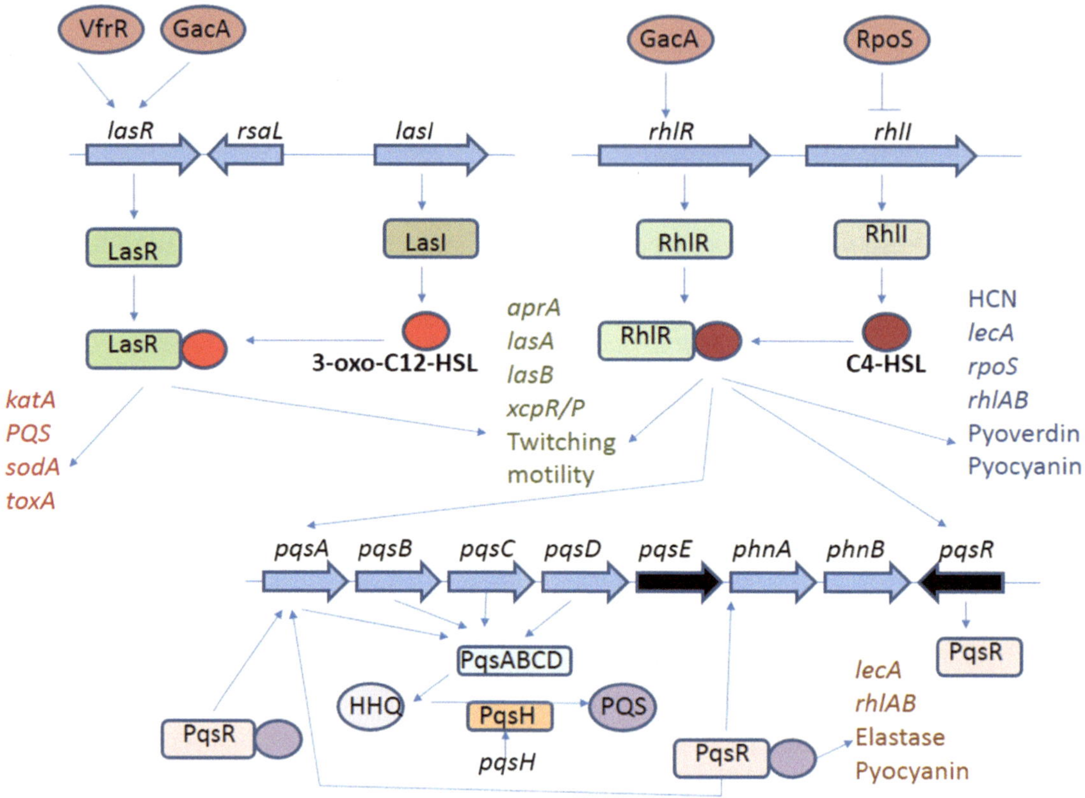

Fig. 2 Regulation of the LasI/R, RhlI/R, and PQS quorum-sensing systems in *Pseudomonas aeruginosa*

reported to be present in culture supernatants (Diggle et al. 2007). PQS regulates the production of virulence determinants including elastase, rhamnolipids, the galactophilic lectin, LecA, and pyocyanin and influences biofilm development. Thus, PQS signaling plays an important role in *P. aeruginosa* pathogenesis. In contrast to the AHLs, when supplied exogenously, PQS overcomes the cell population density-dependent production of *P. aeruginosa* exoproducts (McKnight et al. 2000; Dubern and Diggle 2008). PQS is also produced in the lungs of cystic fibrosis patients infected with *P. aeruginosa* and is required for virulence in eukaryotes. It can also induce apoptosis and decrease viability of eukaryotic cells (Wade et al. 2005).

The QS signaling systems of *P. aeruginosa* are complex in nature and are well regulated in a hierarchical cascade targeting the expression of genes required for growth, survival, virulence, and biofilm formation (Fig. 2).

Quorum-Sensing Regulatory Network in *P. aeruginosa*

Quorum sensing controls a significant proportion of the virulence factors used by *P. aeruginosa* for establishing infection. Additional genes can also influence the QS response. The QS regulator (QscR) represses 3-oxo-C12-HSL-regulated virulence factors and prevents the premature activation of QS cascade within a host and in environments where it is not required. This inhibitory effect is controlled by the global activator protein GacA. RsaL, the product of a gene found between *lasI* and *lasR*, negatively regulates the Las QS circuit. The product of the *vrf* gene is a cAMP receptor homologue and is required for the transcription of *lasR*. RsmA avoids early activation and downregulates production of QS-regulated phenotypes – protease, elastase, and staphylolytic activities – and the production of a cytotoxic lectin, hydrogen cyanide, and pyocyanin. Overexpression of RsmA results in reduction in the expression of the AHL synthase genes *lasI* and *rhlI*. The RpoN is a negative transcriptional regulator of the *lasIR* and *rhlIR* but positively regulates the expression of *rhlI* in minimal media (Willcox et al. 2008). The catabolite repressor homologue Vfr directly induces *lasR* transcription. The stringent response protein RelA, which synthesizes guanosine tetraphosphate (ppGpp) under amino acid starvation conditions on overexpression, causes early activation of several QS-controlled processes and *lasR* and *rhlR* expression. The GacA/GacS two-component regulatory system posttranscriptionally regulates QS through RsmZ and RsmA. In the absence of RsmZ, RsmA represses the synthesis of acyl-HSL signals. RsmA also regulates the production of virulence factors. The anaerobic regulator ANR activates expression of the quorum-controlled hydrogen cyanide biosynthetic genes *hcnABC*; ANR appears to be an important factor in the co-regulation of quorum-controlled genes under oxygen-limiting conditions. The *rsaL* gene, which is directly activated by LasR-3OC12-HSL, encodes an 11-kDa protein which inhibits QS by repressing *lasI*. The stationary-phase sigma factor RpoS can affect the expression of 40 % of QS-regulated genes. The transcriptional regulator VqsR, which is activated by LasR-3-oxo-C12-HSL, is required for AHL production and the expression of QS-controlled genes. Further, the genes required for the synthesis of a direct precursor of PQS (*pqsABCD* and *phnAB*) are activated by the transcriptional regulator MvfR, and it itself is under the control of LasR-3-oxo-C12-HSL. MvfR regulates *rhl*-dependent genes without affecting the production of AHLs or the expression of *lasR* or *rhlR*. MvfR/PQS and *rhl* QS are parallel pathways that converge at the promoters of their target genes. All these suggest that QS in *P. aeruginosa* is highly complex and the QS gene expression is integrated in a highly interconnected network of other regulatory systems (Schuster and Greenberg 2006; Williams and Camara 2009).

Quorum Sensing in *Pseudomonas fluorescens*

A quorum-sensing system has been found in *P. fluorescens* strain NCIMB 10586 (EI-Sayed et al. 2001). Two genes, *mupR* and *mupI*, have been cloned and sequenced and the gene products are identical to LasR/LuxR and LasI/LuxI, respectively. *mupR* encodes a predicted protein of 234 amino acids with a molecular mass of 26 kDa, while mupI produces a 191 amino acid protein with a molecular mass of 21 kDa. The *mupR/mupI* quorum sensing has been found to regulate the mupirocin biosynthetic gene cluster in *P. fluorescens* NCIMB 10586. Although the QS molecule has not been identified in *P. fluorescens* NCIMB 10586, several QS signals have been reported to be present in different *P. fluorescens* strains. *P. fluorescens* F113 produces at least three different AHLs, N-(3-hydroxy-7-cistetradecenoyl)homoserine lactone (3 OH, C14:1-HSL), N-decanoyl homoserine lactone (C10-HSL), and N-hexanoyl homoserine lactone (C6-HSL). C10-HSL has not been previously found as a naturally occurring AHL (Laue et al. 2000). A gene in *P. fluorescens* F113, termed *hdtS*, was capable of directing synthesis of all three AHLs. The HdtS, a 33-KDa protein, does not belong to the known AHL synthase families and is related to the lysophosphatidic acid acyltransferase family. It was concluded that HdtS is from a third protein family capable of AHL biosynthesis. The three AHLs identified in *P. fluorescens* F113 have yet been identified in any other *P. fluorescens* strains (Laue et al. 2000).

Many AHLs have been identified in *P. fluorescens* 2–79 such as N-(3-hydroxyhexanoyl)-L-HSL, N-(3-hydroxyoctanoyl)-LHSL and N-(3-hydroxydecanoyl)-L-HSL, N-octanoyl-L-HSL, and N-hexanoyl-HSL (Shaw et al. 1997). In *P. fluorescens* NCIMB 10586, a compound characterized as cyclo(L-Phe-L-Pro), a diketopiperazine (DKP), was identified. This shows the complexity of the QS system and the existence of cross-talk among bacterial signaling systems in *P. fluorescens* (Cui 2004).

A food isolate *P. fluorescens* strain 395 was capable of producing C4-HSL and 3-oxo-C8-HSL. It was also demonstrated that the alkaline metalloprotease gene in *P. fluorescens* is regulated by the AHL-based quorum-sensing system at a transcriptional level during the late exponential growth phase (Liu et al. 2007).

Quorum Sensing in *Pseudomonas syringae*

Plant pathogen *P. syringae* pv. *syringae* strain B728a possesses an AHL QS system called AhlI/R which produces and responds to 3-oxo-C6-HSL. The two regulators are independently involved in the positive regulation of *ahlI* expression. The GacA/GacS system regulates AHL QS in *P. syringae* through regulation of *ahlI* expression. A TetR family transcriptional regulator designated as AefR positively regulates *ahlI* expression independently from GacA/GacS. The expression of *ahlI* is restored in *gacA* and *aefR* mutants on exogenous addition of 3-oxo-C6-AHL, implying that *ahlI* expression is responsive to AHL through a positive feedback mechanism as in other AHL QS systems. This also indicates that neither regulator is acting through a direct interaction at the promoter sequence. The GacA/GacS system present at the top of the regulatory cascade system affects AHL accumulation in *P. syringae* pv. *tomato* strain DC3000 (Chatterjee et al. 2003; Quinones et al. 2004).

Quorum Sensing in *Pseudomonas putida*

The AHL QS system designated PpuI/R has been identified in two *P. putida* plant-beneficial rhizobacteria. The *ppuI/R* genes of *P. putida* have between them a repressor gene called *rsaL*, which negatively controls *ppuI* and *lasI* gene expression. In *P. putida*, *ppuI* expression is under strong negative transcriptional regulation by RsaL. It plays a major role in keeping the AHL system at very low expression levels, by competing for *ppuI* promoter binding with PpuR/3-oxo-C_{12}-HSL. *ppuI* expression in *P. putida* is positively regulated by the GacA/GacS two-component system and undergoes cross-regulation with the stationary-phase RpoS sigma factor (Bertani and Venturi 2004).

Quorum Sensing in *Pseudomonas aureofaciens* and *Pseudomonas chlororaphis*

The biological control bacterium *Pseudomonas chlororaphis (aureofaciens)* strain 30–84 employs two quorum-sensing (QS) systems: PhzR/PhzI regulates the production of the antibiotics phenazine-1-carboxylic acid, 2-hydroxy-phenazine-1-carboxylic acid, and 2-hydroxy-phenazine, whereas CsaR/CsaI regulates various aspects of the cell surface. The PhzR–PhzI quorum-sensing system regulates phenazine production in a cell density-dependent manner. The *phzR* gene encodes a transcriptional regulator of the phenazine operon, and *phzI* encodes an AHL synthase that directs the synthesis of the signal hexanoyl homoserine lactone (HHL). Upon binding HHL, PhzR becomes activated and induces transcription of the phenazine genes. The GacS–GacA two-component signal transduction system is also involved in controlling phenazine production by regulating transcription of *phzI* and other regulatory elements. The second QS regulatory system, termed CsaR–CsaI (for "cell surface alterations"), is only marginally involved in phenazine regulation. The primary function of the CsaR–CsaI system is the regulation of exoprotease production in conjunction with the PhzR–PhzI system and also the regulation of cell surface properties. CsaI and CsaR are similar to RhlI and RhlR of *P. aeruginosa*. However, *rhlI* and *rhlR* are separated by 181 bp and *rhlI* has its own promoter, whereas *csaR* and *csaI* are separated by only 30 bp, and *csaI* has an RBS but no promoter, implying that *csaI* expression is dependent on *csaR*. The PhzR–PhzI and CsaR–CsaI appear to function independently and do not exist in a hierarchical relationship. However, the AHL produced by PhzI can interact with CsaR and the AHL produced by CsaI can activate PhzR. The two systems are cooperatively involved in the regulation of exoprotease production and colonization of the wheat rhizosphere. The CsaI/CsaR system is also under positive regulation by GacA/GacS (Zhang and Pierson 2001).

In the rhizobacterium *P. chlororaphis* PCL1391, the expression of the antifungal metabolite phenazine-1-carboxamide (PCN) biosynthetic gene cluster is population density dependent and is regulated by the quorum-sensing genes phzI and phzR via synthesis of the autoinducer N-hexanoyl-L-homoserine lactone (C6-HSL). A mutation in the psrA gene (*Pseudomonas* sigma regulator), the gene product of which is a member of the TetR/AcrR family of transcriptional regulators, resulted in increased production of autoinducer molecules and PCN. PsrA also negatively regulates an as yet unidentified AHL QS system(s) in *P. chlororaphis*. PsrA in *P. chlororaphis* is well connected with the RpoS stationary-phase sigma factor, as it positively regulates *rpoS* expression (Chin et al. 2001, 2005). The *psrA* gene is itself positively regulated by the two-component GacA/GacS system in *P. chlororaphis* (Bertani and Venturi 2004; Chin et al. 2005).

Quorum Sensing in *Pseudomonas corrugata* and *Pseudomonas mediterranea*

Pseudomonas corrugata and *Pseudomonas mediterranea* are two closely related phytopathogenic bacteria. *P. corrugata* CFBP 5454 has an *N*-acyl-homoserine lactone (AHL) QS system PcoI/PcoR which produces C6-HSL, 3-oxo-C6-HSL, and C8-HSL and is involved in virulence and regulates lipodepsipeptides LDP production at high population densities. *P. mediterranea* also produces LDPs as well as possessing an AHL-dependent QS system, designated PmeI/PmeR, which is highly homologous to the PcoI/PcoR system of *P. corrugata* producing and responding to C_6-AHL. Downstream of *pmeI* revealed the presence of a homologue of the *rfiA* gene of *P. corrugata* which encodes a transcriptional regulator involved in bacterial virulence. As in other Gram-negative bacteria, the production of AHL signal molecules occurs in a cell density-dependent fashion and requires the expression of the AHL synthase gene, *pcoI*, and the *pcoR* regulator gene. The protein RfiA in *P. corrugate* is an important and novel transcriptional regulator, directly linked to QS by co-transcription with *pcoI* (Licciardello et al. 2009). RfiA and PcoR are required for full virulence in tomato. Mutation of either *pcoR* or *rfiA* drastically reduces virulence in tomato. PmeI and PcoI and PmeR and PcoR have high homology to LuxI and LuxR family proteins of different pathovars of *P. syringae* and other oxidase-negative *Pseudomonas* plant-associated bacterial species other than to oxidase-positive species such as *P. fluorescens* and *P. putida* (Licciardello et al. 2012).

AI-2-Mediated QS in *P. aeruginosa*

P. aeruginosa is unique from the other bacteria because it is does not make its own signaling molecule autoinducer-2 (AI-2). However, there is an increase in the expression of its virulence factor in response to AI-2 produced by other microflora. Although the *P. aeruginosa* QS circuit is AHL signaling based, it is capable of sensing AI-2 and is therefore susceptible to AI-2-mediated QS inhibition. *P. aeruginosa* was able to sense and respond to AI-2 produced by the normal microflora of cystic fibrosis patients, which led to increased virulence factor expression and infection (Duan et al. 2003; Roy et al. 2011).

Additional AHLs and Signaling Molecules in *Pseudomonas*

Apart from the two main AHLs of *P. aeruginosa* (C4-HSL and 3-oxo-C12-HSL), low concentrations of other distinct AHLs, such as 3-oxo-C14-HSL and 3-oxo-C10-HSL, have been detected in *P. aeruginosa*. These AHLs may be synthesized due to the LasI synthase coupling the wrong acyl carrier protein (ACP) to *S*-adenosylmethionine

(SAM) or from the action of a different type of AHL synthase. The two other unrelated AHL synthase families that have been reported are the LuxM synthase family similar to *Vibrio* spp. and the HdtS synthase homologous to those of *P. fluorescens*. The other signaling molecules are diketopiperazines (DKPs), a novel family of cyclic dipeptides, identified from the culture supernatants of various bacteria. These can interfere with the quorum-sensing systems of various bacteria by binding to the LuxR family of receptors and either activating or antagonizing AHL signals. The DKPs present in the supernatants of pseudomonads are cyclo(ΔAla-L-Val) and cyclo(L-Pro-L-Tyr) in *P. aeruginosa* and cyclo(L-Phe-L-Pro) in *P. fluorescens* and *P. alcaligenes* (Sio et al. 2012).

Interkingdom Signaling in *P. aeruginosa*

The 3-oxo-C12-HSL and PQS of *P. aeruginosa* are capable of modulating inflammatory and immune responses in mammals. The QS signal 3-oxo-C12-HSL exerts immune-suppressive or anti-inflammatory effects at concentrations below 10 μM, whereas pro-inflammatory or pro-apoptotic effects are found at much higher concentrations. It attenuates LPS-induced inflammation required for the establishment of chronic *P. aeruginosa* infection. The host environment modulates QS in *P. aeruginosa* either through nonenzymatic or enzymatic destruction of AHLs through lactonolysis. The mammalian paraoxonases (PON1, PON2, PON3) are a unique family of calcium-dependent hydrolases and are known to possess enzymatic activities toward a broad range of substrates including the AHLs (Williams and Camara 2009).

Quorum Quenching in *P. aeruginosa*

The functional analysis of putative acylase genes in the *P. aeruginosa* PAO1 genome, the PA2385 gene, revealed the presence of an acylase that removes the fatty acid side chain from the homoserine lactone (HSL) nucleus of AHL-dependent QS signal molecules. The posttranslational processing of the acylase and the hydrolysis reaction type are similar to those of the beta-lactam acylases, implying that the PA2385 protein is a member of the *N*-terminal nucleophile hydrolase superfamily. The purified acylase was capable of degrading AHLs with side chains ranging in length from 11 to 14 carbons at physiologically relevant low concentrations. The substituent at the 3′ position of the side chain did not affect activity, indicating broad-range AHL quorum-quenching activity. Of the two main AHL signal molecules of *P. aeruginosa* PAO1, *N*-butanoyl-L-homoserine lactone (C4-HSL) and *N*-(3-oxododecanoyl)-L-homoserine lactone (3-oxo-C12-HSL), only 3-oxo-C12-HSL is degraded by the enzyme. The purified protein completely inhibited the accumulation of 3-oxo-C12-HSL and production of the signal molecule 2-heptyl-3-hydroxy-4(1*H*)-quinolone and reduced production of the virulence factors elastase and pyocyanin in *P. aeruginosa* PAO1 cultures. Similar results were observed on overexpressing the PA2385 gene in *P. aeruginosa*. The AHL acylase has in situ quorum-quenching activity, and it enables *P. aeruginosa* PAO1 to modulate its own quorum-sensing-dependent pathogenic potential (Sio et al. 2012).

Conclusion

P. aeruginosa is the most common nosocomial pathogen and is associated with chronic lung disease in cystic fibrosis (CF) patients. The major virulence signaling systems in *P. aeruginosa* are the AHL systems Las and Rhl, which together control the expression of multiple virulence factors in response to cell density. The third group of signaling system, the PQS, connects virulence factor production with adaptation and survival as a strategy to eliminate competition when survival depends on iron availability. Similarly, a number of regulators and more than one AHL-based QS system have been identified in plant growth promoting *P. aureofaciens*/*P. chlororaphis*, root colonizing *P. fluorescens*, and the plant pathogen *Pseudomonas syringae*. Understanding the molecular mechanisms of regulation of QS and the complete signaling integration network in *Pseudomonas* will help in identifying the various drug targets to interrupt pathogenic activities by designing novel antimicrobials.

References

Bai AJ, Rai VR (2011) Bacterial quorum sensing and food industry. Compr Rev Food Sci Food Saf 10:183–193

Bertani I, Venturi V (2004) Regulation of the *N*-acyl homoserine lactone-dependent quorum-sensing system in rhizosphere *Pseudomonas putida* WCS358 and cross-talk with the stationary-phase RpoS sigma factor and the global regulator GacA. Appl Environ Microbiol 70:5493–5502

Cha C, Gao P, Chen YC, Shaw PD, Farrand S (1998) Production of acyl-homoserine lactone quorum-sensing signals by gram-negative plant-associated bacteria. Mol Plant Microbe 11:1119–1129

Chatterjee A, Cui Y, Yang H, Collmer A, Alfano JR, Chatterjee AK (2003) GacA, the response regulator of a two-component system, acts as a master regulator in *Pseudomonas syringae* pv. tomato DC3000 by controlling regulatory RNA, transcriptional activators, and alternate sigma factors. Mol Plant Microbe Interact 16:1106–1117

Chin AWTF, van den Broek D, de Voer G, van der Drift KM, Tuinman S, Thomas-Oates JE, Lugtenberg BJ, Bloemberg GV (2001) Phenazine-1-carboxamide production in the biocontrol strain *Pseudomonas chlororaphis* PCL1391 is regulated by multiple factors secreted into the growth medium. Mol Plant Microbe Interact 14:969–979

Chin AWTF, van den Broek D, Lugtenberg BJ, Bloemberg GV (2005) The *Pseudomonas chlororaphis* PCL1391 sigma regulator psrA represses the production of the antifungal metabolite phenazine-1-carboxamide. Mol Plant Microbe Interact 18:244–253

Choi Y, Park HY, Park SJ, Kim SK, Ha C, Im SJ, Lee JH (2011) Growth phase-differential quorum sensing regulation of anthranilate metabolism in *Pseudomonas aeruginosa*. Mol Cell 32:57–65

Chugani S, Greenberg EP (2010) LuxR homolog-independent gene regulation by acyl-homoserine lactones in *Pseudomonas aeruginosa*. Proc Natl Acad Sci USA 107:10673–10678

Chugani SA, Whiteley M, Lee KM, D'Argenio D, Manoil C, Greenberg EP (2001) QscR, a modulator of quorum-sensing signal synthesis and virulence in *Pseudomonas aeruginosa*. Proc Natl Acad Sci USA 98:2752–2757

Cui X (2004) Regulation of biosurfactant production by quorum sensing in *Pseudomonas fluorescens* 5064, the Cause of Broccoli Head Rot Disease. PhD thesis, The University of Edinburgh

Czajkowski R, Jafra S (2009) Quenching of acyl-homoserine lactone-dependent quorum sensing by enzymatic disruption of signal molecules. Acta Biochim Pol 56:1–16

de Kievit TR, Iglewski BH (2000) Bacterial quorum sensing in pathogenic relationships. Infect Immun 68:4839–4849

Diggle SP, Matthijs S, Wright VI, Fletcher MP, Chhabra SR, Lamvont IL, Kong X, Hider RC, Cornelis P, Camara M (2007) The *Pseudomonas aeruginosa* 4-quinolone signal molecules HHQ and PQS play multifunctional roles in quorum sensing and iron entrapment. Chem Biol 14:87–96

Dong YH, Zhang XF, Soo LHM, Greenberg EP, Zhang LH (2005) The two-component response regulator PprB modulates quorum-sensing signal production and global gene expression in *Pseudomonas aeruginosa*. Mol Microbiol 56:1287–1301

Duan K, Dammel C, Stein J, Rabin H, Surette MG (2003) Modulation of *Pseudomonas aeruginosa* gene expression by host microflora through interspecies communication. Mol Microbiol 50:1477–1491

Dubern JF, Diggle SP (2008) Quorum sensing by 2-alkyl-4-quinolones in *Pseudomonas aeruginosa* and other bacterial species. Mol Biosyst 4:882–888

El-Sayed AK, Hothersall J, Thomas CM (2001) Quorum sensing-dependent regulation of biosynthesis of the polyketide antibiotic mupirocin in *Pseudomonas fluorescens* NCIMB 10586. Microbiology 147: 2127–2139

Gallagher LA, McKnight SL, Kuznetsova MS, Pesci EC, Manoil C (2002) Functions required for extracellular quinolone signaling by *Pseudomonas aeruginosa*. J Bacteriol 184:6472–6480

Ha C, Park SJ, Im SJ, Lee JH (2012) Interspecies signaling through QscR, a quorum receptor of *Pseudomonas aeruginosa*. Mol Cell 33:53–59

Laue BE, Jiang Y, Chhabra SR, Jacob S, Stewart GSAB, Hardman A, Downie JA, O'Gara F, Williams P (2000) The biocontrol strain *Pseudomonas fluorescens* F113 produces the Rhizobium small bacteriocin, N-(3-hydroxy-7-cistetradecenoyl) homoserine lactone, via HdtS, a putative novel *N*-acylhomoserine lactone synthase. Microbiology 146:2469–2480

Lee JH, Lequette Y, Greenberg EP (2006) Activity of purified QscR, a *Pseudomonas aeruginosa* orphan quorum-sensing transcription factor. Mol Microbiol 59:602–609

Lequette Y, Lee JH, Ledgahm F, Lazdunski A, Greenberg EP (2006) A distinct QscR regulon in the *Pseudomonas aeruginosa* quorum-sensing circuit. J Bacteriol 188:3365–3370

Licciardello G, Bertani I, Steindler L, Bella P, Venturi V, Catara V (2009) The transcriptional activator *rfiA* is Quorum Sensing regulated by cotranscription with the *luxI* homolog *pcoI* essential for plant virulence in

Pseudomonas corrugate. Mol Plant Microbe Interact 22:1514–1522

Licciardello G, Cinzia PS, Bertani I, Bella P, Fiore A, Fogliano V, Venturi V, Catara V (2012) *N*-acyl-homoserine-lactone quorum sensing in tomato phytopathogenic *Pseudomonas* spp. is involved in the regulation of lipodepsipeptide production. J Biotechnol 159:274–282

Lintz MJ, Oinuma KI, Wysoczynski CL, Greenberg EP, Churchill MEA (2011) Crystal structure of QscR, a *Pseudomonas aeruginosa* quorum sensing signal receptor. Proc Natl Acad Sci USA 108:15763–15768

Liu M, Wang H, Griffiths MW (2007) Regulation of alkaline metalloprotease promoter by *N*-acyl homoserine lactone quorum sensing in *Pseudomonas fluorescens*. J Appl Microbiol 103:2174–2184

Liu HB, Koh KP, Lee JH, Kim JS, Park S (2009) Characterization of LasR protein involved in bacterial quorum sensing mechanism of *Pseudomonas aeruginosa*. Biotechnol Bioprocess Eng 14:146–154

McKnight SL, Iglewski BH, Pesci EC (2000) The Pseudomonas quinolone signal regulates rhl quorum sensing in *Pseudomonas aeruginosa*. J Bacteriol 182:2702–2708

Park SJ, Liu HB, Park S, Lee JH (2013) Modulation of QscR, a quorum sensing receptor of *Pseudomonas aeruginosa*, by truncation of a signal binding domain. Res Microbiol 164:375–381

Quinones B, Pujol CJ, Lindow SE (2004) Regulation of AHL production and its contribution to epiphytic fitness in *Pseudomonas syringae*. Mol Plant Microbe Interact 17:521–531

Roy V, Adams BL, Bentley WE (2011) Developing next generation antimicrobials by intercepting AI-2 mediated quorum sensing. Enzyme Microb Technol 49:113–123

Schuster M, Greenberg EP (2006) A network of networks: quorum-sensing gene regulation in *Pseudomonas aeruginosa*. Int J Med Microbiol 296: 73–81

Schuster M, Urbanowski ML, Greenberg EP (2004) Promoter specificity in *Pseudomonas aeruginosa* quorum sensing revealed by DNA binding of purified LasR. Proc Natl Acad Sci USA 101:15833–15839

Shaw PD, Ping G, Daly S, Cronan JE, Rhinehart K, Farrand SK (1997) Detecting and characterizing acyl-homoserine lactone signal molecules by thin layer chromatography. Proce Natl Acad Sci USA 94:6036–6041

Sio FC, Otten LG, Cool RH, Diggle SP, Braun PG, Bos R, Daykin M, Camara M, Williams P, Quax WJ (2012) Quorum Quenching by an *N*-Acyl-Homoserine Lactone Acylase from *Pseudomonas aeruginosa* PAO. Infect Immun 74:1673–1682

Turovskiy Y, Kashtanov D, Paskhover B, Chikindas ML (2007) Quorum sensing: fact, fiction, and everything in between. Adv Appl Microbiol 62:191–234

Venturi V (2006) Regulation of quorum sensing in *Pseudomonas*. FEMS Microbiol Rev 30:274–291

Wade SD, Calfee MW, Rocha RE, Ling EA, Engstrom E, Coleman JP, Pesci EC (2005) Regulation of *Pseudomonas* quinolone signal synthesis in *Pseudomonas aeruginosa*. J Bacteriol 187:4372–4380

Willcox MDP, Zhu H, Conibear TCR, Hume EBH, Givskov M, Kjelleberg S, Rice SA (2008) Role of quorum sensing by *Pseudomonas aeruginosa* in microbial keratitis and cystic fibrosis. Microbiology 154:2184–2194

Williams P, Camara M (2009) Quorum sensing and environmental adaptation in *Pseudomonas aeruginosa*: a tale of regulatory networks and multifunctional signal molecules. Curr Opin Microbiol 12:182–191

Winzer K, Falconer C, Garber NC, Diggle SP, Camara M, Williams P (2000) The *Pseudomonas aeruginosa* lectins PA-IL and PA-IIL are controlled by quorum sensing and by RpoS. J Bacteriol 182:6401–6411

Zhang Z, Pierson LS (2001) A second quorum-sensing system regulates cell surface properties but not phenazine antibiotic production in *Pseudomonas aureofaciens*. Appl Environ Microbiol 67:4305–4315

Quorum Sensing in *Escherichia coli*: Interkingdom, Inter- and Intraspecies Dialogues, and a Suicide-Inducing Peptide

Bloom-Ackermann Zohar and Ilana Kolodkin-Gal

Introduction

An emerging theme in microbiology is the ability of bacteria to communicate with one another *via* quorum-sensing signal molecules (Bassler and Losick 2006; Camilli and Bassler 2006; Fuqua et al. 1996; Waters and Bassler 2005). Quorum sensing provides a mechanism for bacteria to monitor one another's presence and to modulate gene expression in response to population density. In the simplest scenario, accumulation of a threshold autoinducer concentration, which is correlated with increasing population density, initiates a signal transduction cascade that culminates in a population-wide alteration in gene expression. Our text brought here is highlighting the recent development in the study of quorum-sensing behaviors in *E. coli. E. coli* is a rod-shaped bacterium from the family *Enterobacteriaceae*. It is able to grow both aerobically and anaerobically, preferably at 37 °C, and can either be nonmotile or motile. Besides being prominent and a fascinating model organism, *Escherichia coli* can be an innocuous resident of the gastrointestinal tract or cause significant diarrheal and extraintestinal diseases (Croxen et al. 2013). Genome sizes of *E. coli* can differ by a million base pairs between commensals and pathogenic variants, and this extra-genetic content can contain virulence and fitness genes. The pathogenic ability of *E. coli* is therefore largely afforded by the flexible gene pool through the gain and loss of genetic material at a number of hot spots throughout the genome (Touchon et al. 2009).

Quite surprisingly, this Gram-negative bacterium, which has been intensively investigated for over 60 years and is the most widely studied prokaryotic model system, is poorly understood and investigated for its social behaviors and more particularly for quorum-sensing pathways. The relative meagerness of data regarding the quorum-sensing pathways participating in the regulation of group behaviors in *E. coli* may also relate to some critical riddles regarding the exact mechanism of pathogenicity, for example, the regulation of attaching and effacing lesions and acid resistance during the persistent cattle infection by enterohemorrhagic *E. coli* (EHEC) that causes severe foodborne disease (Kanamaru et al. 2000).

This review focuses on the major quorum-sensing systems comprehensively studied in *E. coli*. We chose to divide them into the following five categories:

(I) *SdiA-mediated signaling* [SdiA is a LuxR homolog, a receptor for homoserine lactones]

(II) *Indole signaling*, mediated by the self-produced effector indole

(III) *AI-2 signaling*, mediated by an autoinducer produced by the enzyme LuxS

B.-A. Zohar • I. Kolodkin-Gal (✉)
Department of Molecular Genetics, Weizmann Institute of Science, Rehovot, Israel
e-mail: Ilana.Kolodkin-Gal@weizmann.ac.il

V.C. Kalia (ed.), *Quorum Sensing vs Quorum Quenching: A Battle with No End in Sight*,
DOI 10.1007/978-81-322-1982-8_9, © Springer India 2015

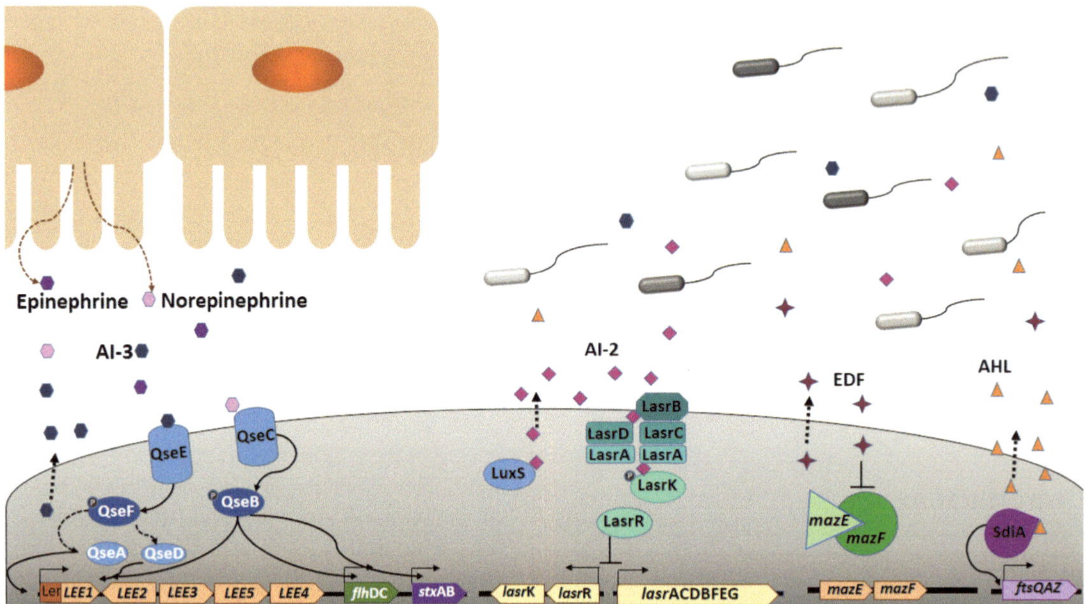

Fig. 1 *Self-produced QS signals as well as signals produced by epithelial cells and neighbor gut bacteria are integrated by Escherichia coli.* The QscEF two-component system integrates self-produced autoinducer names AI-3, and the hormones produced by the mammalian host (epinephrine and norepinephrine) represented as hexagons (Fig. 1). Self and neighbor bacteria-produced AI-2 represented as diamond is sensed by LasR. Self-produced EDF represented by a star is sensed by the toxin MazF, and finally, SdiA senses HSLs produced by neighbor bacteria represented by triangles. The overall gene expression and behavior of *E. coli*, residing in the gut, represents an outcome of integration of all these signals

(IV) *EDF signaling* conveyed by a self-produced peptide that triggers the activation of toxin–antitoxin systems

(V) *AI-3/epinephrine/norepinephrine signaling pathway*, involved in host–bacteria communication

The majority of these signaling systems are involved in interspecies communication, and the AI-3/epinephrine/norepinephrine signaling system is also involved in interkingdom communication (Fig. 1).

SdiA Quorum-Sensing System: Sensing Bacterial Neighbors in the Gastrointestinal Track

In Gram-negative bacteria, the most studied type of quorum-sensing (QS) systems are LuxI/LuxR homologs. The LuxI homolog synthesizes a QS signal molecule, and the LuxR homolog, the signal receptor, binds the signal and responds by regulating gene transcription (Engebrecht and Silverman 1984; Fuqua et al. 1996). LuxI homologs produce a spectrum of related *N*-acyl-homoserine lactone (acyl-HSL) signal molecules. The acyl-HSLs have a conserved homoserine lactone ring connected by an amide linkage to a variably structured acyl side chain. The acyl side chain can vary in length, ranging from 4 to 18 carbons, can be substituted with a carbonyl or hydroxyl group on the third carbon, and may or may not be saturated. Each individual LuxI produces a type of acyl-HSL specific for detection by its cognate LuxR. These autoinducers are synthesized in the cytoplasm during exponential growth and can diffuse passively through the bacterial membrane, accumulating both inside and outside of the cell. When the concentration of AHLs reaches the stimulatory level, the AHLs are bound by LuxR-type protein molecules (Hanzelka and Greenberg 1995). The LuxR–AHL complexes activate the transcription of quorum-sensing-regulated target genes by binding to the appropriate promoters (Hanzelka and Greenberg 1995). Most LuxR-type proteins require binding of AHL as a folding switch that stabilizes them; as in the absence of the signal, they are targeted to degradation (Zhu and Winans 2001).

In a singular exception, one branch of the proteobacteria phylum, within the *Gammaproteobacteria* class, encodes an orphan LuxR homolog named SdiA (Ahmer 2004). This branch includes the medically relevant genus of *Escherichia*, *Salmonella*, *Klebsiella*, *Shigella* (Smith and Ahmer 2003), and *Enterobacter* (Swearingen et al. 2013). In these genomes, there are no acyl-HSL synthase genes, reminiscent to LuxI. Furthermore, it has been experimentally verified that *Escherichia coli* and *Salmonella* do not synthesize acyl-HSLs. While *E. coli* and *Salmonella* do not produce acyl-HSLs, they can sense and respond to a variety of acyl-HSLs produced by other QS bacterial species, a phenomenon described as eavesdropping (Lee et al. 2009).

The *sdiA* (suppressor of cell division inhibitor) gene was first isolated as a positive regulator of the cell division genes *ftsQAZ* (Wang et al. 1991). When expressed from a multi-copy plasmid, SdiA suppresses the expression of a number of chromosomally encoded cell division inhibitors leading to an increase in cell division (Wang et al. 1991). SdiA acts as a positive transcription regulator by controlling the P2 promoter of the *ftsQAZ* operon. The overexpression of the *ftsQAZ* gene cluster increases bacterial cell division and blocks the action of endogenous cell division inhibitors (Wang et al. 1991). The effect of chromosomal *sdiA* on chromosomal *ftsQAZ* expression was not examined, but *sdiA* mutant does not have any apparent cell division defects (Wang et al. 1991). Sitnikov and colleagues explored the hypothesis that SdiA regulates *ftsQAZ* in response to AHL, examining the activation of a plasmid-encoded ftsQ–lacZY fusion by a plasmid-encoded *sdiA* using synthetic AHL (3-oxo-C6-HSL, 3-OH-C4-HSL, or C12-HSL) (Sitnikov et al. 1996). A fourfold activation of the system was observed without AHL, yet upon addition of AHL, there was a further increase in the activation up to 7.5-fold. Experiments using chromosomal *sdiA* or chromosomal fusions were not reported (Sitnikov et al. 1996).

Later, an intriguing finding by Kanamaru and colleagues (2000) linked *sdiA*-mediated signaling to pathogenicity in enterohemorrhagic *E. coli* (EHEC) O157:H7. Overexpression of *sdiA* (using an overexpression vector) caused abnormal cell division and reduced adherence to cultured epithelial cells (Kanamaru et al. 2000). In addition, SdiA reduced expression of virulence factors, encoded on the enterocyte effacement (LEE) pathogenicity island (Lee et al. 2009). The signals activating *sdiA* remained unknown.

Additional link between SdiA-mediated signaling and intestinal colonization of calves by *E. coli* O157:H7 was provided by comparing wild-type and *sdiA* mutant strains. Transcriptome studies of SdiA–AHL signaling in EHEC revealed that SdiA–AHL signaling altered the expression of 49 genes, including the LEE (locus of enterocyte effacement) locus and GAD (glutamic acid decarboxylase) system. LEE is a pathogenicity island, encoding for type III secretion system, and the associated chaperons and effectors required attaching and effacing (AE) lesions in the large intestine. The GAD system is the most efficient acid resistance (AR) mechanism in *E. coli* (Castanie-Cornet et al. 1999). Both are essential for EHEC colonization of cattle and pathogenicity (Dziva et al. 2004; Sheng et al. 2006). Importantly, when oxo-C6-homoserine lactone was added, transcription of the LEE genes was decreased in the WT strain but not in an *sdiA* mutant. These results suggested that AHLs repress transcription of the LEE genes, and, critically, this repression is mediated through SdiA. In addition, Hughes and colleagues demonstrated, by electrophoretic mobility shift assays, that SdiA binds to the promoter of ler (LEE-encoded regulator). Ler (encoded by ler) is the master regulator of the LEE locus. Thus, SdiA in the presence of AHL appears to repress the LEE virulence locus in *E. coli* O157:H7 by regulating the transcription of ler (Hughes et al. 2010).

SdiA has an additional role in controlling biofilm formation. Biofilms are bacterial multicellular communities in close association with surfaces and interfaces, who acquire phenotypic resistance versus environmental insults and antimicrobials (Costerton et al. 1999). Overexpression of *sdiA* in *E. coli* at 37 °C led to an increase in biofilm formation, and an isogenic *sdiA* null mutant showed a threefold decrease in biofilm formation as compared to the wild type (Suzuki et al. 2002). Lee and colleagues found that an *sdiA* mutant of *E. coli* demonstrated a 51-fold increase in biofilm formation as compared to the wild type at moderate temperatures. The presence of SdiA is required to reduce *E. coli* biofilm formation in the presence of AHLs as well as in the presence of indole, which will be discussed later (Lee et al. 2007). It was also demonstrated that SdiA plays a role in decreasing the formation of extracellular matrix components, such as the curli fibers anchoring cell to cell within the biofilm, and regulates motility (Lee et al. 2009).

SdiA was also demonstrated to play a role in antibiotic resistance. Overexpression of *sdiA* confirmed resistance to mitomycin C (MMC) (Wei et al. 2001). The proposed mechanism for this resistance was the activation of efflux pumps by SdiA. It was also noted that, although overexpression of sdiA resulted in increased resistance to MMC, an *sdiA* mutation in the chromosome did not increase sensitivity to MMC (Wei et al. 2001).

In addition to mitomycin C, ectopic expression of *sdiA* can confer resistance to quinolones and chloramphenicol, as well as a less pronounced increase in resistance to kanamycin and tetracycline (Rahmati et al. 2002). This *sdiA*-dependent resistance was found to depend on the drug efflux genes *acrAB*. An *sdiA* mutant was reported to be two- to threefold more sensitive to fluoroquinolones than wild type but was not more sensitive to chloramphenicol, tetracycline, or nalidixic acid (Rahmati et al. 2002).

In *E. coli* K-12 (with chromosomal *sdiA*), the *gadA* gene was upregulated when C6-HSL was added; upregulation was noted at 30 °C but not at 37 °C (Van Houdt et al. 2006). *gadA* encodes glutamate decarboxylase A, involved in acid resistance (i.e., acid resistance system 2 [AR2]), which allows survival of the bacteria at extreme pH values. The addition of C6-HSL increased the tolerance of the wild-type *E. coli* strain to pH 4, whereas the addition of AHL to an isogenic strain with a deleted sdiA did not lead to acid tolerance and resulted in cell death (Van Houdt et al. 2006). Thus, SdiA is involved in acid resistance. The authors did not complement the mutant *E. coli* strain with an active sdiA gene followed by addition of C6-HSL. Lee and colleagues showed that overexpression of SdiA downregulated the genes associated with acid resistance, including *gadA* (Lee et al. 2007).

In summary, SdiA-dependent signaling plays various important roles in regulating social and communal behaviors in pathogenic and nonpathogenic strains of *E. coli.* It is highly unlikely that it is not dedicated to sense a signal with physiological relevance. It was suggested in several studies to sense a stationary-phase-related secondary metabolite, which remains to be elucidated. However, both *E. coli* and *Salmonella* carry functional SdiA, but lack the LuxI homologs, and both inhabit the intestinal environment of humans and many other animals. We suggest that the simplest hypothesis is that *E. coli* and *Salmonella* use SdiA to detect acyl-HSL production of the normal intestinal microbiota.

Indole Signaling

Indole production by *E. coli* during a stationary growth phase was reported in 1897 (Smith 1897). Indole is an aromatic heterocyclic organic compound, consisting of a six-membered benzene ring fused to a five-membered nitrogen-containing pyrrole ring. In bacteria, indole is produced by the *tnaA* gene product, coding for the enzyme tryptophanase, whose major enzymatic reaction is the breakdown of tryptophan into ammonia, pyruvate, and indole. *E. coli* produces very little indole during exponential growth. However, the expression of tryptophanase is strongly upregulated by the stationary-phase sigma factor RpoS, so indole production rises as cells approach the stationary phase (Gaimster et al. 2014).

Indole is highly produced by *E. coli*, and a test for its presence has been regularly used as a diagnostic marker for the identification of *E. coli* (Wang et al. 2001). The identity of indole as a possible extracellular signaling molecule was confirmed by Wang and colleagues (2001) who demonstrated that indole can act as a signaling molecule activating the transcription of *gabT*, *astD*, and *tnaAB* genes (Wang et al. 2001). Activation of the *tnaAB* operon is predicted to induce more indole production potentially creating a positive feedback loop. The other two targets of indole-mediated signaling, *astD* and *gabT*, are involved in pathways that degrade amino acids to pyruvate or succinate providing energy to starving cells. These results led to a speculation that signaling by indole may have a role in adaptation of bacterial cells to a nutrient-poor environment where amino acid catabolism is an important energy source (Wang et al. 2001). Currently, multiple roles have and are been assigned for indole signaling in various aspects of bacterial physiology. The biological function of indole has been extensively studied in the past decade, and diverse functions were reported for the molecule.

Multiple studies have shown that indole increases drug resistance by inducing the intrinsic xenobiotic exporter genes (*mdtEF* and *acrD*), acting via a two-component system (BaeSR and CpxAR) (Hirakawa et al. 2005). Hirakawa and colleagues (2005) proposed a model suggesting that indole first acts on the sensor kinases BaeS and CpxA; the signals are then transmitted to the cognate response regulators which directly bind to different sites in the promoter regions of the exporter genes upregulating their expression. Further evidence for indole-induced drug resistance was introduced when the development of antibiotic-resistant strains was studied. Lee and colleagues found that a few highly resistant mutants rose in the population upon increasing levels of antibiotic treatment. These mutants improved the survival of the population's less-resistant constituents, in part by producing indole (Lee et al. 2010). Following this research, Vega and colleagues examined the hypothesis that indole signaling may trigger the formation of bacterial persisters, a phenomenon in which a subset of an isogenic bacterial population tolerates antibiotic treatment. Incubating *E. coli* cultures with indole prior to treatment with high concentrations of bactericidal antibiotics led to different degrees of persistence to different antibiotics (Vega et al. 2012). Furthermore, the incubation with indole increased persistence to each of the tested antibiotics by at least an order of magnitude indicating that the protective effects of indole are not specific to a single antibiotic and suggesting that indole induces the transition to a persistent state (Vega et al. 2012).

Indole also controls group behavior such as biofilm formation, i.e., a process whereby microorganisms irreversibly attach to and grow on a surface and produce extracellular polymers that facilitate attachment. Using DNA microarrays, Ren and colleagues discovered that genes for the synthesis of indole (*tnaAL*) were induced by a stationary-phase signal (Ren et al. 2004b). Following this work, they revealed that the gene encoding tryptophanase, *tnaA*, was significantly repressed in 6-day-old *E. coli* biofilms in complex medium (Ren et al. 2004a). The differential gene expression of two *E. coli* mutants, *yliH* and *yceP*, both exhibiting increased biofilm formation, demonstrated that indole probably inhibits biofilm formation (Domka et al. 2006). The deletion of each gene leads to biofilms with lower intracellular indole concentrations, resulting in a dramatic increase in biofilm formation. At the same time, the addition of extracellular indole reduced biofilm formation for these mutants (Domka et al. 2006).

In contrast, Di Martino and colleagues reported that indole induces biofilm formation in *E. coli* as *tnaA* deletion resulted in reduction in biofilm formation which was restored by the addition of indole (Martino et al. 2003). To explore this contradiction, Lee and colleagues examined the role of indole in biofilms by performing DNA microarrays on mutants of genes that control indole synthesis in *E. coli* (Lee et al. 2007). They revealed that the effect of exogenous indole is more significant in the presence of glucose, as glucose turns off endogenous indole production resulting in a profound effect of the experimental conditions on biofilm formation. In addition they discovered that indole negatively regulates *E. coli* biofilm formation through SdiA by reducing motility and by influencing acid resistance (Lee et al. 2007). An additional explanation for the discrepancy observed in the effects of indole on biofilm formation was the discovery that indole signaling occurs primarily at low temperatures (below 37 °C) (Lee et al. 2008).

Indole has also been shown to influence biofilm formation of enterohemorrhagic *E. coli* O157:H7 (EHEC) (Bansal et al. 2007) by influencing motility, acid resistance, chemotaxis (i.e., movement of an organism in response to a chemical stimulus), and adherence to HeLa cells. The migration of EHEC to the epithelial cell surface while primarily driven by epinephrine and norepinephrine occurs only in regions of the biofilm where indole concentrations are below the critical threshold. Accordingly, EHEC colonization occurs to a large extent in regions of the gastrointestinal tract that are not colonized by nonpathogenic *E. coli* (regions with low indole levels). Moreover, since *E. coli* O157:H7 itself secretes indole, subsequent colonization will occur only in places that are not already colonized by the pathogen, thereby contributing to further colonization and the spread of infection (Bansal et al. 2007).

The relationship between indole and type III secretion and the formation of A/E lesions in pathogenic *Escherichia coli* (EHEC) O157:H7 was examined by Hirakawa and colleagues (2009). They revealed that indole increases the production and secretion of type III secretion system-mediated translocators, leading to an increase in the formation of A/E lesions in HeLa cells. Addition of indole restored and enhanced the secretion of type III secretion system-mediated translocators as well as the formation of A/E lesions by the *tnaA* deletion mutant EHEC. Taken together, the results reported by Bansal and colleagues and Hirakawa and colleagues may indicate that indole has dual roles in the virulence of EHEC (Bansal et al. 2007; Hirakawa et al. 2009). These observations suggest that the virulence of EHEC is tightly regulated by the concentration of the indole. While addition of 500–600 μM of indole decreased motility, biofilm formation, and attachment to HeLa cells (Bansal et al. 2007), type III secretion-related protein production and virulence phenotypes are stimulated by indole concentrations of 125 μM (Hirakawa et al. 2009). Indole concentration in the enteric site may change dynamically with the amount of indole-producing enteric bacteria the amount of commensal bacteria and the environmental conditions leading to a tight control of EHEC virulence.

In summary, indole is one of most influential cell-to-cell signaling systems that have been identified in *E. coli*. It may be metabolic in nature or true "quorum-sensing" systems meant to coordinate the behavior of microbial populations. It seems to play a cardinal role in regulating *E. coli* social behaviors, persistence, antibiotic resistance, and maybe most importantly virulence.

luxS/AI-2 System

AHLs represent the major class of known bacterial cell–cell signaling molecules. However, the bacterial repertoire of communication was significantly enhanced by the discovery of a family of molecules generically termed autoinducer-2 (AI-2). These families have been found to be widespread in the bacterial world and to facilitate interspecies communication. AI-2s are all derived from a common precursor, 4,5-dihydroxy-2,3-pentanedione (DPD), the product of the LuxS enzyme (Surette et al. 1999). DPD undergoes spontaneous rearrangements to produce a collection of interconverting molecules, some (and perhaps all) of which encode information (Xavier and Bassler 2005a). Presumably, AI-2 interconversions allow bacteria to respond to endogenously produced AI-2 and also to AI-2 produced by other bacterial species in the vicinity. Thus, in contrast with the specific dialogue-based communication channels provided by the HSLs, AI-2 represents a universal language. AI-2, often in conjunction with an AHL or oligopeptide autoinducer, controls a variety of traits in different bacteria ranging from bioluminescence in *V. harveyi* to growth in *Bacillus anthracis* to virulence in

Vibrio cholerae and many other clinically relevant pathogens. *E. coli* carries an active copy of LuxS, the enzyme involved in the metabolism of S-adenosyl methionine (SAM); it converts S-ribosyl homocysteine into homocysteine and 4,5-dihydroxy-2,3-pentanedione (DPD).

In *E. coli*, expression of LuxS demonstrated a fascinating case of coupling the production of quorum-sensing signals to the cell's metabolic state. As expected, the expression of *luxS* increases with bacterial growth (Li et al. 2006; Xavier and Bassler 2005b). However, luxS is also induced at low pH, high osmolarity, and in the presence of a preferred carbon source such as glucose (Ahmer 2004). The synthesis and uptake of AI-2 are subject to catabolite repression through a complex of cyclic AMP (cAMP)–CRP (catabolite regulation protein), which directly stimulates transcription of the *lsr* (luxS regulated) operon and indirectly represses *luxS* expression. The cAMP–CRP complex was shown to bind to a CRP binding site located in the upstream region of the lsr promoter and that mutation in the CRP binding site abolishes this stimulation (Wang et al. 2005a). The expression of *luxS* and the production of AI-2 are regulated at the posttranscriptional level by a small RNA (sRNA) cyaR, by direct binding with complementary sequences in *luxS* mRNA, activating its degradation (De Lay and Gottesman 2009). Given that cyaR is positively regulated by the cAMP–CRP complex, it is thus induced under conditions of low glucose, providing an explanation for the observed increase of LuxS in the presence of glucose (De Lay and Gottesman 2009). A second sRNA, *micA*, was found to affect the length and transcript levels of *luxS* in an RNase III-dependent manner (Udekwu 2010), but whether this regulation affects protein amounts and activity or explains the observed growth dependence of LuxS expression is not yet known (Udekwu 2010).

Alternative pathways for AI-2 formation which are LuxS independent were reported by Li and colleagues, as AI-2 activity was observed from *luxS*-deficient extracts supplied with adenosine (Li et al. 2006). Similar results were observed by Tavender and colleagues using an *E. coli luxS* mutant carrying additional mutations that alter carbon fluxes, suggesting LuxS-independent formation of AI-2, via spontaneous conversion of ribulose-5-phosphate (Tavender et al. 2008). The importance of this route for AI-2 production in *E. coli* may be negligible; it may, however, be responsible for the AI-2-like signals reported for some higher organisms or bacteria lacking *luxS* (Tavender et al. 2008).

While the export mechanisms of AI-2 are not fully understood, the uptake of AI-2 by the Lsr transport system has been extensively studied in multiple organisms (Pereira et al. 2013; Wang et al. 2005a, b; Xavier and Bassler 2005b). The AI-2 uptake system by the *lsrACDBFG* operon was first elucidated by Taga and colleagues in *Salmonella enterica* serovar *typhimurium* (Taga et al. 2001). The receptor for AI-2 is LsrB, a high-affinity substrate-binding periplasmic protein which interacts with the membrane components of an ABC transport system. The transporter comprises a membrane channel formed by two transmembrane proteins, LsrC and LsrD, and an ATPase, LsrA, which provides the energy for the transport of the AI-2 signal (Pereira et al. 2013; Xavier et al. 2007). The transporter proteins are regulated by cyclic AMP/cyclic AMP receptor protein and by the product of *lsrK* and *lsrR*, located immediately upstream of lsr and divergently transcribed in its own lsrRK operon (Li et al. 2007). Following uptake, AI-2 is phosphorylated by the kinase LsrK to produce phospho-AI-2 (P-AI-2); the active molecule then binds and derepresses the lsr repressor LsrR (Wang et al. 2005b). In the absence of P-AI-2, LsrR represses the transcription of both the lsr operon and lsrRK operon, thus regulating its own expression and that of LsrK (Pereira et al. 2013). This positive feedback loop derives the uptake of AI-2, allowing a rapid uptake. The AI-2 is further processed by two additional genes within the *lsr* operon, LsrF and LsrG (Xavier et al. 2007). Recently, it was reported by Pereira and colleagues that phosphoenolpyruvate phosphotransferase system (PTS) is required for Lsr activation and AI-2 internalization (Pereira et al. 2012). The PTS catalyzes transport across the periplasmic membrane of a large range of compounds concomitant with their phosphorylation. The proposed mechanism suggests that the initial uptake of AI-2 is PTS dependent, thereby allowing derepression of the *lsr* operon and the initiation of Lsr-dependent uptake of AI-2 (Pereira et al. 2012).

The uptake of AI-2 would lead to differential gene expression regulating various social behaviors. Multiple studies have used DNA microarrays to show AI-2 controls 166 to 404 genes, including those for chemotaxis, flagellar synthesis, motility, and virulence factors in *E. coli* (DeLisa et al. 2001; Ren et al. 2004b; Sperandio et al. 2001). Gonzalez Barrios and colleagues demonstrated that AI-2 stimulates biofilm formation and motility in five different *E. coli* hosts (ATCC 25404, MG1655, BW25113, DH5α, and JM109) in two different media (Gonzalez Barrios et al. 2006). The induction was dependent upon the presence of the LsrK enzyme, indicating that AI-2 signaling through the Lsr system was responsible for these phenotypes (Gonzalez Barrios et al. 2006). These results further explain previous reports by Sperandio and colleagues and Ren and colleagues indicating that AI-2 controls chemotaxis, flagellar synthesis, and motility in *E. coli* and that the quorum-sensing antagonist furanone was effective in preventing the biofilms of *E. coli* by repressing these same chemotaxis, flagellar synthesis, and motility genes (Ren et al. 2004a, b; Sperandio et al. 2001). Further analysis of *lsrR* and *lsrK* mutant strains using microarrays increased the evidence for Lsr-dependent regulation of biofilm formation and motility (Li et al. 2007).

An additional role for AI-2 is as a chemoattractant (i.e., a chemical agent that induces movement of chemotactic cells) for *E. coli* K-12, a behavior dependent upon both the L-serine receptor (Tsr) and LsrB (Englert et al. 2009; Hegde et al. 2011). It is hypothesized that LsrB binding to AI-2 in the periplasm enables interaction with the periplasmic domain of Tsr; the resulting signaling elicits downstream chemotactic responses, while the LsrC seems to be dispensable for this process (Hegde et al. 2011). Given that AI-2 is produced by many different species of biofilm-forming bacteria, the proposed ecological context in which AI-2 chemotaxis occurs may serve to recruit free-swimming, planktonic bacteria to biofilm (Hegde et al. 2011). Concentration-dependent chemoattraction of enterohemorrhagic *E. coli* (EHEC) by AI-2 has also been observed through the use of *luxS* mutant bacteria (Bansal et al. 2008). In this verotype, AI-2 also regulates pathogen motility and attachment to HeLa cells (Bansal et al. 2008). Similarly, *luxS* mutant of enteropathogenic *E. coli* (EPEC) exhibits reduced motility compared to wild-type bacteria when in the presence of epithelial cells (Girón et al. 2002).

Overall, these studies show that in *E. coli*, *luxS* gene expression and AI-2 signaling are important for proper regulation of biofilm formation, motility, attachment to epithelial cells, and other crucial virulence traits.

Peptide Signaling in *E. coli*, the EDF

In Gram-positive bacteria, autoinducers are short, usually modified peptides processed from precursors. They are involved, for example, in the development of competence and sporulation in *B. subtilis* (Lazazzera 2001; Magnuson et al. 1994; Tortosa and Dubnau 1999) as well as in the virulence response in *S. aureus* (Ji et al. 1997; Lazdunski et al. 2004; Novick 2003) and in biofilm formation in *Streptococcus mutans* (Aspiras et al. 2004). These oligopeptide autoinducers are actively transported out of the cell, and they interact with the external domains of membrane-bound sensor proteins. Signal transduction occurs by a phosphorylation cascade that activates a DNA binding protein that controls transcription of target genes. These autoinducers in Gram-positive bacteria are highly specific because each sensor oligopeptide selects for a given peptide signal (Lyon and Novick 2004; Waters and Bassler 2005). Unique case for peptide based signaling linking Programmed Cell Death and QS exists in *E. coli*, several prokaryotic genetic modules have been described as systems that mediate programmed cell death. Among these is the *E. coli* toxin–antitoxin module *mazEF*, which is the first bacterial programmed cell death system that was described. *mazF* encodes a stable toxin, MazF, and *mazE* encodes a labile antitoxin, MazE, that prevents the lethal effect of MazF. Thus, any stressful condition that prevents the expression of the chromosomally borne *mazEF* module will lead to the reduction of MazE in the cell, permitting toxin MazF to act freely (Engelberg-Kulka et al. 2006). Such conditions include (1) short-term inhibition of transcription and/or translation by antibiotics such as rifampin, chloramphenicol, and spectinomycin (Sat et al. 2001); (2) the overproduction of ppGpp, which inhibits *mazEF* transcription (Aizenman et al. 1996); and (3) DNA damage caused by thymine starvation (Sat et al. 2003) as well as by DNA-damaging agents, such as mitomycin C or nalidixic acid (Hazan et al. 2004). These antibiotics and stressful conditions that are well known to cause bacterial cell death have been found to act through the *mazEF* module. Clearly, a system that causes any given cell to die is not advantageous to that particular cell. On the other hand, the death of an individual cell may be advantageous for the bacterial population, enduring stressful conditions as a whole. *E. coli mazEF*-mediated cell death could only be observed in cultures with high population density. Therefore, it was assumed that cell death probably requires a "quorum sensing," a process in which a secreted autoinducer allows cells to "measure" cell density of the medium. This quorum-sensing process involves a secreted signaling molecule that was designated the *e*xtracellular *d*eath *f*actor (EDF). EDF was determined to be an oligopeptide. Thus, it was different from other molecules so far described to participate in quorum sensing in *E. coli*: AHL, AI-2, and indole. The characterization of the chemical nature of EDF revealed that EDF is a linear pentapeptide NNWNN. Each of the five amino acids in EDF is important for its *mazEF*-mediated killing activity, and the terminal asparagines are the most crucial (Kolodkin-Gal et al. 2007). The quorum-sensing process involved in *mazEF*-mediated cell death and the quorum-sensing peptide EDF are particularly interesting not only because no other peptide has apparently been reported to be involved in quorum sensing in *E. coli* but also because EDF appears to be a distinct type of molecule related to the quorum-sensing peptides of gram-positive bacteria. An additional surprising finding was the source of the pentapeptide. Many self-generated, secreted signaling molecules in bacteria are small molecules, including peptides. So the fact that the "extracellular death factor" is a peptide was no surprise. But the peptide signals described in earlier studies are encoded in small genes that generate prepeptides, which are processed to yield the signaling molecule. The *E. coli* peptide is derived from the degradation of glucose-6-phosphate dehydrogenase, a metabolic enzyme. Although the exact pathway leading to the production of this pentapeptide remains to be defined, it seems reasonable that it is made when the bacterium is under an apparent internal stress. Such stress can be due to the onset of starvation that can generate this signal, priming the population such that some cells can be killed through the programmed cell death pathway when confronted with other stresses (Kolter 2007). The *zwf* product, carrying the sequence NNWDN, may generate the full NNWNN sequence only by subsequent amidation step. Amidation may occur either before or after the cleavage of the precursor by one of *E. coli's* proteases. The involvement of Asn synthetase A, an enzyme that can perform such a modification (Humbert and Simoni 1980), was implied; deleting the gene *asnA* prevented production of active EDF. In the mid-log phase only, the rate-limiting enzyme in the pentose phosphate shunt, glucose-6-phosphate dehydrogenase (G6PD), *E. coli* Zwf, is apparently cleaved between its catalytic and structural domains to release an internal pentapeptide, NNWDN. Amidation of the third asparagine residue is presumed to yield the killer pentapeptide, NNWNN; exposure to recombinant NNWNN, together with a transient stress (e.g., 10 min rifampicin), which activates MazEF, results in significant, but not complete, loss of viability. Other stresses that enforce MazF stabilization or activate RelA to synthesize ppGpp might also activate this pathway (Kolodkin-Gal and Engelberg-Kulka 2008; Kolodkin-Gal et al. 2007).

More recently, the mode of action for EDF was revealed (Belitsky et al. 2011). It was shown that EDF specifically affects the endoribonucleolytic activities of MazF, the mRNA interfering toxin, and of an homologous toxin ChpBK. In vitro, EDF significantly amplified the endoribonucleolytic activities of both MazF and ChpBK. EDF also overcame the inhibitory activities of the antitoxin MazE over the toxin MazF and of the antitoxin ChpBI over the toxin ChpBK. EDF and MazF were directly interacting, and peptide–protein modeling showed parallel contacts between EDF–MazF and MazE–MazF. Thus, the EDF system is a quorum-sensing system of many novel components, both at the level of the peptide autoinducer, derived by proteolysis of metabolic protein, and at the mechanistic level, where once accumulated, the peptide carries catalytic properties.

Very recently, Kumar and colleagues have further demonstrated that the *E. coli mazEF*-mediated cell death could be triggered by QS peptides from the supernatants (SN) of the Gram-positive bacterium *Bacillus subtilis* and the Gram-negative bacterium *Pseudomonas aeruginosa*. A hexapeptide RGQQNE was produced by *Bacillus*, and three peptides, a nonapeptide INEQTVVTK and two hexadecapeptides VEVSDDGSGGNTSLSQ and APKLSDGAAAGYVTKA, were produced by *Pseudomonas aeruginosa*. When added to diluted *E. coli* cultures, each of these peptides acted as an interspecies EDF that triggered *mazE*F-mediated death. Furthermore, though their sequences are very different, each of these EDFs amplified the endoribonucleolytic activity of *E. coli* MazF, probably by interacting with different sites on *E. coli* MazF. Therefore, the group of QS peptides was expended to several additional different peptides. This family provides the first example of quorum-sensing molecules participating in interspecies bacterial cell death. Furthermore, each of these peptides provides the basis of a new class of antibiotics triggering death by acting from outside the cell (Kumar et al. 2013).

AI-3/Epinephrine/Norepinephrine System

Prokaryotes and eukaryotes coexist in both commensal and pathogenic relationships for millions of years, consequently coevolving to sense and respond to each other's signaling molecules (Karavolos et al. 2013). Interkingdom chemical signaling plays an important role in the relationships forged between bacteria and animals. Such interkingdom signaling processes led to high-jacking by bacterial pathogens, such as enterohemorrhagic *Escherichia coli* (EHEC), to activate virulence genes, colonize the host, and initiate the disease process (Clarke et al. 2006; Walters and Sperandio 2006). EHEC colonizes the large bowel and causes a lesion on intestinal epithelial cells (AE lesions). The genes involved in the formation of the AE lesion are encoded within the chromosomal LEE pathogenicity island, which is present in EHEC but absent in commensal and K-12 *E. coli* (Sperandio et al. 2003). The majority of the LEE genes are organized in five operons (LEE1–5), containing a transcriptional activator (Ler) essential for the expression of the LEE genes in addition to the type three secretion system and virulence effectors (Sperandio et al. 2001; Sperandio et al. 2003). Upon reaching the human colon, EHEC senses the autoinducer-3 (AI-3) produced by the microbial gastrointestinal flora and epinephrine and norepinephrine, produced by the host, through histidine sensor kinases (HKs) in their membrane (Clarke et al. 2006). HKs constitute the predominant family of signaling proteins in bacteria, usually acting in concert with a response regulator (RR) protein constituting a two-component system (Hughes et al. 2009). Two HKs, QseC and QseE, both part of the two-component systems (QseCB and QseEF) characterized in *E. coli* have been reported to sense AI-3, epinephrine, and norepinephrine. Upon sensing AI-3, QseC initiates the signaling cascade that will activate the flagella regulon, leading to swimming motility, which may aid EHEC to reach the intestinal epithelial layer. QseC activates transcription of the master regulators of the flagellar regulon, directly through QseB as well as activating AE lesion formation and Shiga toxin expression. While the activation of the flagellar regulon is dependent on QseB's phosphorylation state (Clarke and Sperandio 2005), the expression of the LEE and Shiga toxin genes is not regulated by QseB. As EHEC approaches the epithelium and starts forming AE lesions, it is exposed to epinephrine and/or norepinephrine; AE lesion formation and the commencement of bloody diarrhea may increase EHEC exposure to epinephrine and norepinephrine, further upregulating expression of virulence genes in EHEC (Hughes et al. 2009).

A second two-component system, the QseEF system (Reading et al. 2007), where QseE is the HK and QseF is the RR was shown to also regulate virulence in EHEC. QseE can also respond to the host hormone epinephrine like QseC but, in contrast, does not sense the bacterial signal AI-3. The QseEF system is not involved in the regulation of flagella and motility but plays an important role in activating genes necessary for AE lesion formation (Reading et al. 2007) and also activates expression of Shiga toxin.

Norepinephrine has been reported to induce bacterial growth (Freestone et al. 2000), in addition to a role in activates virulence, e.g., Shiga toxin expression in *E. coli* (Lyte et al. 1996) by an unknown mechanism of induction. Because epinephrine and norepinephrine exert a profound effect on the host physiology and immune system, the ability to sense these hormones by bacteria may facilitate gauging the fitness of the host.

Following their earlier studies (Clarke et al. 2006; Clarke and Sperandio 2005) (Hughes et al. 2009; Walters et al. 2006), the Sperandio group recently discovered a novel two-component signal transduction system, named FusKR, where FusK is the histidine sensor kinase and FusR the response regulator. FusK senses fucose and controls expression of virulence and metabolic genes. This fucose-sensing system is required for robust EHEC colonization of the mammalian intestine (Pacheco et al. 2012). The proposed model is that fucose freed from the mucus layer by a member of the microbiota, *Bacteroides thetaiotaomicron*, inhibits LEE expression, relieving the pathogen from the metabolic burden of expressing the type III secretion system and giving it a competitive growth advantage in the lumen of the gut (Pacheco et al. 2012). Once EHEC approaches the mucosal surface, adrenergic metabolites derepress the LEE, initiating its adherence mechanisms. *B. thetaiotaomicron* was able to repress *ler* expression when incubated with mucin, a derivative of mucus that contains bound fucose, suggesting that the fucosidases produced by *B. thetaiotaomicron* cleave fucose from mucin to directly repress virulence expression of EHEC (Keeney and Finlay 2013).

The existence of multiple evidences for interkingdom communication by bacterial pathogens, relying on the host adrenergic network, may hence reflect the culmination of many millions of years of receptor structural evolution to accommodate these signals and orchestrate the best survival response. These interactions are delicately balanced, and it is clear that in most circumstances, they can benefit the pathogen.

Summary

The existence and importance of quorum-sensing signaling in *E. coli* have been established throughout the last decades. Some scenarios in which this occurs in nature, however, are still largely unknown and many questions remain. First, it is clear that the gastrointestinal tracts of most animals, excluding the bovine rumen, do not contain AHL. It appears that EHEC uses *sdiA* in the bovine rumen to repress the LEE pathogenicity island and to increase acid resistance. What organism(s) are producing the AHLs and what role do the AHLs play in the rumen community? Is detection of bacteria in the rumen the only scenario in which *sdiA* provides a benefit to EHEC? Second, it has been embroiled that LuxS synthesizes AI-2, as well as an autoinducer coined AI-2, which remains structurally unsolved. Thus, LuxS can govern bacterial behaviors such as the secretion of virulence factors, biofilm formation, and swarming motility. Can novel approaches to interrupt LuxS sensing be recognized as next-generation antimicrobials? How does interspecies communication from neighbor gastrointestinal tract resident [both Gram-negative and Gram-positive bacteria] use this pathway to communicate with *E. coli*? And lastly, how prominent is the role of interkingdom communication is in shaping the behaviors of *E. coli* in vivo? Clearly, the study of QS and QS sensing by *Escherichia coli* has only just begun.

References

Ahmer BM (2004) Cell-to-cell signalling in *Escherichia coli* and Salmonella enterica. Mol Microbiol 52: 933–945

Aizenman E, Engelberg-Kulka H, Glaser G (1996) An *Escherichia coli* chromosomal "addiction module" regulated by guanosine [corrected] 3′,5′-bispyrophosphate: a model for programmed bacterial cell death. Proc Natl Acad Sci USA 93:6059–6063

Aspiras MB, Ellen RP, Cvitkovitch DG (2004) ComX activity of Streptococcus mutans growing in biofilms. FEMS Microbiol Lett 238:167–174

Bansal T, Englert D, Lee J, Hegde M, Wood TK, Jayaraman A (2007) Differential effects of epinephrine, norepinephrine, and indole on *Escherichia coli* O157:H7 chemotaxis, colonization, and gene expression. Infect Immun 75:4597–4607

Bansal T, Jesudhasan P, Pillai S, Wood TK, Jayaraman A (2008) Temporal regulation of enterohemorrhagic *Escherichia coli* virulence mediated by autoinducer-2. Appl Microbiol Biotechnol 78:811–819

Bassler BL, Losick R (2006) Bacterially speaking. Cell 125:237–246

Belitsky M, Avshalom H, Erental A, Yelin I, Kumar S, London N, Sperber M, Schueler-Furman O, Engelberg-Kulka H (2011) The *Escherichia coli* extracellular death factor EDF induces the endoribonucleolytic activities of the toxins MazF and ChpBK. Mol Cell 41:625–635

Camilli A, Bassler BL (2006) Bacterial small-molecule signaling pathways. Science 311:1113–1116

Castanie-Cornet MP, Penfound TA, Smith D, Elliott JF, Foster JW (1999) Control of acid resistance in *Escherichia coli*. J Bacteriol 181:3525–3535

Clarke MB, Sperandio V (2005) Transcriptional regulation of flhDC by QseBC and sigma (FliA) in enterohaemorrhagic *Escherichia coli*. Mol Microbiol 57:1734–1749

Clarke MB, Hughes DT, Zhu C, Boedeker EC, Sperandio V (2006) The QseC sensor kinase: a bacterial adrenergic receptor. Proc Natl Acad Sci USA 103: 10420–10425

Costerton JW, Stewart PS, Greenberg EP (1999) Bacterial biofilms: a common cause of persistent infections. Science 284:1318–1322

Croxen MA, Law RJ, Scholz R, Keeney KM, Wlodarska M, Finlay BB (2013) Recent advances in understanding enteric pathogenic *Escherichia coli*. Clin Microbiol Rev 26:822–880

De Lay N, Gottesman S (2009) The Crp-activated small noncoding regulatory RNA CyaR (RyeE) links nutritional status to group behavior. J Bacteriol 191: 461–476

DeLisa MP, Wu CF, Wang L, Valdes JJ, Bentley WE (2001) DNA microarray-based identification of genes controlled by autoinducer 2-stimulated quorum sensing in *Escherichia coli*. J Bacteriol 183:5239–5247

Domka J, Lee J, Wood TK (2006) YliH (BssR) and YceP (BssS) regulate *Escherichia coli* K-12 biofilm formation by influencing cell signaling. Appl Environ Microbiol 72:2449–2459

Dziva F, van Diemen PM, Stevens MP, Smith AJ, Wallis TS (2004) Identification of *Escherichia coli* O157: H7 genes influencing colonization of the bovine gastrointestinal tract using signature-tagged mutagenesis. Microbiology 150:3631–3645

Engebrecht J, Silverman M (1984) Identification of genes and gene products necessary for bacterial bioluminescence. Proc Natl Acad Sci USA 81:4154–4158

Engelberg-Kulka H, Amitai S, Kolodkin-Gal I, Hazan R (2006) Bacterial programmed cell death and multicellular behavior in bacteria. PLoS Genet 2: e135

Englert DL, Manson MD, Jayaraman A (2009) Flow-based microfluidic device for quantifying bacterial chemotaxis in stable, competing gradients. Appl Environ Microbiol 75:4557–4564

Freestone PPE, Lyte M, Neal CP, Maggs AF, Haigh RD, Williams PH (2000) The mammalian neuroendocrine hormone norepinephrine supplies iron for bacterial growth in the presence of transferrin or lactoferrin. J Bacteriol 182:6091–6098

Fuqua C, Winans SC, Greenberg EP (1996) Census and consensus in bacterial ecosystems: the LuxR-LuxI family of quorum-sensing transcriptional regulators. Annu Rev Microbiol 50:727–751

Gaimster H, Cama J, Hernandez-Ainsa S, Keyser UF, Summers DK (2014) The indole pulse: a new perspective on indole signalling in *Escherichia coli*. PLoS One 9:e93168

Girón JA, Torres AG, Freer E, Kaper JB (2002) The flagella of enteropathogenic *Escherichia coli* mediate adherence to epithelial cells. Mol Microbiol 44: 361–379

Gonzalez Barrios AF, Zuo R, Hashimoto Y, Yang L, Bentley WE, Wood TK (2006) Autoinducer 2 controls biofilm formation in *Escherichia coli* through a novel motility quorum-sensing regulator (MqsR, B3022). J Bacteriol 188:305–316

Hanzelka BL, Greenberg EP (1995) Evidence that the N-terminal region of the Vibrio fischeri LuxR protein constitutes an autoinducer-binding domain. J Bacteriol 177:815–817

Hazan R, Sat B, Engelberg-Kulka H (2004) *Escherichia coli* mazEF-mediated cell death is triggered by various stressful conditions. J Bacteriol 186:3663–3669

Hegde M, Englert DL, Schrock S, Cohn WB, Vogt C, Wood TK, Manson MD, Jayaraman A (2011) Chemotaxis to the quorum-sensing signal AI-2 requires the Tsr chemoreceptor and the periplasmic LsrB AI-2-binding protein. J Bacteriol 193:768–773

Hirakawa H, Inazumi Y, Masaki T, Hirata T, Yamaguchi A (2005) Indole induces the expression of multidrug exporter genes in *Escherichia coli*. Mol Microbiol 55:1113–1126

Hirakawa H, Kodama T, Takumi-Kobayashi A, Honda T, Yamaguchi A (2009) Secreted indole serves as a signal for expression of type III secretion system translocators in enterohaemorrhagic *Escherichia coli* O157:H7. Microbiology 155:541–550

Hughes DT, Clarke MB, Yamamoto K, Rasko DA, Sperandio V (2009) The QseC adrenergic signaling cascade in Enterohemorrhagic *E. coli* (EHEC). PLoS Pathog 5:e1000553

Hughes DT, Terekhova DA, Liou L, Hovde CJ, Sahl JW, Patankar AV, Gonzalez JE, Edrington TS, Rasko DA, Sperandio V (2010) Chemical sensing in mammalian host-bacterial commensal associations. Proc Natl Acad Sci USA 107:9831–9836

Humbert R, Simoni RD (1980) Genetic and biomedical studies demonstrating a second gene coding for asparagine synthetase in *Escherichia coli*. J Bacteriol 142:212–220

Ji G, Beavis R, Novick RP (1997) Bacterial interference caused by autoinducing peptide variants. Science 276:2027–2030

Kanamaru K, Kanamaru K, Tatsuno I, Tobe T, Sasakawa C (2000) SdiA, an *Escherichia coli* homologue of quorum-sensing regulators, controls the expression of virulence factors in enterohaemorrhagic *Escherichia coli* O157:H7. Mol Microbiol 38:805–816

Karavolos MH, Winzer K, Williams P, Khan CM (2013) Pathogen espionage: multiple bacterial adrenergic sensors eavesdrop on host communication systems. Mol Microbiol 87:455–465

Keeney KM, Finlay BB (2013) Microbiology: EHEC downregulates virulence in response to intestinal fucose. Curr Biol CB 23:R108–R110

Kolodkin-Gal I, Engelberg-Kulka H (2008) The extracellular death factor: physiological and genetic factors influencing its production and response in *Escherichia coli*. J Bacteriol 190:3169–3175

Kolodkin-Gal I, Hazan R, Gaathon A, Carmeli S, Engelberg-Kulka H (2007) A linear pentapeptide is a quorum-sensing factor required for mazEF-mediated cell death in *Escherichia coli*. Science 318:652–655

Kolter R (2007) Microbiology. Deadly priming. Science 318:578–579

Kumar S, Kolodkin-Gal I, Engelberg-Kulka H (2013) Novel quorum-sensing peptides mediating interspecies bacterial cell death. mBio 4:e00314–00313

Lazazzera BA (2001) The intracellular function of extracellular signaling peptides. Peptides 22:1519–1527

Lazdunski AM, Ventre I, Sturgis JN (2004) Regulatory circuits and communication in Gram-negative bacteria. Nat Rev Microbiol 2:581–592

Lee J, Jayaraman A, Wood TK (2007) Indole is an interspecies biofilm signal mediated by SdiA. BMC Microbiol 7:42

Lee J, Zhang XS, Hegde M, Bentley WE, Jayaraman A, Wood TK (2008) Indole cell signaling occurs primarily at low temperatures in *Escherichia coli*. ISME J 2:1007–1023

Lee J, Maeda T, Hong SH, Wood TK (2009) Reconfiguring the quorum-sensing regulator SdiA of *Escherichia coli* to control biofilm formation via indole and *N*-acylhomoserine lactones. Appl Environ Microbiol 75:1703–1716

Lee HH, Molla MN, Cantor CR, Collins JJ (2010) Bacterial charity work leads to population-wide resistance. Nature 467:82–85

Li J, Wang L, Hashimoto Y, Tsao C-Y, Wood TK, Valdes JJ, Zafiriou E, Bentley WE (2006) A stochastic model of *Escherichia coli* AI-2 quorum signal circuit reveals alternative synthesis pathways. Mol Syst Biol 2:67

Li J, Attila C, Wang L, Wood TK, Valdes JJ, Bentley WE (2007) Quorum sensing in *Escherichia coli* is signaled by AI-2/LsrR: effects on small RNA and biofilm architecture. J Bacteriol 189:6011–6020

Lyon GJ, Novick RP (2004) Peptide signaling in Staphylococcus aureus and other Gram-positive bacteria. Peptides 25:1389–1403

Lyte M, Arulanandam BP, Frank CD (1996) Production of Shiga-like toxins by *Escherichia coli* O157:H7 can be influenced by the neuroendocrine hormone norepinephrine. J Lab Clin Med 128:392–398

Magnuson R, Solomon J, Grossman AD (1994) Biochemical and genetic characterization of a competence pheromone from B. subtilis. Cell 77:207–216

Martino PD, Fursy R, Bret L, Sundararaju B, Phillips RS (2003) Indole can act as an extracellular signal to regulate biofilm formation of *Escherichia coli* and other indole-producing bacteria. Can J Microbiol 49: 443–449

Novick RP (2003) Autoinduction and signal transduction in the regulation of staphylococcal virulence. Mol Microbiol 48:1429–1449

Pacheco AR, Curtis MM, Ritchie JM, Munera D, Waldor MK, Moreira CG, Sperandio V (2012) Fucose sensing regulates bacterial intestinal colonization. Nature 492:113–117

Pereira CS, Santos AJM, Bejerano-Sagie M, Correia PB, Marques JC, Xavier KB (2012) Phosphoenolpyruvate phosphotransferase system regulates detection and processing of the quorum sensing signal autoinducer-2. Mol Microbiol 84:93–104

Pereira CS, Thompson JA, Xavier KB (2013) AI-2-mediated signalling in bacteria. FEMS Microbiol Rev 37:156–181

Rahmati S, Yang S, Davidson AL, Zechiedrich EL (2002) Control of the AcrAB multidrug efflux pump by quorum-sensing regulator SdiA. Mol Microbiol 43:677–685

Reading NC, Torres AG, Kendall MM, Hughes DT, Yamamoto K, Sperandio V (2007) A novel two-component signaling system that activates transcription of an enterohemorrhagic *Escherichia coli* effector involved in remodeling of host actin. J Bacteriol 189:2468–2476

Ren D, Bedzyk LA, Thomas SM, Ye RW, Wood TK (2004a) Gene expression in *Escherichia coli* biofilms. Appl Microbiol Biotechnol 64:515–524

Ren D, Bedzyk LA, Ye RW, Thomas SM, Wood TK (2004b) Differential gene expression shows natural brominated furanones interfere with the autoinducer-2 bacterial signaling system of *Escherichia coli*. Biotechnol Bioeng 88:630–642

Sat B, Hazan R, Fisher T, Khaner H, Glaser G, Engelberg-Kulka H (2001) Programmed cell death in *Escherichia coli*: some antibiotics can trigger mazEF lethality. J Bacteriol 183:2041–2045

Sat B, Reches M, Engelberg-Kulka H (2003) The *Escherichia coli* mazEF suicide module mediates thymineless death. J Bacteriol 185:1803–1807

Sheng H, Lim JY, Knecht HJ, Li J, Hovde CJ (2006) Role of *Escherichia coli* O157:H7 virulence factors in colonization at the bovine terminal rectal mucosa. Infect Immun 74:4685–4693

Sitnikov DM, Shadel GS, Baldwin TO (1996) Autoinducer-independent mutants of the LuxR transcriptional activator exhibit differential effects on the two lux promoters of Vibrio fischeri. Mol Gen Genet MGG 252:622–625

Smith T (1897) A modification of the method for determining the production of indol by bacteria. J Exp Med 2:543–547

Smith JN, Ahmer BM (2003) Detection of other microbial species by Salmonella: expression of the SdiA regulon. J Bacteriol 185:1357–1366

Sperandio V, Torres AG, Giron JA, Kaper JB (2001) Quorum sensing is a global regulatory mechanism in enterohemorrhagic *Escherichia coli* O157:H7. J Bacteriol 183:5187–5197

Sperandio V, Torres AG, Jarvis B, Nataro JP, Kaper JB (2003) Bacteria-host communication: the language of hormones. Proc Natl Acad Sci USA 100:8951–8956

Surette MG, Miller MB, Bassler BL (1999) Quorum sensing in *Escherichia coli*, Salmonella typhimurium, and Vibrio harveyi: a new family of genes responsible for autoinducer production. Proc Natl Acad Sci USA 96:1639–1644

Suzuki K, Wang X, Weilbacher T, Pernestig AK, Melefors O, Georgellis D, Babitzke P, Romeo T (2002) Regulatory circuitry of the CsrA/CsrB and BarA/UvrY systems of *Escherichia coli*. J Bacteriol 184: 5130–5140

Swearingen MC, Sabag-Daigle A, Ahmer BM (2013) Are there acyl-homoserine lactones within mammalian intestines? J Bacteriol 195:173–179

Taga ME, Semmelhack JL, Bassler BL (2001) The LuxS-dependent autoinducer AI-2 controls the expression of an ABC transporter that functions in AI-2 uptake in Salmonella typhimurium. Mol Microbiol 42:777–793

Tavender TJ, Halliday NM, Hardie KR, Winzer K (2008) LuxS-independent formation of AI-2 from ribulose-5-phosphate. BMC Microbiol 8:98

Tortosa P, Dubnau D (1999) Competence for transformation: a matter of taste. Curr Opin Microbiol 2: 588–592

Touchon M, Hoede C, Tenaillon O, Barbe V, Baeriswyl S, Bidet P, Bingen E, Bonacorsi S, Bouchier C, Bouvet O et al (2009) Organised genome dynamics in the *Escherichia coli* species results in highly diverse adaptive paths. PLoS Genet 5:e1000344

Udekwu KI (2010) Transcriptional and post-transcriptional regulation of the *Escherichia coli* luxS mRNA; involvement of the sRNA MicA. PLoS One 5:e13449

Van Houdt R, Aertsen A, Moons P, Vanoirbeek K, Michiels CW (2006) *N*-acyl-L-homoserine lactone signal interception by *Escherichia coli*. FEMS Microbiol Lett 256:83–89

Vega NM, Allison KR, Khalil AS, Collins JJ (2012) Signaling-mediated bacterial persister formation. Nat Chem Biol 8:431–433

Walters M, Sperandio V (2006) Quorum sensing in *Escherichia coli* and Salmonella. Int J Med Microbiol IJMM 296:125–131

Walters M, Sircili MP, Sperandio V (2006) AI-3 synthesis is not dependent on luxS in *Escherichia coli*. J Bacteriol 188:5668–5681

Wang XD, de Boer PA, Rothfield LI (1991) A factor that positively regulates cell division by activating transcription of the major cluster of essential cell division genes of *Escherichia coli*. EMBO J 10: 3363–3372

Wang D, Ding X, Rather PN (2001) Indole can act as an extracellular signal in *Escherichia coli*. J Bacteriol 183:4210–4216

Wang L, Hashimoto Y, Tsao C-Y, Valdes J, Bentley WE, Valdes JJ (2005a) Cyclic AMP (cAMP) and cAMP receptor protein influence both synthesis and uptake of extracellular autoinducer 2 in *Escherichia coli*. J Bacteriol 187(6):2066–2076

Wang L, Li J, March JC, Valdes JJ, Bentley WE (2005b) luxS-dependent gene regulation in *Escherichia coli* K-12 revealed by genomic expression profiling luxS-dependent gene regulation in *Escherichia coli* K-12 revealed by genomic expression profiling. J Bacteriol 187:8350–8360

Waters CM, Bassler BL (2005) Quorum sensing: cell-to-cell communication in bacteria. Annu Rev Cell Dev Biol 21:319–346

Wei Y, Lee JM, Smulski DR, LaRossa RA (2001) Global impact of sdiA amplification revealed by comprehensive gene expression profiling of *Escherichia coli*. J Bacteriol 183:2265–2272

Xavier KB, Bassler BL (2005a) Interference with AI-2-mediated bacterial cell-cell communication. Nature 437:750–753

Xavier KB, Bassler BL (2005b) Regulation of uptake and processing of the quorum-sensing autoinducer AI-2 in *Escherichia coli*. J Bacteriol 187:238–248

Xavier KB, Miller ST, Lu W, Kim JH, Rabinowitz J, Pelczer I, Semmelhack MF, Bassler BL (2007) Phosphorylation and processing of the quorum-sensing molecule autoinducer-2 in enteric bacteria. ACS Chem Biol 2:128–136

Zhu J, Winans SC (2001) The quorum-sensing transcriptional regulator TraR requires its cognate signaling ligand for protein folding, protease resistance, and dimerization. Proc Natl Acad Sci USA 98: 1507–1512

Quorum Sensing in *Acinetobacter baumannii*

Nidhi Bhargava, Prince Sharma, and Neena Capalash

Introduction

Despite all advances in antimicrobials and vaccines development, infectious agents rank among the most common causes of death worldwide (Perez et al. 2007). Microorganisms cover almost all the outer surfaces of the human body and normally do no harm to the host. They exist as commensals and play a vital role in maintaining human body homeostatis. At times these microbes may also result in severe infections in individuals with a compromised immune system. Such microorganisms are called opportunistic. Breach in the first line of defence, i.e., the skin, may allow bacteria to enter the host. Production of various virulence factors by the pathogen may cause host tissue damage, thus leading to establishment of infection in the host (Fig. 1). Factors which predispose such occurrences include increased number of invasive medical procedures, long tenure of hospitalization and capability of the microorganisms to evolve mechanisms of antimicrobial resistance. An exponential increase in drug resistance among pathogens is a major challenge for the management of infectious diseases.

N. Bhargava • N. Capalash (✉)
Department of Biotechnology, Panjab University, Chandigarh 160014, India
e-mail: caplash@pu.ac.in

P. Sharma
Department of Microbiology, Panjab University, Chandigarh 160014, India

Acinetobacter baumannii is a Gram-negative, non-fermentative coccobacillus that is widely distributed in nature and colonizes the human skin. It has recently gained importance in the health-care setup because of its ability to survive in the hospital environment for extended periods of time which makes it a frequent cause of health-care-associated infections. *A. baumannii* shows immunity to disinfectants, can survive desiccation conditions and also resists antibiotics. It is an opportunistic pathogen which has been implicated in a wide spectrum of infections, e.g., nosocomial pneumonia, urinary tract infections, secondary meningitis and superinfections in burn patients (Adams et al. 2011) with varying frequency of occurrence (Gaynes and Edwards 2005). *A. baumannii* has emerged as an important pathogen associated with the most common autosomal genetic disorder with high death toll called cystic fibrosis (Davies and Rubin 2007). The infection rate of *A. baumannii* in populations is alarming and accounts for 20 % of all the organisms isolated from intensive care units (ICUs). The increasing number of *A. baumannii*-related infection cases has led to its comparison with methicillin-resistant *Staphylococcus aureus* (MRSA) and accounts for 5 % mortality in general wards and 54 % mortality in ICUs (Fournier and Richet 2006).

Global surveillance programmes conducted over the last decade have shown an unparallel increase in resistance rates against different classes of antimicrobials among the clinical isolates of *Acinetobacter* spp. (Adams et al. 2011).

V.C. Kalia (ed.), *Quorum Sensing vs Quorum Quenching: A Battle with No End in Sight*,
DOI 10.1007/978-81-322-1982-8_10, © Springer India 2015

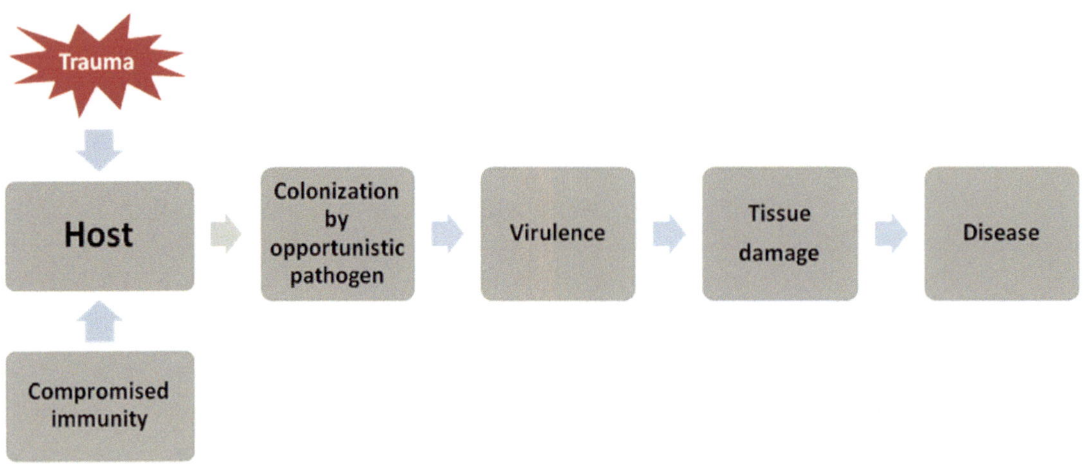

Fig. 1 Infection caused by an opportunistic pathogen

This situation is variable in different countries but has gained worldwide importance, and multiple outbreaks are being reported. *A. baumannii* shows resistance against major antibiotic classes, namely, broad-spectrum β-lactams (Bonnin et al. 2011), third-generation cephalosporins (Meyer et al. 2010), carboxypenicillins (Mammeri et al. 2003) and carbapenems (Poirel and Nordmann 2006). They also produce a wide range of aminoglycoside-inactivating enzymes, and most strains are resistant to fluoroquinolones (Singh et al. 2013). Such resistance pattern is defined as multidrug resistance (MDR) and is a major cause of worry as it is rapidly nullifying the therapeutic armamentarium. The emergence of carbapenem-resistant New Delhi metallo-β-lactamase-producing *A. baumannii* strains has started a new wave of MDR worldwide (Bonnin et al. 2012). Drug resistance in *A. baumannii* is due to both acquired and intrinsic mechanisms. The main underlying mechanisms of resistance to multiple antibiotics in *Acinetobacter* spp. can be summarized as follows:

1. Production of hydrolyzing enzymes, e.g., β-lactam hydrolysis by different kinds of β-lactamases (Classes A to D)
2. Changes in penicillin-binding proteins (PBPs) that prevent action of β-lactams
3. Alterations in the structure and number of porin proteins that result in decreased permeability of antibiotics through the outer membrane of the bacterial cell
4. The activity of efflux pumps that further decrease the concentration of antibiotics within the bacterial cell

MDR as well as the ability to stand environmental stresses makes eradication of *A. baumannii* infection difficult, particularly from hospital settings. A possible remedy to the emergence of *A. baumannii* as a superbug is to adopt anti-virulence strategy targeting its quorum-sensing system with compounds or enzymes holding quorum-quenching potential.

Quorum Sensing in *A. baumannii*

During infections, bacteria interact with each other (intraspecies), with other bacteria (interspecies) and with eukaryotic hosts. Gram-negative bacteria communicate with their neighbours through chemical signals, namely, N-acyl homoserine lactones (AHLs). The density of these signal molecules indicates the number of bacteria in a population. Single-celled bacteria at a minimum population unit, 'quorum', exhibit a complex pattern of multicellular behaviour. Such concerted population response is called quorum sensing (QS) (Diggle et al. 2007).

Table 1 AHLs produced by *Acinetobacter baumannii*

Molecular formula	Structure	Probable AHLs in *A. baumannii*
C16H29NO4	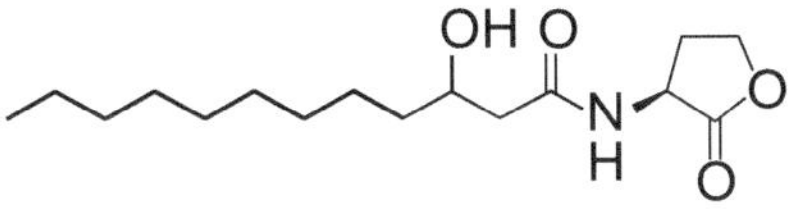	*N*-(3-hydroxy-dodecanoyl)-L-homoserine lactone (3-OH-C12 HSL)
C16H29NO3	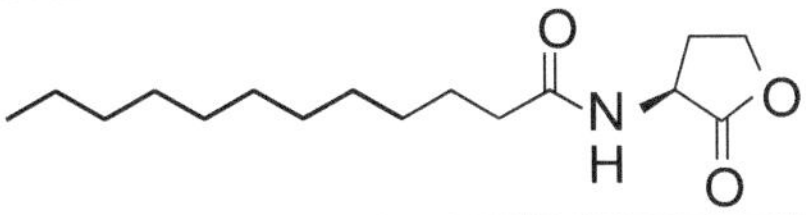	*N*-dodecanoyl-L-homoserine lactone (C-12 HSL)
C14H25NO3	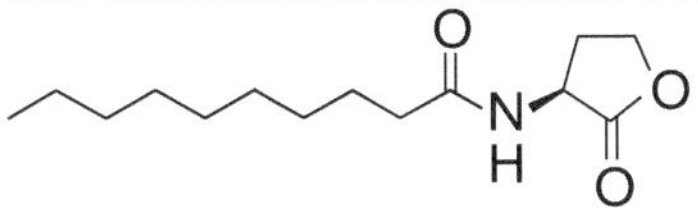	*N*-dodecanoyl-L-homoserine lactone (C-10 HSL)
C18H33NO3	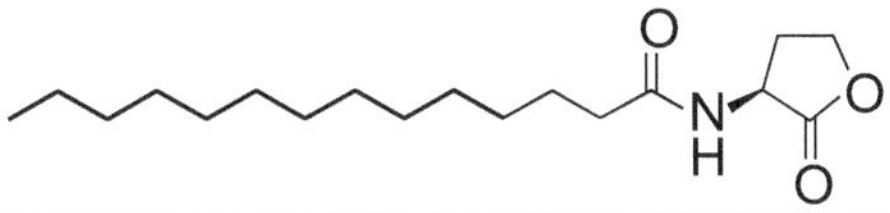	*N*-tetradecanoyl-L-homoserine lactone (C-14 HSL)
C17H29NO4	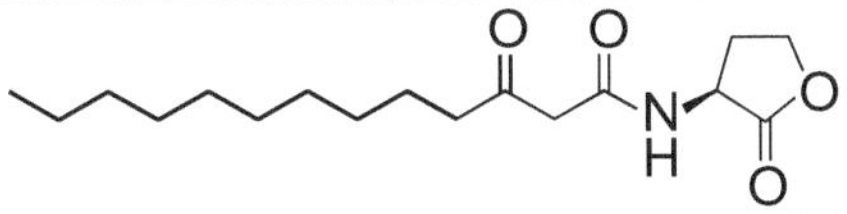	*N*-(3-oxotridecanoyl)-L-homoserine lactone (3-oxo-C13 HSL)
C20H35NO3	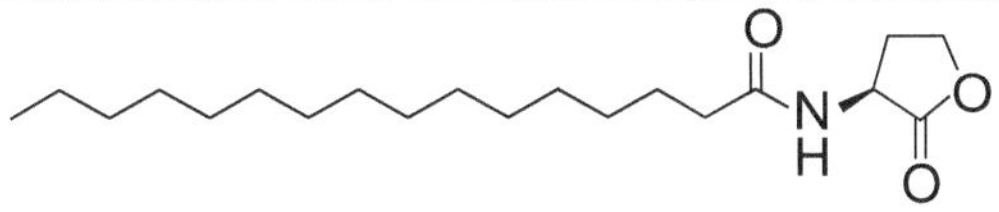	*N*-hexanoyldecanoyl-L-homoserine lactone (C-16 HSL)

Quorum-sensing system in *A. baumannii*, like other Gram-negative bacteria, consists of:

- Signal molecules: AHLs (autoinducers)
- AbaI: AHL synthase
- AbaR: AHL receptor (regulator)

Acyl Homoserine Lactones in *A. baumannii*

Crosstalking between Gram-negative bacteria belonging to the same or different species or genera is carried out through N-acyl homoserine lactones (AHLs) which are hormone like signal molecules. These molecules have a conserved homoserine lactone ring that is acylated with an acyl group at the α-position. AHL may contain a C4–C18 long acyl side chain with either oxo, hydroxy or no substitution at C3 position. Based on the length of the acyl chain, AHLs are classified as short- or long-chain molecules. *A. baumannii* synthesizes more than one long-chain (≥C10) AHLs (Table 1). Other species of *Acinetobacter* like *A. calcoaceticus* also produce one or more than one AHL molecules. However, none of the AHL signals could be specifically assigned to a particular species of this genus (Gonzalez et al. 2009). Since quorum signals are not homogenously distributed among the different strains of *Acinetobacter*, distinction between virulent and non-virulent strains on the basis of quorum-sensing signals is difficult. Transportation of AHLs is fundamental in the process of QS. Movement of AHLs from inside of the cell to the environment and vice versa is by diffusion where AHLs move freely across the bacterial membrane in a density-dependent manner. Since permeability of bacterial membrane is limited only to short-chain AHLs, transportation of long-chain AHLs (>C8) takes place through efflux

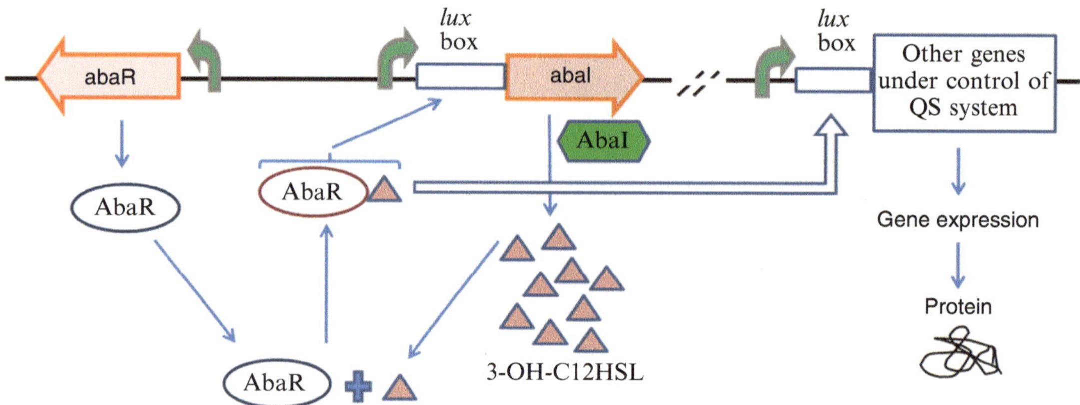

Fig. 2 Quorum-sensing system in *Acinetobacter baumannii*

pumps. However, very little is known about the pumps involved in transportation of AHLs in *Acinetobacter*.

Quorum-sensing system in *A. baumannii* consists of one chromosomally encoded AHL-dependent signalling system, comprising of LuxI and LuxR homologues. *aba*I is responsible for the synthesis of 3-hydroxy-dodecanoyl-(L)-homoserine lactone (3-OH-C12 HSL), and the cognate receptor of this ligand is synthesized by *aba*R. When a threshold concentration is attained within the culture, the AHL molecules which are either present inside the cell or have been transported into it from the environment bind to its cognate receptor, AbaR (Fig. 2). This complex binds to the putative *lux*-box (CTGTAAATTCTTACAG) which is located 67 bp upstream the putative ATG of *aba*I and results in the synthesis of more AHL molecules. Since AHL molecules can induce production of more 'like' molecules themselves, they are referred to as autoinducers. It is suggested that all those virulence genes whose expression is controlled by AHL-mediated quorum-sensing system in *Acinetobacter* are likely to have the *lux*-box sequence in their respective promoter region. Since *aba*I is the only autoinducer synthase deciphered in *A. baumannii* which is responsible for synthesis of 3-OH-C12 HSL, other types of AHLs may probably be synthesized by the organism through the action of acyltransferases (Niu et al. 2008).

Structure of AHL Synthase (AbaI)

Sequence analysis of *A. baumannii* AHL synthase (AbaI) shows that it contains a conserved consensus pattern, [LMFYA]-R-x(3)-F-x(2)-W-x-[LIVM]-x(6,9)-E-x-D-x-[FY]-D, which is characteristic of autoinducer synthase family. The protein contains a single domain and is predicted to perform the function of signal transduction and quorum sensing. LasI, autoinducer synthase in *Pseudomonas aeruginosa*, is the only protein closest (46 % similarity and 27.5 % identity) to AbaI whose X-ray crystal structure (1RO5) is known (Gould et al. 2004). Thus, similarity in the primary structure of LasI and AbaI could form the basis for making predictions about the tertiary structure of AbaI. LasI structure is a three-layered (αβα) sandwich containing eight α helices (31 %) surrounding a highly twisted platform of nine β strands (29 %) (Fig. 3). This platform is composed of antiparallel β-sheet with prominent V-shaped cleft between parallel β-strands, β4 and β5. A six-stranded β-sheet platform is buttressed by three α helices, forming a V-shaped substrate binding cleft that leads to a tunnel passing through the enzyme, which could accommodate the acyl chain of acyl-acyl carrier protein (acyl-ACP). Residues Arg23, Phe27 and Try33 in the N-terminal region of LasI form the substrate binding pocket. Several of these residues, including Met79, Phe105, Thr142 and Thr144, are well conserved

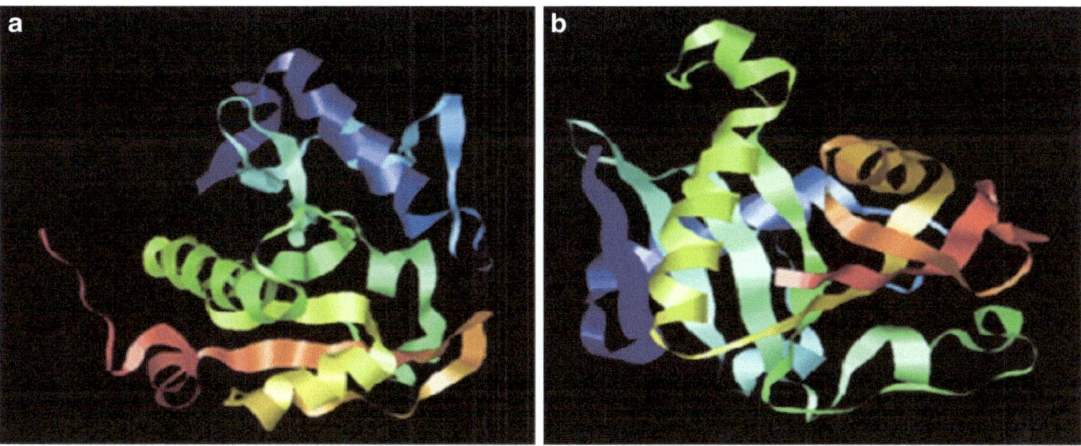

Fig. 3 Predicted tertiary structure of *Acinetobacter baumannii* AHL synthases (AbaI) (**a**) based on *P. aeruginosa* AHL synthases (LasI) (**b**) by protein homology modelling server CPHmodels 3.2

among AHL synthases and appear to position the acyl chain in the proper orientation for catalysis. The conservation of the same residues in AbaI suggests the same role being played by these residues in catalysis of AHL molecule in *Acinetobacter*. Mutational studies reveal that Arg23, Asp42, Asp44, Asp47, Arg70, Glu101, Ser103 and Arg104 residues play a role either in stabilizing the interactions or in catalytic function (Gould et al. 2004). AbaI is predicted to interact with chemical ligands, sulphate and zinc ions and contains two cysteine molecules which may be involved in disulphide bond formation.

Structure of AHL Receptor/Regulator (AbaR)

AHL receptor protein in *A. baumannii*, AbaR, is 198 amino acids in length and is 29.2 % identical and 45 % similar to autoinducer receptor, LasR, of *P. aeruginosa*. Since the structures of autoinducer synthases of *P. aeruginosa* (LasI) and *A. baumannii* (AbaI) are similar, the structure of AbaR too was predicted based on the X-ray crystallographic structure of LasR (3IX3). LasR protein is 239 amino acids in length and is comprised of 50 % helical and 15 % β-sheet structure (Fig. 4). The protein has LuxR-type DNA-binding, helix-turn-helix (HTH) domain of about 65 amino acids, present in transcription regulators of the LuxR/FixJ family of response regulators. The domain is named after *V. fischeri* LuxR, a transcriptional activator for quorum-sensing-controlled luminescence. The DNA-binding HTH domain is usually located in the C-terminal, and the N-terminal contains an autoinducer binding domain or a response regulatory domain. Most LuxR-type regulators act as transcription activators, but some can be repressors or have a dual role for different sites. LuxR-type HTH regulators control a wide variety of activities in various biological processes. Several structures of LuxR-type HTH proteins have been resolved and show that the DNA-binding domain is formed by a four-helix bundle. The HTH is involved in DNA-binding into the major groove, where the N-terminal part of the recognition helix makes most DNA contacts. The regulators bind DNA as multimers (Ducros et al. 2001; Egland and Greenberg 2001; Vannini et al. 2002; Pristovsek et al. 2003).

LuxR-type HTH protein can be activated to form multimers by four different mechanisms:

(a) Regulators belonging to two-component sensory transduction systems in bacteria are activated by phosphorylation, generally on an aspartate residue, by a transmembrane kinase.
(b) Regulators involved in quorum-sensing systems get activated on binding with

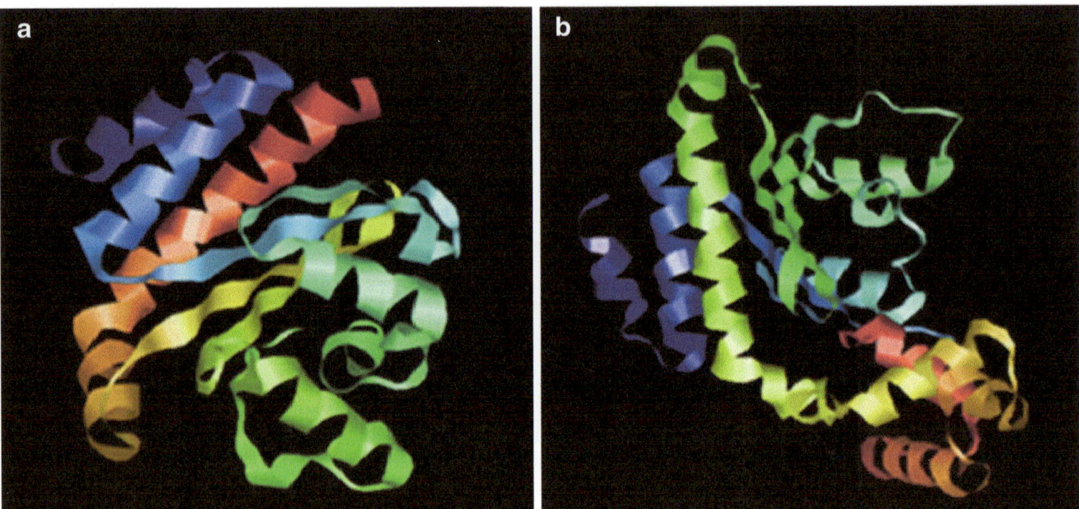

Fig. 4 Predicted tertiary structure of *Acinetobacter baumannii* AbaR (**a**) based on *P. aeruginosa* LasI (**b**) by protein homology modelling server CPHmodels 3.2

autoinducer molecule. AbaR falls under this category with consensus pattern of the LuxR-type HTH domain: [GDC] – x(2) – [NSTAVY] – x(2) – [IV] – [GSTA] – x(2) – [LIVMFYWCT] – x – [LIVMFYWCR] – x(3) – [NST] – [LIVM] – x(2) – {T} – x(2) – [NRHSA] – [LIVMSTA] – x(2) – [KR]. It gets activated and dimerizes on binding to its cognate lactone (3-OH-C12 HSL).

(c) Spore germinating protein (GerE) in *Bacillus subtilis* is regulated by an autonomous effector but lacks the regulatory domain.

(d) Regulators with multiple ligand binding, exemplified by MalT in *E. coli*, activate the maltose operon by binding to maltose and ATP.

Evolution of *A. baumannii* Quorum-Sensing System

A. baumannii was found to be closest to *Bukholderia ambifaria* on the basis of 16S rRNA sequence. Phylogenetic comparison of autoinducer synthase (AbaI) and autoinducer receptor protein (AbaR) sequences of pathogenic and nonpathogenic organisms exhibiting quorum sensing shows that these two proteins of *A. baumannii* ATCC 17978 were most closely related to those of the archaebacterium *Halothiobacillus neapolitanus*. *A. baumannii* and *H. neapolitanus* belong to the order *Pseudomonadales* and *Chromatidales*, respectively. Evolutionary studies on members of the class *Gammaproteobacteria* suggested that *Pseudomonadales* are more recent than *Chromatidales*. This suggested that *aba*I might have been acquired by *A. baumannii* from *H. neapolitanus* and not from *B. ambifaria* with which it shared ancestry (Bhargava et al. 2010).

Pathogenicity and Virulence Factors of *A. baumannii*

Pathogenicity is the ability of microorganisms to cause disease. Virulence is the degree of pathogenicity of a microorganism that can vary among the members of the same species. Virulence depends on various parameters of the microorganism, the host and the interaction between both (Winn et al. 2005). An infection begins when the balance between the host resistance and bacterial pathogenicity is not stable. Although virulence factors responsible for pathogenicity of *Acinetobacter* are still not well defined, a few of them are described below:

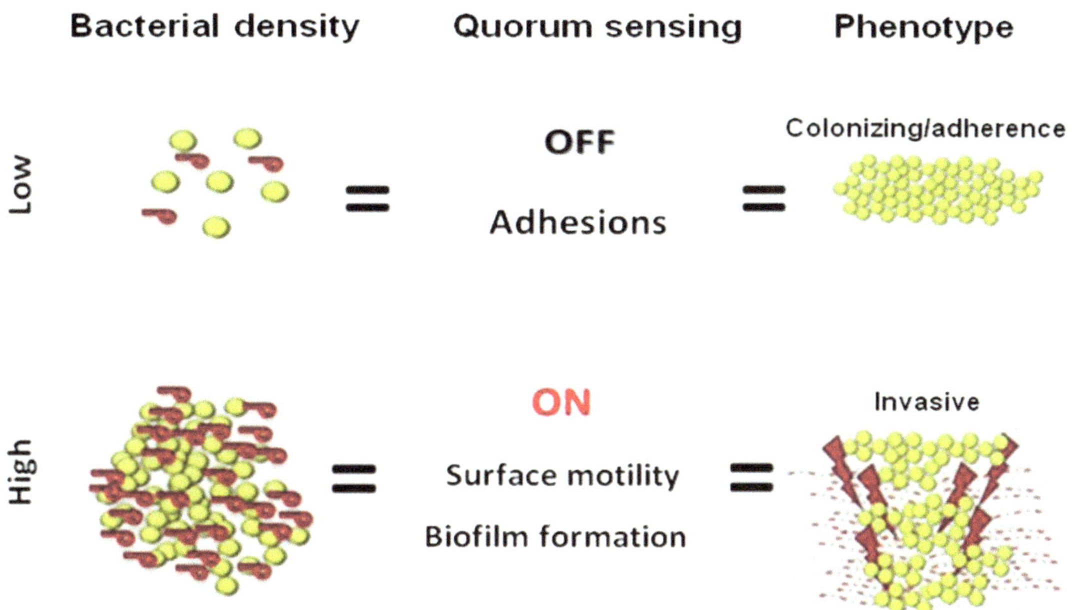

Fig. 5 Quorum-sensing-controlled virulence factors in *Acinetobacter baumannii*

Quorum-Sensing-Regulated Production of Virulence Factors in *A. baumannii*

Quorum sensing in *A. baumannii* is a central mechanism known for the autoinduction of two virulence factors: biofilm formation and surface motility (Fig. 5).

Biofilm: It is a sedentary mode of life exhibited by bacteria. Cells in biofilm, unlike their planktonic counterparts, attach themselves to the substratum irreversibly and exist as a sessile community. The biofilm cells show an altered phenotype with respect to growth rate and gene transcription as compared to the planktonic cells. The biofilm cells are embedded in a matrix of extracellular polymeric substances comprising of polysaccharides, proteins and DNA which makes it 1,000–1,500-fold more resistant to antibiotics as compared to the free living cells.

Formation of biofilm is a complex process which can be described in four stages: (1) attachment, (2) growth, (3) maturity and (4) dispersal. *A. baumannii*, like many other Gram-negative pathogens, also forms biofilm. Biofilm formation is a community phenotype and is associated with quorum sensing. An autoinducer synthase mutant (△*aba*I) of *A. baumannii* strain M2, exhibiting a disrupted QS system, formed biofilm which was impaired in the later stages of development (Niu et al. 2008). Exogenous supplementation of its quorum-sensing molecule (3-OH-C12 HSL) compensated for the deficiency which occurred due to mutation and rescued the phenotype similar to the wild type. AbaI deficiency did not affect the survival of the pathogen but was vital for the formation of mature biofilm.

Surface motility: Bacteria use surface appendages such as flagella and pili to show motility of different types, namely, swimming, swarming, twitching, gliding and sliding. Motility is an important virulence factor as it allows bacteria to penetrate into the host followed by its colonization. Genetically, *A. baumannii* is regarded as nonmotile as genes coding for flagellar assembly responsible for motility are absent in it. However, it demonstrated motility on soft agar (agar concentration ranging from 0.2 to 0.4 %) which was defined as surface motility. In *A. baumannii*, surface motility was found to be quorum sensing dependent as decreased motility of autoinducer synthase mutant (△*aba*I) of *A. baumannii* strain M2 was restored on exogenous addition of 3-OH-C12 HSL (Clemmer et al. 2011). Further, RNA sequencing and transcriptome analysis of *A. baumannii* M2 revealed a subset of the genes (A1S_0112 to A1S_0118) which were activated by quorum sensing. One of the genes, A1S_0119, encoded phosphopantetheine protein transferase which has a function in the production of a lipopeptide, but its role in virulence is not yet determined.

Other Virulence Factors

Hydrolytic enzymes: Many bacteria synthesize enzymes that play an important role in resisting the host immune system. Enzymatic activities, namely, esterases, certain aminopeptidases, ureases and acid phosphatases, have been reported in *A. baumannii* and may also be associated with virulence (Bergogne-Berezin et al. 2008). Esterases might help the pathogen to successfully enter the host and damage host lipid tissues by hydrolyzing short-chain fatty acids at ester linkages, while urease production might help them to colonize the hypochlorhydric or achlorhydric stomach causing inflammation (Rathinavelu et al. 2003).

Lipopolysaccharide (LPS): The outer membrane of *A. baumannii* contains LPS which renders the organism resistant to bactericidal action of host serum (Panilaitis et al. 2002). As a result, bacteria survive in the blood and cause infection. Diamine tetraacetic acid treatment made the pathogen susceptible to antibacterial action of serum as it reduced the LPS content.

Hydrophobicity: Surface hydrophobicity of bacteria protects it from being phagocytozed and plays an important role in attachment to various surfaces such as that of catheters. Surface hydrophobicity of strains obtained from catheter surfaces was higher than that of strains obtained from healthy carrier skin (Boujaafar et al. 1990). Thus, virulent and non-virulent bacteria could be differentiated on the basis of their hydrophobicity.

Siderophore – the iron regulator: *A. baumannii* isolates survive iron-deficient conditions by producing siderophores which solubilize the polymeric ferric oxy-hydroxides into soluble iron (Actis et al. 1993). The ability of bacteria to assimilate iron is known to be related to invasiveness and thus virulence. Other genes responsible for iron uptake in *A. baumannii* are regulated by the Fur protein of *A. baumannii*. Since Fur and Fur-like repressors are known to regulate some virulence-determinant genes in other bacteria (Wooldridge and Williams 1993), similar genes in *A. baumannii* may also regulate a subset of genes with vital role in pathogenesis.

Phospholipases: Phospholipases are group of enzymes produced by *A. baumannii* which cleave phospholipids present in the cell membrane and are linked to virulence.

Phospholipase-C (PLC) and phospholipase-D (PLD) remove glycerophosphate bond and the head group, respectively, from the phospholipid molecule. PLC production increases toxicity of epithelial cells (Camarena et al. 2010), whereas PLD production is linked to serum resistance and epithelial cell invasion (Jacobs et al. 2010).

Quorum-Sensing-Controlled Production of Superoxide Dismutase (SOD) and Catalase Enzymes

A. baumannii produces SOD and catalase to protect itself against toxic products of oxygen reduction, namely, hydrogen peroxide (H_2O_2), superoxide (O_2^*) and hydroxyl (OH^*) radicals. Pathogens are exposed to killing mechanisms by the host including various reactive oxygen species (ROS) generated by respiratory burst response. It is proposed that by the production of SOD and catalase enzymes, the pathogen is able to resist free radicals and persist under oxidative stress.

Autoinducer synthase mutant (Δ*aba*I) of *A. baumannii* strain M2 produced lower levels of antioxidant enzymes, namely, SOD and catalase, as compared to the wild type (unpublished data). Production of these enzymes in the mutant increased on exogenous addition of the quorum-sensing molecule of *A. baumannii* (3-OH-C12 HSL). Treatment of the wild type with a known quorum-quencher salicylic acid interfered with the quorum-sensing system and decreased the SOD and catalase enzyme production. This suggested that production of antioxidant enzymes was controlled by AbaI-mediated quorum-sensing system in *A. baumannii* (Bhargava et al. 2014).

Crosstalking of *A. baumannii* with Other Pathogens

A. baumannii and *P. aeruginosa* have overlapping sites of infection and exist as coinfecting organisms in patients (Dent et al. 2011). None of the two pathogens showed any inhibitory effect on the other, and both could coexist stably in vitro (Bhargava et al. 2012). Comparative genomics analysis of *A. baumannii* with *P. aeruginosa* reveals that they share 65 % orthologs (Gospodarek et al. 2009). Both the organisms exhibit quorum sensing which involves acyl homoserine lactone (AHLs)-mediated regulation.

P. aeruginosa has two AHL-dependent quorum-sensing systems, *las*I-R and *rhl*I-R, and one 2-heptyl-3-hydroxy-4-quinolone (PQS)-dependent quinolone system (Dekimpe and Deziel 2009). *A. baumannii* has only one AHL-dependent AbaI-R-mediated quorum-sensing system (Niu et al. 2008). LasR in *P. aeruginosa* is a transcriptional regulator protein which responds to *N*-(3-oxododecanoyl)-L-homoserine lactone (3-oxo-C12 HSL) generated by *las*I, and AbaR in *A. baumannii* is a transcriptional regulator that responds best to *aba*I-generated *N*-(3-hydroxydodecanoyl)-L-homoserine lactone (3-OH-C12 HSL). Due to high similarity between the receptor proteins of the two organisms, they were probably able to interact with each other's AHL molecules. Such flexibility of AbaR and LasR to C-12 HSL, irrespective of the modification at C3, suggested that *A. baumannii* and *P. aeruginosa* can show two-way communication with each other (Bhargava et al. 2012). It is suggested that interspecies interactions between *A. baumannii* and *P. aeruginosa* could have serious medical implications as both the nosocomial pathogens are associated with overlapping sites of infection and can influence disease courses and therapeutic control measures during infection.

Managing *Acinetobacter* Infections by Inhibiting Quorum Sensing

Quorum sensing has drawn the interest of medical community due to the role it plays in controlling the virulence of *A. baumannii* as it does in other Gram-negative pathogens such as *Yersinia enterocolitica*, *P. aeruginosa* and *Vibrio cholera*. Not much work has been reported on quorum-sensing inhibition of *A. baumannii*. The continued evolution of resistance mechanisms and alarming increase in the prevalence of multiple drug resistance among the emerging bacterial pathogens like *A. baumannii* necessitates the search for new antibacterial agents aimed at new targets. Quorum-sensing inhibition can provide an alternative treatment option when the 'last-resort' drugs such as colistin and tigecycline lose their efficacy. Since QS is not directly involved in essential processes, such as growth of the bacteria, one can reason that unlike antibiotics, the use of quorum-sensing inhibitors (quorum quenchers) will not lead to selective pressure on the bacteria for the development of resistance (Sperandio 2007). Misregulation or inhibition of quorum sensing can be achieved through three mechanisms:

1. Inhibition of AHL synthesis
2. Inhibition of AHL binding to AbaR receptor
3. Quorum quenching enzymes

Inhibition of AHL Synthesis

Methylthioadenosine nucleosidase (MTAN) is a hydrolase found only in bacteria and not in humans which is involved in the synthesis of AHL. It salvages methionine for S-adenosyl methionine (SAM) regeneration and thus is linked to quorum sensing in Gram-negative bacteria. BuT-DADMe-ImmA is an MTAN inhibitor which could suppress AHL synthesis by disruption of SAM supply with high efficacy

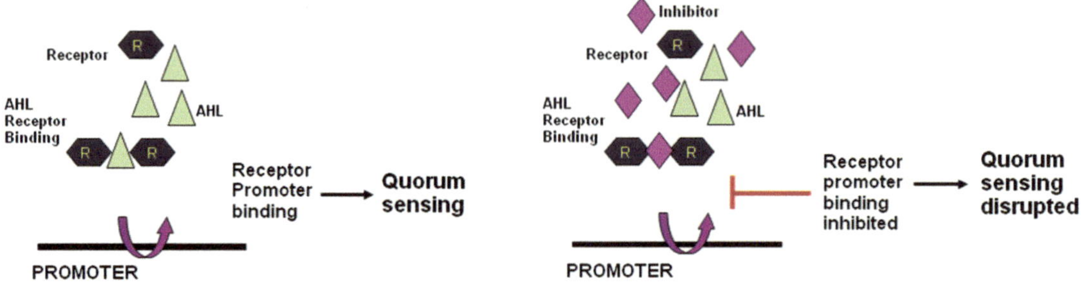

Fig. 6 Quorum quenching by inhibition of AHL binding to its cognate receptor

even at picomolar concentration (Schramm et al. 2008). Inhibition of MTAN could block the synthesis of quorum-sensing molecule without challenging the bacterial survival. Another protein with a vital role in AHL acyl chain formation is enoyl-acyl carrier protein reductase (FabI) and is a potential candidate for quorum-sensing inhibition of pathogenic bacteria. Triclosan at low concentration inhibited FabI to almost 50 % (Hoang and Schweizer 1999). Since enzymes like MTAN and FabI are involved in AHL synthesis, a process conserved in all Gram-negative bacteria exhibiting QS, it is suggested that inhibitors of these enzymes can be used to disrupt QS in *A. baumannii* as well.

Inhibition of AHL Binding to AbaR Receptor

Antagonists are structural analogues or chemicals which inhibit the binding of natural AHLs to AbaR receptor (Fig. 6). AHLs with non-native stereochemistry and chemicals containing aromatic acyl group act as antagonists and can disrupt the AbaR receptor's binding activity for native AHLs in *A. baumannii*. L-Conformation of AHLs is biologically active while the inactive D-conformation is regarded as non-native. D-AHLs were successful in inhibiting quorum-sensing-regulated surface motility and biofilm formation in *A. baumannii*. Recently, it has been found that streptomycin (an aminoglycoside) at subinhibitory concentration acts as antagonist of AHL and prevents the formation of functional 3-OH-C12 HSL-AbaR complexes, thereby obstructing QS in *A. baumannii* (Saroj and Rather 2013). A macrolide like azithromycin, usually used for the treatment of cystic fibrosis patients infected with *P. aeruginosa*, besides showing antibacterial activity also decreases QS which reduces its pathogenicity and could show better clinical outcome (Skindersoe et al. 2008). Thus, it is suggested that these antibiotics could act as signalling agents in a concentration-dependent manner, thereby affecting quorum-sensing-mediated expression of virulence genes in pathogenic bacteria.

Compounds with not only structural but also functional similarity to AHLs are known as agonists. These can bind and activate AbaR to cause QS inhibition in *A. baumannii*. AHLs closely resembling OH-dDHLs (Stacy et al. 2012) and triazole HL (IV-AE) exhibited agonist activity (Stacy et al. 2013).

Quorum-Quenching Enzymes

Gram-negative bacteria share common AHL types. Hence, enzymatic cleavage of these signalling moieties, also known as enzymatic quorum quenching, might be useful in broad-spectrum attenuation of virulence of various pathogens (Czajkowski and Jafra 2009). There are two kinds of quorum-quenching enzymes: (a) those that hydrolyze the lactone ring of AHL and (b) those that break the amide linkages of the AHLs. The former are called lactonases, while the latter are acylases (Fig. 7). Since the

Fig. 7 Quorum-quenching enzymes

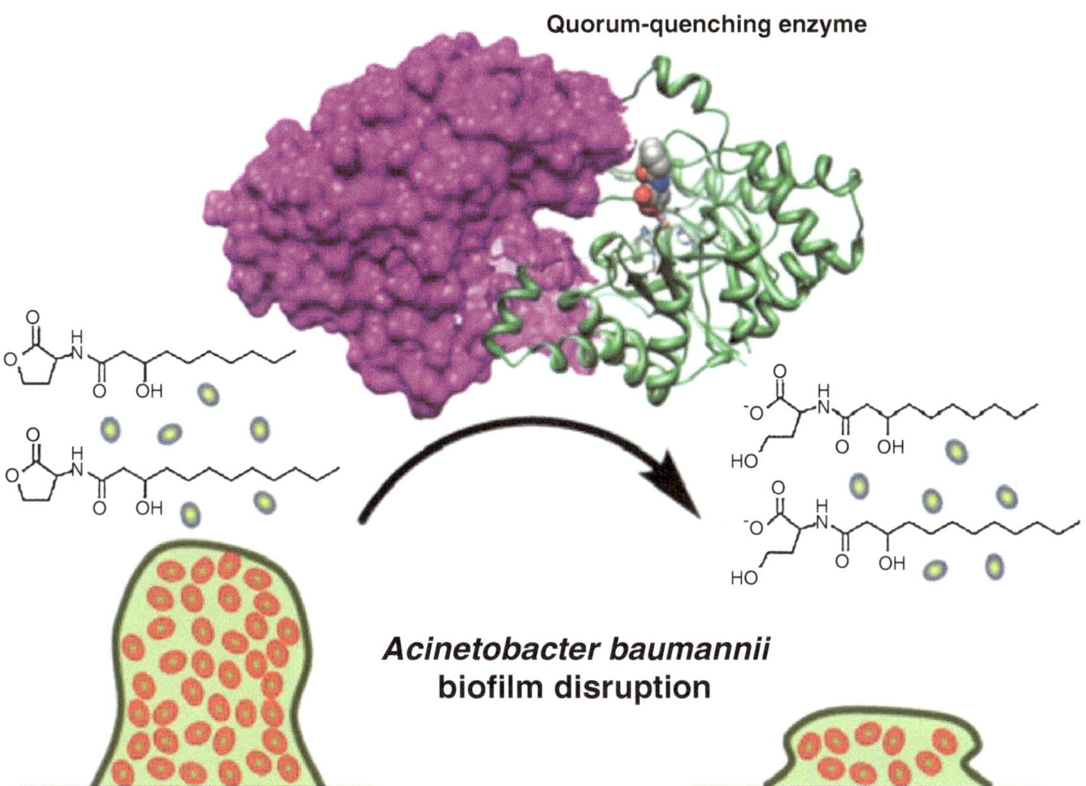

Fig. 8 Lactonase-mediated quorum quenching controls biofilm formation by *Acinetobacter baumannii* (Adapted from Tay and Yew 2013)

substitution groups on acyl chain do not affect the enzymatic activities, it is suggested that enzymes could be more effective in the disruption of a broad range of AHL signals. Lactonase from *Geobacillus kaustophilus* was engineered to enhance its activity against different AHLs (chain length range C6–C12) (Chow et al. 2014). The engineered lactonase was able to disrupt biofilm formation in *A. baumannii* (Fig. 8) due to its broad AHL specificity and could be used to treat infectious diseases caused by various Gram-negative bacteria.

Conclusion

Bacteria have a tendency to adopt more economic processes such as quorum-sensing-controlled virulence which help the organism to conserve resources and avoid wasteful expression of virulence genes which are otherwise ineffective at lower cell densities (Czárán and Hoekstra 2009). The QS system, including autoinducer synthase and its cognate receptor, is regarded as a new target for antimicrobial strategies since, unlike antibiotics, quorum-sensing inhibitors control pathogenesis without placing immediate life-or-death pressure on the targeted pathogen. Currently there is a need to develop such alternate strategies to combat rapidly emerging multidrug resistance. Understanding the role of quorum-sensing-controlled network of genes in *A. baumannii* will suggest new targets and strategies to control infections.

References

Actis LA, Tolmasky ME, Crosa LM, Corsa JH (1993) Effect of iron-limiting conditions on growth of clinical isolates of *Acinetobacter baumannii*. J Clin Microbiol 31:2812–2815

Adams D, Yee L, Rimmer JA, Williams R, Martin H, Ovington C (2011) Investigation and management of an *A. baumannii* outbreak in ICU. Br J Nurs 20: 140–147

Bergogne-Berezin E, Friedman H, Bendinelli M (2008) *Acinetobacter* biology and pathogenesis. Springer Science + Business Media, New York

Bhargava N, Sharma P, Capalash N (2010) Quorum sensing in *Acinetobacter*: an emerging pathogen. Crit Rev Microbiol 36:349–360. doi:10.3109/1040841X.2010.512269

Bhargava N, Sharma P, Capalash N (2012) N-acyl homoserine lactone mediated interspecies interactions between *A. baumannii* and *P. aeruginosa*. Biofouling 28:813–822. doi:10.1080/08927014.2012.714372

Bhargava N, Sharma P, Capalash N (2014) Pyocyanin stimulates quorum sensing-mediate tolerance to oxidative stress and increases persister cells population in Acinetobacter baumannii. Infect Immun pii:IAI.01600-14 [Epub ahead of print]

Bonnin RA, Nordmann P, Pottron A, Lecuyer H, Zahar JR, Piorel L (2011) Carbapenem hydrolyzing GES-type extended spectrum β-lactamase in *Acinetobacter baumannii*. Antimicrob Agents Chemother 55:349–354. doi:10.1128/AAC.00773-10

Bonnin RA, Poirel L, Naas T, Pirs M, Seme K, Schrenzel J, Nordmann P (2012) Dissemination of New Delhi metallo-β-lactamase-1-producing *Acinetobacter baumannii* in Europe. Clin Microbiol Infect 18:E362–E365. doi:10.1111/j.1469-0691.2012.03928.x

Boujaafar N, Freney J, Bouvet PJM, Jeddi M (1990) Cell surface hydrophobicity of 88 clinical strains of *Acinetobacter baumannii*. Res Microbiol 141:477–482. doi:10.1016/0923-2508(90)90073-Y

Camarena L, Bruno V, Euskirchen G, Poggio S, Synder M (2010) Molecular mechanisms of ethanol-induced pathogenesis revealed by RNA-sequencing. PLoS Pathog 6:e1000834. doi:10.1371/journal.ppat.1000834

Chow JY, Yang Y, Tay SB, Chua KL, Yew SW (2014) Disruption of biofilm formation by the human pathogen *Acinetobacter baumannii* using engineered quorum-quenching lactonases. Antimicrob Agents Chemother 58:1802–1805. doi:10.1128/AAC.02410-13

Clemmer KM, Bonomo RA, Rather PN (2011) Analysis of surface motility in *Acinetobacter baumannii*. Microbiology 157:2534–2544. doi:10.1099/mic.0.049791-0

Czajkowski R, Jafra S (2009) Quenching of acyl-homoserine lactone-dependent quorum sensing by enzymatic disruption of signal molecules. Acta Biochim Pol 56:1–16

Czárán T, Hoekstra RF (2009) Microbial communication, cooperation and chatting: quorum sensing drives the evolution of cooperation in bacteria. PLoS One 4(8):e6655. doi:10.1371/journal.pone.0006655

Davies JC, Rubin BK (2007) Emerging and unusual Gram-negative infections in cystic fibrosis. Semin Respir Crit Care Med 28:312–321. doi:10.1055/s-2007-981652

Dekimpe V, Deziel E (2009) Revisiting the quorum-sensing hierarchy in *Pseudomonas aeruginosa* the transcriptional regulator RhlR regulates LasR-specific factors. Microbiology 155:712–723. doi:10.1099/mic.0.022764-0

Dent LL, Marshall DR, Pratap S, Hulette RB (2011) Multidrug resistant *Acinetobacter baumannii*: a descriptive study in a city hospital. BMC Infect Dis 10:196. doi:10.1186/1471-2334-10-196

Diggle SP, Crusz SA, Camara M (2007) Quorum sensing. Curr Biol 17:R907–R910

Ducros VM, Lewis RJ, Verma CS, Dodson EJ, Leonard G, Turkenburg JP, Murshudov GN, Wilkinson AJ, Brannigan JA (2001) Crystal structure of GerE, the ultimate transcriptional regulator of spore formation in *Bacillus subtilis*. J Mol Biol 306:759–771. doi:10.1006/jmbi.2001.4443

Egland KA, Greenberg EP (2001) Quorum sensing in *Vibrio fischeri*: analysis of the LuxR DNA binding region by alanine-scanning mutagenesis. J Bacteriol 183:382–386. doi:10.1128/JB.183.1.382-386.2001

Fournier PE, Richet H (2006) The epidemiology and control of *Acinetobacter baumannii* in health care facilities. Clin Infect Dis 42:692–699. doi:10.1086/500202

Gaynes R, Edwards JR (2005) Overview of nosocomial infections caused by gram-negative bacilli. Clin Infect Dis 41:848–854. doi:10.1086/432803

Gonzalez RH, Dijkshoorn L, Van den Barselaar M, Nudel C (2009) Quorum sensing signal profile of *Acinetobacter* and environmental sources. Rev Argent Microbiol 41:73–78

Gospodarek E, Bogiel T, Zalas-Wiecek P (2009) Communication between microorganisms as a basis for production of virulence factors. Pol J Microbiol 58:191–198

Gould TA, Schwelaer HP, Churchill MEA (2004) Structure of the *Pseudomonas aeruginosa* acyl-homoserine lactone synthase LasI. Mol Microbiol 53:1135–1146. doi:10.1111/j.1365-2958.2004.04211.x

Hoang TT, Schweizer HP (1999) Characterization of *Pseudomonas aeruginosa* enoyl-acyl carrier protein reductase (FabI): a target for the antimicrobial triclosan and its role in acylated homoserine lactone synthesis. J Bacteriol 181:5489–5497

Jacobs AC, Hood I, Boyd KL, Olson PD, Morrison JM, Carson S, Sayood K, Iwen PC, Skaar EP, Dunman PM (2010) Inactivation of phospholipase D diminishes *Acinetobacter baumannii* pathogenesis. Infect Immun 78:1952–1962. doi:10.1128/IAI.00889-09

Mammeri H, Poirel L, Mangeney N, Nordmann P (2003) Chromosomal integration of a cephalosporinase gene from *Acinetobacter baumannii* into *Oligella urethralis* as a source of acquired resistance to β-Lactams. Antimicrob Agents Chemother 47:1536–1542. doi:10.1128/AAC.47.5.1536-1542.2003

Meyer E, Schwab F, Schroeren-Boersch B, Gastmeier P (2010) Dramatic increase of third-generation cephalosporin-resistant *E. coli* in German intensive care units: secular trends in antibiotic drug use and bacterial resistance, 2001 to 2008. Crit Care 14:R113

Niu C, Clemmer KM, Bonomo RA, Rather PN (2008) Isolation and characterization of an autoinducer synthase from *Acinetobacter baumannii*. J Bacteriol 190:3386–3392. doi:10.1128/JB.01929-07

Panilaitis B, Johri A, Blank W, Kaplan D, Fuhrman J (2002) Adjuvant activity of emulsan, a secreted lipopolysaccharide from *Acinetobacter calcoaceticus*. Clin Diagn Lab Immunol 9:1240–1247. doi:10.1128/CDLI.9.6.1240-1247.2002

Perez F, Hujer AM, Hujer KM, Decker BK, Rather NP, Bonomo RA (2007) Global challenge of multidrug-resistant *Acinetobacter baumannii*. Antimicrob Agents Chemother 51:3471–3484. doi:10.1128/AAC.01464-06

Poirel L, Nordmann P (2006) Carbapenem resistance in *Acinetobacter baumannii*: mechanisms and epidemiology. Clin Microbiol Infect 12:826–836. doi:10.1111/j.1469-0691.2006.01456.x

Pristovsek P, Sengupta K, Löhr F, Schäfer B, von Trebra MW, Rüterjans H, Bernhard F (2003) Structural analysis of the DNA-binding domain of the *Erwinia amylovora* RcsB protein and its interaction with the RcsAB box. J Biol Chem 278:17752–17759. doi:10.1074/jbc.M301328200

Rathinavelu S, Zavros Y, Merchant JL (2003) *Acinetobacter lwoffii* infection and gastritis. Microbes Infect 5:651–657. doi:10.1016/S1286-4579(03)00099-6

Saroj SD, Rather PN (2013) Streptomycin inhibits quorum sensing in *Acinetobacter baumannii*. Antimicrob Agents Chemother 57:1926–1929. doi:10.1128/AAC.02161-12

Schramm VL, Gutierrez JA, Cordovano G, Basu I, Guha C, Belbin TJ, Evans GB, Tyler PC, Furneaux RH (2008) Transition state analogues in quorum sensing and sam recycling. Nucleic Acids Symp Ser 52:75–76. doi:10.1093/nass/nrn038

Singh H, Thangaraj P, Chakrabarti A (2013) *Acinetobacter baumannii*: a brief account of mechanisms of multidrug resistance and current and future therapeutic management. J Clin Diagn Res 7:2602–2605. doi:10.7860/JCDR/2013/6337.3626

Skindersoe ME, Alhede M, Phipps R, Yang L, Jensen PO, Rasmussen TB, Bjarnsholt T, Nielsen TT, Hoiby N, Givskov M (2008) Effects of antibiotics on quorum sensing in *Pseudomonas aeruginosa*. Antimicrob Agents Chemother 52:3648–3663. doi:10.1128/AAC.01230-07

Sperandio V (2007) Novel approaches to bacterial infection therapy by interfering with bacteria-to-bacteria signalling. Expert Rev Anti Infect Ther 5:271–276. doi:10.1586/14787210.5.2.271

Stacy DM, Welsh MA, Rather PN, Blackwell HE (2012) Attenuation of quorum sensing in pathogen *Acinetobacter baumannii* using non native N-acyl homoserine lactones. ACS Chem Biol 7:1719–1728. doi:10.1021/cb300351x

Stacy DM, Le Quement ST, Hansen CL, Clausen JW, Tolker-Nielsen T, Brummond JW, Givskov M, Nielsen TE, Blackwell HE (2013) Synthesis and biological evaluation of Triazole-containing N-acyl homoserine lactones as quorum sensing modulators. Org Biomol Chem 11:938–954. doi:10.1039/c2ob27155a

Tay SB, Yew WS (2013) Development of quorum-based anti-virulence therapeutics targeting Gram-negative bacterial pathogens. Int J Mol Sci 14:16570–16599. doi:10.3390/ijms140816570

Vannini A, Volpari C, Gargioli C, Muraglia E, Cortese R, De Francesco R, Neddermann P, Marco SD (2002) The crystal structure of the quorum sensing protein TraR bound to its autoinducer and target DNA. EMBO J 21:4393–4401. doi:10.1093/emboj/cdf459

Winn WC Jr, Allen SD, Janda WM, Koneman EW, Schreckenberger PC, Procop GW, Woods GL (2005) Koneman's color atlas and textbook of diagnostic microbiology. Lippincott Williams & Wilkins, Philadelphia

Wooldridge KG, Williams PH (1993) Iron uptake mechanisms of pathogenic bacteria. FEMS Microbiol Rev 12:325–348. doi: 10.1111/j.1574-6976.1993.tb00026.x

Quorum Sensing Systems in *Aeromonas* spp.

Weihua Chu, Wei Zhu, and Xiyi Zhuang

Introduction

Quorum sensing (QS) is an intercellular communication system by which bacterial cells are capable of indirectly monitoring their own population density through production and exchange of diffusible signal molecules. This enables bacteria to control gene expression dependent on population size and thereby perform coordinated phenotypic changes in a multicellular fashion. At present, QS regulatory systems have been reported for Gram-positive and Gram-negative bacteria. These systems rely on two major components, a small diffusible signalling molecule which accumulates in a population density-dependent manner and a transcriptional activator protein which, in concert with the signalling molecule, activates the expression of relevant genes. For Gram negatives, they employ N-acyl homoserine lactones (acyl HLs, AHLs) as the signalling molecule. AHL synthase encoded by LuxI homolog synthesizes AHL molecules. Short side-chain AHLs diffuse freely across cell membranes, whereas long side-chain AHLs have to use active efflux to partition to the membrane (Pearson et al. 1999). Upon reaching a threshold concentration in the extracellular medium, AHL molecules are perceived by cytoplasmic LuxR family proteins to regulate the downstream processes (Parsek and Greenberg 2000). As a consequence, expression of certain functions, often virulence factors required for pathogenesis, is triggered at high population densities in a coordinated manner.

Aeromonas species are facultative anaerobic Gram-negative bacteria that belong to the family Aeromonadaceae, and to date, there are 24 validly published species names in the genus *Aeromonas* spp. Misidentification of *Aeromonas* as members of the genus *Vibrio* or as an *Escherichia coli* is a continuing problem. Among different *Aeromonas* species, *A. caviae* is most frequently isolated from fecal specimens, followed by *A. hydrophila* and *A. veronii.* These bacteria have a broad host spectrum, with both cold- and warm-blooded animals, including humans, and are known as psychrophilic and mesophilic. In humans, *Aeromonas* spp. cause diarrhea, gastroenteritis, and extraenteritic conditions such as septicemia, wound infection, endocarditis, meningitis, and pneumonia. *Aeromonas* are Gram-negative rods with rounded ends to coccoid, 0.3–1.0 μm in diameter and 1.0–3.5 μm in length, occur singly, in pairs, or short chains. This genus is divided into two well-separated groups: the nonmotile *A. salmonicida* and the motile group. *A. salmonicida* cells are nonmotile and atrichous. Optimum growth temperature is 22–25 °C. Most strains grow at 5 °C. Maximum temperature for growth is usually 35 °C. The following carbohydrates are usually fermented by *A. salmonicida*: arabinose, trehalose,

W. Chu (✉) • W. Zhu • X. Zhuang
Department of Microbiology, China Pharmaceutical University, Nanjing 210009, People's Republic of China
e-mail: chuweihua@cpu.edu.cn

V.C. Kalia (ed.), *Quorum Sensing vs Quorum Quenching: A Battle with No End in Sight*, DOI 10.1007/978-81-322-1982-8_11, © Springer India 2015

galactose, mannose, and dextrin. The following biochemical tests are universally negative for *A. salmonicida*: growth in KCN broth, growth in nutrient broth containing 7.5 % NaCl, urease, ornithine decarboxylase, tetrathionate reductase, and acidification of media containing rhamnose, sorbose, sorbitol, lactose, raffinose, and cellobiose. Some strains produce pigments. The motile *Aeromonas* can be divided into three species: *A. hydrophila*, *A. caviae*, and *A. sobria*. The optimum growth temperature for motile *Aeromonas* species is 28 °C. Some strains can grow at 5 °C. The maximum growth temperature is usually 38–41 °C. Temperature-dependent differences in soluble protein production in some of the *Aeromonas* strains were observed. On nutrient agar, colonies are round, raised, with an entire edge and a smooth surface, translucent, and white to buff (deep yellow) in color. Usually most strains do not have pigment. The following tests are universally positive for motile species: catalase, starch hydrolysis, lecithinase, phosphatase, arginine dihydrolase, hydrolysis of o-nitrophenyl-beta-D-galactopyranoside (ONPG), growth in nutrient broth without NaCl, and fermentation of mannitol, trehalose, fructose, galactose, and dextrin (Martin-Carnahan and Joseph 2005). These *Aeromonas* spp. produce a variety of extracellular products that contribute to the pathogenesis of disease. The virulence factors produced by *Aeromonas* spp. include amylase, chitinase, elastase, aerolysin, nuclease, gelatinase, lecithinase, lipase, and protease. These proteins are known as virulence factors that cause disease in fish and humans (Janda and Abbott 2010; Igbinosa et al. 2012). Their capacity to produce various virulence factors could contribute to the pathogenesis of disease mediated by *Aeromonas* spp. The regulation of virulence determinants by pathogenic bacteria, such as *Aeromonas*, throughout the infection and transmission cycle is an important consideration for the etiology of disease. A number of virulence factors have been identified in *Aeromonas* spp., such as adhesins (e.g., pili), S layers, exotoxins such as hemolysins and enterotoxin, and a repertoire of exoenzymes which digest cellular components such as proteases, amylases, and lipases (Pemberton et al. 1997). The expression of several of these exoproducts is associated with high cell densities in the late exponential/stationary phase, and they therefore represent putative phenotypes for control by quorum sensing (Khajanchi et al. 2009).

Quorum Sensing in *Aeromonas* spp.

Quorum sensing is a mechanism for controlling gene expression in response to an expanding bacterial population. In many Gram-negative bacteria, the diffusible quorum sensing signal molecule is a member of the N-acyl homoserine lactone (AHL) family (Whitehead et al. 2001). Accumulation of this molecule above a threshold concentration, through the activity of a signal generator protein, indicates that the population has reached a minimum population size, and the appropriate target gene(s) is activated via the action of a member of the *LuxR* family of transcriptional activators (Galloway 2011). In general, the signal generator proteins responsible for the synthesis of AHLs belong to the *LuxI* family, the archetypal member of which was originally identified within the *Vibrio* (Photobacterium) *fischeri lux* operon as the gene product responsible for the synthesis of N-(3-oxohexanoyl)-L-homoserine lactone and then bind to its cognate receptor (*LuxR*) which in turn will regulate gene expression. The *LuxR* protein consists of two domains with an AHL binding site within the N-terminal end and a helix-turn-helix DNA binding motif within the C-terminal domain. It is plausible that the expression of various virulence factors of *Aeromonas* spp. could be controlled by quorum sensing. The role of an AHL-dependent, quorum sensing system, based on the *LuxRI* homolog *AsaIR* and *AhyRI* in *A. salmonicida* and *A. hydrophila*, has been reported. As a consequence, expression of certain functions, often virulence factors required for pathogenesis, is triggered at high population densities in a coordinated manner. Autoinducers produced by *Aeromonas* spp. has been reported (Table 1).

A. salmonicida comprises the *LuxIR*-type QS system, termed *AsaIR*, where *AsaI* is the autoinducer synthase and *AsaR* the transcriptional regulator. The QS signals, N-butanoyl-L-homoserine lactone (C4-HSL) and N-hexanoyl-L-homoserine lactone (C6-HSL) (Fig. 1), and the genes involved in production of the AHL synthase (*LuxI*) and AHL receptor (*LuxR*) have been detected in *A. salmonicida* by Swift (Swift et al. 1997). *A. salmonicida* subsp. *salmonicida* also produce three additional AHLs, i.e., N-decanoyl-L-homoserine lactone (C10-HSL), N-hexanoyl-L-homoserine lactone (C6-HSL), and N-3-oxohexanoyl-L-homoserine lactone (3-oxo-C6-HSL) (Swift et al. 1997).

The major signal molecule synthesized by the *ahyI* locus in *A. hydrophila* is N-(butanoyl)-L-homoserine lactone (BHL) (Sha et al. 2005). Downstream of the *ahyI* locus was a gene with homology to *iciA* gene, an inhibitor of chromosome replication in *E. coli*, suggesting that in *Aeromonas* cell division could be linked to quorum sensing. Further, it was noted that both *AhyRI* and BHL were required for the transcription of *ahyI*. Chan et al. determined AHLs production in 22 *Aeromonas* strains isolated from various infected sites from patients (bile, blood, peritoneal fluid, pus, stool, and urine). They found that all isolates produced at least the two principal AHLs, N-butanoyl-L-homoserine lactone (C4-HSL) and N-hexanoyl-L-homoserine lactone (C6-HSL), including ten isolates produced additional AHLs which appeared to have an N-acyl side chain longer than C6 as judged by their Rf values, they also found that *Aeromonas sobria* producing C6-HSL and two additional AHLs with N-acyl side chain longer than C6 (Chan et al. 2011). Myszka and Czaczyk reviewed that N-acyl homoserine lactones (AHLs) produced by Gram-negative bacteria in the natural ecosystems, and they demonstrated that *Aeromonas hydrophila* C-4HSL and AHL can mediate biofilm formation (Myszka and Czaczyk 2012). Using *Chromobacterium violaceum* CV026 as an indicator strain for the detection of C4- and C6-HSLs, Chu et al. found that the fish pathogenic strain Ah-YJ-1 could produce C4- and C6-HSLs (Fig. 2) (Chu et al. 2013).

Table 1 Signal molecule(s) produced by *Aeromonas* spp.

Genus	Signal molecule(s)
Aeromonas hydrophila	C4-HSL, C5-HSL,C6-HSL
Aeromonas salmonicida	C4-HSL, C6-HSL
Aeromonas salmonicida subsp. achromogenes	C4-HSL
Aeromonas sobria	C4-HSL, C6-HSL
Aeromonas media	C4-HSL
Aeromonas veronii	6-carboxy-HHL, C14-HSL

A great number of nutritional and environmental factors are known to influence the growth rate of bacteria. Growth and communication are fundamental processes in biology, and the study of their interaction is of intrinsic interest. Jahid et al. investigated the influence of glucose on the production of AHLs and found that more than 0.05 % glucose significantly impaired ($P < 0.05$) quorum sensing, biofilm formation, protease production, and swarming and swimming motility, whereas when *Aeromonas hydrophila* treated with 0.05 % glucose had activity similar to that of the control (0 % glucose). A stage shift biofilm assay revealed that the addition of glucose (2.5 %) inhibited initial biofilm formation but not in later stages (Jahid et al. 2013).

Jangid et al. (2012) investigated the quorum sensing control in *Aeromonas veronii* MTCC 3249, originally isolated as *A. culicicola* from

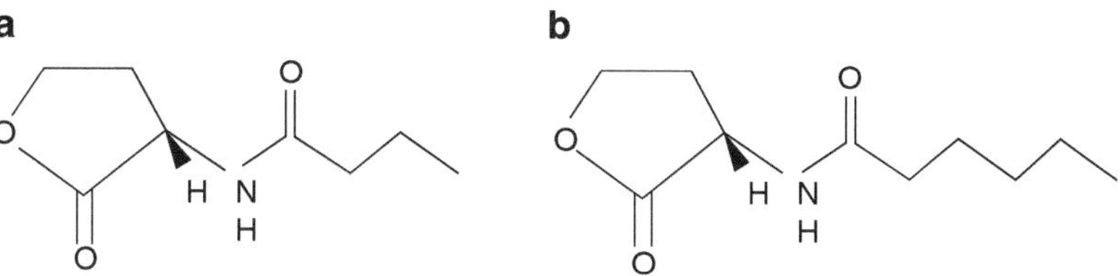

Fig. 1 Autoinducers produced by *Aeromonas* spp. (**a**) N-butanoyl-L-homoserine lactone (C4-HSL), BHL; (**b**) N-hexanoyl-L-homoserine lactone (C6-HSL), HHL

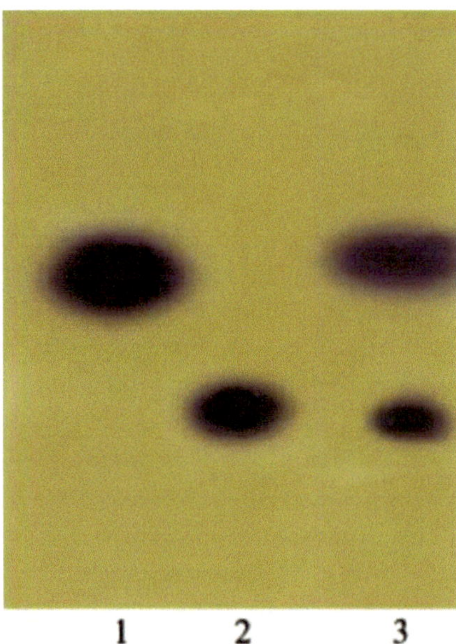

Fig. 2 Detection of AHLs by thin layer chromatography (TLC) with the *Chromobacterium violaceum* CV026 strain used as a biosensor. Lane 1, N-butanoyl-L-homoserine lactone; lane 2, N-hexanoyl-L-homoserine lactone; lane 3, *Aeromonas hydrophila* YJ-1 extract

the midgut of *Culex quinquefasciatus*. Based on biosensor assays, they found that *A. veronii* MTCC 3249 produced multiple acyl homoserine lactones (AHLs) with increasing cell density. The *luxRI* gene homologs, *acuR* (*A. culicicola* transcriptional regulator) and acuI (*A. culicicola* autoinducer), were successfully amplified by inverse PCR. Sequence analysis indicated *acuRI* were divergent from all known quorum sensing gene homologs in *Aeromonas* spp.

Role of Quorum Sensing in *Aeromonas* spp. Virulence

Quorum sensing controls processes that include bioluminescence, virulence, and biofilm formation, in various bacterial species. In general, quorum sensing regulates processes that are effective only when a population of bacteria acts in a coordinated manner but not when the bacteria act as individuals. *Aeromonas* spp. produce a wide range of virulence factors. These virulence factors are expressed differently, depending on environmental and metabolic aspects of its current habitat. Production of virulence factors by *Aeromonas* spp. is under the hierarchical control of two pairs of LuxI/LuxR homologs, *AsaIR* and *AhyIR* in *A. salmonicida* and *A. hydrophila*. Both *AsaI* and *AhyI* are autoinducer synthases that catalyze the formation of HSLs, N-butanoyl-L-homoserine lactone (C4-HSL) or N-hexanoyl-L-homoserine lactone (C6-HSL). A link between quorum sensing and virulence has been established and reported for *Aeromonas* spp.

Several of the virulence factors expressed by *A. salmonicida*, such as α-hemolysin, glycerophospholipid-cholesterol acyltransferase, lipase, and serine protease, are associated with high cell densities and may therefore potentially be controlled by QS. It has been shown that the autoinducer synthase *AsaI* plays a role in the virulence of *A. salmonicida* subsp. *achromogenes*. A knockout mutant of *AsaI* was constructed by allelic exchange which did not produce any detectable QS signals, and its virulence in fish was significantly impaired, as LD50 of the *AsaI*-deficient mutant was 20-fold higher than that of the isogenic wt strain, and the mean day to death of the mutant was significantly prolonged. In addition to this, the expression of two virulence factors (the toxic protease AsaP1 and a cytotoxic factor) was reduced in the mutant. *AsaP1* production was also inhibited by synthetic quorum sensing inhibitors. (Schwenteit et al. 2011).

In *A. hydrophila*, the involvement of quorum sensing in pathogenicity could be demonstrated. An *ahyR* mutant was highly attenuated relative to the wild-type strain. The analysis of exoenzyme activity revealed that the *ahyR* mutant could not produce exoproteases, amylases, hemolysins, and Dnases, while the wild-type strain of *A. hydrophila* had a high level of exoenzyme activity. The S layer of *A. hydrophila* could not be detected in the mutant either (Bi et al. 2007; Chu et al. 2011). Khajanchi et al.'s research showed that N-acyl homoserine lactones involved in quorum sensing control the type VI secretion system, biofilm formation, protease production, and in vivo virulence. They constructed a double knockout Δ*ahyRI* mutant of *A. hydrophila* SSU. The results suggest that AHL-mediated QS modulates the virulence of *A. hydrophila* SSU by regulating the T6SS, metalloprotease production, and biofilm formation (Khajanchi et al. 2009).

The effect of quorum sensing on the virulence of *A. hydrophila* AH-1 N towards burbot larvae was investigated by Natrah et al. using quorum sensing mutants (N-(butyryl)-L-homoserine lactone production and receptor mutants). A challenge with these mutants resulted in higher survival of burbot larvae when compared to a challenge with the wild type, and the addition of the signal molecule N-butyryl-L-homoserine lactone restored the virulence of the quorum sensing production mutant (Natrah et al. 2012).

A. hydrophila produces both a serine protease and a metalloprotease, and in the ahyI-negative strain, both of these proteases were produced in reduced amounts. However, their production was restored after exogenous addition of C4-HSL (Swift et al. 1999). On the other hand, mutation in the ahyR gene resulted in the loss of protease activity which could not be restored by the addition of C4-HSL. However, studies of Vivas et al. (2004) reported no correlation between production of AHLs and protease in vitro in an aroA live vaccine strain of *A. hydrophila*. The presence of the C4-HSL in *A. hydrophila* biofilm development has also been reported (Swift et al. 1999). The *ahyI* mutant that could not produce C4-HSL failed to form mature biofilms. A mutation in the *ahyR* locus increased the coverage of the available surface to around 80 %, with no obvious effect upon biofilm microcolony formation (Lynch et al. 2002). Bacteria in biofilms are more resistant to host defenses and antimicrobial agents and could express more virulent phenotypes as a result of gene activation through bacterial communication (quorum sensing) or gene transfer (Swift et al. 1997). Recent studies indicated a correlation between the T3SS and Act of *A. hydrophila* and the production of lactones (Sha et al. 2005). More in-depth studies are needed to definitively establish the role of iron acquisition and quorum sensing in *Aeromonas*-associated infections.

Quorum Sensing Inhibitors: New Way to Combat with *Aeromonas* spp. Disease

Bacteria communicate with one another using chemical signalling molecules. They release, detect, and respond to the accumulation of these molecules, which are called autoinducers. Detection of autoinducers allows bacteria to distinguish between low and high cell population density and to control gene expression in response to changes in cell numbers. This process allows a population of bacteria to coordinately control gene expression of the entire community. Many bacterial behaviors are regulated by quorum sensing, including symbiosis, virulence, antibiotic production, and biofilm formation. So the disruption of quorum sensing has been suggested as a new strategy for combating disease spread by pathogenic bacteria. There are two major strategies for the control of bacterial infection, either to kill the organism or to attenuate its virulence such that it fails to adapt to the host environment and is readily cleared by the innate host defenses. The discovery that bacteria employ quorum sensing molecules, to regulate the production of secondary metabolites and virulence determinants, now offers a novel target for animal diseases control (Defoirdt 2013). Quorum sensing inhibitors (QSIs) are compounds that antagonize bacterial QS systems without affecting growth of the bacteria. Different quorum sensing-disrupting pathways have been reported in controlling *Aeromonas* disease. Effective QSIs to *Aeromonas* spp. include vanillin, garlic extract, clove oil, brominated furanones and brominated thiophenones, antagonistic acyl homoserine lactones, and signal molecule-degrading enzymes and also bacteria which can produce AHL-degrading enzymes.

Sub-MICs of clove oil on *A. hydrophila* WAF-38 virulence factors (exoprotease, EPS production) and biofilm formation were studied by Husain, and they found that substantial reduction in protease and EPS production and inhibition in the form of biofilm was observed with bacteria treated with clove oil (Husain et al. 2013).

A novel *Bacillus* sp. strain, designated quorum sensing inhibitor-1 (QSI-1) by the researchers, was isolated and identified from the intestine of the fish, *Carassius auratus gibelio*. The isolated QSI-1 is capable of utilizing AHL molecules as the sole source of energy and possesses at least one type of AHL-degrading enzymes capable of degrading the AHLs of fish pathogen, *Aeromonas hydrophila*, thus significantly decreasing the production of its extracellular proteases, in turn increasing the survival rate of infected fish. Thus, perhaps this QSI-1 strain can be used as a probiotic in aquaculture (Chu et al. 2010).

In a study by Ponnusamy et al. (2009), vanillin from vanilla beans showed significant inhibition in short-chain and long-chain AHL molecules and repressed the biofilm formation by *A. hydrophila* on polystyrene surface.

A nonhalogenated, commercially available 2(5H)-furanone was found to inhibit the AHL molecules with varying chain lengths and significantly reduced the biofilm mass of *A. hydrophila* isolated from a biologically fouled RO membrane on polystyrene surface, which suggested that 2(5H)-furanone could be used as potential QS inhibitor compounds that reduced the biofouling on RO membranes (Ponnusamy et al. 2010). Feng et al. reviewed the quorum sensing in water and wastewater treatment biofilms, and they found that AHL-mediated communication widely exists in Gram-negative bacteria in water and is very important for biofilm formation, and they also reviewed that QS inhibitors can be used to control of membrane biofouling. They reviewed that the biofilm formation on RO membrane surface with two selected biofouling bacteria *Aeromonas hydrophila* and *Pseudomonas putida*, which had been isolated from a biofouled RO membrane system, was suppressed in the presence of Acylase I at a concentration of 60 ug ml^{-1} (Feng et al. 2013).

Cao et al. cloned and expressed the AHL-degrading enzyme AiiAAI96 from *Bacillus*, and supplementation of AiiAAI96 into fish feed by oral administration, results showed that AiiAAI96 significantly attenuated *A. hydrophila* infection in zebra fish (Cao et al. 2012).

Conclusion

It is becoming apparent that QS is nearly always involved in *Aeromonas* spp. virulence factors production and biofilm formation and is essential for its pathogenicity. So the QS of *Aeromonas* spp. has the potential as therapeutic targets for control of infections caused by these organisms. It is a new way to combat with bacteria diseases in aquaculture instead of antibiotics.

References

Bi ZX, Liu YJ, Lu CP (2007) Contribution of *AhyR* to virulence of *Aeromonas hydrophila* J-1. Res Vet Sci 83:150–156

Cao YA, He SX, Zhou ZG, Zhang MC, Mao W, Zhang HT, Yao B (2012) Orally administered thermostable N-acyl homoserine lactones from *Bacillus* sp strain AI96 attenuates *Aeromonas hydrophila* infection in zebrafish. Appl Environ Microbiol 78:1899–1908. doi:10.1128/AEM.06139-11

Chan KG, Puthucheary SD, Chan XY, Yin WF, Wong CS, Too WS, Chua KH (2011) Quorum sensing in *Aeromonas* species isolated from patients in Malaysia. Curr Microbiol 62(1):167–172

Chu WH, Lu F, Zhu W, Kang CT (2010) Isolation and characterization of new potential probiotic bacteria based on Quorum sensing system. J Appl Microbiol 110:202–208

Chu WH, Jiang Y, Liu YW, Zhu W (2011) Role of the quorum sensing system in biofilm formation and virulence of *Aeromonas hydrophila*. Afr J Microbiol Res 5(32):5819–5825

Chu WH, Liu YW, Jiang Y, Zhu W, Zhuang XY (2013) Production of N-acyl Homoserine Lactones and virulence factors of waterborne *Aeromonas hydrophila*. Ind J Microbiol 53(3):264–268

Defoirdt T (2013) Antivirulence therapy for animal production: filling an arsenal with novel weapons for sustainable disease control. PLoS Pathog 9(10):e1003603. doi:10.1371/journal.ppat.1003603

Feng L, Wu ZY, Yu X (2013) Quorum sensing in water and wastewater treatment biofilms. J Environ Biol 34:437–444

Galloway WR, Hodgkinson JT, Bowden SD, Welch M, Spring DR (2011) Quorum sensing in Gram-negative bacteria: small-molecule modulation of AHL and AI-2 quorum sensing pathways. Chem Rev 111(1):28–67

Husain FM, Ahmad I, Asif M, Tahseen Q (2013) Influence of clove oil on certain quorum-sensing-regulated functions and biofilm of *Pseudomonas aeruginosa* and *Aeromonas hydrophila*. J Biosci 38: 835–844. doi: 10.1007/s12038-013-9385-9

Igbinosa IH, Igumbor EU, Aghdasi F, Tom M, Okoh AI (2012) Emerging *Aeromonas* species infections and their significance in public health. Scientific World Journal 2012:625023. doi:10.1100/2012/625023

Jahid IK, Lee NA, Kim A, Ha SD (2013) Influence of glucose concentrations on biofilm formation, motility, exoprotease production, and quorum sensing in *Aeromonas hydrophila*. J Food Prot 76(2): 239–247

Janda JM, Abbott SL (2010) The genus *Aeromonas*: taxonomy, pathogenicity, and infection. Clin Microbiol Rev 23(1):35–73

Jangid K, Parameswaran PS, Shouche YS (2012) A variant quorum sensing system in *A. veronii* MTCC 3249. Sensors 12: 3814–3830. doi:10.3390/s120403814

Khajanchi BK, Sha J, Kozlova EV, Erova TE, Suarez G, Sierra JC, Popov VL, Horneman AJ, Chopra AK (2009) N-Acylhomoserine lactones involved in quorum sensing control the type VI secretion system, biofilm formation, protease production, and in vivo virulence in a clinical isolate of *Aeromonas hydrophila*. Microbiology 155(11):3518–3531

Lynch MJ, Swift S, Kirke DF, Keevil CW, Dodd CE, Williams P (2002) The regulation of biofilm development by quorum sensing in *Aeromonas hydrophila*. Environ Microbiol 4(1):18–28

Martin-Carnahan A, Joseph SW (2005) Genus I. Aeromonas *Stanier* 1943, 213AL. In: Brenner DJ, Krieg NR, Staley JT, Garrity GM (ed) Bergey's manual of systematic bacteriology, 2nd edn, vol 2, part B. Springer, New York, NY, p 557–578

Myszka K, Czaczyk K (2012) N-Acylhomoserine Lactones (AHLs) as phenotype control factors produced by gram-negative bacteria in natural ecosystems. Pol J Environ Stud 21(1):15–21

Natrah FM, Alam MI, Pawar S, Harzevili AS, Nevejan N, Boon N, Sorgeloos P, Bossier P, Defoirdt T (2012) The impact of quorum sensing on the virulence of *Aeromonas hydrophila* and *Aeromonas salmonicida* towards burbot (Lota lota L.) larvae. Vet Microbiol 159(1–2):77–82

Parsek MR, Greenberg EP (2000) Acyl-homoserine lactone quorum sensing in Gram-negative bacteria: a signaling mechanism involved in associations with higher organisms. Proc Natl Acad Sci U S A 97:8789–8793

Pearson KG, Fouad K, Misiaszek JE (1999) Adaptive changes in motor activity associated with functional recovery following muscle denervation in walking cats. J Neurophysiol 82:370–381

Pemberton JM, Stephen PK, Radomir S (1997) Secreted enzymes of *Aeromonas*. FEMS Microbiol Lett 152:1–10

Ponnusamy K, Paul D, Kweon JH (2009) Inhibition of quorum sensing mechanism and *Aeromonas hydrophila* biofilm formation by Vanillin. Environ Eng Sci 26(8):1359–1363. doi:10.1089/ees.2008.0415

Ponnusamy K, Paul D, Kim YS, Kweon JH (2010) 2(5H)-Furanone: a prospective strategy for biofouling-control in membrane biofilm bacteria by quorum sensing inhibition. Braz J Microbiol 41:227–234

Schwenteit J, Gram L, Nielsen KF, Fridjonsson OH, Bornscheuer UT, Givskov M, Gudmundsdottir BK (2011) Quorum sensing in *Aeromonas salmonicida* subsp. *achromogenes* and the effect of the autoinducer synthase AsaI on bacterial virulence. Vet Microbiol. 27;147(3–4):389–397.

Sha J, Pillai L, Fadl AA, Galindo CL, Erova TE, Chopra AK (2005) The type III secretion system and cytotoxic enterotoxin alter the virulence of *Aeromonas hydrophila*. Infect Immun 73:6446–6457

Swift S, Karlyshev AV, Fish L, Durant EL and others (1997) Quorum sensing in *A. hydrophila* and *A. salmonicida*: identification of the LuxRI homologs AhyRI and AsaRI and their cognate N-acylhomoserine lactone signal molecules. J Bacteriol 179: 5271–5281

Swift S, Lynch MJ, Fish L, Kirke DF, Tomás JM, Stewart GS, Williams P (1999) Quorum sensing-dependent regulation and blockade of exoprotease production in *Aeromonas hydrophila*. Infect Immun 67(10): 5192–5199

Vivas J, Razquin BE, López-Fierro P, Naharro G, Villena A (2004) Correlation between production of acyl homoserine lactones and proteases in an *Aeromonas hydrophila* aroA live vaccine. Vet Microbiol 101(3):167–176

Whitehead NA, Barnard AML, Slater H, Simpson NJL, Salmond GPC (2001) Quorum sensing in Gram-negative bacteria. FEMS Microbiol Rev 25: 365–404

Rhizobial Extracellular Signaling Molecules and Their Functions in Symbiotic Interactions with Legumes

Walter Giordano

Introduction

Rhizobia are soil bacteria that fix nitrogen after becoming established inside root nodules of legume plants. Rhizobia require a plant host; they are not capable of fixing nitrogen independently. Many legume species respond to inoculation with rhizobia by developing root nodules. The development of a root nodule in which rhizobia convert atmospheric nitrogen into ammonia requires the exchange of specific signaling molecules between the host plant and its microsymbionts. Flavonoids released by plant roots or seeds act as chemoattractants to rhizobia, and certain flavonoids have been shown to induce transcription of rhizobial (*nod*) genes. The products of these genes synthesize Nod factor, a lipochitooligosaccharide molecule whose effects include root hair deformation, root cortical cell division, and nodule morphogenesis (Downie and Walker 1999).

"Quorum sensing" (QS) is a mechanism whereby bacteria sense population density and regulate gene expression, leading to activation of specific phenotypes in the population. This process depends on the accumulation in the environment of signaling molecules termed autoinducers. Many Gram-negative bacteria use N-acyl homoserine lactones (AHLs) as signaling molecules. Some have been reported to use other fatty acid derivatives such as 3-hydroxypalmitic acid methyl ester and cis-unsaturated fatty acids. In contrast, many Gram-positive bacteria use amino acids or modified peptides as signaling molecules. Both Gram-positive and Gram-negative bacteria use isomers of methyl-2,3,3,4-tetrahydroxytetrahydrofuran (AI-2 autoinducer) as signals. Signaling molecules belonging to other structural classes (indole and its derivatives, quinolones, (S)-3-hydroxytridecan-4-one, and cyclic dipeptides) have also been described (Ryan and Dow 2008; Li and Nair 2012).

AHL molecules from different species are chemically distinct, although their basic structures are similar. The molecules consist of a homoserine lactone (HSL) ring, covalently linked via an amide bond to an acyl side chain (ranging from 4 to 18 carbons) which may be saturated or unsaturated. The carbon at the 3 position of the *N*-linked acyl chain may contain a hydrogen-, oxo-, or hydroxyl substitution. This variability, in combination with the ability of most bacteria to produce more than one type of AHL, provides a mechanism for specificity in QS communication and for the ability of bacteria to distinguish their own AHLs from those produced by other species. The identification of AHL-based QS systems in a variety of bacteria and the understanding of how these systems work have been facilitated by the use of AHL bioreporters. There are many such reporter systems. The two most widely used are those in

W. Giordano (✉)
Departamento de Biología Molecular, Universidad Nacional de Río Cuarto, Ruta 36 Km 601, X5804BYA Río Cuarto, Córdoba, Argentina
e-mail: wgiordano@exa.unrc.edu.ar

V.C. Kalia (ed.), *Quorum Sensing vs Quorum Quenching: A Battle with No End in Sight*,
DOI 10.1007/978-81-322-1982-8_12, © Springer India 2015

Chromobacterium violaceum and *Agrobacterium tumefaciens*, which are often applied for visualization of AHLs separated by thin-layer chromatography (Sanchez-Contreras et al. 2007).

The role of QS in rhizobia-legume symbioses has been studied extensively in many species and has been documented in symbiotic nodulation (Yang et al. 2009), bacterial growth and nodule formation (Cao et al. 2009), symbiotic nitrogen fixation (Daniels et al. 2002), and successful symbiosis establishment (Marketon et al. 2002). On the other hand, many rhizobia with mutations of QS genes are able to establish effective symbioses with their legume hosts, indicating that the role of QS mechanisms is primarily to optimize bacteria/host interactions. Our study of peanut-nodulating rhizobia strains suggested that, depending on the bacterial strain, QS mechanisms may either (i) induce dispersion of cellular aggregates and thereby allow individual bacteria to colonize the peanut root or (ii) promote autoaggregation and thereby improve overall bacterial survival in soil (Nievas et al. 2012).

AHL molecules may also provide a means for rhizobia to communicate with the legume host. Certain higher plants, including legumes, can synthesize AHL-mimicking compounds that may activate or disrupt rhizobial communication and thus affect rhizobial/host symbiosis (Gao et al. 2003). Conversely, the legume *Medicago truncatula* is able to perceive rhizobial AHL signals, with resulting changes in gene expression in the plant (Mathesius et al. 2003). A recent study indicates that research models such as *Arabidopsis thaliana* perceive AHLs in diverse manners (Zarkani et al. 2013).

Rhizobia are a highly diverse group of nitrogen-fixing symbiotic bacteria belonging to the α- and β-subclasses of Proteobacteria. Rhizobial genera include *Rhizobium*, *Sinorhizobium*, *Mesorhizobium*, *Bradyrhizobium*, *Azorhizobium*, *Allorhizobium*, *Methylobacterium* (α-rhizobia), *Burkholderia*, and *Cupriavidus* (β-rhizobia). Recent studies have revealed other rhizobial genera, some of which have arisen through lateral transfer of symbiotic genes. At present, 13 genera and 98 species of rhizobia (α- and β-proteobacteria) are known (www.rhizobia.co.nz/taxonomy/rhizobia). An integrated view of QS systems in rhizobia is presented in this chapter. Available data on QS signaling molecules in rhizobia are summarized in Table 1. The following sections present data for the four best-studied genera: *Sinorhizobium*, *Mesorhizobium*, *Rhizobium*, and *Bradyrhizobium*.

Sinorhizobium

Sinorhizobium (*Ensifer*) *meliloti* is best known for its ability to establish nitrogen-fixing symbioses with plant hosts in the genera *Medicago*, *Melilotus*, and *Trigonella*. The signaling and regulatory events in these symbioses have been well studied. Less is known regarding *S. meliloti* behaviors outside the host that affect the symbioses, e.g., bacterial QS signaling in the rhizosphere (Gao et al. 2012).

The Sin QS system of *S. meliloti* involves an AHL synthase, SinI, and at least two LuxR-type regulators, SinR and ExpR. SinR appears to be independent of AHLs for its control of *sinI* expression, whereas ExpR is almost completely dependent on AHLs. The *S. meliloti* genes *sinI*, *sinR*, and *expR* are all essential for QS regulation in this system, and their expression is dependent on not only the presence of AHLs but also the level of AHLs supplied in growth culture of strains incapable of producing AHLs (McIntosch et al. 2009). The *sinI* gene encodes an AHL synthase that catalyzes synthesis of several long-chain AHLs, including oxo-C14-HL, oxo-C16:1-HL, and C16:1-HL (Teplitski et al. 2003). Upstream of *sinI* is *sinR*, which encodes a transcription regulator that controls activity of the *sinI* promoter. The *sinR* promoter responds to environmental cues such as nutrient limitation by increasing *sinR* transcription (McIntosch et al. 2009). The SinR protein is necessary for activation of the *sinI* promoter, most likely through a SinR binding site, and increased SinR production therefore results in increased *sinI* expression.

The *expR* gene of the Sin system is disrupted by an insertion element in *S. meliloti* strains Rm1021 and Rm2011 (Pellock et al. 2002),

Table 1 Summary of rhizobial QS signaling molecules

Rhizobial species	Signaling molecules	Phenotypes	References
S. meliloti			
Rm1021	3-O-C14-HSL, C16:1-HSL, 3-O-C16:1-HSL, 3-O-C16-HSL, C18-HSL, C12-HSL	EPSs production, swarming	Marketon et al. (2002), Teplitski et al. (2003), Gao et al. (2005), Hoang et al. (2008), Gao et al. (2012)
Rm41	3-O-C8-HSL	Plasmid transfer	Marketon and Gonzalez (2002)
Mesorhizobium			
M. loti R7A	Unknown acyl-HSLs	Symbiosis island transfer	Ramsay et al. (2009)
M. loti NZP2213	3-O-C6-HSL, C8-HSL, C10-HSL, C12-HSL	Nodulation efficiency	Yang et al. (2009)
M. tianshanense	Unknown acyl-HSLs	Legume nodulation	Cao et al. (2009)
M. huakuii	Unknown acyl-HSLs	Biofilm formation, nodulation	Wang et al. (2004), Gao et al. (2006)
Rhizobium			
R. Etli CNPAF512	3-OH-slc-HSL, short-chain acyl-HSLs	Nitrogen fixation, symbiosome development, growth inhibition	Rosemeyer et al. (1998), Daniels et al. (2002)
R. leguminosarum A34	3-OH-C14:1-HSL	Growth inhibition, polysaccharide degradation	Lithgow et al. (2000), Edwards et al. (2009)
	C6-HSL, C7-HSL, C8-HSL	Nodulation efficiency	Cubo et al. (1992); Rodelas et al. (1999)
	3-O-C8-HSL, C8-HSL	Plasmid transfer	Wilkinson et al. (2002), Danino et al. (2003)
Rhizobium sp.			
NGR234	3-O-C8-HSL	Plasmid transfer	He et al. (2003)
Bradyrhizobium			
B. japonicum USDA110	Bradyoxetin	*Nod* gene control	Loh and Stacey (2001), Loh et al. (2002)
	Isovaleryl-HSLs	*BjaI* gene control	Lindemann et al. (2011)
B. japonicum USDA110/290 and *B. elkanii*	Acyl-HSLs like autoinducers	Not reported	Pongsilp et al. (2005)
Bradyrhizobium sp.	Acyl-HSLs like autoinducers	Autoaggregation, biofilm formation, motility	Nievas et al. (2012)

which have been most intensively studied. Under nonstarvation conditions in the laboratory, these strains produce detectable quantities of the symbiotically important exopolysaccharide (EPS) succinoglycan (also known as EPS I) but does not produce galactoglucan (EPS II). The presence of a functional *expR* open reading frame (ORF) on a plasmid or in the genome is sufficient to promote the production of symbiotically active EPS II, e.g., in strain Rm8530, which has an intact *expR*, is termed *expR*+, and has a mucoid phenotype. Restoration of the *expR* gene in strains Rm1021 and Rm2011 confers bacterial QS capability and a notable increase in production of EPS I and EPS II (Pellock et al. 2002). The DNA binding activity of ExpR depends upon the presence of AHLs. The ExpR-AHL complex regulates many promoters (the precise number is not known) throughout the genome. Binding has been demonstrated for the promoters of genes controlling EPS II production (*wgeA* and *wgaA*), genes related to EPS I production (*exoI* and *exsH*), genes controlling flagellum production (*visNR*), and the Sin system genes *sinR* and *sinI* (McIntosch et al. 2009). Because of the importance of EPSs in bacterial attachment, rhizobial cell surface components in combination with bacterial functional signals are essential for this process (Bogino et al. 2013). Accordingly, strains Rm1021 and Rm2011 did not form organized biofilms (Rinaudi and Giordano 2010) and displayed poorly autoaggregative phenotypes (Sorroche et al. 2010).

The amount of AHLs in the environment presumably determines the predominance of positive vs. negative feedback mechanisms and eventually leads to an equilibrium state between the mechanisms at high population densities (McIntosh et al. 2009). Recent studies confirmed the presence of previously detected ExpR-DNA binding sites in *S. meliloti* and identified several additional sites, some of which regulate genes not previously known to be members of the ExpR/AHL regulon. The activities of ExpR/AHL-dependent promoters were titrated against AHL levels, with varying effects on AHL sensitivity. The findings suggest a type of temporal expression program whereby the activity of each promoter is subject to a specific range of AHL concentrations. Genes responsible for EPS production are activated at lower AHL concentrations than those required for repression of genes that control flagellum production. Several features of ExpR-regulated promoters determine their response to AHLs. The location of the ExpR-binding site relative to the relevant transcription start site within each promoter region determines whether ExpR/AHL activates or represses promoter activity. The strength of the response depends on the AHL concentration. This differential sensitivity to AHLs provides a bacterial colony with a transcription control program that is both dynamic and precise (Charoenpanich et al. 2013).

Rm1021 and Rm41 (the parent strain of AK631) are two independently isolated *S. meliloti* strains that are commonly used for studying various aspects of symbiosis with alfalfa (*M. sativa*). Both of these strains display autoinducer activity but have different AHL patterns. TraR, a QS transcriptional activator found in the family Rhizobiaceae, is regulated negatively by the antiactivator TraM via formation of a TraR-TraM heterocomplex (Zheng et al. 2012). The *traR*/*traM* locus homologous to the *tra* system of *A. tumefaciens* (Lang and Faure 2014) is part of a QS system unique to AK631 strain and is involved in regulating conjugal plasmid transfer in the presence of a second QS system, *sinR*/*sinI*, that is present in both AK631 and Rm1021 (Marketon and González 2002). The *traR*/*traM* QS regulators may also be involved in other functions (e.g., host range specificity) uniquely in AK631. The *sinRI* locus, common to Rm1021 and AK631, may regulate components of symbiosis or the free-living state in these strains (Marketon and González 2002).

Disruption of *sinI* expression eliminates these AHLs, and *sinR* disruption results in basal AHL levels. The same *sinI* and *sinR* mutations lead to reduced numbers of pink nodules in nodulation assays and a slight delay in appearance of these nodules, indicating a role of QS in symbiosis. The *sinI* and *sinR* mutants are still capable of producing several short-chain AHLs, one of which was identified as octanoyl homoserine lactone. Marketon et al. (2002) proposed that these short-chain AHLs belong to another QS system in Rm1021, which is termed *mel* system, for "*S. meliloti*."

Mesorhizobium

Mesorhizobium is a moderately fast-growing genus that fixes nitrogen in symbiotic association with legumes. *M. loti* is able to form determinant-type globular nodules and perform nitrogen fixation in several *Lotus* species. The genome sequence of *M. loti* contains ten predicted LuxR homologs and six predicted AHL synthases (Kaneko et al. 2000).

A number of bacterial processes are regulated by exchanges of chemical signals that permit a bacterial community (monospecies and multispecies biofilms) to coordinate its responses to novel environmental challenges or opportunities (Burmølle et al. 2014). Such responses include pathogenesis, symbiosis, antibiotic production, motility, genetic competence, and biofilm formation. Diffusible signals have been suggested to help a bacterial community take a census of its population size; this is a QS phenomenon. The abovementioned chemical signals consist of a set of diffusible AHLs. The key regulatory components of these signaling systems are LuxI-type proteins (which act as AHL synthases) and LuxR-type proteins (which act as AHL receptors and AHL-dependent transcription factors). The TraI and TraR proteins of *A. tumefaciens* are members of this family. The symbiosis island ICE*Ml*Sym of *M. loti* R7A is an integrative and conjugative element (ICE) that carries genes required for nitrogen-fixing symbiosis with *Lotus* species. ICE*Ml*Sym encodes homologs (TraR, TraI1, TraI2) of proteins that regulate plasmid transfer by QS in *A. tumefaciens* (Ramsay et al. 2009). Horizontal transfer is activated by TraR and AHLs. In wild-type cultures, the ICE is excised at low frequency. Ramsay et al. (2013) demonstrated recently that QseM, a widely conserved ICE-encoded protein, is an antiactivator of TraR.

Various isolates of *M. tianshanense* that form nodules on various types of licorice plants produce several different AHL molecules. Root hair attachment and nodulation in this species are controlled by *mrtI/R* QS genes (Cao et al. 2009).

M. huakuii is best known as a symbiont of *Acacia* and *Astragalus*, but was recently shown to form nodules in *Thermopsis* spp. (Ampomah and Huss-Danell 2011). Many AHLs were detected in this species, some at extremely low concentrations (Zhu et al. 2003). To investigate QS regulation in *M. huakuii*, the *Agrobacterium* QS regulator TraR was heterologously expressed in the bacterium. AHL production in supernatant was lower in the resulting strains than in wild type, but intracellular AHL levels were similar, suggesting that AHLs in *M. huakuii* can be bound to intracellular TraR proteins and thus become unavailable for the bacterium's own QS systems. TraR-overexpressing *M. huakuii* formed thinner biofilms than did wild type, suggesting a role of QS in biofilm formation (Wang et al. 2004).

Rhizobium

The genus *Rhizobium* is currently known to include 30 rhizobial species and 11 nonrhizobial species (www.rhizobia.co.nz/taxonomy/rhizobia). *R. etli* establishes a nitrogen-fixing symbiosis by infecting the roots of its leguminous host *Phaseolus vulgaris* (common bean). *R. etli* has multiple AHL synthase genes. The *cinR* and *cinI* genes are required for normal symbiotic nitrogen fixation and swarming and for normal expression levels of *raiI*, which encodes another AHL synthase. The expression of *raiI* in *R. etli* is regulated by RaiR (Rosemeyer et al. 1998; Daniels et al. 2002). QS genes similar to the *cinI/R* genes have been identified in *R. etli* and *M. tianshanense* (*cinI/R* and *mrtI/R*, respectively). Despite high sequence similarities, the roles of *cin* in *R. etli* and *mrt* in *M. tianshanense* are different from their roles in *R. leguminosarum*. In *R. etli*, a *cinI* mutation increased lag phase, slowed growth, and led to abnormal symbiosome development and nitrogen fixation (Daniels et al. 2002). It is possible that the observed symbiotic phenotypes resulted from growth problems in the mutants. The *cin* locus in *R. etli* is required for normal swarming (Daniels et al. 2004), and induction of *cinI* and *cinR* leads to enhanced expression of RaiI-made AHLs. RaiR is involved in the restriction of nodule number. Rosemeyer et al. (1998) found that in vitro mutation of *raiI* led to increases in nodule number and nitrogenase activity, although no significant increase in nitrogen fixation was observed *in planta*.

R. leguminosarum has three biovars: bv. *viciae* (which nodulates peas, vetch, and lentils), bv. *trifolii* (which nodulates clover), and bv. *phaseoli* (which nodulates *Phaseolus*). bv. *viciae* has received the most research attention. Four LuxI-type AHL synthase genes have been identified to date in isolates of bv. *viciae* (Wisniewski-Dye and Downie 2002).

R. leguminosarum A34 contains *cin*, *rai*, *rhi*, and *tra* QS genes. *cinI* and *cinR* are located on the chromosome and are on top of a regulatory cascade that induces production of

RaiI-, RhiI-, and TraI-made AHLs (Lithgow et al. 2000; Wisniewski-Dye and Downie 2002). CinI-made 3-hydroxy-C14:1-HSL was originally termed "small bacteriocin" because it was found to inhibit growth of *Rhizobium* strains carrying the symbiotic plasmid pRL1JI (Schripsema et al. 1996). This compound was also involved in adaptation to stationary phase, because cultures entering stationary phase at high population density showed no loss of viability over long periods, whereas cultures entering stationary phase at low population density did. The addition of 3-hydroxy-C14:1-HSL to cultures at low population density reversed such loss of viability (Thorne and Williams 1999). Mutation of *cinI* or *cinR* did not cause growth difficulties under laboratory conditions, and pea nodulation was normal (Lithgow et al. 2000).

The *traI* and *traR* genes on the symbiotic plasmid pRL1JI are homologous to those in *A. tumefaciens*. Expression of *traR* and *traI* is induced by CinI-made AHLs and results in recipient-induced plasmid transfer. The crucial factor in this process is the presence of BisR, a LuxR-type regulator encoded on pRL1JI that can act as either an inducer or repressor (Danino et al. 2003). In strains carrying pRL1JI (donor strains), BisR represses *cinI* expression, thus preventing synthesis of CinI-made 3-hydroxy-C14:1-HSLs (Wilkinson et al. 2002). In strains that do not carry pRL1JI (recipient strains), such repression does not occur, and CinI therefore produces 3-hydroxy-C14:1-HSLs. When a recipient strain and donor strain come into close proximity, BisR in the donor strain senses the 3-hydroxy-C14:1-HSLs produced by the recipient strain, and the activated BisR then induces *traR* expression (Wilkinson et al. 2002). TraR is activated by TraI-made AHLs and induces expression of plasmid transfer genes, thus initiating conjugation of the symbiotic plasmid to the recipient strain (Danino et al. 2003). The bivalent mode of action of BisR (activator and repressor) is thus responsible for a regulatory mechanism that allows a recipient strain to induce plasmid transfer in the presence of a possible donor strain. This mechanism leads to very high conjugation frequencies and prevents the waste of energy that would occur if unnecessary plasmid transfer took place; i.e., plasmid transfer is initiated only in the presence of recipient strains that do not carry a BisR-containing plasmid (Danino et al. 2003).

Lithgow et al. (2000) showed that CinR induces production of RhiI-made AHLs, which are present on the symbiotic plasmid. The *rhiI* and *rhiR* genes were first identified in *R. leguminosarum* based on the high expression level of RhiA protein, which was not produced by strains lacking the *nod-nif* gene region. RhiR regulates expression of the *rhiABC* genes in response to RhiI-made C6-, C7-, and C8-HSLs. *rhiA* encodes a protein of unknown function that is highly expressed in the rhizosphere (Cubo et al. 1992). Mutation of *rhiA* or *rhiR* caused a decrease in nodule number in strains that were already nodulation compromised (Cubo et al. 1992). Mutations in the *cin* and *rhi* QS systems also affect biofilm formation, rhizosphere growth, and symbiotic interactions (Russo et al. 2006; Edwards et al. 2009). The *cinS* gene, which is co-transcribed with the AHL synthase gene *cinI*, is required for full induction of *rhiR* and *raiR*, whose products (in combination with their partner AHL synthases) regulate other genes in a QS-regulated hierarchy (Frederix et al. 2011). Expression of the *rhi* genes is inhibited by the presence of flavonoids; this is a *nod*-dependent effect mediated via *rhiR* expression. RhiA is present in all strains of bv. *viciae* but is absent in bv. *trifolii* and *phaseoli*, suggesting that this protein may function to optimize interactions between bv. *viciae* and its hosts (peas, vetch lentils).

Rhizobium sp. strain NGR234 is a unique α-proteobacterium that forms nitrogen-fixing nodules with a greater variety of legumes than any other microsymbiont. Transfer of the symbiotic plasmid of this strain apparently has the potential to be under QS regulation because the *traI* and *traR* genes are on the plasmid pNGR234a. TraI synthesizes 3-oxo-C8-HSL, and two other AHLs have been detected in a *traI* mutant, indicating that the corresponding synthase(s) is encoded elsewhere in the genome (He et al. 2003). NGR234 carries at least six loci linked to the quenching of QS signals as well as one gene (*ngrI*) that may encode a novel type of autoinducer I molecule (Schmeisser et al. 2009).

Bradyrhizobium

This genus comprises a diverse group of soil microorganisms having the ability to establish associations with legume (e.g., soybean, peanut) and nonlegume plants (e.g., *Parasponia*). Detectable amounts of AHLs were found in 20 % of *B. japonicum* and *B. elkanii* strains examined with an AHL detection bioassay (Pongsilp et al. 2005). In *B. japonicum* strain USDA110, nodulation genes are expressed in a population density-dependent manner under regulation by bradyoxetin, a factor in the growth medium. Bradyoxetin was shown to be an iron chelator and is therefore not a typical QS regulation signaling molecule (Loh and Stacey 2001; Loh et al. 2002). It acts as a NolA inducer leading to *nod* gene suppression, and NwsB activity is modulated in a cell density-dependent manner. The observed production of AHLs in *B. japonicum* indicates that nodulation and other biological processes are modulated by different autoinducers and different global regulator families to coordinate cellular physiology (Westenberg 2002; Pongsilp et al. 2005; Lindemann et al. 2011).

Production of AHL signaling molecules has been demonstrated in native strains of soybean-nodulating *Bradyrhizobium* (Pongsilp et al. 2005). From several strains analyzed, only 22 % were capable of producing AHL molecules when *A. tumefaciens* NT1 (pZLR4) (Cha et al. 1998) was used as a biosensor to detect autoinducer production. All strains positive for autoinducer activity belonged to the species *B. japonicum* or *B. elkanii*. The *luxI-luxR* genes responsible for AHL synthesis and the biological functions regulated by AHLs were not identified.

The strain USDA110 of *B. japonicum* has genes homologous to *luxI-luxR*, termed *bjaI-bjaR*. No AHLs derived from BjaI have yet been detected. Lindemann et al. (2011) reported the synthesis of a novel signaling molecule catalyzed by BjaI synthase in USDA110. The molecule consists of a branched-chain fatty acyl-HSL, identified as isovaleryl-HSL (IV-HSL). The $BjaR_1$ regulator, a LuxR homolog, has high affinity for IV-HSL. In analogy to results in other AHL QS systems, *bjaI* expression was upregulated by IV-HSL addition. USDA110 is able to respond to AHLs but not to synthesize them. The high sensitivity and low specificity of the $BjaR_1$ regulator to AHLs may reflect a strategy of *B. japonicum* to both avoid detection of AHLs by other bacterial strains and detect QS signals from other microorganisms, in order to acquire a competitive advantage in the rhizospheric microniche (Lindemann et al. 2011). This QS system has been well described for *B. japonicum*, but the biological functions of AHLs in the system has not revealed.

Soil bacteria of the genus *Bradyrhizobium* form symbiotic relationships with peanut (*Arachis hypogaea* L.) root cells and fix atmospheric nitrogen (Bogino et al. 2006). In comparison to well-studied models of rhizobia-legume interaction such as *S. meliloti*-alfalfa, *R. leguminosarum*-bean, and *B. japonicum*-soybean, the symbiotic *Bradyrhizobium* sp.-peanut interaction is poorly understood. Novel AHL signaling molecules have been recently detected and identified in *Bradyrhizobium* strains that are phylogenetically related to peanut-nodulating strains (Ahlgren et al. 2011; Lindemann et al. 2011). We showed recently that addition of various types and concentrations of AHLs to cellular cultures of peanut-nodulating *Bradyrhizobium* sp. affects physiological processes related to bacterial survival, particularly autoaggregation, biofilm formation, and motility (Nievas et al. 2012). However, the symbiotic or other functional roles of these autoinducers in these strains remain essentially unknown.

Conclusions

The interactions between root-associated bacteria and plants encompass many levels. We have summarized here the effects of bacterial QS molecules on the symbiotic interactions between rhizobia and members of the Fabaceae family. QS is a mechanism whereby bacteria regulate their gene expression in a population density-dependent manner. Many aspects of nodulation are also regulated by QS. The best-studied QS signaling systems involve AHLs as signaling molecules. Not surprisingly, the genetic determinants of AHL production and perception are usually integrated in complex regulatory networks and affect numerous aspects of bacterial lifestyles.

Rhizobia have a variety of QS regulatory systems that affect or regulate plasmid transfer, symbiotic interactions, surface polysaccharides, growth inhibition, and stationary-phase adaptation. Several rhizobial species studied to date have been to establish effective symbioses even when genes for AHL synthases and certain AHL receptors are mutated. Rhizobia may therefore not be as dependent as are many bacterial pathogens on QS for successful interactions with their hosts. Findings to date indicate that QS-regulated functions in the bacterium and host serve primarily to optimize various interactions between the partners. The manners in which host plants perceive AHLs appear to be diverse; AHL-producing bacteria therefore have great potential for agricultural applications.

QS signals have been detected in many species of legume-nodulating rhizobia. Besides the role of rhizobial QS signals in symbiotic interactions with legumes, these signals play important roles in microbial cross-communication. Under natural conditions, bacteria typically occur as a mixture of species and have therefore developed many ways to communicate with each other and to "listen in" on other conversations. It has been proposed that bacteria use AI-2 for interspecies communication and that the variable chemical nature of AHLs allows intraspecies communication. In this regard, certain LuxR-type regulators may interact with non-cognate AHL molecules, and such interactions may lead to unwanted activation or inhibition of QS.

It is likely that several additional genes regulated by the above-described systems remain to be identified. Although the regulatory mechanism of one of the most complex QS cascades has been well elucidated, many of the functions regulated by QS genes are poorly understood. Molecular genetic studies of the systems responsible for production of signaling molecules are essential for understanding the mechanisms whereby rhizobia communicate with each other and interact symbiotically with the host plant.

Acknowledgments This review and studies by the author's group described herein were supported by grants from the Secretaría de Ciencia y Técnica (UNRC), Agencia Nacional de Promoción Científica y Tecnológica (ANPCyT), and Consejo Nacional de Investigaciones Científicas y Técnicas of the República Argentina (CONICET). WG is a Career Member of CONICET. Thanks to Dr. S. Anderson for English editing of the manuscript.

References

Ahlgren NA, Harwood CS, Schaefer AL, Giraud E, Greenberg P (2011) Aryl-homoserine lactone quorum sensing in stem-nodulating photosynthetic bradyrhizobia. Proc Natl Acad Sci U S A 108:7183–7188

Ampomah OY, Huss-Danell K (2011) Nodulation of *Thermopsis lupinoides* by a *Mesorhizobium huakuii* strain with a unique *nodA* gene in Kamchatka, Russia. Appl Environ Microbiol 77:5513–5516

Bogino P, Banchio E, Rinaudi L, Cerioni G, Bonfiglio C, Giordano W (2006) Peanut (*Arachis hypogaea*) response to inoculation with *Bradyrhizobium* sp. in soils of Argentina. Ann Appl Biol 148:207–212

Bogino P, Oliva MM, Sorroche FG, Giordano W (2013) The role of bacterial biofilms and surface components in plant-bacterial associations. Int J Mol Sci 14:15838–15859

Burmølle M, Ren D, Bjarnsholt T, Sørensen S (2014) Interactions in multispecies biofilms: do they actually matter? Trends Microbiol 22:84–91

Cao H, Yang M, Zheng H, Zhang J, Zhong Z, Zhu J (2009) Complex quorum-sensing regulatory systems regulate bacterial growth and symbiotic nodulation in *Mesorhizobium tianshanense*. Arch Microbiol 191:283–289

Cha CE, Gao O, Chen YC, Shaw PD, Farrand SK (1998) Production of acyl-homoserine lactone quorum-sensing signals by gram-negative plant-associated bacteria. Mol Plant Microbe Interact 11:1119–1129

Charoenpanich P, Meyer S, Becker A, McIntosh M (2013) Temporal expression program of quorum sensing-based transcription regulation in *Sinorhizobium meliloti*. J Bacteriol 195:3224–3236

Cubo MT, Economou A, Murphy G, Johnston AW, Downie JA (1992) Molecular characterization and regulation of the rhizosphere-expressed genes *rhiABCR* that can influence nodulation by *Rhizobium leguminosarum* biovar *viciae*. J Bacteriol 174:4026–4035

Daniels R, De Vos DE, Desair J et al (2002) The *cin* quorum sensing locus of *Rhizobium etli* CNPAF512 affects growth and symbiotic nitrogen fixation. J Biol Chem 277:462–468

Daniels R, Vanderleyden J, Michiels J (2004) Quorum sensing and swarming migration in bacteria. FEMS Microbiol Rev 28:261–289

Danino VE, Wilkinson A, Edwards A, Downie JA (2003) Recipient-induced transfer of the symbiotic plasmid pRL1JI in *Rhizobium leguminosarum* bv. *viciae* is regulated by a quorum-sensing relay. Mol Microbiol 50:511–525

Downie JA, Walker SA (1999) Plant responses to nodulation factors. Curr Opin Plant Biol 2:483–489

Edwards A, Frederix M, Wisniewski-Dye F, Jones J, Zorreguieta A, Downie JA (2009) The *cin* and *rai* quorum-sensing regulatory systems in *Rhizobium leguminosarum* are coordinated by ExpR and CinS, a small regulatory protein coexpressed with CinI. J Bacteriol 191:3059–3067

Frederix M, Edwards A, McAnulla C, Downie JA (2011) Co-ordination of quorum-sensing regulation in *Rhizobium leguminosarum* by induction of an anti-repressor. Mol Microbiol 81:994–1007

Gao M, Teplitski M, Robinson JB, Bauer WD (2003) Production of substances by *Medicago truncatula* that affect bacterial quorum sensing. Mol Plant Microbe Interact 16:827–834

Gao M, Chen H, Eberhard A et al (2005) *sinI*- and *expR*-dependent quorum sensing in *Sinorhizobium meliloti*. J Bacteriol 187:7931–7944

Gao YJ, Zhong ZT, Sun KJ, Wang H, Zhu J (2006) The quorum sensing system in a plant bacterium *Mesorhizobium huakuii* affects growth rate and symbiotic nodulation. Plant Soil 286:53–60

Gao M, Coggin A, Yagnik K, Teplitski M (2012) Role of specific quorum-sensing signals in the regulation of exopolysaccharide II production within *Sinorhizobium meliloti* spreading colonies. PLoS One 7(8):e42611

He X, Chang W, Pierce DL, Seib LO, Wagner J, Fuqua C (2003) Quorum sensing in *Rhizobium* sp. strain NGR234 regulates conjugal transfer (*tra*) gene expression and influences growth rate. J Bacteriol 185:809–822

Hoang HH, Gurich N, Gonzalez JE (2008) Regulation of motility by the ExpR/Sin quorum-sensing system in *Sinorhizobium meliloti*. J Bacteriol 190:861–871

Kaneko T, Nakamura Y, Sato S et al (2000) Complete genome structure of the nitrogen-fixing symbiotic bacterium *Mesorhizobium loti*. DNA Res 7:381–406

Lang J, Faure D (2014) Functions and regulation of quorum-sensing in *Agrobacterium tumefaciens*. Front Plant Sci 5:14. doi:10.3389/fpls.2014.00014

Li Z, Nair SK (2012) Quorum sensing: how bacteria can coordinate activity and synchronize their response to external signals? Protein Sci 21:1403–1417

Lindemann A, Pessi G, Schaefer AL et al (2011) Isovaleryl-homoserine lactone, an unusual branched-chain quorum-sensing signal from the soybean symbiont *Bradyrhizobium japonicum*. Proc Natl Acad Sci U S A 108:16765–16770

Lithgow JK, Wilkinson A, Hardman A, Rodelas B, Wisniewski-Dye F, Williams P, Downie JA (2000) The regulatory locus *cinRI* in *Rhizobium leguminosarum* controls a network of quorum-sensing loci. Mol Microbiol 37:81–97

Loh J, Stacey G (2001) Feedback regulation of the *Bradyrhizobium japonicum* nodulation genes. Mol Microbiol 41:1357–1364

Loh J, Carlson RW, York WS, Stacey G (2002) Bradyoxetin, a unique chemical signal involved in symbiotic gene regulation. Proc Natl Acad Sci U S A 99: 14446–14451

Marketon MM, González JE (2002) Identification of two quorum-sensing systems in *Sinorhizobium meliloti*. J Bacteriol 148:3466–3475

Marketon MM, Gronquist MR, Eberhard A, Gonzalez JE (2002) Characterization of the *Sinorhizobium meliloti* *sinR/sinI* locus and the production of novel *N*-acyl homoserine lactones. J Bacteriol 184:5686–5695

Mathesius U, Mulders S, Gao M, Teplitski M, Caetano-Anolles G, Rolfe BG, Bauer WD (2003) Extensive and specific responses of a eukaryote to bacterial quorum-sensing signals. Proc Natl Acad Sci U S A 100: 1444–1449

McIntosh M, Meyer S, Becker A (2009) Novel *Sinorhizobium meliloti* quorum sensing positive and negative regulatory feedback mechanisms respond to phosphate availability. Mol Microbiol 74: 1238–1256

Nievas F, Bogino P, Sorroche F, Giordano W (2012) Detection, characterization, and biological effect of quorum-sensing signaling molecules in peanut-nodulating bradyrhizobia. Sensors-Basel 12:2851–2873

Pellock BJ, Teplitski M, Boinay RP, Bauer WD, Walker GC (2002) A LuxR homolog controls production of symbiotically active extracellular polysaccharide II by *Sinorhizobium meliloti*. J Bacteriol 184: 5067–5076

Pongsilp N, Triplett EW, Sadowsky MJ (2005) Detection of homoserine lactone like quorum sensing molecules in *Bradyrhizobium* strains. Curr Microbiol 51: 250–254

Ramsay JP, Sullivan JT, Jambari N et al (2009) A LuxRI-family regulatory system controls excision and transfer of the *Mesorhizobium loti* strain R7A symbiosis island by activating expression of two conserved hypothetical genes. Mol Microbiol 73: 1141–1155

Ramsay JP, Major AS, Komarovsky VM et al (2013) A widely conserved molecular switch controls quorum sensing and symbiosis island transfer in *Mesorhizobium loti* through expression of a novel antiactivator. Mol Microbiol 87:1–13

Rinaudi L, Giordano W (2010) An integrated view of biofilm formation in Rhizobia. FEMS Microbiol Lett 304:1–11

Rodelas B, Lithgow JK, Wisniewski-Dye F et al (1999) Analysis of quorum-sensing dependent control of rhizosphere-expressed (*rhi*) genes in *Rhizobium leguminosarum* bv. *viciae*. J Bacteriol 181: 3816–3823

Rosemeyer V, Michiels J, Verreth C, Vanderleyden J (1998) *luxI*- and *luxR*-homologous genes of *Rhizobium etli* CNPAF512 contribute to synthesis of autoinducer molecules and nodulation of *Phaseolus vulgaris*. J Bacteriol 180:815–821

Russo DM, Williams A, Edwards A et al (2006) Proteins exported via the PrsD-PrsE type I secretion system and the acidic exopolysaccharide are involved in biofilm formation by *Rhizobium leguminosarum*. J Bacteriol 188:4474–4486

Ryan RP, Dow JM (2008) Diffusible signals and interspecies communication in bacteria. Microbiology 154:1845–1858

Sanchez-Contreras M, Wolfgang BD, Mengsheng G, Robinson JB, Downie A (2007) Quorum sensing regulation in rhizobia and its role in symbiotic interactions with legumes. Philos Trans R Soc B 362:1149–1163

Schmeisser C, Liesegang H, Krysciak D et al (2009) *Rhizobium* sp. Strain NGR234 possesses a remarkable number of secretion systems. Appl Environ Microbiol 75:4035–4045

Schripsema J, de Rudder KE, van Vliet TB et al (1996) Bacteriocin small of *Rhizobium leguminosarum* belongs to the class of *N*-acyl-L-homoserine lactone molecules, known as autoinducers and as quorum sensing co-transcription factors. J Bacteriol 178: 366–371

Sorroche F, Rinaudi L, Zorreguieta A, Giordano W (2010) EPS II-dependent autoaggregation of *Sinorhizobium meliloti* planktonic cells. Curr Microbiol 61:465–470

Teplitski M, Eberhard A, Gronquist MR, Gao M, Robinson JB, Bauer WD (2003) Chemical identification of N-acyl homoserine lactone quorum-sensing signals produced by *Sinorhizobium meliloti* strains in defined medium. Arch Microbiol 180: 494–497

Thorne SH, Williams HD (1999) Cell density-dependent starvation survival of *Rhizobium leguminosarum* bv. *phaseoli*: identification of the role of an *N*-acyl homoserine lactone in adaptation to stationary-phase survival. J Bacteriol 181:981–990

Wang H, Zhong Z, Cai T, Li S, Zhu J (2004) Heterologous overexpression of quorum-sensing regulators to study cell-density-dependent phenotypes in a symbiotic plant bacterium *Mesorhizobium huakuii*. Arch Microbiol 182:520–525

Westenberg DJ (2002) Evidence of AHL autoinducer production by the soybean symbiont *Bradyrhizobium japonicum*. In: Finan et al (eds) Nitrogen fixation: global perspectives. CABI Publishing/Oxford University Press, Cary

Wilkinson A, Danino V, Wisniewski-Dye F, Lithgow JK, Downie JA (2002) *N*-acyl-homoserine lactone inhibition of rhizobial growth is mediated by two quorum-sensing genes that regulate plasmid transfer. J Bacteriol 184:4510–4519

Wisniewski-Dye F, Downie JA (2002) Quorum-sensing in *Rhizobium*. Antonie Van Leeuwenhoek 81:397–407

Yang M, Sun K, Zhou L, Yang R, Zhong Z, Zhu J (2009) Functional analysis of three AHL autoinducer synthase genes in *Mesorhizobium loti* reveals the important role of quorum sensing in symbiotic nodulation. Can J Microbiol 55:210–214

Zarkani AA, Stein E, Röhrich CR, Schikora M, Evguenieva-Hackenberg E, Degenkolb T, Vilcinskas A, Klug G, Kogel KH, Schikora A (2013) Homoserine lactones influence the reaction of plants to rhizobia. Int J Mol Sci 14:17122–17171

Zheng Z, Fuqua C, Chen L (2012) The quorum sensing transcriptional regulator TraR has separate binding sites for DNA and the anti-activator. Biochem Biophys Res Commun 418:396–401

Zhu J, Chai Y, Zhong Z, Li S, Winans SC (2003) *Agrobacterium* bioassay strain for ultrasensitive detection of N-acylhomoserine lactone-type quorum sensing molecules: detection of autoinducers in *Mesorhizobium huakuii*. Appl Environ Microbiol 69:6949–6953

Quorum Sensing Systems in Clostridia

Charles Darkoh and Godfred Ameyaw Asiedu

Introduction

The genus *Clostridium* consists of bacteria that are obligate anaerobes, Gram-positive, rod shaped, sporeformers, and catalase negative. This genus is one of the largest prokaryotic genera in the phylum *Firmicutes* (*Clostridia* is also a class in the phylum), containing over 300 species. All clostridial species form endospores with a strict fermentative type of metabolism. These anaerobes have two forms of growth stages in their life cycle: the vegetative stage, where they divide and proliferate during favorable conditions, and the sporulation stage, in which the vegetative cells form spores under unfavorable conditions and remain dormant until suitable conditions return. The spores are non-metabolizing and highly resistant to heat, desiccation, radiation, oxidation, and many other microbial control agents. The majority of clostridial species do not grow under aerobic conditions, and the vegetative cells are killed upon exposure to oxygen. These anaerobes are ancient bacteria that colonize and thrive in nearly all of the natural anaerobic habitats where organic compounds are present such as soils, aquatic sediments, and the intestinal tracts of animals.

Clostridia are well known for their clinical importance and biotechnological potential. The bacteria in this genus are capable of fermenting a wide variety of organic compounds. They produce butyric acid, acetic acid, butanol, acetone, and large amounts of gas (CO_2 and H_2) as end products of sugar fermentation. A variety of foul-smelling compounds are formed during the fermentation of amino acids and fatty acids. These microbes also produce a wide variety of extracellular enzymes to break down large biological molecules such as proteins, lipids, collagen, and cellulose in the environment into fermentable metabolites. In anaerobic clostridial infections, these enzymes also play a role in host invasion and pathology. Therefore, the clostridial species play an important role in natural biodegradation and carbon cycling, with significant applications in biotechnology and medicine. The nonpathogenic *Clostridium acetobutylicum* has garnered widespread interest because of its role in biobutanol production (Jones and Woods 1986; Lee et al. 2008). As the genome sequence data and DNA mutagenesis tools become available, recombinant strains with superior biobutanol-producing ability could be engineered (Durre 2007, 2008). Furthermore, the use of clostridial spores in cancer therapy has also been described for *Clostridium sporogenes* and *Clostridium novyi* (Minton 2003; Minton et al. 1995; Theys et al. 2006; Mengesha et al. 2010).

C. Darkoh (✉)
Division of Epidemiology, Human Genetics, and Environmental Sciences, Center for Infectious Diseases, The University of Texas School of Public Health, 1200 Herman Pressler, Suite E715, Houston, TX 77030, USA
e-mail: Charles.darkoh@uth.tmc.edu

G.A. Asiedu
Faculty of Law, Humanities and the Arts, University of Wollongong, Australian National Center for Ocean Resources and Security, Wollongong, NSW, Australia

V.C. Kalia (ed.), *Quorum Sensing vs Quorum Quenching: A Battle with No End in Sight*,
DOI 10.1007/978-81-322-1982-8_13, © Springer India 2015

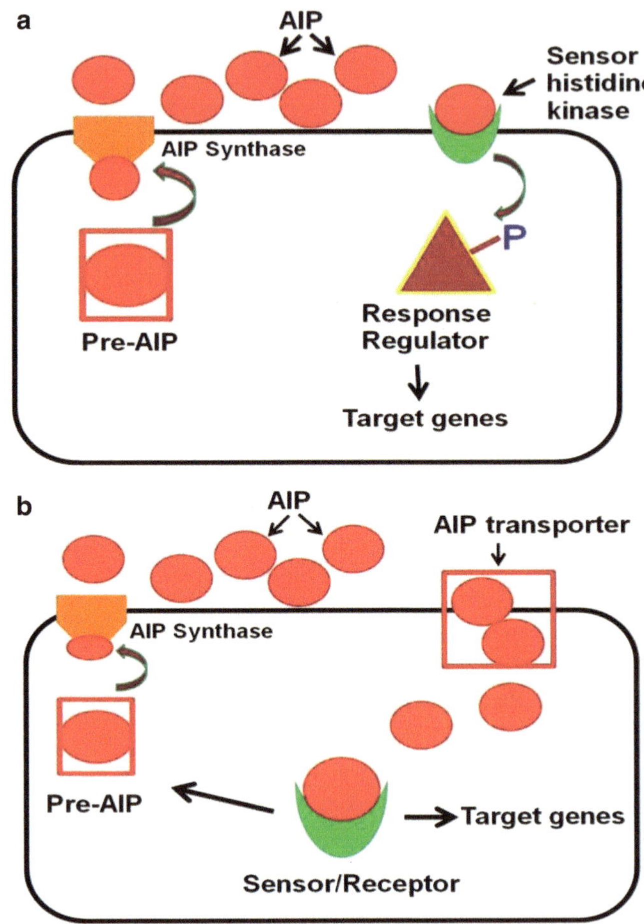

Fig. 1 Autoinducing peptide (AIP)-mediated quorum sensing in Gram-positive bacteria by (**a**) two-component signaling or (**b**) an AIP-binding transcription factor. (**a**) The autoinducer pre-peptide is processed by the transmembrane protein AIP synthase, leading to the release of mature AIP into the extracellular *milieu*. A sensor histidine kinase protein senses the extracellular AIP, which activates its ATPase activity leading to the phosphorylation of the response regulator. Phosphorylated response regulator modulates the transcription of its target genes. (**b**) The extracellular AIP is transported into the cell by a transmembrane transporter, which is then sensed by a cytoplasmic receptor leading to the modulation of its target genes

Nevertheless, the pathogenic *Clostridia* have overshadowed the beneficial properties of the nonpathogenic ones. The pathogenic *Clostridia* cause disease by releasing potent toxins, notably the enterotoxins from *Clostridium perfringens,* cytotoxins from *Clostridium difficile*, and severe neurotoxins from *Clostridium botulinum* and *Clostridium tetani.* Tetanus and botulinum neurotoxins, produced by *C. tetani* and *C. botulinum*, respectively, are two of the most harmful natural compounds known to man. Both produce neurotoxins that are lethal in very low concentrations. The estimated lethal dose of *C. botulinum* neurotoxin type A for a human of 70 kg is approximately 70 μg orally and 0.7–0.9 μg if inhaled (Arnon et al. 2001). *C. perfringens* produces a number of extracellular toxins and enzymes associated with gas gangrene and clostridial myonecrosis in humans and animals (Collie et al. 1998). *C. botulinum* is widely used in the cosmetics industry and it is important in other medical applications. However, *C. botulinum* has long been associated with food poisoning and concern for the use of the lethal toxins in bioterrorism (Schantz and Johnson 1992). The most recent publicized pathogenic *Clostridium* is *C. difficile*, the causative agent of pseudomembranous colitis and antibiotic-associated diarrhea, which has been wreaking havoc in the elderly and the immunocompromised.

Quorum Sensing in Gram-Positive *Clostridia*

Gram-positive and Gram-negative bacteria utilize different kinds of quorum sensing systems. Figure 1 shows the two paradigmatic quorum sensing signaling networks that have been postulated in Gram-positive bacteria such as *Clostridia* (Rutherford and Bassler 2012). Quorum sensing in Gram-positive bacteria occurs through a range of different signals, but all systems involve a two-component regulatory signal transduction system consisting of a membrane-bound receptor/sensor histidine kinase protein and an intracellular response regulator (Sturme et al. 2002). Gram-positive bacteria mainly utilize peptides, called autoinducing peptides (AIPs), as quorum signaling molecules. Once produced in the cell, the AIPs are processed and secreted. When the extracellular concentration of the AIP is high, which occurs at high cell density, it binds to a cognate membrane-bound two-component histidine kinase receptor. Typically, binding activates the kinase activity of the receptor, which autophosphorylates and transfers a phosphate to its cognate cytoplasmic response regulator. The phosphorylated response regulator activates the transcription of the genes in the quorum sensing regulon (Fig. 1a). In other types of Gram-positive quorum sensing systems, the AIPs are transported from the extracellular milieu back into the cell cytoplasm where they interact with specific receptors and transcription factors to modulate gene expression (Fig. 1b).

Quorum sensing in Gram-positive bacteria depends on three principles: production, sensing, and response to the AIPs. In the majority of the Gram-positive bacteria, the AIPs are small diffusible oligopeptides that are sensed by membrane-bound two-component signal transduction systems (Havarstein et al. 1995; Ji et al. 1995; Solomon et al. 1996). The AIPs are diverse in sequence and structure and are genetically encoded as autoinducer pre-peptides, which are processed posttranslationally into fully functional AIPs (Thoendel and Horswill 2010; Thoendel et al. 2011; Okada et al. 2005). Specialized transporters secrete the AIPs extracellularly due to the impermeability of the oligopeptides to the cell membrane. The functionally processed AIPs, which range in size from 5 to 17 amino acids, are posttranslationally modified into either a cyclic or linear molecule (Bouillaut et al. 2008; Okada et al. 2005; Mayville et al. 1999). Extracellular AIPs are detected via membrane-bound two-component sensor kinases (Flamez et al. 2007). The binding of the AIP to the sensor kinases leads to autophosphorylation at conserved histidines. The phosphoryl group is subsequently passed from the histidine to a conserved aspartate on a cognate cytoplasmic response regulator protein. This phosphorylation activates the response regulator, leading to the modulation of expression of the quorum sensing-regulated genes. The autoinducer pre-peptide, its transporter, histidine kinase receptor, and response regulator are typically encoded in an operon (Peterson et al. 2000). In some bacteria, the phosphorylation of the response regulator activates the expression of the operon, resulting in an autoinducing feedforward loop that synchronizes the quorum sensing response. Two main quorum sensing systems have been identified in *Clostridia*, namely, the accessory gene regulator (agr) quorum-sensing system and the LuxS quorum-sensing system. These systems are briefly described in the sections that follow.

The Accessory Gene Regulator (agr) Quorum Sensing System in *Clostridia*

The accessory gene regulator *(agr)* system is not well characterized in *Clostridia*, but has been extensively studied in *Staphylococcus aureus*. In *S. aureus*, it regulates the temporal expression of cell surface colonization factors and various toxins, which contribute to virulence (Novick 2003; Novick and Geisinger 2008; Novick and Muir 1999). The *agr* locus in *S. aureus* is approximately 3.5 kilo base pairs and consists of two divergent operons, RNAII and RNAIII, which are transcribed from nonoverlapping promoters, P2 and P3, respectively. The P2 operon contains four open reading frames: *agrA, agrC, agrD,* and *agrB. The agrA* and *agrC* form a two-component system for signal sensing and transduction (Balaban and Novick 1995; Novick et al. 1995). The P3-generated transcript, RNAIII, is the actual intracellular effector of the *agr* response that modulates the expression of the agr-regulated genes (Novick et al. 1993). It also encodes a staphylococcal toxin (Novick et al. 1995). The P2 operon is doubly autocatalytic since its transcription requires P2 operon products and is therefore suited to the need for swift expression of proteins required when overall growth is coming to a halt (Morfeldt et al. 1988). The specific roles of each of the agr quorum-sensing system components are outlined below.

AgrD is the pre-peptide which is synthesized within the cell during growth, processed, and released extracellularly. The processed pre-peptide becomes the AIP, which activates the agr quorum-sensing system. The transmembrane protein AgrB is involved in processing the pre-peptide and secretion of the posttranslationally modified AIP (Ji et al. 1995). The AgrD protein is processed at both the N- and C-terminal ends. A cyclization reaction occurs, where a thioester linkage is formed between a conserved cysteine residue and the C-terminal carboxyl group, in order to form the mature thiolactone AIP (Lyon and Novick 2004; Novick 2003).

AgrB is a transmembrane protein that is crucial for the generation and secretion of the AIP. This protein is required for the proteolytic modification of the autoinducer pre-peptide encoded by *agrD* and transport of the fully functional AIP outside the cell (Sturme et al. 2002). Accordingly, the *agrBD* genes are responsible for the generation and processing of AIP.

AgrC is the transmembrane sensor kinase of a two-component system, whose N- terminal interacts with the AIP (Lina et al. 1998). The activation of AgrC then occurs via a phosphorylation event. The C-terminal was originally proposed to autophosphorylate at a conserved histidine residue upon stimulation by the AIP (Ji et al. 1995; Novick et al. 1995). However, it was later observed that AgrC is partially phosphorylated in the absence of the AIP. This suggested that AgrC may be partially activated constitutively and becomes fully activated upon binding to the AIP. The autophosphorylation of AgrC occurs as a *trans* process (Novick 2003). The cytoplasmic domain of AgrC is dimeric containing a four-helix bundle that aids in the sensing and binding of the AIP.

AgrA is the response regulator protein required for the activation of the P2 and P3 promoters and it is phosphorylated by AgrC. AgrA binds with high affinity to the RNAIII *agr* intergenic region, and this binding is localized to a pair of direct repeats in the P2 and P3 promoter regions (Koenig et al. 2004). The phosphorylation of AgrA activates the system, which subsequently results in the modulation of AgrA-regulated genes. The *agr* genes are constitutively expressed at low cell densities, but when the AIP level reaches a threshold concentration, the *agr* response is activated (Ji et al. 1995). The signaling transduction cascade produced by the AgrA and AgrC two-component system results in an increase in the transcription of RNAII and RNAIII in a positive feedback manner (Yarwood and Schlievert 2003).

Homologues of the *S. aureus agr* genes are present in several clostridial species. It is evident that the *agr*-based quorum sensing system plays an important role in the pathogenesis of some toxigenic strains of *Clostridia* such as toxin production in *C. botulinum* and *C. perfringens,* as discussed later in this section. Some species have two copies of the agr autoinducer pre-peptide signal generation and processing genes, *agrBD*. For instance, *C. botulinum* contains two homologues of the *agrBD* genes, named *agrBD1* and *agrBD2* (Cooksley et al. 2010). The role of the agr quorum-sensing system in the pathogenesis and virulence of *Clostridia* is discussed later in this section.

The LuxS Quorum Sensing System

The LuxS quorum-sensing system utilizes a signaling molecule designated as autoinducer-2 (AI-2), which is synthesized by enzymes encoded by the *luxS* family of genes. It has been demonstrated to be involved in quorum signaling in both Gram-negative and Gram-positive bacteria (Vendeville et al. 2005; Doherty et al. 2006). The signaling molecule utilized by the LuxS quorum-sensing system (AI-2) is a by-product of the activated methyl cycle, which recycles S-adenosyl-L-methionine (SAM). SAM is the main methyl donor in archaea, prokaryotes, and eukaryotes that supplies methyl groups to biomolecules such as DNA and proteins. In the activated methyl cycle, SAM is converted to S-adenosyl-L-homocysteine (SAH), which is toxic to cells. In eukaryotes, SAH is converted to nontoxic homocysteine in a single step (Winzer et al. 2002). However, in archaea and prokaryotes, SAH is detoxified in a two-step process, first by the Pfs enzyme (S-adenosylhomocysteine nucleosidase) to generate adenine and S-ribosylhomocysteine (SRH). SRH is finally converted by LuxS (S-ribosylhomocysteinase) to 4,5-dihydroxy-2,3-pentanedione (DPD) and homocysteine (Winzer et al. 2002; Schauder and Bassler 2001). DPD undergoes spontaneous cyclization, and it can be sensed and modified by different bacterial species as a signaling molecule that controls gene expression.

AI-2 appears to mediate both intraspecies and interspecies communication in bacteria (Schauder et al. 2001; Curry et al. 2013). AI-2/*luxS*-mediated regulation is important in controlling different activities in a variety of bacterial species (Curry et al. 2013). These activities include biofilm production in *Streptococcus mutans*, *Salmonella enterica* serovar *Typhimurium*, and *Vibrio cholerae* (Prouty et al. 2002; Hammer and Bassler 2003; Merritt et al. 2003; Yoshida et al. 2005); motility in *Campylobacter jejuni*, enterohemorrhagic *Escherichia coli* (EHEC), and enteropathogenic *E. coli* (Sperandio et al. 2001; Elvers and Park 2002; Giron et al. 2002); and expression of virulence factors in *Aggregatibacter actinomycetemcomitans*, *EHEC*, *Porphyromonas gingivalis*, and *V. cholerae* (Sperandio et al. 1999; Chung et al. 2001; Fong et al. 2001; Ohtani et al. 2002; Zhu et al. 2002).

Despite the proposed role of LuxS in quorum sensing, it is speculated that LuxS is in fact a metabolic by-product produced during the activated methyl cycle (Winzer et al. 2003). Indeed, Doherty et al. (2006) demonstrated that the inactivation of LuxS in *S. aureus* had no effect on virulence-associated traits. It is possible that the phenotypic changes observed in some bacteria following the inactivation of LuxS may be due to the absence of the AI-2 signal molecule, but they could also be attributed to the disruption of the activated methyl cycle (Winzer et al. 2003).

Quorum Sensing-Mediated Regulation of Toxin Production in *Clostridia*

The pathogenic *Clostridia* are an important group of anaerobic bacteria that cause various neurotoxic, cytopathic, and enterotoxic diseases in humans and animals. The ability of these anaerobes to cause disease is mediated by the various toxins (also known as large clostridial toxins) that they produce during colonization and infection. The most common clostridial species that are known to cause serious disease in mammals are *C. difficile*, *C. botulinum*, *C. tetani*, *C. perfringens*, and *C. sordellii*. *C. difficile* is an increasingly important nosocomial pathogen, which is considered the leading cause of antibiotic-associated diarrhea and pseudomembranous colitis in the developed world (Carter et al. 2011a, 2012). The toxins produced by *C. botulinum* and *C. tetani* are neurotoxigenic and result in flaccid and convulsive paralysis, respectively (Montecucco et al. 1994; Montecucco and Schiavo 1994). *C. perfringens* is the causative agent of a myriad of different diseases ranging from gas gangrene and necrotic enteritis to food poisoning (Rood 1998; Uzal and McClane 2011), depending on the types of toxins produced by the infecting strain. *C. sordellii* is an emerging pathogen that causes a range of diseases including myonecrosis, sepsis, and shock (Carter et al. 2011a). In addition to these species, *C. chauvoei*, *C. colinum*, *C. novyi*, *C. septicum*, and *C. spiroforme* also cause severe diseases in numerous domestic and livestock animals (Songer 1996, 1998, 2010). Even though these anaerobes cause different diseases, in most cases the symptoms of disease are associated with the production of potent exotoxins that result in damage to the host tissues (Stevens et al. 2012). The pathogenic *Clostridia* produce some of the most potent toxins known to mankind such as the *C. botulinum* neurotoxin, which is the most potent toxin ever identified (Arnon et al. 2001; Schechter and Arnon 2000; Katona 2012).

Most of the clostridial toxins have been purified and subjected to extensive structural and functional analysis. However, challenges in genetically manipulating clostridial species have limited our understanding of the mechanisms and factors that control the production of these toxins. Certainly, the construction of genetically defined mutants in many clostridial species remains a challenge, and techniques for successful genetic manipulation of these bacteria are still in the developmental stages. The limited methods developed so far have facilitated genetic manipulation of some of these species (Lyristis et al. 1994; Chen et al. 2005; O'Connor et al. 2006; Heap et al. 2007; Bradshaw et al. 2010; Carter et al. 2011a) and have led to the understanding of some of the genes and regulatory pathways that control toxin production at the molecular level. The agr quorum-sensing system has been demonstrated to control toxin production in a number of clostridial species, and this is discussed in the sections that follow.

Clostridium perfringens

Infection by *C. perfringens* results in gastrointestinal distress, but this species is also the causative agent for gas gangrene. *C. perfringens* is an important medical pathogen that causes both histotoxic infections (e.g., traumatic gas gangrene) and infections originating in the intestines (e.g., human food poisoning). This bacterium produces four types of toxins, namely, alpha-, beta-, epsilon-, and iota-toxins. Clinical and environmental isolates of *C. perfringens* are commonly assigned to one of five biotypes (from A to E) based on the type of the toxins they produce. Infections due to *C. perfringens* show evidence of tissue necrosis, bacteremia, emphysematous cholecystitis, and gas gangrene, which is also known as clostridial myonecrosis.

Deep wounds create anaerobic conditions and provide an ideal environment for the germination of the environmentally ubiquitous *C. perfringens* endospores. Germinating *C. perfringens* produces several proteases and toxins to digest host tissues, which subsequently supply the necessary amino acids and nutrients for which *C. perfringens* lacks the necessary biosynthetic pathways (Gray et al. 2013). This in situ exotoxin-mediated release of nutrients rapidly accelerates bacterial growth and further complicates the treatment of *C. perfringens* infection. In addition, strains of *C. perfringens* may also produce a number of other toxins such as neuraminidase and enterotoxin (CPE). The CPE is one of the most important toxins produced by this bacterium, although it is not part of the toxinotyping classification system. The CPE-producing type A strains are responsible for *C. perfringens* type A food poisoning, which is the second most commonly identified bacterial foodborne disease in the United States, where nearly a million cases occur annually (Li et al. 2011; Scallan et al. 2011a, b).

Site-directed mutagenesis and random mutagenesis of the *agrB* or *agrBD* null mutants of strain 13, a *cpe*-negative and non-sporulating *C. perfringens* type A strain, has established that the Agr-like system regulates the production of alpha-toxin (CPA) and perfringolysin O (PFO), two chromosomally encoded toxins expressed during vegetative growth. This Agr-like system-mediated regulation of toxin production was shown to involve a diffusible molecule, consistent with a QS mechanism (Ohtani et al. 2009; Vidal et al. 2009). Recently, Li et al. (2011) reported that the *agr* locus also regulates CPE, beta-toxin (CPB2), and PFO expression by *C. perfringens* strain F5603 in DS sporulation medium. Using real-time PCR, they were able to detect the mRNA transcripts of *cpe*, *cpb2*, *cpa*, and *pfo* genes in F5603 wild type that was cultured in DS medium for 5 h. However, they could not detect the *cpe*, *cpb2*, or *pfo* gene transcripts under the same conditions in an *agrB* null mutant during the 5 h incubation period. The loss of toxin gene transcription in DS cultures of the *agrB* mutant was reversible by complementation.

The production of epsilon-toxin and beta2-toxins was lower in Δ*agrB* type B strains compared to wild-type bacteria in in vitro assays using enterocyte-like $CaCo_2$ cells (Chen et al. 2011;

Chen and McClane 2012). A*grB* has also been reported to control in vivo toxicity and colitis in a rabbit intestinal loop model with a type C strain (Vidal et al. 2012). The group A strain *agrBD* has also been demonstrated to regulate expression of genes on a plasmid (Li et al. 2011), a novel result previously unheralded in the *agr* literature. The cognate receptor for *C. perfringens* AIP has not been conclusively identified, but in type A strains the 2CRS *virRS*, which regulates the expression of several toxins and virulence factors through the regulatory RNA molecule VR-RNA (Ohtani et al. 2010; Shimizu et al. 2002), may be involved in *agr*-mediated quorum sensing and pathogenesis. However, the link between *agr* and *vir* is unclear in other strain types, as VirRS does not appear to be involved in *agrBD*-mediated toxin production in type B or D strains of *C. perfringens* (Chen et al. 2011; Chen and McClane 2012).

Clostridium botulinum

Clostridium botulinum neurotoxins are classified into seven structurally related but antigenically distinct serogroups (serogroups A–G). These toxins form heterocomplexes with other nontoxic proteins and are the most potent toxins known to humans (Schantz and Johnson 1992). *C. botulinum* is a heterogeneous species consisting of four physiologically and phylogenetically distinct groups (Cooksley et al. 2010). In humans and various animal species, the toxin induces a potentially fatal condition known as botulism. Botulism intoxication in healthy adults is not normally due to infection with *C. botulinum*, although colonization of the gut is the accepted mechanism for infant botulism. In adults, consuming preserved food initially contaminated with *C. botulinum* spores leads to the ingestion of the botulism neurotoxin, of which there are several types. Symptoms of botulism include progressive, descending, symmetrical paralysis that affects the musculature innervated by the cranial nerves at the initial stage, before spreading to the rest of the body (Schantz and Johnson 1992). Naturally occurring forms of botulism include foodborne, intestinal, and wound botulism. The most common form of disease caused by this anaerobe is foodborne botulism and occurs when food or drink containing the preformed neurotoxin is ingested (Cooksley et al. 2010). In humans, members of group I (proteolytic) and group II (non-proteolytic) *C. botulinum* are responsible for most of the cases observed. In contrast, intestinal and wound forms of botulism are a consequence of in vivo production of the *C. botulinum* toxins following infection. In humans, these forms of botulism are mostly caused by group I *C. botulinum.* Intestinal botulism is usually observed in infants and occurs after the ingestion of spores and subsequent colonization of the gastrointestinal tract by the vegetative cells. Wound botulism manifests as a consequence of localized tissue infection. Due to its extremely high potency, security agencies fear that it might be an attractive weapon for bioterrorists (Villar et al. 2006). *C. botulinum* has therefore developed renewed strategic importance post-September 11, 2001 terrorist attacks, particularly in the United States.

All the recently completed genomes of *C. botulinum* hold putative homologues of the *agr* quorum-sensing system. In silico analysis of the 10 completed *C. botulinum* genomes shows the presence of two sets of genes encoding homologues of the AgrB and AgrD proteins. These two putative *agr* loci are located in close proximity to each other on the chromosome. Consequently, the *C. botulinum* agr quorum-sensing systems have been designated *agrBD1* and *agrBD2*. The flanking regions of both loci are conserved among sequenced group I *C. botulinum* strains and two *C. sporogenes* strains (Cooksley et al. 2010). The AgrD1 and AgrD2 sequences from group I *C. botulinum* and *C. sporogenes* strains share 17 of 44 amino acids (38.6 % identity). Additionally, the AgrD1 sequences of these strains are almost identical. All group I *C. botulinum* sequences share the same putative AIP-encoding region, which differs from that of the two *C. sporogenes* strains in only one of the five putative amino acids that are expected to form the cyclic AIP peptide. However, the agrD2 sequences have some degree of variability. The variation in these sequences occurs primarily in the region that contains the putative AIP ring structure, but these sequences are 77 % similar. Based on the putative AIP encoded, the currently available *agrD2* genes form five strain clusters, consisting of (1) the two *C. sporogenes* strains; (2) strain NCTC 2916, strain ATCC 19397, A strain Hall, A2 strain Kyoto, and A3 strain Loch Maree; (3) strain ATCC 3502; (4) Ba4 strain 657 and strain Bf; and (5) strains Langeland and Okra. The third putative *agrD* gene in the Loch Maree strain is 90 % identical to the third *agrD* homologue found in *C. sporogenes* ATCC 15579, with no differences in the putative AIP-containing region. This analysis suggests that group I *C. botulinum* strains are likely to produce the same AIP, but may differ with respect to the AIP2 signal.

Cooksley et al. (2010) characterized the Agr system in *C. botulinum* ATCC 3502 (group I) using *C. sporogenes* NCIMB 10696 as a model system for initial experiments, as it is considered a non-toxigenic version of group I (proteolytic) *C. botulinum* (Carter et al. 2009). They reported that the production of neurotoxins was affected in agrD mutants. They generated both agrD1 and agrD2 mutants in *C. botulinum* strain 3502 and assayed for the botulinum neurotoxin in the wild type and its *agrD* mutants by using an ELISA-based method. The assay demonstrated that extracellular neurotoxin accumulation peaked after entry into the stationary phase and that the production of the neurotoxin was negatively affected in both mutants. The largest reduction occurred in the *agrD2* mutant, where the levels of toxin produced were considerably depressed throughout the period assessed (up to 72 h). In the case of the *agrD1* mutant, whereas the levels of toxin produced were evidently reduced up to and including the 24 h time point, the quantities of toxin present in late-stationary-phase cultures (from 48 h onward) had returned to wild-type levels. Thus, the inactivation of either *agrD1* or *agrD2* resulted in lower botulinum neurotoxin production compared to the wild-type strain, with the latter mutant displaying the greatest reduction (Cooksley et al. 2010). These results provide evidence that toxin production in *C. botulinum* involves the agr quorum-sensing system.

Clostridium difficile

The incidence of *Clostridium difficile* infection (CDI) has been increasing such that it is now the most common definable cause of hospital-acquired and antibiotic-associated diarrhea in the United States (DuPont 2011). The number of cases occurring in the community and nonhospital healthcare facilities makes CDI the most common form of bacterial diarrhea, with the total cost of treatment estimated between 1 and 3.2 billion US dollars annually (Kyne et al. 2002; O'Brien et al. 2007; Wilkins and Lyerly 2003; DuPont 2011). This bacterium has been identified as the cause of 10–25 % of the cases of antibiotic-associated diarrhea, 50–75 % of antibiotic-associated colitis, and 90–100 % of pseudomembranous colitis (Bartlett 2002; Elliott et al. 2007). Morbidity and mortality resulting from *C. difficile*-associated diseases have also increased significantly (McDonald et al. 2005; Redelings et al. 2007; O'Brien et al. 2007).

Pathogenic strains of *C. difficile* possess a 19.6 kb pathogenicity locus, which is composed of the *tcdR, tcdB, tcdE, tcdA*, and *tcdC* genes. This locus is responsible for the production of toxins A and B that are encoded by *tcdA* and *tcdB*, respectively. The toxins are essential for *C. difficile* pathogenesis (Lyerly et al. 1985; Rupnik et al. 2001; Geric et al. 2004; Voth and Ballard 2005; Kuehne et al. 2010); strains that are unable to produce either of these toxins are not associated with disease. Both toxins have similar enzymatic cleavage activities (Dillon et al. 1995; Just et al. 1995a, b) and are cytotoxic to cultured cells; however, toxin B is 100–1,000-fold more potent than toxin A (von Eichel-Streiber et al. 1996; Just and Gerhard 2004; Voth and Ballard 2005). During infection, these toxins are released into the intestinal lumen where they bind to surface receptors on colonic epithelial cells via their receptor-binding domain and are then internalized by host cells via receptor-mediated endocytosis (Ho et al. 2005; Dingle et al. 2008). The acidic environment within endosomes activates the autocatalytic cysteine protease activity of the toxins, resulting in cleavage and release of the N-terminal glucosyltransferase domain into the cytosol of the mammalian host (Hofmann et al. 1997; Egerer et al. 2007; Reineke et al. 2007; Pfeifer et al. 2003; Rupnik et al. 2005). The glucosyltransferase domain monoglucosylates low-molecular-weight GTPases of the Rho family in the cytosol using cellular uridine diphosphoglucose (UDP-glucose) as the glucose donor (Just et al. 1995a; Just and Gerhard 2004). This monoglucosylation interrupts the normal function of the Rho GTPases, leading to a variety of deleterious effects including apoptosis, cell rounding, actin cytoskeleton dysregulation, and altered cellular signaling (Hofmann et al. 1997; Just and Gerhard 2004; Genth et al. 2008; Just et al. 1995a; Huelsenbeck et al. 2009). Cellular intoxication by the toxins also induces the release of inflammatory mediators from epithelial cells, phagocytes, and mast cells, resulting in colonic inflammation and accumulation of neutrophils (Thelestam and Chaves-Olarte 2000). Early detection of *C. difficile* during infection is critical to prevent further damage by these toxins.

C. difficile overpopulates the human gastrointestinal tract after the normal gut microflora has been reduced by antibiotic therapy (Pothoulakis and Lamont 2001; Lawley et al. 2012). Treatment of CDI has been hampered by increased virulence of the causative strains, sporulation, recurrence of the infection, resistance to multiple antibiotics, and the lack of drugs that preserve or restore the colonic bacterial flora. Specifically, a large number of *C. difficile* isolates show an alarming pattern of resistance to the majority of antibiotics currently used in hospitals and outpatient settings (Bishara et al. 2006; Aspevall et al. 2006; Shubeita et al. 1987; Pelaez et al. 2002). As a result of the dwindling number of antibiotics available to effectively clear this infection and prevent recurrence, there is new interest in identifying and developing alternative nonantibiotic treatments, either as stand-alone therapies or as therapies designed to augment the efficacy of the currently used antibiotic regimens. Nonantibiotic treatment options that have been proposed include infusion of stool from healthy donors (Aas et al. 2003), adjunctive use of monoclonal antibodies specific to the toxins (Lowy et al. 2010), probiotics (McFarland et al. 1994), and the use of non-toxigenic *C. difficile* strains to outcompete the toxigenic strains (Borriello and Barclay 1985; Wilson and Sheagren 1983; Sambol et al. 2002). As the toxins are the main agents of CDI pathogenesis, the inhibition of either toxin production or toxin activity, which directly targets the cause of tissue injury and illness, is a promising therapeutic approach to combat this major public health problem. This strategy requires understanding the molecular regulatory mechanisms that control toxin production, which remain largely unknown. Such knowledge will be critical in developing innovative strategies for the prevention and treatment of CDI. Furthermore, the disruption of the regulatory mechanisms that control the expression of the toxins and other virulence factors (but do not affect cell growth) is less likely to generate selective pressure on the bacterium; this approach may thus minimize the risk of resistance development and promote intestinal homeostasis.

The *tcdR* gene of the pathogenicity locus encodes an RNA polymerase sigma factor that controls transcription from the toxin promoters and from its own promoter (Mani and Dupuy 2001; Delon et al. 1999). Genetic and biochemical evidence suggests that TcdC negatively regulates toxin production by disrupting the capability of TcdR-containing RNA polymerase holoenzyme to recognize the *tcdA* and *tcdB* promoters (Mani and Dupuy 2001; Delon et al. 1999). Thus, it has been proposed that *tcdC* encodes a negative regulator of toxin production (Hundsberger et al. 1997; Carter et al. 2011b; Matamouros et al. 2007). Evidence to support this hypothesis includes the inverse transcription pattern of *tcdC* in relation to the toxin genes and the emergence of epidemic strains (NAP1/027 strains) with deletions or frameshift mutations in the *tcdC* gene that produce high toxin levels (Carter et al. 2011b; Curry et al. 2007; Matamouros et al. 2007; Hundsberger et al. 1997; McDonald et al. 2005; Warny et al. 2005; Jarraud et al. 2000). Moreover, all NAP1/027 isolates from the 1980s and 1990s, like those from recent outbreaks, carry *tcdC* mutations (McDonald et al. 2005; Loo et al. 2005). These reports highlight the importance of *tcdC* in the pathogenesis of *C. difficile*.

However, Cartman and coworkers (Giraud et al. 1999) found no association between toxin production and the *tcdC* genotype when they deleted the *tcdC* gene by allelic exchange. Furthermore, the restoration of the Δ117 frameshift mutation and the spontaneous 18-nucleotide deletion in the *tcdC* gene of some virulent *C. difficile* strains such as R20291 did not alter toxin production (Giraud et al. 1999). These data suggest that the regulation of *C. difficile* toxin synthesis is more complex than previously thought and that the mechanism may involve other key players yet to be discovered.

AI-2/LuxS-mediated regulation is important in controlling different activities in a variety of bacterial species (Curry et al. 2013). The transcription of the toxin genes in *C. difficile* has been reported to be affected by a universal AI-2 signaling molecule involved in quorum signaling (Goldstein et al. 2013). *C. difficile* produces AI-2 that is able to induce bioluminescence in a *Vibrio harveyi* reporter strain (Goldstein et al. 2013). AI-2-containing cell-free supernatants from the mid-log phase of *C. difficile* and *E. coli* DH5α producing recombinant *C. difficile* LuxS increased the transcript levels of TcdA, TcdB, and TcdE (Goldstein et al. 2013). However, Carter et al. reported that AI-2 has no significant effect on either the timing of toxin production or the amount of toxins produced by *C. difficile* (Carter et al. 2005). Therefore, it is still not clear whether AI-2/LuxS is involved in *C. difficile* toxin production. Moreover, no other known quorum signaling system has been conclusively shown to influence the production of these toxins.

Toxin synthesis in *C. difficile* occurs during the stationary phase, where the cell density is optimum. We hypothesized that *C. difficile* toxin synthesis is regulated by a quorum sensing mechanism. Two lines of evidence supported this hypothesis. First, a small (<1 KDa), diffusible, heat-stable toxin-inducing activity accumulates in the medium of high-density *C. difficile* cells. This conditioned medium when incubated with low-density log-phase cells causes them to produce toxin early (2–4 h instead of 12–16 h) and at elevated levels when compared with cells grown in fresh medium (Darkoh 2012). These data suggested that *C. difficile* cells extracellularly release an inducing molecule during growth that is able to activate toxin synthesis prematurely and demonstrates for the first time that toxin synthesis in *C. difficile* is regulated by quorum signaling. Second, this toxin-inducing activity was partially purified from high-density stationary-phase culture supernatant fluid by High performance liquid chromatography (HPLC) and confirmed to induce early toxin synthesis, even in *C. difficile* virulent strains that overproduce the toxins. Mass spectrometry analysis of the purified toxin-inducing fraction from HPLC revealed a cyclic peptide with a mass between 600 and 700 Da. It is anticipated that the identification of this toxin-inducing compound will advance our understanding of the mechanism involved in the quorum-dependent regulation of *C. difficile* toxin synthesis. This finding should lead to the development of even more sensitive tests to diagnose *C. difficile* infections and may lead to the discovery of promising novel therapeutic targets that could be harnessed for the treatment of *C. difficile* infections.

In line with our finding, a recent study by Martin et al. (2013) showed that insertional inactivation of *agrA*, a response regulator located in the *agr2* locus of strain R20291, resulted in a two-fold reduction in *tcdA* gene expression and a colonization defect in a mouse model of *C. difficile* infection (Martin et al. 2013). These results suggest that the *agr2* quorum sensing system might play a role in regulating the virulence response of this organism. Since similar *agrBD* homologues appear widespread among *Clostridia*, it is likely that future research will show that similar quorum sensing systems play an important role in controlling virulence factor expression in many toxigenic clostridial species, which may provide important targets for therapeutic development.

Quorum Sensing-Mediated Regulation of Sporulation in *Clostridia*

Sporulation and germination in clostridial species have not been well characterized. However, the physiological and morphological changes that occur during sporulation in the well-studied *Bacillus* species are similar to that of clostridial species (Lofland et al. 2013). Sporulation is initiated under conditions of nutrient limitation or other unfavorable conditions, when the cells can no longer maintain vegetative growth. In *Bacillus* species, signal transduction systems that control sporulation initiation have been broadly studied and consist of extended variants of two-component signal transduction systems (sporulation phosphorelay) (Aronsson et al. 1985). Sporulation-associated sensor histidine kinases sense various environmental and cellular cues when the cells can no longer sustain vegetative growth. As a result, a specific histidine located in the catalytic domain of sensor histidine kinases becomes autophosphorylated (Kyne et al. 2001). The phosphoryl group is then transferred to an aspartate on the Spo0F response regulator, and this leads to the activation of the Spo0A response regulator transcription factor via the Spo0B phosphotransferase (Kyne et al. 2001). The phosphorylation of the active site Spo0A aspartate promotes binding to a specific target sequence (the "0A box") in or near the promoters of genes under Spo0A control, leading to either the activation or repression of these genes (Kyne et al. 2001).

The phosphoryl groups on sensor kinases activate two-component systems and phosphorelays. Sensor kinases generally consist of an N-terminal signal input domain and a catalytic C-terminal kinase domain containing the dimerization and histidine phosphotransfer sub-domain and an ATP-binding sub-domain (Kyne et al. 2001; Aronsson et al. 1985). The Spo0F in the sporulation phosphorelay can be phosphorylated by multiple sensor kinases (Leung et al. 1991). In *B. subtilis,* there are five sensor kinases, KinA to KinE, with conserved active sites that influence sporulation (Aboudola et al. 2003). One sensor kinase is normally responsive to a single specific signal ligand; hence, incorporating several kinases increases the variety of the signals that can be sensed and allows various different signals to influence Spo0A activation.

In *Bacillus* and *Clostridium* species, Spo0A is highly conserved (Lofland et al. 2013). The regions upstream of genes most likely controlled by Spo0A in *Clostridia* contain "0A boxes," suggesting that the mechanism of Spo0A-mediated gene control is similar to that of *Bacillus*. In *Clostridium* and *Bacillus* species, the most important residues of Spo0A that mediate interaction with the nucleotides of the "0A box" are highly conserved (Dionne et al. 2013). However, protein sequence homology analysis indicates that the genomes of sequenced *Clostridium* species do not appear to encode homologues of Spo0F and Spo0B (Leung et al. 1991). The general consensus has been that the sporulation initiation pathway in *Clostridium* species does not involve a multicomponent phosphorelay but utilizes a two-component system to sense signals via sporulation-associated sensor kinases that phosphorylate Spo0A directly (Gens et al. 2013; Guerrero et al. 2013). Activated Spo0A regulates several post-exponential-phase phenomena (Lofland et al. 2013). To sustain their pathogenic lifestyle, bacterial sporeformers appear to have evolved to allow Spo0A to control virulence as well as survival responses (Gens et al. 2013). For instance, sporulation and toxin production in *Clostridium perfringens* involve Spo0A, which binds to the putative "0A box" upstream of the Cpe cytotoxin (Ogami et al. 2013).

Upon sporulation initiation, an asymmetrically placed division septum is formed that divides the cell into two unequal compartments (the mother cell and a forespore). Each of these two compartments contains one copy of the chromosome. The larger mother cell compartment engulfs the forespore, leading to maturation (Hilbert and Piggot 2004). The process of maturation involves the addition of a peptidoglycan cortex and several layers of proteins that coat around the forespore. The mother cell finally lyses and the spore is released into the environment (Henriques and Moran 2007). The spore is metabolically dormant when released from the mother cell but is resistant to various harsh environmental conditions, such as high temperature, oxygen, pH, alcohols, etc. Under suitable conditions, the spores germinate and grow as vegetative cells. Germination in *B. subtilis* can be artificially induced by different compounds such as L-alanine or a mixture of asparagine, glucose, fructose, and potassium ions (Kelly et al. 2012). The *B. subtilis* receptors that have been identified to be involved in sensing these environmental signals are GerA, GerB, and GerK. A large amount of calcium dipicolinate is released subsequent to sensing of the germinant, leading to hydration of the core, degradation of the cortex, and resumption of metabolism (Setlow 2003). Homologues of GerA, GerB, and GerK have been identified in many *Bacillus* and clostridial species except *C. difficile*. This suggests that *C. difficile* spores respond to a different kind of environmental signal for germination (Sebaihia et al. 2006). However, germination and outgrowth of the spores have not been well studied. Cholate derivatives of bile salts, such as taurocholate and the amino acid glycine, act as co-germinants of *C. difficile* spores (Kelly et al. 2012) and improve the germination of *C. difficile* spores from environmental surfaces and stool samples (Weese et al. 2000; Bliss et al. 1997). Lysozyme and thioglycolate also improve the colony formation of *C. difficile* spores (Mattila et al. 2012; Brandt et al. 2012). Unfortunately, the mechanism by which these molecules stimulate the germination of *C. difficile* spores is unknown.

Clostridium difficile

The formation of spores by *C. difficile* is a significant impediment in overcoming hospital-acquired *C. difficile*-associated diseases and recurrence. The spores contribute to the survival of this bacterium after treatment of surfaces with antiseptics and antibiotic therapy, which disrupts the colonic microflora and precipitates *C. difficile* infection, colonization, and overgrowth in the intestinal tract (Montoya and Detorres 2013). To date, no conclusive evidence exist that links sporulation and toxin production in *C. difficile*. Furthermore, many reports describing direct correlations between toxin production, sporulation, and stationary-phase events (Karlsson et al. 2008; Rupnik and Poxton 2013; Yakob et al. 2013) have been subsequently disputed by other reports demonstrating negative correlations or no link at all (Fitzpatrick 2013; Piacenti and Leuthner 2013). These ambiguities have been made difficult by differences in the strains and the growth media used in these studies, as well as the lack of effective *C. difficile* genetic tools to generate mutants in key genes to enable definite analysis of the pathways involved. Furthermore, no quorum sensing system has been conclusively associated with sporulation in *C. difficile*.

Clostridium perfringens

To investigate whether a functional *agr* locus is required for *C. perfringens* sporulation, Li et al. (2011) incubated wild-type F5603, the isogenic *agrB* null mutant, and the *agr* complementing strains in DS sporulation medium. They observed that wild-type F5603 formed refractive spores with sporulation efficiency between 60 and 70 % after 8 h of incubation at 37 °C. In contrast, the *agrB* null mutant formed only a trace amount of spores, barely detectable by phase-contrast microscopy with less than 1 % sporulation efficiency. They concluded that the observed effect was specifically due to the inactivation of the *agr* locus, since the complementing strain showed similar sporulation efficiency as that of the wild-type parent. Furthermore, they also tested the sporulating abilities of F5603, the *agrB* null mutant, and the complementing strain by quantitating the formation of heat-resistant spores. Their results demonstrated that wild-type F5603 with a functional agr locus produced a significantly larger number of spores than the isogenic *agrB* null mutants. Complementation of the *agrB* null mutants restored spore formation. These results together demonstrated that the *agr* locus is needed for efficient formation of spores in *C. perfringens*.

Clostridium botulinum

Cooksley et al. (2010) generated both *agrD1* and *agrD2* mutants in *C. botulinum* strain 3502 and investigated these mutants for sporulation defects. They observed that both *C. botulinum agrD* mutants showed a reduction in sporulation. The most noticeable effect was observed for cultures of the *agrD1* mutant, with a reduction in spore content of about 1,000-fold. In contrast, only a 70-fold reduction was observed for cultures of the *agrD2* mutant. These results showed that the agr quorum-sensing system is involved in sporulation in *C. botulinum*.

Quorum Sensing-Mediated Control of Biofilm Formation in *Clostridia*

Bacteria form biofilm to enable them to assume a temporary multicellular lifestyle to promote their survival under adverse conditions as a group. A biofilm is a complex, multifaceted, and dynamic developmental process that involves a community of microorganisms attached to a surface. The transition from planktonic growth to biofilm generally occurs in response to environmental changes. This involves several regulatory networks, which decode environmental signals into rigorous gene expression, thereby mediating the spatial and temporal reorganization of the bacterial cell (Pratt and Kolter 1998; Kuchma and O'Toole 2000; O'Toole et al. 2000a, b; Leid et al. 2009; Monds and O'Toole 2009). This cellular reprogramming equips the bacteria with an arsenal of properties that enable their survival in unfavorable conditions and alters the expression of surface molecules, nutrient utilization, and virulence factors (Klebensberger et al. 2009; Schleheck et al. 2009; Zhang and Mah 2008). The ability to grow as a biofilm can facilitate survival of bacteria in the environment, enhance persistence, and promote infection.

Within the biofilm, bacteria are insulated in a self-produced extracellular matrix, which accounts for approximately 90 % of the biomass (Flemming and Wingender 2010; Tielen et al. 2010). The matrix is composed of extracellular polymeric substances that, along with carbohydrate-binding proteins (Diggle et al. 2006; Branda et al. 2006), flagella, pili, other adhesive fibers (Zogaj et al. 2001; Cegelski et al. 2009), and extracellular DNA (Thomas et al. 2009; Guiton et al. 2009; Vilain et al. 2009), act as a stabilizing scaffold for the three-dimensional biofilm structure. In the matrix, nutrients are trapped by the resident bacteria, and water is efficiently retained through H-bond interactions with hydrophilic polysaccharides (Conrad et al. 2003; Flemming and Wingender 2010). The bacteria within biofilm secrete enzymes that modify the composition of the extracellular polymeric substances in response to changes in nutrient availability, thereby altering biofilm architecture to the specific environment (Gjermansen et al. 2005; Sauer et al. 2004). The structural components of the matrix provide a highly hydrated, robust structure with high tensile strength that keeps bacteria in close proximity, enabling cell-to-cell interactions and DNA exchange (Flemming and Wingender 2010; Xiao and Koo 2010) while protecting the biomass from drying, predation, oxidizing molecules, radiation, and other damaging agents (Flemming and Wingender 2010; Mai-Prochnow et al. 2008; Walters et al. 2003). The resilient nature of biofilm is also partly attributed to the presence of environmental gradients within the biomass, which give rise to "division of labor" within the community with subpopulations of bacteria showing differential gene expression in response to local conditions such as nutrient and oxygen availability (Lewis 2005).

It has been reported that there are metabolically inactive nondividing persister cells present within biofilms, which are tolerant to a number of antibiotics despite the fact that they are genetically identical to the rest of the bacterial population (Lewis 2005). These persister cells are believed to be responsible for the reseeding of biofilms on cessation of antibiotic treatment in clinical settings (Lewis 2005). The biofilm matrix protects the bacterial community from the host innate immune defenses and antibiotic treatments (Cerca et al. 2006, 2007; Leid et al. 2009; Jesaitis et al. 2003). Inter-bacterial interactions can promote the spread of drug resistance markers and other virulence factors (Vuong et al. 2004). As a result, biofilm-forming pathogens persist, establishing chronic and refractory infections such as upper respiratory infections, urinary tract infections, periodontitis, and other device-associated infections. Particularly in immunocompromised

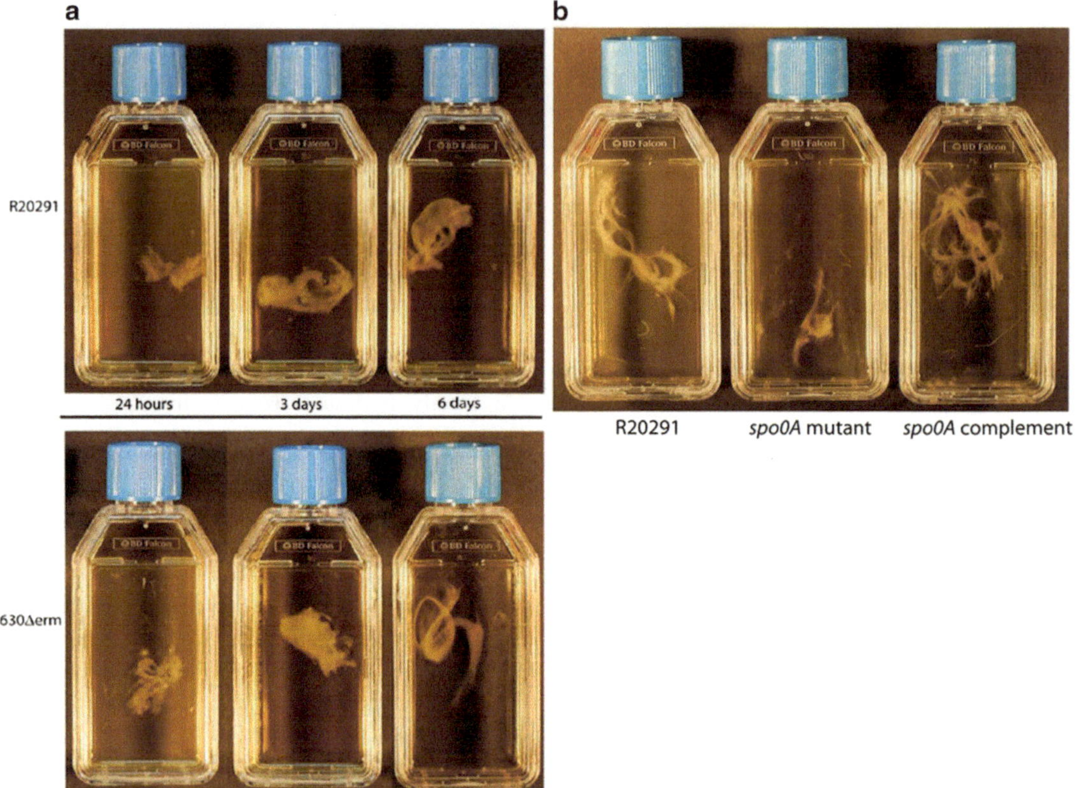

Fig. 2 (**a**) *C. difficile* biofilm formation from 24 h to 6 days in the hypervirulent strain R20291 and the erythromycin-sensitive strain 630 Δerm. (**b**) Comparison of biofilm formation in wild-type strain R20291, sporulation mutant (R20291_*spo0A*::CT), and complemented sporulation mutant (Figure taken from Dawson et al. 2012)

patients, the manifestation of infections by opportunistic biofilm-forming pathogens can be devastating, leading to severe symptoms and, in many instances, death.

Biofilm formation has been characterized in a very few clostridial species that colonize the gut. This could be attributed to the difficulties in cultivating these anaerobes and genetic manipulation of such bacteria. For these species, the involvement of quorum sensing in biofilm formation has only been demonstrated in *C. difficile*.

Clostridium difficile

The ability to adhere and form biofilm influences the capability of *C. difficile* to colonize and establish an infection. Biofilm formation in *C. difficile* was first characterized by Dawson et al. (2012) in 2012. They established that the *C. difficile* hypervirulent strain R20291 and the erythromycin-sensitive strain 630 form polymicrobial aggregates in vitro on abiotic surfaces, which were attached to each other by an extracellular polymeric matrix (Fig. 2a). The matrix provides the scaffold bonding together vegetative cells and spores, as well as forming a protective barrier for vegetative cells against oxygen stress. Dawson et al. (2012) also reported that the master regulator of sporulation, Spo0A, may play a key role in biofilm formation, as genetic inactivation of spo0A in strain R20291 exhibited decreased biofilm formation (Fig. 2b).

Ethapa et al. (2013) also recently examined biofilm formation in *C. difficile* strains R20291 and 630 by allowing the biofilm to form on glass slides in BHIS medium [Brain Heart Infusion medium supplemented with 0.1 % (wt/vol) L-cysteine and yeast extract (5 mg/ml)] containing 0.1 M glucose for 1 or 3 days. The biofilm was then evaluated by bacterial viability and biofilm thickness by Live/Dead staining using Syto 9 and propidium iodide. Syto 9 stains the live bacteria green, whereas propidium iodide stains the dead bacteria red. They reported that the majority of bacteria in the biofilms formed by both *C. difficile* strains R20291 and 630 were alive with a minor number of dead cells after the first day. For strain 630, they observed that the Live/Dead staining of the biofilm was not homogeneous compared to that of R20291 after day 1 or day 3. The maximum thickness of the biofilm was 30 μm (72 h). They concluded based on their microscopic analysis that *C. difficile* biofilms are structured, with several layers of largely live bacteria encased within a dense matrix.

Quorum sensing has been demonstrated to be involved in the formation of biofilms in several bacterial species. The role of the LuxS quorum-sensing system in biofilm formation has been established in various other bacteria (Lombardia et al. 2006; Ohtani et al. 2002; Vendeville et al. 2005) but unknown in *C. difficile* until Ethapa et al.'s (Ethapa et al. 2013) study. Using a *C. difficile luxS* mutant, they observed a dramatic defect in biofilm formation. The examination of the *luxS* mutant showed that it is unable to form even a bacterial monolayer on glass surface. Biofilm formation defect of this mutant was complemented by episomal expression of the full-length *luxS* gene under the control of the native promoter (*luxS*-C). However, complementation did not completely restore the mutant to the wild-type phenotype, but resulted in the formation of several layers of bacteria in the complemented strain. Although the growth curves for the mutant and complemented strains were similar, they suggested that it is possible that expressing *luxS* episomally may be toxic for the bacteria. These results suggested that the LuxS quorum-sensing system may be involved in biofilm formation in *C. difficile*. Nonetheless, they also reported that in addition to LuxS, virulence-associated proteins, Cwp84, and flagella are all required for maximal biofilm formation by vegetative *C. difficile* cells, underscoring the need for further studies to elucidate the role of quorum sensing and biofilm formation in *Clostridia*.

Concluding Remarks

Recent development of new genetic technologies has facilitated the study of the pathogenic *Clostridia* at the molecular level and has made possible the regulatory cascades and environmental factors involved in toxin production, sporulation, and biofilm formation in these bacteria to be elucidated. The regulatory pathways involved are complex and multifaceted, and further research is needed to provide more insight into the involvement of quorum sensing in the pathogenesis of these anaerobes. It is also evident that a lot of work needs to be done to uncover the effect of the Agr and LuxS quorum-sensing systems in other clostridial species. Targeting nonvital, pro-pathogenic quorum sensing mechanisms in clostridial species offers one of the best approaches to combat bacterial virulence without engendering resistance. Presumably, therapies that affect bacterial behavior will not be as susceptible to resistance as are the targets of traditional antibiotics that result in outright killing of bacteria or inhibition of their growth. Thus, therapeutics that interfere with small molecule-controlled pathways could have longer functional shelf lives than second- and third-generation antibiotics. The marked degree of homology and distinct functional features shared between *agr* family members and orthologues across the *Firmicutes* offer the prospect of developing a small suite of therapies that can inhibit pathogenesis in numerous community-acquired and nosocomial pathogens. Importantly, with the development of new antibiotics in decline, understanding these regulatory networks may identify important new targets for the development of novel antimicrobial or anti-virulence compounds that act against the pathogenic *Clostridia* and that may be efficacious in treating diseases caused by this group of bacteria.

References

Aas J, Gessert CE, Bakken JS (2003) Recurrent *Clostridium difficile* colitis: case series involving 18 patients treated with donor stool administered via a nasogastric tube. Clin Infect Dis 36:580–585

Aboudola S, Kotloff KL, Kyne L, Warny M, Kelly EC, Sougioultzis S, Giannasca PJ, Monath TP, Kelly CP (2003) *Clostridium difficile* vaccine and serum immunoglobulin G antibody response to toxin A. Infect Immun 71:1608–1610

Arnon SS, Schechter R, Inglesby TV, Henderson DA, Bartlett JG, Ascher MS, Eitzen E, Fine AD, Hauer J, Layton M, Lillibridge S, Osterholm MT, O'Toole T, Parker G, Perl TM, Russell PK, Swerdlow DL, Tonat K (2001) Botulinum toxin as a biological weapon: medical and public health management. JAMA 285: 1059–1070

Aronsson B, Mollby R, Nord CE (1985) Antimicrobial agents and *Clostridium difficile* in acute enteric disease: epidemiological data from Sweden, 1980–1982. J Infect Dis 151:476–481

Aspevall O, Lundberg A, Burman LG, Akerlund T, Svenungsson B (2006) Antimicrobial susceptibility pattern of *Clostridium difficile* and its relation to PCR ribotypes in a Swedish university hospital. Antimicrob Agents Chemother 50:1890–1892

Balaban N, Novick RP (1995) Translation of RNAIII, the *Staphylococcus aureus* agr regulatory RNA molecule, can be activated by a 3′-end deletion. FEMS Microbiol Lett 133:155–161

Bartlett JG (2002) Clinical practice. Antibiotic-associated diarrhea. N Engl J Med 346:334–339

Bishara J, Bloch Y, Garty M, Behor J, Samra Z (2006) Antimicrobial resistance of *Clostridium difficile* isolates in a tertiary medical center, Israel. Diagn Microbiol Infect Dis 54:141–144

Bliss DZ, Johnson S, Clabots CR, Savik K, Gerding DN (1997) Comparison of cycloserine-cefoxitin-fructose agar (CCFA) and taurocholate-CCFA for recovery of *Clostridium difficile* during surveillance of hospitalized patients. Diagn Microbiol Infect Dis 29:1–4

Borriello SP, Barclay FE (1985) Protection of hamsters against *Clostridium difficile* ileocaecitis by prior colonisation with non-pathogenic strains. J Med Microbiol 19:339–350

Bouillaut L, Perchat S, Arold S, Zorrilla S, Slamti L, Henry C, Gohar M, Declerck N, Lereclus D (2008) Molecular basis for group-specific activation of the virulence regulator PlcR by PapR heptapeptides. Nucleic Acids Res 36:3791–3801

Bradshaw M, Marshall KM, Heap JT, Tepp WH, Minton NP, Johnson EA (2010) Construction of a nontoxigenic *Clostridium botulinum* strain for food challenge studies. Appl Environ Microbiol 76:387–393

Branda SS, Chu F, Kearns DB, Losick R, Kolter R (2006) A major protein component of the *Bacillus subtilis* biofilm matrix. Mol Microbiol 59:1229–1238

Brandt LJ, Aroniadis OC, Mellow M, Kanatzar A, Kelly C, Park T, Stollman N, Rohlke F, Surawicz C (2012) Long-term follow-up of colonoscopic fecal microbiota transplant for recurrent *Clostridium difficile* infection. Am J Gastroenterol 107:1079–1087

Carter GP, Purdy D, Williams P, Minton NP (2005) Quorum sensing in *Clostridium difficile*: analysis of a luxS-type signalling system. J Med Microbiol 54: 119–127

Carter AT, Paul CJ, Mason DR, Twine SM, Alston MJ, Logan SM, Austin JW, Peck MW (2009) Independent evolution of neurotoxin and flagellar genetic loci in proteolytic *Clostridium botulinum*. BMC Genomics 10:115

Carter GP, Awad MM, Hao Y, Thelen T, Bergin IL, Howarth PM, Seemann T, Rood JI, Aronoff DM, Lyras D (2011a) TcsL is an essential virulence factor in *Clostridium sordellii* ATCC 9714. Infect Immun 79:1025–1032

Carter GP, Douce GR, Govind R, Howarth PM, Mackin KE, Spencer J, Buckley AM, Antunes A, Kotsanas D, Jenkin GA, Dupuy B, Rood JI, Lyras D (2011b) The anti-sigma factor TcdC modulates hypervirulence in an epidemic BI/NAP1/027 clinical isolate of *Clostridium difficile*. PLoS Pathog 7:e1002317

Carter GP, Rood JI, Lyras D (2012) The role of toxin A and toxin B in the virulence of *Clostridium difficile*. Trends Microbiol 20:21–29

Cegelski L, Pinkner JS, Hammer ND, Cusumano CK, Hung CS, Chorell E, Aberg V, Walker JN, Seed PC, Almqvist F, Chapman MR, Hultgren SJ (2009) Small-molecule inhibitors target *Escherichia coli* amyloid biogenesis and biofilm formation. Nat Chem Biol 5:913–919

Cerca N, Jefferson KK, Oliveira R, Pier GB, Azeredo J (2006) Comparative antibody-mediated phagocytosis of *Staphylococcus epidermidis* cells grown in a biofilm or in the planktonic state. Infect Immun 74: 4849–4855

Cerca N, Maira-Litran T, Jefferson KK, Grout M, Goldmann DA, Pier GB (2007) Protection against *Escherichia coli* infection by antibody to the *Staphylococcus aureus* poly-N-acetylglucosamine surface polysaccharide. Proc Natl Acad Sci USA 104:7528–7533

Chen J, Mcclane BA (2012) Role of the Agr-like quorum-sensing system in regulating toxin production by *Clostridium perfringens* type B strains CN1793 and CN1795. Infect Immun 80:3008–3017

Chen Y, Mcclane BA, Fisher DJ, Rood JI, Gupta P (2005) Construction of an alpha toxin gene knockout mutant of *Clostridium perfringens* type A by use of a mobile group II intron. Appl Environ Microbiol 71: 7542–7547

Chen J, Rood JI, Mcclane BA (2011) Epsilon-toxin production by *Clostridium perfringens* type D strain CN3718 is dependent upon the agr operon but not the VirS/VirR two-component regulatory system. mBio 2. doi: 10.1128/mBio.00275-11. e00275-11 [pii]. mBio.00275-11 [pii]

Chung WO, Park Y, Lamont RJ, Mcnab R, Barbieri B, Demuth DR (2001) Signaling system in *Porphyromonas gingivalis* based on a LuxS protein. J Bacteriol 183:3903–3909

Collie RE, Kokai-Kun JF, Mcclane BA (1998) Phenotypic characterization of enterotoxigenic *Clostridium perfringens* isolates from non-foodborne human gastrointestinal diseases. Anaerobe 4:69–79

Conrad A, Suutari MK, Keinanen MM, Cadoret A, Faure P, Mansuy-Huault L, Block JC (2003) Fatty acids of lipid fractions in extracellular polymeric substances of activated sludge flocs. Lipids 38:1093–1105

Cooksley CM, Davis IJ, Winzer K, Chan WC, Peck MW, Minton NP (2010) Regulation of neurotoxin production and sporulation by a Putative agrBD signaling system in proteolytic *Clostridium botulinum*. Appl Environ Microbiol 76:4448–4460

Curry SR, Marsh JW, Muto CA, O'Leary MM, Pasculle AW, Harrison LH (2007) tcdC genotypes associated with severe TcdC truncation in an epidemic clone and other strains of *Clostridium difficile*. J Clin Microbiol 45:215–221

Curry SR, Muto CA, Schlackman JL, Pasculle AW, Shutt KA, Marsh JW, Harrison LH (2013) Use of multilocus variable number of tandem repeats analysis genotyping to determine the role of asymptomatic carriers in *Clostridium difficile* transmission. Clin Infect Dis 57(8):1094–1102

Darkoh C (2012) Regulation of toxin synthesis by *Clostridium difficile*. University of Texas Graduate School of Biomedical Sciences dissertations and theses (Open Access), paper, 250pp

Dawson LF, Valiente E, Faulds-Pain A, Donahue EH, Wren BW (2012) Characterisation of *Clostridium difficile* biofilm formation, a role for Spo0A. PLoS One 7:e50527

Delon A, Levasseur LM, Giraudon M, Bouquet S, Couet W (1999) Antagonistic interaction between the convulsant activities of pefloxacin and its main metabolite norfloxacin in rats. Pharm Res 16:1894–1897

Diggle SP, Stacey RE, Dodd C, Camara M, Williams P, Winzer K (2006) The galactophilic lectin, LecA, contributes to biofilm development in *Pseudomonas aeruginosa*. Environ Microbiol 8:1095–1104

Dillon ST, Rubin EJ, Yakubovich M, Pothoulakis C, Lamont JT, Feig LA, Gilbert RJ (1995) Involvement of Ras-related Rho proteins in the mechanisms of action of *Clostridium difficile* toxin A and toxin B. Infect Immun 63:1421–1426

Dingle T, Wee S, Mulvey GL, Greco A, Kitova EN, Sun J, Lin S, Klassen JS, Palcic MM, Ng KK, Armstrong GD (2008) Functional properties of the carboxy-terminal host cell-binding domains of the two toxins, TcdA and TcdB, expressed by *Clostridium difficile*. Glycobiology 18:698–706

Dionne LL, Raymond F, Corbeil J, Longtin J, Gervais P, Longtin Y (2013) Correlation between *Clostridium difficile* bacterial load, commercial real-time pcr cycle thresholds, and results of diagnostic tests based on enzyme immunoassay and cell-culture cytotoxicity assay. J Clin Microbiol 51:3624–3630

Doherty N, Holden MT, Qazi SN, Williams P, Winzer K (2006) Functional analysis of luxS in *Staphylococcus aureus* reveals a role in metabolism but not quorum sensing. J Bacteriol 188:2885–2897

Dupont HL (2011) The search for effective treatment of *Clostridium difficile* infection. N Engl J Med 364: 473–475

Durre P (2007) Biobutanol: an attractive biofuel. Biotechnol J 2:1525–1534

Durre P (2008) Fermentative butanol production: bulk chemical and biofuel. Ann N Y Acad Sci 1125: 353–362

Egerer M, Giesemann T, Jank T, Satchell KJ, Aktories K (2007) Auto-catalytic cleavage of *Clostridium difficile* toxins A and B depends on cysteine protease activity. J Biol Chem 282:25314–25321

Elliott B, Chang BJ, Golledge CL, Riley TV (2007) *Clostridium difficile*-associated diarrhoea. Intern Med J 37:561–568

Elvers KT, Park SF (2002) Quorum sensing in Campylobacter jejuni: detection of a luxS encoded signalling molecule. Microbiology 148:1475–1481

Ethapa T, Leuzzi R, Ng YK, Baban ST, Adamo R, Kuehne SA, Scarselli M, Minton NP, Serruto D, Unnikrishnan M (2013) Multiple factors modulate biofilm formation by the anaerobic pathogen *Clostridium difficile*. J Bacteriol 195:545–555

Fitzpatrick LR (2013) Probiotics for the treatment of *Clostridium difficile* associated disease. World J Gastrointest Pathophysiol 4:47–52

Flamez C, Ricard I, Arafah S, Simonet M, Marceau M (2007) Two-component system regulon plasticity in bacteria: a concept emerging from phenotypic analysis of *Yersinia pseudotuberculosis* response regulator mutants. Adv Exp Med Biol 603: 145–155

Flemming HC, Wingender J (2010) The biofilm matrix. Nat Rev Microbiol 8:623–633

Fong KP, Chung WO, Lamont RJ, Demuth DR (2001) Intra- and interspecies regulation of gene expression by *Actinobacillus actinomycetemcomitans* LuxS. Infect Immun 69:7625–7634

Gens KD, Elshaboury RH, Holt JS (2013) Fecal microbiota transplantation and emerging treatments for *Clostridium difficile* infection. J Pharm Pract 26: 498–505

Genth H, Dreger SC, Huelsenbeck J, Just I (2008) *Clostridium difficile* toxins: more than mere inhibitors of Rho proteins. Int J Biochem Cell Biol 40: 592–597

Geric B, Rupnik M, Gerding DN, Grabnar M, Johnson S (2004) Distribution of *Clostridium difficile* variant toxinotypes and strains with binary toxin genes among clinical isolates in an American hospital. J Med Microbiol 53:887–894

Giraud MN, Motta C, Romero JJ, Bommelaer G, Lichtenberger LM (1999) Interaction of indomethacin and naproxen with gastric surface-active phospholipids: a possible mechanism for the gastric toxicity of nonsteroidal anti-inflammatory drugs (NSAIDs). Biochem Pharmacol 57:247–254

Giron JA, Torres AG, Freer E, Kaper JB (2002) The flagella of enteropathogenic *Escherichia coli* mediate adherence to epithelial cells. Mol Microbiol 44: 361–379

Gjermansen M, Ragas P, Sternberg C, Molin S, Tolker-Nielsen T (2005) Characterization of starvation-induced dispersion in *Pseudomonas putida* biofilms. Environ Microbiol 7:894–906

Goldstein EJ, Citron DM, Tyrrell KL, Merriam CV (2013) Comparative in-vitro activity of SMT19969, a new antimicrobial agent, against *Clostridium difficile* and 350 Gram-positive and negative aerobic and anaerobic intestinal flora isolates. Antimicrob Agents Chemother 57:4872–4876

Gray B, Hall P, Gresham H (2013) Targeting agr- and agr-Like quorum sensing systems for development of common therapeutics to treat multiple gram-positive bacterial infections. Sensors (Basel) 13:5130–5166

Guerrero DM, Becker JC, Eckstein EC, Kundrapu S, Deshpande A, Sethi AK, Donskey CJ (2013) Asymptomatic carriage of toxigenic *Clostridium difficile* by hospitalized patients. J Hos Infect 85:155–158

Guiton PS, Hung CS, Kline KA, Roth R, Kau AL, Hayes E, Heuser J, Dodson KW, Caparon MG, Hultgren SJ (2009) Contribution of autolysin and Sortase a during *Enterococcus faecalis* DNA-dependent biofilm development. Infect Immun 77:3626–3638

Hammer BK, Bassler BL (2003) Quorum sensing controls biofilm formation in *Vibrio cholerae*. Mol Microbiol 50:101–104

Havarstein LS, Coomaraswamy G, Morrison DA (1995) An unmodified heptadecapeptide pheromone induces competence for genetic transformation in *Streptococcus pneumoniae*. Proc Natl Acad Sci USA 92: 11140–11144

Heap JT, Pennington OJ, Cartman ST, Carter GP, Minton NP (2007) The ClosTron: a universal gene knock-out system for the genus Clostridium. J Microbiol Methods 70:452–464

Henriques AO, Moran CP Jr (2007) Structure, assembly, and function of the spore surface layers. Annu Rev Microbiol 61:555–588

Hilbert DW, Piggot PJ (2004) Compartmentalization of gene expression during *Bacillus subtilis* spore formation. Microbiol Mol Biol Rev 68:234–262

Ho JG, Greco A, Rupnik M, Ng KK (2005) Crystal structure of receptor-binding C-terminal repeats from *Clostridium difficile* toxin A. Proc Natl Acad Sci USA 102:18373–18378

Hofmann F, Busch C, Prepens U, Just I, Aktories K (1997) Localization of the glucosyltransferase activity of *Clostridium difficile* toxin B to the N-terminal part of the holotoxin. J Biol Chem 272:11074–11078

Huelsenbeck SC, May M, Schmidt G, Genth H (2009) Inhibition of cytokinesis by Clostridium difficile toxin B and cytotoxic necrotizing factors–reinforcing the critical role of RhoA in cytokinesis. Cell Motil Cytoskeleton 66:967–975

Hundsberger T, Braun V, Weidmann M, Leukel P, Sauerborn M, Von Eichel-Streiber C (1997) Transcription analysis of the genes tcdA-E of the pathogenicity locus of *Clostridium difficile*. Eur J Biochem 244:735–742

Jarraud S, Lyon GJ, Figueiredo AM, Lina G, Vandenesch F, Etienne J, Muir TW, Novick RP (2000) Exfoliating-producing strains define a fourth agr specificity group in *Staphylococcus aureus*. J Bacteriol 182:6517–6522

Jesaitis AJ, Franklin MJ, Berglund D, Sasaki M, Lord CI, Bleazard JB, Duffy JE, Beyenal H, Lewandowski Z (2003) Compromised host defense on *Pseudomonas aeruginosa* biofilms: characterization of neutrophil and biofilm interactions. J Immunol 171:4329–4339

Ji G, Beavis RC, Novick RP (1995) Cell density control of staphylococcal virulence mediated by an octapeptide pheromone. Proc Natl Acad Sci USA 92:12055–12059

Jones DT, Woods DR (1986) Acetone-butanol fermentation revisited. Microbiol Rev 50:484–524

Just I, Gerhard R (2004) Large clostridial cytotoxins. Rev Physiol Biochem Pharmacol 152:23–47

Just I, Selzer J, Wilm M, Von Eichel-Streiber C, Mann M, Aktories K (1995a) Glucosylation of Rho proteins by *Clostridium difficile* toxin B. Nature 375:500–503

Just I, Wilm M, Selzer J, Rex G, Von Eichel-Streiber C, Mann M, Aktories K (1995b) The enterotoxin from *Clostridium difficile* (ToxA) monoglucosylates the Rho proteins. J Biol Chem 270:13932–13936

Karlsson S, Burman LG, Akerlund T (2008) Induction of toxins in *Clostridium difficile* is associated with dramatic changes of its metabolism. Microbiology 154:3430–3436

Katona P (2012) Botulinum toxin: therapeutic agent to cosmetic enhancement to lethal biothreat. Anaerobe 18:240–243

Kelly CR, De Leon L, Jasutkar N (2012) Fecal microbiota transplantation for relapsing *Clostridium difficile* infection in 26 patients: methodology and results. J Clin Gastroenterol 46:145–149

Klebensberger J, Birkenmaier A, Geffers R, Kjelleberg S, Philipp B (2009) SiaA and SiaD are essential for inducing autoaggregation as a specific response to detergent stress in *Pseudomonas aeruginosa*. Environ Microbiol 11:3073–3086

Koenig RL, Ray JL, Maleki SJ, Smeltzer MS, Hurlburt BK (2004) *Staphylococcus aureus* AgrA binding to the RNAIII-agr regulatory region. J Bacteriol 186: 7549–7555

Kuchma SL, O'Toole GA (2000) Surface-induced and biofilm-induced changes in gene expression. Curr Opin Biotechnol 11:429–433

Kuehne SA, Cartman ST, Heap JT, Kelly ML, Cockayne A, Minton NP (2010) The role of toxin A and toxin B in *Clostridium difficile* infection. Nature 467:711–713

Kyne L, Warny M, Qamar A, Kelly CP (2001) Association between antibody response to toxin A and protection against recurrent *Clostridium difficile* diarrhoea. Lancet 357:189–193

Kyne L, Hamel MB, Polavaram R, Kelly CP (2002) Health care costs and mortality associated with nosocomial diarrhea due to *Clostridium difficile*. Clin Infect Dis 34:346–353

Lawley TD, Clare S, Walker AW, Stares MD, Connor TR, Raisen C, Goulding D, Rad R, Schreiber F, Brandt C, Deakin LJ, Pickard DJ, Duncan SH, Flint HJ, Clark TG, Parkhill J, Dougan G (2012) Targeted restoration of the intestinal microbiota with a simple, defined bacteriotherapy resolves relapsing *Clostridium difficile* disease in mice. PLoS Pathog 8:e1002995

Lee SY, Park JH, Jang SH, Nielsen LK, Kim J, Jung KS (2008) Fermentative butanol production by Clostridia. Biotechnol Bioeng 101:209–228

Leid JG, Kerr M, Selgado C, Johnson C, Moreno G, Smith A, Shirtliff ME, O'Toole GA, Cope EK (2009) Flagellum-mediated biofilm defense mechanisms of *Pseudomonas aeruginosa* against host-derived lactoferrin. Infect Immun 77:4559–4566

Leung DY, Kelly CP, Boguniewicz M, Pothoulakis C, Lamont JT, Flores A (1991) Treatment with intravenously administered gamma globulin of chronic relapsing colitis induced by *Clostridium difficile* toxin. J Pediatr 118:633–637

Lewis K (2005) Persister cells and the riddle of biofilm survival. Biochemistry (Mosc) 70:267–274

Li J, Chen J, Vidal JE, Mcclane BA (2011) The Agr-like quorum-sensing system regulates sporulation and production of enterotoxin and beta2 toxin by *Clostridium perfringens* type A non-food-borne human gastrointestinal disease strain F5603. Infect Immun 79: 2451–2459

Lina G, Jarraud S, Ji G, Greenland T, Pedraza A, Etienne J, Novick RP, Vandenesch F (1998) Transmembrane topology and histidine protein kinase activity of AgrC, the agr signal receptor in *Staphylococcus aureus*. Mol Microbiol 28:655–662

Lofland D, Josephat F, Partin S (2013) Fecal transplant for recurrent *Clostridium difficile* infection. Clin Lab Sci 26:131–135

Lombardia E, Rovetto AJ, Arabolaza AL, Grau RR (2006) A LuxS-dependent cell-to-cell language regulates social behavior and development in *Bacillus subtilis*. J Bacteriol 188:4442–4452

Loo VG, Poirier L, Miller MA, Oughton M, Libman MD, Michaud S, Bourgault AM, Nguyen T, Frenette C, Kelly M, Vibien A, Brassard P, Fenn S, Dewar K, Hudson TJ, Horn R, Rene P, Monczak Y, Dascal A (2005) A predominantly clonal multi-institutional outbreak of *Clostridium difficile*-associated diarrhea with high morbidity and mortality. N Engl J Med 353: 2442–2449

Lowy I, Molrine DC, Leav BA, Blair BM, Baxter R, Gerding DN, Nichol G, Thomas WD Jr, Leney M, Sloan S, Hay CA, Ambrosino DM (2010) Treatment with monoclonal antibodies against *Clostridium difficile* toxins. N Engl J Med 362:197–205

Lyerly DM, Saum KE, Macdonald DK, Wilkins TD (1985) Effects of *Clostridium difficile* toxins given intragastrically to animals. Infect Immun 47:349–352

Lyon GJ, Novick RP (2004) Peptide signaling in *Staphylococcus aureus* and other Gram-positive bacteria. Peptides 25:1389–1403

Lyristis M, Bryant AE, Sloan J, Awad MM, Nisbet IT, Stevens DL, Rood JI (1994) Identification and molecular analysis of a locus that regulates extracellular toxin production in *Clostridium perfringens*. Mol Microbiol 12:761–777

Mai-Prochnow A, Lucas-Elio P, Egan S, Thomas T, Webb JS, Sanchez-Amat A, Kjelleberg S (2008) Hydrogen peroxide linked to lysine oxidase activity facilitates biofilm differentiation and dispersal in several gram-negative bacteria. J Bacteriol 190:5493–5501

Mani N, Dupuy B (2001) Regulation of toxin synthesis in *Clostridium difficile* by an alternative RNA polymerase sigma factor. Proc Natl Acad Sci USA 98: 5844–5849

Martin MJ, Clare S, Goulding D, Faulds-Pain A, Barquist L, Browne HP, Pettit L, Dougan G, Lawley TD, Wren BW (2013) The agr locus regulates virulence and colonization genes in *Clostridium difficile* 027. J Bacteriol 195:3672–3681

Matamouros S, England P, Dupuy B (2007) *Clostridium difficile* toxin expression is inhibited by the novel regulator TcdC. Mol Microbiol 64:1274–1288

Mattila E, Uusitalo-Seppala R, Wuorela M, Lehtola L, Nurmi H, Ristikankare M, Moilanen V, Salminen K, Seppala M, Mattila PS, Anttila VJ, Arkkila P (2012) Fecal transplantation, through colonoscopy, is effective therapy for recurrent *Clostridium difficile* infection. Gastroenterology 142:490–496

Mayville P, Ji G, Beavis R, Yang H, Goger M, Novick RP, Muir TW (1999) Structure-activity analysis of synthetic autoinducing thiolactone peptides from *Staphylococcus aureus* responsible for virulence. Proc Natl Acad Sci USA 96:1218–1223

Mcdonald LC, Killgore GE, Thompson A, Owens RC Jr, Kazakova SV, Sambol SP, Johnson S, Gerding DN (2005) An epidemic, toxin gene-variant strain of *Clostridium difficile*. N Engl J Med 353:2433–2441

Mcfarland LV, Surawicz CM, Greenberg RN, Fekety R, Elmer GW, Moyer KA, Melcher SA, Bowen KE, Cox JL, Noorani Z et al (1994) A randomized placebo-controlled trial of *Saccharomyces boulardii* in combination with standard antibiotics for *Clostridium difficile* disease. JAMA 271:1913–1918

Mengesha A, Wei JZ, Zhou SF, Wei MQ (2010) Clostridial spores to treat solid tumours – potential for a new therapeutic modality. Curr Gene Ther 10:15–26

Merritt J, Qi F, Goodman SD, Anderson MH, Shi W (2003) Mutation of luxS affects biofilm formation in *Streptococcus mutans*. Infect Immun 71:1972–1979

Minton NP (2003) Clostridia in cancer therapy. Nat Rev Microbiol 1:237–242

Minton NP, Mauchline ML, Lemmon MJ, Brehm JK, Fox M, Michael NP, Giaccia A, Brown JM (1995) Chemotherapeutic tumour targeting using clostridial spores. FEMS Microbiol Rev 17:357–364

Monds RD, O'Toole GA (2009) The developmental model of microbial biofilms: ten years of a paradigm up for review. Trends Microbiol 17:73–87

Montecucco C, Schiavo G (1994) Mechanism of action of tetanus and botulinum neurotoxins. Mol Microbiol 13:1–8

Montecucco C, Papini E, Schiavo G (1994) Bacterial protein toxins penetrate cells via a four-step mechanism. FEBS Lett 346:92–98

Montoya M, Detorres O (2013) Antimicrobial selection and its impact on the incidence of *clostridium difficile*-associated diarrhea. J Pharm Pract 26:483–487

Morfeldt E, Janzon L, Arvidson S, Lofdahl S (1988) Cloning of a chromosomal locus (exp) which regulates the expression of several exoprotein genes in *Staphylococcus aureus*. Mol Gen Genet MGG 211:435–440

Novick RP (2003) Autoinduction and signal transduction in the regulation of staphylococcal virulence. Mol Microbiol 48:1429–1449

Novick RP, Geisinger E (2008) Quorum sensing in staphylococci. Annu Rev Genet 42:541–564

Novick RP, Muir TW (1999) Virulence gene regulation by peptides in staphylococci and other Gram-positive bacteria. Curr Opin Microbiol 2:40–45

Novick RP, Ross HF, Projan SJ, Kornblum J, Kreiswirth B, Moghazeh S (1993) Synthesis of staphylococcal virulence factors is controlled by a regulatory RNA molecule. EMBO J 12:3967–3975

Novick RP, Projan SJ, Kornblum J, Ross HF, Ji G, Kreiswirth B, Vandenesch F, Moghazeh S (1995) The agr P2 operon: an autocatalytic sensory transduction system in *Staphylococcus aureus*. Molecular Gen Genet MGG 248:446–458

O'Brien JA, Lahue BJ, Caro JJ, Davidson DM (2007) The emerging infectious challenge of *Clostridium difficile*-associated disease in Massachusetts hospitals: clinical and economic consequences. Infect Control Hosp Epidemiol 28:1219–1227

O'Connor JR, Lyras D, Farrow KA, Adams V, Powell DR, Hinds J, Cheung JK, Rood JI (2006) Construction and analysis of chromosomal *Clostridium difficile* mutants. Mol Microbiol 61:1335–1351

O'Toole G, Kaplan HB, Kolter R (2000a) Biofilm formation as microbial development. Annu Rev Microbiol 54:49–79

O'Toole GA, Gibbs KA, Hager PW, Phibbs PV Jr, Kolter R (2000b) The global carbon metabolism regulator Crc is a component of a signal transduction pathway required for biofilm development by *Pseudomonas aeruginosa*. J Bacteriol 182:425–431

Ogami N, Yoshida J, Ishimaru T, Kikuchi T, Matsubara N, Ueno T, Asano I (2013) Is *Clostridium difficile* infection influenced by antimicrobial use density in wards? Jpn J Antibiot 66:87–95

Ohtani K, Hayashi H, Shimizu T (2002) The luxS gene is involved in cell-cell signalling for toxin production in *Clostridium perfringens*. Mol Microbiol 44:171–179

Ohtani K, Yuan Y, Hassan S, Wang R, Wang Y, Shimizu T (2009) Virulence gene regulation by the agr system in *Clostridium perfringens*. J Bacteriol 191:3919–3927

Ohtani K, Hirakawa H, Tashiro K, Yoshizawa S, Kuhara S, Shimizu T (2010) Identification of a two-component VirR/VirS regulon in *Clostridium perfringens*. Anaerobe 16:258–264

Okada M, Sato I, Cho SJ, Iwata H, Nishio T, Dubnau D, Sakagami Y (2005) Structure of the *Bacillus subtilis* quorum-sensing peptide pheromone ComX. Nat Chem Biol 1:23–24

Pelaez T, Alcala L, Alonso R, Rodriguez-Creixems M, Garcia-Lechuz JM, Bouza E (2002) Reassessment of *Clostridium difficile* susceptibility to metronidazole and vancomycin. Antimicrob Agents Chemother 46:1647–1650

Peterson S, Cline RT, Tettelin H, Sharov V, Morrison DA (2000) Gene expression analysis of the *Streptococcus pneumoniae* competence regulons by use of DNA microarrays. J Bacteriol 182:6192–6202

Pfeifer G, Schirmer J, Leemhuis J, Busch C, Meyer DK, Aktories K, Barth H (2003) Cellular uptake of *Clostridium difficile* toxin B Translocation of the N-terminal catalytic domain into the cytosol of eukaryotic cells. J Biol Chem 278:44535–44541

Piacenti FJ, Leuthner KD (2013) Antimicrobial stewardship and *Clostridium difficile*-associated diarrhea. J Pharm Pract 26:506–513

Pothoulakis C, Lamont JT (2001) Microbes and microbial toxins: paradigms for microbial-mucosal interactions II. The integrated response of the intestine to *Clostridium difficile* toxins. Am J Physiol Gastrointest Liver Physiol 280:G178–G183

Pratt LA, Kolter R (1998) Genetic analysis of *Escherichia coli* biofilm formation: roles of flagella, motility, chemotaxis and type I pili. Mol Microbiol 30:285–293

Prouty AM, Schwesinger WH, Gunn JS (2002) Biofilm formation and interaction with the surfaces of gallstones by Salmonella spp. Infect Immun 70:2640–2649

Redelings MD, Sorvillo F, Mascola L (2007) Increase in *Clostridium difficile*-related mortality rates, United States, 1999–2004. Emerg Infect Dis 13:1417–1419

Reineke J, Tenzer S, Rupnik M, Koschinski A, Hasselmayer O, Schrattenholz A, Schild H, Von Eichel-Streiber C (2007) Autocatalytic cleavage of *Clostridium difficile* toxin B. Nature 446:415–419

Rood JI (1998) Virulence genes of *Clostridium perfringens*. Annu Rev Microbiol 52:333–360

Rupnik M, Poxton IR (2013) *Clostridium difficile* – special collection. J Med Microbiol 62:1368

Rupnik M, Brazier JS, Duerden BI, Grabnar M, Stubbs SL (2001) Comparison of toxinotyping and PCR ribotyping of *Clostridium difficile* strains and description of novel toxinotypes. Microbiology 147:439–447

Rupnik M, Pabst S, Von Eichel-Streiber C, Urlaub H, Soling HD (2005) Characterization of the cleavage site and function of resulting cleavage fragments after limited proteolysis of *Clostridium difficile* toxin B (TcdB) by host cells. Microbiology 151:199–208

Rutherford ST, Bassler BL (2012) Bacterial quorum sensing: its role in virulence and possibilities for its control. Cold Spring Harb Perspect Med 2

Sambol SP, Merrigan MM, Tang JK, Johnson S, Gerding DN (2002) Colonization for the prevention of *Clostridium difficile* disease in hamsters. J Infect Dis 186: 1781–1789

Sauer K, Cullen MC, Rickard AH, Zeef LA, Davies DG, Gilbert P (2004) Characterization of nutrient-induced dispersion in *Pseudomonas aeruginosa* PAO1 biofilm. J Bacteriol 186:7312–7326

Sebaihia M et al (2006) The multidrug-resistant human pathogen Clostridium difficile has a highly mobile, mosaic genome. Nat Genet 38:779–786

Scallan E, Griffin PM, Angulo FJ, Tauxe RV, Hoekstra RM (2011a) Foodborne illness acquired in the United States–unspecified agents. Emerg Infect Dis 17:16–22

Scallan E, Hoekstra RM, Angulo FJ, Tauxe RV, Widdowson MA, Roy SL, Jones JL, Griffin PM (2011b) Foodborne illness acquired in the United States–major pathogens. Emerg Infect Dis 17:7–15

Schantz EJ, Johnson EA (1992) Properties and use of botulinum toxin and other microbial neurotoxins in medicine. Microbiol Rev 56:80–99

Schauder S, Bassler BL (2001) The languages of bacteria. Genes Dev 15:1468–1480

Schauder S, Shokat K, Surette MG, Bassler BL (2001) The LuxS family of bacterial autoinducers: biosynthesis of a novel quorum-sensing signal molecule. Mol Microbiol 41:463–476

Schechter R, Arnon SS (2000) Extreme potency of botulinum toxin. Lancet 355:237–238

Schleheck D, Barraud N, Klebensberger J, Webb JS, Mcdougald D, Rice SA, Kjelleberg S (2009) Pseudomonas aeruginosa PAO1 preferentially grows as aggregates in liquid batch cultures and disperses upon starvation. PLoS One 4:e5513

Setlow P (2003) Spore germination. Curr Opin Microbiol 6:550–556

Shimizu T, Yaguchi H, Ohtani K, Banu S, Hayashi H (2002) Clostridial VirR/VirS regulon involves a regulatory RNA molecule for expression of toxins. Mol Microbiol 43:257–265

Shubeita HE, Sambrook JF, Mccormick AM (1987) Molecular cloning and analysis of functional cDNA and genomic clones encoding bovine cellular retinoic acid-binding protein. Proc Natl Acad Sci USA 84:5645–5649

Solomon JM, Lazazzera BA, Grossman AD (1996) Purification and characterization of an extracellular peptide factor that affects two different developmental pathways in *Bacillus subtilis*. Genes Dev 10:2014–2024

Songer JG (1996) Clostridial enteric diseases of domestic animals. Clin Microbiol Rev 9:216–234

Songer JG (1998) Clostridial diseases of small ruminants. Vet Res 29:219–232

Songer JG (2010) Clostridia as agents of zoonotic disease. Vet Microbiol 140:399–404

Sperandio V, Mellies JL, Nguyen W, Shin S, Kaper JB (1999) Quorum sensing controls expression of the type III secretion gene transcription and protein secretion in enterohemorrhagic and enteropathogenic *Escherichia coli*. Proc Natl Acad Sci USA 96:15196–15201

Sperandio V, Torres AG, Giron JA, Kaper JB (2001) Quorum sensing is a global regulatory mechanism in enterohemorrhagic *Escherichia coli* O157:H7. J Bacteriol 183:5187–5197

Stevens DL, Aldape MJ, Bryant AE (2012) Life-threatening clostridial infections. Anaerobe 18: 254–259

Sturme MH, Kleerebezem M, Nakayama J, Akkermans AD, Vaugha EE, de Vos WM (2002) Cell to cell communication by autoinducing peptides in gram-positive bacteria. Antonie Van Leeuwenhoek 81:233–243

Thelestam M, Chaves-Olarte E (2000) Cytotoxic effects of the *Clostridium difficile* toxins. Curr Top Microbiol Immunol 250:85–96

Theys J, Pennington O, Dubois L, Anlezark G, Vaughan T, Mengesha A, Landuyt W, Anne J, Burke PJ, Durre P, Wouters BG, Minton NP, Lambin P (2006) Repeated cycles of Clostridium-directed enzyme prodrug therapy result in sustained antitumour effects in vivo. Br J Cancer 95:1212–1219

Thoendel M, Horswill AR (2010) Biosynthesis of peptide signals in gram-positive bacteria. Adv Appl Microbiol 71:91–112

Thoendel M, Kavanaugh JS, Flack CE, Horswill AR (2011) Peptide signaling in the staphylococci. Chem Rev 111:117–151

Thomas VC, Hiromasa Y, Harms N, Thurlow L, Tomich J, Hancock LE (2009) A fratricidal mechanism is responsible for eDNA release and contributes to biofilm development of *Enterococcus faecalis*. Mol Microbiol 72:1022–1036

Tielen P, Rosenau F, Wilhelm S, Jaeger KE, Flemming HC, Wingender J (2010) Extracellular enzymes affect biofilm formation of mucoid *Pseudomonas aeruginosa*. Microbiology 156:2239–2252

Uzal FA, Mcclane BA (2011) Recent progress in understanding the pathogenesis of *Clostridium perfringens* type C infections. Vet Microbiol 153:37–43

Vendeville A, Winzer K, Heurlier K, Tang CM, Hardie KR (2005) Making 'sense' of metabolism: autoinducer-2, LuxS and pathogenic bacteria. Nat Rev Microbiol 3:383–396

Vidal JE, Chen J, Li J, Mcclane BA (2009) Use of an EZ-Tn5-based random mutagenesis system to identify a novel toxin regulatory locus in *Clostridium perfringens* strain 13. PLoS One 4:e6232

Vidal JE, Ma M, Saputo J, Garcia J, Uzal FA, Mcclane BA (2012) Evidence that the Agr-like quorum sensing system regulates the toxin production, cytotoxicity and pathogenicity of *Clostridium perfringens* type C isolate CN3685. Mol Microbiol 83:179–194

Vilain S, Pretorius JM, Theron J, Brozel VS (2009) DNA as an adhesin: *Bacillus cereus* requires extracellular DNA to form biofilms. Appl Environ Microbiol 75:2861–2868

Villar RG, Elliott SP, Davenport KM (2006) Botulism: the many faces of botulinum toxin and its potential for bioterrorism. Infect Dis Clin North Am 20: 313–327, ix

Von Eichel-Streiber C, Boquet P, Sauerborn M, Thelestam M (1996) Large clostridial cytotoxins–a family of glycosyltransferases modifying small GTP-binding proteins. Trends Microbiol 4:375–382

Voth DE, Ballard JD (2005) *Clostridium difficile* toxins: mechanism of action and role in disease. Clin Microbiol Rev 18:247–263

Vuong C, Voyich JM, Fischer ER, Braughton KR, Whitney AR, Deleo FR, Otto M (2004) Polysaccharide intercellular adhesin (PIA) protects *Staphylococcus epidermidis* against major components of the human innate immune system. Cell Microbiol 6:269–275

Walters MC 3rd, Roe F, Bugnicourt A, Franklin MJ, Stewart PS (2003) Contributions of antibiotic penetration, oxygen limitation, and low metabolic activity to tolerance of *Pseudomonas aeruginosa* biofilms to ciprofloxacin and tobramycin. Antimicrob Agents Chemother 47:317–323

Warny M, Pepin J, Fang A, Killgore G, Thompson A, Brazier J, Frost E, Mcdonald LC (2005) Toxin production by an emerging strain of *Clostridium difficile* associated with outbreaks of severe disease in North America and Europe. Lancet 366:1079–1084

Weese JS, Staempfli HR, Prescott JF (2000) Isolation of environmental *Clostridium difficile* from a veterinary teaching hospital. J Vet Diagn Invest 12: 449–452

Wilkins TD, Lyerly DM (2003) *Clostridium difficile* testing: after 20 years, still challenging. J Clin Microbiol 41:531–534

Wilson KH, Sheagren JN (1983) Antagonism of toxigenic *Clostridium difficile* by nontoxigenic *C. difficile*. J Infect Dis 147:733–736

Winzer K, Hardie KR, Burgess N, Doherty N, Kirke D, Holden MT, Linforth R, Cornell KA, Taylor AJ, Hill PJ, Williams P (2002) LuxS: its role in central metabolism and the in vitro synthesis of 4-hydroxy-5-methyl-3(2H)-furanone. Microbiology 148:909–922

Winzer K, Hardie KR, Williams P (2003) LuxS and autoinducer-2: their contribution to quorum sensing and metabolism in bacteria. Adv Appl Microbiol 53:291–396

Xiao J, Koo H (2010) Structural organization and dynamics of exopolysaccharide matrix and microcolonies formation by *Streptococcus mutans* in biofilms. J Appl Microbiol 108:2103–2113

Yakob L, Riley TV, Paterson DL, Clements AC (2013) *Clostridium difficile* exposure as an insidious source of infection in healthcare settings: an epidemiological model. BMC Infect Dis 13:376

Yarwood JM, Schlievert PM (2003) Quorum sensing in Staphylococcus infections. J Clin Invest 112: 1620–1625

Yoshida A, Ansai T, Takehara T, Kuramitsu HK (2005) LuxS-based signaling affects *Streptococcus mutans* biofilm formation. Appl Environ Microbiol 71: 2372–2380

Zhang L, Mah TF (2008) Involvement of a novel efflux system in biofilm-specific resistance to antibiotics. J Bacteriol 190:4447–4452

Zhu J, Miller MB, Vance RE, Dziejman M, Bassler BL, Mekalanos JJ (2002) Quorum-sensing regulators control virulence gene expression in *Vibrio cholerae*. Proc Natl Acad Sci USA 99:3129–3134

Zogaj X, Nimtz M, Rohde M, Bokranz W, Romling U (2001) The multicellular morphotypes of *Salmonella typhimurium* and *Escherichia coli* produce cellulose as the second component of the extracellular matrix. Mol Microbiol 39:1452–1463

Quorum-Sensing Systems in Enterococci

Ravindra Pal Singh and Jiro Nakayama

Introduction

Enterococcus is a genus of Gram-positive bacteria that is ubiquitous in natural ecosystems, plants, and animals. Some species of *Enterococcus* are present in the normal gastrointestinal bacterial community. However, others notably *Enterococcus faecalis* and *Enterococcus faecium* often cause opportunistic infections such as bacteremia, endocarditis, urinary tract infections, posttreatment endodontic infections, and endophthalmitis (Arias et al. 2010). Bacteria often use quorum sensing (QS) systems to control the expression of certain virulence genes and establish infection efficiently (Waters and Bassler 2005).

Similar to those in other Gram-positive bacteria, known QS systems in enterococci are primarily mediated by peptide molecules which allow specific cell-cell communication within species (Horinouchi et al. 2010). These molecules have been given various names such as 'pheromone', 'autoinducing peptide (AIP)', and 'quormone'. A pheromone is a type of communication signal between individuals, which are cells in the case of bacteria. The sex pheromone of *E. faecalis* induces conjugative transfer of a specific plasmid, resembles insect or animal pheromones in this sense (Dunny et al. 1978). On the contrary, the terms AIP and quormone are more indicative of autoinducers that orchestrate the phenotype of certain groups and correspond to the same species, strains, or clones in the case of bacteria (Waters and Bassler 2005). Gelatinase biosynthesis-activating pheromone (GBAP) is a typical AIP molecule that controls the expression of two pathogenicity related extracellular proteases in *E. faecalis* (Nakayama et al. 2001). GBAP is a peptide of 11 amino acids with a lactone ring. Given this cyclic structure and mode of action that triggers the histidine kinase receptor involved in two-component regulatory systems, the GBAP-mediated QS system is recognized as a cognate QS system, which is frequently encoded in the genomes of *Firmicutes* including staphylococci, clostridia, and listeria as well as enterococci (Wuster and Babu 2008). Conversely, the QS system regulating cytolysin production is unique in terms of sensor structure. These three types of peptide-mediated QS are discussed in this chapter.

Sex Pheromone/Inhibitor System Controlling Plasmid Transfer

In enterococci, a number of virulence factors and antibiotic-resistance genes are encoded on plasmid. Among several enterococcal plasmids, one large group is known to transfer at high frequency on the order of 10^{-3} per donor (or higher)

R.P. Singh • J. Nakayama (✉)
Laboratory of Microbial Technology, Department of Bioscience and Biotechnology, Faculty of Agriculture, Graduate School, Kyushu University, 6-10-1 Hakozaki, Higashi-ku, Fukuoka 812-8581, Japan
e-mail: nakayama@agr.kyushu-u.ac.jp

V.C. Kalia (ed.), *Quorum Sensing vs Quorum Quenching: A Battle with No End in Sight*, DOI 10.1007/978-81-322-1982-8_14, © Springer India 2015

within a few hours in broth suspensions. This type of plasmid encodes a unique system that produces a series of components required for conjugation in response to sex pheromone peptides secreted from plasmid-receiving cells. One of the main events induced by sex pheromones is sexual clumping between plasmid-donating and plasmid-receiving cells, which is mediated by an adhesion protein called an aggregation substance that is expressed on the surface of plasmid-donating cells. This sexual clumping enables a high frequency of plasmid transfer in the liquid phase. These sex pheromone plasmids encode genes for cytolysin, bacteriocins, and tetracycline and vancomycin resistance (Clewell et al. 2014). Thus far, this system appears to be unique to enterococci and may be related to the frequent occurrence of drug-resistant enterococci as well as other mobile genetic elements that can confer resistance (Clewell et al. 2014). Co-mobilization of the latter, including plasmids and transposons, with sex pheromone plasmids, is a possible risk for the transfer of antibiotic-resistance genes to other species. Beyond gene transfer, sexual behaviour is associated with adhesion and biofilm production in hosts (Chuang et al. 2009). The sexual clumping of *E. faecalis* facilitates fibrin adhesion (Hirt et al. 2000), increases biofilm formation on *ex vivo* cardiac valves (Chuang-Smith et al. 2010), and enhances vegetation formation in a rabbit model of endocarditis (Chuang et al. 2009).

Sex pheromones are hydrophobic oligopeptides, and the fact that they are encoded in a leader peptide moiety of lipoproteins suggests that they are processed products of the leader after the mature proteins are secreted (Cook and Federle 2013). Excreted sex pheromones are internalized into recipient cells by oligopeptide permease and may also be imported via natural membrane diffusion; once internalized, they bind to intracellular receptors (Leonard et al. 1996; Nakayama et al. 1998). Sex pheromone peptides are internalized into donor cells in the same manner. However, donor cells prevent self-induction by producing sex pheromone inhibitors that are antagonistic peptides of sex pheromone (Nakayama et al. 1998; Chatterjee et al. 2013; Clewell et al. 2014). Thus, the activation of either sex pheromone or inhibitor is determined by the ratio and distance between donor and recipient cells in nature and subsequently determines the potential 'mates' within the same population (Chatterjee et al. 2013). The system also ensures that conjugation does not occur between cells harbouring the same type of conjugative plasmid. This pheromone/inhibitor system appears to be effective among complex microbiota to transfer virulence, even within a specific population such as enterococci.

To date, five pheromone/inhibitor sets have been discovered (Horinouchi et al. 2010). Interestingly, each plasmid responds specifically to the corresponding pheromone and is blocked specifically by the corresponding inhibitor (Nakayama et al. 1995). In this context, pheromone and inhibitor designations follow plasmid nomenclature: pheromones are indicated by 'c', which stands for clumping-inducing agent and inhibitors are indicated by 'i' for inhibitor (e.g. cPD1 and iPD1 are pheromone and inhibitor of plasmid pPD1, respectively). The fact that sex pheromone precursors are encoded on the chromosome implies that *E. faecalis* has the potential to produce all sets of pheromones. However, once *E. faecalis* acquires a plasmid, it shuts down the activity of the corresponding pheromone via a pheromone shutdown system in addition to producing pheromone inhibitor. Pheromone shutdown is achieved via a plasmid-encoded protein (TraB for pAD1 and pPD1 and PrgY for pCF10) (Nakayama et al. 1995).

The tetracycline-resistant plasmid pCF10 is one of the best-studied sex pheromone plasmids (Dunny 2013). Therefore, detailed understanding of the mechanism of pheromone signalling herein uses examples from the pCF10 system (Fig. 1). The corresponding pheromone, cCF10, is expressed continuously as part of lipoprotein CcfA (Antiporta and Dunny 2002). At least three peptidases are involved in the processing of cCF10: one is a signal peptidase II involved in the cutoff of the leader moiety, the second is an Eep metalloprotease involved in the removal

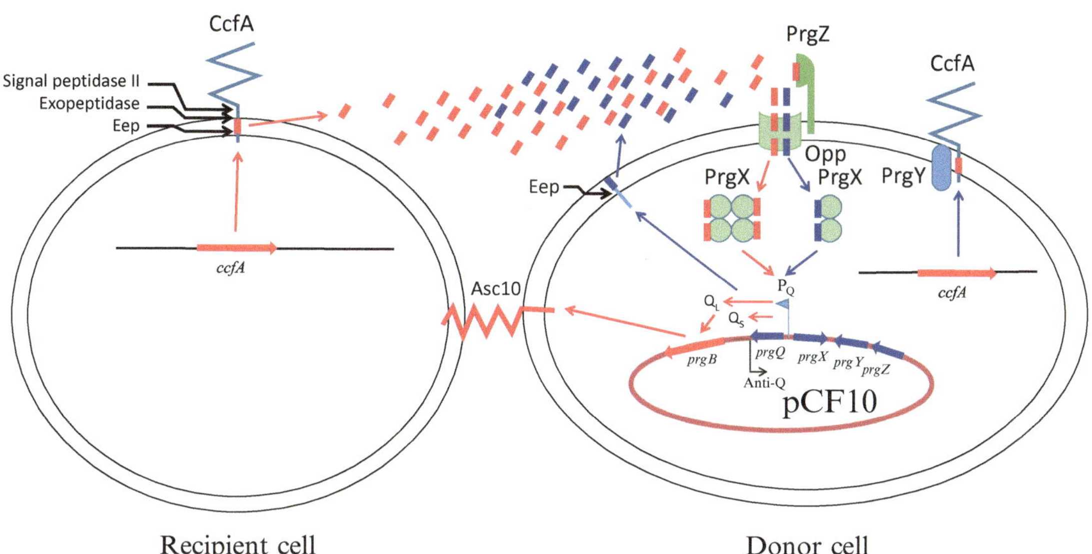

Fig. 1 Processing of conjugative peptide signalling in *Enterococcus faecalis*. The cCF10/iCF10-controlled conjugative transfer of pCF10 in *E. faecalis*. A lipoprotein of the pheromone precursor is translated from a chromosomal gene, *ccfA*, and processed by three peptidases signal peptidase II, Eep, and an exopeptidase resulting in mature cCF10. The inhibitor precursor is translated from *prgQ* encoded by pCF10 and processed by Eep, resulting in mature iCF10. The pheromone and inhibitor are captured by PrgZ or OppA and then imported through oligopeptide permease encoded by the *opp* gene cluster on the chromosome. Imported cCF10 and iCF10 competitively bind to an intracellular receptor, PrgX. The binding of iCF10 stabilizes the tetrameric form of PrgX that binds upstream of the P_Q promoter and blocks transcription. When cCF10 is in abundance, it outcompetes iCF10 at the binding site of PrgX and destabilizes PrgX tetramers, resulting in greater expression of Q_S as well as extended transcripts such as Q_L and other longer RNAs including aggregation substance gene *asc10*, required for conjugation. Both Q_S and Q_L determine the pre-iCF10 polypeptide. The opposing and partially overlapping *prgX* encodes the PrgX protein and a small RNA, Anti-Q, which promotes the termination of *prgQ* transcription. PrgY, which functions in pheromone shutdown, is localized in the cell membrane and directly inactivates self-produced cCF10

of N-terminal extensions, and the third is an exopeptidase involved in the removal of three residues on the C-terminal. Finally, the seven-residue cCF10 is generated and secreted outside the cells (Chandler and Dunny 2008). The corresponding inhibitor, iCF10, is encoded by a small open reading frame (ORF), *prgQ*, encoding 22 amino acid residues. The mature iCF10 sequence corresponds to the seven C-terminal residues of this *prgQ*. The length and hydrophobic nature of the PrgQ peptide resembles the leader peptide of CcfA, suggesting a similar or the same machinery for the secretion and maturation of the inhibitor peptide. Indeed, iCF10 is generated in an Eep-dependent manner (see Fig. 1) (Chandler and Dunny 2008). The pheromone and inhibitor are imported by oligopeptide permease encoded by the *opp* gene cluster on the *E. faecalis* chromosome. pCF10 encodes an accessory protein, PrgZ, which supports the transporter function of Opp (Leonard et al. 1996). PrgZ shows sequence similarity to OppA, which is an oligopeptide-binding protein associated with the Opp complex. cCF10 is thought to be captured by PrgZ and may be also mediated by OppA on the cell membrane and subsequently passed to the membrane transporter complex of peptide permease. There is also another possibility for pheromone intake. Nakayama et al. (1999) have demonstrated that an *opp* knockout mutant of *Escherichia coli* can import a tritium-labelled cPD1 corresponding to wild-type *E. coli*. Thus, it can be presumed that peptide sex pheromones are permeable through the cytoplasmic membrane and can be internalized without any transporter owing to their highly hydrophobic nature.

Imported cCF10 and iCF10 bind directly to an intracellular receptor, PrgX. The binding of iCF10 stabilizes the tetrameric form of PrgX, which binds upstream of the P_Q promoter and blocks transcription (Chatterjee et al. 2013). A crystallography study has suggested that the PrgX-cCF10 complex opens a 70-bp pCF10 DNA loop required for conjugation repression (Kozlowicz et al. 2006). When cCF10 is abundant, it outcompetes iCF10 at the binding site of PrgX and destabilizes the PrgX tetramers, resulting in increased expression of Q_S and extended transcripts such as Q_L as well as longer RNAs including aggregation substance gene *asc10*, which is required for conjugation (Shi et al. 2005). Both Q_S and Q_L determine the pre-iCF10 polypeptide. The opposing and partially overlapping *prgX* encodes PrgX protein and a small RNA, Anti-Q, which promotes the termination of *prgQ* transcription. The unique organization of these two operons provides several layers of both co-transcriptional and post-transcriptional regulation that allow the system to function as a sensitive biological switch.

The gene of a negative regulator, *prgY*, resides between *prgZ* and *prgX* and encodes a protein that functions in pheromone shutdown (Chandler et al. 2005; Chandler and Dunny 2008). Combined genetic and physiological studies suggest that PrgY function is Eep independent and involves direct interaction with cCF10. In another words, PrgY localized in the cell membrane interacts directly with cCF10, after which, cCF10 is sequestered or inactivated as it is released from the membrane (Chandler et al. 2005; Chandler and Dunny 2008), and residual cCF10 escaping PrgY is captured by iCF10. Despite an understanding of these details, the complete molecular mechanism for pheromone shutdown by PrgY remains unclear.

The mechanisms for specific recognition of pheromone and inhibitor peptides have been well studied in another pheromone plasmid, pPD1. Nakayama et al. (1998) used radio-labelled cPD1 and recombinant TraA protein corresponding to PrgX of pCF10 to demonstrate on the molecular level that TraA of pPD1 shows highly specific affinity to cPD1 and iPD1 but not to other pheromones and inhibitors, suggesting that TraA plays a central role in the specific recognition of pheromone and inhibitor peptides. On the contrary, pheromone-binding proteins encoded by *traC* of pPD1, which corresponds to PrgZ of pCF10, can be replaced by TraC encoded by pAD1. Furthermore, the pheromone shutdown function encoded by *traB*, which corresponds to *prgY* of cCF10, cannot be complemented by *traB* of pAD1, suggesting that pheromone shutdown occurs specifically with each corresponding plasmid. The conjugative plasmid appears to have evolved to develop a specific recognition system for the peptide pheromone/inhibitor and pheromone shutdown, which enables recipient-targeted transfer (Nakayama and Suzuki 1997).

GBAP

Gelatinase is an extracellular metalloprotease that liquefies gelatine and collagen and is considered a virulence factor in *E. faecalis*. In addition to having proteolytic activity in the extracellular matrix, gelatinase digests a broad range of host substrates such as fibrinogen, fibrin, endothelin-1, bradykinin, LL-37, and complement components C3, C3a, and C5a (Schmidtchen et al. 2002; Waters et al. 2003; Park et al. 2008; Thurlow et al 2010).

Gelatinase production in *E. faecalis* occurs in the late log to early stationary phases, but the addition of conditioned medium shifts its proteolytic phase to the mid-log phase, suggesting the presence of an autoinducer which mediates QS. Nakayama et al. (2001) succeeded in purifying the autoinducer and identified it as a cyclic peptide called GBAP that induces gelatinase biosynthesis at nanomolar concentrations. The results of Edman degradation and mass analysis have revealed that GBAP is a peptide of 11 amino acid residues with an intramolecular lactone bridge between the hydroxyl group of the third serine residue and the α-carboxyl group of C-terminal methionine. Specifically, GBAP is composed of an exocyclic tail consisting of two N-terminal

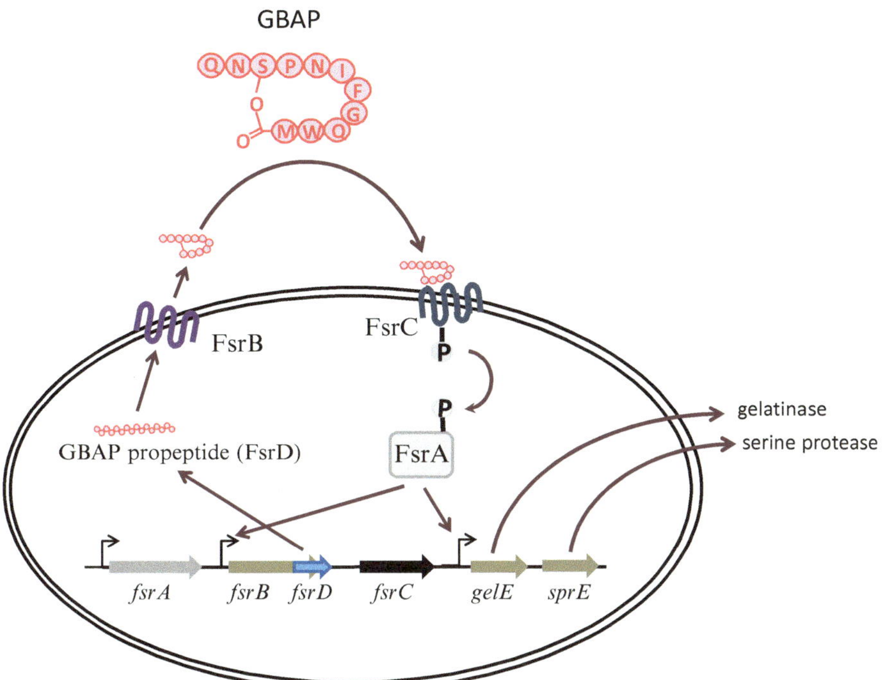

Fig. 2 The *fsr* quorum-sensing system in *Enterococcus faecalis*. After translation, FsrD is processed, cyclized, and secreted by FsrB. Eventually, the mature form of gelatinase biosynthesis-activating pheromone (GBAP) is secreted and accumulates outside the cells. GBAP binds to FsrC and triggers the phosphorylation of FsrA. Phosphorylated FsrA binds to the promoters of *fsrBDC* and *gelE-sprE*, resulting in the upregulation of these operons

amino acids and a macrocyclic moiety consisting of nine amino acids, as shown in Fig. 2. The ring moiety is crucial for induction activity, and alanine scanning on the peptide has suggested that two aromatic residues (Phe-7 and Trp-10) are essential for receptor binding of GBAP (Nishiguchi et al. 2009).

The amino acid sequence of GBAP was found within the C-terminal of a deduced 242-residue protein encoded by *fsrB* in the *E. faecalis* genome (Nakayama et al. 2001). Later, a GBAP precursor, FsrD, was demonstrated to be translated from an in frame methionine codon at Met-210 of *fsrB* (Nakayama et al. 2006). It indicated the presence of a small ORF termed *fsrD* which independently encodes the AIP precursor similar to the encoding of staphylococcal AIP. Genes encoding two component regulatory proteins, FsrA and FsrC, are present upstream and downstream of *fsrBD*, respectively, suggesting that the GBAP signal is received by the FsrC histidine protein kinase receptor and then transmitted to response regulator FsrA through phosphorylation. This type of QS system, in which a cyclic AIP control gene is expressed through a two-component regulatory system, is often found in low-GC Gram-positive bacteria. One such system, the *agr* QS system of staphylococci, has been widely studied (Novick 2003).

Downstream of the gelatinase gene, *gelE*, and another extracellular protease gene, *sprE*, is encoded. According to the amino acid sequence, SprE is deduced as serine protease. These two protease genes constitute operons and are under the control of the *fsr* QS system. A number of studies in animal models have shown the potential roles of the *fsr* system and its controlled proteases in *E. faecalis* virulence. For instance, in a mouse peritonitis model, *sprE*, *fsrA*, and *fsrB* mutants showed delayed time of killing of the host compared with that of the wild-type strain (Qin et al. 2000). A rabbit endophthalmitis model, an exquisitely sensitive infection system for investigating QS, suggests a pleiotropic influence of the *fsr* locus in the virulence of infection (Engelbert et al. 2004). Using a rabbit model of enterococcal endocarditis, Thurlow et al. (2010) demonstrated that GelE, not SprE, is the principal mediator of pathogenesis in endocarditis through the suppression of neutrophil migration via breakdown of anaphylatoxin complement C5a. Proteolytic activity of cell surface proteins is also known to modulate morphology and adhesion properties, which may directly or indirectly associate with biofilm formation by *E. faecalis* cells. Hancock and Perego (2004) have demonstrated that GelE, is involved in biofilm formation of *E. faecalis*. Waters et al. (2003) have demonstrated that GelE clears the bacterial cell surface of misfolded proteins, and disruption of GelE production increases cell chain length in *E. faecalis*. Pinkston et al. (2011) have demonstrated that the display of Ace, a collagen-binding protein in *E. faecalis*, is modulated by *fsr* QS. Ace is a microbial surface component recognizing adhesive matrix (MSCRAMM) that is under consideration as an immunotherapy target.

Tissue translocation of *E. faecalis* is also known to be controlled by the *fsr* QS system. Specially, GelE is important for the translocation of *E. faecalis* across polarized human enterocyte-like T84 cells (Zeng et al. 2005). Nakayama et al. (2013) have developed a GBAP antagonist, ZBzl-YAA5911, and demonstrated its in vivo efficacy in suppressing translocation of the *E. faecalis* from the aqueous humour into the vitreous cavity, significantly reducing retinal damage. GelE and SprE may also modify the cell wall affinity of autolysin, which releases DNA as a component of the biofilm matrix. Finally, transcriptomic analysis has been performed in a comprehensive investigation of genes under the control of *fsr* QS (Bourgogne et al. 2006). The results identified a variety of genes as newly recognized Fsr-regulated targets, including factors other than gelatinase that are important in biofilm (BopD), genes predicted to encode surface proteins, and proteins implicated in several metabolic pathways.

Notably, gelatinase is not produced by all clinical isolates despite its role as a possible virulence factor (Nakayama et al. 2002; Roberts et al. 2004). The ratio of gelatinase-positive isolates was 51 % in a study by Nakayama et al. (2002) and 60 % in a study by Roberts et al. (2004). The latter found that a similar ratio of isolates from healthy volunteers was also gelatinase positive. Nakayama et al. (2002) have found that the majority of gelatinase-negative phenotypes could be attributed to a lack not of *gelE* but *fsr*. Genetic analysis revealed that most of gelatinase-negative strains of *E. faecalis* have intact *gelE-sprE* gene clusters but display a 23.9-kb deletion covering the entire regions of *fsrA*, *fsrB*, and *fsrD* and part of the 5′-end of *fsrC*. Roberts et al. (2004) found this 23.9-kb deletion in 14 of 80 endocarditis isolates tested. A study by Strzelecki et al. (2011) found that among 153 isolates, 140 had *gelE*, whereas only 53 % produced gelatinase. These results imply that the gelatinase-negative phenotype is primarily determined not by the lack of the gelatinase gene but by the absence of a QS gene. The fact that the 23.9-kb deletion corresponds to a chromosomal region ranging from inside an ORF (*ef1841*) to inside *fsrC* indicates that it is not an insertion but instead a deletion in the chromosome. The high frequency of deletion suggests an evolutionary benefit associated with the loss of gelatinase production. Among mixed populations of gelatinase-positive and gelatinase-negative strains, *fsr*-negative mutants may act as 'QS cheaters', which do not respond to the GBAP signal and do not contribute to the rest of the population but rather take advantage of it (Thomas et al. 2009).

The QS inhibitor-targeting *fsr* system has attracted interest for its potential to attenuate the virulence of *E. faecalis* without bactericidal effects, which could lead to the emergence of drug-resistant strains. The mechanism of action of ZBzl-YAA5911 is an example of effective in vivo suppression of virulence expression in *E. faecalis*. As described above, ZBzl-YAA5911 attenuates the translocation of *E. faecalis* from the aqueous humour into the vitreous cavity by inhibiting the expression of gelatinase, consequently reducing retinal damage. Details and other examples of QS inhibitors targeting the *fsr* system are described in Chap. 25.

Subunit-Mediated Regulation of Cytolysin Production

Cytolysin, one of the best-characterized virulence factors of *E. faecalis*, shows cytotoxic activity against both eukaryotic and prokaryotic cells. It efficiently lyses red blood cells and can be considered a hemolysin. It also shows bacteriolytic activity similar to bacteriocin and may have a function in complex microbiota in nature. Cytolysin is a complex molecule consisting of two lantibiotic peptides (Coburn and Gilmore 2003). The biosynthetic gene cluster of cytolysin is organized by eight ORFs: *cylR1, cylR2*, $cylL_L$, $cylL_S$, *cylM, cylB, cylA*, and *cylI* (Coburn and Gilmore 2003). The $cylL_L$ and $cylL_S$ encode a large subunit ($CylL_L$) and a small subunit ($CylL_s$) of cytolysin, respectively. These subunits are post-translationally modified by CylM, secreted and trimmed by CylB, and further processed by CylA. Eventually, active cytolysin consisting of two mature subunit peptides is generated.

Interestingly, $CylL_S$ also functions as an autoinducer for a two-component regulatory system consisting of CylR1 and CylR2. CylR1-CylR2 is a novel two-component regulatory protein that differs from the cognate family, which consists of a conserved histidine kinase and response regulator. Instead, CylR2 has a small helix-turn-helix DNA-binding motif, and CylR1 appears to be a transmembrane protein (Haas et al. 2002). In vitro culture of *E. faecalis* has demonstrated that $CylL_L$ and $CylL_S$ form an oligomeric cytolysin complex that is an inactive form of AIP. However, when target animal cells get close to *E. faecalis* cells, $CylL_L$ is captured by the animal cells and free $CylL_S$ molecules increase. Consequently, $CylL_S$ triggers the CylR1-CylR2 two-component regulatory system and activates transcription of the cytolysin operon (Coburn et al. 2004). Specifically, this system appears to sense both the quorum of bacterial cells and the distance to target cells.

Conclusion

Enterococci are ubiquitous in nature and commensal in the digestive tracts of animals. Some enterococcal species such as *E. faecalis* and *E. faecium* cause opportunistic infection using a small set of virulence genes, including cytolysin and gelatinase. In addition, these species frequently carry mobile genetic elements such as plasmids and transposons which often encode antibiotic-resistance genes. Sex pheromone plasmids are particularly effective at inducing resistance owing to their highly transmissible properties. Pheromone/inhibitor systems enable efficient transconjugation between donor and recipient cells even in complex microbiota involving a number of other species. Cytolysin is often encoded by a pheromone plasmid and is itself regulated by QS using the cytolysin component $CylL_S$. The $CylL_S$-$CylL_L$ system is involved in sensing the distance between bacterial and animal cells. The nonessential virulence factor gelatinase is also regulated by QS using the cyclic peptide autoinducer GBAP. The QS function is encoded in the *fsr* gene cluster, a cognate QS system often found in low-GC Gram-positive bacteria in which AIP activates target gene expression via a two-component regulatory system. However, the *fsr* system is somewhat different from other cognate systems that employ thiolactone peptide instead of lactone and regulate a number of genes dispersed throughout the genome via regulatory RNA expression. GBAP directly regulates its downstream operon encoding gelatinase and serine protease.

Opinion

As antibiotic-resistant enterococci are frequently observed not only at clinical sites but also in the natural environment, enterococci are sometimes recognized as a reservoir of antibiotic-resistance genes. Although sex pheromone plasmids can be transferred within enterococci, they sometimes integrate transposons, which may generate the capability to rearrange genomes and raise the risk of transfer to other bacterial species including pathogenic ones. Interestingly, sex pheromone plasmids use the processed products of chromosomally encoded proteins. Contrary, sex pheromone inhibitors are produced from plasmid-encoded propeptides without other functions, suggesting that sex pheromone plasmids are extra-chromosomal elements that have evolved to gain an efficient transferring system using chromosomal factors.

Two virulence-related extracellular proteases, GelE and SprE, have been demonstrated to degrade not only gelatin and collagen but also protein on the surface of bacteria. These proteins could associate with adhesion or biofilm formation property of bacterial cell, and a variety of host proteins, which could be involved in the immune system. It suggests *fsr* QS system to strongly link with virulence of *E. faecalis*. However, it should be noted that GelE/SprE is not essential for enterococci infection as *fsr*-negative *E. faecalis* are frequently isolated from patients. Are they 'QS cheater' in the enterococcal community? It is still in question how *fsr*-induced proteases function at infectious locations.

Acknowledgements This research was supported in part by Grants-in-Aid for Scientific Research (B) No. 24380050 and Grant-in-Aid for JSPS Fellows No. 25–03389.

References

Antiporta M, Dunny G (2002) *ccfA*, the genetic determinant for the cCF10 peptide pheromone in *Enterococcus faecalis*. J Bacteriol 184:1155–1162

Arias CA, Contreras GA, Murray BE (2010) Management of multi-drug resistant enterococcal infections. Clin Microbiol Infect 16:555–562

Bourgogne A, Hilsenbeck SG, Dunny GM, Murray BE (2006) Comparison of OG1RF and an isogenic *fsrB* deletion mutant by transcriptional analysis: the Fsr system of *Enterococcus faecalis* is more than the activator of gelatinase and serine protease. J Bacteriol 188:2875–2884

Chandler JR, Dunny GM (2008) Characterization of the sequence specificity determinants required for processing and control of sex pheromone by the intramembrane protease Eep and the plasmid-encoded protein PrgY. J Bacteriol 190:1172–1183

Chandler JR, Hirt H, Dunny GM (2005) A paracrine peptide sex pheromone also acts as an autocrine signal to induce plasmid transfer and virulence factor expression *in vivo*. Proc Natl Acad Sci USA 102:15617–15622

Chatterjee A, Cook LC, Shu CC, Chen Y, Manias DA, Ramkrishna D, Dunny GM, Hu WS (2013) Antagonistic self-sensing and mate-sensing signaling controls antibiotic resistance transfer. Proc Natl Acad Sci USA 110:7086–7090

Chuang O, Schlievert P, Wells CL, Manias DA, Tripp TJ, Dunny GM (2009) Multiple functional domains of *Enterococcus faecalis* aggregation substance Asc10 contribute to endocarditis virulence. Infect Immun 77:539–548

Chuang-Smith O, Wells C, Henry-Stanley M, Dunny GM (2010) Acceleration of *Enterococcus faecalis* biofilm formation by aggregation substance expression in an *ex vivo* model of cardiac valve colonization. PLoS One 5:e15798

Clewell DB, Weaver KE, Dunny GM, Coque TM, Francia MV, Hayes F (2014) Extrachromosomal and mobile elements in enterococci: transmission, maintenance, and epidemiology. In: Clewell DB, Shankar N, Ike Y, Gilmore M (eds) Enterococci: from commensals to leading causes of drug resistant infection [Internet]. Massachusetts Eye and Ear Infirmary, Boston

Coburn PS, Gilmore MS (2003) The *Enterococcus faecalis* cytolysin: a novel toxin active against eukaryotic and prokaryotic cells. Cell Microbiol 5:661–669

Coburn PS, Pillar CM, Jett BD, Haas W, Gilmore MS (2004) *Enterococcus faecalis* senses target cells and in response expresses cytolysin. Science 306:2270–2272

Cook LC, Federle MJ (2013) Peptide pheromone signaling in *Streptococcus* and *Enterococcus*. FEMS Microbiol Rev 38:473–492

Dunny GM (2013) Enterococcal sex pheromones: signaling, social behavior, and evolution. Annu Rev Gen 47:457–482

Dunny GM, Brown BL, Clewell DB (1978) Induced cell aggregation and mating in *Streptococcus faecalis*: evidence for a bacterial sex pheromone. Proc Natl Acad Sci USA 75:3479–3483

Engelbert M, Mylonakis E, Ausubel FM, Calderwood SB, Gilmore MS (2004) Contribution of gelatinase, serine protease, and *fsr* to the pathogenesis of *Enterococcus faecalis* endophthalmitis. Infect Immun 72:3628–3633

Haas W, Shepard BD, Gilmore MS (2002) Two-component regulator of *Enterococcus faecalis* cytolysin responds to quorum-sensing autoinduction. Nature 415:84–87

Hancock LE, Perego M (2004) The *Enterococcus faecalis fsr* two-component system controls biofilm development through production of gelatinase. J Bacteriol 186:5629–5639

Hirt H, Erlandsen SL, Dunny GM (2000) Heterologous inducible expression of *Enterococcus faecalis* pCF10 aggregation substance Asc10 in *Lactococcus lactis* and *Streptococcus gordonii* contributes to cell hydrophobicity and adhesion to fibrin. J Bacteriol 182:2299–2306

Horinouchi S, Ueda K, Nakayama J, Ikeda T (2010) Cell-to-cell communications among microorganisms. In: Mander L, Liu HW (eds) Comprehensive natural products II, vol 4. Elsevier Science, Amsterdam, pp 283–337

Kozlowicz BK, Shi K, Gu ZY, Ohlendorf DH, Earhart CA, Dunny GM (2006) Molecular basis for control of conjugation by bacterial pheromone and inhibitor peptides. Mol Microbiol 62:958–969

Leonard BA, Podbielski A, Hedberg PJ, Dunny GM (1996) *Enterococcus faecalis* pheromone binding protein, PrgZ, recruits a chromosomal oligopeptide permease system to import sex pheromone cCF10 for induction of conjugation. Proc Natl Acad Sci USA 93:260–264

Nakayama J, Suzuki A (1997) Genetic analysis of plasmid-specific pheromone signaling encoded by pPD1 in *Enterococcus faecalis*. Biosci Biotechnol Biochem 61:1796–1799

Nakayama J, Yoshida K, Kobayashi H, Isogai A, Clewell DB, Suzuki A (1995) Cloning and characterization of a region of *Enterococcus faecalis* plasmid pPD1 encoding pheromone inhibitor (*ipd*), pheromone sensitivity (*traC*), and pheromone shutdown (*traB*) genes. J Bacteriol 177:5567–5573

Nakayama J, Takanami Y, Horii T, Sakuda S, Suzuki A (1998) Molecular mechanism of peptide-specific pheromone signaling in *Enterococcus faecalis*: functions of pheromone receptor TraA and pheromone-binding protein TraC encoded by plasmid pPD1. J Bacteriol 180:449–456

Nakayama J, Horii T, Suzuki A (1999) Bacterial peptide pheromone is imported where it directly binds to an intracellular receptor. In: Proceedings of the 1st international peptide symposium 1997, in press. Kluwer Academic Publishers, BV, Dordrecht, pp 215–217

Nakayama J, Cao Y, Horii T, Sakuda S, Akkermans AD, de Vos WM, Nagasawa H (2001) Gelatinase biosynthesis-activating pheromone: a peptide lactone that mediates a quorum sensing in *Enterococcus faecalis*. Mol Microbiol 41:145–154

Nakayama J, Kariyama R, Kumon H (2002) Description of a 23.9-kilobase chromosomal deletion containing a region encoding *fsr* genes which mainly determines the gelatinase-negative phenotype of clinical isolates of *Enterococcus faecalis* in urine. Appl Environ Microbiol 68:3152–3155

Nakayama J, Chen S, Oyama N, Nishiguchi K, Azab EA, Tanaka E, Kariyama R, Sonomoto K (2006) Revised model for *Enterococcus faecalis fsr* quorum-sensing system: the small open reading frame *fsrD* encodes the gelatinase biosynthesis-activating pheromone propeptide corresponding to staphylococcal AgrD. J Bacteriol 188:8321–8326

Nakayama J, Yokohata R, Sato M, Suzuki T, Matsufuji T, Nishiguchi K, Kawai T, Yamanaka Y, Nagata K, Tanokura M, Sonomoto K (2013) Development of a peptide antagonist against fsr quorum sensing of Enterococcus faecalis. ACS Chem Biol 8:804–811

Nishiguchi K, Nagata K, Tanokura M, Sonomoto K, Nakayama J (2009) Structure-activity relationship of gelatinase biosynthesis-activating pheromone of *Enterococcus faecalis*. J Bacteriol 191:641–650

Novick RP (2003) Autoinduction and signal transduction in the regulation of staphylococcal virulence. Mol Microbiol 48:1429–1449

Park SY, Shin YP, Kim CH, Park HJ, Seong YS, Kim BS, Seo SJ, Lee IH (2008) Immune evasion of *Enterococcus faecalis* by an extracellular gelatinase that cleaves C3 and iC3b. J Immunol 181:6328–6336

Pinkston KL, Gao P, Diaz-Garcia D, Sillanpää J, Nallapareddy SR, Murray BE, Harvey BR (2011) The Fsr quorum-sensing system of *Enterococcus faecalis* modulates surface display of the collagen-binding MSCRAMM Ace through regulation of *gelE*. J Bacteriol 193:4317–4325

Qin X, Singh KV, Weinstock GM, Murray BE (2000) Effects of *Enterococcus faecalis fsr* genes on production of gelatinase and a serine protease and virulence. Infect Immun 68:2579–2586

Roberts JC, Singh KV, Okhuysen PC, Murray BE (2004) Molecular epidemiology of the *fsr* locus and of gelatinase production among different subsets of *Enterococcus faecalis* isolates. J Clin Microbiol 42:2317–2320

Schmidtchen A, Frick IM, Andersson E, Tapper H, Björck L (2002) Proteinases of common pathogenic bacteria degrade and inactivate the antibacterial peptide LL-37. Mol Microbiol 46:157–168

Shi K, Brown CK, Gu ZY, Kozlowicz BK, Dunny GM, Ohlendorf DH, Earhart CA (2005) Structure of peptide sex pheromone receptor PrgX and PrgX/pheromone complexes and regulation of conjugation in *Enterococcus faecalis*. Proc Natl Acad Sci USA 102:18596–18601

Strzelecki J, Hryniewicz W, Sadowy E (2011) Gelatinase-associated phenotypes and genotypes among clinical isolates of *Enterococcus faecalis* in Poland. Polish J Microbiol 60:287–292

Thomas VC, Hiromasa Y, Harms N, Thurlow L, Tomich J, Hancock LE (2009) A fratricidal mechanism is responsible for eDNA release and contributes to biofilm envelopment of *Enterococcus faecalis*. Mol Microbiol 72:1022–1036

Thurlow LR, Thomas VC, Narayanan S, Olson S, Fleming SD, Hancock LE (2010) Gelatinase contributes to the pathogenesis of endocarditis caused by *Enterococcus faecalis*. Infect Immun 78:4936–4943

Waters CM, Bassler BL (2005) Quorum sensing: cell-to-cell communication in bacteria. Annu Rev Cell Dev Biol 21:319–346

Waters CM, Antiporta MH, Murray BE, Dunny GM (2003) Role of the *Enterococcus faecalis* GelE protease in determination of cellular chain length, supernatant pheromone levels, and degradation of fibrin and misfolded surface proteins. J Bacteriol 185:3613–3623

Wuster A, Babu MM (2008) Conservation and evolutionary dynamics of the *agr* cell-to-cell communication system across firmicutes. J Bacteriol 190:743–746

Zeng J, Teng F, Murray BE (2005) Gelatinase is important for translocation of *Enterococcus faecalis* across polarized human enterocyte-like T84 cells. Infect Immun 73:1606–1612

Quorum-Sensing Systems in *Bacillus*

Lalit K. Singh*, Neha Dhasmana*, and Yogendra Singh

Introduction

Since centuries bacteria were thought to be unicellular organisms. The discovery of bacterial communication through small molecules has asserted that bacteria can efficiently coordinate intraspecies as well as interspecies. The bacteria become more benefitted and suitable of behaving like a multicellular organism to adopt new modes of growth in limited nutrient supply. Under adverse conditions, single bacterial cell has less chance to survive in isolation; consequently bacterial language has been developed during evolution to communicate with its neighbours through self-generated signals (Bassler and Losick 2006). These signalling small molecules are called as pheromones or autoinducers. These autoinducers sense a critical bacterial density in population (Kievit and Iglewski 2000; Williams et al. 2007).

Communication in bacterial population is also observed in symbiotic bacteria which use these pheromones while interacting with their host. Bacteria are associated with the lives of *Homo sapiens* as well as live stocks concerning that they cause deadly diseases, can be in symbiotic relationship or are employed in various food processing industries (Steidle et al. 2001). *Bacillus* genera belong to the Firmicutes phylum of bacterial kingdom and include various significant organisms, namely, *B. anthracis*, the aetiological agent of anthrax; *B. cereus* that causes diarrhoeal symptoms; and *B. thuringiensis*, the insect pathogen. Numerous pathogens are known to cause upper respiratory and urinary tract infections, given their ability to communicate intraspecies and thus form multicellular organization called biofilms. Enhanced antibiotic resistance is also observed when these pathogenic microorganisms form clinically more interesting recalcitrant biofilms (Davey and O'toole 2000; Donlan and Costerton 2002). Furthermore, *B. subtilis* is known to colonize the *Arabidopsis thaliana* roots through matrix-enclosed multicellular communities in which matrix production is triggered by plant exopolysaccharides and reported to be beneficial for plant growth (Beauregard et al. 2013).

*Author contributed equally with all other contributors.

L.K. Singh
Allergy and Infectious Diseases, CSIR-Institute of Genomics and Integrative Biology, Delhi University Campus, Mall Road, Delhi 110007, India

N. Dhasmana • Y. Singh (✉)
Allergy and Infectious Diseases, CSIR-Institute of Genomics and Integrative Biology, Delhi University Campus, Mall Road, Delhi 110007, India

Academy of Scientific & Innovative Research (AcSIR), 2, Rafi Marg, Anusandhan Bhawan, New Delhi 110001, India
e-mail: ysingh@igib.res.in

V.C. Kalia (ed.), *Quorum Sensing vs Quorum Quenching: A Battle with No End in Sight*,
DOI 10.1007/978-81-322-1982-8_15, © Springer India 2015

Quorum-Sensing-Mediated Biofilm Formation

Bacillus subtilis is the most extensively studied organism in *Bacillus* genera. *B. subtilis* is a gram-positive, endospore-forming, rod-shaped and aerophilic bacteria (Kumar et al. 2013). It secretes diffusible oligopeptides for communication with the neighbouring cells; the process referred as quorum sensing (QS) (Kalia and Purohit 2011; Kalia 2013). Neighbouring bacteria perceive and broadcast the signals in its vicinity consequently causing behavioural modifications in the bacterial population (Parsek and Greenberg 2005; Bassler and Losick 2006; Mehta et al. 2009; Boyle et al. 2013; Vlamakis et al. 2013). Biofilms are the cocoons made up of extracellular polymeric substances (EPS) in which bacteria thrive under adverse environmental conditions (Flemming et al. 2007). The biofilm matrix consists of exopolysaccharides, proteins, enzymes and extracellular DNA along with pili and flagella. Biofilm provides adhesion that facilitates the initial step of colonization and imparts protection against the innate host defence. Following adhesion, biofilm also maintains the moisture content and absorption of nutrients for the better survival throughout the infectious phase or under environmental stress (Flemming and Wingender 2010; Li and Tian 2012; Kostakioti et al. 2013). Biofilm formation precisely describes the transition between unicellular bacteria to the partial multicellular organism (Aguilar et al. 2007; Shank et al. 2011). It is an ingenious plan of nature to sweep over the nutritional stress using coordinated biological pathways (Lemon et al. 2008; Shank et al. 2011). Bacteria within the biofilm display heterogeneity at phenotypic and genotypic level, and the social behaviour is governed through paracrine signalling (Stewart and Franklin 2008; López et al. 2009; Monds and O'Toole 2009; Kalia et al. 2011; Kalia 2013).

Competence and sporulation stimulating factor (CSF) is the master regulator of QS-mediated biofilm formation, competence and sporulation (Waters and Bassler 2005). The CSF is conserved in *Bacillus* spp.; however, the protein sequence shows the polymorphism. CSF is secreted in its precursor form (40 amino acids) encoded by *phrC* gene. The N-terminal of pre-CSF contains the guiding sequence for secretion. Membrane-associated serine proteases cleave the C-terminal of precursor CSF and release mature CSF (Miller and Bassler 2001; López et al. 2009). Mature CSF is a pentapeptide secreted into extracellular milieu, which regulates the competence factors ComA, ComK, ComS and ComX expression. The competence factor ComA activates the surfactin operon (*srfA-D*). Additionally, the ComK interacts with ComP and switches on the *srfA-D* operon, which ensures the cell to be competent and able to produce the exopolysaccharides. Surfactin-producing cells are capable to show the competency as ComS is synthesized, while the ComK shows the alternative cascade regulation; only a fraction of cells producing surfactin displays the competency (Shank and Kolter 2011). The pathways of competency and biofilm formation are similar in *B. subtilis* and in other organisms, for example, *Streptococcus pneumoniae*. Although the mechanism of action of QS molecules is well established, how the QS molecules are synthesized in the cell is relatively less understood (Turovskiy et al. 2007; orthington et al. 2012).

Quorum-Sensing-Mediated Cannibalism

Bacillus undergoes sporulation process in high stress condition such as nutritional imbalance or heat stress (Fig. 1). Sporulation is highly energy-driven and time-investing process in which vegetative cell differentiates into a dormant structure. Among different subpopulations under starvation conditions, a discrete set of bacterial cells secretes two toxin peptides, namely, Sdp (sporulation-delaying factor) and Skf (sporulation-killing factor) whose function is to delay sporulation and kill the siblings, respectively (Lamsa et al. 2012). The Skf peptide toxin kills their neighbouring sensitive cells

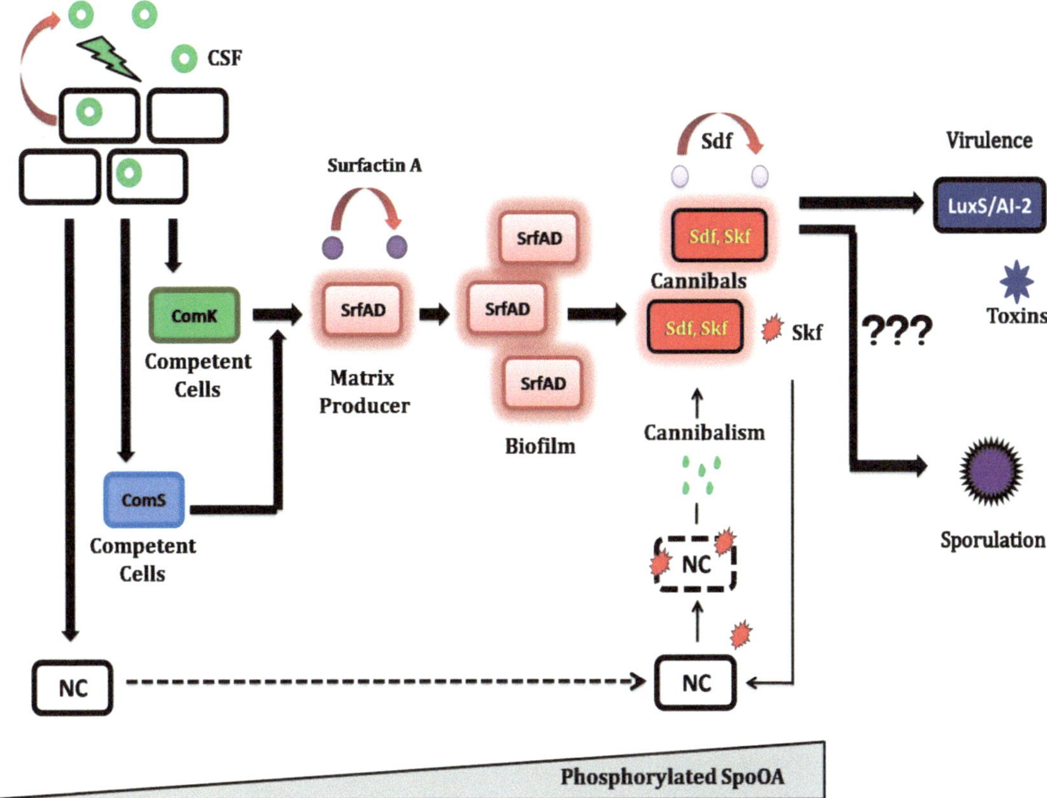

Fig. 1 Quorum-sensing systems in *Bacillus*: Heterogeneous bacterial population undergoes competence with the trigger of the colony and sporulation stimulating factor (CSF) during low phosphorylated Spo0A levels. The CSF acts differently on the cells, giving rise to distinct subpopulations of non-competent cells (NC), ComS and ComK expressing cells. The ComK majorly regulates the expression of surfactin-encoding operon (*srfA-D*), which triggers onset of matrix production by ComK and ComS expressing competent cells. Under persistent nutritional limitations, the matrix producers generate quorum-sensing (QS) signal through releasing sporulation-killing factor (Skf) which causes the killing of neighbouring non-competent cells, hence releasing nutrients to be used for prolonged survival and matrix production, the process known as cannibalism. Additionally, the matrix producers generate other signals known as a sporulation-delaying factor (Sdf), which results in enhanced survival in vegetative phase of the bacterial life cycle. At early stationary phase, cells generate QS molecules, i.e. autoinducer AI-2 signals, and regulate various virulence factors like toxin synthesis and secretion, S-layer formation, etc. At higher levels of phosphorylated Spo0A signals, sporulation is triggered simultaneously in the whole bacterial population. The QS regulators behind this highly synchronized onset of sporulation process are still unknown

(non-competent cells) and utilizes the nutrients released as food to overcome the nutrient limitations. The procedure of eating their siblings is termed as cannibalism and is described to be transitory in nature (Schultz et al. 2009; Shank and Kolter 2011). The sporulation master regulator Spo0A governs the expression of Sdp and Skf proteins. The low level of phosphorylated Spo0A regulates the *sdpABC, srf* operon consequently causing matrix production (López and Kolter 2009). The same subpopulation forms the biofilm, out of which some cells display cannibalistic behaviour. Both biofilm formation and cannibalism are reported to be triggered by the QS signal molecule, surfactin. Higher level of phosphorylated Spo0A promotes the cell into the committed sporulation phase of the bacterial life cycle (Fujita and Losick 2005).

Cannibalism is analogous to programmed cell death, in which cells not required for development of bacterial community are removed (López et al. 2009; Li and Tian 2012). In *B. subtilis* paracrine signalling through QS molecules regulates the three major life events in the life cycle, e.g. biofilm formation, cannibalism and sporulation. Similarly the other *Bacillus* spp. like *Lactococcus lactis* secrete the cannibalism toxin nisin that functions as the antimicrobial peptide (Williams et al. 2007). The nisin treatment with *B. subtilis* kills the cells, those that are unable to produce the cannibalism toxin, consequently giving rise to stronger biofilm that are competing for survival in limited nutrients. It has been reported that *B. subtilis* also kills the neighbouring bacteria through the Skf cannibalism toxin (Turovskiy et al. 2007). Further investigations are needed to see the role of QS molecules and cannibalism toxins in multispecies biofilm formation.

Quorum-Sensing-Mediated Virulence

QS governs the major phenomenon in the bacterial alternating life stages. Interestingly, the virulence of pathogenic members of *Bacillus* is also reported to be governed by the societal communication (Fig. 1). The pathogenic clade of *Bacillus* species includes *B. anthracis*, *B. cereus* and *B. thuringiensis*. *B. thuringiensis* is an insect pathogen, which is a parasite of the economically significant crop, cotton. The paracrine signalling peptide PlcR (34 kDa or 48aa) is secreted by the *B. thuringiensis* and imported by neighbouring bacterial cells through oligopeptide permease. The PlcR interacts with PapR and binds to the DNA, and this ternary interaction is known to cause pleiotropic effects, including secretion of toxins (Kievit and Iglewski 2000). The PlcR regulator is also secreted by another pathogen *B. cereus* causing diarrhoeal and nausea symptoms upon food poisoning. PlcR is known to express at the onset of stationary phase that regulates the synthesis of various toxins like enterotoxins, cytotoxins and hemolysins. Also, the deletion of *plcR* causes abolished virulence in animal model systems (Grenha et al. 2012). Thus, PlcR plays central role in controlling the virulence in *B. thuringiensis* and *B. cereus* through QS (Atkinson and Williams 2009; Hong et al. 2012; Rutherford and Bassler 2012).

Anthrax is a zoonotic disease prevalent in developing countries, and recurrent outbreaks of the disease are reported across the world. *B. anthracis* is the aetiological agent of anthrax and secretes the tripartite toxin, namely, lethal factor, oedema factor and protective antigen encoded by *lef, cya* and *pag* genes, respectively. The toxin genes are present on the pathogenic islands of extrachromosomal plasmid pXO1. Small molecular weight protein AI-2 secreted by *B. anthracis* is known to modulate the pathogenicity. Jones and co-workers have investigated the role of *luxS* in the secretion of toxins from *B. anthracis*. Using microarray, they demonstrated that *B. anthracis* LuxS regulates the AI-2-dependent toxin secretion, bacterial growth and S-layer protein expression. The concentration of autoinducer AI-2 is found to be directly proportional to the bacterial cell density and toxin secretion (Kievit and Iglewski 2000; Turovskiy et al. 2007; Jones et al. 2010).

Conclusion

Bacillus exhibits the alternative life phase, i.e. vegetative cells and spore. Furthermore, it displays temporary multicellular organism behaviour that facilitates bacteria to survive in different environmental conditions. Under nutritional stress, bacteria show matrix production and cannibalism. However, persistent environmental stress or nutritional imbalance leads to the sporulating fate of vegetative cells in the biofilms. *Bacillus* is also reported to express the toxins at critical bacterial cell density. Sporulation is a highly energy-driven process in which the active vegetative cells undergo cellular differentiation process and form a dormant structure which can be viable over years. The decision to enter sporulation is reported to be critical for bacterial cell. Therefore, to avoid the sporulation, bacterium secretes proteins that are toxic to neighbouring cells, and this process of feeding on the siblings is known as cannibalism. The bacteria utilize the degraded cell components very efficiently for delayed survival under starving conditions. Consequently, cannibalistic behaviour assists the bacterium to maintain the vegetative phase of its life cycle. Despite that, prolonged starvation leads its entry into the sporulation phase.

These major life events in *Bacillus* are temporally interdependent and are founded on precise intercellular communications in the bacterial population through its own molecular language, i.e. QS. The QS-mediated biofilm formation and consequent cannibalism behaviour are well elucidated. However, the QS modulators involved in sporulation are still not known.

Acknowledgements The authors wish to thank the director of CSIR-Institute of Genomics and Integrative Biology (IGIB), CSIR-GENCODE (BSC0123), Government of India, for providing the necessary funds and facilities. Authors are also thankful to the Academy of Scientific and Innovative Research (AcSIR), New Delhi. LKS is thankful to UGC and ND is thankful to CSIR for granting Senior Research Fellowships. We highly acknowledge Dr. V. C. Kalia and Mr. Prasun Kumar from CSIR-IGIB, Delhi, India, for their critical comments in the manuscript.

References

Aguilar C, Vlamakis H, Losick R, Kolter R (2007) Thinking about *Bacillus subtilis* as a multicellular organism. Curr Opin Microbiol 10:638–643. doi:10.1016/j.mib.2007. 09.006

Atkinson S, Williams P (2009) Quorum sensing and social networking in the microbial world. J R Soc Interface 6:959–978. doi:10.1098/rsif.2009.0203

Bassler BL, Losick R (2006) Bacterially speaking. Cell 125:237–246. doi:10.1016/j.cell.2006.04.001

Beauregard PB, Chai Y, Vlamakis H, Losick R, Kolter R (2013) *Bacillus subtilis* biofilm induction by plant polysaccharides. Proc Natl Acad Sci 110:E1621–E1630. doi:10.1073/pnas.1218984110

Boyle KE, Heilmann S, Ditmarsch DV, Xavier JB (2013) Exploiting social evolution in biofilms. Curr Opin Microbiol 16:207–212. doi:10.1016/j.mib.2013.01.003

Davey ME, O'toole GA (2000) Microbial biofilms: from ecology to molecular genetics. Microbiol Mol Biol Rev 64:847–867

Donlan RM, Costerton JW (2002) Biofilms: survival mechanisms of clinically relevant microorganisms. Clin Microbiol Rev 15:167–193. doi:10.1128/CMR.15.2.167-193.2002

Flemming HC, Wingender J (2010) The biofilm matrix. Nat Rev Microbiol 8:623–633. doi:10.1038/nrmicro2415

Flemming HC, Neu TR, Wozniak DJ (2007) The EPS matrix: "The house of biofilm cells". J Bacteriol 189:7945–7947. doi:10.1128/JB.00858-07

Fujita M, Losick R (2005) Evidence that entry into sporulation in *Bacillus subtilis* is governed by a gradual increase in the level and activity of the master regulator Spo0A. Genes Dev 19:2236–2244. doi:10.1101/gad.1335705

Grenha R, Slamti L, Nicaise M, Refes Y, Lereclus D, Nessler S (2012) Structural basis for the activation mechanism of PlcR virulence regulator by the quorum-sensing signal peptide PapR. Proc Natl Acad Sci 110:1047–1052. doi:10.1073/pnas.1213770110

Hong KW, Koh CL, Sam CK, Yin WF, Chan KG (2012) Quorum quenching revisited-from signal decays to signalling confusion. Sensors 12:4661–4696. doi:10.3390/s120404661

Jones MB, Peterson SN, Benn R, Braisted JC, Jarrahi B, Shatzkes K, Ren D, Wood TK, Blaser MJ (2010) Role of luxS in *Bacillus anthracis* growth and virulence factor expression. Virulence 1:72–83. doi:10.4161/viru.1.2.10752

Kalia VC (2013) Quorum sensing inhibitors: an overview. Biotechnol Adv 31:224–245. doi:10.1016/j.biotechadv.2012.10.004

Kalia VC, Purohit HJ (2011) Quenching the quorum sensing system: potential antibacterial drug targets. Crit Rev Microbiol 37:121–140. doi:10.3109/1040841X. 2010.532479

Kalia VC, Raju SC, Purohit HJ (2011) Genomic analysis reveals versatile organisms for quorum quenching enzymes: acyl-homoserine lactone-acylase and lactonase. Open Microbiol J 5:1–13. doi:10.2174/1874285801105010001

Kievit TR, Iglewski BH (2000) Bacterial quorum sensing in pathogenic relationships. Infect Immun 68:4839–4849. doi:10.1128/IAI.68.9.4839-4849.2000

Kostakioti M, Hadjifrangiskou M, Hultgren SJ (2013) Bacterial biofilms: development, dispersal and therapeutic strategies in the dawn of the postantibiotic era. Cold Spring Harb Perspect Med 3:a010306. doi:10.1101/cshperspect.a010306

Kumar P, Patel SKS, Lee JK, Kalia VC (2013) Extending the limits of *Bacillus* for novel biotechnological applications. Biotechnol Adv 31:1543–1561. doi:10.1016/j.biotech adv.2013.08.007

Lamsa A, Liu WT, Dorrestein PC, Pogliano K (2012) The *Bacillus subtilis* cannibalism toxin SDP collapses the proton motive force and induces autolysis. Mol Microbiol 84:486–500. doi:10.1111/j.1365-2958.2012.08038

Lemon KP, Earl AM, Vlamakis HC, Aguilar C, Kolter R (2008) Biofilm development with an emphasis on *Bacillus subtilis*. Curr Top Microbiol Immunol 322:1–16

Li YH, Tian X (2012) Quorum sensing and bacterial social interactions in biofilms. Sensors 12:2519–2538. doi:10.3390/s120302519

López D, Kolter R (2009) Extracellular signals that define distinct and coexisting cell fates in *Bacillus subtilis*. FEMS Microbiol Rev 34:134–149. doi:10.1111/j.1574-6976.2009.00199.x

López D, Vlamakis H, Losick R, Kolter R (2009) Cannibalism enhances biofilm development in *Bacillus subtilis*. Mol Microbiol 74:609–618. doi:10.1111/j.1365-2958.2009.06882.x

Mehta P, Goyal S, Long T, Bassler BL, Wingreen NS (2009) Information processing and signal integration in bacterial quorum sensing. Mol Syst Biol 5:325. doi:10.1038/msb. 2009.79

Miller MB, Bassler BL (2001) Quorum sensing in bacteria. Annu Rev Microbiol 55:165–199. doi:10.1146/annurev.micro.55.1.165

Monds RD, O'Toole GA (2009) The developmental model of microbial biofilms: ten years of a paradigm up for review. Trends Microbiol 17:73–87. doi:10.1016/j.tim.2008.11.001

Parsek MR, Greenberg EP (2005) Sociomicrobiology: the connection between quorum sensing and biofilms. Trends Microbiol 13:27–33. doi:10.1016/j.tim.2004.11.007

Rutherford ST, Bassler BL (2012) Bacterial quorum sensing: Its role in virulence and possibilities for its control. Cold Spring Harb Perspect Med 2:a012427. doi:10.1101/cshperspect.a012427

Schultz D, Wolynes PG, Ben Jacob E, Onuchic JN (2009) Deciding fate in adverse times: sporulation and competence in *Bacillus subtilis*. Proc Natl Acad Sci 106:21027–21034. doi:10.1073/pnas.0912185106

Shank EA, Kolter R (2011) Extracellular signaling and multicellularity in *Bacillus subtilis*. Curr Opin Microbiol 14:741–747. doi: 10.1016/j.mib.2011.09.016

Shank EA, Klepac-Ceraj V, Collado-Torres L, Powers GE, Losick R, Kolter R (2011) Interspecies interactions that result in *Bacillus subtilis* forming biofilms are mediated mainly by members of its own genus. Proc Natl Acad Sci 108:E1236–E1243. doi:10.1073/pnas.1103630108

Steidle A, Sigl K, Schuhegger R, Ihring A, Schmid M, Gantner S, Stoffels M, Riedel K, Givskov M, Hartmann A, Langebartels C, Eberl L (2001) Visualization of N-acylhomoserine lactone-mediated cell-cell communication between bacteria colonizing the tomato rhizosphere. Appl Environ Microbiol 67:5761–5770. doi:10.1128/AEM.67.12.5761-5770.2001

Stewart PS, Franklin MJ (2008) Physiological heterogeneity in biofilms. Nat Rev Microbiol 6:199–210. doi:10.1038/nrmicro1838

Turovskiy Y, Kashtanov D, Paskhover B, Chikindas ML (2007) Quorum sensing: fact, fiction and everything in between. Adv Appl Microbiol 62:191–234. doi:10.1016/S0065-2164(07)62007-3

Vlamakis H, Chai Y, Beauregard P, Losick R, Kolter R (2013) Sticking together: building a biofilm the *Bacillus subtilis* way. Nat Rev Microbiol 11:157–168. doi:10.1038/nrmicro2960

Waters CM, Bassler BL (2005) Quorum sensing: cell-to cell communication in bacteria. Annu Rev Cell Dev Biol 21:319–346. doi:10.1146/annurev.cellbio.21.012704.131001

Williams P, Winzer K, Chan WC, Cámara M (2007) Look who's talking: communication and quorum sensing in the bacterial world. Philos Trans R Soc Lond B Biol Sci 362:1119–1134. doi:10.1098/rstb.2007.2039

Worthington RJ, Richards JJ, Melander C (2012) Small molecule control of bacterial biofilms. Org Biomol Chem 10:7457–7474. doi:10.1039/c2ob25835h

Part III

Detectors for Quorum Sensing Signals

Quorum Sensing Biosensors

Navneet Rai, Rewa Rai, and K.V. Venkatesh

Introduction

The microbial world frequently faces abrupt changes in the surrounding environment. For survival in such a dynamic environment, it is advantageous to take decision of future action in advance at the population level. Bacteria do so by a phenomenon called quorum sensing (QS), which is sensing the presence of other bacteria in the surrounding environment and taking the decision at a certain threshold cell density. In QS active state, bacteria induce a set of target genes at population level. QS among bacteria is mediated by autoinducers (AIs), whose concentration is a function of cell density, and above a threshold concentration of cells, AIs mediate QS. There are three classes of AIs. (a) *AI-1:* chemically these are acyl homoserine lactone (AHL). These are of several types, based on their acyl side chain and substitutions at 3rd C in the acyl side chain. AHL-type quorum sensing takes place primarily in Gram-negative bacteria and is involved primarily in intraspecies signaling. (b) *AI-2:* chemically it is a furanosyl borate diester. AI-2 was first identified in marine bacterium *V. harveyi.* AI-2 is produced and recognized by both Gram-positive and Gram-negative bacteria and is involved in inter- as well as intraspecies QS. (c) *Autoinducing peptide (AIP):* Gram-positive bacteria typically use modified oligopeptide autoinducers as QS molecules. AIPs are recognized by a membrane-bound histidine kinase receptor. AIP-bound receptor activates a phosphorylation cascade that alters the activity of a DNA-binding transcriptional regulatory protein termed as a response regulator protein, which subsequently alters the expression of target genes. QS regulates a diverse range of physiological processes, including bioluminescence, virulence determinants, plasmid conjugal transfer, sporulation, competence, and biofilm formation (Rutherford and Bassler 2012).

Since QS molecules regulate several physiological processes, including virulence gene expression and biofilm formation which are directly linked to human and plants, it is important to know the nature and type of AI molecules. AI molecules are frequently detected using QS biosensor strains. During the last decade, several biosensors have been developed to detect and quantify AI molecules. Most of the biosensors are designed to sense either AI-1 or AI-2. These biosensors are efficient in detecting presence of AI(s) in bacteria, as well as in the screening of QS inhibitors (Tello et al. 2013; Anbazhagan et al. 2012). A bacterial biosensor is a genetically

N. Rai (✉)
Genome Center, University of California Davis, Davis, CA, USA
e-mail: nrai.iitb@gmail.com

R. Rai
Department of Chemistry, Indian Institute of Technology Delhi, Hauz Khas, New Delhi, India

K.V. Venkatesh
Department of Chemical Engineering, Indian Institute of Technology Bombay, Powai, Mumbai, India

V.C. Kalia (ed.), *Quorum Sensing vs Quorum Quenching: A Battle with No End in Sight*,
DOI 10.1007/978-81-322-1982-8_16, © Springer India 2015

recombinant bacterium, which harbors a QS gene circuit of interest coupled with an easily detectable and quantifiable reporter gene circuit. Reporter gene generates different classes of readable output, including fluorescence, color pigments, and luminescence. QS molecules can also be screened using several analytical methods, including HPLC and mass spectrometry (Rutherford and Bassler 2012). But, in comparison to chemical-based methods, biosensors are easy to use, inexpensive to construct, and provide real-time data in physiologically relevant environmental conditions. Most of the quorum sensing biosensors express a reporter gene from a QS-responsive promoter. This promoter gets activated when a complex of AI and QS transcriptional activator binds with the promoter. Some biosensors respond to a very specific type of AI, and some respond to a group of AIs. Regularly used biosensors are mentioned in Table 1. In following sections, we will first discuss several methods for the detection of AHLs and subsequently biosensors available for the detection of different classes of AIs.

Methods to Screen AI Producers

Several qualitative and quantitative methods have been developed during the last decade to detect and characterize AIs. Selection of suitable screening method depends on the type of study being carried out. Choice of biosensors also plays a role in screening of AHL producers. Some biosensors are very specific to a particular type of AI, and some respond to a range of AIs. So, it is recommended to use different classes of biosensors, if the physical and chemical properties of AI produced by the tester strain are unknown. In the following sections, we will discuss commonly used detection techniques, along with their merits and demerits.

T-Streak Plate Method

T-streak plate is an easy and rapid method to qualitatively screen AI producers. Here, biosensor and tester strains are streaked on agar plate forming a T shape. If a biosensor strain responds to AI produced by tester strain, it produces a particular type of visible output depending on reporter gene present in the biosensor. Biosensors like *Chromobacterium violaceum* strain CV026 and biosensors with reporter *lacZ* are compatible with this method (McClean et al. 1997). *C. violaceum* CV026 produces a visible purple pigment violacein. *lacZ* encodes enzyme β-galactosidase, which breaks X-gal into a blue color product, which can be visually detected. As it was mentioned above, T streak is rapid, easy to perform, and does not require any instrument, but it reports only the presence or absence of AI in the tester strain.

Thin-Layer Chromatography (TLC)-Agarose Overlay Method

This method is better than T streak in the fact that it reports the size and the naïvely indicates the type of AI present in the tester strain. This method is also more sensitive. Supernatants are loaded on C18-reverse phase TLC plate, and separation is performed using organic solvents because AIs (AI-1 and AI-2) are organic in nature. Along with supernatant, AI standards are also loaded on to the TLC plate. Once AIs are separated, plates are dried to remove organic solvents, and then these plates are overlaid with an agarose suspension containing a suitable biosensor strain. Biosensor strain produces visible output if AI is present. It has been noted that a combination of specific AI and specific biosensor creates a specific type of visible spots on the TLC plate, including circular and tear-shaped spots. Though TLC-based method does not pinpoint the type of AI present in the tester strain, the R_fs of samples can be calculated, which in conjunction with AI standards and shape of visible spots can be helpful in predicting the presence of specific AI in the tester strain, which can further be confirmed by mass spectrometry (Anbazhagan et al. 2012). Biosensors compatible with TLC-based method are also compatible with this method (McClean et al. 1997).

Table 1 List of biosensors for the QS signal detection

AHL biosensors						
Plasmid-based biosensors						
Plasmid	*Host strain*	*Activator*	*Promoter*	*AHL detected*	*Reporter*	*References*
pAL105	*E. coli* JLD271	LasR	P*lasI*	3-Oxo-C12-AHL	*luxCDABE*	Lindsay and Ahmer (2005)
pAL101	*E. coli* JLD271	RhlR	P*rhlI*	C4-AHL	*luxCDABE*	Lindsay and Ahmer (2005)
pCF218, pMV26	*A. tumefaciens*	TraR	P*traI*	High sensitive: 3-oxo-C6- to 3-oxo-C12-AHL Less sensitive: C4- to C12-AHL	*luxCDABE*	Chambers et al. (2005) and Sokol et al. (2003)
pSB1075	*E. coli* JM109	LasR	P*lasI*	3-Oxo-C12- to 3-oxo-C16-AHL, C12-to C16-AHL,C14:1- to C18:1-AHL, 3-oxo-C16:1- to 3-oxo-C18-AHL	*luxCDABE*	Winson et al. (1998) and Savka et al. (2011)
pSB406	*E. coli* JM109	RhlR	P*rhlI*	C4- to C12-AHL, 3-oxo-C4- to 3-oxo-C14-AHL	*luxCDABE*	Winson et al. (1998)
pSB536	*E. coli* JM109	AhyR	P*ahyI*	C4-AHL	*luxCDABE*	Swift et al. (1997)
pSB401	*E. coli* JM109	LuxR	P*luxI*	C4- to C12-AHL, 3-oxo-C4- to 3-oxo-C14-AHL	*luxCDABE*	Winson et al. (1998)
pSB403	Broad host range	LuxR	P*luxI*	C4- to C12-AHL, 3-oxo-C4- to 3-oxo-C14-AHL	*luxCDABE*	Winson et al. (1998)
pHV2001	*E. coli* VJS533	LuxR	P*luxI*	3-Oxo-C6-AHL	*luxCDABE*	Pearson et al. (1994)
pKDT17	*E. coli*	LasR	P*lasI*	3-Oxo-C12-AHL	*lacZ*	Pearson et al. (1994)
pZLR4	*A. tumefaciens* NT1	TraR	P*traI*	3-Oxo- C8-AHL	*lacZ*	Farrand et al. (2002)
pCF218, pCF372	*A. tumefaciens* WCF47	TraR	P*traI*	3-Oxo-C4- to 3-oxo-C12-AHLs, C5- to C10-AHLs	*lacZ*	Zhu et al. (1998)
pJZ384, pJZ410, pJZ372	*A. tumefaciens* KYC55	TraR	P*traI*	3-Oxo-C4- to 3-oxo-C18-AHLs, C4- to C18-AHLs	*lacZ*	Zhu et al. (2003)
pSF105, pSF107	*P. fluorescens* 1855	PhzR	P*phzA*	3-OH-C6-AHL, C6-AHL, 3-OH-C8-AHL	*lacZ*	Khan et al. (2005)
pUCP18 P*rsaL*::*lux*	*P. aeruginosa* PA14	LasR	P*rsaL*	3-Oxo-C12-AHL	*luxCDABE*	Massai et al. (2011)
pMS402 P*rsaL*::*lux*	*P. aeruginosa* PA14	LasR	P*rsaL*	3-Oxo-C12-AHL	*luxCDABE*	Massai et al. (2011)
pUCGMAT1 – 4	*E. coli*	AhlR	P*ahlI*	3-Oxo-C6-AHL	*mcherry*	Deng et al. (2014)
pREC-FF	*E. coli* MG1655 K12Z1	LuxR	P*luxI*	3-Oxo-C6-AHL	*cfp*	Rai et al. (2012)
Chromosomally integrated biosensors						
Strain		*Activator*	*Promoter*	*AHL detected*	*Reporter*	*References*
S. meliloti sinI::*lacZ*		SinR	P*sinI*	3-Oxo-C14-AHL, 3-oxo-C16-AHL, C14-AHL, C16-AHL, C16:1-AHL, 3-oxo-C16:1-AHL	*lacZ*	Llamas et al. (2004)
P. aeruginosa PA14-R3		LasR	P*rsaL*	3-Oxo-C12-AHL	*luxCDABE*	Massai et al. (2011)
C. violaceum CV026		CviR	P*vioA*	C4- to C8-AHL	*vioABCD*	McClean et al. (1997)
P. aeruginosa M71LZ		LasR	P*lasI*	3-Oxo-C12-AHL	*lacZ*	Dong et al. (2005)

Calorimetric Assay

Calorimetric assays are sensitive and can be used both for the qualitative and quantitative detections of AHLs. This method is compatible with biosensors using *lacZ* as a reporter gene. Here, a biosensor is grown in the extract of tester strain, supplemented with *ortho*-nitrophenyl-β-galactoside (ONPG). β-galactosidase, an enzymatic product of *lacZ*, converts ONPG into galactose and a yellow product, *ortho*-nitrophenol. *ortho*-nitrophenol can be quantified using a spectrophotometer at 420 nm, by performing the Miller assay. If AHL present in the extract of tester strain is known, AHL standard curve can be generated by growing biosensor in known concentrations of synthetic AHL. Standard curve can be used to calculate the concentration of AHL in the extract of tester strain. This method does not provide any information of the size and type of AHL(s) present in the extract of tester strain (Pearson et al. 1994).

Luminescence Assay

This method is compatible with biosensors using luminescence genes, *luxCDABE*, as a reporter. Luminescence is detected by a luminometer. This assay is performed in liquid extracts of tester strain containing a biosensor strain. If AI is unknown, this method will provide only qualitative information, but if AI is known, this method can be used to quantitate the AI present in extracts, by plotting a standard curve. Similar to bioluminescence method, this method also does not provide information of the size and nature of AHL, but this method is highly sensitive (Massai et al. 2011).

Fluorescence Assay

Fluorescence-based methods can be used for the detection and quantification of AIs, both at the population and single cell levels, using a fluorimeter, fluorescence microscope, and cytometer. Single cell measurements can be useful to trace out the variability in response at the single cell level. Biosensors using florescence reporters, such as *gfp*, *cfp*, are reported to be sensitive (Rai et al. 2012).

Biosensors for the Detection of AI-1 (AHL)

AHLs mediate QS in Gram-negative bacteria. AHLs are synthesized by a family of LuxI homologue proteins in several bacteria. LuxI uses appropriately charged acyl carrier protein as the major acyl chain donor and S-adenosyl methionine that provides the homoserine lactone moiety to synthesize AHL. At low cell density, bacterial cells produce basal level of LuxI, which in turn produces basal level of AHL. AHL accumulates in proportion to the increasing bacterial population. At a certain threshold concentration, AHL molecules bind to a transcriptional activator protein LuxR. AHL-LuxR complex forms dimers or multimers and subsequently activates the target genes (Rutherford and Bassler 2012). AHL biosensors principally contain similar types of sensing gene network topology: (1) a QS transcriptional activator (LuxR homologue) is expressed either from a constitutive or inducible promoter, and (2) a reporter gene is expressed from the cognate promoter of LuxR homologue. Sensing mechanism of AHL biosensors can be summarized as follows: once above the threshold concentration, a particular type of AHL binds to its transcriptional activator protein, and then this complex binds to its cognate promoter and generates a readable output (Fig. 1a). Recognition and binding of transcriptional activator to AHL is very specific and depends on the length of the acyl side chain of AHL and modification in the side chain at position C3, which may contain either hydroxyl or oxo groups. A wide range of biosensors have been designed, differing in sensitivity and specificity toward AHLs (Table 1) (Steindler and Venturi 2007).

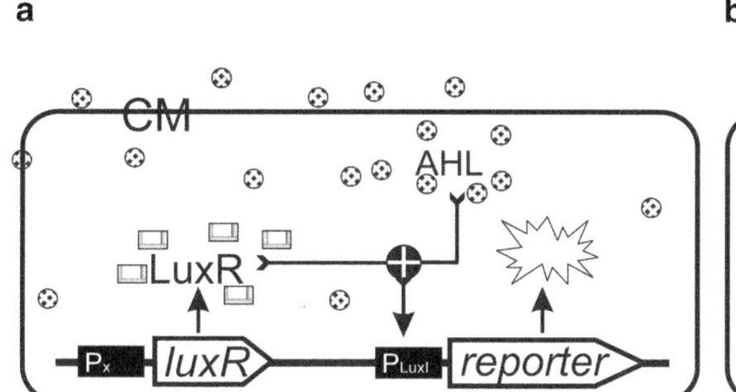

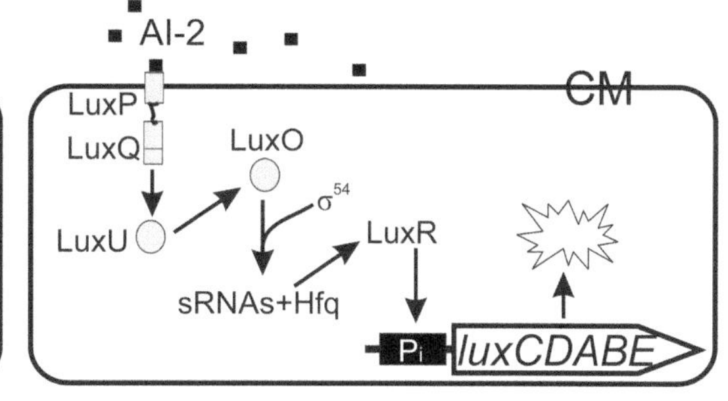

Fig. 1 Generalized signaling flow of biosensors for detecting AI-1 (AHL) and AI-2. (**a**) Design principle of biosensor to detect AI-1(AHL). (**b**) Design principle of biosensor to detect AI-2. CM, cell membrane; AHL, acyl homoserine lactone; AI-2, autoinducer-2; P_x, an arbitrary promoter for *luxR* expression; P_{LuxI} and P_i, cognate promoters of LuxR

Engineered *E. coli*-Based Biosensors

E. coli is a Gram-negative bacterium and is routinely used in labs. *E. coli* does not contain an AHL synthase gene, so is unable to make its own AHL, which makes it a suitable host for the detection of AHLs. Several plasmids with AHL sensing gene circuits have been integrated into various strains of *E. coli* and have been used to measure AHLs (Table 1). Recently, *E. coli* carrying plasmid pSB401 was used to detect and quantify AHLs present in supernatants of *Burkholderia* sp. strain A9 (Anbazhagan et al. 2012). This biosensor detects a wide range of AHLs and produces bioluminescence which is measured using a luminometer. Plasmid pSB401 was constructed by fusing *Photorhabdus luminescens* bioluminescence genes *luxCDABE* with *V. fischeri luxRI'* DNA fragment, containing a *luxR* transcriptional activator and *luxI* promoter. It can detect AHLs ranging from C4 to C14, with a different level of sensitivity. This system responded best toward its cognate AHL, i.e, C6-AHL, and least toward C4-AHL. *E. coli* harboring pSB401 produces a high dynamic range of 3log (Winson et al. 1998). In the same study, pSB406 and pSB1075 were also constructed using reporter genes *luxCDABE*. pSB406 contains *P. aeruginosa rhlRI'::luxCDABE* fusion, and pSB1075 contains *P. aeruginosa lasRI'::luxCDABE* fusion. Similar to pSB401, pSB406 and pSB1075 also detect a wide range of AHLs, generating varying levels of bioluminescence in the following order: pSB401 > pSB406 > pSB1075 (Winson et al. 1998).

Similarly, several other biosensors, by implementing *luxCDABE*, were developed to detect short- and long-chain AHLs. Plasmids pAL103 and pAL105 were developed by fusion of *luxR* and P*luxI*::*luxCDABE* and *lasR* and *lasI*::*luxCDABE*, respectively. *E. coli* carrying pAL103 was able to detect 10 nM 3-oxo-C6-AHL, and *E. coli* carrying pAL105 was able to detect 10 nM 3-oxo-C12-AHL. It has been reported that in some cases *ΔsdiA E. coli* can be a better host to detect AHLs. SdiA is a homologue of LuxR, and in some cases, it interferes with quorum sensing machinery and thus influences the output. It has been reported that SdiA influences RhlI/RhlR QS system of *Pseudomonas aeruginosa*. Lindsay and Ahmer (2005) used *sdiA* mutant *E. coli* to measure the response of plasmid pAL101 (Lindsay and Ahmer 2005). pAL101 was created by fusing *rhlR* activator with P*rhlI*::*luxCDABE*. This biosensor detects 1 μM C4- or 10 μM 3-oxo-C6-AHL. pAL101 generates 913-fold change in the output at 100 μM C4 and 53-fold change in the output at 1 μM C4-AHL. When this plasmid was expressed in *E. coli* strain containing wild-type *sdiA*, it generated only 2 to 3-fold change in response. In *E. coli* with natural *sdiA*, *sdiA* is constitutively expressed, which leads to activation of the cognate promoter even in the absence of AHL, and thus causes higher basal level expression. Basal level in wild-type *E. coli* was found to be equal to what was observed for *sdiA* mutant *E. coli* treated with 100 μM of C4-AHL (Lindsay and Ahmer 2005). Plasmid pSB536 was created using quorum sensing components *ahyR* and P*ahyI* of *Aeromonas hydrophila*. AhyR is a LuxR homologue, and P*ahyI* is a cognate *ahyI* (a *luxI* homologue) promoter. Plasmid pSB536 was constructed by fusing P*ahyI* with bioluminescence genes *luxCDABE*. This biosensor works best and as expected in *sdiA*-deleted *E. coli*. This biosensor efficiently detects C4-AHL, but provides a dynamic range only of fourfold at 100 μM C4-AHL, making it less suitable candidate for biosensing (Lindsay and Ahmer 2005; Swift et al. 1997). These studies also demonstrate the necessity of proper selection of host strain before going for any biosensing measurements.

Apart from bioluminescence-based biosensor, fluorescence-based sensors have also been created. Rai et al. (2012) created an AHL-responsive network (Rec-FF) on a high copy number plasmid pSB1A2. Rec-FF was created by fusing *V. fischeri luxI* promoter with gene encoding cyan florescent protein and *luxR* with a tetracycline promoter. *E. coli* K12Z1 strain carrying the above plasmids responded as predicted at different levels of exogenously supplied natural or synthetic C6-AHL. This plasmid was able to detect 1 μM of synthetic C6-AHL (Rai et al. 2012). Fluorescence-based biosensors have advantages over other reporter methods such as luminescence, because responses of fluorescence-based biosensors can be monitored at the single cell level, when a proper detection method is used, such as fluorescence microscope or flow cytometer, while luminescence-based detection methods reveal QS response at the population level.

Engineered *Pseudomonas* sp.-Based Biosensors

P. aeruginosa is an opportunistic human pathogen. It regulates its virulence gene expression implicating QS system. *P. aeruginosa* contains two LuxI/LuxR-type QS systems: (a) LasI/LasR and (b) RhlI/RhlR. In LasI/LasR system, LasI synthesizes N-3-oxo-dodecanoyl-homoserine lactone (OdDHL), which binds to the transcriptional activator LasR, leading to the upregulation of cognate promoters. In RhlI/RhlR system, RhlI synthesizes N-butanoyl-homoserine lactone (BHL), which then binds to RhlR, leading to the expression from cognate promoters (de Kievit et al. 1999). *Pseudomonas*-based biosensors have been constructed by mutating autoinducer synthase gene (*lasI* or *rhlI*) and expressing reporter such as *lacZ* from QS-responsive promoter. *P. aeruginosa* PAO1 M71LZ strain has been used to detect long-chain AHL such as 3-oxo-C10- and 3-oxo-C12-AHL. M71LZ is a *lasI* deletion mutant and contains a chromosomally integrated fusion of *rsaL* promoter and *lacZ* (Anbazhagan et al. 2012). Studies have demonstrated that LasI/LasR QS system directly regulates expression of *rsaL*, making this system a suitable biosensor (de Kievit et al. 1999). The presence of AHL is quantified by measuring the level of lacZ and by performing β-galactosidase assay.

Later, a biosensor strain was developed to sense primarily 3-OH-C6-AHL, by engineering PhzI/PhzR QS system of *P. fluorescens* 2–79 (Khan et al. 2005). PhzI/PhzR QS system regulates the expression of the *phzABCDEFG* operon. Products of this operon synthesize chorismate and convert it into an antimicrobial compound phenazine-1-carboxylic acid. Autoinducer synthase PhzI synthesizes six different AHLs including 3-OH-C6-AHL, 3-OH-C8-AHL, and 3-OH-C10-AHL. *P. fluorescens* 2–79-based biosensor carrying engineered PhzI/PhzR QS system is most responsive toward 3-OH-C6-AHL (Swift et al. 1997). Biosensing networks of *P. fluorescens* 2–79-based biosensor are present on two plasmids pSF105 and pSF107. Plasmid pSF105 expresses *phzR* from *trc* promoter. Plasmid pSF107 contains PhzR-regulated divergent dual promoter *phzR-phzA* (*phzR-phzA* intergenic region). In dual promoter *phzR-phzA*, transcription toward *phzR* controls expression of reporter *uidA*, while transcription toward *phzA* controls expression of reporter *lacZ*. PhzA-lacZ reporter module of pSB107 demonstrated highest sensitivity for 3-OH-C6-AHL, which was tenfold higher than 3-OH-C8-AHL and C6-AHL. Recently, *P. aeruginosa*-based sensitive biosensors, PA14-R1, PA14-R2, and PA14-R3, were developed to detect 3-oxo-C12-AHL (Massai et al. 2011). PA14-R1 is a *lasI* mutant strain of *P. aeruginosa* PA14 carrying high copy number plasmid pUCP18 P*rsaL*::*luxCDABE*. PA14-R2 is identical to PA14-R1, excluding that it carries QS-responsive gene network on low copy number plasmid pMS402. PA14-R3 is engineered *lasI* mutant *P. aeruginosa* PA14, carrying a chromosomal copy of P*rsaL*::*luxCDABE* fusion (Massai et al. 2011). All three biosensors are reported to be sensitive to 3-oxo-C12-AHL, having sensitivity in the orders of nM. In PA14-R1 and PA14-R2 bioluminescence is proportional to the log concentration of 3-oxo-C12-AHL in the range of 2 nM to 1 μM. PA14-R3 appeared to be most sensitive and had broader detection range. PA14-R3 shows proportionality between bioluminescence and 3-oxo-C12-AHL in the range of 1.4 nM to 3 μM log concentration of 3-oxo-C12-AHL. PA14-R3 is highly sensitive to 3-oxo-C12-AHL with a detection limit of 10 pM, while it also detects C4-AHL with a lower detection limit of 10 μM (Massai et al. 2011).

Engineered *Agrobacterium tumefaciens*-Based Biosensors

A. tumefaciens is a Gram-negative pathogenic bacterium. It infects plants and causes crown gall disease. During infection, a DNA fragment, called T-DNA, which is present on tumor-inducing (Ti) plasmid of bacteria, is copied and transferred into plants, where it is integrated into chromosomes of plants and causes the crown gall. Ti plasmid also contains several genes involved in the regulation and transfer of T-DNA. Regulation of Ti copy number and conjugative transfer of Ti plasmid into other bacteria are regulated by LuxI/LuxR-type quorum sensing, TraI/TraR. TraI is an autoinducer synthase and synthesizes 3-oxo-C8-AHL, and TraR is a transcriptional activator. *A. tumefaciens* is frequently used as a biosensor because it detects a wide range of AHLs with a high level of sensitivity. *A. tumefaciens*-based biosensors have been implemented with reverse phase TLC and HPLC for the qualitative detection of AHLs. Several biosensors use reporter *lacZ*, making them suitable for the quantitative measurements. *A. tumefaciens* strain NT1 carrying plasmid pZLR4 is one of the most commonly used biosensors. NT1 does not harbor Ti plasmid and thus does not produce native AHL. Plasmid pZLR4 contains *traR* and one of *tra* operons, where *traG* has been fused with reporter *lacZ* (Farrand et al. 2002).

Later, two biosensors, *A. tumefaciens* A136 (pCF218)(pMV26) and *A. tumefaciens* A136 (pCF218)(pCF372), were developed and used to detect different types of AHLs (Chambers et al. 2005; Zhu et al. 1998). Both biosensors are TraI negative and contain plasmid pCF218, which overexpresses TraR transcriptional activator. Plasmid pMV26 contains P*traI* promoter region from −143 to +68 fused with reporter *luxCDABE*. Plasmid pCF372 contains P*traI* promoter region from −143 to +359 fused with reporter *lacZ*. *A. tumefaciens* A136 (pCF218)(pCF372) was used to detect AHLs separated by TLC, while *A. tumefaciens* A136 (pCF218)(pMV26) was used to detect and quantify AHLs in the column fractions. Luminescence-based assays were performed with biosensor *A. tumefaciens* A136 (pCF218)(pMV26), and it was noticed that this biosensor is highly sensitive toward a range of AHLs including C4-AHL (25 μM), C6-AHL (250 nM), C8-AHL (0.25 nM), C10-AHL (25 nM), C12-AHL (250 nM), 3-oxo-C6-AHL (20 pM), 3-oxo-C8-AHL (0.2 pM), 3-oxo-C10-AHL (0.02 pM), and 3-oxo-C12-AHL (0.02 pM) (Chambers et al. 2005). Apart from the above-mentioned biosensors, an ultrasensitive *A. tumefaciens* KYC55 (pJZ372)(pJZ384)(pJZ410) biosensor has also been developed to detect a wide range of AHLs (Zhu et al. 2003). This biosensor was made ultrasensitive by overexpressing *traR* from T7 promoter. *A. tumefaciens* strain KYC55, which lacks Ti plasmid and consequently does not produce

TraI, was used as a host. Biosensor networks were cloned on three plasmids, pJZ372, pJZ384, and pJZ410. Plasmid pJZ384 contains a fusion of T7 promoter and traR, plasmid pJZ410 contains T7 RNA polymerase, and plasmid pJZ372 contains P*traI*::*lacZ* fusion, derived from the plasmid pCF372 (Zhu et al. 2003). This strain demonstrated better sensitivity and range compared to a closely related biosensor *A. tumefaciens* strain WCF47 (pCF218)(pCF372), where *traR* is expressed from tetracycline-inducible promoter (P*tet*) (Zhu et al. 1998). Since WCF47 uses *lacZ* reporter gene, it can detect and quantify AHLs using both TLC-based qualitative and Miller assay-based quantitative methods.

Engineered *C. violaceum*-Based Biosensors

C. violaceum is a Gram-negative bacterium. This bacterium possesses CviI/CviR-type quorum sensing and uses QS to regulate cell density-dependent production of water-insoluble violacein, a purple pigment. McClean et al. (1997) developed AHL-negative mutant, *C. violaceum* strain CV026, by mini-T5 mutagenesis in *cviI*. CV026 is unable to synthesize its own AHL. CV026 has been widely used as a biosensor to screen AHL-producing bacteria, using agar plate and thin-layer chromatography-agarose overlay methods. CV026 detects a wide range of AHLs ranging from C4- to C8-AHL. Limitation with CV026 is that this does not detect any of 3-hydroxy-AHLs. QS system of CV026 is not induced by AHL molecules with acyl side chains, ranging from C10 to C14. But these larger AHL molecules can indirectly be detected by the QS inhibition assay. Larger AHL molecules act as antagonists of smaller molecules. Fully induced CV026, in the presence of small AHLs, when is treated with large AHLs, demonstrate reduction in the level of induction (McClean et al. 1997).

In one of the latest studies, CV026 was used to detect small-chain AHLs C6-AHL and C8-AHL using TLC, in soil and air living bacterium *Burkholderia* sp. strain A9. Extracts of *Burkholderia* sp. strain A9, along with AHL standards, were loaded and separated on C18-TLC plates; when CV026 was overlaid, two spots were developed. These spots were matching with R_fs of C6-AHL and C8-AHL. Spots were further reconfirmed by Triple Quadrupole LC/MS (Chen et al. 2013).

Anbazhagan et al. (2012) screened large AHL molecules from the biofilm forming clinical isolates of *Acinetobacter* sp. by performing well-diffusion-based inhibition assay and later characterization by mass spectrometry. *Acinetobacter* sp. are multidrug-resistant, Gram-negative opportunistic bacteria causing several deadly diseases. *Acinetobacter* sp. form a biofilm, and it has been observed that the development of biofilm in several bacteria is controlled by quorum sensing. In this study, first, CV026 violacein induction assay was performed with 30 biofilms forming isolates of *Acinetobacter* sp., but none of the samples induced production of violacein indicating that *Acinetobacter* sp. lacks short-chain AHLs. Later, to screen possible presence of long-chain AHLs, CV026 violacein inhibition assay was performed on a plate where CV026 has previously been grown in the presence of synthetic short-chain AHLs. The presence of short AHLs resulted in purple colonies. When isolates were added on this plate, 7 out of 30 isolates inhibited development of purple color, indicating the presence of long-chain AHLs. Later, AHLs present in isolates were quantified by well-diffusion method. When mass spectrometry was performed with these 7 isolates, 5 isolates produced C10-AHL and the remaining two C12-AHL. This study describes the versatility of CV026, in detection of large arrays of AHLs (Anbazhagan et al. 2012).

Biosensors for the Detection of AI-2

AI-2 molecules mediate inter- and intraspecies communications, both in Gram-positive and in Gram-negative bacteria. AI-2 is synthesized in two enzymatic steps: one is catalyzed by Pfs enzyme and other by LuxS enzyme. S-adenosyl methionine is used as a methyl donor to produce adenosyl homocysteine, which is subsequently hydrolyzed by a nucleosidase, Pfs, to yield adenine and S-ribosyl homoserine. Subsequently, S-ribosyl homocysteine is converted to 4,5-dihydroxy-2,3-pentanedione (DPD), by LuxS. DPD forms a cyclic molecule, which undergoes rearrangement to form AI-2 (Rutherford and Bassler 2012).

Detection and quantification of AI-2 molecules are performed predominantly using either *Vibrio harveyi* BB170 (AI-1$^-$ and AI-2$^+$) or *V. harveyi* MM32 (AI-1$^-$ and AI-2$^-$) (Fig. 1b). Density-dependent AI-2 detection systems of *Vibrio* strains have been thoroughly studied. AI-2 signal detection cascade is comprised of a periplasmic AI-2 binding protein LuxP and a phosphorelay cascade consisting of several kinases. The first member of this phosphorelay cascade is a hybrid sensor kinase LuxQ, *which* contains N-terminal periplasmic membrane-bound sensory domain and a C-terminal intracellular response regulator domain. Signal from LuxQ is transferred to the other phosphorelay protein LuxU and then to the LuxO. LuxO belongs to highly conserved σ^{54} dependent transcriptional regulators. LuxO contains three conserved domains: (1) response regulator domain, (2) the σ^{54} activation domain, and (3) a HTH (helix-turn-helix) motif that is involved in direct DNA binding. At a low cell density, LuxQ protein autophosphorylates and signal is transferred from its conserved aspartate residue to the histidine residue of LuxU, and then LuxU phosphorylates the aspartate residue of the response regulator LuxO. Active LuxO increases the expression of the small regulatory RNAs (sRNAs). These sRNAs form a complex with a sRNA chaperone protein Hfq. This complex destabilizes the mRNA of the quorum sensing regulator molecule LuxR, resulting in inhibition of Lux operon expression. When cell density is high, AI-2 molecules present in the periplasmic space bind to the LuxP protein, converting LuxQ from kinase to phosphatase, reversing the phosphorelay pathway. Subsequently, at higher cell density sRNAs are not expressed and *luxR* mRNAs are not degraded, leading to high translation of *luxR* mRNA and giving a quorum sensing-mediated luminescence in *V. harveyi* (LaSarre and Federle 2013).

Luminescence generated by *V. harveyi* is measured using a luminometer. Briefly, in BB170 and MM32 bioassays, first of all, tester bacteria are grown in the growth media, and then bacteria are filtered out using 0.22 μM filter. Supernatant is used to measure the presence of AI-2. Different bacteria produce AI-2 at different phase of growth, so if unknown bacteria are being tested for AI-2, it is recommended to harvest the supernatant at different time points of growth. In some bacterial species like *V. harveyi*, the supernatant is stable and can be preserved at −20 °C, and in some cases like *E. coli*, it is unstable and refrigeration is not advised. Biosensors are grown overnight in AB medium at 30 °C; later these cells are diluted to 1:5000 in sterile AB medium and mixed with the supernatant of the tester strain. Ideally, nine volumes of diluted biosensor is mixed with one volume of tester supernatant and is incubated at 30 °C. Luminescence is measured at 490 nm every 30 min to 6 h. When BB170 is used as a biosensor, BB170 growing in the supernatants of *V. harveyi* BB152 (AI-1$^-$, AI-2$^+$) and *V. harveyi* MM77 (AI-1$^-$, AI-2$^-$) is always used as positive and negative controls, respectively. It should be noted that BB170 is an AI-2-positive strain and thus always produces basal level of AI-2, while MM32 is AI-2 negative, and hence MM32 senses and responds only to exogenous AI-2 (Taga and Xavier 2011).

Conclusions and Future Directions

QS enables bacteria to communicate with each other to synchronously express target genes at high cell density. Latest studies have demonstrated that QS systems of pathogens get compromised when they are treated with AI analogues, making such pathogens less infectious. So, it is important to determine if a suspected bacterium produces AI or not. Biosensors are fast and cost-effective sensing systems to detect AIs and AI analogues in surrounding environments. These are able to detect a wide range of AHLs ranging from C4 to C18. One of the limitations of biosensors is that a single biosensor cannot detect all types of AHLs, so if an unknown strain is being screened for possible AHL production, different groups of biosensors, sensing different types of AHLs, should be used. Biosensors derived from *C. violaceum* and *A. tumefaciens* are preferably used for the screening of unknown AHL producer, as these sense a broad range of AHLs. Responses of several biosensors including *V. harveyi* BB170 are influenced by the composition of growth medium and also by the genotype of the biosensor strain (Lindsay and Ahmer 2005). So, these limitations should be considered during the selection of biosensor and during interpreting screening results. Though biosensors have some drawbacks, their ease of use and versatility make them a preferable choice for the screening of new AI-producing strains. Once AI producers are screened, further characterization of AI should be done by using sophisticated techniques like mass spectrometry and HPLC.

References

Anbazhagan D, Mansor M, Yan GO, Md Yusof MY, Hassan H, Sekaran SD (2012) Detection of quorum sensing signal molecules and identification of an autoinducer synthase gene among biofilm forming clinical isolates of *Acinetobacter* spp. PLoS One 7(7):e36696

Chambers CE, Visser MB, Schwab U, Sokol PA (2005) Identification of N-acylhomoserine lactones in mucopurulent respiratory secretions from cystic fibrosis patients. FEMS Microbiol Lett 244(2):297–304

Chen JW, Koh CL, Sam CK, Yin WF, Chan KG (2013) Short chain N-acyl homoserine lactone production by soil isolate *Burkholderia* sp. strain A9. Sensors (Basel) 13(10):13217–13227

de Kievit T, Seed PC, Nezezon J, Passador L, Iglewski BH (1999) RsaL, a novel repressor of virulence gene expression in *Pseudomonas aeruginosa.* J Bacteriol 181(7):2175–2184

Deng X, Zhuang G, Ma A, Yu Q, Zhuang X (2014) Construction of a dual fluorescence whole-cell biosensor to detect N-acyl homoserine lactones. J Environ Sci 26(2):415–422

Dong YH, Zhang XF, Soo HM, Greenberg EP, Zhang LH (2005) The two-component response regulator PprB modulates quorum-sensing signal production and global gene expression in *Pseudomonas aeruginosa.* Mol Microbiol 56(5):1287–1301

Farrand SK, Qin Y, Oger P (2002) Quorum-sensing system of *Agrobacterium plasmids*: analysis and utility. Methods Enzymol 358:452–484

Khan SR, Mavrodi DV, Jog GJ, Suga H, Thomashow LS, Farrand SK (2005) Activation of the phz operon of *Pseudomonas fluorescens* 2–79 requires the LuxR homolog PhzR, N-(3-OH-Hexanoyl)-L-homoserine lactone produced by the LuxI homolog PhzI, and a cis-acting phz box. J Bacteriol 187(18):6517–6527

LaSarre B, Federle MJ (2013) Exploiting quorum sensing to confuse bacterial pathogens. Microbiol Mol Biol Rev 77(1):73–111

Lindsay A, Ahmer BM (2005) Effect of *sdiA* on biosensors of N-acylhomoserine lactones. J Bacteriol 187(14):5054–5058

Llamas I, Keshavan N, Gonzalez JE (2004) Use of *Sinorhizobium meliloti* as an indicator for specific detection of long-chain N-acyl homoserine lactones. Appl Environ Microbiol 70(6):3715–3723

Massai F, Imperi F, Quattrucci S, Zennaro E, Visca P, Leoni L (2011) A multitask biosensor for micro-volumetric detection of N-3-oxo-dodecanoyl-homoserine lactone quorum sensing signal. Biosens Bioelectron 26(8):3444–3449

McClean KH, Winson MK, Fish L, Taylor A, Chhabra SR, Camara M, Daykin M, Lamb JH, Swift S, Bycroft BW, Stewart GS, Williams P (1997) Quorum sensing and *Chromobacterium violaceum*: exploitation of violacein production and inhibition for the detection of N-acylhomoserine lactones. Microbiology 143(Pt 12):3703–3711

Pearson JP, Gray KM, Passador L, Tucker KD, Eberhard A, Iglewski BH, Greenberg EP (1994) Structure of the autoinducer required for expression of *Pseudomonas aeruginosa* virulence genes. Proc Natl Acad Sci U S A 91(1):197–201

Rai N, Anand R, Ramkumar K, Sreenivasan V, Dabholkar S, Venkatesh KV, Thattai M (2012) Prediction by promoter logic in bacterial quorum sensing. PLoS Comput Biol 8(1):e1002361

Rutherford ST, Bassler BL (2012) Bacterial quorum sensing: its role in virulence and possibilities for its control. Cold Spring Harb Perspect Med 2(11)

Savka MA, Le PT, Burr TJ (2011) LasR receptor for detection of long-chain quorum-sensing signals: identification of N-acyl-homoserine lactones encoded by the avsI locus of *Agrobacterium vitis*. Curr Microbiol 62(1):101–110

Sokol PA, Sajjan U, Visser MB, Gingues S, Forstner J, Kooi C (2003) The CepIR quorum-sensing system contributes to the virulence of *Burkholderia cenocepacia* respiratory infections. Microbiology 149(Pt 12):3649–3658

Steindler L, Venturi V (2007) Detection of quorum-sensing N-acyl homoserine lactone signal molecules by bacterial biosensors. FEMS Microbiol Lett 266(1):1–9

Swift S, Karlyshev AV, Fish L, Durant EL, Winson MK, Chhabra SR, Williams P, Macintyre S, Stewart GS (1997) Quorum sensing in *Aeromonas hydrophila* and *Aeromonas salmonicida*: identification of the LuxRI homologs AhyRI and AsaRI and their cognate N-acylhomoserine lactone signal molecules. J Bacteriol 179(17):5271–5281

Taga ME, Xavier KB (2011) Methods for analysis of bacterial autoinducer-2 production. Curr Protoc Microbiol. Chapter 1:Unit 1C.1

Tello E, Castellanos L, Duque C (2013) Synthesis of cembranoid analogues and evaluation of their potential as quorum sensing inhibitors. Bioorg Med Chem 21(1):242–256

Winson MK, Swift S, Fish L, Throup JP, Jorgensen F, Chhabra SR, Bycroft BW, Williams P, Stewart GS (1998) Construction and analysis of *luxCDABE*-based plasmid sensors for investigating N-acyl homoserine lactone-mediated quorum sensing. FEMS Microbiol Lett 163(2):185–192

Zhu J, Beaber JW, More MI, Fuqua C, Eberhard A, Winans SC (1998) Analogs of the autoinducer 3-oxooctanoyl-homoserine lactone strongly inhibit activity of the TraR protein of *Agrobacterium tumefaciens*. J Bacteriol 180(20):5398–5405

Zhu J, Chai Y, Zhong Z, Li S, Winans SC (2003) Agrobacterium bioassay strain for ultrasensitive detection of N-acylhomoserine lactone-type quorum-sensing molecules: detection of autoinducers in *Mesorhizobium huakuii*. Appl Environ Microbiol 69(11): 6949–6953

Caenorhabditis elegans as an In Vivo Non-Mammalian Model System to Study Quorum Sensing in Pathogens

Sajal Sarabhai, Neena Capalash, and Prince Sharma

Introduction

One of the most alarming problems in the current times is the continuous emergence of multiple drug resistance in the most pathogenic bacterial species (Dzidic and Bedekovic 2003). Even the World Health Organization has regarded antibiotic resistance as the major challenge of the future. Since 1980, no new drug has been discovered, and bacteria are becoming resistant to all the currently available antibiotics. Therefore, alternative methods to combat the bacterial sternness are being search continuously. In 1970, Nealson et al. identified a cell-density-dependent expression of bioluminescence in *Vibrio fischeri* which lives in the photophore (or light-producing organ) of the Hawaiian bobtail squid as the mutualistic symbiont. The mechanism was termed as quorum sensing (QS) comprising of signalling molecules (autoinducers or pheromones) which are produced in a cell-density-dependent fashion. At a certain threshold concentration, signalling molecules interact with their cognate receptors and regulate the transcription of target genes. In simpler words, bacteria require a minimum number of cells or "quorum" for gene expression and could "sense" this density by the levels of signal molecules. Later on, quorum sensing was found to exists in both bacterial (Gram positive and negative) and fungal species. The genes which fall under the regulation of quorum sensing are specific to the microbial species, e.g. in case of pathogenic bacteria, QS genes regulate the expression of virulence genes, while in environmental species, genes required for nutrients uptake are expressed (Smith and Iglewski 2003).

The first quorum-sensing inhibitor was identified from the marine macroalga *Delisea pulchra*, which produces halogenated furanones and was able to inhibit quorum-sensing-regulated colonization of marine bacterial species on its surface (Rasmussen et al. 2000). Under laboratory conditions, when halogenated furanone was tested against *E. coli* with *luxR* construct of *V. fischeri*, there was accelerated degradation of transcriptional regulator LuxR protein on binding with halogenated furanone (Manefield et al. 2002). Also, halogenated furanone was able to inhibit the expression of PluxI-GFP in *E. coli*. As quorum sensing contributes significantly to virulence gene expression in pathogenic bacteria, therefore interference in the cell–cell communication has emerged as an attractive strategy to battle the infections.

S. Sarabhai • P. Sharma (✉)
Department of Microbiology, Panjab University, Chandigarh 160014, India
e-mail: princess@pu.ac.in

N. Capalash
Department of Biotechnology, Panjab University, Chandigarh 160014, India

V.C. Kalia (ed.), *Quorum Sensing vs Quorum Quenching: A Battle with No End in Sight*,
DOI 10.1007/978-81-322-1982-8_17, © Springer India 2015

Importance of an In Vivo Model System to Study the Effectiveness of Quorum-Sensing Inhibition

To understand the host–pathogen interactions at both cellular and molecular level, an appropriate in vivo model is required. In vitro model is used simply to describe the effect of QS inhibitors (QSI) on a certain bacterial population and to collect data for proving the hypothesis. The precise effect of the QSI at the site of infection can be analysed only in an in vivo model system, specially developed to study virulence of a particular bacterium. In most of the in vitro studies, dealing with host pathogenesis, we ignore the immune responses of the host evoked during bacterial invasion. Also, the toxicity, effective dose and metabolism of the QSIs are not even dealt with, which is disadvantageous when dealing with its commercial value. For the last 10–15 years, more than 100 synthetic and natural compounds with QSI activity have been identified. However, their efficacy in vivo still remains a question. Most of the QS inhibitors either have not been tested in in vivo model or their efficacy were not as good as observed by in vitro analysis. Garlic has undergone clinical trials in cystic fibrosis patients with *P. aeruginosa* infections but was not as effective as in in vitro analysis and in vivo burnt mouse model (Smyth et al. 2010). Quorum-sensing-regulated pathogenesis models have been very well established in both invertebrate (e.g. *Drosophila melanogaster*, *C. elegans*, *Galleria mellonella*) and vertebrate model systems (rodents, mice, guinea pig). This chapter will focus on using an invertebrate model *C. elegans* to study and validate the efficacy of QS inhibitors in vivo.

C. elegans

The nematode *C. elegans* was isolated from mushroom compost near Bristol (designated as N2 strain), England, by L. N. Stainland and was further cultured by W. L. Nicholas. Subsequently, the strain was identified by Gunther and Nigon and was given to Brenner in 1965 who later developed *C. elegans* as a model to study cell cycle.

C. elegans is a eukaryotic, multicellular 1 mm long and self-reproducing hermaphrodite with a life cycle of approximately 3.5 days at 20 °C. It lends itself for easy culture and propagation on *E. coli* strain OP50 as feed. It is able to get internally fertilized and lays eggs through its uterus. Under favourable environmental conditions, hatched larvae develop through four stages designated as L1–L4 (Fig. 1), and hundreds of progeny can be generated rapidly in petri dishes at 25 °C. When conditions are stressful, e.g. in case of food insufficiency or high temperature, *C. elegans* enters an alternative third larval stage called the dauer state. Dauer larvae are stress resistant and do not age. This is the preferred stage to store worms in liquid nitrogen or deep freezer for long time. The genome is 9.7×10^7 bp in size with gene density of 1gene/5 kb. It has a small chromosome number (2n = 6 named I, II, III, IV, V and X) and a mitochondrial genome (Corsi 2006).

C. elegans Pathogenesis Model: An Edge over Others

C. elegans has emerged as a cost-effective, amenable, genetically tractable in vivo model. It proved to be effective in studying various biological processes including apoptosis, cell signalling, cell cycle, cell polarity, gene regulation, metabolism, ageing and sex determination and host–microbial pathogenesis interaction studies (Riddle et al. 1997). As compared to other in vivo model systems, *C. elegans* is relatively easy to handle and manipulate with panoply of genetic and molecular tools available for its study (**see at the end, *C. elegans* information databases**). Caenorhabditis Genetic Centre (CGC) in Minnesota, USA, encourages scientists to use *C. elegans* for in vivo research. There is no ethical clearance involved in using *C. elegans* as in vivo model. Therefore, it has an edge over other higher vertebrate in vivo model systems. *C. elegans* reproduces rapidly (small generation time) which allows drug

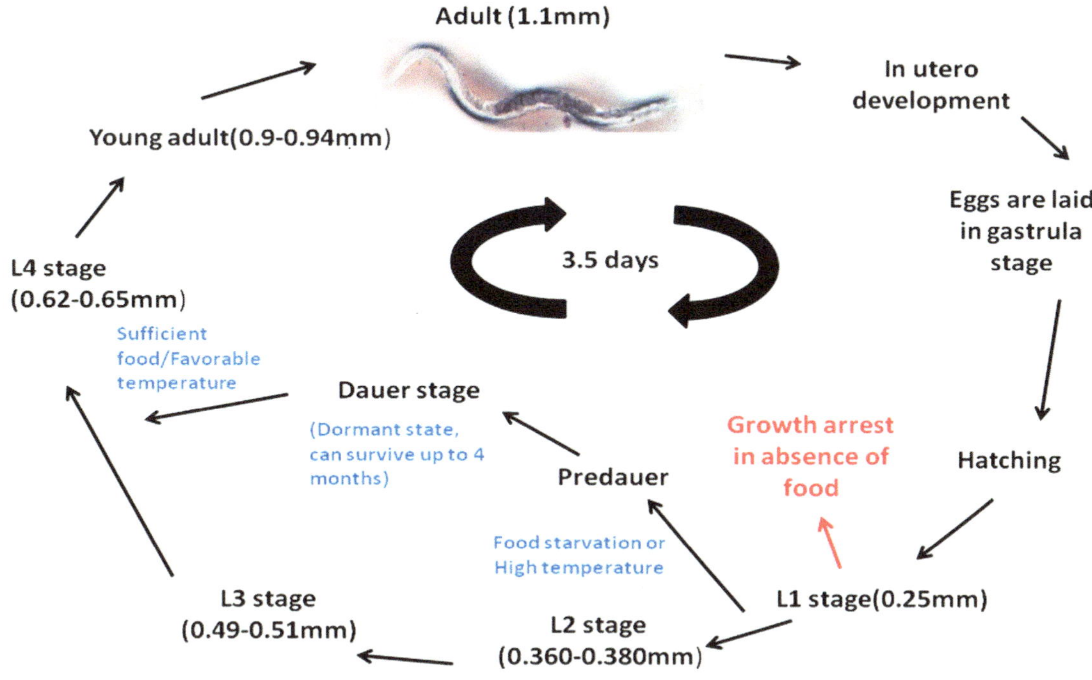

Fig. 1 Life cycle of *C. elegans* N2 at 21 °C

efficacy to be studied easily, while in higher vertebrates like mouse model, it takes fairly long time to reproduce which limits the study. Being small in size, hundreds of worms can fit in the 96-well microtitre plate, and scoring can be easily done by checking the viability of worm. This provides an easy high-throughput screening which is impossible in higher vertebrate models (Moy et al. 2009).

The transparent body of *C. elegans* allows visualization of fluorescence-tagged molecules and proteins, and their fate can be determined. The delivery of anti-infective compounds in the model system offers a major challenge when pharmacodynamic properties are to be evaluated. Delivery in *C. elegans* is simple. Anti-infective compounds are either ingested by worms or absorbed through the skin or sensory neuronal endings and spread throughout the body after absorption from intestinal cavity. As almost any type of microbe can be ingested by *C. elegans*, therefore, it has emerged successfully as a model to study interactions and interference during host–pathogen coexistence. A comparison between *C. elegans* and other model system is given in Table 1.

C. elegans as a Model to Study Quorum Sensing (QS) and QS Inhibition (QSI) in Microbial Pathogens

Host–pathogen interaction studies are well established in *C. elegans* involving both bacterial and fungal species. In case of bacteria, both Gram-positive and Gram-negative species utilize QS for colonization and pathogenesis in worms. Some of the important pathogens causing disease in *C. elegans* and their survival by anti-infectives are described below.

Pseudomonas aeruginosa

P. aeruginosa is a soil bacterium that is pathogenic to both plants and animals. It is the common part of the human microflora, but when immune system gets compromised during traumatic surgeries, burns and chemotherapy, it causes lethal infections (Oncul et al. 2009; Senturk et al. 2012). It accounts for 80 %

Table 1 A comparison of different in vivo model systems

Parameters	*D. discoideum*	*C. elegans*	*D. melanogaster*	*G. mellonella*	*Bombyx mori*	*Danio rerio*	*Mus musculus*	Cell lines
Size	2–4 mm	1 mm	2.5 mm	2 cm	5 cm	6.4 cm	10 cm	0.1 mm
Favourable temperature	21–25 °C	15–25 °C	18–29 °C	25–37 °C	27 °C	29 °C	37 °C	37 °C
Generation time	12 h	4 days	10 days	30 days	40–60 days	3–4 months	10 weeks	24–48 h
Handling	Very easy	Very easy	Very easy	Easy	Easy	Easy	Difficult	Easy
Laboratory space	Minor	Minor	Minor	Minor	Minor	Minor	Major	Minor
Maintenance cost	Low	Low	Low	Low	Low	Low	High	High
High-throughput screening	Yes	Yes	Yes	Yes	Yes	Yes	No	Yes
Experimental results	Days	Days	Days	Days	Days	Days	Months	Days
Innate immunity	Absent	Present	Present	Present	Present	Present	Present	Absent
Adaptive immunity	Absent	Absent	Absent	Absent	Absent	Present	Present	Absent
Clinical relevance	Potential	Potential	Potential	Potential	Potential	Confirmed	Confirmed	Potential
Ethical clearance	Not required	Not required	Not required	Not required	Not required	Required	Required	Not required
Genome matching	40 %	40–60 %	60 %	60 %	60 %	60–80 %	75–90 %	100 %
Genetic malleability	Very easy	Very easy	Very easy	Easy	Easy	Difficult	Very Difficult	Difficult
Quorum quenching experimentation	Very easy	Very easy	Very easy	Easy	Easy	Difficult	Difficult	Easy

deaths in cystic fibrosis patients, and the clinical management is difficult due to multiple drug resistance. Attenuation of *P. aeruginosa* virulence by interfering with QS proved to be promising therapy as studied through in vitro experimentation (O'Loughlin et al. 2013). Besides using conventional in vivo model systems, the *P. aeruginosa* pathogenesis model was also established in *C. elegans*, and depending on the bacterial strain and culture conditions, a total of five types of worm killing (four agar-based and one liquid-based killing) are reported. *P. aeruginosa* strain P14 was first identified by Ausubel and his co-workers in 1997. It is a clinical strain and is able to cause disease in both plants and animals using shared subsets of virulence factors. P14 strain showed two types of *C. elegans* killing, namely, "fast killing" and "slow killing". When grown on high-osmolarity, nutrient-rich (glucose, peptone, sorbitol) medium, P14 releases certain diffusible exoproducts (Mahajan-Miklos et al. 1999) which cause fast killing of worms. Approximately, 50 % of total *C. elegans* population (LT_{50}) died within 4 h of bacterial exposure. During mutagenesis analysis for finding the reason behind the worm killing, seven Tn*phoA* mutants failed to cause fast killing. The genes disrupted in these mutants were involved in synthesis of phenazine molecules which lie under regulation of quorum sensing in *P. aeruginosa*. Three types of phenazines are reported in P14, namely, 1-hydroxyphenazine, phenazine-1-carboxylic acid and pyocyanin, and their toxicity to *C. elegans* depends on the pH of the medium (Cezairliyan et al. 2013). Phenazine-1-carboxylic acid is the major phenazine responsible for fast killing, while pyocyanin and 1-hydroxyphenazine are produced in low concentration by P14, insufficient to kill worms. Strain PAO1 secretes toxic metabolites when grown on brain–heart infusion medium which causes cessation of pharyngeal pumping in *C. elegans*. Worms become paralyzed, fail to move and die within 2–4 h, a phenomenon termed as paralytic killing. A quorum-sensing mutant of PAO1 was unable to cause paralytic killing of *C. elegans*, and pyocyanin was found to be the responsible candidate (Darby et al. 1999). Later on Gallagher and Manoil (2001) proposed cyanide (*hcn*) as the sole molecule responsible for paralytic killing. *P. aeruginosa* produces cyanide maximally in the late-exponential and early-stationary phases, and the production is known to be under QS regulation. The addition of exogenous hydrogen cyanide alone showed similar death kinetics in *C. elegans* as that of wild-type *P. aeruginosa* strain. Cyanide, being a known inhibitor of cytochrome oxidase, shuts down the mitochondrial respiration in neuromuscular tissues and causes rapid and dramatic paralytic killing of nematodes by *P. aeruginosa* PAO1. However, the role of cyanide in human infections is still not known.

"Slow killing" in *C. elegans* is mediated by both PAO1 and P14 strains when grown in minimal medium. Bacteria colonize the intestine of worm, and infections develop within 12 h which is also visualized by accumulation of GFP-tagged bacterial population. Continuous exposure of *C. elegans* to *P. aeruginosa* results in cessation of pharyngeal pumping, and worms became immobile and die. L4-stage worms are more susceptible than adults, and LT_{50} was 38 and 48 h, respectively. Slow-killing defective mutants (*gacA*$^-$, *lasR*$^-$ and *lemA*$^-$) of *P. aeruginosa* were unable to colonize/proliferate in worm gut. Since LasR and GacA are involved in QS, this suggests that QS plays role in slow killing of worms (Tan et al. 1999).

Exposure of *P. aeruginosa* P14 to aqueous extracts of *Conocarpus erectus*, *Callistemon viminalis* and *Bucida buceras* effectively reduced both the fast and slow killing of *C. elegans* (Adonizio et al. 2008). Similarly, these extracts were able to reduce paralytic killing of worms infected with PAO1. As these plant extracts showed reduced production of QS-regulated extracellular virulence factors in *P. aeruginosa*, hence, increase in worm survival was observed. Similarly, Rudrappa and Bais (2008) showed reduction in worm killing when fed on bacteria treated with 3 μg/ml of QSI curcumin. *C. elegans* has remarkable chemosensory network, and due to this there could be a possibility that worms did not ingest bacteria exposed to curcumin. However, on

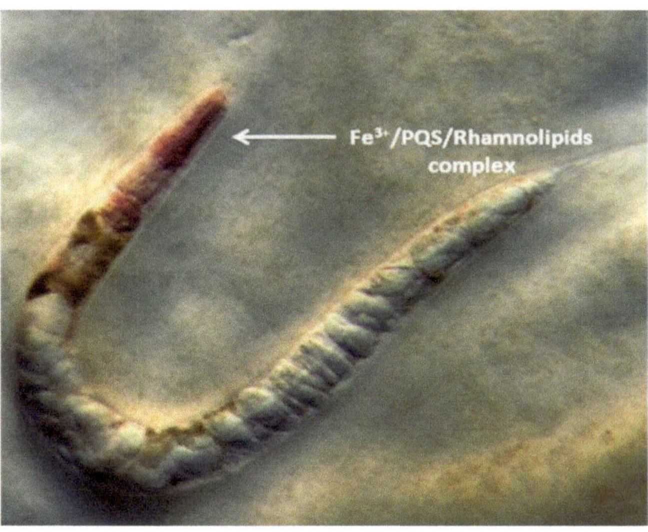

Fig. 2 Red death in *C. elegans* (Courtesy of Dr. Alverdy; http://www.newswise.com/articles/research-could-lead-to-new-non-antibiotic-drugs-to-counter-hospital-infections)

determining the total bacterial load in the worm gut, the number was found to be same when fed on either curcumin-treated or curcumin-untreated PAO1. The anti-infective potential of clove oil (1.6 % v/v) (Husain et al. 2013), AHL analogue meta-bromo-thiolactone (O'Loughlin et al. 2013), ellagic acid derivatives of *Terminalia chebula* (Sarabhai et al. 2013), furanone (Maeda et al. 2012), garlic extracts and 4-nitropyridine-N-oxide (Rasmussen et al. 2005) was found to be effective in rescuing *C. elegans* killing by *P. aeruginosa* due to their QS inhibitory potential. Another approach for QS inhibition is the use of signal-degrading enzymes, namely, acylases and lactonases which degrade N-acyl homoserine lactone molecules in Gram-negative bacteria. These enzymes were effective in reducing the QS-regulated virulence gene expression of *P. aeruginosa* in vitro. When worms were allowed to feed on acylase (PvdQ) overproducing *P. aeruginosa* PAO1, reduction in paralytic killing was observed as compared to their wild type. Same was also seen in case where purified acylase was added to the *C. elegans–P. aeruginosa* killing model (Papaioannou et al. 2009).

Besides being the primary lung pathogen, *P. aeruginosa* is also detected in the intestines of 20 % populations. Zaborin et al. (2009) observed red colour in the worm gut infected with PAO1 and called it "red death". The medium used for PAO1 growth was phosphate depleted which mimicked physiological state of human intestines during injuries (Murono et al. 2003). In PAO1, depletion of phosphate induces PhoB-regulated MvfR expression which consequently enhances production of QS signal molecules, especially 2-heptyl-3-hydroxy-4-quinolone (PQS). Increased PQS molecules chelate extracellular iron which in turn increases pyoverdin production in PAO1. PQS forms red complex with Fe^{3+} in the intestinal tract of *C. elegans* and is responsible for their red death (Fig. 2).

"Liquid killing" of *C. elegans* reveals altogether a new array of virulence factors responsible for worm death by *P. aeruginosa* which were different from the classic pathways. In liquid minimal medium, there is competition between the Fe^{3+} acquisition by both worms and bacteria. PAO1 secretes pyoverdin which chelates all the available Fe^{3+}. This stabilizes the HIF-1, a transcriptional factor which induces hypoxic response in *C. elegans* and eventually results in death of worms (Epstein et al. 2001). Pyoverdin also stimulates production of proteases and exotoxins which assist in pathogenesis of PAO1 in worms (Kirienko et al. 2013). Table 2 summarizes the resemblance of *C. elegans–P. aeruginosa* killing models with the mammalian infections.

In both "red death" and "liquid killing" *C. elegans–P. aeruginosa* pathogenesis model systems, quorum sensing plays an important role in

Table 2 *C. elegans–P. aeruginosa* model system mimicking human infections

Killing models	Worm killing	Infection sate in humans	Reference
Slow killing	QS-regulated virulence factors like pyocyanin, proteases, elastase, etc., required by *P. aeruginosa* for successful infection in *C. elegans*	Same virulence factors are required by *P. aeruginosa* for successful in humans leading to nosocomial infections	Tan et al. (1999)
Fast killing	Phenazine-1-carboxylic acid is the major toxin released by *P. aeruginosa* P14 causing worm death	Phenanzine-1-carboxylic acid caused toxicity in cystic fibrosis patients and inflammation in acute wounds infection which gets exaggerated under acidic condition	Mahajan-Miklos et al. (1999), Cezairliyan et al. (2013)
Lethal paralysis	Cyanide (HCN) produced under quorum-sensing regulation, causes paralysis of neuromuscular tissues leading to asphyxia and worm death	The role cyanide in *P. aeruginosa* infection is still not clear, but it helps bacteria to persist and dominate in human infection site	Darby et al. (1999)
Red death	Depletion of phosphate induces *Pho* regulon which perhaps enhance the production of phenazines forming a red coloured complex in worm intestine	Phosphate depletion occurred during major traumatic surgeries creating a perfect environment for commensal *P. aeruginosa* to become virulent	Zaborin et al. (2009)
Liquid killing	Depletion of iron liquid medium induces various hypoxic responses which causes worn death	Microaerophilic condition in lungs of cystic fibrosis patients mimicked the exact condition. It induces various inflammatory responses which damages host cells	Kirienko et al. (2013)

the development of disease. As both the models are in the initial stages of study, no QS inhibitors have been tested as yet for their efficacy in reducing the worm killing.

Staphylococcus aureus

S. aureus is a Gram-positive bacterium causing a variety of diseases including skin infections, food poisoning, toxic shock syndrome and septic shock. Nosocomial infections with methicillin-resistant *S. aureus* (MRSA) strains are the most serious risk factors associated with hospitalization (Lowry 1998). *S. aureus* colonizes intestine of the worm assisted by various extracellular virulence factors expressed under QS regulation. Mutation of the quorum-sensing global virulence regulators *agr* and *sarA* of *S. aureus* reduced worm killing. *S. aureus* is capable of infecting all the developmental stages (L1–L4 and adults) of *C. elegans*. Exposure of worm to bacteria for only 8 h is sufficient for causing infection (Sifri et al. 2003).

Other virulence determinants like α-hemolysin and V-proteases are important for pathogenesis in both human host and *C. elegans*. *S. aureus* induces *dar* phenotype (deformed anal region) in *C. elegans* which is dependent on β-catenin and ERK-MAPK signalling system. In higher vertebrates, β-catenin activates NF-κB-mediated immune responses. Although NF-κB is not present in *C. elegans* system, its downstream genes are well conserved. Hence, *C. elegans* model can be very well studied for pathogenesis development and resistance development against *S. aureus*. Anti-infective potential of 37 natural extracts and 29 synthetic compounds against *S. aureus* pathogenicity to *C. elegans* was assessed by liquid killing assay. Approximately, 14 natural and synthetic compounds were able to enhance the survival of worms infected with *S. aureus* by 2.4–4 fold. However, these anti-infective compounds have not been tested for their anti-QS activity, but these compounds could be further tested for the mechanism and toxicity through which they are able to attenuate virulence of *S. aureus* in *C. elegans* (Kong et al. 2014).

Burkholderia cenocepacia

B. cenocepacia is the secondary inhabitant of cystic fibrosis lungs and plays a major role in clinical outcomes of nosocomial infections. Two types of killing have been reported for *B. cenocepacia* in *C. elegans*. Slow killing occurs in worms fed on bacteria growing in standard minimal medium. Death occurred due to gut colonization followed by infection. When bacterium was allowed to grow under high-osmolarity conditions, worms were killed within 24 h with no detectable bacterium in the gut. Conditioned medium with diffusible products of *B. cenocepacia* was able to kill worm within 4 h. However, the nature of the toxin has not been detected. Both types of worm killing by *B. cenocepacia* required *cep* quorum-sensing regulatory system (Kothe et al. 2003). An effective QS inhibitor reduced the killing of worms infected with *B. cenocepacia*. The heterologous expression of QS signal-degrading enzyme (*aiiA* lactonase) of *Bacillus* sp. strain 240B1 in *B. cepacia* attenuated bacterial virulence. Even the levels of N-acyl homoserine lactone signalling molecules were less as compared to wild type. Combination of antibiotic (tobramycin) and QS inhibitor (baicalin hydrate) enhanced the worm survival significantly ($p < 0.01$) fed on *B. cenocepacia* strains LMG18828 and LMG 16656. Also the combinatorial action of antibiotic and QSI was more effective in rescuing worm killing than either of them used alone (Brackman et al. 2011).

Escherichia coli

E. coli OP50 strain is the food for *C. elegans* N2 and is used for the propagation of worms under laboratory conditions. It is completely digested by the worm and shows no gut colonization (Aballay et al. 2000). Another strain of *E. coli*, i.e. O157:H7, is a causative agent of hemorrhagic colitis and hemolytic uremic syndrome (HUS), and its virulence expression is regulated via quorum-sensing (QS) signalling pathway (Sperandio et al. 2003). Broccoli extract (BE) suppressed autoinducer signalling synthesis and bacterial motility in *E. coli* O157:H7 in vitro. Worms fed on *E. coli* strain O157:H7 developed gut infection leading to death. However, BE treatment of bacterial strain attenuated its virulence in vivo and worms survived longer than those fed solely on the pathogenic bacteria (Lee et al. 2011), indicating the therapeutic potential of broccoli in controlling infections.

Candida albicans

It is a commensal fungus found on skin and mucosal surfaces of virtually all humans and causes candidiasis as a hospital-acquired infection. Both the yeast and hyphal forms of fungus are capable of killing worms where the former colonizes the surfaces and facilitate dissemination while the latter causes host invasion and tissue destruction. Hyphae formation is essential for the expression of fungal virulence, and the switch over of yeast cell to hypha and vice versa depends on subtle levels of QS signalling molecules, namely, tyrosol and farnesol. Tyrosol initiates while farnesol inhibits filamentation (Chen et al. 2004; Albuquerque and Casadevall 2012). To date, a direct role of these QS signals in *C. elegans* killing has not been established, but as filamentation is important for host pathogenesis, therefore, these QS signals may be playing certain role in *C. elegans* killing. Filastatin, an antifungal agent, reduced worm killing by inhibiting morphogenesis of *C. albicans*, a plausible indication of the importance of fungal morphology (Fazly et al. 2013). *C. elegans*–*C. albicans* pathogenesis model has been used to study coinfections. *C. elegans* infected with *C. albicans* showed inhibition in hyphal growth when coinfected with *A. baumannii* or *P. aeruginosa* (Peleg et al. 2008) which may be due to interactions between the QS signalling molecules. This model mimicked the polymicrobial interaction during coinfection in mammalian host tissues. *C. elegans* has also been used for high-throughput screening of the antifungal compounds from the chemical library of 3,228 compounds. Twelve compounds were found to be effective in reducing *C. elegans* killing by *C. albicans* in liquid medium. The model also helped in assessing the relative minimum inhibitory concentrations, effective concentration in vivo and toxicity of tested compounds (Okoli et al. 2009).

Significance and Future Prospects

C. elegans is susceptible to a wide range of bacteria and fungi, and the key phenotype for successful microbial pathogenesis is worm killing. Quorum sensing being the primary regulator of microbial pathogenesis in *C. elegans*, a novel QS inhibitor, could be easily screened as the one which is able to reduce worm killing. These high-throughput screening assays using *C. elegans* virtually bridge the gap that intrinsically exist between in vitro and in vivo approaches providing physiologically relevant data derived from whole animal setting. The ease of genetic malleability of *C. elegans* (e.g. RNAi silencing) has provided a tool to identify the plausible drug targets during host–microbial interactions. The virulence factors of *P. aeruginosa* required for pathogenicity in burnt mouse model and worms are same and lie under QS regulation. Therefore, it would be cost-effective to validate QS inhibitory potential of compounds using *C. elegans* rather than going to higher vertebrates. However, *C. elegans* models do suffer from certain limitations that must be considered during experimentation. First, although 60–80 % genes are homologous in worms and human genome, *C. elegans* fails to represent exact physiology of host–microbe interactions as encountered by humans. However, even the best vertebrate model like mouse cannot predict the safety and exact target of anti-infective compounds. Secondly, *C. elegans* lacks cell-mediated immune system as found in humans and depends solely on the secretions of antimicrobial peptides. This also represents the major limitation in accessing the therapeutic potential of the tested compounds. Despite some critical disadvantages, *C. elegans* model provides enormous wealth of genetic resources available for the study. *C. elegans*, being microscopically small in size, is becoming one of the most popular emerging in vivo model systems for testing quorum-sensing inhibitory potential of the compounds with results comparable to higher model organisms. In the near future, *C. elegans* will definitely become a popular model to study the trickiest pathways in human–pathogen interaction studies and eventually the ways to inhibit this interaction.

C. elegans information databases

Databases	Link	Information
WormBase	www.wormbase.org	*C. elegans* genome
Textpresso	www.textpresso.org	*C. elegans* literature
ORFeome	http://worfdb.dfci.harvard.edu	Predicted ORFs
Worm Interactome	http://elegans.bcgsc.ca/home/sage.html	Gene expression during development along with FACS sorted cells and tissues
SAGE data	http://elegans.bcgsc.ca/home/sage.hmtl	Gene expression during development along with FACS sorted cells and tissues
Structural genomics	http://sgce.cbse.uab.edu/index.php	Protein expression, crystallization, structure modelling
WormAtlas	http://www.wormatlas.org	Behavioural anatomical literature
EST by Yuji Kohara	http://www.ddbj.nig.ac.jp	EST in *C. elegans* genome

References

Aballay A, Yorgey P, Ausubel FM (2000) *Salmonella typhimurium* proliferates and establishes a persistent infection in the intestine of *Caenorhabditis elegans*. Curr Biol 10:1539–1542. doi:10.1016/S0960-9822(00)00830-7

Adonizio A, Ausubel FM, Mathee K (2008) Attenuation of *Pseudomonas aeruginosa* virulence by medicinal plants in a *Caenorhabditis elegans* model system. J Med Microbiol 57:809–813. doi:10.1099/jmm.0.47802-0

Albuquerque P, Casadevall A (2012) Quorum sensing in fungi – a review. Med Mycol 50:337–345. doi:10.3109/13693786.2011.652201

Ausubel FM, Rahme LG, Tan MW, Le L, Wong SM, Tompkins RG, Calderwood SB (1997) Use of model plant hosts to identify *Pseudomonas aeruginosa* virulence factors. Proc Natl Acad Sci USA 94: 13245–13250

Brackman G, Cos P, Maes L, Nelis HJ, Coenye T (2011) Quorum sensing inhibitors increase the susceptibility of bacterial biofilms to antibiotics *in vitro* and *in vivo*. Antimicrob Agents Chemother 55:2655–2661. doi:10.1128/AAC.00045-11

Cezairliyan B, Vinayavekhin N, Grenfell-Lee D, Yuen GJ, Saghatelian A, Ausuble FM (2013) Identification of *Pseudomonas aeruginosa* phenazines that kill *Caenorhabditis elegans*. PLoS Pathog 9:e1003101. doi:10.1371/journal.ppat.1003101

Chen H, Fujita M, Feng Q, Clardy J, Fink GR (2004) Tyrosol is a quorum-sensing molecule in *Candida albicans*. Proc Natl Acad Sci USA 101:5048–5052. doi:10.1073/pnas.0401416101

Corsi AK (2006) A biochemists guide to *C. elegans*. Anal Biochem 359:1–17. doi:10.1016/j.ab.2006.07.033

Darby C, Cosma CL, Thomas JH, Manoil C (1999) Lethal paralysis of *Caenorhabditis elegans* by *Pseudomonas aeruginosa*. Proc Natl Acad Sci U S A 96:15202–15207. doi:10.1073/pnas.96.26.15202

Dzidic S, Bedekovic V (2003) Horizontal gene transfer-emerging multidrug resistance in hospital bacteria. Acta Pharmacol Sin 24:519–526

Epstein AC, Gleadle JM, McNeill LA, Hewitson KS, O'Rourke J, Mole DR, Mukherji M, Metzen E, Wilson MI, Dhanda A, Tian YM, Masson N, Hamilton DL, Jaakkola P, Barstead R, Hodgkin J, Maxwell PH, Pugh CW, Schofield CJ, Ratcliffe PJ (2001) *C. elegans* EGL-9 and mammalian homologs define a family of dioxygenases that regulate HIF by prolyl hydroxylation. Cell 107:43–54. doi:10.1016/S0092-8674(01)00507-4

Fazly A, Jain C, Dehner AC, Issi L, Lilly EA, Ali A, Cao H, Fidel PL Jr, Rao RP, Kaufman PD (2013) Chemical screening identifies filastatin, a small molecule inhibitor of *Candida albicans* adhesion, morphogenesis, and pathogenesis. Proc Natl Acad Sci USA 110:13594–13599. doi:10.1073/pnas.1305982110

Gallagher LA, Manoil C (2001) *Pseudomonas aeruginosa* PAO1 kills *Caenorhabditis elegans* by cyanide poisoning. J Bacteriol 183:6207–6214. doi:10.1128/JB.183.21.6207-6214.2001

Husain FM, Ahmad I, Asif M, Tahseen Q (2013) Influence of clove oil on certain quorum-sensing-regulated functions and biofilm of *Pseudomonas aeruginosa* and *Aeromonas hydrophila*. J Biosci 38:835–844. doi:10.1007/s12038-013-9385-9

Kirienko NV, Kirienko DR, Larkins-Ford J, Wählby C, Ruvkun G, Ausubel FM (2013) *Pseudomonas aeruginosa* disrupts *Caenorhabditis elegans* iron homeostasis, causing a hypoxic response and death. Cell Host Microbe 13:406–416. doi:10.1016/j.chom.2013.03.003

Kong C, Yehye WA, Rahman NA, Tan MW, Nathan S (2014) Discovery of potential anti-infectives against *Staphylococcus aureus* using a *Caenorhabditis elegans* infection model. BMC Complement Altern Med 14:4. doi:10.1186/1472-6882-14-4

Köthe M, Antl M, Huber B, Stoecker K, Ebrecht D, Steinmetz I, Eberl L (2003) Killing of *Caenorhabditis elegans* by *Burkholderia cepacia* is controlled by the *cep* quorum-sensing system. Cell Microbiol 5:343–351. doi:10.1046/j.1462-5822.2003.00280.x

Lee KM, Lim J, Nam S, Yoon MY, Kwon YK, Jung BY, Park Y, Park S, Yoon SS (2011) Inhibitory effect of broccoli extract on *Escherichia coli* O157:H7 quorum sensing and *in vivo* virulence. FEMS Microbiol Lett 321:67–74. doi:10.1111/j.1574-6968

Lowry F (1998) *Staphylococcus aureus* infections. N Engl J Med 339:520–532

Maeda T, García-Contreras R, Pu M, Sheng L, Garcia LR, Tomás M, Wood TK (2012) Quorum quenching quandary: resistance to antivirulence compounds. ISME J 6:493–501. doi:10.1038/ismej.2011

Mahajan-Miklos S, Tan MW, Rahme LG, Ausubel FM (1999) Molecular mechanisms of bacterial virulence elucidated using a *Pseudomonas aeruginosa*-*Caenorhabditis elegans* pathogenesis model. Cell 96:47–56

Manefield M, Rasmussen TB, Henzter M, Andersen JB, Steinberg P, Kjelleberg S, Givskov M (2002) Halogenated furanones inhibit quorum sensing through accelerated LuxR turnover. Microbiology 148: 1119–1127

Moy TI, Conery AL, Ford LJ, Wu G, Mazitschek R, Casadei G, Lewis K, Carpenter AE, Ausuble FM (2009) High-throughput screen for novel antimicrobials using a whole animal infection model. ACS Chem Biol 4:527–533. doi:10.1021/cb900084v

Murono K, Hirano Y, Koyano S, Ito K, Fujieda K (2003) Molecular comparison of bacterial isolates from blood with strains colonizing pharynx and intestine in immuno-compromised patients with sepsis. J Med Microbiol 52:527–530. doi:10.1099/jmm.0.05076-0

Nealson KH, Platt T, Hastings JW (1970) The cellular control of the synthesis and activity of the bacterial luminescent system. J Bacteriol 104:313–322

Okoli I, Coleman JJ, Tampakakis E, An WF, Holson E, Wagner F, Conery AL, Larkins-Ford J, Wu G, Stern A, Ausubel FM, Mylonakis E (2009) Identification of antifungal compounds active against *Candida albicans* using an improved high-throughput *Caenorhabditis elegans* assay. PLoS One 4:e7025. doi:10.1371/journal.pone.0007025

O'Loughlin CT, Miller LC, Siryaporn A, Drescher K, Semmelhack MF, Bassler LB (2013) A quorum-sensing inhibitor blocks *Pseudomonas aeruginosa* virulence and biofilm formation. Proc Natl Acad Sci USA 110:17981–17986. doi:10.1073/pnas.1316981110

Oncul O, Ulkur E, Acar A, Turhan V, Yeniz E, Karacaer Z, Yildiz F (2009) Prospective analysis of nosocomial infections in a burn care unit, Turkey. Indian J Med Res 130:758–764

Papaioannou E, Wahjudi M, Nadal-Jimenez P, Koch G, Setroikromo R, Quax WJ (2009) Quorum-quenching acylase reduces the virulence of *Pseudomonas aeruginosa* in a *Caenorhabditis elegans* infection model. Antimicrob Agents Chemother 53:4891–4897. doi:10.1128/AAC.00380-09

Peleg AY, Tampakakis E, Fuchs BB, Eliopoulos GM, Moellering RC Jr, Mylonakis E (2008) Prokaryote-eukaryote interactions identified by using *Caenorhabditis elegans*. Proc Natl Acad Sci USA 105:14585–14590. doi:10.1073/pnas.0805048105

Rasmussen TB, Manefield M, Andersen JB, Eberl L, Anthoni U, Christophersen C, Steinberg P, Kjelleberg S, Givsko VM (2000) How *Delisea pulchra* furanones affect quorum sensing and swarming motility in *Serratia liquefaciens* MG1. Microbiology 12:3237–32344

Rasmussen TB, Bjarnsholt T, Skindersoe ME, Hentzer M, Kristoffersen P, Kote M, Nielsen J, Eberl L, Givskov M (2005) Screening for quorum-sensing inhibitors (QSI) by use of a novel genetic system, the QSI selector. J Bacteriol 187:1799–1814

Riddle DL, Blumenthal T, Meyers BJ, Priess JR (1997) *C. elegans* II edition. Cold Spring Harbor Laboratory, New York

Rudrappa T, Bais HP (2008) Curcumin a known phenolic from *Curcuma longa* attenuates the virulence of *Pseudomonas aeruginosa* PAO1 in whole plant and animal pathogenicity models. J Agric Food Chem 56:1955–1962. doi:10.1021/jf072591j

Sarabhai S, Sharma P, Capalash N (2013) Ellagic acid derivatives from *Terminalia chebula* Retz. downregulate the expression of quorum sensing genes to attenuate *Pseudomonas aeruginosa* PAO1 virulence. PLoS One 8:e53441

Senturk S, Ulusoy S, Bosgelmez-Tinaz G, Yagci A (2012) Quorum sensing and virulence of *Pseudomonas aeruginosa* during urinary tract infections. J Infect Dev Ctries 6:501–507. doi:10.3855/jidc.2543

Sifri CD, Begun J, Ausubel FM, Calderwood SB (2003) *Caenorhabditis elegans* as a model host for *Staphylococcus aureus* pathogenesis. Infect Immun 71:2208–2219. doi:10.1128/IAI.71.4.2208-2217.2003

Smith RS, Iglewski BH (2003) *P. aeruginosa* quorum-sensing systems and virulence. Curr Opin Microbiol 6:56–60. doi:10.1016/S1369-5274(03)00008-0

Smyth AR, Cifelli PM, Ortori CA, Righetti K, Lewis S, Erskine P, Holland ED, Givskov M, Williams P, Cámara M, Barrett DA, Knox A (2010) Garlic as an inhibitor of *Pseudomonas aeruginosa* quorum sensing in cystic fibrosis–a pilot randomized controlled trial. Pediatr Pulmonol 45:356–362. doi:10.1002/ppul.21193

Sperandio V, Torres AG, Jarvis B, Nataro JP, Kaper JB (2003) Bacteria–host communication: *the language of hormones*. Proc Natl Acad Sci USA 100:8951–8956. doi:10.1073/pnas.1537100100

Tan MW, Rahme LG, Sternberg JA, Tompkins RG, Ausuble FM (1999) *Pseudomonas aeruginosa* killing of *Caenorhabditis elegans* used to identify *P. aeruginosa* virulence factors. Proc Natl Acad Sci USA 96:2408–2413. doi:10.1073/pnas.96.5.2408

Zaborin A, Romanowski K, Gerdes S, Holbrook C, Francios L, Jason L, Valeriy P, Diggle SP, Wike A, Righetti K, Morozova I, Babrowski T, Liu DC, Zaborina O, Alverdy JC (2009) Red death in *Caenorhabditis elegans* caused by *Pseudomonas aeruginosa* PAO1. Proc Natl Acad Sci USA 106:6327–6332. doi:10.1073/pnas.0813199106

Strategies for Silencing Bacterial Communication

Kristina Ivanova, Margarida M. Fernandes, and Tzanko Tzanov

Introduction

Bacteria are the oldest living organisms on Earth, and regardless of their simple structural features as single prokaryotic cells, they possess the capacity to cooperate as multicellular organisms and efficiently perform a wide range of collective behaviours (Kalia 2013; Benneche et al. 2008; Kalia and Purohit 2011; Choudhary and Schmidt-Dannert 2010; Parker and Sperandio 2009). Virulence factor expression, bioluminescence, sporulation, dispersal, foraging and biofilm formation are just some examples of the coordinated behaviour that bacteria use as a survival strategy in hostile environments. This process is based on a mechanism of bacterial communication, recently known as quorum sensing (QS). It is now known that bacteria use chemical language to 'talk' to each other. They secrete small diffusible hormonelike molecules, called autoinducers (AIs), into the extracellular environment, and upon reaching a threshold concentration, these AIs are recognised by receptors situated in the intracellular compartment of gram-positive bacteria or in the cell membrane of gram-negative bacteria (Kalia and Purohit 2011; Choudhary and Schmidt-Dannert 2010; Parker and Sperandio 2009). This mechanism allows bacteria to transduce stimuli from the environment into gene expression, enabling a single cell to sense the number of bacteria in their environment and thus turn on collective behaviours when they are in sufficient number to do so.

Although microorganisms live in perfect balance with the human body, supporting our immune and digestive systems, the diversity of QS-controlled phenotypes can have devastating effects on human health (Relman 2012). When the balance is broken and the normal flora grows beyond its typical ranges, due to lower immunity, heart diseases, chemotherapy, etc., bacteria can act as opportunistic pathogens causing life-threatening infections (Baron 1996). The discovery of antibiotics in the beginning of the twentieth century allowed bacteria-mediated infection treatment. However, their intensive and extended use and misuse have led to the emergence and spread of multidrug-resistant bacterial species (Rodríguez-Rojas et al. 2013; Kalia 2013).

Alternative approaches providing new perspectives for development of novel pharmaceutical drugs capable of preventing bacteria-associated infections have been sought. Silencing bacterial communication processes by targeting distinct components of the QS systems has attracted significant attention in the recent years as a novel anti-infective strategy. While conventional antibiotics kill or inhibit bacteria by targeting their structure or survival processes, interrupting the QS pathways is thought to induce

K. Ivanova • M.M. Fernandes • T. Tzanov (✉)
Group of Molecular and Industrial Biotechnology, Department of Chemical Engineering, Universitat Polìtecnica de Catalunya, Rambla Sant Nebridi 22, 08222 Terrassa, Spain
e-mail: tzanko.tzanov@upc.edu

V.C. Kalia (ed.), *Quorum Sensing vs Quorum Quenching: A Battle with No End in Sight*,
DOI 10.1007/978-81-322-1982-8_18, © Springer India 2015

less selective pressure on bacteria, reducing the risk of resistance development (Rasko and Sperandio 2010). The QS disruption affects the ability of the systems to control genes required for the host invasion and infection establishment; thus, it is a promising strategy to prevent bacterial pathogenesis and biofilm formation, among others (Rasko and Sperandio 2010; Parker and Sperandio 2009). This review provides a brief overview of the different strategies aiming at interruption of bacterial cell-to-cell communication process. QS signal synthesis, inactivation and receptor blockage are among the various targets of QS pathways that are discussed in detail. In addition, the efflux pumps of QS-regulated bacterial behaviour as promising novel anti-virulence and anti-biofilm strategies are also outlined.

Quorum-Sensing Mechanisms in Gram-Negative and Gram-Positive Bacteria

Quorum-sensing mechanisms differ greatly from organism to organism, depending on the characterisation of the bacterium, either gram positive or gram negative; the type of signalling molecules; and the signal transduction system used (Wang et al. 2008). In gram-negative bacteria, acyl-homoserine lactone (AHL)-type molecules serve as the main signalling molecules for their intraspecies communication (Fig. 1) (Galloway et al. 2012). These molecules consist of a homoserine lactone ring and a fatty acid side chain, varying in length (C4 to C18), saturation and oxidation at C3 position (Fig. 1b) (Watson et al. 2002; Yang et al. 2012). AHLs are produced by intracellular synthases (LuxI homologues of *V. fischeri*), by an acylation reaction of S-adenosyl-methionine (SAM), a source for homoserine lactone formation, and acyl-acyl carrier protein (ACP), a source of fatty acid side chain. The signals accumulate in the surrounding environment and in high concentration levels (threshold level); the AHL binds to the cognate cytoplasmic transcriptional regulators, LuxR-type proteins, inducing a target gene expression (Fig. 1a) (Choudhary and Schmidt-Dannert 2010; Atkinson and Williams 2009). This results in the induction of the quorum response and the positive regulation of an autoinduce synthetase by a signal generator which will provide more AIs for further QS responses.

The QS signalling systems in some gram-negative bacteria are usually more complicated than described above. Extensive studies have been carried out to understand the QS pathways of pathogenic bacteria *P. aeruginosa*, which exploit three types of QS systems associated with different virulence phenotypes (Tay and Yew 2013). Two of these QS systems (known as *las* and *rhl*) are AHL based, i.e. use 3-oxo-C12-HSL and C4-HSL to regulate the production and secretion of various virulence factors including elastase, exotoxin A, pyocyanin and biofilm formation (Smith and Iglewski 2003; Fux, Costerton et al. 2005). The third one is based on 2-heptyl-3-hydroxy-4(1H)-quinolone (PQS) and connects the other two *las* and *rhl* systems (Moore et al. 2014). Despite this complex system presented in *P. aeruginosa*, most pathogenic gram-negative species possess a single AI-based QS system, consisting of an AI synthase and a transcription regulator.

Gram-positive bacteria, on the other hand, use peptides or cyclic oligopeptide signals, called autoinducing peptides (AIPs), as the primary means of signalling and a two-component signal transduction system as the main quorum mechanism (Fig. 2) (Kalia and Purohit 2011). AIP signals are synthesised in bacterial cells as precursors (pro-AIP) and excreted through specialised membrane-bound AIP transporters. During the pro-AIP transportation through the membrane, the transporters process the signal in sizes that vary from 5 to 17 amino acids (Rutherford and Bassler 2012; Parker and Sperandio 2009). In contrast to gram-negative bacteria, gram-positive bacteria recognise the signal via a membrane-associated two-component histidine kinase receptor. At threshold concentration levels in the extracellular environment, the peptide signal binds to and activates a sensor protein, histidine kinase, located in the cell membrane of the bacterium. The activation of the histidine

a **Signal transduction system**

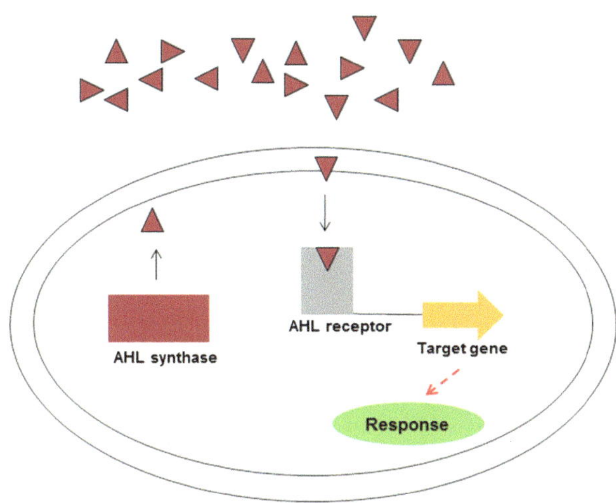

b **QS signals**

= Acyl-homoserine lactones (AHLs)

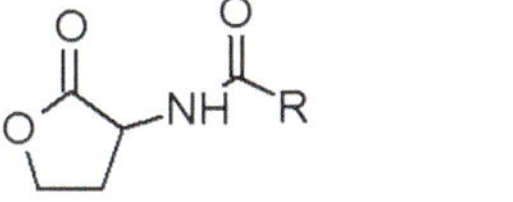

$R=C_3H_7$
P. aeruginosa; A. hydrophila; A. salmonicida (Smith and Iglewski 2003; S. Swift 1997)

$R=C_3H_6OH$
V. harveyi (Cao and Meighen 1989)

$R= C_5H_{11}$
C.violaceum;P.aureofaciens; Y. pseudotuberculosis (Zhang and Pierson 2001; McClean et al. 1997; Atkinson et al. 1999)

$R=C_7H_{15}$
V. fischeri; *B. cenocepacia* (Lupp and Ruby 2005; Sokol et al. 2003)

$R_1=C_3H$
V.fischeri;E.carotovora;E. chrysanthemi (Lupp and Ruby 2005; Welch et al. 2005; Nasser et al. 1998)

$R_1=C_5H_7$
A.tumefaciens (Haudecoeur and Faure 2010)

$R_1=C_7H_{15}$
P.putida; Y. pestis (Fekete et al. 2010; Kirwan et al. 2006)

$R_1=C_9H_{19}$
P. aeruginosa (Smith and Iglewski 2003)

Fig. 1 (**a**) Quorum-sensing signal transduction system in gram-negative bacteria consisting of AHL synthase (LuxI) and cognate intracellular AHL receptor (LuxR); (**b**) chemical structure of AHL signalling molecules secreted and detected by gram-negative bacteria

kinase leads to a phosphorylation of cognate response-regulating protein, which modulates the target gene expression, thus leading to wide-range changes in bacterial behaviour (Fig. 2a) (Rutherford and Bassler 2012).

Besides being able to communicate with members of their own species through AHL and AIP signals (intraspecies communication), many bacteria can also communicate with members of other species (interspecies communication)

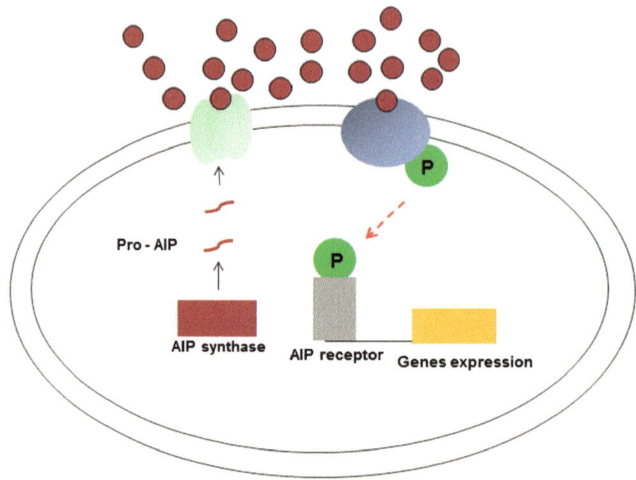

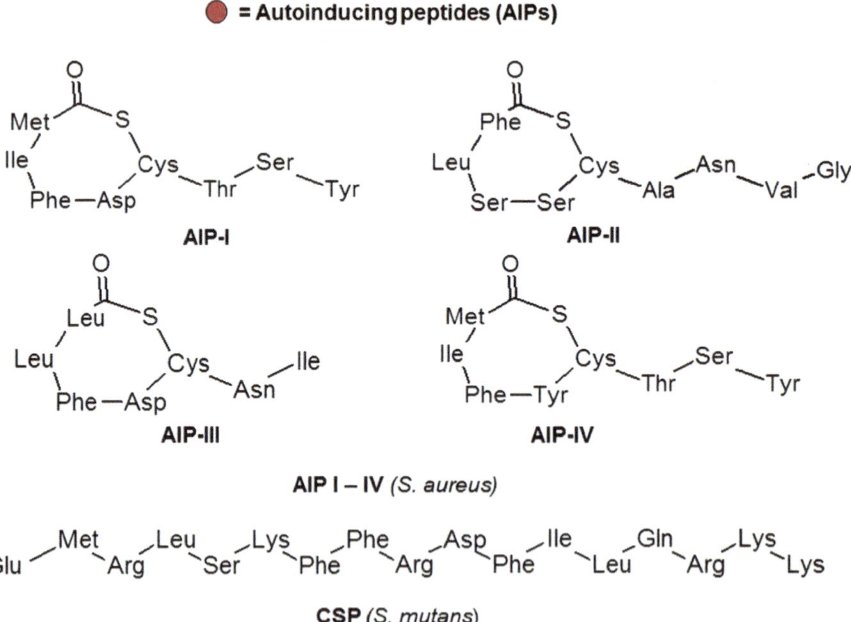

Fig. 2 (**a**) Quorum-sensing system in gram-positive bacteria; (**b**) chemical structure of AIPs produced by *S. aureus* and *S. mutants*

using a signal molecule, known as autoinducer-2 (AI-2) (Gamby et al. 2012; Parker and Sperandio 2009; Galloway et al. 2010). AI-2 signalling molecules are derived from 4,5-dihydroxy-2,3-pentanedione (DPD), which is produced by an enzyme, known as LuxS. The linear DPD is very unstable and can undergo spontaneous cyclisation forming a variety of derivatives that interconvert and exist in equilibrium in solution (known as AI-2 pool). Each bacterial strain recognises specific DPD derivatives; however, the interconversion of the molecules within the

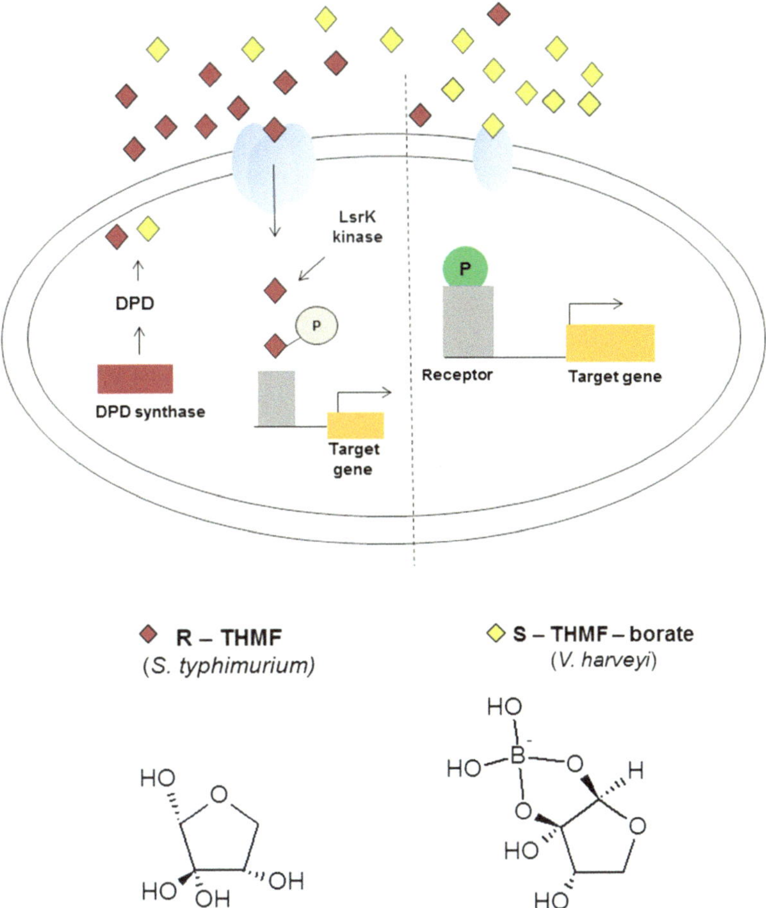

Fig. 3 Quorum-sensing systems in *S. typhimurium* and *V. harveyi* using either R-THMF (*red diamonds*) or S-THMF-borate (*yellow diamonds*) (Miller et al. 2004)

AI-2 pool allows bacteria to respond to their own AI-2 as well as to those produced by other bacterial species (Galloway et al. 2010; Geske et al. 2008). Although genome sequencing studies have revealed the presence of *luxS* gene homologues in many pathogens, including *E. coli*, *Helicobacter*, *Neisseria*, *Porphyromonas*, *Proteus*, *Salmonella*, *E. faecalis*, *S. pyogenes* and *S. aureus,* only a minority of bacterial species have been shown to possess the receptors that sense and respond to the AI-2 (Hentzer and Givskov 2003; Sintim et al. 2010). Three protein receptors involved in AI-2 signalling have been characterised till now: (i) LuxP in bacteria *V. harveyi* that recognises (2S,4S)-2-methyl-2,3,3,4-tetrahydroxytetrahydrofuran borate ((S)-THMF-borate); (ii) LsrB in *S. typhimurium* responding to (2R,4S)-2-methyl-2,3,3,4-tetrahydroxytetrahydrofuran((R)-THMF); and (iii) RbsB in *A. actinomycetemcomitans*, which signal structure is still unknown (Galloway et al. 2010; Pereira et al. 2013). Nevertheless, the studies on the use of small molecules able to interfere with the AI-2 signalling have been focused on the best characterised interspecies LuxP- and LsrB-based systems of *V. harveyi* and *S. typhimurium*, respectively (Fig. 3) (Galloway et al. 2010).

Quorum-Sensing Inhibition Targets/Strategies

Bacterial coordinated behaviours such as virulence factor production and formation of biofilms are controlled by QS mechanisms and may be related to human life-threatening diseases. For that reason, QS has been considered as an attractive target for the development of new anti-infective and therapeutic agents (Rasko and Sperandio 2010; Ivanova, Fernandes et al. 2013). Although different QS systems have been identified in bacteria, all of them function with the same principle that involves signal production, followed by their accumulation in the extracellular environment and finally signal detection by the specific receptor. In this review different quorum-sensing inhibition strategies, such as (i) inhibition of QS signal biosynthesis, (ii) QS signal degradation in the extracellular environment, (iii) inhibition of QS signal detection by receptor blockage and (iv) disruption of efflux pumps, targeting these QS components are highlighted.

Silencing Bacterial Communication by Targeting the Biosynthesis of QS Signals

Interfering with AI synthesis in bacteria is an attractive strategy for inhibition of QS pathways, by limiting the signal accumulation in the cells surrounding. Strategies involve the inhibition of AHL and AIP production on gram-negative and gram-positive bacteria for their intraspecies communication, as well as inhibition of AI-2 production, responsible for bacterial interspecies communication. Despite that this approach is less used than signal degradation and inactivation, the inhibition of signal production is feasible and effective both in vitro and in vivo for several types of autoinducers.

Blockage of Acyl-Homoserine Lactone Production in Gram-Negative Bacteria

Three different types of AHLs producing the enzymes (AHL synthases) LuxI, HdtS and LuxM have been identified. Among all, the Lux I-type synthases are the best understood and their identical homologues have been found in many bacterial species (Churchill and Chen 2010). The majority of the investigated AHL synthase inhibitors are structural analogues of the SAM. Compounds such as L/D-S-adenosylhomocysteine, sinefungin and butyryl-SAM have been proved to suppress the AHL synthesis and inhibit the first step in QS signalling in vitro of pathogenic *P. aeruginosa* (Parsek et al. 1999; Hentzer and Givskov 2003). It is relevant to emphasise that SAM is used by other enzymes in living systems and its analogues could lead to side effects (Ni et al. 2009b; Rasmussen and Givskov 2006).

Recently, an analogue of *B. glumae* AHL synthase, called J8-C8, has been demonstrated to inhibit QS signal production by occupying the binding site for the ACP of the cognate substrate. Moreover, the synergistic effect of this analogue, with 5′-methylthioadenosine a by-product of the AI synthesis, was also demonstrated (Chung et al. 2011). Calfee et al. have shown that methylated analogue of anthranilate, a PQS precursor, can inhibit the signal synthesis and consequently prevent the virulence factor production by *P. aeruginosa* (Calfee et al. 2001). Moreover, halogenated analogues of anthranilate have been shown to attenuate bacterial pathogenesis affecting QS-regulated pathways in *P. aeruginosa* (Lesic et al. 2007).

Blockage of Autoinducing Peptide Production in Gram-Positive Bacteria

Ribosomes and peptidases, enzymes responsible for the synthesis of AIPs in gram-positive bacteria, are also involved in the growth and survival of bacterial cells (Kulanthaivel et al. 2004). For this reason, little work has been performed on the blockage of AIPs. By inhibiting these enzymes a bactericidal activity would take place in addition to quorum-quenching activity, which theoretically increases the pressure on bacteria to develop resistance (LaSarre and Federle 2013; Hentzer and Givskov 2003). Nevertheless, the mechanistic understanding of signal synthesis has been widely studied (O'Loughlin et al. 2013; De Kievit 2009; Peeters et al. 2008), which will eventually lead, in the near future, to the development of strategies for the inhibition of signal production as a potential anti-virulence approach.

Targeting the AI-2 Synthases

Interfering with the production of the universal interspecies bacterial signal, AI-2, has been proposed as an alternative approach that may provide inhibition of QS-regulated pathways in broader bacterial spectrum, thereby attenuating polymicrobial pathogenesis. The LuxS-type enzyme uses an S-ribosyl-L-homocysteine (SHR) as a substrate (Galloway et al. 2010). The SHR is derived from SAM, which initially is converted to a toxic intermediate SAH by methyltransferases, thereby is hydrolysed by the 5′-methylthioadenosine/S-adenosylhomocysteine nucleosidase (MTAN) enzyme (known as Pfs nucleosidase). At the end, SHR is cleaved by the LuxS resulting in DPD production and L-homocysteine (Amara et al. 2010; Galloway et al. 2010).

Targeting the direct inhibition of DPD synthesis is a valuable tool for disruption of interspecies communications leading to modified bacterial behaviours. Zhou and co-workers have reported two SHR analogues, S-anhydroribosyl-L-homocysteine and S-homoribosyl-L-cysteine, capable of inhibiting the SHR hydrolysation and deactivating the LuxS synthases (Alfaro et al. 2004). Moreover, series of stable SRH and 2 ketone intermediate analogues have been reported by Pei and co-workers. Kinetic studies demonstrated that these analogues act as reversible and competitive inhibitors of LuxS synthases interfering the cleavage of SRH (Shen et al. 2006). Recently, two analogues of SRH that act as modest competitive inhibitors of LuxS have also been reported by Malladi et al. (Malladi et al. 2011).

Moreover, several naturally occurring brominated furanones were demonstrated to be able to covalently modify and inhibit the LuxS enzyme in a concentration-dependent manner (Zang et al. 2009). Lately, several brominated furanone derivatives were synthesised by Benneche and co-workers and investigated for their ability to interfere with QS process using the reporter bacterial strain *V. harveyi*. *V. harveyi* lacks the receptor for AHL signal, thus responds only to intermicrobial communication via the AI-2 signal. The presence of these furanone derivatives was shown to reduce bioluminescence in the *V. harveyi*, suggesting their possible role in the interruption of interspecies bacterial communication. Moreover, the most effective furanones were shown to reduce *S. epidermidis* biofilm by 57 % without affecting their growth (Benneche et al. 2008).

Another promising target in inhibition of AI-2 signal synthesis is the enzyme MTAN present only in bacteria, which means that the strategies would be bacteria specific without causing harm to human cells (Benneche et al. 2008). As already mentioned, this enzyme is involved in the DPD production by converting the SAH to SRH in the SRH biosynthesis (Galloway et al. 2010). Analogues of MTAN such as 5′-methylthio-DADMe-immucillin-A, 5′-ethylthio-DADMe-immucillin-A and 5′-butylthio-DADMe-immucillin-A have been shown to inhibit the AI-2 synthesis of *V. cholerae* in a dose-dependent manner through binding to the MTAN. Moreover, MT-DADMe-ImmA and BuT-DADMe-ImmA were also demonstrated as potent non-toxic inhibitors of AI-2 synthesis in the pathogenic *E. coli*. The major concern related to this approach is that by targeting the MTAN enzyme in AI-2 synthesis, the overexpression of quorum-sensing pathway (or alternative pathways) might overcome the effect of developed MTAN inhibitors (Guo et al. 2013).

Silencing Bacterial Communication by Targeting the Extracellular QS Signals

Enzymatic Inactivation/Degradation of Autoinducers

Altering the AIs while present in the extracellular environment is a promising approach to attenuate bacterial pathogenicity that is believed to provide less selective pressure on the population, therefore reducing the risk of drug resistance development. The absence of membrane diffusion barriers makes the development of the antibacterial agents easier since there is no need to enter the cell to reach to their target. Several enzymes, normally produced by various bacterial species including *Bacillus* sp. 240B1, *Bacillus* strain COT1, *A. tumefaciens, B. thuringiensis, Arthrobacter* sp. IBN110, *B. thuringiensis*, *B. cereus*, *B. mycoides, V. paradox*, *Anabaena*,

Ralstonia and *Rhodococcus*, have been reported to degrade the AIs and consequently inhibit the QS-regulated gene expression (Roche et al. 2004; Hong et al. 2012).

Dong et al. reported an AHL-inactivating enzyme, lactonase, encoded by *aiiA* gene of *Bacillus* sp. that hydrolyses the AHL lactone ring. The *aiiA* gene expression in plants has been demonstrated to increase the resistance to the plant pathogen *E. carotovora* due to the enzymatic degradation of 3-oxo-C6-HSL (Dong et al. 2001). There are also some evidences about the synergistic effect of quorum-quenching enzymes, such as lactonase and antibiotic-like ciprofloxacin and gentamicin, which was demonstrated to increase the susceptibility of *P. aeruginosa* biofilms (Kiran et al. 2011).

Another quorum-quenching (QQ) enzyme, the acylase, produced by several species including *V. paradoxus,* soil *P. aeruginosa* PAI-A and *Ralstonia* sp., has been proved to interfere with the QS process by degrading the AHLs through the cleavage of the amide bond between the lactone ring and fatty acid side chain (Lin et al. 2003; Hong et al. 2012). Lin et al. demonstrated that acylase expression in pathogenic *P. aeruginosa* PAO1 efficiently suppressed the production of QS-regulated virulence factors (elastase and pyocyanin), decreased its swarming motility and attenuated the pathogenesis of *C. elegans* (Lin et al. 2003). Later, Huang et al. showed that the soil *P. aeruginosa* PAI-A and several clinical isolates were able to degrade long AHLs using them as a sole energy source for growth in selective conditions (Huang et al. 2003). Human acylase, isolated from the kidney, was shown to degrade in vitro several models of AHLs including C4-HSL and C8-HSL. Thereafter, in vitro biofilm inhibition tests against *P. aeruginosa* revealed the potential of this enzyme to prevent surface attachment through QS interruption (Xu et al. 2003).

Oxidoreductases, another group of enzymes capable of confusing bacteria in their communication process, have been reported as a part of the defence mechanisms of *R. erythropolis* and *Burkholderia* sp. *GG4* (Hong et al. 2012; Uroz et al. 2005). The oxidoreductases modify the AHLs altering the C3 keto group of the fatty acid side chain resulting in receptor-AHL-binding interaction failure and further inhibition of QS-regulated gene expression (Hong et al. 2012; Uroz et al. 2005). Chan et al. demonstrated the anti-infective properties of an oxidoreductase produced by *Burkholderia* sp. *GG4* against pathogenic *E. carotovora.* The enzyme modulated the 3-oxo-C6-HSL signalling molecule and attenuated the *E. carotovora* pathogenesis (Hong et al. 2012; Chan et al. 2011).

Recently, another family of human enzymes, the paraoxonases (PONs), has also been reported as QQ agents. PONs exhibit lactonase activity and are capable of disrupting AHL-mediated QS systems of gram-negative bacteria (Camps et al. 2011). Teiber et al. reported that PONs, especially PON 2, efficiently degraded 3-oxo-C12-HSL, a key QS molecule produced by several bacterial species including *P. aeruginosa*, *Burkholderia*, *Yersinia*, *Serratia* and *Aeromonas* (Teiber et al. 2008). It is believed that PONs have a key role in innate human defence mechanisms inactivating the AHL signals of pathogenic *P. aeruginosa* (Teiber et al. 2008).

Interestingly, the AHL signals may be also deactivated at higher pH values by opening of the lactone ring (Kociolek 2009). For instance, one of the early defence mechanisms of the plants against pathogenic bacteria *E. carotovora* is based on increasing of the pH around the infected site and further attenuation of the infection due to the QS disruption (Byers et al. 2002).

Another QQ enzyme that has been reported is 2,4-dioxygenase, Hod, able to disrupt the quinolone-based signalling pathway in *P. aeruginosa.* Hod converts the PQS to N-octanoylanthranilic acid and carbon monoxide preventing its accumulation in the extracellular environment. Despite the fact that Hod is susceptible to exoproteases and could be inhibited by the PQS precursor 2-heptyl-4(1H)-quinolone, the enzyme significantly reduced the expression of key *P. aeruginosa* virulence factors including lectin A, pyocyanin and rhamnolipids and inhibited the growth and tissue damage in a plant leaf infection model (Pustelny et al. 2009).

Antibodies

In addition to the QQ enzymes, antibodies have also been reported to act as a scavenger of the QS signals (Tay and Yew 2013; Kalia 2013). A few organisms possess orphan receptors for QS molecules that might be used to quench the QS signalling. Mammalian adaptive immune system generates antibodies in response to the presence of antigens. It is believed that because of their non-proteinaceous nature and low molecular weight, the AI signals (AHLs and DPD) should not elicit a human immune response. However, bacterial AHL molecules have been reported as small molecule toxins exerting apoptosis and modulation of NF-κB activity on mammalian cells (Kalia 2013; Pustelny et al. 2009; Kaufmann et al. 2006). Marin et al. demonstrated a monoclonal catalytic antibody, XYD-11G2, able to neutralise 3-oxo-C12-HSL and prevent pyocyanin production in *P. aeruginosa* (De Lamo Marin et al. 2007). Another monoclonal antibody, termed RS2-1G9, was also shown to effectively sequester 3-oxo-C12-HSL suppressing the QS signalling in *P. aeruginosa*, but not able to recognise short chain 3-oxo-AHLs. The effect of active immunisation with 3-oxo-C12-HSL-BSA conjugate on acute infection in mice was also investigated. The 3-oxo-C12-HSL-BSA conjugate produced significant amount of specific antibody in the mice leading to the increased survival and ability to prevent infection occurrence (Kaufmann et al. 2008).

Although less reported, the monoclonal antibodies have been also demonstrated to be capable of inactivating the AIPs and disrupt the QS pathways. Several AIP-mediated QS systems in gram-positive bacteria have been studied including the *agr* system of pathogenic *S. aureus* and *agr*-like system of *E. faecalis*. The agr quorum-sensing systems use thiolactone-containing peptides (AIP), as a specific signalling molecules, and have been identified in many staphylococci (Rasko and Sperandio 2010; Ni et al. 2009b). Janda and co-workers designed hepten AP4-5 that elicits the generation of anti-autoinducer monoclonal antibody, termed AP4-24 H11, thus efficiently inhibiting QS in vitro through the sequestration of the AIP-4 produced by the human pathogen *S. aureus*. The AP4-24H11 was able to inhibit the *S. aureus* pathogenicity in an abscess formation mouse model in vivo providing complete protection against a lethal *S. aureus* challenge (Park et al. 2007).

Active Uptake of AI Signalling Molecules by Beneficial Bacteria

A variety of bacterial species have the ability to sequester and process the QS molecules presented in their environment, thereby interfering with the cell-cell communication by acting as competing bacteria (Xavier and Bassler 2005). Certain bacteria, including the members from the Enterobacteriaceae (like the commensal *E. coli* K12 and the pathogens *E. coli* O157), *S. typhimurium*, *S. meliloti* and the pathogen *B. anthracis*, have been shown to be able to control the concentration (or availability) of AI-2 using a QS machinery, known as Lsr. These bacteria remove AI-2 signal from the surrounding environment, thereby eliminating the ability of others in the community to utilise the AI-2 signals and regulate their behaviours. Xavier and Bassler have shown that *E. coli* is able to modulate the QS in *V. harveyi* by uptaking the AI-2 signals (Xavier and Bassler 2005). When both species were cultured together, the production of bioluminescence by *V. harveyi* was diminished by 18 %. On the other hand, when *V. harveyi* was cocultured with *E. coli* mutant containing a constitutively derepressed LsrK, a decrease of 90 % of bioluminescence was observed (Xavier and Bassler 2005). This nature-inspired approach for inhibiting the QS-regulated gene expression is believed to be a broad-spectrum anti-infective strategy to control multidrug resistance development. However, application of LsrK as multispecies anti-infective agent has not been examined in vivo.

Passive Control of QS Signalling

Recently, molecularly imprinted polymers (MIPs) have been proposed as novel class of AHL sequesters capable of attenuating bacterial pathogenesis and virulence factor production. The first-generation MIP with affinity towards a signal molecule of *V. fischeri*,

N-(β-ketocaproyl)-L-homoserine lactone (3-oxo-C6-HSL), was reported by Piletska et al. (2010). Based on computational modelling, the authors designed different polymers, which were shown to form complex with the signalling molecule resulting in its elimination from the extracellular environment, thereby disrupting the QS-regulated bioluminescence production and biofilm formation of *V. fischeri*. Lately, the authors synthesised MIP targeting other signalling molecule, 3-oxo-C12-HSL, involved in the *P. aeruginosa* pathogenesis. After screening a library of 20 functional monomers able to form energetically favourable complexes with the target molecule, the most promising MIPs efficiently reduced the biofilm formation by 80 % (Piletska et al. 2011).

Silencing Bacterial Communication by Targeting and Blocking the Signal Receptors

Targeting the AI receptors is the most intuitive approach to silence bacterial QS process and control bacteria-mediated infections. Numerous QS inhibitors (QSIs) have been developed as analogues of native signals to disrupt the QS pathways in bacteria by interaction with the receptor. The QSIs competitively interact with the signal receptor generating an inactive signal-receptor complex and blocking the downstream signalling (Ni et al. 2009b). However, not all identified QSIs have demonstrated antagonistic activities; other activities such as agonistic, no activity and both have been also addressed (Ni et al. 2009b). This section outlines recent advances on the development of QSI able to interfere with QS process by the signal-receptor interactions.

Inhibitors Targeting AHL Receptors in Gram-Negative Bacteria

The discovery of potent QSIs targeting the LuxR type of receptor proteins has been extensively studied as an alternative strategy to control bacteria-mediated infections through QS silencing. Different strategies have been used in the development of efficient antagonists capable of blocking the AHL binding to the cognate response regulator that are based on the design and synthesis of (i) AHL analogues, (ii) structurally unrelated AHLs and (iii) natural QS inhibitors.

AHL Analogues

The ability of AHLs to regulate QS signalling is strongly dependent on their geometry and chirality (Chan and Chua 2005; Kalia 2013). Various modifications in the AHL acyl side chain and/or lactone ring have been reported as an effective strategy for designing AHLs capable of blocking the ligand-receptor interaction (Atkinson et al. 1999). The length of acyl side chain was found to be a crucial structural feature for AHL-binding capacity (Atkinson et al. 1999). Thus, the incorporation of functional groups in the acyl chain has been a widely used approach to block the interaction between the ligand and receptor. By introducing one methylene unit in the AHL side chain, an activity loss of 50 % was found, while the incorporation of two induced a greater effect inducing 90 % of loss in binding capacity (Hentzer and Givskov 2003; Kalia 2013).

Geske et al. synthesised AHL analogues by introducing various bulky groups in the acyl side chain that were capable of blocking the TraR, LasR and LuxR receptors in *A. tumefaciens*, *P. aeruginosa* and *V. fischeri*, respectively (Geske et al. 2008). For example, 4-substituted N-phenoxyacetyl and 3-substituted N-phenylpropionyl L-homoserine lactones were identified as potent antagonists of TraR and LuxR, respectively. Replacing of carboxamide bond in native AHL by sulphonamide led to the formation of inactive ligand-receptor complex in V. *fischeri*, resulting in the reduction of its bioluminescence production (Olsen et al. 2002). Kim et al. also synthesised structural analogues of QS signal molecule of *A. tumefaciens*, 3-oxo-C8-HSL, by replacing the carboxamide bond with a nicotinamide or sulphonamide bond that were further able to block the TraR response regulator and inhibit the QS signalling in the reporter bacterial strain. Moreover, these compounds have shown ability to affect biofilm formation in *P. aeruginosa* reducing the number of bacterial cell attached to the surface (Kim et al. 2009b).

AHL analogues in which the amide group was replaced by a reverse amide or reverse sulphonamides have been studied as QS modulators. The QS inhibition activity of the synthesised analogues was evaluated by measuring the decrease of the bioluminescence normally induced by the 3-oxo-C6-HSL in *Vibrio fischeri*. Both reverse amide and reverse sulphonamide AHL analogues demonstrated antagonistic activity in the concentration range (1–200 μM) depending on the length of the fatty acid chain. These results suggested that α-amidolactone motif in AHLs plays a key role on the efficient binding to the transcriptional LuxR-type proteins (Tal-Gan et al. 2013). Ishida et al. synthesised several acyl cycloalkylamide derivatives with variety of acyl chain lengths, Cn-CPA ($n = 1 - 10$), as potential QS inhibitors interfering with LasR and RhlR QS systems of *P. aeruginosa*. The most effective inhibitor, N-decanoyl cyclopentylamide (C10-CPA, acyl side chain length of 10 carbons), was shown to efficiently prevent the elastase, pyocyanin and rhamnolipid virulence factor production and biofilm formation of *P. aeruginosa* PAO1 (Ishida et al. 2007). Later the antagonistic activity of these analogues, Cn-CPA, was also demonstrated on QS systems in *V. fischeri* and *S. marcescens* (Morohoshi, Shiono et al. 2007).

Alternative strategy in synthesis of AHL analogues is based on modifications in the lactone ring, while maintaining the native side chain intact (Galloway et al. 2010). By substituting the head part of 3-oxo-C12-HSL with different aromatic rings, an inhibition of the QS-regulated process of *P. aeruginosa* was observed. The QS inhibition capacity in vivo of these synthetic analogues was assessed by measuring the β-galactosidase activity in a recombinant *E. coli* reporter strain carrying a LasR receptor protein, which was reduced in more than 50 % (Kim et al. 2009a). This investigation revealed the importance of the head part of 3-oxo-C12-HSL in active ligand-receptor complex formation, suggesting it as a good target for development of *P. aeruginosa* inhibitors (Galloway et al. 2010; Kim et al. 2009a).

N-Decanoyl-L-homoserine benzyl, an AHL analogue, was reported as a potent anti-infective agent against *P. aeruginosa,* inhibiting the QS-regulated protease and elastase activities, rhamnolipid production and swarming motility without affecting bacterial growth (Yang et al. 2012). Moreover, the synergistic effect of this synthetic compound with several antibiotics including tobramycin, gentamicin, cefepime and meropenem was also demonstrated using a microdilution checkerboard method and time-kill assay (Yang et al. 2012). It is relevant to note that homoserine lactone moiety is unstable in alkaline conditions and can be degraded by mammalian lactonases; thus, the efficacy of pharmacological agents based on these structural features is of particular concern. Moreover, most of the AHL receptors demonstrate very high specificity to the cognate signal molecule hindering the rational design of efficient QS inhibitor.

Structurally Unrelated AHLs

The limited in vivo efficacy of AHL-based analogues implies the need for identification of structurally unrelated AHL signals as alternative compounds capable of inhibiting the QS pathways. Several chemical compounds such as 4-nitropyridine-N-oxide (4-NPO), p-benzoquinone, 2,4,5-tribromo-imidazole and 3-nitro-benzene-sulphonamide were identified by QSI screens as agents able to interfere with QS signalling. Among them the most effective was 4-NPO found to block the LuxR in *E. coli* and LasR/RhlR QS systems in *P. aeruginosa* (Rasmussen and Givskov 2006; Rasmussen et al. 2005a).

Interestingly, Skindersoe et al. proved that antibiotics such as azithromycin (AZM), ceftazidime (CFT) and ciprofloxacin (CPR) in subinhibitory concentration affect the QS process attenuating *P. aeruginosa* virulence (Skindersoe et al. 2008). Moreover, various synthetic furanones were also shown to affect a wide range of QS-regulated behaviours such as biofilm formation and swarming motility in gram-negative bacteria (Kalia and Purohit 2011).

Natural QS Inhibitors

Naturally occurring compounds such as penicillic acid (PA) and patulin, produced by *Penicillium* species, were found to efficiently antagonise the *las* and *rhl* QS systems in *P. aeruginosa* (Rasmussen and Givskov 2006; Rasmussen et al. 2005b; Koch et al. 2005). In addition, due to the synergistic effect of patulin and tobramycin, the *P. aeruginosa* bacterial strain in biofilm mode of growth was more susceptible to the antibiotic (Kociolek 2009).

Numerous QSIs have been extracted from other natural sources such as plants, herbs and fungi and demonstrated to be capable of silencing bacterial QS process (Rasmussen and Givskov 2006). The mode of action of these agents in terms of their targets in QS pathways, e.g. the QS signal itself, the receptor or the synthase, is not well understood. Rasmussen et al. reported a number of plants and herbal extracts including garlic, carrot, bean, water lily, chamomile, habanero and propolis capable to suppress the QS-regulated virulence gene expression (Rasmussen et al. 2005a). Among all, garlic extract that contains at least three different inhibitors of QS pathway was demonstrated as the most potent QSI, and together with tobramycin, a synergistic effect was observed by increasing the biofilm susceptibility (Rasmussen et al. 2005a). Other extracts of several south Florida medicinal plant species including *B. buceras*, *C. erectus* and *C. viminalis* were also found to attenuate *P. aeruginosa* PAO1 pathogenesis inhibiting virulence factor production (Fux et al. 2005).

Various plant-derived polyphenols such as (−)-epigallocatechin gallate (EGCG), ellagic acid, tannic acid and pyrogallol have also been shown to affect the QS circuits in bacteria (Kociolek 2009). EGCG significantly inhibited the pathogenic *E. coli* O157:H7 by suppressing the QS-regulated gene expression (Kang-Mu et al. 2009). Extracts from *Camellia sinensis* (Green tea, GT) also have been demonstrated as QS modulators reducing the virulence factor expression in *P. aeruginosa* PAO1 (Mihalik et al. 2008). Halogenated furanones (fimbrolides) are one of the most investigated groups of natural QSIs isolated from the red macroalga *D. pulchra* (Hentzer and Givskov 2003). The marine algae produce halogenated furanone compounds as secondary metabolites that interfere with AHL-based QS signalling systems. Pure samples of natural furanones were shown to disrupt QS-regulated processes such as swarming motility of *S. liquefaciens* and bioluminescence production of *V. fischeri* and *V. harveyi* (Galloway et al. 2010; Givskov et al. 1996). The furanone (5Z)-4-bromo-5-(bromomethylene)-3-butyl-2(5H) produced by *D. pulchra* also inhibited pathogenic phenotypes of *E. coli* such as swarming motility and biofilm formation. Although these furanones did not demonstrate activity on pathogenic *P. aeruginosa* species, synthetic derivatives showed improved survival time of mice infected with lethal *P. aeruginosa* (Wu et al. 2004). It has to be pointed that halogenated furanones have the disadvantage of being extremely reactive and thus may be toxic to human cells, which limits their potential use as QSI (Chenia 2013).

Targeting AIP Receptors

Gram-positive bacteria possess two-component QS systems that contain a membrane-bound histidine kinase receptor and a responsive transcriptional regulator (Fig. 2a) (Ni et al. 2009b). Understanding the mechanisms of receptor activation by QS molecules led to the development of new QSIs that shut down the entire QS pathway and thus attenuate bacterial pathogenesis. So far, the most intriguing anti-QS strategy is to target the inhibition of membrane sensor histidine kinase receptors by AIP antagonists (Ni et al. 2009b).

Several AIP-mediated QS systems have been well studied, including agr system of pathogenic *S. aureus* and agr-like system of *E. faecalis* that utilise structurally similar AIP signals. The *agr* quorum-sensing system of *S. aureus* uses four different types of thiolactone-containing peptides (AIPs I–IV) to control wide-range bacterial behaviours (Fig. 2b) (Ni et al. 2009b; Rasko and Sperandio 2010). This system consists of AgrC receptor (transmembrane histidine kinase) that specifically interacts with the cognate AIP resulting in an active complex that phosphorylates the intracellular response regulator, AgrA, promoting the autoinduction of the agr system and upregulation of RNA III transcription (Tal-Gan et al. 2013). In *S. aureus* four agr groups using distinct AIP and AgrC sequences have been identified. The AgrC receptor of each group responds to its own AIP molecule by activating the agr regulon, but this activation is cross inhibited by the other heterologous AIPs (the other AIP groups). This activity suggested to provide a competitive advantage of each group when establishing a specific infection (Tal-Gan et al. 2013). To date, the QS in *S. aureus* is a unique case wherein the native QS signals function as QS inhibitors against AgrC receptor of strains belonging to different agr groups (other groups of *S. aureus* strains) (Tal-Gan et al. 2013; LaSarre and Federle 2013).

Early studies on AgrC inhibition used native AIPs capable of cross inhibiting the other three non-cognate receptors. For instance, Otto et al. demonstrated that a synthetic *S. epidermidis* AIP is a competent inhibitor of the *S. aureus* agr system. Moreover, the AIP derivatives, in which the N-terminus or the cyclic bond structure was changed, were synthesised and acted as effective quorum-sensing blockers that suppressed the expression of virulence factors, such as δ-toxin and α-toxin in *S. aureus* (Otto et al. 1999).

In terms of designing AIP analogues capable of inhibiting both the cognate and non-cognate AgrC receptors, changes in the amino acid residue tail or removal of the tail are suggested as efficient methods for developing QSIs that target the AgrC groups (LaSarre and Federle 2013). Combination of various structural changes in the AIP is known to provide antagonistic activity and has been shown to have a potential for generating stronger inhibitors against all *agr* groups (LaSarre and Federle 2013; Lyon et al. 2002). Lyon et al. described a truncated AIP-II peptide lacking the tail as a global inhibitor of *S. aureus* virulence able to inhibit all four *agr* groups including its own cognate receptor. Moreover, the truncated analogue inhibited δ-toxin production with different strengths in *S. aureus*, *S. epidermidis* and *S. warneri*, suggesting its use as a global inhibitor of staphylococcal *agr* activation (Lyon et al. 2000). Another truncated version of AIP-I that lacked an exocyclic tail and had an aspartic acid (D) to alanine (A) mutation in the macrocyclic core, called tAIP-I D2A, was also shown to be a potent inhibitor of four AgrC receptors (groups I–IV) (Lyon et al. 2002). Wright et al. demonstrated in vivo using mouse dermonecrosis that the administration of antagonistic AIP delayed *agr* gene expression and further prevented the formation of ulcers and abscesses in the mice (Wright et al. 2005).

To date, most research attention has been focused on the non-native analogues of AIP-I and AIP-II group antagonist of AgrC. Recently, a set of synthetic AIP-III analogues has been reported as inhibitors of all four AgrC receptors in *S. aureus*. These AIP analogues also blocked the QS-regulated virulence trait, e.g. hemolysis, in wild-type *S. aureus* groups I–IV at picomolar concentrations. Moreover, four of the best identified AgrC inhibitors were also capable of attenuating the production of toxic shock syndrome toxin-1 (also under the control of QS) by over 80 % at nanomolar concentrations in a wild-type *S. aureus* group III strain (Tal-Gan et al. 2013). Small cyclic dipeptides, cyclo(L-Phe-L-Pro) and cyclo(L-Tyr-L-Pro), produced by *L. reuteri* were demonstrated to interfere with the agr QS system repressing the expression of exotoxin toxic shock syndrome toxin-1 in *S. aureus* (Li et al. 2011).

Recently, an analogue of *S. pneumoniae* signalling peptide has been demonstrated to decrease the virulence factor production in vitro. *S. pneumoniae* utilises a linear peptide consisting of 17 amino acids (CSP), which is recognised at the cell surface by the histidine kinase, known as ComD. Alanine substitution of the first amino acid generated a peptide (CSP-E1A) that could block the ability of the native CSP to bind and stimulate ComD downregulation signalling. Co-injection of the inhibitor peptide with S. *pneumoniae* in a mouse lung infection model led to improved mouse survival compared to a non-treated control. In addition, co-injection of the inhibitor blocked the bacteria's ability to transform into an antibiotic-resistant strain within the mouse when DNA encoding antibiotic resistance was supplied simultaneously (Zhu and Lau 2011).

Anti-QS-sensing strategies in principle are promising approach for development of new therapeutics. However, due to the specificity of QS signalling in distinct pathogens, the development of anti-QS drugs requires very careful investigation and usage (Rasko and Sperandio 2010).

Targeting AI-2 Receptors

The development of small molecules capable of inhibiting the AI-2 QS system (interspecies communication) is an attractive approach for prevention of wide-range bacterial behaviours from occurring (LaSarre and Federle 2013; Guo et al. 2013). With the concept of molecular mimicking and the goal of identifying antagonist of AI-2 receptor proteins, Wang and co-workers found a few sulphone compounds using high-throughput virtual screening able to antagonise LuxP receptor in *V. harveyi* (Peng et al. 2009; Ni et al. 2008a). The authors suggested the key role of the sulphone group of these compounds in the interaction with the LuxP receptor. The same group found a series of phenylboronic acids and aromatic polyols (mimic the S-THMF-borate signal molecules) to be capable of inhibiting the bioluminescence production in *V. harveyi* with IC50 at low or sub-micromolar concentrations (Ni et al. 2008a, b, 2009a). Recently, Wang and co-workers used the DOCK6 program for the virtual screening of AI-2-mediated quorum-sensing inhibitors targeting LuxP crystal structure. After various in silico and in vitro evaluation, seven inhibitors of AI-2 QS system in *Vibrio harveyi* were successfully identified (Zhu et al. 2012). Brackman et al. found nucleoside analogues as antagonists against *Vibrio* species. From the screening of a small panel of nucleoside analogues, an adenosine derivative with a p-methoxyphenylpropionamide moiety at C-3′ emerged as a promising inhibitory agent. This compound, as well as a truncated analogue lacking the adenine base, blocked the signal transduction pathway at the level of LuxP in *V. harveyi* and consequently reduced bioluminescence production. Moreover, the most active nucleoside analogue (designated LMC-21) decreased the *Vibrio* species starvation response, biofilm formation in *V. anguillarum*, *V. vulnificus* and *V. cholerae* and also reduced the pigment and protease production in *V. anguillarum* (Brackman et al. 2009).

Structural analogues of well-known natural QS inhibitors have also been demonstrated to efficiently affect the AI-2 pathways in *V. harveyi*. For instance, the thiazolidinediones (resemble the QSIs such as N-acylaminofuranones and/or acyl-homoserine lactone signalling molecules), dioxazaborocanes (structurally resemble the oxazaborolidine derivatives) and brominated thiophenones were shown to block the AI-2 QS in *V. harveyi* (Brackman et al. 2013).

Another approach for the disruption of interspecies QS signalling is modulating the native AI-2 signal. C1-substituted alkyl-DPD derivatives (propyl-133 and butyl-DPD) were found to inhibit the AI-2 signalling in *S. typhimurium* (Defoirdt et al. 2012). The activity of these multiple C1-alkyl-DPD derivatives in *V. harveyi* depended on the assay conditions; they were found to function as either antagonists or synergistic agonists (LaSarre and Federle 2013; Defoirdt et al. 2012; Lowery et al. 2008). However, carbocyclic analogues of DPD did not display significant quenching activities in *S. typhimurium* or *V. harveyi* (Tsuchikama et al. 2011; Guo et al. 2013). Generally, AI-2 QS antagonists can exhibit two different types of activities being antagonists or synergistic agonists (LaSarre and Federle 2013). For example, propyl-DPD and other analogues with longer alkyl chain substitutions were all antagonists of AI-2 QS in *E. coli*, but only butyl-DPD had antagonistic activity in *S. typhimurium* (LaSarre and Federle 2013; Gamby et al. 2012).

The main concern in the developing of AI-2 QSI is the assessment of their stability and efficacy in vivo. Guo et al. recently demonstrated that a bis-ester-protected form of isobutyl-DPD can effectively inhibit the AI-2 QS signalling in *E. coli* comparably to the unprotected inhibitor alleviating instability issues of AI-2 molecules (Lowery et al. 2009; Guo et al. 2012).

Silencing Bacterial Communication by Targeting the Efflux Pumps

Bacteria can evolve resistance towards noxious agents such as antibiotics, through their extrusion to the environment before reaching the target using transport proteins, called efflux pumps (Rodríguez-Rojas et al. 2013; Webber and Piddock 2003). Two different types of efflux pumps have been identified in bacteria, depending on the energy source they use to export the antibiotic: (i) ABC transporters that use ATP for the expulsion of the noxious agent to the environment and (ii) proton-motive-force-dependent transporters that use energy from their cellular metabolism. Recent evidences suggested that the expression of efflux pumps in bacteria is QS mediated and thus could be used in the control of bacteria-mediated infections (Leonard Amaral 2012). For instance, a link between QS and active efflux pump has been demonstrated in *P. aeruginosa*, in which the expression of MexAB-OprM pump was enhanced by the QS signals. As already described, *P. aeruginosa* has two distinct QS systems, *las* and *rhl*, that control the synthesis of 3-oxo-C12-HSL and C4-HSL, respectively. Exogenous addition of C4-HSL QS signals was shown to enhance the expression of MexAB-OprM pump, whereas 3-oxo-C12-HSL had little or no effect, which means that *rhl* QS system in *P. aeruginosa* is actively involved in the efflux pump expression (Sawada et al. 2004). Targeting inhibition of QS mechanisms can thus have an effect on efflux pumps and therefore provide an effective strategy to fight bacterial antibiotic resistance and QS-regulated processes simultaneously.

Interestingly, several compounds such as phenothiazines and trifluoromethyl ketones (TFs) that inhibit various energy-dependent activities such as swarming and swimming were demonstrated to affect both the efflux pump systems of bacteria and their QS systems (Leonard Amaral 2012; Spengler et al. 2004; Molnar et al. 2004; Wolfart et al. 2004). Varga et al. investigated a series of TFs known to inhibit the proton motive force for their effects on the efflux pump of a QS-responding bacterium, for its subsequent effect on the response to a QS signal and its direct inhibition of the response to a QS signal. Some of them were proved to be effective inhibitors of the QS response by the reporter strain *C. violaceum* CV026, as well as inhibit the efflux pumps of both CV026 and *E. coli*, demonstrating that TFs have QS-inhibitory properties that are mediated through the inhibition of efflux pumps that extrude the noxious QS signal before it reaches its intended target (Varga et al. 2012).

Conclusion

QS has been widely used as a term to describe density-dependent bacterial cell-cell communication process since its initial discovery. The mechanistic understanding of this QS system in many bacteria has been growing since then and is now considered as a viable target to alleviate or prevent infection. This new approach has been met with great optimism among research community due to the antibiotic resistance problem that currently impedes treatment of infections caused by many bacterial pathogens. In fact, QS-based strategies are a way of scrambling the wires of bacterial communication before it starts. As a consequence, bacteria will never know that they have achieved the threshold population density necessary to establish an infection in human body. The main advantage of this approach is that it attenuates virulence, exerting less selective pressure on bacteria and reducing the risk of resistance development, due to the fact that QS is not an essential metabolic process for bacterial growth. Despite being a very appealing approach and many enzymes and chemicals have been proved to possess QSI activity, only a few of them have been tested for in vivo efficacy. Nevertheless, some questions still need to be answered in order to evaluate their potential as therapeutic agents, such as whether these agents can elicit an immune response in humans, whether degradation of AHLs would affect the commensal gut microflora of humans to the extent that this may be harmful to the host and whether bacteria can gain resistance to the QSIs.

Acknowledgements This work was supported by the European project "Novel approaches for prevention of biofilms formed on medical indwelling devices, e.g. catheters" (FP7-27840). K.I. would like to thank Generalitat de Catalunya for providing her the Ph.D. grant (FI_B1 00188) and M.M.F. acknowledges the Marie Curie Intra-European Fellowship (IEF) NanoQuench (FP7-331416).

References

Alfaro JF, Zhang T, Wynn DP, Karschner EL, Zhou ZS (2004) Synthesis of LuxS inhibitors targeting bacterial cell-cell communication. Org Lett 6(18):3043–3046. doi:10.1021/ol049182i

Amara N, Krom BP, Kaufmann GF, Meijler MM (2010) Macromolecular inhibition of Quorum sensing: enzymes, antibodies, and beyond. Chem Rev 111(1):195–208. doi:10.1021/cr100101c

Atkinson S, Williams P (2009) Quorum sensing and social networking in the microbial world. J R Soc Interface 6(40):959–978

Atkinson S, Throup JP, Stewart GS, Williams P (1999) A hierarchical quorum-sensing system in Yersinia pseudotuberculosis is involved in the regulation of motility and clumping. Mol Microbiol 33(6): 1267–1277

Baron S (ed) (1996) Medical microbiology, 4th edn. The University of Texas, Galveston

Benneche T, Hussain Z, Aamdal Scheie A, Lonn-Stensrud J (2008) Synthesis of 5-(bromomethylene)furan-2(5H)-ones and 3-(bromomethylene)isobenzofuran-1(3H)-ones as inhibitors of microbial quorum sensing. New J Chem 32(9):1567–1572. doi:10.1039/b803926g

Brackman G, Celen S, Baruah K, Bossier P, Van Calenbergh S, Nelis HJ, Coenye T (2009) AI-2 quorum-sensing inhibitors affect the starvation response and reduce virulence in several Vibrio species, most likely by interfering with LuxPQ. Microbiology 155(12):4114–4122. doi:10.1099/mic.0.032474-0

Brackman G, Al Quntar AAA, Enk CD, Karalic I, Nelis HJ, Van Calenbergh S, Srebnik M, Coenye T (2013) Synthesis and evaluation of thiazolidinedione and dioxazaborocane analogues as inhibitors of AI-2 quorum sensing in Vibrio harveyi. Bioorg Med Chem 21(3):660–667. doi: http://dx.doi.org/10.1016/j.bmc.2012.11.055

Byers JT, Lucas C, Salmond GPC, Welch M (2002) Nonenzymatic turnover of an Erwinia carotovora quorum-sensing signaling molecule. J Bacteriol 184(4):1163–1171

Calfee MW, Coleman JP, Pesci EC (2001) Interference with Pseudomonas quinolone signal synthesis inhibits virulence factor expression by Pseudomonas aeruginosa. Proc Natl Acad Sci 98(20):11633–11637. doi:10.1073/pnas.201328498

Camps J, Pujol I, Ballester F, Joven J, Simó JM (2011) Paraoxonases as potential antibiofilm agents: their relationship with quorum-sensing signals in gram-negative bacteria. Antimicrob Agents Chemother 55(4):1325–1331

Cao JG, Meighen EA (1989) Purification and structural identification of an autoinducer for the luminescence system of Vibrio harveyi. J Biol Chem 264(36):21670–21676

Chan YY, Chua KL (2005) The Burkholderia pseudomallei BpeAB-OprB Efflux Pump: expression and impact on quorum sensing and virulence. J Bacteriol 187(14):4707–4719. doi:10.1128/jb.187.14.4707-4719.2005

Chan K-G, Atkinson S, Mathee K, Sam C-K, Chhabra SR, Cámara M, Koh C-L, Williams P (2011) Characterization of N-acylhomoserine lactone-degrading bacteria associated with the Zingiber officinale (ginger) rhizosphere: co-existence of quorum quenching and quorum sensing in Acinetobacter and Burkholderia. BMC Microbiol 11(1):51

Chenia HY (2013) Anti-quorum sensing potential of crude Kigelia Africana fruit extracts. Sensors (Basel, Switzerland) 13(3):2802–2817

Choudhary S, Schmidt-Dannert C (2010) Applications of quorum sensing in biotechnology. Appl Microbiol Biotechnol 86(5):1267–1279. doi:10.1007/s00253-010-2521-7

Chung J, Goo E, Yu S, Choi O, Lee J, Kim J, Kim H, Igarashi J, Suga H, Moon JS, Hwang I, Rhee S (2011) Small-molecule inhibitor binding to an N-acyl-homoserine lactone synthase. Proc Natl Acad Sci 108(29):12089–12094. doi:10.1073/pnas.1103165108

Churchill MEA, Chen L (2010) Structural basis of acyl-homoserine lactone-dependent signaling. Chem Rev 111(1):68–85. doi:10.1021/cr1000817

De Kievit TR (2009) Quorum sensing in Pseudomonas aeruginosa biofilms. Environ Microbiol 11(2):279–288. doi:10.1111/j.1462-2920.2008.01792.x

De Lamo Marin S, Xu Y, Meijler MM, Janda KD (2007) Antibody catalyzed hydrolysis of a quorum sensing signal found in Gram-negative bacteria. Bioorg Med Chem Lett 17(6):1549–1552. doi:http://dx.doi.org/10.1016/j.bmcl.2006.12.118

Defoirdt T, Benneche T, Brackman G, Coenye T, Sorgeloos P, Scheie AA (2012) A quorum sensing-disrupting brominated thiophenone with a promising therapeutic potential to treat luminescent vibriosis. PLoS One 7(7):e41788. doi:10.1371/journal.pone.0041788

Dong YH, Wang LH, Xu JL, Zhang HB, Zhang XF, Zhang LH (2001) Quenching quorum-sensing-dependent bacterial infection by an N-acyl homoserine lactonase. Nature 411(6839):813–817

Fekete A, Kuttler C, Rothballer M, Hense BA, Fischer D, Buddrus-Schiemann K, Lucio M, Müller J, Schmitt-Kopplin P, Hartmann A (2010) Dynamic regulation of N-acyl-homoserine lactone production and degradation in Pseudomonas putida IsoF. FEMS Microbiol Ecol 72(1):22–34. doi:10.1111/j.1574-6941.2009.00828.x

Fux CA, Costerton JW et al (2005) Survival strategies of infectious biofilms. Trends Microbiol 13(1):34–40

Galloway WRJD, Hodgkinson JT, Bowden SD, Welch M, Spring DR (2010) Quorum sensing in gram-negative bacteria: small-molecule modulation of AHL and AI-2 quorum sensing pathways. Chem Rev 111(1):28–67. doi:10.1021/cr100109t

Galloway WRJD, Hodgkinson JT, Bowden S, Welch M, Spring DR (2012) Applications of small molecule activators and inhibitors of quorum sensing in Gram-negative bacteria. Trends Microbiol 20(9):449–458. doi:http://dx.doi.org/10.1016/j.tim.2012.06.003

Gamby S, Roy V, Guo M, Smith JAI, Wang J, Stewart JE, Wang X, Bentley WE, Sintim HO (2012) Altering the communication networks of multispecies microbial systems using a diverse toolbox of AI-2 analogues. ACS Chem Biol 7(6):1023–1030

Geske GD, Mattmann ME, Blackwell HE (2008) Evaluation of a focused library of N-aryl l-homoserine lactones reveals a new set of potent quorum sensing modulators. Bioorg Med Chem Lett 18(22):5978–5981. doi:http://dx.doi.org/10.1016/j.bmcl.2008.07.089

Givskov M, de Nys R, Manefield M, Gram L, Maximilien R, Eberl L, Molin S, Steinberg PD, Kjelleberg S (1996) Eukaryotic interference with homoserine lactone-mediated prokaryotic signalling. J Bacteriol 178(22):6618–6622

Guo M, Gamby S, Nakayama S, Smith J, Sintim HO (2012) A pro-drug approach for selective modulation of AI-2-mediated bacterial cell-to-cell communication. Sensors 12(3):3762–3772

Guo M, Gamby S, Zheng Y, Sintim H (2013) Small molecule inhibitors of AI-2 signaling in bacteria: state-of-the-art and future perspectives for anti-quorum sensing agents. Int J Mol Sci 14(9): 17694–17728

Haudecoeur E, Faure D (2010) A fine control of quorum-sensing communication in Agrobacterium tumefaciens. Commun Integr Biol 3(2):84–88

Hentzer M, Givskov M (2003) Pharmacological inhibition of quorum sensing for the treatment of chronic bacterial infections. J Clin Invest 112(9):1300–1307. doi:10.1172/jci20074

Hong K-W, Koh C-L, Sam C-K, Yin W-F, Chan K-G (2012) Quorum quenching revisited—from signal decays to signalling confusion. Sensors 12(4): 4661–4696

Huang JJ, Han JI, Zhang LH, Leadbetter JR (2003) Utilization of acyl-homoserine lactone quorum signals for growth by a soil pseudomonad and pseudomonas aeruginosa PAO1. Appl Environ Microbiol 69(10):5941–5949

Ishida T, Ikeda T, Takiguchi N, Kuroda A, Ohtake H, Kato J (2007) Inhibition of quorum sensing in Pseudomonas aeruginosa by N-acyl cyclopentylamides. Appl Environ Microbiol 73(10):3183–3188

Ivanova K, Fernandes MM et al (2013) Current advances on bacterial pathogenesis inhibition and treatment strategies. Microbial pathogens and strategies for combating them: science, technology and education. Mendez-Vilas A (ed), vol 1, pp 322–336. Formatex Research Center, Badajoz

Kalia VC (2013) Quorum sensing inhibitors: an overview. Biotechnol Adv 31(2):224–245. doi:http://dx.doi.org/10.1016/j.biotechadv.2012.10.004

Kalia VC, Purohit HJ (2011) Quenching the quorum sensing system: potential antibacterial drug targets. Crit Rev Microbiol 37(2):121–140

Kang-Mu LEE, Wan-Seok KIM, Jeesun LIM, Sunyoung NAM, Youn MIN, Seong-Won NAM, Younghoon KIM, Sae-Hun KIM, Park W, Park S (2009) Antipathogenic properties of green tea polyphenol epigallocatechin gallate at concentrations below the MIC against enterohemorrhagic escherichia coli O157:H7. J Food Prot 72(2):325–331

Kaufmann GF, Sartorio R, Lee S-H, Mee JM, Altobell LJ, Kujawa DP, Jeffries E, Clapham B, Meijler MM, Janda KD (2006) Antibody interference with N-acyl homoserine lactone-mediated bacterial quorum sensing. J Am Chem Soc 128(9):2802–2803. doi:10.1021/ja0578698

Kaufmann GF, Park J et al (2008) The quorum quenching antibody RS2-1G9 protects macrophages from the cytotoxic effects of the *Pseudomonas aeruginosa* quorum sensing signalling molecule N-3-oxo-dodecanoyl-homoserine lactone. Mol Immunol 45(9):2710–2714

Kim C, Kim J, Park H-Y, Lee J-H, Park H-J, Kim C, Yoon J (2009a) Structural understanding of quorum-sensing inhibitors by molecular modeling study in Pseudomonas aeruginosa. Appl Microbiol Biotechnol 83(6):1095–1103. doi:10.1007/s00253-009-1954-3

Kim C, Kim J, Park H-Y, Park H-J, Kim C, Yoon J, Lee J-H (2009b) Development of inhibitors against TraR quorum-sensing system in Agrobacterium tumefaciens by molecular modeling of the ligand-receptor interaction. Mol Cells 28(5):447–453. doi:10.1007/s10059-009-0144-6

Kiran S, Sharma P, Harjai K, Capalash N (2011) Enzymatic quorum quenching increases antibiotic susceptibility of multidrug resistant Pseudomonas aeruginosa. Iran J Microbiol 3(1):1–12

Kirwan JP, Gould TA, Schweizer HP, Bearden SW, Murphy RC, Churchill MEA (2006) Quorum-sensing signal synthesis by the Yersinia pestis acyl-homoserine lactone synthase YspI. J Bacteriol 188(2):784–788

Koch B, Liljefors T, Persson T, Nielsen J, Kjelleberg S, Givskov M (2005) The LuxR receptor: the sites of interaction with quorum-sensing signals and inhibitors. Microbiology 151(11):3589–3602. doi:10.1099/mic.0.27954-0

Kociolek MG (2009) Quorum-sensing inhibitors and biofilms. Anti-Infect Agent Med Chem 8(4):315–326. doi:10.2174/187152109789760117

Kulanthaivel P, Kreuzman AJ et al (2004) Novel lipoglycopeptides as inhibitors of bacterial signal peptidase I. J Biol Chem 279(35):36250–36258

LaSarre B, Federle MJ (2013) Exploiting quorum sensing to confuse bacterial pathogens. Microbiol Mol Biol Rev 77(1):73–111. doi:10.1128/mmbr.00046-12

Leonard Amaral JM (2012) Inhibitors of efflux pumps of Gram-negative bacteria inhibit Quorum Sensing. Open J Pharmacol 3

Lesic B, Lépine F, Déziel E, Zhang J, Zhang Q, Padfield K, Castonguay M-H, Milot S, Stachel S, Tzika AA, Tompkins RG, Rahme LG (2007) Inhibitors of pathogen intercellular signals as selective anti-infective compounds. PLoS Pathog 3(9):e126. doi:10.1371/journal.ppat.0030126

Li J, Wang W, Xu SX, Magarvey NA, McCormick JK (2011) Lactobacillus reuteri-produced cyclic dipeptides quench agr-mediated expression of toxic shock syndrome toxin-1 in staphylococci. Proc Natl Acad Sci 108(8):3360–3365. doi:10.1073/pnas.1017431108

Lin YH, Xu JL, Hu J, Wang LH, Leong Ong S, Renton Leadbetter J, Zhang LH (2003) Acyl-homoserine lactone acylase from Ralstonia strain XJ12B represents a novel and potent class of quorum-quenching enzymes. Mol Microbiol 47(3):849–860

Lowery CA, Park J, Kaufmann GF, Janda KD (2008) An unexpected switch in the modulation of AI-2-based quorum sensing discovered through synthetic 4,5-dihydroxy-2,3-pentanedione analogues. J Am Chem Soc 130(29):9200–9201. doi:10.1021/ja802353j

Lowery CA, Abe T, Park J, Eubanks LM, Sawada D, Kaufmann GF, Janda KD (2009) Revisiting AI-2 quorum sensing inhibitors: direct comparison of Alkyl-DPD analogues and a natural product fimbrolide. J Am Chem Soc 131(43):15584–15585. doi:10.1021/ja9066783

Lupp C, Ruby EG (2005) Vibrio fischeri uses two quorum-sensing systems for the regulation of early and late colonization factors. J Bacteriol 187(11): 3620–3629

Lyon GJ, Mayville P, Muir TW, Novick RP (2000) Rational design of a global inhibitor of the virulence response in Staphylococcus aureus, based in part on localization of the site of inhibition to the receptor-histidine kinase, AgrC. Proc Natl Acad Sci 97(24):13330–13335. doi:10.1073/pnas.97.24.13330

Lyon GJ, Wright JS, Muir TW, Novick RP (2002) Key determinants of receptor activation in the agr autoinducing peptides of Staphylococcus aureus. Biochemistry 41(31):10095–10104. doi:10.1021/bi026049u

Malladi VLA, Sobczak AJ, Meyer TM, Pei D, Wnuk SF (2011) Inhibition of LuxS by S-ribosylhomocysteine analogues containing a [4-aza]ribose ring. Bioorg Med Chem 19(18):5507–5519. doi:http://dx.doi.org/10.1016/j.bmc.2011.07.043

McClean KH, Winson MK, Fish L, Taylor A, Chhabra SR, Camara M, Daykin M, Lamb JH, Swift S, Bycroft BW, Stewart GS, Williams P (1997) Quorum sensing and Chromobacterium violaceum: exploitation of violacein production and inhibition for the detection of N-acylhomoserine lactones. Microbiology 143(Pt 12):3703–3711

Mihalik K, Chung DW, Crixell SH, McLean RJC, Vattem DA (2008) Quorum sensing modulators of Pseudomonas aeruginosa characterized in Camellia sinensis. Asian J Trad Med 3, 12–23

Miller ST, Xavier KB, Campagna SR, Taga ME, Semmelhack MF, Bassler BL, Hughson FM (2004) Salmonella typhimurium recognizes a chemically distinct form of the bacterial quorum-sensing signal AI-2. Mol Cell 15(5):677–687

Molnar A, Wolfart K, Kawase M, Motohashi N, Molnar J (2004) Effect of a trifluoromethyl ketone on the motility of proton pump-deleted mutant of Escherichia coli strain and its wild-type. Vivo 18(4):505–507

Moore JD, Gerdt JP, Eibergen NR, Blackwell HE (2014) Active efflux influences the potency of quorum sensing inhibitors in pseudomonas aeruginosa. ChemBioChem 15(3):435–442. doi:10.1002/cbic.201300701

Morohoshi T, Shiono T et al (2007) Inhibition of quorum sensing in Serratia marcescens AS-1 by synthetic analogs of N-acylhomoserine lactone. Appl Environ Microbiol 73(20):6339–6344

Nasser W, Bouillant ML, Salmond G, Reverchon S (1998) Characterization of the Erwinia chrysanthemi expI-expR locus directing the synthesis of two N-acyl-homoserine lactone signal molecules. Mol Microbiol 29(6):1391–1405

Ni N, Chou H-T, Wang J, Li M, Lu C-D, Tai PC, Wang B (2008a) Identification of boronic acids as antagonists of bacterial quorum sensing in Vibrio harveyi. Biochem Biophys Res Commun 369(2):590–594. doi:http://dx.doi.org/10.1016/j.bbrc.2008.02.061

Ni N, Choudhary G, Li M, Wang B (2008b) Pyrogallol and its analogs can antagonize bacterial quorum sensing in Vibrio harveyi. Bioorg Med Chem Lett 18(5):1567–1572. doi:http://dx.doi.org/10.1016/j.bmcl.2008.01.081

Ni N, Choudhary G, Peng H, Li M, Chou H-T, Lu C-D, Gilbert ES, Wang B (2009a) Inhibition of quorum sensing in Vibrio harveyi by boronic acids. Chem Biol Drug Des 74(1):51–56. doi:10.1111/j.1747-0285.2009.00834.x

Ni N, Li M, Wang J, Wang B (2009b) Inhibitors and antagonists of bacterial quorum sensing. Med Res Rev 29(1):65–124. doi:10.1002/med.20145

O'Loughlin CT, Miller LC, Siryaporn A, Drescher K, Semmelhack MF, Bassler BL (2013) A quorum-sensing inhibitor blocks Pseudomonas aeruginosa virulence and biofilm formation. Proc Natl Acad Sci 110(44):17981–17986. doi:10.1073/pnas.1316981110

Olsen JA, Severinsen R, Rasmussen TB, Hentzer M, Givskov M, Nielsen J (2002) Synthesis of new 3- and 4-substituted analogues of acyl homoserine lactone quorum sensing autoinducers. Bioorg Med Chem Lett 12(3):325–328. doi:http://dx.doi.org/10.1016/S0960-894X(01)00756-9

Otto M, Süßmuth R, Vuong C, Jung G, Götz F (1999) Inhibition of virulence factor expression in Staphylococcus aureus by the Staphylococcus epidermidis agr pheromone and derivatives. FEBS Lett 450(3):257–262

Park J, Jagasia R et al (2007) Infection control by antibody disruption of bacterial quorum sensing signaling. Chem Biol 14(10):1119–1127

Parker CT, Sperandio V (2009) Cell-to-cell signalling during pathogenesis. Cell Microbiol 11(3):363–369. doi:10.1111/j.1462-5822.2008.01272.x

Parsek MR, Val DL, Hanzelka BL, Cronan JE, Greenberg EP (1999) Acyl homoserine-lactone quorum-sensing signal generation. Proc Natl Acad Sci U S A 96(8):4360–4365

Peeters E, Nelis H, Coenye T (2008) Comparison of multiple methods for quantification of microbial biofilms grown in microtiter plates. J Microbiol Methods 72(2):157–165

Peng H, Cheng Y, Ni N, Li M, Choudhary G, Chou HT, Lu C-D, Tai PC, Wang B (2009) Synthesis and evaluation of new antagonists of bacterial quorum sensing in Vibrio harveyi. ChemMedChem 4(9):1457–1468. doi:10.1002/cmdc.200900180

Pereira CS, Thompson JA, Xavier KB (2013) AI-2-mediated signalling in bacteria. FEMS Microbiol Rev 37(2):156–181. doi:10.1111/j.1574-6976.2012.00345.x

Piletska EV, Stavroulakis G, Karim K, Whitcombe MJ, Chianella I, Sharma A, Eboigbodin KE, Robinson GK, Piletsky SA (2010) Attenuation of Vibrio fischeri quorum sensing using rationally designed polymers. Biomacromolecules 11(4):975–980. doi:10.1021/bm901451j

Piletska EV et al (2011) Passive control of quorum sensing: prevention of *Pseudomonas aeruginosa* biofilm formation by imprinted polymers. Biomacromolecules 12(4):1067–1071

Pustelny C, Albers A, Büldt-Karentzopoulos K, Parschat K, Chhabra SR, Cámara M, Williams P, Fetzner S (2009) Dioxygenase-mediated quenching of quinolone-dependent quorum sensing in Pseudomonas aeruginosa. Chem Biol 16(12):1259–1267

Rasko DA, Sperandio V (2010) Anti-virulence strategies to combat bacteria-mediated disease. Nat Rev Drug Discov 9(2):117–128

Rasmussen TB, Givskov M (2006) Quorum sensing inhibitors: a bargain of effects. Microbiology 152(4):895–904

Rasmussen TB, Bjarnsholt T, Skindersoe ME, Hentzer M, Kristoffersen P, Köte M, Nielsen J, Eberl L, Givskov M (2005a) Screening for quorum-sensing inhibitors (QSI) by use of a novel genetic system, the QSI selector. J Bacteriol 187(5):1799–1814

Rasmussen TB, Skindersoe ME, Bjarnsholt T, Phipps RK, Christensen KB, Jensen PO, Andersen JB, Koch B, Larsen TO, Hentzer M, Eberl L, Hoiby N, Givskov M (2005b) Identity and effects of quorum-sensing inhibitors produced by Penicillium species. Microbiology 151(5):1325–1340. doi:10.1099/mic.0.27715-0

Relman DA (2012) Microbiology: learning about who we are. Nature 486(7402):194–195

Roche DM, Byers JT, Smith DS, Glansdorp FG, Spring DR, Welch M (2004) Communications blackout? Do N-acylhomoserine-lactone-degrading enzymes have any role in quorum sensing? Microbiology 150(7):2023–2028

Rodríguez-Rojas A, Rodríguez-Beltrán J, Couce A, Blázquez J (2013) Antibiotics and antibiotic resistance: a bitter fight against evolution. Int J Med Microbiol 303(6–7):293–297. doi:http://dx.doi.org/10.1016/j.ijmm.2013.02.004

Rutherford ST, Bassler BL (2012) Bacterial quorum sensing: its role in virulence and possibilities for its control. Cold Spring Harbor Perspect Med 2(11)

Swift S, Karlyshev AV, Fish L, Durant EL, Winson MK, Chhabra SR, Williams P, Macintyre S, Stewart GS (1997) Quorum sensing in Aeromonas hydrophila and Aeromonas salmonicida: identification of the LuxRI homologs AhyRI and AsaRI and their cognate N-acylhomoserine lactone signal molecules. Research Repository. http://eprints.kingston.ac.uk/4461/

Sawada I, Maseda H, Nakae T, Uchiyama H, Nomura N (2004) A quorum-sensing autoinducer enhances the mexAB-oprM efflux-pump expression without the MexR-mediated regulation in Pseudomonas aeruginosa. Microbiol Immunol 48(5):435–439. doi:10.1111/j.1348-0421.2004.tb03533.x

Shen G, Rajan R, Zhu J, Bell CE, Pei D (2006) Design and synthesis of substrate and intermediate analogue inhibitors of S-ribosylhomocysteinase. J Med Chem 49(10):3003–3011. doi:10.1021/jm060047g

Sintim HO, Smith JAI, Wang J, Nakayama S, Yan L (2010) Paradigm shift in discovering next-generation anti-infective agents: targeting quorum sensing, c-di-GMP signaling and biofilm formation in bacteria with small molecules. Future Med Chem 2(6):1005–1035. doi:10.4155/fmc.10.185

Skindersoe ME, Alhede M, Phipps R, Yang L, Jensen PO, Rasmussen TB, Bjarnsholt T, Tolker-Nielsen T, Høiby N, Givskov M (2008) Effects of antibiotics on quorum sensing in Pseudomonas aeruginosa. Antimicrob Agents Chemother 52(10):3648–3663

Smith RS, Iglewski BH (2003) P. aeruginosa quorum-sensing systems and virulence. Curr Opin Microbiol 6(1):56–60. doi:http://dx.doi.org/10.1016/S1369-5274(03)00008-0

Sokol PA, Sajjan U, Visser MB, Gingues S, Forstner J, Kooi C (2003) The CepIR quorum-sensing system contributes to the virulence of Burkholderia cenocepacia respiratory infections. Microbiology 149(Pt 12): 3649–3658

Spengler G, Molnár A, Klausz G, Mándi Y, Kawase M, Motohashi N, Molnár J (2004) The antimotility action of a trifluoromethyl ketone on some gram-negative bacteria. Acta Microbiol Immunol Hung 51(3):351–358. doi:10.1556/AMicr.51.2004.3.12

Tal-Gan Y, Stacy DM, Foegen MK, Koenig DW, Blackwell HE (2013) Highly potent inhibitors of quorum sensing in Staphylococcus aureus revealed through a systematic synthetic study of the Group-III autoinducing peptide. J Am Chem Soc 135(21):7869–7882. doi:10.1021/ja3112115

Tay S, Yew W (2013) Development of quorum-based anti-virulence therapeutics targeting gram-negative bacterial pathogens. Int J Mol Sci 14(8):16570–16599

Teiber JF, Horke S, Haines DC, Chowdhary PK, Xiao J, Kramer GL, Haley RW, Draganov DI (2008) Dominant role of paraoxonases in inactivation of the Pseudomonas aeruginosa quorum-sensing signal N-(3-oxododecanoyl)-L-homoserine lactone. Infect Immun 76(6):2512–2519

Tsuchikama K, Lowery CA, Janda KD (2011) Probing autoinducer-2 based quorum sensing: the biological consequences of molecules unable to traverse equilibrium states. J Org Chem 76(17):6981–6989. doi:10.1021/jo200882k

Uroz S, Chhabra SR, Cámara M, Williams P, Oger P, Dessaux Y (2005) N-acylhomoserine lactone quorum-sensing molecules are modified and degraded by Rhodococcus erythropolis W2 by both amidolytic and novel oxidoreductase activities. Microbiology 151(10):3313–3322

Varga ZG, Armada A, Cerca P, Amaral L, Mior Ahmad Subki MA, Savka MA, Szegedi E, Kawase M, Motohashi N, Molnar J (2012) Inhibition of quorum sensing and efflux pump system by trifluoromethyl ketone proton pump inhibitors. Vivo 26(2):277–285

Wang W, Morohoshi T, Ikeda T, Chen L (2008) Inhibition of Lux quorum-sensing system by synthetic N-acyl-L-homoserine lactone analogous. Acta Biochim Biophys Sin 40(12):1023–1028. doi:10.1111/j.1745-7270.2008.00490.x

Watson WT, Minogue TD, Val DL, von Bodman SB, Churchill MEA (2002) Structural basis and specificity of acyl-homoserine lactone signal production in bacterial quorum sensing. Mol Cell 9(3):685–694. doi:http://dx.doi.org/10.1016/S1097-2765(02)00480-X

Webber MA, Piddock LJV (2003) The importance of efflux pumps in bacterial antibiotic resistance. J Antimicrob Chemother 51(1):9–11. doi:10.1093/jac/dkg050

Welch M, Dutton JM, Glansdorp FG, Thomas GL, Smith DS, Coulthurst SJ, Barnard AML, Salmond GPC, Spring DR (2005) Structure activity relationships of Erwinia carotovora quorum sensing signaling molecules. Bioorg Med Chem Lett 15(19):4235–4238

Wolfart K, Molnar A, Kawase M, Motohashi N, Molnar J (2004) Effects of trifluoromethyl ketones on the motility of Proteus vulgaris. Biol Pharma Bull 27(9): 1462–1464

Wright JS, Jin R, Novick RP (2005) Transient interference with staphylococcal quorum sensing blocks abscess formation. Proc Natl Acad Sci U S A 102(5):1691–1696. doi:10.1073/pnas.0407661102

Wu H, Song Z et al (2004) Synthetic furanones inhibit quorum-sensing and enhance bacterial clearance in *Pseudomonas aeruginosa* lung infection in mice. J Antimicrob Chemother 53(6):1054–1061

Xavier KB, Bassler BL (2005) Interference with AI-2-mediated bacterial cell-cell communication. Nature 437(7059):750–753

Xu F, Byun T, Dussen H-J, Duke KR (2003) Degradation of N-acylhomoserine lactones, the bacterial quorum-sensing molecules, by acylase. J Biotechnol 101(1):89–96. doi:http://dx.doi.org/10.1016/S0168-1656(02)00305-X

Yang Y-X, Xu Z-H, Zhang Y-Q, Tian J, Weng L-X, Wang L-H (2012) A new quorum-sensing inhibitor attenuates virulence and decreases antibiotic resistance in Pseudomonas aeruginosa. J Microbiol 50(6):987–993. doi:10.1007/s12275-012-2149-7

Zang T, Lee BWK, Cannon LM, Ritter KA, Dai S, Ren D, Wood TK, Zhou ZS (2009) A naturally occurring brominated furanone covalently modifies and inactivates LuxS. Bioorg Med Chem Lett 19(21):6200–6204. doi:http://dx.doi.org/10.1016/j.bmcl.2009.08.095

Zhang Z, Pierson LS (2001) A second quorum-sensing system regulates cell surface properties but not phenazine antibiotic production in Pseudomonas aureofaciens. Appl Environ Microbiol 67(9):4305–4315

Zhu L, Lau GW (2011) Inhibition of competence development, horizontal gene transfer and virulence in Streptococcus pneumoniae by a modified competence stimulating peptide. PLoS Pathog 7(9):e1002241. doi:10.1371/journal.ppat.1002241

Zhu P, Peng H, Ni N, Wang B, Li M (2012) Novel AI-2 quorum sensing inhibitors in Vibrio harveyi identified through structure-based virtual screening. Bioorg Med Chem Lett 22(20):6413–6417. doi:http://dx.doi.org/10.1016/j.bmcl.2012.08.062

Part IV

Natural Quorum Sensing Inhibitors

Silencing Bacterial Communication Through Enzymatic Quorum-Sensing Inhibition

Manuel Romero, Celia Mayer, Andrea Muras, and Ana Otero

Introduction

The development of antibiotic resistance by some pathogenic bacteria is a serious global problem, giving rise to multiresistance strains wherein treatment is longer and frequently ineffective and forcing the pharmaceutical industry to develop new generations of more potent antibiotics, as well as exploring new strategies to fight bacterial infections. Since a great number of human, animal, plant, and aquatic pathogens use cell-to-cell bacterial communication or "quorum sensing" (QS) systems to regulate the production of virulence factors (Bruhn et al. 2005; Williams et al. 2007), interference with these communication systems constitutes an interesting and promising strategy for the control of infectious bacterial diseases (Dong et al. 2007). Of special significance is the role of QS processes in the formation and differentiation of bacterial biofilms (Williams et al. 2007), since these organized structures are involved in antibiotic resistance in several important pathogens and constitute the basis of other important biological problems such as fouling, corrosion, etc. The mechanisms causing the inactivation of QS communication systems have been generally termed as "quorum-sensing inhibition" (QSI). One of the major interests of the QSI approach is that it blocks the expression of the virulence factors, but it does not exert direct selective pressure as it does not directly affect the pathogen survival, therefore reducing potential emergence of resistances. Furthermore, this strategy in combination with antibiotics may be an interesting alternative in the treatment of infectious diseases by multiresistant pathogens such as *Pseudomonas aeruginosa*, in which QS control on virulence mechanisms has been described (Jimenez et al. 2012).

Although the inactivation of all QS systems is equally interesting, the bacterial gram-negative signals *N*-acyl-homoserine lactones (AHLs), as the first-discovered and most-characterized pathway, have been the main target of this area of research. Among the AHL-based QSI systems, the *blockage of AHL synthesis* could be the most effective communication interception system; however, only a scarce number of studies have explored this possibility (Parsek et al. 1999; Pechere 2001). Interference with bacterial communication can also be achieved by *decreasing the concentration of active AHL* in the media. AHLs suffer spontaneous lactonization due to high pH values (Yates et al. 2002), but this can also occur by enzymatic mechanisms. Several bacteria with the capacity to degrade AHL signals have been described and could therefore be

M. Romero
School of Life Sciences, Centre for Biomolecular Sciences, University of Nottingham, NG7 2RD Nottingham, UK

C. Mayer • A. Muras • A. Otero (✉)
Departamento de Microbioloxía e Parasitoloxía, Universidade de Santiago de Compostela, Facultade de Bioloxía (CIBUS), 15782 Santiago de Compostela, Spain
e-mail: anamaria.otero@usc.es

V.C. Kalia (ed.), *Quorum Sensing vs Quorum Quenching: A Battle with No End in Sight*,
DOI 10.1007/978-81-322-1982-8_19, © Springer India 2015

used as QS quenchers and their enzymes could have great interest for their biotechnological applications. Inhibition of the signal efflux to prevent its accumulation in the environment has also been proposed as QSI strategy (Dong et al. 2007). QS signal transduction can be inhibited by antagonists able to compete or interfere with AHL and receptor union. The competitive antagonists would be similar enough to AHL to allow their union to the receptor but would not activate later signal transduction; noncompetitive antagonists show low or no structural similitude with AHLs and will join to different parts of the receptor (Ni et al. 2009).

Given that bacterial species that employ quorum sensing (QS) have important competition advantages in the ubiquitous interactions with other prokaryotes and eukaryotes, it is rational that competitors have evolved mechanisms to disarm their QS systems. Such mechanisms to avoid bacterial colonization and competence through QS interception have been found in nature. "Natural" QSI mechanisms described so far include the production of inhibitors or antagonists of signal reception by marine algae (Givskov et al. 1996), invertebrates (Skindersoe et al. 2008), terrestrial plants (Gao et al. 2003), and bacteria (Teasdale et al. 2009) or the enzymatic inactivation of signals found in mammalian cells (Camps et al. 2011), plants (Delalande et al. 2005), and bacteria (Dong and Zhang 2005). Besides QSI produced by competitors, different studies have developed synthetic signal antagonists and antibodies capable of sequestering QS signals (Kaufmann et al. 2006; Ni et al. 2009). Arguably, among the different possibilities, enzymatic degradation has been the most deeply explored strategy to date.

Bacterial AHLases

Up to now, numerous bacterial species with enzymatic QSI activity have been identified belonging to five phyla of the *Bacteria* kingdom: *Actinobacteria*, *Firmicutes*, *Proteobacteria*, *Bacteroidetes*, and *Cyanobacteria*, although the activity has not been characterized in many of them (bacteria with characterized enzymatic QSI activity can be found in Table 1). The wide distribution of enzymatic QSI activity among bacteria suggests that blocking bacterial communication is important to confer a competitive advantage in bacterial populations. Three main groups of AHL inactivation enzymes have been identified so far and classified according to the AHL cleavage/modification mechanism. The metalloproteins *AHL lactonases* hydrolyze the ester bond of the homoserine lactone ring (HSL) to produce the corresponding acyl-homoserines (Dong et al. 2000, 2001) (Fig. 1). These are the enzymes with the broadest range of AHL specificity as many of them degrade all signals independently of the size and substitutions in the acyl side chain (Liu et al. 2008; Otero et al. 2013). The second group are the acylases that cleave the AHL amide bond generating the corresponding free fatty acid and homoserine lactone ring (Leadbetter and Greenberg 2000; Lin et al. 2003) (Fig. 1). Unlike lactonases, acylases show more substrate specificity, probably due to recognition of the acyl chain of the signal as was shown for the *Pseudomonas* acylase PvdQ (Bokhove et al. 2010). The third group of known AHL-inactivating enzymes are the *oxidoreductases* that, in contrast to lactonase and acylase activities, do not degrade the signal but oxidize or reduce the acyl chain of the AHLs to modify them. These reactions do not degrade the signals, but the modifications affect the specificity and therefore the union between the signal and the receptor. However, the role of QSI enzymes as signal quenchers has been questioned as have been related to the catabolism of other molecules not related to QS such as γ-butyrolactone (Khan and Farrand 2009) or penicillin G (Park et al. 2005). We cannot therefore disregard the possibility that QSI enzymes have additional functions beyond AHL inactivation that would increase their ecological significance, including the mere obtainment of energy from the metabolism of the molecule, as demonstrated by the relative success of the enrichment methods for the isolation of strains with enzymatic QSI activity (Leadbetter and Greenberg 2000).

Table 1 Characterized prokaryotic QSI enzymes

Species/origin	Enzyme	Enzyme family	Target signal/activity	Reference
Bacillus sp.	AiiA	Metallo-β-lactamase	AHL lactonase	Dong et al. (2001)
Agrobacterium tumefaciens	AttM AiiB	Metallo-β-lactamase Metallo-β-lactamase	AHL lactonase AHL lactonase	Zhang et al. (2002) Carlier et al. (2003)
Klebsiella pneumoniae	AhlK	Metallo-β-lactamase	AHL lactonase	Park et al. (2003)
Arthrobacter sp.	AhlD	Metallo-β-lactamase	AHL lactonase	Park et al. (2003)
Mesorhizobium loti	MLR6805	Metallo-β-lactamase	AHL lactonase	Funami et al. (2005)
Rhizobium sp.	DlhR QsdR1	Dienelactone hydrolase Metallo-β-lactamase	AHL lactonase AHL lactonase	Krysciak et al. (2011)
Chryseobacterium sp.	AidC	Metallo-β-lactamase	AHL lactonase	Wang et al. (2012)
Solibacillus silvestris	AhlS	Metallo-β-lactamase	AHL lactonase	Morohoshi et al. (2012)
Tenacibaculum sp.	Aii20J	Metallo-β-lactamase	AHL lactonase	Otero et al. (2013)
Rhodococcus erythropolis	QsdA	Phosphotriesterase	AHL lactonase	Uroz et al. (2008)
Geobacillus kaustophilus	GKL	Phosphotriesterase	AHL lactonase	Chow et al. (2010)
Sulfolobus solfataricus	SsoPox	Phosphotriesterase	AHL lactonase	Merone et al. (2005)
Mycobacterium avium	MCP	Phosphotriesterase	AHL lactonase	Chow et al. (2009)
Microbacterium testaceum	AiiM	α/β hydrolase	AHL lactonase	Wang et al. (2010)
Ochrobactrum sp.	AidH AiiO	α/β hydrolase α/β hydrolase	AHL lactonase AHL acylase	Mei et al. (2010) Czajkowski et al. (2011)
Pseudoalteromonas byunsanensis	QsdH	GDSL-like hydrolase	AHL lactonase	Huang et al. (2012)
Soil metagenome	QlcA	Metallo-β-lactamase	AHL lactonase	Riaz et al. (2008)
Soil metagenome	BpiB01 BpiB04 BpiB05 BpiB07 BpiB09	– Glycosyl hydrolase – Dienelactone hydrolase NADP-dependent dehydrogenase/reductase	AHL lactonase AHL lactonase AHL lactonase AHL lactonase AHL oxidoreductase	Bijtenhoorn et al. (2011a, b), Schipper et al. (2009)
Ralstonia sp.	AiiD	Ntn hydrolase	AHL acylase	Lin et al. (2003)
Pseudomonas aeruginosa	PvdQ QuiP HacB	Ntn hydrolase Ntn hydrolase Ntn hydrolase	AHL acylase AHL acylase AHL acylase	Huang et al. (2003) Huang et al. (2006) Wahjudi et al. (2011)
Pseudomonas syringae	HacA	Ntn hydrolase	AHL acylase	Shepherd and Lindow (2009)
	HacB	Ntn hydrolase	AHL acylase	
Brucella melitensis	AibP	Ntn hydrolase	AHL acylase	Terwagne et al. (2013)
Streptomyces sp.	AhlM	Ntn hydrolase	AHL acylase	Park et al. (2005)
Kluyvera citrophila	KcPGA	Ntn hydrolase	AHL acylase	Mukherji et al. (2014)
Anabaena sp.	AiiC	Ntn hydrolase	AHL acylase	Romero et al. (2008)
Shewanella sp.	Aac	Ntn hydrolase	AHL acylase	Morohoshi et al. (2008)
Bacillus megaterium	CYP102A1	Cytochrome P450	AHL oxidoreductase	Chowdhary et al. (2007)
Pseudomonas sp.	CarAB	Carbamoyl phosphate synthase	DSF	Newman et al. (2008)
Arthrobacter nitroguajacolicus	Hod	Dioxygenase	PQS	Pustelny et al. (2009)
E. coli	LsrK	Kinase	AI-2	Roy et al. (2010)
	LsrG	–	AI-2	Marques et al. (2011)

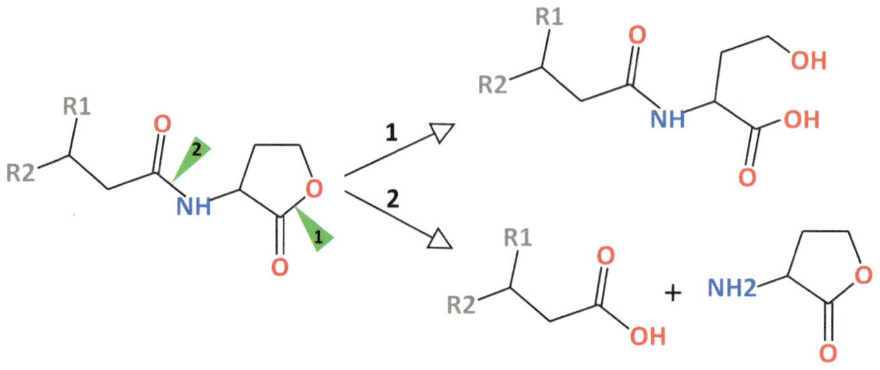

Fig. 1 AHL degradation mechanisms of lactonases (*1*) and acylases (*2*)

Lactonases

The first enzymatic degradation of AHLs was described in *Bacillus* sp. 240B1, detected after a screening of QSI activity in soil and laboratory isolates. Genomic DNA from strain 240B1 was then cloned in a library to identify the gene responsible for the AHL-inactivation activity. The gene responsible for the activity was named *aiiA* (*a*uto*i*nducer *i*nactivator), characterized as an AHL lactonase belonging to the metallo-β-lactamase superfamily and containing the Zn-binding motif "HXDH ~ H ~ D" that is conserved in several groups of metallohydrolases (Dong et al. 2001). This lactonase was later shown to be present in various species of the genus *Bacillus*, with the different AiiA homologues sharing 90 % identity at the amino acid sequence level (Lee et al. 2002; Molina et al. 2003).

In addition to *Bacillus*, other bacteria were shown to produce AHL lactonases belonging to the metallo-β-lactamase superfamily (Table 1, Fig. 2), including *Agrobacterium tumefaciens* (AttM and AiiB), *Klebsiella pneumoniae* (AhlK), *Arthrobacter* sp. (AhlD), *Mesorhizobium loti* (MLR6805), *Rhizobium* sp. (DlhR and QsdR1), *Chryseobacterium* sp. (AidC), *Solibacillus silvestris* (AhlS), and *Tenacibaculum* sp. (Aii20J) (Carlier et al. 2003; Funami et al. 2005; Krysciak et al. 2011; Morohoshi et al. 2012; Otero et al. 2013; Park et al. 2003; Wang et al. 2012; Zhang et al. 2002). All these lactonases belong to the same clade, presenting the conserved Zn-binding domain which has been proven to be essential for AHL-lactonase activity (Liu et al. 2008) (Fig. 2) even though they share low identity (<30 %) at the amino acidic sequence level. It should be noted that the identity of the AHL lactonases in this group does not correlate with phylogenetical proximity. For example, Aii20J, the lactonase from *Tenacibaculum*, a member of the CFB phylum, is located in a very distant branch from another member of this group, *Chryseobacterium* sp. (Fig. 2), which may indicate an active horizontal gene transfer of these genes. In an attempt to elucidate the role of the AHL-lactonase AiiA in *B. thuringiensis*, Park et al. (2008) performed competence assays using *B. thuringiensis aiiA* mutants and the AHL-producer plant pathogen *Pectobacterium carotovorum* (*Erwinia carotovora*) in a pepper root system. Results showed that the survival rate of the *aiiA* mutant significantly decreased over time compared with that of wild type suggesting that AiiA plays an important role in rhizosphere competence of *B. thuringiensis*. A remarkable case is the number of ORFs linked to AHL-QSI activity present in the nitrogen-fixing microsymbiont *Rhizobium* sp. strain NGR234. NGR234 possesses at least five different genes codifying for AHLases, two located on the plasmid pNGR234b (*dlhR* and *qsdR1*) that were demonstrated to function as AHL lactonases and three in the chromosome (*qsdR2*, *aldR*, and *hitR-hydR*). The affluence of QSI enzymes in NGR234 suggests that this is an important characteristic for rhizosphere colonization and soil growth, as it was shown that a NGR234 strain overexpressing *dlhR* and

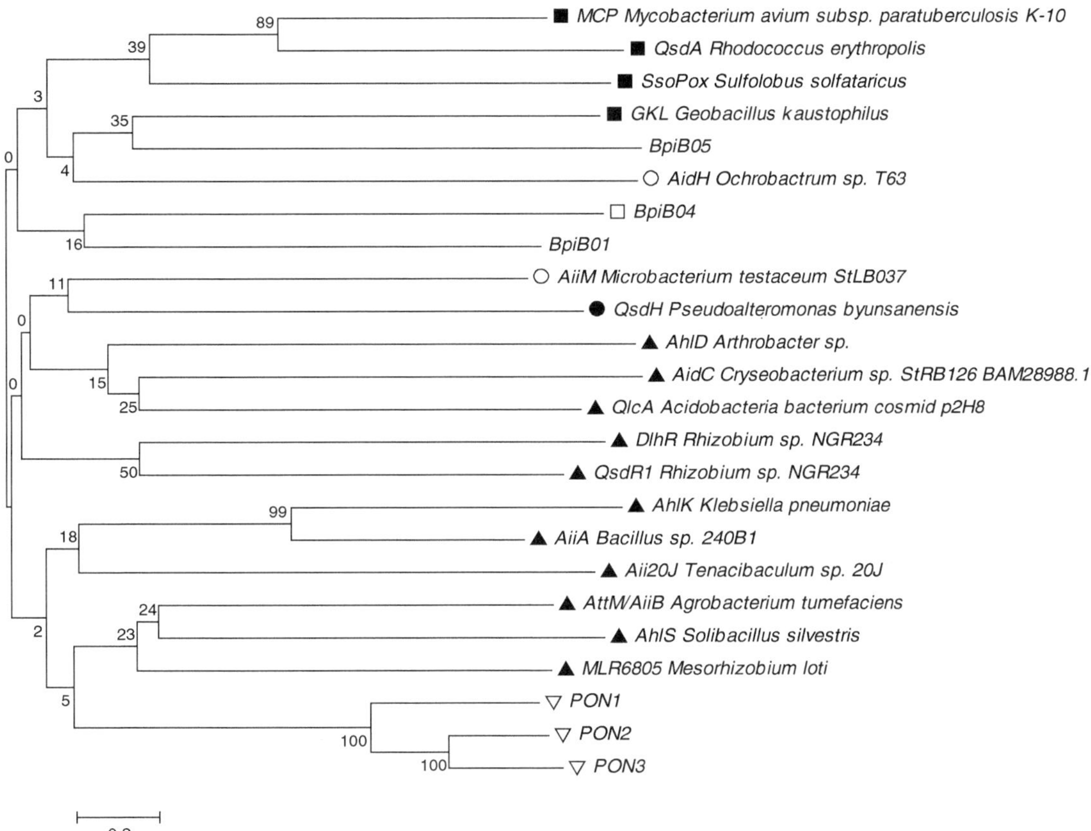

Fig. 2 Phylogenetic tree based on amino acid sequences of Aii20J from *Tenacibaculum* sp. 20 J (Otero et al. 2013), AhlD from *Arthrobacter* sp. (accession number AAP57766.1, Park et al. 2003), AhlK from *K. pneumoniae* (AAO47340.1, Park et al. 2003), AhlS from *Solibacillus silvestris* (BAK54003.1, Morohoshi et al. 2012), AidC from *Chryseobacterium* sp. strain StRB126 (BAM28988.1, Wang et al. 2012), AidH from *Ochrobactrum* sp. T63 (ACZ73823.1, Mei et al. 2010), AiiA from *Bacillus* sp. Strain 240B1 (AAF62398.1, Dong et al. 2000), AiiM from *Microbacterium testaceum* strain StlB037 (YP_004225655.1, Morohoshi et al. 2011), AttM/AiiB from *A. tumefaciens* (AAL13075.1, Zhang et al. 2002), QlcA from *Acidobacteria bacterium cosmid* p2H8 (ABV58973.1, Riaz et al. 2008), QsdA from *Rhodococcus erythropolis* (AAT06802.1, Uroz et al. 2008), DlhR and QsdR1 from *Rhizobium* sp. Strain NGR234 (ACP23660.1, Krysciak et al. 2011), MLR6805 from *Mesorhizobium loti* (BAB53031.1, Funami et al. 2005), GKL from *Geobacillus kaustophilus* (YP_147359.1, Chow et al. 2010), SsoPox from *Sulfolobus solfataricus* (AAW47234.1, Merone et al. 2005), QsdH from *Pseudoalteromonas byunsanensis* (AFV15299.1, Huang et al. 2012), MCP from *Mycobacterium avium* subsp. *paratuberculosis* strain k-10 (NP_962602.1, Chow et al. 2009), BpiB01, BpiB04, and BpiB05 from soil metagenomes (ABU51084.1, ABU51107.1, ABU51109.1; Schipper et al. 2009), and PON1, PON2, and PON3 from mouse serum (P52430.2, Q62086.2, Q62087.2; Camps et al. 2011). The enzymes belonging to the metallo-β-lactamase superfamily are indicated with filled triangles. PTE members are shown with filled squares, α/β hydrolase family members with unfilled circles, GDSL-like hydrolase family members with filled circles, glycosyl hydrolase family members with unfilled squares, and paraoxonases with unfilled triangles. The dendrogram was constructed by the neighbor-joining method using the MEGA 5.1 program. The scale bar represents 0.2 substitutions per amino acid position

qsdR1 genes was less effective in cowpea plants root colonization (Krysciak et al. 2011).

Despite most of the QSI systems are thought to be involved in bacterial competitive behaviors, AHL degradation enzymes can also function as QS self-regulators as was shown in the plant pathogen *Agrobacterium tumefaciens*, causative agent of the crown gall disease. This bacterium produces two AHL-lactonases AttM (BlcC) and AiiB (Carlier et al. 2003; Zhang et al.

2002), both enzymes fascinatingly modulate QS-regulated functions of the bacterium, such as conjugation of Ti plasmid and seriousness of plant symptoms, by controlling the levels of the own signal *N*-3-oxooctanoyl-homoserine lactone (OC8-HSL) (Haudecoeur and Faure 2010). In a proposed model, the infected plant receives the T-DNA from *A. tumefaciens* that integrates in the plant genome, inducing cell proliferation and tumor formation. Emerging tumor accumulates opines that trigger OC8-HSL production, but also the lactonase AiiB; in addition, plant-produced GABA and salicylic acid stimulate AttM expression; thus, signal levels remain low. Later, growing tumor starts to accumulate proline which antagonize GABA intake by *A. tumefaciens*. This may allow the OC8-HSL levels overcome the degradation capacity of the bacterium and the signal release in bacterial cell environment, triggering horizontal transfer of Ti plasmid and the seriousness of infection symptoms. Therefore, *A. tumefaciens* lactonases appear to contribute to bacterial fitness within the plant tumor and may also function in inter-kingdom signaling between the plant and the bacterium (Haudecoeur and Faure 2010).

The strain *Arthrobacter* sp. IBN110 was the first AHL-lactonase-producing bacterium obtained with AHL enrichment culture approach. This soil isolate was able to grow in media amended with *N*-3-oxohexanoyl-homoserine lactone (OC6-HSL) as the sole carbon source, and the enzyme catalyzing the AHL degradation was then identified and named AhlD, another member of the metallo-β-lactamase superfamily. Mass spectrometry analysis with the signal *N*-hexanoyl-homoserine lactone (C6-HSL) showed that AhlD hydrolyzed the AHL to generate one product presumably an acyl-homoserine (Park et al. 2003). Interestingly, a concurrent report showed that the isolate *Arthrobacter* sp. VAI-A, also obtained by enrichment methods, could only utilize the AHL breakdown products, HSL and acyl-homoserines, rather than AHL as the sole carbon and nitrogen source and could grow on AHLs only when cocultured with the AHL degrading bacteria *Variovorax paradoxus*. This result suggests the existence of enzymatic complementation activities of both bacteria and could explain why none of the three compounds accumulate in soil environments (Flagan et al. 2003).

On the other hand, Funami et al. (2005) studying a 4-pyridoxolactonase (MLR6805), another member of the metallo-β-lactamase superfamily involved in the degradation pathway of pyridoxine in the nitrogen-fixing symbiotic bacterium *Mesorhizobium loti* MAFF303099, discovered that this enzyme could also degrade AHL signals. But authors suggest that the natural substrate of MLR6805 is 4-pyridoxolactone because the adjacent gene (*mlr6806*) of the 4-pyridoxolactonase encodes for a pyridoxamine-pyruvate aminotransferase. Also closely located genes as *mll6785* and *mlr6788* encode for a pyridoxine 4-oxidase and a 3-hydroxy-2-methylpyridine-5-carboxylic acid oxygenase. Thus, the enzymes involved in the degradation for pyridoxine could be clustered in *M. loti* MAFF303099 genome.

Despite the metallo-β-lactamases representing the most abundant cluster among the enzymes capable of hydrolyzing the lactone ring of AHL signals, several completely unrelated families of enzymes with AHL-lactonase activity have been described (Table 1, Fig. 2). The first AiiA-unrelated lactonase discovered was QsdA, identified by screening of a genomic library of *Rhodococcus erythropolis* strain W2. QsdA is related to phosphotriesterases (PTE) and confers the bacteria the ability to inactivate AHLs with acyl side chains ranging from C6 to C14, with or without substitution at carbon 3 (Uroz et al. 2008). Members of the PTE family are also metal-dependent proteins and were first described for their ability to cleave phosphotriester bonds but were later shown to be promiscuous enzymes, harboring lactonase or amidohydrolase activities (Roodveldt and Tawfik 2005). Interestingly, thermophilic bacteria from the genus *Geobacillus* also produce thermostable AHL-lactonase members of the PTE metalloproteins. The thermostability of these enzymes makes them attractive because of their increased potential biotechnological applications. GKL, the thermostable lactonase from *G. kaustophilus*, exhibits substrate preference for medium- to long-chain AHLs (≥8 carbons) (Chow et al. 2010). Besides members of the *Eubacteria* domain, another thermostable lactonase of the PTE family was discovered in the archaeon *Sulfolobus solfataricus* and named SsoPox. This lactonase was related to organophosphate hydrolysis, but later studies showed that it also degrades AHLs (Merone et al. 2005). *Mycobacterium avium* subsp. *paratuberculosis* K-10 has been reported to produce a PTE amidohydrolase too. K-10 lactonase was named MCP and shown to degrade medium- to long-chain AHLs (Chow et al. 2009).

The third family of AHL lactonases discovered was the α/β hydrolase fold family which includes the lactonase AiiM from *Microbacterium testaceum* StLB037 and AidH from *Ochrobactrum anthropi* ATCC 49188, both cloned from genomic libraries of plant leaf and soil isolate, respectively (Mei et al. 2010; Wang et al. 2010). Mutational analyses indicated that the G-X-Nuc-X-G motif or the histidine residue conserved in α/β hydrolases is critical for the AHL-degradation activity (Mei et al. 2010).

Special mention requires the lactonase QsdH detected in a *Pseudoalteromonas byunsanensis strain* 1A01261 genomic library screening. Interestingly, this AHL-degrading enzyme is a GDSL-like hydrolase located in the N-terminus of a multidrug efflux transporter protein of the resistance-nodulation-cell division (RND) family. QsdH locates in the first periplasmic loop of the pump and possibly could degrade AHLs that influx or efflux through the inner membrane (Huang et al. 2012). GDSL hydrolases have multifunctional properties derived from broad substrate specificity and with potential for use in the hydrolysis and synthesis of important ester compounds of pharmaceutical, food, biochemical, and biological interest (Akoh et al. 2004).

Soil metagenomic libraries analyses further extended the diversity of the lactonase family proteins by allowing the identification of the novel AHL-lactonase QlcA, a zinc-dependent metallohydrolase distantly related to the other AHL lactonases discovered (Riaz et al. 2008), and the lactonases BpiB01, BpiB04, and BpiB05 showing no significant similarities at their deduced amino acid sequences with other AHL lactonases (Bijtenhoorn et al. 2011b; Schipper et al. 2009).

AHL-lactonase activity is also present in mammalian tissues as paraoxonases (PONs), enzymes that were related to hydrolysis of organophosphates, were shown to degrade AHLs. There are three PON members: PON1, PON2, and PON3, sharing about 65 % identity at the amino acidic sequence level. While PON2 is an intracellular enzyme expressed in different tissues, *PON1* and *PON3* genes are expressed in the liver and kidney, and their protein products can be found in circulation bound to high-density lipoprotein (HDL). In addition to different distribution, these enzymes also display different efficiency against AHLs as PON2 was shown to have the highest lactonase activity of the three members (Camps et al. 2011).

Acylases

The second type of AHL-QSI enzymes was discovered in a soil isolate classified as *Variovorax paradoxus* (Leadbetter and Greenberg 2000). This *Betaproteobacterium* was shown to be able to use different AHLs as source of energy and nitrogen; moreover, the *V. paradoxus* molar growth yields correlated with the length of the acyl chain of the signal. Even though the enzyme responsible for AHL inactivation was not identified, authors showed that the enzyme responsible breaks the amide bond between the HSL and the acyl side chain (Leadbetter and Greenberg 2000). The first gene encoding for an AHL-acylase was cloned from *Ralstonia* sp. XJ12B and called *aiiD* (Lin et al. 2003). This strain was isolated among other bacteria in a biofilm collected from an experimental water treatment system and found to inactivate AHLs. AiiD acylase was also active when cloned in *E. coli* and, opposite to AHL lactonases, AiiD degradation products could not revert to active forms. Thus, acylases completely inactivate AHL signals unlike the chemical decomposition that occurs under alkaline conditions and the biological decomposition catalyzed by AHL lactonases to generate acyl-homoserines, which are stable for weeks or months at pH values of 5 to 6 (Schaefer et al. 2000) and could revert to active forms at lower pHs (Dong et al. 2001). AiiD shows similarity with Ntn (N-terminal nucleophile) hydrolases like cephalosporin and penicillin acylases; however, this QSI enzyme could not degrade penicillin G or ampicillin suggesting that AiiD has probably evolved to serve as a dedicated AHL-acylase (Lin et al. 2003).

The Ntn-hydrolases are activated autocatalytically by posttranslational cleavage of the inactive precursor to generate an active two-subunit form (Oinonen and Rouvinen 2000). This seems to be the case for AiiD, as Lin et al. (2003) identified the well-conserved domains (signal peptide followed by an α-subunit, an spacer sequence, and a β-subunit) and residues important for autocatalytic and AHL degradation activities in the sequence of the acylase. Furthermore, authors suggest a membrane or periplasmic location of AiiD which would explain the release of HSL into the medium by *V. paradoxus* (Leadbetter and Greenberg 2000). The external localization of the acylase could prevent HSL accumulation into the cells as HSL was shown to be toxic for bacteria (Zakataeva et al. 1999).

Other close homologues of AiiD have been identified in members of the genus *Ralstonia* (up to 83 % identity at the amino acid sequence level) and *Deinococcus radiodurans* (52 %) and 38–40 % identities with homologues of in *Pseudomonas* spp. (Lin et al. 2003). The first study describing AHL-QSI activity in *Pseudomonas* allowed the isolation of a soil pseudomonad capable of growth in media supplemented with *N*-3-oxododecanoyl-homoserine lactone (OC12-HSL) (Huang et al. 2003). Moreover, the human opportunistic pathogen *P. aeruginosa* PAO1, which produces both OC12-HSL and *N*-butanoyl-homoserine lactone (C4-HSL) (signals not detected in the soil isolate), also exhibited growth on AHLs with side chains longer than 8 carbons. But in contrast to soil *Pseudomonas* isolate, PAO1 showed no increase in molar growth yield depending on AHL acyl chain lengths, and the growth on AHLs started only after very long lag phases (Huang et al. 2003). Authors also cloned *pvdQ* (*pa2385*) gene, a previously identified homologue of AiiD in *P. aeruginosa* PAO1 (Lin et al. 2003), and showed that was effectively catalyzing the inactivation of long acyl chain AHLs and release of HSL. Crystal structure of PvdQ revealed that this acylase has a typical α/β heterodimeric Ntn-hydrolase fold but an unusually large hydrophobic-binding pocket that can recognize acyl chains of 12 carbons, which could explain its preference to degrade long acyl side-chain AHLs (Bokhove et al. 2010). PAO1 constitutively expressing *pvdQ* did not accumulate OC12-HSL when grown in rich media; nevertheless, *pvdQ* mutants of PAO1 still grew on media supplemented with OC12-HSL suggesting that other AHL-QSI enzymes could be present in this bacterium. The second *P. aeruginosa* PAO1 acylase QuiP (PA1032) was identified 3 years later. But contrary to *pvdQ*, *quiP* mutant of PAO1 was unable to grow in media supplemented with *N*-decanoyl-homoserine lactone (C10-HSL) and complementation of *quiP* restored growth defect. Moreover, when compared to wild type, a PAO1 strain engineered to constitutively express *quiP* exhibited a reduction in OC12-HSL levels (Huang et al. 2006).

Another member of the genus *Pseudomonas*, *P. syringae* B728a, was described to produce two Ntn-hydrolase acylases termed HacA and HacB (Shepherd and Lindow 2009). While HacA degrades AHLs with acyl chains longer than 8 carbons, HacB was found to inactivate short and long AHLs. Surprisingly, disruption of *hacA*, *hacB*, and both genes had no effect on OC6-HSL levels produced by *P. syringae*. When compared with PAO1's putative Ntn-hydrolases, HacA shared 55 % identity at the amino acid sequence level with PvdQ, while HacB showed 67 % identity with PA0305. Despite initial studies with a *pa0305* mutant showed no effect in growth on AHL signals (Huang et al. 2006), this gene was later described to codify the third AHL-acylase known in PAO1 and named HacB (Wahjudi et al. 2011). HacB displays broader substrate specificity than PvdQ and QuiP; however, in spite of the sequence similarity with its homologue from *P. syringae*, PA0305 is unable to degrade C4-HSL (Wahjudi et al. 2011).

The fact that *Pseudomonas* members could display AHL-based QS systems together with AHL degradation activity may indicate that these bacteria could fine-tune its own QS systems. To date, it is not clear if QSI activity in *Pseudomonas* modulates communication processes by recycling AHLs once quorum is achieved, or just use AHL breakdown products as carbon source. But the fact that PAO1 has two AHL QS systems, LasR/LasI and RhlR/RhlI, that produce and sense OC12-HSL and C4-HSL, respectively, and that these systems operate hierarchically, being RhlR/RhlI transcriptional and posttranslational regulated by LasR/LasI system (Latifi et al. 1996) suggests that QSI processes in *P. aeruginosa* PAO1 could modulate AHLs ratios to fine-tune the expression of virulent factors in this bacterium (Roche et al. 2004). Also, the long lag phase observed when *P. aeruginosa* is grown on AHLs implies that this bacterium is not adapted to AHL metabolism (Huang et al. 2003). Similarly, the AHL-acylase AibP discovered in the intracellular pathogen *Brucella melitensis* was shown to degrade endogenous AHL activity within *B. melitensis* in vitro and during macrophage infection impairing AHL accumulation. Moreover, a deletion of *aibP* in *B. melitensis* resulted in an increased expression of virulence determinants of the bacterium suggesting that QS and QSI systems in *Brucella* act modulating its virulence (Terwagne et al. 2013).

Despite the above cited, AHL-acylases were defined as AHL degraders only; a new Ntn-hydrolase discovered in a *Streptomyces* sp. added complexity to ascertain the role of these QSI enzymes, since *Streptomyces* sp. acylase AhlM was found to degrade a wide range of AHL sizes but also penicillin G by deacylation, showing that AhlM has a broad substrate specificity (Park et al. 2005). Moreover, the penicillin G acylase KcPGA from *Kluyvera citrophila*, used to deacylate penicillin to obtain the essential beta-lactam nucleus for semi-synthetic penicillins, was reported to cleave AHLs with acyl chains between 6 and 8 carbons in size with our without oxo substitutions (Mukherji et al. 2014).

Other AHL-acylase activities were later found in the cyanobacterium *Anabaena* sp. PCC7120 (AiiC, Romero et al. 2008), *Shewanella* sp. (Aac, Morohoshi et al. 2008), *Ochrobactrum* sp. (AiiO, Czajkowski et al. 2011), *Comamonas* sp., and *Rhodococcus erythropolis* (Uroz et al. 2005, 2007). More yet-to-be-identified enzymes responsible for AHL hydrolysis were described in *Acinetobacter* sp., *Bosea* sp., *Sphingomonas* sp., *Delftia acidovorans*, *Marinobacterium* sp., *Glaciecola* sp., *Hyphomonas* sp., *Alteromonas* sp., *Oceanobacillus* sp., *Phaeobacter* sp., *Halomonas taeanensis*, *Roseovarius aestuarii*, *Salinicola salarius*, *Olleya marilimosa*, *Maribacter* sp., *Flaviramulus basaltis*, *Muricauda olearia*, *Salinimonas lutimaris*, *Thalassomonas agariperforans*, *Marivita byusanensis*, *Novosphingobium tardaugens*, *Colwellia aestuarii*, *Rhodobacter ovatus*, *Lysinibacillus* sp., *Serratia* sp., *Myroides* sp., *Chryseobacterium* sp., and *Nocardioides kongjuensis* (D'Angelo-Picard et al. 2005; Jafra et al. 2006; Kang et al. 2004; Ma et al. 2013; Rashid et al. 2011; Romero et al. 2011, 2012a; Tait et al. 2009; Tang et al. 2013; Yoon et al. 2006).

Oxidoreductases

In addition to lactonase and acylase activities, the acyl chain of AHLs can suffer oxidase or reductase activities that modify them. These reactions do not degrade the signals but the modifications could affect the specificity and therefore the union between the signal and the receptor. This activity was first discovered in *Rhodococcus erythropolis* W2 (Uroz et al. 2005). Examination of the mechanism by which *R. erythropolis* modifies AHLs revealed that 3-oxo substituted signals, with acyl side chains ranging from C8 to C14, were converted to their corresponding 3-hydroxy derivatives. This oxidoreductase activity was also observed for other 3-oxo-substituted substrates without HSL or AHLs with aromatic compounds in the acyl chain (Uroz et al. 2005). However, the gene responsible for this activity has yet to be identified. The second report studying oxidase activity on AHLs was done with CYP102A1, a cytochrome P450 from *Bacillus megaterium*, which was previously shown to oxidize fatty acids. CYP102A1 was able to efficiently oxidize AHLs and their lactonolysis products at the ω-1, ω-2, and ω-3 carbons of the acyl chain, similar to this enzyme's well-known activity on fatty acids (Chowdhary et al. 2007). Interestingly, authors suggested that AHL oxidation could make the acyl-homoserines more water soluble and membrane permeable, to avoid lactonolysis (it should be noted that *Bacillus* spp. were shown to produce AHL lactonases) to accumulate inside the cell and prevent spontaneously conversion of oxo-acyl-homoserine lactones into tetramic acid compounds highly toxic to *Bacillus* spp. (Chowdhary et al. 2007). An AHL-enrichment-based study to detect QSI activity in the rhizosphere of ginger allowed the isolation of *Burkholderia* sp. GG4 with 3-oxo-substituted AHLs reducing activity to generate their corresponding 3-hydroxy forms. But, unlike *R. erythropolis* strain W2 (Uroz et al. 2005), in *Burkholderia* sp. GG4 AHL-reducing activity is not followed by signal degradation (Chan et al. 2011). Soil metagenome also allowed the detection of an AHL-oxidoreductase denominated BpiB09. BpiB09 was described as a NADP-dependent short-chain dehydrogenase/reductase (SDR) able to inactivate OC12-HSL. Although no chemical data was made available, authors speculate that aside OC12-HSL reduction, BpiB09 could also interfere with AHL synthesis by reducing the signal precursor 3-oxo-acyl-ACP, as it was found an almost complete lack of *lasI* and *rhlI* transcription in a *P. aeruginosa* PAO1 strain expressing BpiB09 (Bijtenhoorn et al. 2011a).

Enzymatic Inactivation of Other QS Systems

Besides AHL-QSI, a small number of studies have focused in the discovery of novel enzymes targeting the other cell-cell communication systems. The main explanation for this is the modest attention that has been given to other QS systems. An example we can cite, although not discussed in relation to QS, is the oxidation of the QS signal used by the yeast *Candida albicans* farnesol (Ramage et al. 2002) by the Cytochrome P450 of *Bacillus megaterium* (Murataliev et al. 2004). The oxidation of farnesol has been shown in vitro, but the potential physiological significance of this reaction has not been explored.

The diffusible signal factor (DSF) is a family of chemically diverse fatty acid signals required for expression of virulence factors in *Xanthomonas* spp., *Xylella fastidiosa*, *Stenotrophomonas maltophilia*, *Burkholderia cepacia*, and *P. aeruginosa* (Deng et al. 2011). A screening with 1,093 bacteria, isolated from plants infected with DSF-producing pathogens, for enzymatic QSI activity against DSF of *Xanthomonas campestris* yielded nine strains capable of rapidly inactivate DSF, including species from *Bacillus*, *Paenibacillus*, *Microbacterium*, *Staphylococcus*, and *Pseudomonas*. Despite the low percentage of DSF-inactivating bacteria initially observed, this percentage could be significantly increased with longer incubation times as authors showed that common lab bacteria like *E. coli* DH10β could also degrade DSF when incubated for 24 h or more (Newman et al. 2008). A random transposon mutagenesis of the isolate G, identified as *Pseudomonas* sp., allowed the detection of the genes *carA* and *carB* as involved in DSF inactivation and previously described as the subunits of the heterodimeric complex responsible for the synthesis of carbamoyl phosphate, precursor for arginine, and pyrimidine biosynthesis (Llamas et al. 2003). Moreover, complementation of *carAB* mutants restored both pyrimidine prototrophy and DSF inactivation ability of the strain G mutant. While the *carAB* mechanism of DSF inactivation remains unknown, the fact that DSF signal incubated with *Pseudomonas* sp. strain G was altered in mobility in a thin-layer chromatography assay and the signaling activity was significantly diminished suggests that DSF signals could be modified. Interestingly, mutation of *carAB* showed that this operon is also necessary for DSF inactivation in *E. coli*. Therefore, this operon seems to be required for DSF modification although showing diverse inactivation rates in different bacteria (Newman et al. 2008).

Besides the AHL-QS system, *P. aeruginosa* also regulates virulence in a concerted manner using Alkyl quinolone (AQ) molecules as autoinducers. Although *P. aeruginosa* produces over 50 different AQ congeners, a 2-heptyl-3-hydroxy-4(1H)-quinolone is the major AQ in *P. aeruginosa* and named the *Pseudomonas* quinolone signal (PQS) (Deziel et al. 2004). Due to the structural similarity of the dioxygenase Hod (1H-3-hydroxy-4-oxoquinaldine 2,4-dioxygenase) of *Arthrobacter nitroguajacolicus* substrate and PQS, Pustelny et al. (2009) examined if this enzyme could inactive PQS, thus capable of quenching AQ signaling in *P. aeruginosa*. Hod could in fact perform a 2,4-dioxygenolytic cleavage of PQS with formation of *N*-octanoylanthranilic acid and carbon monoxide, consistent with the natural substrate of the enzyme. On the other hand, authors observed a decay of Hod activity by proteolytic cleavage by extracellular *P. aeruginosa* proteases, a competitive inhibition by the PQS precursor 2-heptyl-4(1H)-quinolone and the PQS binding to rhamnolipids reduced the efficiency of Hod as a quorum-quenching agent (Pustelny et al. 2009).

In the case of the interspecies QS signal AI-2 ((S)-4,5-dihydroxy-2,3-pentanedione, DPD), two mechanisms that disrupt communication have been described that exploit enzymes to process this signal in enteric bacteria. In *E. coli* and *Salmonella typhimurium*, the LuxS protein is responsible for the synthesis of AI-2, which accumulates in the media and then is sensed and internalized by the *lsrABCD* operon in the bacterial cell (Xavier et al. 2007). Once in the cell, AI-2 is phosphorylated (P-DPD) by the kinase LsrK. This phosphorylation serves to trap and activate the signal within the cell, which in turn binds to LsrR derepressing the *lsr* operon transcription and potentially other QS-regulated genes (Xue et al. 2009). Therefore, using the Lsr system, these bacteria remove AI-2 from the media and, thus, eliminate the ability of others to utilize AI-2 to regulate their behaviors. Since *lsrK* mutant strains were unable to trap and activate AI-2 (Pereira et al. 2012), Roy et al. (2010) explored the use of purified LsrK to phosphorylate AI-2 and thereby restricting its transport into the cell via binding the Lsr complex. LsrK was shown to successfully quench AI-2 responses both in pure and mixed cultures of *E. coli*, *S. typhimurium*, and *V. harveyi*. Moreover, AI-2-quenching by LsrK was also observed in cocultures of an AI-2-generating *E. coli* and AI-2-sensing *S. typhimurium*.

Because P-DPD is the active form of AI-2, any enzymes involved in P-DPD modification are of interest as potential inhibitors of Lsr expression. The enzyme LsrG of enteric bacteria is one of them, as it was shown to catalyze the modification of P-DPD to its isomer 3,4,4-trihydroxy-2-pentanone-5-phosphate (P-TPO) in *E. coli*. Moreover, a *lsrG* mutant strain accumulate 10 times more P-DPD than wild-type cell levels, and the overexpression of *lsrG* in this mutant prevented this accumulation suggesting that LsrG could be a useful target for AI-2-based quorum-quenching therapies (Marques et al. 2011).

Use of Bacterial QSI Enzymes as Anti-pathogenic Tools

Due to the high prevalence of QS systems among plant, animal, and human pathogens (Williams et al. 2007), biotechnological applications of the interception of bacterial communication as a promising method to control pathogenic bacteria have attracted attention from numerous researchers. Rationally, the majority of AHL-degrading enzymes derived from soil and plant isolates were assessed for their capacity to interfere with plant diseases caused by pathogens with demonstrated QS-regulated virulence. Thereby, a significant decrease in infection symptoms have been observed in potato, eggplant, Chinese cabbage, carrot, celery, cauliflower, and tobacco plants infected with the AHL-dependent pathogen *Pectobacterium carotovorum* (*Erwinia carotovora*) expressing the AHL-lactonase AiiA from *Bacillus* sp. (Dong et al. 2000). Similar results were achieved with the causative agent of the fire blight disease *P. amylovora* (*E. amylovora*) expressing AiiA in apple leaves (Molina et al. 2005). Besides AiiA, heterologous expression of other AHL lactonases in *P. carotovorum* was also effective for bacterial biocontrol in plants. Examples we can cite are the expression of the lactonases AttM and AiiB from *A. tumefaciens* (Carlier et al. 2003), the soil metagenomic-derived lactonase QlcA (Riaz et al. 2008), the lactonase AiiM from *M. testaceum* (Wang et al. 2010), AidH from *Ochrobactrum* sp. (Mei et al. 2010), and AhlS from *Solibacillus silvestris* (Morohoshi et al. 2012) to protect potato tubers, radish, or Chinese cabbage. Co-inoculation of AiiA-expressing bacteria and plant pathogens has also been tried with positive results in potato, tomato, cactus, and rice (Cho et al. 2007; Dong et al. 2004; Lee et al. 2002; Molina et al. 2003; Zhao et al. 2008; Zhu et al. 2006). Moreover, transgenic plants expressing the lactonase AiiA like *Amorphophallus konjac*, tobacco, potato, and Chinese cabbage resulted in plants resistant to *P. carotovorum* (Ban et al. 2009; Dong et al. 2001; Vanjildorj et al. 2009), demonstrating the potential of disease prevention approach by quorum-quenching processes. As shown by Chan et al. (2011), AHL-oxidoreductase QSI could also be used in bacterial biocontrol as *Burkholderia* sp. strain GG4 attenuates potato tubers maceration caused by *P. carotovorum*. Regarding non-AHL QSI bacterial enzymes, it was shown that DSF-degrading strains are capable of disease control when co-inoculated with *X. campestris* or *X. fastidiosa* in mustard, grape, and cabbage leaves. Results showed a decrease in the severity of black rot and Pierce's disease, and this attenuation was *carB* dependent (Newman et al. 2008). Using a lettuce leaf model of *P. aeruginosa* infection, Pustelny et al. (2009) demonstrated that loss of PQS by co-injection of Hod enzyme resulted in attenuation of virulence *in planta*. Besides these strategies, biostimulants for the growth of AHL-degrading bacteria in the plant rhizosphere has been proposed as new approach for bacterial biocontrol. Studies on molecules structurally related to AHLs have shown growth promotion of AHL-QSI bacteria like *R. erythropolis* and biocontrol activity of *P. atrosepticum* in potato (Cirou et al. 2012).

The effectiveness of enzymatic QSI strategies has been also demonstrated in animal infection models. Lin et al. (2003) showed that heterologous expression of the acylase AiiD from *Ralstonia* sp. markedly reduced *P. aeruginosa* virulence in the nematode *Caenorhabditis elegans*, adding evidence that AHL-mediated quorum-sensing systems are important for regulation of virulence gene expression and pathogenesis. Furthermore, it was shown that more than 75 % of *C. elegans* worms exposed to *P. aeruginosa* PAO1 overexpressing the acylase PvdQ survived and continued to grow when using this strain as a food source (Papaioannou et al. 2009). Additionally, in a slow-killing assay (monitoring the survival of the nematodes during 4 days), a *pvdQ* deletion mutant was more virulent than wild-type strain, confirming the role of PvdQ as a virulence-reducing factor. In the same study, purified PvdQ protein was also added to nematodes infected with wild-type PAO1, and this resulted in reduced pathogenicity and increased the life span of the nematodes (Papaioannou et al. 2009). On the contrary, a *hacB* mutant of *P. aeruginosa* PAO1 showed a slightly increased capacity to kill *C. elegans* compared to wild-type and HacB overproducing strains (Wahjudi et al. 2011). These results show that despite similar specificity of signal degradation, QSI enzymes may show different effectiveness in infection assays. Again, AHL oxidoreductases could also be considered as tools for QSI strategies since the metagenome-derived NADP-dependent reductase BpiB09 was also shown to protect *C. elegans* when infected with PAO1 expressing this enzyme (Bijtenhoorn et al. 2011a).

Special attention has been paid to the feasibility of the application of QSI strategies to control diseases in aquatic organisms, for which the use of antibiotics is highly restricted and there is a need of sustainable alternatives for maintaining a healthy microbial environment. In this sense, encouraging results have been obtained with different enzymatic QSI strategies. In example, the use of bacterial consortium with AHL-degradation activity isolated from the digestive tract of Pacific white shrimp *Penaeus vannamei* was shown to neutralize the negative effect of *Vibrio harveyi* on the rotifer *Brachionus plicatilis* cultures (Tinh et al. 2007). Moreover, Nhan et al. (2010) demonstrated that bacterial consortia with AHL-degrading activity, isolated from European and Asian sea bass gut, were capable of protecting the giant freshwater prawn *Macrobrachium rosenbergii* larvae from *V. harveyi* infections, when AHL-degrading consortia were added both directly into the larval-rearing water and via enriched *Artemia nauplii* used for larval feeding. Interestingly, the consortia also improved the health of non-challenged larvae, indicating that AHL degradation may also protect shrimp larvae against deleterious bacteria that are naturally associated with larval prawn rearing (Nhan et al. 2010). These conclusions are further supported by previous studies that showed that AHL-degrading bacteria reduced the negative effect in survival of *M. rosenbergii* and turbot larvae when AHLs were exogenously added to the cultures (Baruah et al. 2009; Tinh et al. 2008). QSI enzymes were also demonstrated to have potential for use in aquaculture field, as when the recombinant purified lactonase AiiA from *Bacillus* was co-injected with the fish pathogen *Aeromonas hydrophila* in carp, recombinant AiiA decreased the mortality rate and delayed the mortality time of fish (Chen et al. 2010). Moreover, feed supplemented with *Bacillus*-expressing *aiiA* or purified thermostable AiiA showed protective effects against *A. hydrophila* in carp and zebra fish intraperitoneally or bath challenged, respectively (Cao et al. 2012; Chu et al. 2011).

Previous examples show that QS signal degradation could be successfully used for blocking bacterial infections. In addition, the interest of these mechanisms is that they do not affect directly the survival of the pathogen but the expression of virulence factors, and so they do not exert direct selective pressure reducing potential emergence of resistances.

Conclusion

The discovery of multiple and different bacterial signaling systems has changed our vision on the control that bacteria exert on their behavior in relation to the environment. Now we know that bacteria coordinate efforts to achieve benefits for the whole population, including antibiotics biosynthesis, motility, swarming, aggregation, plasmid conjugal transfer, luminescence, or biofilm formation (Williams et al. 2007). Antibiotic-resistant bacteria have become a global threat to public health, forcing researches to seek for novel therapeutics that exert less selection pressure on pathogens and act as anti-pathogenic drugs. QS, as key regulatory system of virulence determinants, has become a good example of potential new anti-pathogenic target for novel drug design. In the past decade, numerous enzymes capable of inactivating QS signals have been described but, although the inactivation of all QS systems is equally interesting, most known QSI enzymes degrade or modify AHLs, the gram-negative QS signal molecules. Despite the specificity of these enzymes as signal quenchers remains unclear, enzymatic QSI has been explored as a novel anti-pathogenic therapy to control bacterial infections with positive results in plants, the nematode infection model, and in aquaculture field.

Opinion

An increasing amount of evidence is being accumulated regarding the feasibility of the use of QSI strategies to fight bacterial infections. Researchers have focused their attention in the interference with AHL-mediated QS systems, which is reflected in the increasing number of patent applications that are related to this strategy (Romero et al. 2012b). The effectiveness of bacterial enzymatic QSI mechanisms has been mainly applied to control infections in plants, like the transgenic expression of the AHL-lactonase AiiA from *Bacillus* in tobacco and potato plants, or the protection of infections by gram-negative pathogens in aquaculture. The advantage of enzymatic degradation of QS signals is that the use of enzymes with a wide range of degradation activity will allow the interference with a higher number of signaling systems while antagonists are usually species-specific. However, in spite of the considerable activity that is being developed in the field and the interest of the strategy for medical and biotechnological applications, more robust proofs of concept and medical trials are surely required before we can see enzymatic QSI-related products in the market.

Acknowledgment M.R. was financially supported by Fundación Barrié de la Maza. This work has been partially supported by a research project from Fundación Ramón Areces (CIVP16A1814).

References

Akoh CC, Lee GC, Liaw YC, Huang TH, Shaw JF (2004) GDSL family of serine esterases/lipases. Prog Lipid Res 43:534–552

Ban H, Chai X, Lin Y, Zhou Y, Peng D, Zou Y, Yu Z, Sun M (2009) Transgenic *Amorphophallus konjac* expressing synthesized acyl-homoserine lactonase (*aiiA*) gene exhibit enhanced resistance to soft rot disease. Plant Cell Rep 28:1847–1855

Baruah K, Cam DTV, Dierckens K, Wille M, Defoirdt T, Sorgeloos P, Bossier P (2009) In vivo effects of single or combined N–acyl homoserine lactone quorum sensing signals on the performance of *Macrobrachium rosenbergii* larvae. Aquaculture 288:233–238

Bijtenhoorn P, Mayerhofer H, Muller-Dieckmann J, Utpatel C, Schipper C, Hornung C, Szesny M, Grond S, Thurmer A, Brzuszkiewicz E, Daniel R, Dierking K, Schulenburg H, Streit WR (2011a) A novel metagenomic short-chain dehydrogenase/reductase attenuates *Pseudomonas aeruginosa* biofilm formation and virulence on *Caenorhabditis elegans*. PLoS ONE 6: e26278

Bijtenhoorn P, Schipper C, Hornung C, Quitschau M, Grond S, Weiland N, Streit WR (2011b) BpiB05, a novel metagenome-derived hydrolase acting on N-acylhomoserine lactones. J Biotechnol 155: 86–94

Bokhove M, Nadal Jimenez P, Quax WJ, Dijkstra BW (2010) The quorum-quenching N-acyl homoserine lactone acylase PvdQ is an Ntn-hydrolase with an unusual substrate-binding pocket. Proc Natl Acad Sci U S A 107:686–691

Bruhn JB, Dalsgaard I, Nielsen KF, Buchholtz C, Larsen JL, Gram L (2005) Quorum sensing signal molecules (acylated homoserine lactones) in gram-negative fish pathogenic bacteria. Dis Aquat Organ 65:43–52

Camps J, Pujol I, Ballester F, Joven J, Simó JM (2011) Paraoxonases as potential antibiofilm agents: their relationship with quorum-sensing signals in Gram-negative bacteria. Antimicrob Agents Chemother 55:1325–1331

Cao Y, He S, Zhou Z, Zhang M, Mao W, Zhang H, Yao B (2012) Orally administered thermostable N-acyl homoserine lactonase from *Bacillus* sp. strain AI96 attenuates *Aeromonas hydrophila* infection in zebrafish. Appl Environ Microbiol 78:1899–1908

Carlier A, Uroz S, Smadja B, Fray R, Latour X, Dessaux Y, Faure D (2003) The Ti plasmid of *Agrobacterium tumefaciens* harbors an *attM*-paralogous gene, *aiiB*, also encoding N-acyl homoserine lactonase activity. Appl Environ Microbiol 69:4989–4993

Chan KG, Atkinson S, Mathee K, Sam CK, Chhabra SR, Cámara M, Koh CL, Williams P (2011) Characterization of N-acylhomoserine lactone-degrading bacteria associated with the *Zingiber officinale* (ginger) rhizosphere: co-existence of quorum quenching and quorum sensing in *Acinetobacter* and *Burkholderia*. BMC Microbiol 11:51

Chen RD, Zhou ZG, Cao YA, Bai YG, Yao B (2010) High yield expression of an AHL-lactonase from *Bacillus* sp B546 in *Pichia pastoris* and its application to reduce *Aeromonas hydrophila* mortality in aquaculture. Microb Cell Fact 9:39

Cho HS, Park SY, Ryu CM, Kim JF, Kim JG, Park SH (2007) Interference of quorum sensing and virulence of the rice pathogen *Burkholderia glumae* by an engineered endophytic bacterium. FEMS Microbiol Ecol 60:14–23

Chow JY, Wu L, Yew WS (2009) Directed evolution of a quorum-quenching lactonase from *Mycobacterium avium* subsp. *paratuberculosis* K-10 in the amidohydrolase superfamily. Biochemistry 48:4344–4353

Chow JY, Xue B, Lee KH, Tung A, Wu L, Robinson RC, Yew WS (2010) Directed evolution of a thermostable quorum-quenching lactonase from the amidohydrolase superfamily. J Biol Chem 285:40911–40920

Chowdhary PK, Keshavan N, Nguyen HQ, Peterson JA, Gonzalez JE, Haines DC (2007) *Bacillus megaterium* CYP102A1 oxidation of acyl homoserine lactones and acyl homoserines. Biochemistry 46:14429–14437

Chu W, Lu F, Zhu W, Kang C (2011) Isolation and characterization of new potential probiotic bacteria based on quorum-sensing system. J Appl Microbiol 110: 202–208

Cirou A, Mondy S, An S, Charrier A, Sarrazin A, Thoison O, DuBow M, Faure D (2012) Efficient biostimulation of native and introduced quorum-quenching *Rhodococcus erythropolis* populations is revealed by a combination of analytical chemistry, microbiology, and pyrosequencing. Appl Environ Microbiol 78:481–492

Czajkowski R, Krzyżanowska D, Karczewska J, Atkinson S, Przysowa J, Lojkowska E, Williams P, Jafra S (2011) Inactivation of AHLs by *Ochrobactrum* sp. A44 depends on the activity of a novel class of AHL acylase. Environ Microbiol Rep 3:59–68

D'Angelo-Picard C, Faure D, Carlier A, Uroz S, Raffoux A, Fray R, Dessaux Y (2005) Bacterial populations in the rhizosphere of tobacco plants producing the quorum-sensing signals hexanoyl-homoserine lactone and 3-oxo-hexanoyl-homoserine lactone. FEMS Microbiol Ecol 51:19–29

Delalande L, Faure F, Raffoux A, Uroz S, D'Angelo-Picard C, Elasri M, Carlier A, Berruyer R, Petit A, Williams P, Dessaux Y (2005) N-hexanoyl-l-homoserine lactone a mediator of bacterial quorum-sensing regulation exhibits plant-dependent stability and may be inactivated by germinating *Lotus corniculatus* seedlings. FEMS Microbiol Ecol 52: 13–20

Deng Y, Wu J, Tao F, Zhang LH (2011) Listening to a new language: DSF-based quorum sensing in Gram-negative bacteria. Chem Rev 111:160–173

Déziel E, Lépine F, Milot S, He J, Mindrinos MN, Tompkins RG, Rahme LG (2004) Analysis of *Pseudomonas aeruginosa* 4-hydroxy-2-alkylquinolines (HAQs) reveals a role for 4-hydroxy-2-heptylquinoline in cell-to-cell communication. Proc Natl Acad Sci U S A 101:1339–1344

Dong YH, Zhang LH (2005) Quorum sensing and quorum-quenching enzymes. J Microbiol 43:101–109

Dong YH, Xu JL, Li XZ, Zhang LH (2000) AiiA, an enzyme that inactivates the acylhomoserine lactone quorum-sensing signal and attenuates the virulence of *Erwinia carotovora*. Proc Natl Acad Sci U S A 97:3526–3531

Dong YH, Wang LH, Xu JL, Zhang HB, Zhang XF, Zhang LH (2001) Quenching quorum-sensing-dependent bacterial infection by an N-acyl homoserine lactonase. Nature 411:813–817

Dong YH, Zhang XF, Xu JL, Zhang LH (2004) Insecticidal *Bacillus thuringiensis* silences *Erwinia carotovora* virulence by a new form of microbial antagonism, signal interference. Appl Environ Microbiol 70: 954–960

Dong YH, Wang LH, Zhang LH (2007) Quorum-quenching microbial infections: mechanisms and implications. Phil Trans R Soc B 362:1201–1211

Flagan S, Ching WK, Leadbetter JR (2003) *Arthrobacter* strain VAI-A utilizes acyl-homoserine lactone inactivation products and stimulates quorum signal biodegradation by *Variovorax paradoxus*. Appl Environ Microbiol 69:909–916

Funami J, Yoshikane Y, Kobayashi H, Yokochi N, Yuan B, Iwasaki K, Ohnishi K, Yagi T (2005) 4-Pyridoxolactonase from a symbiotic nitrogen-fixing bacterium *Mesorhizobium loti*: cloning, expression, and characterization. Biochim Biophys Acta 1753:234–239

Gao MS, Teplitski M, Robinson JB, Bauer WD (2003) Production of substances by Medicago truncatula that affect bacterial quorum sensing. Mol Plant Microbe Interact 16:827–834

Givskov M, de Nys R, Manefield M, Gram L, Maximilien R, Eberl L, Molin S, Steinberg PD, Kjelleberg S (1996) Eukaryotic interference with homoserine lactone-mediated prokaryotic signalling. J Bacteriol 178:6618–6622

Haudecoeur E, Faure D (2010) A fine control of quorum-sensing communication in *Agrobacterium tumefaciens*. Commun Integr Biol 3:84–88

Huang JJ, Han JI, Zhang LH, Leadbetter JR (2003) Utilization of acyl-homoserine lactone quorum signals for growth by a soil pseudomonad and *Pseudomonas aeruginosa* PAO1. Appl Environ Microbiol 69: 5941–5949

Huang JJ, Petersen A, Whiteley M, Leadbetter JR (2006) Identification of QuiP, the product of gene PA1032, as the second acyl-homoserine lactone acylase of *Pseudomonas aeruginosa* PAO1. Appl Environ Microbiol 72:1190–1197

Huang W, Lin Y, Yi S, Liu P, Shen J, Shao Z, Liu Z (2012) QsdH, a novel AHL lactonase in the RND-type inner membrane of marine *Pseudoalteromonas byunsanensis* strain 1A01261. PLoS ONE 7:e46587

Jafra S, Przysowa J, Czajkowski R, Michta A, Garbeva P, van der Wolf JM (2006) Detection and characterization of bacteria from the potato rhizosphere degrading N-acyl-homoserine lactone. Can J Microbiol 52: 1006–1015

Jimenez PN, Koch G, Thompson JA, Xavier KB, Cool RH, Quax WJ (2012) The multiple signaling systems regulating virulence in *Pseudomonas aeruginosa*. Microbiol Mol Biol Rev 76:46–65

Kang BR, Lee JH, Ko SJ, Lee YH, Cha JS, Cho BH, Kim YC (2004) Degradation of acyl-homoserine lactone molecules by *Acinetobacter* sp. strain C1010. Can J Microbiol 50:935–941

Kaufmann GF, Sartorio R, Lee SH, Mee JM, Altobell LJ 3rd, Kujawa DP, Jeffries E, Clapham B, Meijler MM, Janda KD (2006) Antibody interference with N-acyl homoserine lactone-mediated bacterial quorum sensing. J Am Chem Soc 128:2802–2803

Khan SR, Farrand SK (2009) The BlcC (AttM) lactonase of *Agrobacterium tumefaciens* does not quench the quorum-sensing system that regulates Ti plasmid conjugative transfer. J Bacteriol 191: 1320–1329

Krysciak D, Schmeisser C, Preuss S, Riethausen J, Quitschau M, Grond S, Streit WR (2011) Involvement of multiple loci in quorum quenching of autoinducer I molecules in the nitrogen-fixing symbiont *Rhizobium* (*Sinorhizobium*) sp. strain NGR234. Appl Environ Microbiol 77:5089–5099

Latifi A, Foglino M, Tanaka K, Williams P, Lazdunski A (1996) A hierarchical quorum-sensing cascade in *Pseudomonas aeruginosa* links the transcriptional activators LasR and RhlR (VsmR) to expression of the stationary-phase sigma factor RpoS. Mol Microbiol 21:1137–1146

Leadbetter JR, Greenberg EP (2000) Metabolism of acyl-homoserine lactone quorum-sensing signals by *Variovorax paradoxus*. J Bacteriol 182:6921–6926

Lee SJ, Park SY, Lee JJ, Yum DY, Koo BT, Lee JK (2002) Genes encoding the N-acyl homoserine lactone-degrading enzyme are widespread in many subspecies of *Bacillus thuringiensis*. Appl Environ Microbiol 68:3919–3924

Lin YH, Xu JL, Hu J, Wang LH, Ong SL, Leadbetter JR, Zhang LH (2003) Acyl-homoserine lactone acylase from *Ralstonia* strain XJ12B represents a novel and potent class of quorum-quenching enzymes. Mol Microbiol 47:849–860

Liu D, Momb J, Thomas PW, Moulin A, Petsko GA, Fast W, Ringe D (2008) Mechanism of the quorum-quenching lactonase (AiiA) from *Bacillus thuringiensis*. 1. Product-bound structures. Biochemistry 47:7706–7714

Llamas I, Suarez A, Quesada E, Bejar V, del Moral A (2003) Identification and characterization of the *carAB* genes responsible for encoding carbamoylphosphate synthetase in *Halomonas eurihalina*. Extremophiles 7:205–211

Ma A, Lv D, Zhuang X, Zhuang G (2013) Quorum quenching in culturable phyllosphere bacteria from tobacco. Int J Mol Sci 14:14607–14619

Marques JC, Lamosa P, Russell C, Ventura R, Maycock C, Semmelhack MF, Miller ST, Xavier KB (2011) Processing the interspecies quorum-sensing signal autoinducer-2 (AI-2): characterization of phospho-(S)-4,5-dihydroxy-2,3-pentanedione isomerization by LsrG protein. J Biol Chem 286:18331–18343

Mei GY, Yan XX, Turak A, Luo ZQ, Zhang LQ (2010) AidH, an alpha/beta-hydrolase fold family member from an *Ochrobactrum* sp. strain, is a novel N-acylhomoserine lactonase. Appl Environ Microbiol 76:4933–4942

Merone L, Mandrich L, Rossi M, Manco G (2005) A thermostable phosphotriesterase from the archaeon *Sulfolobus solfataricus*: cloning, overexpression and properties. Extremophiles 9:297–305

Molina L, Constantinescu F, Michel L, Reimmann C, Duffy B, Defago G (2003) Degradation of pathogen quorum-sensing molecules by soil bacteria: a preventive and curative biological control mechanism. FEMS Microbiol Ecol 45:71–81

Molina L, Rezzonico F, Defago G, Duffy B (2005) Autoinduction in *Erwinia amylovora*: evidence of an acyl-homoserine lactone signal in the fire blight pathogen. J Bacteriol 187:3206–3213

Morohoshi T, Nakazawa S, Ebata A, Kato N, Ikeda T (2008) Identification and characterization of N-acylhomoserine lactone-acylase from the fish intestinal *Shewanella* sp. strain MIB015. Biosci Biotechnol Biochem 72:1887–1893

Morohoshi T, Wang W-Z, Someya N, Ikeda T (2011) Genome sequence of Microbacterium testaceum StLB037, an N-acylhomoserine lactone-degrading bacterium isolated from potato leaves. J Bacteriol 193:2072–2073

Morohoshi T, Tominaga Y, Someya N, Ikeda T (2012) Complete genome sequence and characterization of the N-acylhomoserine lactone-degrading gene of the potato leaf-associated *Solibacillus silvestris*. J Biosci Bioeng 113:20–25

Mukherji R, Varshney NK, Panigrahi P, Suresh CG, Prabhune A (2014) A new role for penicillin acylases: degradation of acyl homoserine lactone quorum sensing signals by *Kluyvera citrophila* penicillin G acylase. Enzyme Microb Technol 56:1–7

Murataliev MB, Trinh LN, Moser LV, Bates RB, Feyereisen R, Walker FA (2004) Chimeragenesis of the fatty acid binding site of cytochrome P450BM3. Replacement of residues 73–84 with the homologous residues from the insect cytochrome P450 CYP4C7. Biochemistry 43:1771–1780

Newman KL, Chatterjee S, Ho KA, Lindow SE (2008) Virulence of plant pathogenic bacteria attenuated by degradation of fatty acid cell-to-cell signaling factors. Mol Plant Microbe Interact 21:326–334

Nhan DT, Cam DTV, Wille M, Defoirdt T, Bossier P, Sorgeloos P (2010) Quorum quenching bacteria protect *Macrobrachium rosenbergii* larvae from *Vibrio harveyi* infection. J Appl Microbiol 109:1007–1016

Ni N, Li M, Wang J, Wang B (2009) Inhibitors and antagonists of bacterial quorum sensing. Med Res Rev 29:65–124

Oinonen C, Rouvinen J (2000) Structural comparison of Ntn-hydrolases. Protein Sci 9:2329–2337

Otero A, Romero M, Mayer C (2013) Péptido con actividad inhibidora de quorum sensing, polinucleótido que codifica dicho péptido y sus aplicaciones. Patent application OEPM P201331060

Papaioannou E, Wahjudi M, Nadal-Jimenez P, Koch G, Setroikromo R, Quax WJ (2009) Quorum-quenching acylase reduces the virulence of *Pseudomonas aeruginosa* in a *Caenorhabditis elegans* infection model. Antimicrob Agents Chemother 53: 4891–4897

Park SY, Lee SJ, Oh TK, Oh JW, Koo BT, Yum DY, Lee JK (2003) AhlD, an N-acylhomoserine lactonase in *Arthrobacter* sp., and predicted homologues in other bacteria. Microbiology 149:1541–1550

Park SY, Kang HO, Jang HS, Lee JK, Koo BT, Yum DY (2005) Identification of extracellular N-acylhomoserine lactone acylase from a *Streptomyces* sp. and its application to quorum quenching. Appl Environ Microbiol 71:2632–2641

Park SJ, Park SY, Ryu CM, Park SH, Lee JK (2008) The role of AiiA, a quorum-quenching enzyme from *Bacillus thuringiensis*, on the rhizosphere competence. J Microbiol Biotechnol 18:1518–1521

Parsek MR, Val DL, Hanzelka BL, Cronan JE Jr, Greenberg EP (1999) Acyl homoserine-lactone quorum-sensing signal generation. Proc Natl Acad Sci U S A 96:4360–4365

Pechere JC (2001) New perspectives on macrolide antibiotics. Int J Antimicrob Agents 1:93–97

Pereira CS, Santos AJ, Bejerano-Sagie M, Correia PB, Marques JC, Xavier KB (2012) Phosphoenolpyruvate phosphotransferase system regulates detection and processing of the quorum sensing signal autoinducer-2. Mol Microbiol 84:93–104

Pustelny C, Albers A, Buldt-Karentzopoulos K, Parschat K, Chhabra SR, Camara M, Williams P, Fetzner S (2009) Dioxygenase-mediated quenching of quinolone-dependent quorum sensing in *Pseudomonas aeruginosa*. Chem Biol 16:1259–1267

Ramage G, Saville SP, Wickes BL, Lopez-Ribot JL (2002) Inhibition of *Candida albicans* biofilm formation by farnesol, a quorum-sensing molecule. Appl Environ Microbiol 68:5459–5463

Rashid R, Morohoshi T, Someya N, Ikeda T (2011) Degradation of N-acylhomoserine lactone quorum sensing signaling molecules by potato root surface-associated *Chryseobacterium* strains. Microbes Environ 26: 144–148

Riaz K, Elmerich C, Moreira D, Raffoux A, Dessaux Y, Faure D (2008) A metagenomic analysis of soil bacteria extends the diversity of quorum-quenching lactonases. Environ Microbiol 10:560–570

Roche DM, Byers JT, Smith DS, Glansdorp FG, Spring DR, Welch M (2004) Communications blackout? Do N-acylhomoserine-lactone-degrading enzymes have any role in quorum sensing? Microbiology 150: 2023–2028

Romero M, Diggle SP, Heeb S, Camara M, Otero A (2008) Quorum quenching activity in Anabaena sp. PCC7120: identification of AiiC, a novel AHL-acylase. FEMS Microbiol Lett 280:73–80

Romero M, Martin-Cuadrado AB, Roca-Rivada A, Cabello AM, Otero A (2011) Quorum quenching in cultivable bacteria from dense marine coastal microbial communities. FEMS Microbiol Ecol 75:205–217

Romero M, Acuna L, Otero A (2012a) Patents on quorum quenching: interfering with bacterial communication as a strategy to fight infections. Recent Pat Biotechnol 6:2–12

Romero M, Martin-Cuadrado AB, Otero A (2012b) Determination of whether quorum quenching is a common activity in marine bacteria by analysis of cultivable bacteria and metagenomic sequences. Appl Environ Microbiol 78:6345–6348

Roodveldt C, Tawfik DS (2005) Shared promiscuous activities and evolutionary features in various members of the amidohydrolase superfamily. Biochemistry 44:12728–12736

Roy V, Fernandes R, Tsao CY, Bentley WE (2010) Cross species quorum quenching using a native AI-2 processing enzyme. ACS Chem Biol 5:223–232

Schaefer AL, Hanzelka BL, Parsek MR, Greenberg EP (2000) Detection, purification and structural elucidation of acylhomoserine lactone inducer of *Vibrio fischeri* luminescence and other related molecules. Methods Enzymol 305:288–301

Schipper C, Hornung C, Bijtenhoorn P, Quitschau M, Grond S, Streit WR (2009) Metagenome-derived clones encoding two novel lactonase family proteins involved in biofilm inhibition in *Pseudomonas aeruginosa*. Appl Environ Microbiol 75:224–233

Shepherd RW, Lindow SE (2009) Two dissimilar N-acyl-homoserine lactone acylases of *Pseudomonas syringae* influence colony and biofilm morphology. Appl Environ Microbiol 75:45–53

Skindersoe ME, Ettinger-Epstein P, Rasmussen TB, Bjarnsholt T, de Nys R, Givskov M (2008) Quorum sensing antagonism from marine organisms. Mar Biotechnol 10:56–63

Tait K, Williamson H, Atkinson S, Williams P, Cámara M, Joint I (2009) Turnover of quorum sensing signal molecules modulates cross-kingdom signalling. Environ Microbiol 11:1792–1802

Tang K, Zhang Y, Yu M, Shi X, Coenye T, Bossier P, Zhang XH (2013) Evaluation of a new high-throughput method for identifying quorum quenching bacteria. Sci Rep 3:2935

Teasdale ME, Liu J, Wallace J, Akhlaghi F, Rowley DC (2009) Secondary metabolites produced by the marine bacterium *Halobacillus salinus* that inhibit quorum sensing-controlled phenotypes in gram-negative bacteria. Appl Environ Microbiol 75:567–572

Terwagne M, Mirabella A, Lemaire J, Deschamps C, De Bolle X, Letesson JJ (2013) Quorum sensing and self-quorum quenching in the intracellular pathogen *Brucella melitensis*. PLoS ONE 8:e82514

Tinh NT, Asanka Gunasekara RA, Boon N, Dierckens K, Sorgeloos P, Bossier P (2007) N-acyl homoserine lactone-degrading microbial enrichment cultures isolated from *Penaeus vannamei* shrimp gut and their probiotic properties in *Brachionus plicatilis* cultures. FEMS Microbiol Ecol 62:45–53

Tinh NTN, Yen VHN, Dierckens K, Sorgeloos P, Bossier P (2008) An acyl homoserine lactone–degrading microbial community improves the survival of first feeding turbot larvae (*Scophthalmus maximus* L.). Aquaculture 285:56–62

Uroz S, Chhabra SR, Camara M, Williams P, Oger P, Dessaux Y (2005) N-Acylhomoserine lactone quorum-sensing molecules are modified and degraded by *Rhodococcus erythropolis* W2 by both amidolytic and novel oxidoreductase activities. Microbiology 151:3313–3322

Uroz S, Oger P, Chhabra SR, Camara M, Williams P, Dessaux Y (2007) N-acyl homoserine lactones are degraded via an amidolytic activity in *Comamonas* sp. strain D1. Arch Microbiol 187:249–256

Uroz S, Oger PM, Chapelle E, Adeline MT, Faure D, Dessaux Y (2008) A *Rhodococcus qsdA*-encoded enzyme defines a novel class of large-spectrum quorum-quenching lactonases. Appl Environ Microbiol 74:1357–1366

Vanjildorj E, Song SY, Yang ZH, Choi JE, Noh YS, Park S, Lim WJ, Cho KM, Yun HD, Lim YP (2009) Enhancement of tolerance to soft rot disease in the transgenic Chinese cabbage (*Brassica rapa* L. ssp. *pekinensis*) inbred line, Kenshin. Plant Cell Rep 28: 1581–1591

Wahjudi M, Papaioannou E, Hendrawati O, van Assen AH, van Merkerk R, Cool RH, Poelarends GJ, Quax WJ (2011) PA0305 of *Pseudomonas aeruginosa* is a quorum quenching acylhomoserine lactone acylase belonging to the Ntn hydrolase superfamily. Microbiology 157:2042–2055

Wang WZ, Morohoshi T, Ikenoya M, Someya N, Ikeda T (2010) AiiM, a novel class of N-acylhomoserine lactonase from the leaf-associated bacterium *Microbacterium testaceum*. Appl Environ Microbiol 76: 2524–2530

Wang WZ, Morohoshi T, Someya N, Ikeda T (2012) AidC, a novel N-acylhomoserine lactonase from the potato root-associated *Cytophaga-Flavobacteria-Bacteroides* (CFB) group bacterium *Chryseobacterium* sp. strain StRB126. Appl Environ Microbiol 78:7985–7992

Williams P, Winzer K, Chan WC, Cámara M (2007) Look who's talking: communication and quorum sensing in the bacterial world. Phil Trans R Soc B 362: 1119–1134

Xavier KB, Miller ST, Lu W, Kim JH, Rabinowitz J, Pelczer I, Semmelhack MF, Bassler BL (2007) Phosphorylation and processing of the quorum-sensing molecule autoinducer-2 in enteric bacteria. ACS Chem Biol 2:128–136

Xue T, Zhao L, Sun H, Zhou X, Sun B (2009) LsrR-binding site recognition and regulatory characteristics in *Escherichia coli* AI-2 quorum sensing. Cell Res 19:1258–1268

Yates EA, Philipp B, Buckley C, Atkinson S, Chhabra SR, Sockett RE, Goldner M, Dessaux Y, Camara M, Smith H, Williams P (2002) N-acylhomoserine lactones undergo lactonolysis in a pH-, temperature-, and acyl chain length-dependent manner during growth of *Yersinia pseudotuberculosis* and *Pseudomonas aeruginosa*. Infect Immun 70:5635–5646

Yoon JH, Lee JK, Jung SY, Kim JA, Kim HK, Oh TK (2006) *Nocardioides kongjuensis* sp. nov., an N-acylhomoserine lactone-degrading bacterium. Int J Syst Evol Microbiol 56:1783–1787

Zakataeva NP, Aleshin VV, Tokmakova IL, Troshin PV, Livshits VA (1999) The novel transmembrane *Escherichia coli* proteins involved in the amino acid efflux. FEBS Lett 452:228–232

Zhang HB, Wang LH, Zhang LH (2002) Genetic control of quorum-sensing signal turnover in *Agrobacterium tumefaciens*. Proc Natl Acad Sci U S A 99: 4638–4643

Zhao C, Zeng H, Yu Z, Sun M (2008) N-Acyl homoserine lactonase promotes prevention of *Erwinia* virulence with zwittermicin A-producing strain *Bacillus cereus*. Biotechnol Bioeng 100: 599–603

Zhu C, Yu Z, Sun M (2006) Restraining *Erwinia* virulence by expression of N-acyl homoserine lactonase gene *pro3A-aiiA* in *Bacillus thuringiensis* subsp *leesis*. Biotechnol Bioeng 95:526–532

Fungal Quorum Sensing Inhibitors

Rohit Sharma and Kamlesh Jangid

Quorum Sensing in Fungi

Adaptation to various ecosystems is important to all microbes for growth and survival. Pathogenic microbes, especially bacteria, have developed mechanisms of synchronised expression of virulence genes with increasing population density through quorum sensing, enabling their survival in various host organisms (Casadevall and Pirofski 1999; San-Blas et al. 2000). Due to its direct significance to humans, quorum sensing has been studied in great detail in pathogens and has significantly contributed to our understanding of microbe-host interactions (Gonzalez and Marketon 2003; Shiner et al. 2005; Von Bodman et al. 2003). Quorum sensing regulation is now known to exist in numerous bacterial genera, with few, such as *Aeromonas*, harbouring it ubiquitously amongst all known species (Jangid et al. 2007). In many cases, quorum sensing allows the pathogen to delay or evade host detection until it reaches an effective population (Hogan 2006). In others, it allows developing resistance to biocides and heavy metals through the formation of biofilms. For instance, *Pseudomonas aeruginosa* forms biofilms that are 1,000-fold more resistant than their planktonic counterpart (Rasmussen and Givskov 2006b). The development of resistance towards antibiotics and various other agents amongst bacteria is an increasing concern, and the inhibition of quorum sensing in such bacteria is the most sought-after target to overcome this problem.

Just like bacteria, fungi are known to biocommunicate between themselves or with other organisms through signalling compounds to regulate numerous phenotypic and genotypic changes (Fig. 1) (Witzany 2010, 2012). Communication in fungi is known to regulate complex morphological changes, such as the regulation of fruiting body formation, shape and colour, as well as genetic regulation, namely, bioluminescence in fungi (secretion of luciferase) and fairy ring formation in some mushrooms (species of *Clitocybe*, *Tricholoma*, *Agaricus*, etc.). Amongst other mechanisms, filamentation, growth, pathogenicity, mating and physiological changes in response to nutrient availability and light are also signal mediated (Alspaugh et al. 2000; Bell-Pederson et al. 1996; Borges-Walmsley and Walmsley 2000; Dechant and Peter 2008; Leeder et al. 2011; Witzany 2012). In response to nutrient/substrate availability, fungi secrete several extracellular enzymes, such as cellulase, lipase and protease, and form hyphal or spores of different shape, size and colour. Fungi also use extracellular signalling substances during communication between different mating-type hyphal (pheromone type), increased hygrophobicity, dikaryotic hyphal formation, germination of basidiospores, etc.

R. Sharma • K. Jangid (✉)
Microbial Culture Collection, National Centre for Cell Science, University of Pune Campus, Ganeshkhind, Pune 411007, Maharashtra, India
e-mail: jangidk@nccs.res.in; jangidk@gmail.com

V.C. Kalia (ed.), *Quorum Sensing vs Quorum Quenching: A Battle with No End in Sight*,
DOI 10.1007/978-81-322-1982-8_20, © Springer India 2015

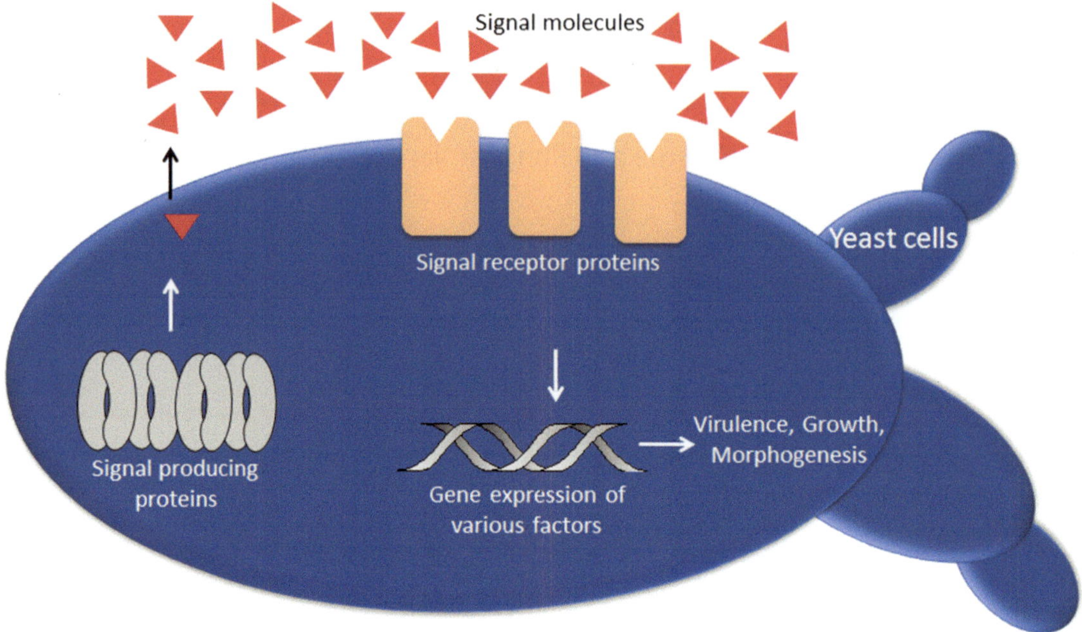

Fig. 1 General mechanism of quorum sensing in fungi. Similar to bacteria, signal-producing proteins are involved in the synthesis of signal molecules which are then detected by the signal receptor proteins which later on regulate the expression of various genes

Several signal molecules play an important role in fungal physiology and genetics. Mitogen-activated protein kinase (MAPK) signalling regulates cell integrity, cell wall construction, pheromones/mating and osmoregulation (Dohlman and Slessareva 2006; Yu et al. 2008). Cyclic adenosine monophosphate (cAMP)/PKA system is involved in fungal development and virulence; calcium-calmodulin-calcineurin are involved in the survival of fungi under oxidative stress, high temperature and membrane perturbation; and rapamycin is involved in the control of fungal growth and proliferation (Fernandes et al. 2005). The GTPase family of proteins is known to activate both of the above signal pathways in *Candida albicans* (yeast) morphogenesis (Han et al. 2011). Several other mechanisms and pathways involved in morphological changes of *C. albicans* are also under signalling control (Alonso-Monge et al. 2009; Cassola et al. 2004; Cheetham et al. 2007; Cloutier et al. 2003; Eisman et al. 2006; Fu et al. 2002; Leberer et al. 2001; Leng et al. 2001; Rocha et al. 2001). Aromatic alcohols (tryptophol and phenylethyl alcohol) and various other volatile and non-volatile compounds are involved in different signalling mechanisms of fungi, such as farnesol in regulating morphogenesis in yeasts, volatile oxylipins in regulating sporulation and mycotoxin production, etc. (Borrego and Kolomiets 2012; Leeder et al. 2011). Similarly, molecules known as fungal quorum sensors are also released by fungi for biocommunication during growth, morphogenesis, virulence, etc. These have been studied mostly in *C. albicans*, *Histoplasma capsulatum*, *Saccharomyces cerevisiae*, *Ceratocystis ulmi* and *Neurospora crassa* (Albuquerque and Casadevall 2012; Chen and Fink 2006; Hornby et al. 2001, 2004; Kügler et al. 2000; Roca et al. 2005).

Fungal pathogens are involved in some severe infections. Some of these infections are so severe that populations of many organisms are now on the verge of extinction due to fungi, namely, frogs (due to *Botryotrichum*), eastern massasauga rattlesnake (*Chrysosporium*), loggerhead turtle eggs (*Fusarium solani*), climbing perch fish (*Aphanomyces invadans*), little brown bat

(*Geomyces destructans*), European crayfish plague (*Aphanomyces astaci*), etc. (Rex 2012). While the role of fungal communication is not well studied in these cases, it is very likely that quorum sensing might play a role in such cases wherein the establishment of high-density populations is required for such disastrous effects on the host. Communication in fungi is probably one of the reasons why fungi have been able to infect humans, animals and plants alike as it likely involves a well-coordinated action of several virulence factors. However, this needs further validation.

History and Evolution of Quorum Sensing in Fungi

By the end of the twentieth century, the phenomenon of quorum sensing was well established in many bacteria. Its near-universal presence amongst eubacteria probably led researchers to extend the exploration towards eukaryotes as well. If it was true, it was going to open a new insight into the mycelial-yeast-form dimorphism, pathogenesis and pathogenicity of fungi and other roles of these compounds in regulatory mechanisms. Sometimes regarded as a "dual personality" of fungi, i.e. the same organism behaving differently with change in conditions, dimorphism was thought to be an evolutionary process in certain fungi: from soil-specific saprotrophic mycelial form to human- or animal-specific pathogenic yeast form capable of growing easily at 37 °C. Temperature and other environmental factors, namely, carbon source, nitrogen source, glucose levels, pH and cell population density, were known to affect this transition from mycelial to yeast form (Nickerson et al. 2006; Román et al. 2009).

However, the concept of quorum sensing in fungi probably emerged much earlier. Filamentation in *C. albicans* was inhibited at high cell density and in the presence of a supernatant from stationary-phase cultures of *C. albicans* (Lingappa et al. 1969). The quorum sensing molecules (QSMs) inhibiting mycelial state were identified as tryptophol and phenylethyl alcohol. Later on Hazen and Cutler (1979 and 1983) first isolated an auto-regulatory substance, named MARS (morphogenic auto-regulatory substance) from *C. albicans* that prevented the transition from yeast to mycelial form. Although the exact mechanism for its action could not be elucidated, it gave a direction towards auto-regulatory substances. In 2000, Kügler et al., while working on the yeast form of *H. capsulatum*, showed that a cell wall-bound polysaccharide controls the expression of yeast-form-specific genes and production of certain other molecules, suggesting that the mode of function may be in the form of quorum sensing compound (as in bacteria). A year later, Hornby et al. (2001) proved that the transition from mycelial to yeast form was indeed cell density dependent. An initial inoculum of $>10^6$ cells/ml favoured the mycelial form, whereas $<10^6$ cells/ml favoured the yeast form. They demonstrated that *farnesol*, an isoprenoid (15-carbon sesquiterpene, 3,7,11-trimethyl-2,6,10-dodecatrien-1-ol) QSM produced by *C. albicans*, prevented germ tube formation (i.e. mycelial form of fungi) by inhibiting the germ tube factors, namely, L-proline, N-acetylglucosamine and serum. Recently, various other factors have been discovered that trigger yeast to filament transition. With the help of gene mutagenesis and transcriptional studies during the past several years, various signalling pathways, transcriptional factors and other regulatory components have been identified that regulate the transition from yeast to mycelial form (Gow 2009; Han et al. 2011). Han et al. (2011) compiled an extensive list of 31 signal pathway genes, 13 transcription factor genes, 11 receptor genes, 5 metabolic genes, 7 cell wall-related genes and 6 cell cycle-associated genes that make modifications in cell wall, cell polarity, metabolism, cell cycle, virulence factors and morphogenesis.

The evolution of quorum sensing regulation in fungi has two hypotheses. Fungi have been living in different habitats wherein they must coordinate or in other words communicate between different cells. Certain groups of fungi have a dolipore septum through which different cells communicate internally between themselves. However, fungi need to communicate externally amongst themselves for its survival. The first hypothesis stems from this survival criterion. Conditions might have forced certain fungi to live inside a host (animal/human) sometime during the evolution and for which they developed the dimorphic phase through QSMs. The second hypothesis originates from the fact that fungi and bacteria have been living together in terrestrial as well as aquatic ecosystem for the past several million years and while living together, fungi must have adopted the phenomenon from bacteria for competition. Similar views have been expressed by Gould (2012) while discussing polyketide (which has a role in signalling) suggesting the uptake of such genes by fungi from bacteria (or cyanobacteria) while living in close association or together (as in lichens). It could have been a mechanism to evade attack by other organisms, namely, bacteria, by forming biofilms preventing them from various external chemicals. The transition of fungi from soil to humans or animals may be of escapism also, something very similar to an animal living inside a zoo or protected area who is safe from competition and gets sufficient food and protection from predation, i.e. a well-crafted life. Similarly, entering inside a human, animal or plant may be an escape from a highly competitive environment, escape from predation (protists in case of fungi), and may provide sufficient food without competition. For instance, *Cryptococcus* produces drying-digestion-resistant coats (like polysaccharide) to prevent itself from predation by protists (like amoeba) or human macrophages (Frazer 2013).

Clearly, the evolution of quorum sensing system is energy efficient as the organism (fungi or bacteria) only expresses virulence genes when they are in sufficient number, thus ensuring a high probability of success. Most QSMs whether produced by bacteria or fungi are either alcohol or alkaloid based, i.e. volatile in nature. This is an important factor for ensuring that the signal reaches to each and every cell of the colony. Pamela Ronald, Cornell University, describing bacteria as invaders has defined the mechanism of coordination between bacteria and plant as "Bacteria talk, plants listen" (Summers 2012). It has been found that plants, animals, humans and flies contain certain immune receptors against these QSMs to protect themselves (Gómez-Gómez and Boller 2000; Lee et al. 2009). On the other hand, Matson (2011) isolated several fungi from an "extreme environment", like a dishwasher from the Netherlands, China and Slovenia, and none was considered to be highly pathogenic (causing infection in only immune-compromised patients) except *Exophiala* which in rare cases infected cystic fibrosis patients. In our view, apart from being non-virulent, it is probably due to the inability of these fungi to form QSM. However, this needs further validation.

Fungal quorum sensing systems have been studied for their various effects, namely, oxidative stress, biofilm formation, filamentation, drug efflux, etc. (Albuquerque and Casadevall 2012). In fact, *C. albicans* is one of the most well-studied pathogenic fungi for the role of quorum sensing in dimorphism (Chen and Fink 2006; Han et al. 2011; Hogan 2006; Nickerson et al. 2006). While most studies have focused on the use of QSMs as autoinducers to enhance growth or for virulence, the role of QSMs in plant-fungi symbiosis and algae-fungi symbiosis (lichen) is also an important area of study. Very few studies have looked at the role of quorum sensing in biofilm formation by fungi (Kumamoto 2002; Ramage et al. 2005). Studies on *farnesol* have shown that when biofilms were exposed to it, genes related to drug resistance, cell wall maintenance and iron transport and heat shock proteins were influenced (Cao et al. 2005). Studies have shown that farnesol can act as an interspecies (*C. albicans* farnesol affects mycelial development in *C. dubliniensis*) and inter-kingdom (*C. albicans* farnesol reduces *Pseudomonas* quinolone signal and phycocyanin in *P. aeruginosa*) quorum sensing molecule (Fig. 2) (Atkinson and Williams 2009; Cugini et al. 2007). Apart from QSMs, various fungi, such as *Uromyces phaseoli*, *Glomerella cingulata* and *Ceratocystis ulmi*, are known to produce different auto-regulatory substances which affect the cellular morphology and physiology of the organisms which in turn affect their behaviour in the ecosystem (Hornby et al. 2004; Lingappa and Lingappa 1969;

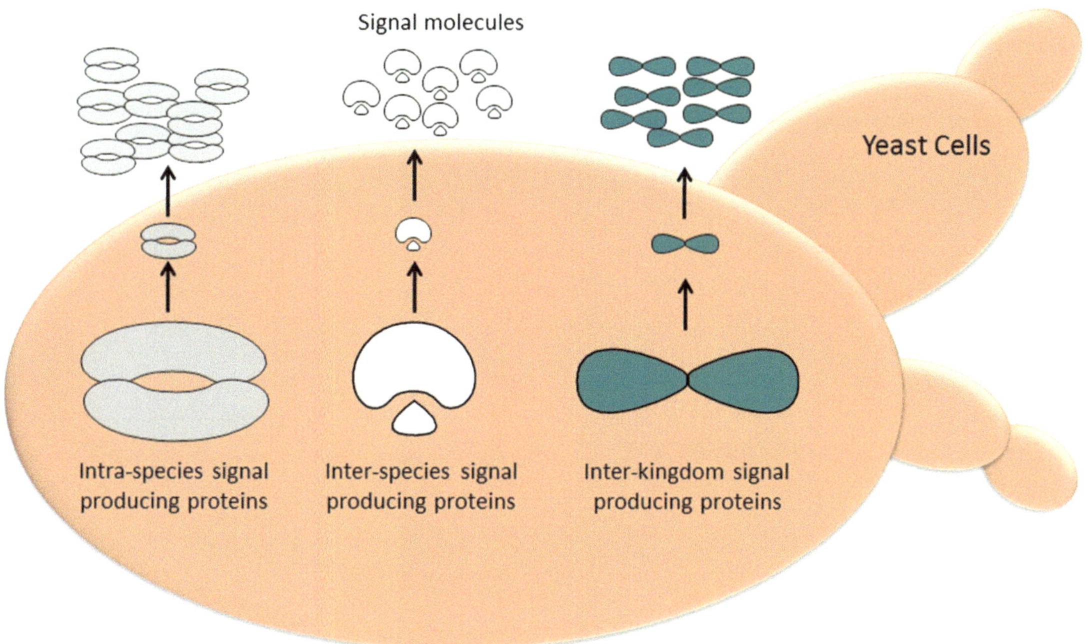

Fig. 2 Specific and non-specific quorum sensing regulation in fungi. Just like bacteria, fungi do not produce a single QSM but different molecules with different recognition spectra

Macko et al. 1970). Thus, quorum sensing in fungi has increased our understanding of the regulation of numerous complex mechanisms in multicellular organisms.

Fungal Quorum Sensing Molecules (QSMs)

Various kinds of QSMs are known from fungi with farnesol being the most well studied (Fig. 3). It blocks the transition from yeast to mycelial form in *C. albicans* but does not stop hyphal that have already developed (Mosel et al. 2005; Navarathna et al. 2005). Farnesol inhibits mycelial formation in the range of 1–50 μm and is produced in higher amounts at temperature range from 37 to 40 °C (Mosel et al. 2005). It is produced aerobically from farnesyl pyrophosphate (an intermediate in the sterol biosynthesis) by enzymatic activity and is independent of the carbon source used for growth (Dumitru et al. 2004; Hornby et al. 2001, 2003). The exact mechanism of farnesol-mediated quorum sensing regulation of morphological changes is still largely unknown. However, Han et al. (2011) had discussed metabolic basis of various pathways and QSMs involved in *C. albicans*. Farnesol inhibits MAP kinase cascades (related to hyphal growth and virulence through nutrient starvation) via suppression of HST7 and CPH1 gene expression (Sato et al. 2004). These are homologues of *S. cerevisiae* STE12 and STE7 and are morphogenetic regulators of *C. albicans*. Thus, farnesol suppresses the hyphal formation by inhibiting the starvation signal which in turn is induced by MAP kinase cascade. Farnesol is also shown to upregulate (TUP1) or downregulate (CRK1 and PDE2) morphogenesis-related genes (Cao et al. 2005) or suppress the RAS1-CDC35 pathway repressing the hyphal formation (induced by cAMP/PKA pathway) (Davis-Hanna et al. 2008). In other studies, various other mechanisms of inhibition of hyphal formation involving farnesol have been proposed (Kebaara et al. 2008). Apart from inhibiting filament formation in *Candida* (*C. dubliniensis*, *C. tropicalis*) (Henriques et al. 2007; Zibafar et al. 2009), farnesol also induces apoptosis in fungi, such as *S. cerevisiae*, *Aspergillus nidulans* and

Source of QSM	Compound	Structure	Reference
Candida albicans	Farnesol#	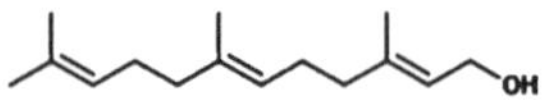	Hornby et al. 2001; Cugini et al. 2007
C. albicans, *Saccharomyces cereviseae*	Tryptophol	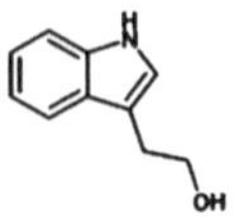	Chen et al. 2004
C. albicans, *S. cereviseae*	Tyrosol#	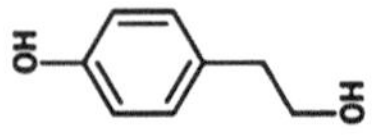	Chen et al. 2004
C. albicans, *S. cereviseae*	Phenylethanol	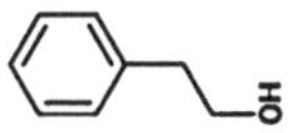	Chen et al. 2004
C. albicans	Farnesoic acid	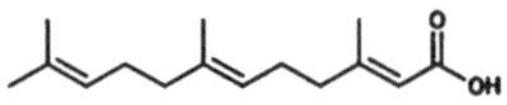	Hogan 2006
Uromyces phaseoli	Dimethoxycinnamate	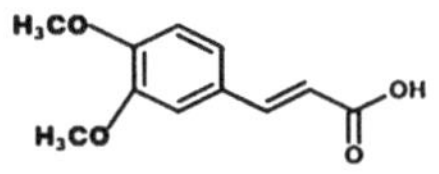	Hogan 2006
Zygomycetes	Trisporic acid	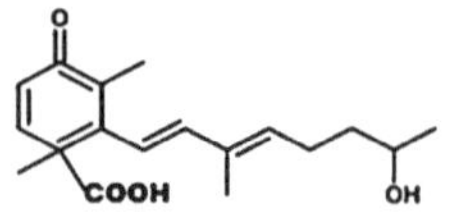	Hogan 2006

Fig. 3 Structures of some known fungal QSMs. #Farnesol and tyrosol are also known quorum sensing inhibitors (QSIs). See main text

Penicillium expansum (Derengowski et al. 2009; Fairn et al. 2007; Liu et al. 2009; Semighini et al. 2006). A different strain of *C. albicans* (ATCC 10231) produces *farnesoic acid* instead of farnesol, which also inhibited filament formation (Oh et al. 2001). However, lately, it has been found that farnesol has higher activity and a different structure (Hornby and Nickerson 2004; Nickerson et al. 2006). In addition, there are other substances which suppress hyphal development like tryptophol and phenylethyl alcohol.

There are QSMs which act opposite to that of farnesol (Fig. 4). Aromatic alcohol (tyrosol) is also found in *C. albicans*; however, it promotes hyphal formation (Hogan 2006). Microarray studies have shown that it affects DNA replication, chromosome segregation and cell cycle regulation (Chen et al. 2004). Other metabolites such as estradiol (mammalian metabolite) and di-butyryl cAMP are also considered to be involved in biocommunication (Han et al. 2011). Additionally, there are certain factors, enzymes, etc. involved in the carbon metabolism (which usually is considered conserved) that may affect morphogenesis. The reader is advised to refer to Han et al. (2011) for more details on this.

Fungal Quorum Sensing Inhibitors (QSIs)

In the past several years, research has focused on searching for a mechanism to inhibit quorum sensing signalling in microbes. In order to understand the inhibition of quorum sensing regulation, it is important to quickly review the general regulation mechanism. Regardless of the signalling compound, the general mechanism involves the synthesis of the signal which is secreted to the exterior of the cell either by diffusion or active transportation. After reaching threshold levels, the signalling compound binds to the receptor, which then binds to the promoter sequence acting as a transcriptional regulator of

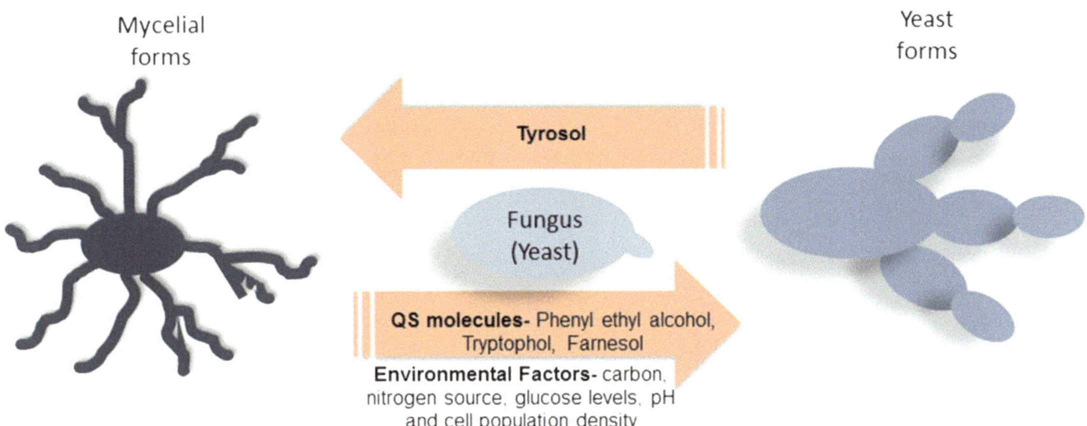

Fig. 4 The effect of quorum sensing molecules on the dimorphic transition of *C. albicans*. Tyrosol accelerates hyphal formation, while farnesol, tryptophol and phenylethyl alcohol suppress hyphal development (Adapted from Han et al. (2011))

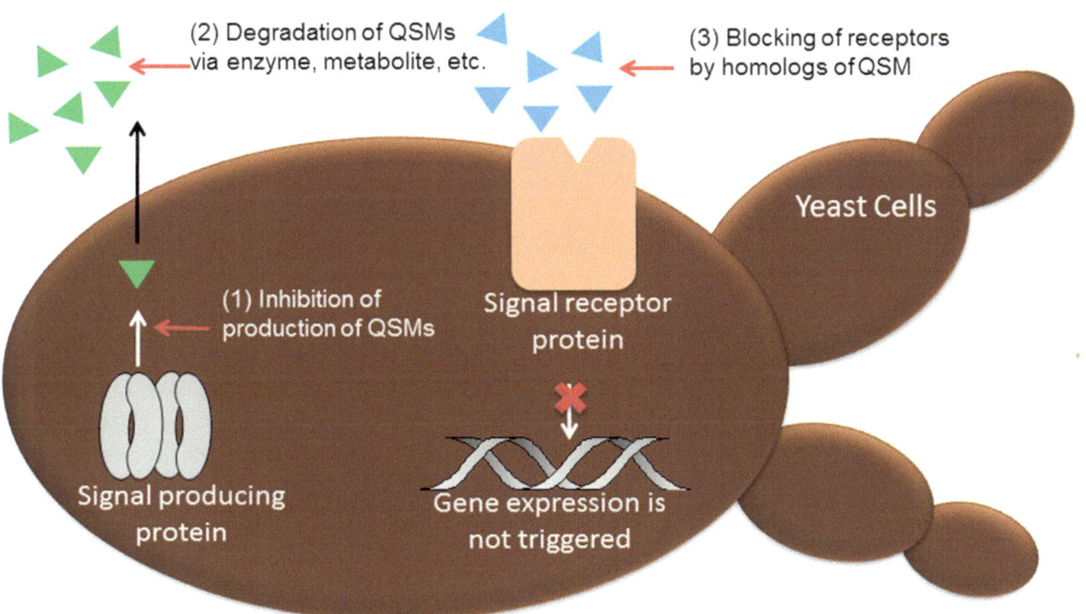

Fig. 5 Three strategies of fungal QSIs to control quorum sensing mechanism of pathogenic organisms

target gene. This suggests that there are three ways by which the quorum sensing signalling can be stopped/hampered (Fig. 5). First is stopping the production of signal molecules. Second, the signal molecules can be degraded by enzyme(s). Third, if the signal molecule is not allowed to bind to the receptor molecule, it will not act as a regulator to the gene promoter. In either case, it would disrupt the regulation circuit.

Fungi are almost cosmopolitan and have been living on various habitats and interacting with other groups of organisms, namely, microbes, plants, animals, etc. They have been occupying extreme niches also. During evolution organisms that live in association with each other in nature have developed mechanisms to tackle each other by producing chemicals, enzymes and metabolites. Rhizosphere and mycorrhizal fungi

Source of QSI	Compound	Structure	Effective against organism	Quorum sensing activity	Reference
Tremella fuciformis (White jelly mushroom)	Extract from fruiting body	--	*Ch. violaceum*	Violacein production	Kalia, 2013; Zhu and Sun, 2008
Auricularia auricular (Jelly fungi)	Natural pigments	--	*Ch. violaceum*	Violacein production	Kalia, 2013; Zhu et al., 2011
Penicillium coprobium	Patulin	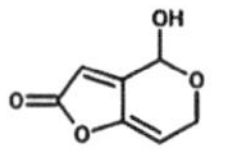	*P. aeruginosa*	Biofilm	Rasmussen et al., 2005b
Pe. radicola	Penicillic acid	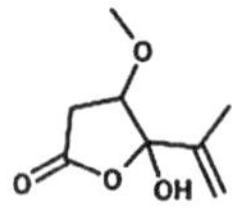	*P. aeruginosa*	Biofilm	Rasmussen et al., 2005b
Delisea pulchra (Australian alga)	Furanone	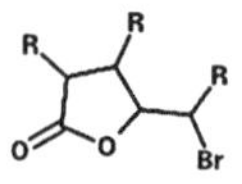	--	--	Rasmussen et al., 2005a

Fig. 6 Structures of some fungal QSIs

interact closely with bacteria in soil. Due to their close interaction, fungi have developed or evolved natural mechanism to combat bacterial population. These may be for nutrition, pathogenicity, space, etc. They are also known to produce various secondary metabolites (Frisvad et al. 1998, 2008), including mycotoxins (Pitt 2000), enzymes and other chemicals which have the potential to act as QSIs (Fig. 6). However, not much information is available on fungal QSIs. The Australian alga (*Delisea pulchra*) produces halogenated furanones, which prevent bacterial growth (Maximilien et al. 1998). QSIs which act on signalling molecules have been known as a potential strategy for bacterial disease control. The inhibition of quorum sensing signalling by inhibiting N-acyl homoserine lactone (AHL) production or degradation is reported for fungi (Fig. 7) (Dong and Zhang 2005; Uroz and Heinonsalo 2008). Even many medicinal and dietary plants are known to produce QSIs for AHL-based regulation (Adonizio et al. 2008; Zhu and Sun 2008). Recently, Teplitski et al. (2000) have demonstrated that several plants produce molecules which imitate the structures of AHL produced by bacteria. *Pisum sativum* (pea) produces molecules which either inhibit the AHL production or its recognition by receptor molecules. For further details on plant-derived natural products which can be used as QSIs, refer to Koh et al. (2013).

There are two types of QSIs reported so far in bacteria, furanones and RNA III inhibiting peptides (RIP). Furanones are the most studied inhibitory molecules and can act on both AHL and AI2 quorum sensing systems. Furanones either increase the susceptibility of biofilms to antibiotics (Pan and Ren 2009) or facilitate the immune system (Wu et al. 2004). However, the toxic nature of furanones has been a concern for the same. The RIP in its amide form (YSPWTNF-NH_2) is known to repress virulence, biofilm formation and antibiotic resistance in *Staphylococcus* (Balaban et al. 2003, 2007; Chen and Wen 2011). In the presence of RIP, TRAP (quorum sensing signal molecule) is not phosphorylated, and *agr* is not activated at the mid-exponential phase, preventing the production of RNA III (which upregulates the toxin production) and in turn reducing toxins produced. Because not many fungal QSIs are known, a detailed screening of fungi from various habitats like endophytes, marine fungi, may help in finding chemicals which have QSI potential.

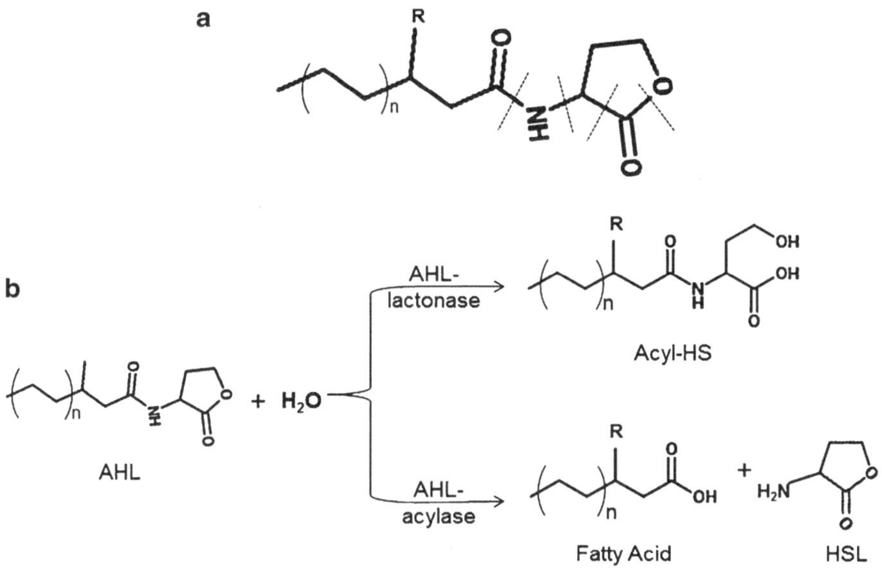

Fig. 7 Enzymatic degradation of AHL. (**a**) The AHL structure and its possible enzyme cleavage sites. (**b**) The corresponding degradation mechanisms of AHL-lactonase and AHL-acylase (Adapted from Dong and Zhang (2005))

Fungi mediated inhibition of QSM Production

Quorum Sensors as QSIs

Candida albicans

Fungi are known to produce several QSMs with farnesol and tyrosol being the two most prominent of them. Farnesol, which is produced by most dimorphic yeasts, has a dramatic effect on their morphogenesis (Shirtliff et al. 2009; Weber et al. 2010). However, farnesol also possesses antimicrobial properties against *Fusarium graminearum* (Semighini et al. 2006), non-albicans *Candida* species (Weber et al. 2010), *Paracoccidioides brasiliensis* (Derengowski et al. 2009), *Staphylococcus aureus*, *St. epidermidis* (Cerca et al. 2012) and other bacteria (Brilhante et al. 2012; Pammi et al. 2011). When used in combination with antibiotics, farnesol acts as an adjuvant against *St. epidermidis* (Pammi et al. 2011) affecting its biofilm matrix composition (Gomes et al. 2011b). It increases cell death when used in combination with tetracycline or rifampicin (Gomes et al. 2011a). Farnesol has also been shown to interfere with the metabolic pathway in *St. aureus* (Kaneko et al. 2011). It also induces apoptosis in *A. nidulans* (Semighini et al. 2006). Farnesol produced by *C. albicans* inhibits its biofilm formation (Ramage et al. 2002), thus protecting mice against oral candidiasis (Hisajima et al. 2008). High concentration of farnesol inside the biofilm inhibits newly formed cells inside the biofilm. A similar study conducted on *C. tropicalis* and *C. parapsilosis* also showed that high concentration of farnesol inhibits biofilm formation (Laffey and Butler 2005; Zibafar et al. 2009). *Cryptococcus neoformans* are also known to be inhibited by farnesol (Cordeiro et al. 2012). Thus, these can be used as a therapeutic agent and also adjuvant to antifungal drugs.

Secondary Metabolites

Tremella fuciformis

Tremella belongs to the family *Tremellaceae* (*Basidiomycota*) commonly called as "jelly fungi" as they form gelatinous fruit bodies. In nature they generally colonise dead woods but are also now artificially cultivated. Their anamorphs are mostly yeasts. *Tremella fuciformis* is commonly called as white-snow fungus or white-ear fungus due to its soft texture, earlike shape and white colour. These are very popular in China due to their edibility and medicinal importance (for their effects in tuberculosis, cancer, immune-modulation, diabetes, high blood pressure and common cold) (Cho et al. 2006, 2007). *T. fuciformis* is known to contain heteroglucan, triterpenes and other compounds (Cho et al. 2006; Zhu et al. 2006). Zhu and Sun (2008) studied the effect of *T. fuciformis* on the inhibition of quorum sensing in bacteria using the mutant *Chromobacterium violaceum* CVO26 (produced from wild-type *Ch. violaceum* ATCC 31532) which is violacein negative. The mutant is devoid of violacein synthase CViI and requires the addition of N-hexanoyl homoserine lactone (HHL) in the medium to produce violacein, which is also a potential antibiotic. Qualitative and quantitative analysis of the effect of *T. fuciformis* extracts on violacein production showed that although it did not have an effect on growth, it inhibited violacein production at different concentrations (0.2–0.8 %). Thus, crude extract or compounds isolated from *T. fuciformis* extracts can be used as QSIs due to its non-toxic nature and effectiveness.

Auricularia auricula

Auricularia belongs to *Auriculariales* (*Basidiomycota*), also commonly called as "jelly fungi". They are generally found on dead wood. Although species of *Auricularia* are edible, they are not very popular. *Auricularia* produces several exopolysaccharides which have potential antibacterial, antitumor and immune-modulatory effects (Li and Dong 2010). It is also used in traditional Ayurvedic preparations for sore throat, sore eyes, jaundice, etc. *Au. auricula* produces several dark-coloured pigments (namely, melanin, melanoid, pheomelanin) (Dastager et al. 2006). Zhu et al. (2011) tested the effect of pigments (hydrochloric infusions of dried fruit bodies) of *Auricularia auricula* against quorum sensing in *Ch. violaceum* CVO26 as test organism. Many heterocyclic compounds are used in the biosynthesis of these pigments like dopaquinone, leucodopachrome and cysteinyldopas. These compounds contain carboxyl and amino group which readily transforms to carbonyl. These compounds inhibit the AHL-regulated signalling system by binding to the active site of receptor proteins. Other studies on the inhibitory effect of extracts of *Au. auricula* on biofilms of *Escherichia coli* showed that the rate of inhibition was 73 % (Li and Dong 2010). The identification of the active compounds and mechanism by which these groups of compounds modify the signalling system of bacteria needs further investigation.

Ganoderma lucidum

Ganoderma, commonly called as "reishi mushroom", is a member of *Ganodermataceae* (*Basidiomycota*). Species of *Ganoderma* are hard, polypore mushrooms which grow on forest trees as pathogens. It is also mentioned in the old Chinese description of medicines. Presently, it has found its use as immune-modulatory, antimicrobial, antidiabetic, antioxidant, etc. and in normalising blood pressure (Silva 2003; Wachtel-Galor et al. 2004). Zhu et al. (2011) used methanolic extracts of *Ganoderma lucidum* to test its effect on the quorum sensing signalling in bacteria and found promising results. Further studies on the active metabolites and the mechanism involved are required.

Phellinus igniarius

Phellinus is a member of *Hymenochaetaceae* (*Basidiomycota*). It is considered to be a plant pathogen and has several medicinal properties, namely, antioxidant, anticancer, antidiabetes, cholesterol reduction, etc. (Lung et al. 2010). Zhu et al. (2012) showed that fermentation metabolites of ten different *Phellinus igniarius* strains inhibited violacein production in *Ch. violaceum* CVO26.

Endophytic Fungi (*Fusarium graminearum* and *Lasiodiplodia* sp.)

Endophytic fungi that live inside a plant host hold a lot of potential for metabolites with antibacterial properties. Recently, Rajesh and Rai (2013) have tested different endophytic isolates from *Ventilago madraspatana* plant for the production of various enzymes and metabolites which can

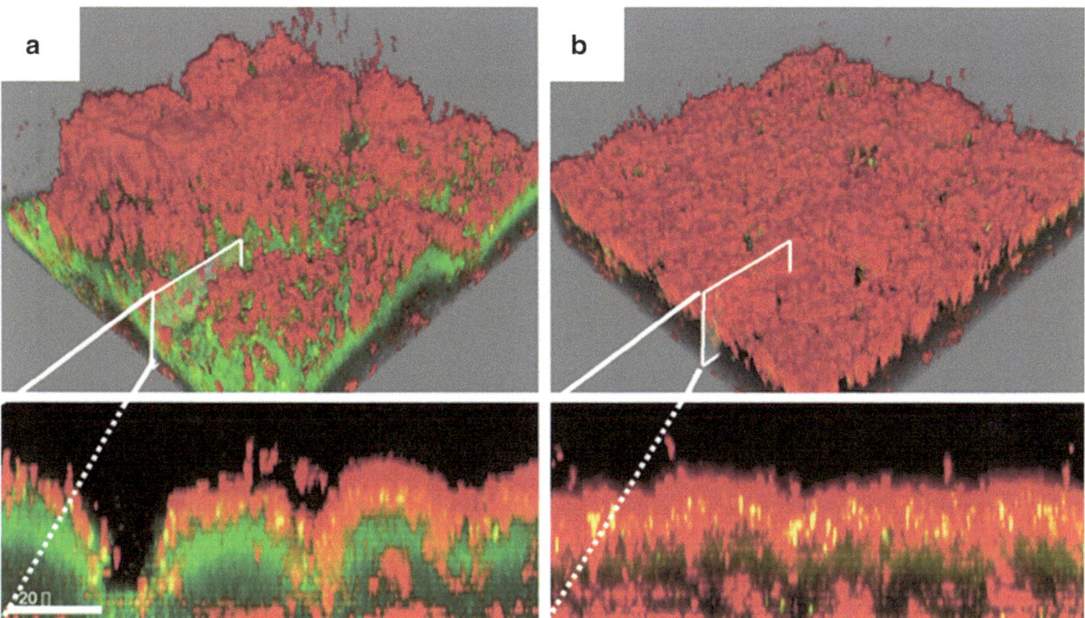

Fig. 8 Effect of QSI on bacteria. Furanone-treated *P. aeruginosa* biofilms are less tolerant to tobramycin. Scanning confocal laser photomicrographs of *P. aeruginosa* PAO1 biofilms grown in the absence (**a**) or the presence (**b**) of 10 μM C-30. After 3 days, the biofilms were exposed to 100 μg/ml tobramycin for 24 h. Bacterial viability was assayed by staining using the LIVE/DEAD *Bac*Light Bacterial Viability Kit. *Red areas* are dead bacteria; *green areas* are live bacteria (Reproduced with permission from Hentzer and Givskov (2003))

act as QSIs. Following the experimental strategy of Ueda et al. (2007) and Zhu et al. (2011), they demonstrated that out of 15 fungal isolates, extracts of *Fusarium graminearum* and *Lasiodiplodia* sp. contained metabolites which inhibit violacein production in *Ch. violaceum* CVO26 without affecting its growth. Thus, these two organisms hold potential to help in the formulation of new antibacterial drugs in the future.

Mycotoxins

Penicillium species

Fungi have been known to produce several mycotoxins which destroy food and degrade postharvest crops, medicinal plants, fruits, vegetables, etc. These are metabolites with different nature, namely, alkaloids. These may cause sickness and food poisoning in humans and animals even in low concentrations. *Aspergillus*, *Penicillium* and *Fusarium* are some common fungi which are known to produce such toxins. Aflatoxin, produced by *A. flavus*, is a well-known mycotoxin noted to cause sickness in humans and animals. Ochratoxin A is produced by *A. ochraceus* which is distributed on various nuts and dry fruits. Fumonisins, trichothecene toxins and zearalenone, produced by *F. moniliforme* and *F. graminearum*, are some other common mycotoxins (Abbott 2002; Pitt 2000). Rasmussen et al. (2005b) studied the effect of two mycotoxins, patulin (produced by *Penicillium coprobium*) and penicillic acid (produced by *Pe. radicola*) for their QSI activity in *P. aeruginosa*. Both patulin and penicillic acid targeted the RhlR and LasR proteins, thereby inhibiting quorum sensing-regulated genes in the bacterium up to 49 % and 34 %, respectively. Moreover, *P. aeruginosa* infection in mouse is cleared rapidly when treated with patulin. The biofilm of *P. aeruginosa* was also found to be susceptible to tobramycin antibiotic in the presence of patulin (Fig. 8).

Enzymes of Fungal Origin Degrading QSMs and Biofilms

The successful invasion of a host by microbe depends on the expression of virulence factor through quorum sensing signalling or biofilm formation. Once the bacterium invades a host, it attaches to the surface of internal lining and forms a slimy structure (biofilms) for its protection. It provides extra protection to the microbes from outside effect (antibiotic, biomolecules, enzymes, etc.). Almost 80 % of the disease infections are caused by bacteria that form biofilms (Davies 2003). Biofilms also facilitate the development of chronic infections of *P. aeruginosa* by protecting the pathogen from host immune infections (Costerone et al. 1999). Fungi are known to produce several extracellular enzymes, namely, cellulases, proteases, amylases, etc., which can be used for degrading bacterial biofilms. Enzymes (AHL-lactonase, AHL-acylase, oxidoreductase) with QSI properties have been already found in bacteria (Fig. 7) and reviewed in depth by Chen et al. (2013). However, very few reports are currently available on fungal QSI via degradation or interruption. Since mycorrhizal (both ecto and endo) and beneficial root-associated fungi are known to protect plants from diseases (caused by various microbes), these will surely have the potential for such biomolecules.

Trichosporon loubieri

Trichosporon is a genus of anamorphic fungi (family *Trichosporonaceae*) with common occurrence in soil with few species inhabiting natural human skin flora. In a first report of a basidiomycete yeast, *Trichosporon loubieri* was found to produce enzyme-based QSI degrading various AHLs with different *N*-acyl side chains ranging from C4 to C10 with or without oxo group substitution at C3 position (Wong et al. 2013). Based on the re-lactonisation experiment, it was confirmed that the degradation was via lactonase enzyme. Further, heat-treated *Tr. loubieri* did not degrade the AHLs, demonstrating that the enzymatic activity is heat sensitive.

Root-Associated Fungi (*Phialocephala fortinii, Ascomycete* Isolate, *Meliniomyces variabilis*, Unidentified Isolate)

Root-associated fungi are known to produce several chemicals to inhibit other groups of organisms. In a study to test different root-associated fungi for the inhibition of QSMs, Uroz and Heinonsalo (2008) showed that *Phialocephala fortinii* (strains JH1, JH8), unidentified *Basidiomycota* mycorrhizal isolate (JH7) and *Meliniomyces variabilis* (JH23) completely inhibited AHL molecules. Interestingly, the AHL degrading ability of JH1 was due to a bacterium associated with the fungus (*Rhodococcus erythropolis* Myc1), a probable endosymbiont. The results concluded that lactonase and another unidentified enzyme were responsible for the activity (Uroz and Heinonsalo 2008). However, the exact mechanism of action is still unknown.

Other Enzymes Degrading Biofilms

Apart from bacterial enzymes, fungal enzymes are also known to degrade the biofilms produced by bacteria. The enzymes produced by *A. niger*, *Trichoderma viride* and *Penicillium* species are known to degrade the biofilms of *P. aeruginosa* (Gautam et al. 2013). The compositions of most of the biofilms are lesser known. However, the composition of planktonic extracellular polysaccharide and others is known. Since a number of fungal enzymes are able to degrade the plant cell wall, it was thought that enzymes produced by various fungi can be used to degrade biofilms. Orgaz et al. (2006) studied the effect of various enzymes for their potential to degrade biofilms of *Pseudomonas* and found that fungi grown on pectin (*Trichoderma viride* and *A. niger* are rich in pectinesterases and poor in proteinase) and gum arabic (from *A. niger* high in proteinase) degraded more biofilm (84 %, 60 % and 65 %, respectively). The study concluded that both proteolytic and carbohydrate- degrading activities are good for biofilm removal.

Resistance to Fungal QSIs

The research on QSIs began with the concept and interest that disruption of quorum sensing system will be an effective anti-infective strategy and it will restrict the growth of pathogenic bacteria. With high likelihood of preventing the development of resistance against the prevalent antibiotics, these were regarded as the best alternative to the discovery of new antibiotics (Kalia et al. 2013). However, there are possibilities that organisms may devise mechanisms to develop resistance to QSIs (Defoirdt et al. 2010). This concept came to light when researchers observed a variation in the expression of core quorum sensing genes between different bacterial species and different strains (possibly virulent and avirulent strains) of *Vibrio* and *Pseudomonas*. It is very likely that if this variation better supports fitness and health of bacteria, natural selection would help the bacteria to spread the QSI resistance further on. Later on, Maeda et al. (2012) showed that *P. aeruginosa* develops resistance when grown on adenosine as sole carbon source. They observed growth of previously resistant cells within four sequential dilutions after transposon mutagenesis in the presence of added QSI (brominated furanone C-30). While resistant phenotypes are known, no clear mechanism is known for such resistance. It is believed to be similar to that of antibiotic resistance: restricted availability, inactivation or even modification of the target. Relatively recently, several studies have been initiated that have revealed some intriguing aspects about the development of resistance to QSIs (Decho et al. 2010; Defoirdt et al. 2011; Mattmann and Blackwell 2010). It has been found that bacteria might develop resistance against QSIs when bacteria interact with antibiotics and mutation occurs in efflux pump without any interaction with QSI (García-Contreras et al. 2013).

While some information is available on the development of resistance to QSIs in bacteria, there is no information about it in fungi. But, given the nature of mechanisms being discovered, it is very likely that resistance to QSIs in fungi will soon be discovered, albeit it needs further exploration.

Applications of Various Fungal QSIs

In 1928, when the first antibiotic was discovered (penicillin), it was thought that it will help in controlling microbial infections. During the past few decades, various antibiotics and there derivatives were discovered which have helped mankind. With the discovery of antibiotics came an imminent problem of antibiotic resistance. The life cycle of microbes, genetics, physiology and ability of an organism to adapt and evolve have led to the resistance against many compounds including antibiotics. Although antibiotics have played a key role in the greater life survival and expectancy of humans, the treatment of microbial infections has become difficult with the emergence of antibiotic-resistant strains of bacteria. Many researchers are now focusing on searching for new antibacterial compounds. They achieve this by disrupting the pathogen's ability to sense its population and trigger its virulence factors. However, relatively recently, there has been an alarming increase in the drug-resistant microbes. The reason includes the extensive use of non-specific drugs and over-the-counter availability of many generic antibiotics. Thus, those diseases which were once treatable have now become untreatable.

Microbial resistance to known antimicrobials and limitation in the discovery of novel antimicrobial compounds have led to the adoption of new strategies. Relatively recently, researchers have focused on using QSIs and enzymes for inhibiting production and degradation of QSMs (Adak et al. 2011; Hentzer and Givskov 2003; Hentzer et al. 2003; Njoroge and Sperandio 2009). The effective use of QSIs in controlling the spread of infection has recently been demonstrated in mice. When mice are infected with virulent bacteria, they grow and

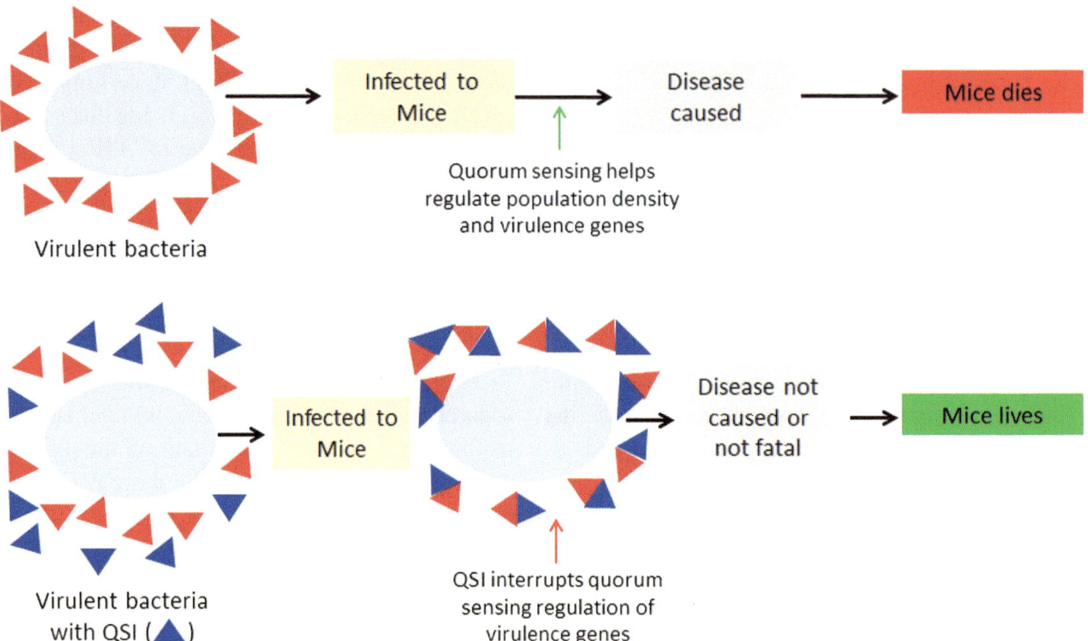

Fig. 9 Biotechnological application of fungal QSIs (Adapted from Bonnie Bassler's lecture on "How bacteria 'talk' in 2009")

multiply, and QSMs are released, resulting in bacterial population to reach a threshold level; the virulence gene expresses and causes disease, killing mice. However, when mice are infected with bacteria along with quorum sensing imitating molecules or QSIs (with little structural change), the bacterial population does not recognise quorum sensing signalling, and its virulence gene remains unexpressed or under-expressed and mice survive (Fig. 9). Thus, these QSIs will help us to develop new sets of antibiotics and also help us to get rid of multidrug resistance as it does not kill the bacteria and only prevent them from coordinating amongst them. Thus, the possibility of selection of resistant bacteria becomes less.

In addition, probiotics and QSIs produced by bacteria have been studied in controlling pathogenic bacteria (Cotar 2013). In animal pulmonary diseases, such QSIs have proved to be beneficial and cleared the infection, thereby reducing the mortality (Rasmussen and Givskov 2006b). Fungi (both macro and micro) with antimicrobial compounds hold a lot of potential. Fungi obtained as endophytes have been studied by workers for antimicrobial properties and they found promising results. Similarly, mushrooms have also shown good antimicrobial properties. When studied in detail, these may be producing QSIs against pathogenic microbes. Fungal QSI has been isolated and screened from fungal extracts (Persson et al. 2005; Rasmussen et al. 2005a, b). The screening of 50 *Penicillium* spp. showed that 66 % of them produced secondary metabolites with QSI activity (Rasmussen and Givskov 2006a, b). Unfortunately penicillic acid and patulin (two fungal QSIs) have been limited in their use to biotech industry due to their mycotoxin nature. Many mushrooms are known for medical importance and used by tribals for antimicrobial properties and documented in ethnomedicinal publications. Hence, the screening of edible mushrooms is an important aspect while searching for fungal-based QSIs which holds a lot of pharmaceutical potential. However, there have been very few studies on the screening of QSIs from fungi as natural resources.

Biofilm formation on glass, steel and plastic surfaces of medical equipment has become a problem for critical surgeries. Various pathogens have been isolated from artificial hip prosthesis, prosthetic heart valve, intrauterine device, central venous catheter, artificial voice prosthesis, etc. (*P. aeruginosa*, *St. aureus*, *Klebsiella pneumonia*, *Enterococcus* sp., *C. albicans*, etc.) (Jain et al. 2007). The QSI enzymes that are found in organisms other than bacteria have the potential to degrade the biofilms of microbes and provide protection against microbial infection. In addition, many fungal enzymes are commercially used in fruit processing industries and waste water treatments (Orgaz et al. 2006). Earlier work on the same has resulted in promising results (Hahn et al. 2001; Johansen et al. 1997). Biofilms treated with furanones are easily eradicated by tobramycin. Many other researchers have expressed their hope in QSIs as an alternative to conventional methods of controlling bacterial infections (Finch et al. 1998; Hentzer and Givskov 2003; Smith and Iglewski 2003). The biofilms of microbes are also known to degrade polymeric compounds which may involve various mechanisms (Mohan and Srivastava 2010). The fungal enzymes hold a lot of potential to protect deterioration of the same.

Spoilage of nonfrozen vegetables by microbes has been an area of concern for food industry. Growth of bacteria to high density is one of the reasons for destruction of perishable crops and vegetables. Spoilage of sprouted beans is one such example where bacteria grow, release enzymes and cause soft rot of vegetables accounting for 10–50 % of loss (Rasch et al. 2007). Studies have found very high concentration of AHLs in sprout bean spoilage (Rasch et al. 2005). As also discussed earlier, QSIs patulin and penicillic acid have antimicrobial effect on *P. aeruginosa*. Hence, these QSIs have a lot of potential to be used as food preservatives. However, their side effects must first be evaluated in great detail.

Treatment of rhizospheric bacterial diseases is also another aspect where the use of fungal QSIs may hold potential for biotechnological exploitation. Production and secretion of QSI and its application against root plant pathogens are likely to help in crop protection.

Conclusion and Our Opinion on the Scope of Research on Fungal QSIs

With the discovery of crosstalk, even unicellular organisms are now being reinvestigated at the population level because they are behaving as multicellular forms (multiple cells) working in coordination. In comparison to bacteria, fungi harbour a greater diversity in forms and functions necessitating much concerted regulation of their functions. Most information on quorum sensing in fungi is based on studies in yeasts like *C. albicans*, *S. cerevisiae* and *H. capsulatum* (yeast phase), amongst which *C. albicans* has been the most extensively studied due to its importance as a human pathogen. Amongst mycelial fungi, quorum sensing has been studied in *Ce. ulmi* and *Neurospora crassa*. While we now know that quorum sensing in fungi is not uncommon, there is dearth of information on QSMs in fungi, except farnesol and tyrosol, produced by *C. albicans*. Similarly, fungi also produce several secondary metabolites, mycotoxins and signalling molecules which can act as QSIs and help in controlling the spread of pathogenic bacteria or fungi. In similar potential, research is required to test various edible mushroom extracts for fungal QSIs, those which are cultivable in lab and those which are not yet cultured but have antimicrobial properties. Several mushrooms which are being used by tribals or reported to be ethnomedically important should be used for studies, namely, *Russula*, *Cantharellus*, *Lactarius*, *Lentinus* and *Daedalea*. Using edible wild mushrooms for such screening will help to exclude the possibility of toxic nature of such compounds. Endophytic fungi that are known to produce several metabolites and antibiotic compounds (cephalosporin derived from *Acremonium*) also hold great potential for fungal QSIs.

Multidrug resistance has become a problem for humans to control various pathogens and disease outbreaks. With the advent in research on QSIs, researchers are now targeting the crosstalk mechanisms amongst the bacteria and other pathogenic microbes to improve the efficacy of existing antibiotics and reduce their dosage. The molecules imitating the species-specific QSM can be used for a specific disease and others not so specific may be used as broad-spectrum antibiotics. Although research in this area is in infancy and still restricted to the laboratory, the results are promising. The restriction of bacterial growth by affecting quorum sensing signalling is better than bactericidal or bacteriostatic and will find its application in medicine, agriculture and food technology in the future. We believe that the time is not far when inhibition of quorum sensing will replace antibiotic treatment for microbial disease control in humans, animals and plants.

Acknowledgements The authors thank the Department of Biotechnology (DBT) for funding the Microbial Culture Collection (MCC) wide grant letter no. BT/PR10054/NDB/52/94/2007.

References

Abbott SP (2002) Mycotoxins and indoor molds. Indoor Environ Connect 3(4):14–24

Adak S, Upadrasta L, Kumar SPJ, Soni R, Banerjee R (2011) Quorum quenching – an alternative antimicrobial therapeutics. In: Méndez-Vilas A (ed) Science against microbial pathogens: communicating current research and technological advances. Formatex Research Center, Badajoz

Adonizio A, Kong KF, Mathee K (2008) Inhibition of QS-controlled virulence factor production in *Pseudomonas aeruginosa* by South Florida plant extracts. Antimicrob Agents Chemother 52:198–203

Albuquerque P, Casadevall A (2012) Quorum sensing in fungi – a review. Med Mycol 50:337–345

Alonso-Monge R, Román E, Arana DM, Pla J, Nombela C (2009) Fungi sensing environmental stress. Clin Microbiol Infect 15:17–19

Alspaugh JA, Cavallo LM, Perfect JR, Heitman J (2000) RAS1 regulates filamentation, mating and growth at high temperature of *Cryptococcus neoformans*. Mol Microbiol 36:352–365

Atkinson S, Williams P (2009) Quorum sensing and social networking in the microbial world. J R Soc Interface 6:969–978

Balaban N, Giacometti A, Cirioni O, Gov M, Hirshberg M, Koyfman N, Mathews HR, Nhan RT, Singh B, Uzeil O (2003) Use the quorum- sensing inhibitor RNA III-inhibiting peptide to prevent biofilm formation in vivo by drug resistant *Staphylococcus epidermidis*. J Infect Dis 187:625–630

Balaban N, Cirioni O, Giacometti A, Ghiselli R, Braunstein JB, Silvestri C, Mocchengiani F, Saba V, Scalise G (2007) Treatment of *Staphylococcus aureus* biofilm infection by the quorum- sensing inhibitor RIP. Antimicrob Agents Chemother 51:2226–2229

Bell-Pederson D, Dunlap JC, Loros JJ (1996) Distinct cis-acting elements mediate clock, light, and developmental regulation of the *Neurospora crassa* eas (ccg2) gene. Mol Cell Biol 16:513–521

Borges-Walmsley MI, Walmsley AR (2000) cAMP signalling in pathogenic fungi: control of dimorphic switching and pathogenicity. Trends Microbiol 8: 133–141

Borrego EJ, Kolomiets MV (2012) Lipid-mediated signalling between fungi and plants. In: Witzany G (ed) Biocommunication of fungi. Springer, Dordrecht

Brilhante RSN, Valente LGA, Rocha MFG, Bandeira TJPG, Cordeiro RA, Lima RAC, Leite, JJG, Ribeiro JF, Pereira JF, Castelo-Branco DSCM, Sidrim JJC (2012) Sesquiterpene farnesol contributing to increased susceptibility to b-lactams in strains of *Burkholderia pseudomallei*. Antimicrob Agents Chemother. http://dx.doi.org/10.1128/AAC.05885-11

Cao YY, Cao YB, Xu Z, Ying K, Li Y, Xie Y, Zhu ZY, Chen WS, Jiang YY (2005) cDNA microarray analysis of differential gene expression in *Candida albicans* biofilm exposed to farnesol. Antimicrob Agents Chemother 49:584–589

Casadevall A, Pirofski LA (1999) Host-pathogen interactions: redefining the basic concepts of virulence and pathogenicity. Infect Immun 67:3703–3713

Cassola A, Parrot M, Silberstein S, Magee BB, Passeron S, Giasson L, Cantore ML (2004) *Candida albicans* lacking the gene encoding the regulatory subunit of protein kinase a displays a defect in hyphal formation and an altered localization of the catalytic subunit. Eukaryotic Cell 3:190–199

Cerca N, Gomes F, Pereira S, Teixeira P, Oliveira R (2012) Confocal laser scanning microscopy analysis of *S. epidermidis* biofilms exposed to farnesol, vancomycin and rifampicin. BMC Res Notes 5:244

Cheetham J, Smith DA, Dantas ADS, Doris KS, Patterson MJ, Bruce CR, Quinn J (2007) A single MAPKKK regulates the Hog1 MAPK pathway in the pathogenic fungus *Candida albicans*. Mol Biol Cell 18:4603–4614

Chen H, Fink GR (2006) Feedback control of morphogenesis in fungi by aromatic alcohols. Genes Dev 20: 1150–1161

Chen L, Wen Yu-mei (2011) The role of bacterial biofilm in persistent infections and control strategies. Int J Oral Sci 3:66–73. doi:10.4248/IJOS11022

Chen H, Fujita M, Feng Q, Clardy J, Fink GR (2004) Tyrosol is a quorum-sensing molecule in *Candida albicans*. Proc Natl Acad Sci U S A 101:5048–5052

Chen F, Gao Y, Chen X, Yu Z, Li X (2013) Quorum quenching enzymes and their application in degrading signal molecules to block quorum sensing-dependent infection. Int J Mol Sci 14:17477–17500. doi:10.3390/ijms140917477

Cho EJ, Oh JY, Chang HY, Yun JW (2006) Production of exopolysaccharides by submerged mycelial culture of a mushroom *Tremella fuciformis*. J Biotechnol 127: 129–140

Cho EJ, Hwang HJ, Kim SW, Oh JY, Baek YM, Choi JW, Bae SH, Yun JW (2007) Hypoglycemic effects of exopolysaccharides produced by mycelial cultures of two different mushrooms *Tremella fuciformis* and *Phellinus baumii* in ob/ob mice. Appl Microbiol Biotechnol 75:1257–1265

Cloutier M, Castilla R, Bolduc N, Zelada A, Martineau P, Bouillon M, Magee BB, Passeron S, Giasson L, Cantore ML (2003) The two isoforms of the cAMP dependent protein kinase catalytic subunit are involved in the control of dimorphism in the human fungal pathogen *Candida albicans*. Fungal Genet Biol 38:133–141

Cordeiro RA, Nogueira GC, Brilhante RSN, Teixeira CEC, Mourão CI, Castelo-Branco D de Souza CM, Paiva MAN, Ribeiro JF, Monteiro AJ, Sidrim JJC, Rocha MFG (2012) Farnesol inhibits in vitro growth of the *Cryptococcus neoformans* species complex with no significant changes in virulence-related exoenzymes. Vet Microbiol 159:375–380

Costerone JW, Stewart PS, Greenberg EP (1999) Bacterial biofilms: a common cause of persistent infections. Science 284:1318–1322

Cotar AI (2013) Quorum sensing inhibitors as anti-pathogenic drugs in the fight against Pseudomonas aeruginosa infections. Clin Microbiol 2(4):e111. doi:10.4172/2327-5073.1000e111

Cugini C, Calfee MW, Farrow JM, Morales DK, Pesci EC, Hogan DA (2007) Farnesol, a common sesquiterpene, inhibits PQS production in *Pseudomonas aeruginosa*. Mol Microbiol 65:896–906

Dastager SG, Li WJ, Dayanand A, Tang SK, Tian XP, Zhi XY, Xu LH, Jiang CL (2006) Separation, identification and analysis of pigment (melanin) production in *Streptomyces*. Afr J Biotechnol 5:1131–1134

Davies D (2003) Understanding biofilm resistance to antibacterial agents. Nat Rev Drug Discov 2:114–122. doi:10.1038/nrd1008

Davis-Hanna A, Piispanen AE, Stateva LI, Hogan DA (2008) Farnesol and dodecanol effects on the *Candida albicans* Ras1-cAMP signalling pathway and the regulation of morphogenesis. Mol Microbiol 67:47–62

Dechant R, Peter M (2008) Nutrient signals driving cell growth. Curr Opin Cell Biol 20:678–687

Decho AW, Norman RS, Visscher PT (2010) Quorum sensing in natural environments: emerging views from microbial mats. Trends Microbiol 18:73–80. doi:10.1016/j.tim.2009.12.008

Defoirdt T, Boon N, Bossier P (2010) Can bacteria evolve resistance to quorum sensing disruption? PLoS Pathog 6:e1000989. doi:10.1371/journal.ppat.1000989

Defoirdt T, Sorgeloos P, Bossier P (2011) Alternatives to antibiotics for the control of bacterial disease in aquaculture. Curr Opin Microbiol 14:251–258. doi:10.1016/j.mib.2011.03.004

Derengowski LS, De-Souza-Silva C, Braz SV, Mello-De-Sousa TM (2009) Antimicrobial effect of farnesol, a *Candida albicans* quorum sensing molecule, on *Paracoccidioides brasiliensis* growth and morphogenesis. Ann Clin Microbiol Antimicrob 8:13. doi:10.1186/1476-0711-8-13

Dohlman HG, Slessareva JE (2006) Pheromone signaling pathways in yeast. Sci Signal 364:cm6. doi:10.1126/stke.3642006cm6

Dong Y-H, Zhang L-H (2005) Quorum sensing and quorum-quenching enzymes. J Microbiol 43:101–109

Dumitru R, Hornby JM, Nickerson KW (2004) Defined anaerobic growth medium for studying *Candida albicans* basic biology and resistance to eight antifungal drugs. Antimicrob Agents Chemother 48:2350–2354

Eisman B, Alonso-Monge R, Román E, Arana D, Nombela C, Pla J (2006) The Cek1 and Hog1 mitogen-activated protein kinases play complementary roles in cell wall biogenesis and chlamydospore formation in the fungal pathogen *Candida albicans*. Eukaryot Cell 5:347–358

Fairn GD, MacDonald K, McMaster CR (2007) A chemogenomic screen in *Saccharomyces cerevisiae* uncovers a primary role for the mitochondria in farnesol toxicity and its regulation by the Pkc1 pathway. J Biol Chem 282:4868–4874

Fernandes L, Araujo MAM, Amaral A, Reis VCB, Martins NF, Felipe MS (2005) Cell signalling pathways in *Paracoccidioides brasiliensis*- inferred from comparisons with other fungi. Genet Mol Res 4:216–231

Finch RG, Pritchard DI, Bycroft BW, Williams P, Stewart GS (1998) Quorum sensing: a novel target for anti-infective therapy. J Antimicrob Chemother 42(5): 569–571

Frazer J (2013) Accident of evolution allows fungi to thrive in our bodies. Scientific American Web Exclusive. http://www.scientificamerican.com/article/fungal-infection-accident-of-evolution-may-thrive-in-our-bodies. Accessed 11 May 2014

Frisvad JC, Thrane U, Filtenborg O (1998) Role and use of secondary metabolites in fungal taxonomy. In: Frisvad JC, Bridge PD, Arora DK (eds) Chemical fungal taxonomy. Marcel Dekker, New York

Frisvad JC, Andersen B, Thrane U (2008) The use of secondary metabolite profiling in chemotaxonomy of filamentous fungi. Mycol Res 112(2):231–240. doi:10.1016/j.mycres.2007.08.018

Fu Y, Ibrahim AS, Sheppard DC, Chen YC, French SW, Cutler JE, Filler SG, Edwards JE Jr (2002) *Candida albicans* Als1p: an adhesin that is a downstream effector of the EFG1 filamentation pathway. Mol Microbiol 44:61–72

García-Contreras R, Maeda T, Wood TK (2013) Resistance to quorum-quenching compounds. Appl Environ Microbiol 79(22):6840–6846. doi:10.1128/AEM.02378-13

Gautam CK, Srivastav AK, Bind S, Madhav M, Shanthi V (2013) An insight into biofilm ecology and its applied aspects. Int J Pharm Pharm Sci 5(4):69–73

Gomes FIA, Leite B, Teixeira P, Cerca N, Azeredo J, Oliveira R (2011a) Farnesol as antibiotics adjuvant in *Staphylococcus epidermidis* control *in vitro*. Am J Med Sci 341(3):191–195. doi:10.1097/MAJ.0b013e3181fcf138

Gomes FIA, Teixeira P, Cerca N, Azeredo J, Oliveira R (2011b) Effect of farnesol on structure and composition of *Staphylococcus epidermidis* biofilm matrix. Curr Microbiol 63:354–359

Gómez-Gómez L, Boller T (2000) FLS2: an LRR receptor–like Kinase involved in the perception of the bacterial elicitor Flagellin in *Arabidopsis*. Mol Cell 5:1003–1011

Gonzalez JE, Marketon MM (2003) Quorum sensing in nitrogen fixing rhizobia. Microbiol Mol Biol Rev 67:574–592

Gould SE (2012) Fungi that steal genes from bacteria. Scientific American Blog. http://blogs.scientificamerican.com/lab-rat/2012/08/12/fungi-that-steal-genes-from-bacteria. Accessed 11 May 2014

Gow NAR (2009) Fungal morphogenesis: some like it hot. Curr Biol 19:621–629

Hahn Berg IC, Kalfas S, Malmsten M, Arnebrant T (2001) Proteolytic degradation of oral biofilms *in vitro* and *in vivo*: potential of proteases originating from *Euphausia superba* for plaque control. Eur J Oral Sci 109:316–324

Han T-L, Cannon RD, Villas-Bôas SG (2011) The metabolic basis of *Candida albicans* morphogenesis and quorum sensing. Fungal Genet Biol 48:747–763

Hazen KC, Cutler JE (1979) Autoregulation of germ tube formation by *Candida albicans*. Infect Immun 24:661–666

Hazen KC, Cutler JE (1983) Isolation and purification of morphogenic autoregulatory substance produced by *Candida albicans*. J Biochem 94:777–783

Henriques M, Martins M, Azeredo J, Oliveira R (2007) Effect of farnesol on *Candida dubliniensis* morphogenesis. Lett Appl Microbiol 44:199–205. doi:10.1111/j.1472-765X.2006.02044.x

Hentzer M, Givskov M (2003) Pharmacological inhibition of quorum sensing for the treatment of chronic bacterial infections. J Clin Invest 112(9):1300–1307. doi:10.1172/JCI200320074

Hentzer M, Eber L, Nielsen J, Givskov M (2003) Quorum sensing: a novel target for the treatment of biofilm infections. Biodrugs 17(4):241–250

Hisajima T, Maruyama N, Tanabe Y, Ishibashi H, Yamada T, Makimura K, Nishiyama Y, Funakoshi K, Oshima H, Abe S (2008) Protective effects of farnesol against oral candidiasis in mice. Microbiol Immunol 52:327–333. doi:10.1111/j.1348-0421.2008.00044.x

Hogan DA (2006) Talking to themselves: autoregulation and quorum sensing in fungi. Eukaryot Cell 5(4):613–619. doi:10.1128/EC.5.4.613-619

Hornby JM, Nickerson KW (2004) Enhanced production of farnesol by *Candida albicans* treated with four azoles. Antimicrob Agents Chemother 48:2305–2307

Hornby JM, Jensen EC, Lisec AD, Tasto JJ, Jahnke B, Shoemaker R, Dussault P, Nickerson W (2001) Quorum sensing in the dimorphic fungus *Candida albicans* is mediated by farnesol. Appl Environ Microbiol 67:2982–2992

Hornby JM, Kebaara BW, Nickerson KW (2003) Farnesol biosynthesis in *Candida albicans*: cellular response to sterol inhibition by zaragozic acid B. Antimicrob Agents Chemother 47:2366–2369

Hornby JM, Jacobitz-Kizzier SM, McNeel DJ, Jensen EC, Treves DS, Nickerson KW (2004) Inoculum size effect in dimorphic fungi: extracellular control of yeast-mycelium dimorphism in *Ceratocystis ulmi*. Appl Environ Microbiol 70:1356–1359

Jain A, Gupta Y, Agrawal R, Khare P, Jain SK (2007) Biofilms—a microbial life perspective: a critical review. Crit Rev Ther Drug Carr Syst 24(5):393–443

Jangid K, Kong R, Patole MS, Shouche YS (2007) *luxRI* homologs are universally present in the genus *Aeromonas*. BMC Microbiol 7:93

Johansen C, Falholt P, Gram L (1997) Enzymatic removal and disinfection of bacterial biofilm. Appl Environ Microbiol 63:3724–3728

Kalia VC (2013) Quorum sensing inhibitors: an overview. Biotechnol Adv 31:224–245

Kalia VC, Wood TK, Kumar P (2013) Evolution of resistance to quorum-sensing inhibitors. Microb Ecol. doi:10.1007/s00248-013-0316-y

Kaneko M, Togashi N, Hamashima H, Hirohara M, Inoue Y (2011) Effect of farnesol on mevalonate pathway of *Staphylococcus aureus*. J Antibiot 64:547–549

Kebaara BW, Langford ML, Navarathna DHMLP, Dumitru R, Nickerson K, Atkin AL (2008) *Candida albicans* Tup1 is involved in farnesol-mediated inhibition of filamentous-growth induction. Eukaryot Cell 7:980–987

Koh C-L, Sam C-K, Yin W-F, Tan LY, Krishnan T, Chong YM, Chan K-G (2013) Plant-derived natural products as sources of anti-quorum sensing compounds. Sensors 13:6217–6228. doi:10.3390/s130506217

Kügler S, Sebghati TS, Eissenberg LG, Goldman WE (2000) Phenotypic variation and intracellular parasitism by *Histoplasma capsulatum*. PNAS 97(16):8794–8798

Kumamoto CA (2002) *Candida* biofilms. Curr Opin Microbiol 5:608–611

Laffey SF, Butler G (2005) Phenotype switching affects biofilm formation by *Candida parapsilosis*. Microbiology 151(4):1073–1081. doi:10.1099/mic.0.27739-0

Leberer E, Harcus D, Dignard D, Johnson L, Ushinsky S, Thomas DY, Schröppel K (2001) RAS links cellular morphogenesis to virulence by regulation of the MAP kinase and cAMP signalling pathways in the pathogenic fungus *Candida albicans*. Mol Microbiol 42:673–687

Lee S-W, Han S-W, Sririyanum M, Park C-J, Seo Y-S, Ronald PC (2009) A Type I–secreted, sulfated peptide triggers XA21-mediated innate immunity. Science 326:850–853. doi:10.1126/science.1173438

Leeder AC, Palma-Guerrero J, Glass NL (2011) The social network: deciphering fungal language. Nat Rev Microbiol 9:440–451

Leng P, Lee PR, Wu H, Brown AJP (2001) Efg1, a morphogenetic regulator in *Candida albicans*, is a sequence-specific DNA binding protein. J Bacteriol 183:4090–4093

Li B, Dong MS (2010) Inhibition effect of extract from *Auricularia auricular* on quorum sensing and biofilm formation of bacteria. Food Sci 31:140–143

Lingappa BT, Lingappa Y (1969) Role of auto-inhibitors on mycelial growth and dimorphism of *Glomerella cingulata*. J Gen Microbiol 56:35–45

Lingappa BT, Prasad M, Lingappa Y, Hunt DF, Biemann K (1969) Phenethyl alcohol and tryptophol: auto antibiotics produced by the fungus *Candida albicans*. Science 163:192–194

Liu P, Deng B, Long CA, Min X (2009) Effect of farnesol on morphogenesis in the fungal pathogen *Penicillium expansum*. Ann Microbiol 59:33–38

Lung MY, Tsai JC, Huang PC (2010) Antioxidant properties of edible basidiomycete *Phellinus igniarius* in submerged cultures. J Food Sci 75(1):18–24. doi:10.1111/j.1750-3841.2009.01384.x

Macko V, Staples RC, Gershon H, Renwick JA (1970) Self inhibitor of bean rust uredospores: methyl 3,4-dimethoxycinnamate. Science 170:539–540

Maeda T, García-Contreras R, Pu M, Sheng L, Garcia LR, Tomás M, Wood TK (2012) Quorum quenching quandary: resistance to anti virulence compounds. ISME J 6:493–501. doi:10.1038/ismej.2011.122

Matson J (2011) Unwanted housemates: dishwashers provide habitat for "extremotolerant" fungi. Scientific American Blog. http://blogs.scientificamerican.com/observations/2011/06/21/unwanted-housemates-dishwashers-provide-habitat-for-extremotolerant-fungi. Assessed 11 May 2014

Mattmann ME, Blackwell HE (2010) Small molecules that modulate quorum sensing and control virulence in *Pseudomonas aeruginosa*. J Org Chem 75:6737–6746. doi:10.1021/j0101237e

Maximilien R, de Nys R, Holmström C, Gram L, Givskov M, Crass K, Kjelleberg S, Steinberg PD (1998) Chemical mediation of bacterial surface colonisation by secondary metabolites from the red alga *Delisea pulchra*. Aquat Microb Ecol 15:233–246

Mohan SK, Srivastava T (2010) Microbial deterioration and degradation of polymeric materials. J Biochem Technol 2(4):210–215

Mosel DD, Dumitru R, Hornby JM, Atkin AL, Nickerson KW (2005) Farnesol concentrations required to block germ tube formation in *Candida albicans* in the presence and absence of serum. Appl Environ Microbiol 71:4938–4940

Navarathna DHMLP, Hornby JM, Hoerrmann N, Parkhurst AM, Duhamel GE, Nickerson KW (2005) Enhanced pathogenicity of *Candida albicans* pre-treated with subinhibitory concentrations of fluconazole in a mouse model of disseminated candidiasis. J Antimicrob Chemother 56:1156–1159

Nickerson KW, Atkin AL, Hornby JM (2006) Quorum sensing in dimorphic fungi: farnesol and beyond. Appl Environ Microbiol 72(6):3805–3813. doi:10.1128/AEM.02765-05

Njoroge J, Sperandio V (2009) Jamming bacterial communication: new approaches for the treatment of infectious diseases. EMBO Mol Med 1:201–210. doi:10.1002/emmm.200900032

Oh K-B, Miyazawa H, Naito T, Matsuoka H (2001) Purification and characterization of an auto regulatory substance capable of regulating the morphological transition in *Candida albicans*. Proc Natl Acad Sci U S A 98:4664–4668

Orgaz B, Kivesb J, Pedregosa AM, Monistrol IF, Laborda F, Jośe CS (2006) Bacterial biofilm removal using fungal enzymes. Enzym Microb Technol 40:51–56

Pammi M, Liang R, Hicks J, Barrish J, Versalovic J (2011) Farnesol decreases biofilms of *Staphylococcus epidermidis* and exhibits synergy with nafcillin and vancomycin. Pediatr Res 70:578–583

Pan JC, Ren DC (2009) Quorum sensing inhibitors: a patent review. Expert Opin Ther Patents 19:1581–1601

Persson T, Hansen TH, Rasmussen TB, Skindersoe ME, Givskov M, Nielsen J (2005) Rational design and synthesis of new quorum sensing inhibitors derived from acylated homoserine lactones and natural product from garlic. Org Biomol Chem 3:253–262

Pitt JI (2000) Toxigenic fungi and mycotoxins. Br Med Bull 56(1):184–192

Rajesh PS, Rai RV (2013) Hydrolytic enzymes and quorum sensing inhibitors from endophytic fungi of *Ventilago madraspatana* Gaertn. Biocatal Agric Biotechnol 2:120–124

Ramage G, Saville SP, Wickes BL, López-Ribot JL (2002) Inhibition of *Candida albicans* biofilm formation by farnesol, a quorum-sensing molecule. Appl Environ Microbiol 68(11):5459–5463. doi:10.1128/AEM.68.11.5459-5463

Ramage G, Saville SP, Thomas DP, Lopez-Ribot JL (2005) *Candida* biofilms: an update. Eukaryot Cell 4:633–638

Rasch M, Andersen JB, Nielsen KF, Flodgaard LR, Christensen H, Givskov M, Gram L (2005) Involvement of bacterial quorum sensing signals in spoilage of bean sprouts. Appl Environ Microbiol 71:3321–3330

Rasch M, Rasmussen TB, Andersen JB, Persson T, Nielsen J, Givskov M, Gram L (2007) Well-known quorum sensing inhibitors do not affect bacterial quorum sensing-regulated bean sprout spoilage. J Appl Microbiol 102:826–837. doi:10.1111/j.1365-2672.2006.03121.x

Rasmussen TB, Givskov M (2006a) Quorum-sensing inhibitors as anti-pathogenic drugs. Int J Med Microbiol 296:149–161. doi:10.1016/j.ijmm.2006.02.005

Rasmussen TB, Givskov M (2006b) Quorum sensing inhibitors: a bargain of effects. Microbiology 152:895–904. doi:10.1099/mic.0.28601-0

Rasmussen TB, Bjarnsholt T, Skindersoe ME, Hentzer M, Kristoffersen P, Köte M, Nielsen J, Eberl L, Givskov M (2005a) Screening for Quorum-Sensing Inhibitors (QSI) by use of a novel genetic system, the QSI selector. J Bacteriol 187:1799–1814. doi:10.1128/JB.187.5.1799-1814.2005

Rasmussen TB, Skindersoe ME, Bjarnsholt T, Phipps RK, Christensen KB, Jensen PO, Andersen JB, Koch B, Larsen TO, Hentzer M, Eberl L, Hoiby N, Givskov M (2005b) Identity and effects of quorum-sensing inhibitors produced by *Penicillium* species. Microbiology 151:1325–1340

Rex E (2012) Invasive fungi wreak havoc on species worldwide [slide show]. Scientific American News. http://www.scientificamerican.com/article/fungi-wreak-havoc-across-species-globally-slide-show. Assessed 11 May 2014

Roca MG, Arlt J, Jeffree CE, Read ND (2005) Cell biology of conidial anastomosis tubes in *Neurospora crassa*. Eukaryot Cell 4:911–919

Rocha CRC, Schröppel K, Harcus D, Marcil A, Dignard D, Taylor BN, Thomas DY, Whiteway M, Leberer E (2001) Signaling through adenylyl cyclase is essential for hyphal growth and virulence in the pathogenic fungus *Candida albicans*. Mol Biol Cell 12: 3631–3643

Román E, Alonso-Monge R, Gong Q, Li D, Calderone R, Pla J (2009) The Cek1 MAPK is a short-lived protein regulated by quorum sensing in the fungal pathogen *Candida albicans*. FEMS Yeast Res 9:942–955

San-Blas G, Travassos LR, Fries BC, Goldman DL, Casadevall A, Carmona AK, Barros TF, Puccia R, Hostetter MK, Shanks SG, Copping VM, Knox Y, Gow NA (2000) Fungal morphogenesis and virulence. Med Mycol 38(suppl 1):79–86

Sato T, Watanabe T, Mikami T, Matsumoto T (2004) Farnesol, a morphogenetic auto regulatory substance in the dimorphic fungus *Candida albicans*, inhibits hyphae growth through suppression of a mitogen-activated protein kinase cascade. Biol Pharm Bull 27:751–752

Semighini CP, Hornby JM, Dumitru R, Nickerson KW, Harris SD (2006) Farnesol-induced apoptosis in *Aspergillus nidulans* reveals a possible mechanism for antagonistic interactions between fungi. Mol Microbiol 59(3):753–764

Shiner EK, Rumbaugh KP, Williams SC (2005) Inter-kingdom signaling: deciphering the language of acyl homoserine lactones. FEMS Microbiol Rev 29: 935–947

Shirtliff ME, Krom BP, Meijering RA, Peters BM, Zhu J, Scheper MA, Harris ML, Jabra-Rizk MA (2009) Farnesol induced apoptosis in *Candida albicans*. Antimicrob Agents Chemother 53:2392–2401

Silva D (2003) *Ganoderma lucidum* (Reishi) in cancer treatment. Integr Cancer Ther 2(4):358–364. doi:10.1177/1534735403259066

Smith RS, Iglewski BH (2003) *P. aeruginosa* quorum-sensing systems and virulence. Curr Opin Microbiol 6:56–60

Summers (2012) Bacteria talk, plants listen: the discovery of plant immune receptors, an interview with Pamela Ronald. Scientific American Blog. http://blogs.scientificamerican.com/guest-blog/2012/04/22/bacteria-talk-plants-listen-the-discovery-of-plant-immune-receptors-an-interview-with-dr-pamela-ronald. Assessed 11 May 2014

Teplitski M, Robinson JB, Bauer WD (2000) Plants secrete substances that mimic bacterial N-acyl homoserine lactone signal activities and affect population density-dependent behaviors in associated bacteria. Mol Plant-Microbe Interact 13:637–648

Ueda M, KuboT MK, Nakamura K (2007) Purification and characterization of fibrinolytic alkaline protease from *Fusarium* sp. BLB. Appl Microbiol Biotechnol 74:331–338

Uroz S, Heinonsalo J (2008) Degradation of N -acyl homoserine lactone quorum sensing signal molecules by forest root-associated fungi. FEMS Microbiol Ecol 65:271–278. doi:10.1111/j.1574-6941.2008.00477.x

Von Bodman SB, Bauer WD, Coplin DL (2003) Quorum sensing in plant-pathogenic bacteria. Annu Rev Phytopathol 41:455–482

Wachtel-Galor S, Tomlinson B, Benzie IFF (2004) *Ganoderma lucidum* ('Lingzhi'), a Chinese medicinal mushroom: biomarker responses in a controlled human supplementation study. Br J Nutr 91(2):263–269. doi:10.1079/BJN20041039

Weber K, Schulz B, Ruhnke M (2010) The quorum-sensing molecule E, E-farnesol- its variable secretion and its impact on the growth and metabolism of Candida species. Yeast 27:727–739

Witzany G (2010) Uniform categorization of biocommunication in bacteria, fungi and plants. World J Biol Chem 1(5):160–180

Witzany G (2012) Introduction: key levels of biocommunication in fungi. In: Witzany G (ed) Biocommunication of fungi. Springer, Dordrecht

Wong C-S, Koh C-L, Sam C-K, Chen J-W, Chong YM, Yin W-F, Chan K-G (2013) Degradation of bacterial quorum sensing signaling molecules by the microscopic yeast *Trichosporon loubieri* isolated from tropical wetland waters. Sensors 13:12943–12957. doi:10.3390/s131012943

Wu H, Song Z, Hentzer M, Andersen JB, Molin S, Givskov M, Høiby N (2004) Synthetic furanones inhibit quorum-sensing and enhance bacterial clearance in *Pseudomonas aeruginosa* lung infection in mice. J Antimicrob Chemother 53:1054–1061

Yu RC, Pesce CG, Colman-Lerner A, Lok L, Pincus D, Serra E, Holl M, Benjamin K, Gordon A, Brent R (2008) Negative feedback that improves information transmission in yeast signalling. Nature 456: 755–761

Zhu H, Sun SJ (2008) Inhibition of bacterial quorum sensing-regulated behaviours by *Tremella fuciformis* extract. Curr Microbiol 57:418–422. doi:10.1007/s00284-008-9215-8

Zhu H, Wang TW, Sun SJ, Shen YL, Wei DZ (2006) Chromosomal integration of the Vitreoscilla hemoglobin gene and its physiological actions in *Tremella fuciformis*. Appl Microbiol Biotechnol 72: 770–776

Zhu H, Liu W, Tian B, Liu H, Ning S (2011) Inhibition of quorum sensing in the opportunistic pathogenic bacterium *Chromobacterium violaceum* by an extract from fruiting bodies of lingzhi or reishi medicinal mushroom, *Ganoderma lucidum* (w.Curt.:Fr.) p. Karst. (higher basidiomycetes). Int J Med Mushroom 13: 559–564

Zhu H, Liu W, Wang S, Tian B, Zhang S (2012) Evaluation of anti-quorum-sensing activity of fermentation metabolites from different strains of a medicinal mushroom, *Phellinus igniarius*. Chemotherapy 58:195–199

Zibafar E, Hashemi SJ, Zaini F, Zeraati H, Rezaie S, Kordbacheh P (2009) Inhibitory effect of farnesol on biofilm formation by *Candida tropicalis*. Daru 17(1):19–23

Marine Organisms as Source of Quorum Sensing Inhibitors

Fohad Mabood Husain and Iqbal Ahmad

Introduction

More than half of all prescribed drugs have their origin in natural sources and the diversity of chemical structures produced by marine organisms provides a potential source of novel pharmaceuticals (Harper et al. 2001; Paterson and Anderson 2005; Marris 2006). While there are few natural products derived from marine organisms in clinical use, the potential applications of these compounds are broad (Marris 2006). Marine natural products are in clinical or preclinical evaluation against severe pain, cancers, allergy/asthma, inflammation, Alzheimer's disease, HIV, and tuberculosis, with the majority of these natural products being sponge or tunicate derived (Proksch et al. 2002; Newman and Cragg 2004a, b). They also have potential application as anti-infective therapies (Fenical and Jensen 2006). The development of novel anti-infectives is particularly critical because of the increasing number of multidrug resistant bacteria (Tenover and Hughes 1996; Reacher et al. 2000; Ahmad et al. 2008). One strategy to develop new mechanisms to combat bacterial infections is through the development of antipathogenic drugs that function by attenuating the bacteria with respect to virulence. Because this can be achieved without affecting the growth of bacteria, this principle is less likely to impose a selective pressure for resistance (Hentzer and Givskov 2003; Hentzer et al. 2003; Aqil et al. 2006). Many pathogenic bacteria synchronize their expression of virulence genes by cell-to-cell communication systems called quorum sensing (QS) which enables bacteria to monitor cell density. QS relies on low molecular weight signal molecules that are capable of activating transcriptional regulators in order to couple gene expression with population density. In Gram-negative bacteria, QS is mediated mostly by N-acyl homoserine lactones (AHLs). QS systems function by means of two proteins: the AHL synthetase (the LuxI-homologue) responsible for the AHL signal generation and the LuxR homologue, which is a dual-function receptor-response protein recognizing and binding the AHL molecules, allowing interactions with QS-controlled promoter sequences. The signal molecules are produced constitutively at a basal low level by the LuxI homologue and are distributed in the cell and the surrounding environment either actively or by diffusion. The concentration of AHL molecules increases with increasing cell density, and reaching a certain critical level, they bind to the LuxR homologues, creating active LuxR–AHL complexes. This complex functions as a transcription factor that binds to the promoter region of the QS-regulated genes, where it triggers gene expression

F.M. Husain • I. Ahmad (✉)
Department of Agricultural Microbiology, Aligarh Muslim University, Aligarh 202002, Uttar Pradesh, India
e-mail: fahadamu@gmail.com; ahmadiqbal8@yahoo.co.in

V.C. Kalia (ed.), *Quorum Sensing vs Quorum Quenching: A Battle with No End in Sight*, DOI 10.1007/978-81-322-1982-8_21, © Springer India 2015

(Fuqua et al. 1994), including virulence traits (Toder et al. 1991; Gambello et al. 1993). It has been established that for some opportunistic pathogens, such as *Pseudomonas aeruginosa*, it is essential to stay "silent" and delay production of virulence factors until the cell population has reached a level to overwhelm the host's defenses. QS-regulated virulence factors and biofilm have been reported in a variety of bacteria pathogenic to humans, animals, as well as plants (Williams 2007; Kalia 2013). The role of QS in virulence and pathogenicity is most widely explored and studied in *Pseudomonas aeruginosa.* Virulence factors like LasB elastase, LasA protease, pyocyanin production, rhamnolipid production, swarming motility, and biofilm are controlled by QS (Miller and Bassler 2001; LaSarre and Federle 2013). Therefore, QS offers the invading bacteria with an advantage during the establishment of infections. A strategy for the development of novel anti-infectives is to block the bacterial communication pathway. In contrast to the sophisticated mammalian immune system plants, fungi and marine invertebrates rely on physical and chemical defense strategies to protect them against bacterial pathogens. This provides a logical source to search for natural products with the ability to inhibit QS in natural libraries of chemically rich organisms (Natrah et al. 2011). A proof of this concept is the halogenated furanones derived from the marine macro alga *Delisea pulchra* (Givskov et al. 1996), which inhibits QS in vitro as well as in vivo experimental settings (Manefield et al. 2000; Hentzer et al. 2003; de Nys et al. 2006). The treatment of the opportunistic pathogen *P. aeruginosa* in a mouse pulmonary infectious model with furanone-derived drugs greatly accelerates eradication of the pathogen (Hentzer et al. 2003). Furanones also reduce the mortality of vibriosis in rainbow trout challenged with pathogenic *Vibrio anguillarum* (Rasch et al. 2004). The use of QS inhibitors (QSIs) has a wide range of applications, from veterinary treatment in aquaculture to medical treatments in humans. Thus, given the importance of QS in virulence and pathogenicity development of pathogenic bacteria, it is attractive how other organisms and certain compounds interfere with the bacterial QS. In this chapter we have reviewed the current progress in the discovery and reports on QSI compounds from marine organisms and their potential applications.

QS Inhibitors from Marine Origin

Quorum sensing pathways reported so far are complex and varied; associated signals are also of different types (Lasarre and Federle 2013). The process of QS can be disrupted by different mechanisms: (1) reducing the activity of AHL cognate receptor protein or AHL synthase, (2) inhibiting the production of QS signal molecules, (3) degradation of the AHL, and (4) mimicking the signal molecules primarily by using synthetic compounds as analogs of signal molecules (Hentzer and Givskov 2003). The compounds reported so far as anti-QS agents act on either one or other above targets. The wide spread of anti-QS compounds from various sources that act against *P. aeruginosa* infection, which includes halogenated furanones from marine red macroalgae *Delisea pulchra* (Manefield et al. 1999; Hentzer et al. 2002), N-acyl homoserine lactone acylase from *Streptomyces species* (Park et al. 2005), N-acyl and homoserine lactonases from *Bacillus* species (Dong et al. 2002), has been reported. However, reports from crude products might be a combination of direct and indirect mechanisms (Adonizio et al. 2008; Zahin et al. 2010; Khan et al. 2009; Husain et al. 2013). Although the marine environment is considered to be a rich source of both biologically and chemically diverse bioactive compounds, the AHL inhibitors discovered from marine resources especially from marine microbes is meager. In recent years, the QSI potential of a few marine microbes has been reported. In the subsequent sections, a detailed account of the QS inhibitors of marine origin is discussed.

Quorum Sensing Inhibition by Algae

Eukaryotes such as algae, protozoa, and fungi live in close proximity with both pathogenic and beneficial bacteria in the aquatic environment. Thus, it is not surprising that eukaryotes have developed different defense mechanisms to interact with bacteria, e.g., by producing secondary metabolites (Rasmussen et al. 2005) impacting QS (Dudler and Elberl 2006).

Marine plants are important potential candidates to interfere with QS (Kjelleberg and Steinberg 2001). The first QS inhibitor was isolated from the red macroalga (also known as seaweed), *Delisea pulchra*, which showed a high antifouling activity (Givskov et al. 1996). The antifouling activity apparently is caused by a broad range of secondary metabolites, halogenated furanones found at the surface of the alga (Dworjanyn and Steinberg 1999). These halogenated furanones are similar in structure to AHL, except that furanones have a furan ring instead of a homoserine lactone ring. Among the earliest investigations on the effect of this alga on bacteria is the addition of *D. pulchra* crude extract to cultures of the human pathogen *Proteus mirabilis* (Gram et al. 1996), where the extract was found to inhibit swarming motility. Although not as strong as the crude extract, the halogenated furanones isolated from the sample also decreased the swarming velocity. The most well-studied natural compound to date is probably (5Z)-4-bromo-5-(bromomethylene)-3-butyl-2(5 H)-furanone (Fig. 1). This compound is shown to have high inhibitory activities in several biological assays of AHL-controlled expression in different Gram-negative bacteria (Rasmussen et al. 2000; Hentzer et al. 2003) and also blocks AI-2 signaling (Ren et al. 2001). The results triggered the chemical synthesis of several furanone analogs as QS inhibitors such as (5Z)-4-bromo-5-(bromomethylene)- 2(5H)-furanone (Fig. 1).

Using a gnotobiotic model system, Defoirdt et al. (2006) demonstrated that natural and synthetic brominated furanones are able to protect brine shrimp (*Artemia franciscana*) from pathogenic isolates belonging to the species *V. harveyi*, *V. campbellii*, and *V. parahaemolyticus*, respectively, through the disruption of AI-2 QS. The natural furanone was also found to counteract the negative effect of different pathogenic *V. harveyi* strains in the rotifer *Brachionus plicatilis* (Tinh et al. 2007a). Moreover, Manefield et al. (2000) showed that the natural furanone blocked the luminescence and toxin T1 production (both of which are QS-regulated) of *V. harveyi* that were pathogenic to farmed shrimp. This compound also decreased death in rainbow trout infected with *V. anguillarum* (Rasch et al. 2004). These data showed that furanones could act as anti-infective compounds in different aquatic host-microbe systems.

The antagonistic effects of furanones are attributed to their ability to destabilize *V. fischeri* LuxR-type proteins, thus reducing the amount of the protein to act as an AHL-mediated regulator (Manefield et al. 2002). In *V. harveyi*, the natural furanone was found to directly target the QS master regulator LuxR (which is not homologous to *V. fischeri* LuxR), obstructing the ability of LuxR to bind to target gene promoter sequences, probably by inducing certain molecular structure modifications (Defoirdt et al. 2007). In addition, findings by Zang et al. (2009) demonstrated that the molecules also disrupt AI-2 synthesis by covalently modifying and inactivating the LuxS enzyme. The furanones were shown to bind to thiol groups in cysteine residues in the LuxS protein and can therefore be hypothesized to be nonspecific. The apparent specificity in activity toward QS-regulated phenotypes might be due to the fact that relatively small changes in the activity of QS regulatory genes have large effects on the phenotypes regulated by QS. Interestingly, another seaweed belonging to the same family, *Bonnemaisonia hamifera*, also showed antifouling activities (Nylund et al. 2005). The chemical structure of the algal metabolites responsible for the inhibition activity is still unknown, although it was speculated to be polyhalogenated 2-heptanones. In addition to this, Kim et al. (2007) discovered three new AHL antagonists in the red alga

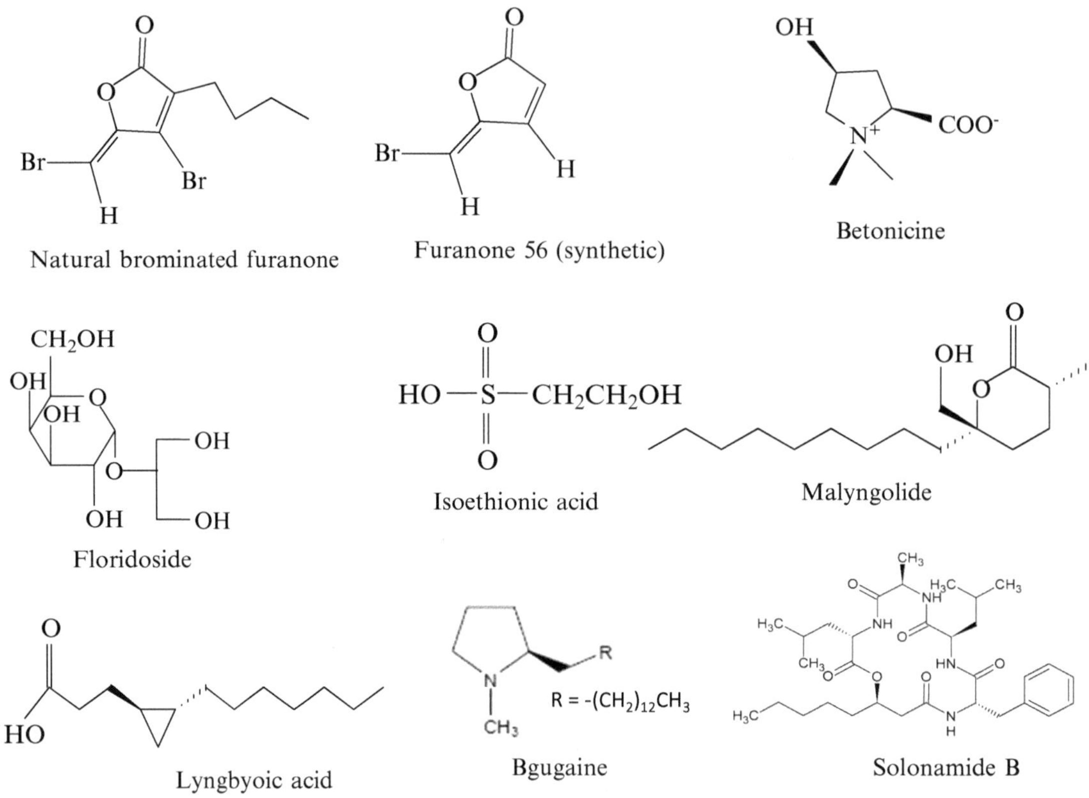

Fig. 1 Inhibitors of QS from marine origins

Ahnfeltiopsis flabelliformis through bioactivity-guided fractionation. The metabolites from the polar active fractions were identified as α-Dgalactopyranosyl-(1–2)-glycerol(floridoside), betonicine, and isethionic acid (Fig. 1). Due to the inavailability of other pure compounds, only commercial isethionic acid was further tested. However, the compound did not show any QSI activity and it was speculated that it worked synergistically with other compounds to block QS. Minor antagonist activity was also observed in algae of the families Caulerpaceae, Rhodomelaceae, and Galaxauraceae (Skindersoe et al. 2008). Some compounds secreted by *Chlamydomonas reinhardtii* were found to mimic bacterial signals and interfere with their QS system (Teplitski et al. 2004). Another enzyme bromoperoxidase produced by algae (*Laminaria digitata*) can also be exploited as QSI to deactivate AHL signal-3OC6HSL by oxidation process (Borchardt et al. 2001). QS inhibitor-like compounds were shown to be produced by 12 % of the 96 epibiotic bacteria associated with brown algae, *Colpomenia sinuosa* using *Serratia rubidaea* JCM14263 as biosensor (Kanagasabhapathy et al. 2009). An interesting feature of this indicator organism is its ability to tolerate high NaCl concentrations, which is necessary when marine organisms are the source material.

Asparagopsis taxiformis, a marine macroalgae demonstrated antiquorum, sensing activity in *Chromobacterium violaceum* CV026 reporter strain. The activity was confirmed using the sensor strain, *Serratia liquefaciens* MG44, having green fluorescent protein (gfp). The ICR-FT/MS data indicated that the expected active compound responsible for the QS inhibitory activity could be 2-dodecanoyloxyethanesulfonate ($C_{14}H_{27}O_5S$) (Jha et al. 2013).

Quorum Sensing Inhibition by Marine Bacteria

Various bacteria have been reported to antagonize/inhibit QS of other bacteria by mechanisms like the production of lactonase, acylase enzymes, and long chain AHLs (Kalia 2013). However, in some cases the mechanism remains unexplored. The role of marine bacteria is also evident in inhibiting QS. Tinh et al. (2007b) isolated microbial communities from the *Penaeus vannamei* shrimp gut and were grown in the presence of a mixture of different short-chain AHLs as sole carbon (and nitrogen) source, eventually resulting in three enrichment cultures. The enrichment cultures were shown to degrade *V. harveyi* HAI-1 in vitro and to improve the growth rate of rotifers challenged to pathogenic *V. harveyi*. Similar studies were conducted by Cam et al. (2009) using AHL degraders from the gut of European seabass, *Dicentrarchus labrax* L., and Asian seabass, *Lates calcarifer*. Two Gram-negative strains with different roles in QS were isolated from the intestinal microbial flora of ayu fish, *Plecoglossus altivelis*, with *Aeromonas* sp. strain MIB015 as AHL producer and *Shewanella* sp. strain MIB010 as AHL degrader, respectively (Morohoshi et al. 2005). The AHL degraders disintegrated 1 μM of synthetic HHL in 3 days and interfered with the exoprotease activity of *Aeromonas*. Further investigation showed that the activity was due to AHL-acylases encoded by the *aac* gene (Morohoshi et al. 2008). The expression of this gene was also found to quench AHL production in the fish pathogen *V. anguillarum*, thus disrupting QS-regulated biofilm formation (Morohoshi et al. 2008). Bacteria can also act as antagonists by releasing molecules that can block QS. Marine actinomycetes screened by You et al. (2007) inhibited the biofilm formation of different *Vibrio* species. Thirty-five out of 88 actinomycetes inhibited the biofilm formation of *V. harveyi*, *Vibrio vulnificus*, and *V. anguillarum* without affecting their growth. The best strain, strain A66 (identified as *Streptomyces albus*), dispersed the biofilm structure and this may have been linked to the inactivation of the AHL QS system. As indigenous species of the marine environment, Actinomycetes predominantly from the genera of *Streptomyces*, *Micromonospora*, and *Salinispora* could offer interesting options for probiotics in aquaculture (Das et al. 2008).

A Gram-positive strain isolated from seagrass communities, *Halobacillus salinus* C42, inhibited the bioluminescence of *V. harveyi* in a co-cultivation assay through the diffusion of small molecules identified as phenethylamine compounds, with 2,3-methyl-N-(2′-phenylethyl)-butyramide being the most effective one. The compounds blocked the expression of several QS-controlled phenotypes in Gram-negative bacteria such as violacein pigment in CV026, fluorescence in AHL reporter strain JB525, and planktonic luminescence in *V. harveyi*. The structural similarities of the compounds with the AHL suggest that they might compete with AHLs for receptor binding (Teasdale et al. 2009).

Nithya et al. (2010) screened bacteria isolated from Palk Bay sediments for anti-QS activity and found eleven strains inhibiting QS signals in *Chromobacterium violaceum* (ATCC12472) and *C. violaceum* CV026. The marine bacterial strain S8-07 identified as *Bacillus pumilus* reduced the accumulation of N-acyl homoserine lactone (AHLs) and showed significant inhibition of LasA protease (76 %), LasB elastase (84 %), caseinase (70 %), pyocyanin (84 %), and pyoverdin and biofilm formation (87 %) in *Pseudomonas aeruginosa* PAO1. Strain S8-07 also showed highly significant reduction (90 %) in prodigiosin, secreted caseinase (92 %), hemolytic activity (73 %), and biofilm formation (61 %) in *Serratia marcescens*. Acidification assay and HPLC analysis revealed that the degradation of AHL was due to acylase activity of S8-07. Another marine bacterial strain *Bacillus* sp. SS4 isolated from the Point Calimere coastal region showed potential quorum sensing inhibitory (QSI) activity in a concentration-dependent manner (0.5–2 mg/ml) against the AHL-mediated violacein production in *C. violaceum* (33–86 %) and biofilm formation (33–88 %), total protease (20–65 %), LasA protease (59–68 %), LasB elastase (36–68 %), pyocyanin (17–86 %), and pyoverdin productions in PAO1 (Musthafa et al. 2011). In another study, the bacterium exhibited a concentration-dependent (50–200 μg/ml) reduction in prodigiosin production in *S. marcescens* to a level of 40–87 %. The azocasein-degrading proteolytic activity of *S. marcescens* was also reduced by up to 60 % (Musthafa et al. 2012).

Five bacterial strains belonging to the genera *Marinobacter* and *Halomonas* and one archaeal strain belonging to the genus *Haloterrigena* were isolated from a hypersaline microbial mat. Purification of QS inhibitory dichloromethane extracts of *Marinobacter* sp. SK-3 resulted in isolation of four related diketopiperazines (DKPs). *Cyclo*(L-Pro-L-Phe) and *cyclo*(L-Pro-L-*iso*Leu) inhibited QS-dependent production of violacein by *C. violaceum* CV017. *Cyclo*(L-Pro-L-Phe), *cyclo*(L-Pro-L-Leu), and *cyclo*(L-Pro-L-*iso*Leu) reduced QS-dependent luminescence of the reporter *E. coli* pSB401 induced by 3-oxo-C6-HSL (Abed et al. 2013).

Quorum Sensing Inhibition by Other Marine Organisms

Other organisms with the potential to be inhibit QS systems include sponges and aquatic invertebrate and the compounds produced by these organisms. The North Sea bryozoan *Flustra foliacea* releases brominated alkaloids that reduce the signal intensities of different QS biosensors with 20–50 % at a concentration of 20 mg/L. Furthermore, the metabolites also inhibit QS-regulated phenotypes, such as protease in P. aeruginosa (Peters et al. 2003). A strong QS inhibition has also been observed for the sponge *Luffariella variabilis* in LuxR-regulated systems. The secondary metabolites manoalide, manoalide monoacetate, and secomanoalide were found to be responsible for the inhibitory activity present in this sponge (Skindersoe et al. 2008). Other AHL-like molecules are the tumonoic acids from the marine cyanobacterium *Blennothrix cantharidosmum*. Although no significant QSI activity was detected in a green fluorescent protein AHL detection assay, some of the compounds were shown to moderately inhibit the bioluminescence of wild-type *V. harveyi* with the highest inhibition by tumonoic acid F (Clark et al. 2008). Dobretsov et al. (2011) reported the screening of seventy-eight natural products from chemical libraries containing compounds from marine organisms (sponges, algae, fungi, tunicates, and cyanobacteria) and terrestrial plants, for the inhibition of bacterial quorum sensing (QS) using a reporter strain *Chromobacterium violaceum* CV017. Twenty-four percent of the tested compounds inhibited QS of the reporter without causing toxicity. QS-dependent luminescence of the LasR-based reporter, which is normally induced by N-3-oxo-dodecanoyl-L-homoserine lactone, was reduced by demethoxy encecalin and hymenialdisin at concentrations >6.6 μM and 15 μM, respectively. Hymenialdisin, demethoxy encecalin, microcolins A and B, and kojic acid inhibited responses of the LuxR-based reporter induced by N-3-oxo-hexanoyl-L homoserine lactone at concentrations >0.2 μM, 2.2 μM, 1.5 μM, 15 μM, and 36 μM, respectively. A few compounds have been reported from marine organisms such as cyanobacteria with abilities to inhibit QS gene expressions. Malyngolide and lyngbyoic acid (Fig. 1) isolated from *Lyngbya majuscula* were able to inhibit violacein production in *C. violaceum* and elastase and pyocyanin production in *P. aeruginosa*, respectively (Dobretsov et al. 2010; Kwan et al. 2011). Two other compounds-malyngamide C and 8-epi-malyngamide isolated from *L. majuscula* were able to inhibit QS activities in P. aeruginosa (Kwan et al. 2010).

The quorum-sensing inhibitory (QSI) activity of marine sponges collected from Palk Bay, India, was evaluated against acyl homoserine lactone (AHL)-mediated violacein production in *Chromobacterium violaceum* (ATCC 12472), CV026, and virulence gene expressions in clinical isolate *Serratia marcescens* PS1. Out of 29 marine sponges tested, the methanol extracts of *Aphrocallistes bocagei*, *Haliclona* (*Gellius*) *megastoma*, and *Clathria atrasanguinea* inhibited the AHL mediated violacein production in *C. violaceum* (ATCC 12472) and CV026. Further, these sponge extracts inhibited the AHL-dependent prodigiosin pigment, virulence enzymes such as protease, hemolysin production, and biofilm formation in *S. marcescens* PS1 (Annapoorani et al. 2012).

(R)-Bgugaine (Fig. 1) is a natural pyrrolidine alkaloid from *Arisarum vulgare* (a plant native of Mediterranean coasts of Morocco and Spain) was found to exhibit significant effect on motility, particularly swarming as well as swimming motilities in *P. aeruginosa*. The compound also showed significant reduction in rhamnolipid production by 42 %, pyocyanin production by 37 %, and Las A production by 34 % (Majik et al. 2013). Recently, Nielsen et al. (2014) reported that a cyclodepsipeptide termed Solonamide B (Fig. 1) isolated from the marine bacterium, *Photobacterium halotolerans* strongly reduces expression of RNAIII, the effector molecule of the *agr* quorum sensing system. They demonstrated that Solonamide B interferes with the binding of *S. aureus* autoinducing peptides (AIPs) to sensor histidine kinase, AgrC, of the *agr* two-component system.

Potential Application and Limitations

The compounds of marine origin identified as QSIs need to be tested in vivo in suitable animal models to test their therapeutic efficacy. In a mouse pulmonary infection model, the furanone inhibited quorum sensing of the *P. aeruginosa* and promoted their clearance by the mouse immune response Hentzer et al. (2003). Further, Wu et al. (2004) reported that furanones successfully interfered with N-acyl homoserine lactone and suppressed bacterial quorum sensing in lungs, which resulted in decreases in expression of green fluorescent protein. Furanones accelerated lung bacterial clearance and reduced the severity of lung pathology. In a lethal *P. aeruginosa* lung infection, treatment with furanone significantly prolonged the survival time of the mice. Several other reports on the in vivo activity of the furanones have been published, but application of these furanones for treatment of human patients is limited due to instability of the furanones (Rasmussen and Givskov 2006). Other reported compounds need to be investigated in search of safe, stable, and nontoxic QS inhibitors for therapeutic use.

Conclusion

The discovery that several marine organisms and their secondary metabolites can interfere with QS inhibitors, together with the findings that QS controls the virulence of many pathogenic bacteria, suggests that the active compounds derived from marine organisms could be used to control infections caused by bacteria. In order to utilize these organisms in practice, in vivo studies in suitable animal models is needed for a better understanding of the processes involved. It is very important to explore in depth the virulence mechanisms of the pathogens and particularly the connection with QS. There is also a need to investigate the precise relationship between the diverse organisms present in the marine environment and the effect of QS on this interaction. A thorough understanding of the mechanism of action is needed to uncover the therapeutic potential of these agents. Further combinational approaches of disease treatment by exploiting synergistic interactions of QSIs with other anti-infective compounds should be explored.

Acknowledgment The authors acknowledge the financial support provided by the Department of Science and Technology (DST), New Delhi, India, for awarding INSPIRE fellowship to FMH.

References

Abed RMM, Dobretsov S, Al-Fori M, Gunasekera SP, Sudesh K, Paul VJ (2013) Quorum-sensing inhibitory compounds from extremophilic microorganisms isolated from a hypersaline cyanobacterial mat. J Ind Microbiol Biotechnol 40:759–772

Adonizio A, Kong KF, Mathee K (2008) Inhibition of quorum sensing-controlled virulence factor production in *Pseudomonas aeruginosa* by south Florida plant extracts. Antimicrob Agents Chemother 52:198–203

Ahmad I, Aqil F, Ahmad F, Zahin M, Musarrat J (2008) Quorum sensing in bacteria: Potential in plant health protection. In: Ahmad I et al (eds) Plant–bacteria interactions. Wiley-VCH Verlag GmbH & Co. KGa, Weinheim, pp 129–153. ISBN 978-3-527-31901-5

Annapoorani A, Jabbar AKKA, Pandian SK, Ravi AV (2012) Inhibition of quorum sensing mediated virulence factors production in urinary pathogen *Serratia marcescens* PS1 by marine sponges. Ind J Microbiol 52(2):160–166

Aqil F, Ahmad I, Owais M (2006) Targeted screening of bioactive plant extracts and phytocompounds against problematic groups of Multidrug resistant bacteria. In: Modern Phytomedicine: turning medicinal plants into drugs. Wiley VCH, Weinheim, pp 173–197

Borchardt SA, Allain EJ, Michels JJ, Stearns GW, Kelly RF, McCoy WF (2001) Reaction of acylated homoserine lactone bacterial signaling molecules with oxidized halogen antimicrobials. Appl Environ Microbiol 67(7):3174–3179

Cam DTV, Nhan DT, Ceuppens S, Hao NV, Dierckens K, Wille M, Sorgeloos P, Bossier P (2009) Effect of N-acyl homoserine lactone degrading enrichment cultures on *Macrobrachium rosenbergii* larviculture. Aquaculture 294:5–13

Clark BR, Engene N, Teasdale ME, Rowley DC, Matainaho T, Valeriote FA, Gerwick WH (2008) Natural products chemistry and taxonomy of the marine cyanobacterium *Blennothrix cantharidosmum*. J Nat Prod 71:1530–1537

Das S, Ward LR, Burke C (2008) Prospects of using marine actinobacteria as probiotics in aquaculture. Appl Microbiol Biotechnol 81:419–429

de Nys R, Givskov M, Kumar N, Kjelleberg S, Steinberg PD (2006) Furanones. Prog Mol Subcell Biol 42:55–86

Defoirdt T, Crab R, Wood TK, Sorgeloos P, Verstraete W, Bossier P (2006) Quorum sensing-disrupting brominated furanones protect the gnotobiotic brine shrimp *Artemia franciscana* from pathogenic *Vibrio harveyi*, *Vibrio campbelii* and *Vibrio parahaemolyticus* isolates. Appl Environ Microbiol 72(9):6419–6423

Defoirdt T, Miyamoto CM, Wood TK, Meighen EA, Sorgeloos P, Verstraete W, Bossier P (2007) The natural furanone (5Z)-4bromo-5-(bromomethylene)-3-butyl-2(5H)-furanone disrupts quorum sensing-regulated gene expression in *Vibrio harveyi* by decreasing the DNA-binding activity of the transcriptional regulator protein luxR. Environ Microbiol 9(10):2486–2495

Dobretsov S, Teplitski M, Alagely A, Gunasekara SP, Paul VJ (2010) Malyngolide from the cyanobacterium *Lyngbya majuscula* interferes with quorum sensing circuitry. Environ Microbiol Rep 2:739–744

Dobretsov S, Teplitski M, Bayer M, Gunasekera S, Proksch P, Paul VJ (2011) Inhibition of marine biofouling by bacterial quorum sensing inhibitors. Biofouling 27:893–905

Dong YH, Gusti AR, Zhang Q, Xu JL, Zhang LH (2002) Identification of quorum-quenching N-acyl homoserine lactonases from *Bacillus* species. Appl Environ Microbiol 68:1754–1759

Dudler R, Elberl E (2006) Interactions between bacteria and eukaryotes via small molecules. Curr Opin Biotechnol 17:268–273

Dworjanyn SA, Steinberg P (1999) Localisation and surface quantitation of secondary metabolites in the red alga *Delisea pulchra*. Mar Biol 133:727–736

Fenical W, Jensen PR (2006) Developing a new resource for drug discovery: marine actinomycete bacteria. Nat Chem Biol 2:666–673

Fuqua WC, Winans SC, Greenberg EP (1994) Quorum sensing in bacteria-the LuxR-LuxI family of cell density responsive transcriptional regulators. J Bacteriol 176:269–275

Gambello MJ, Kaye S, Iglewski BH (1993) LasR of *Pseudomonas aeruginosa* is a transcriptional activator of the alkaline protease gene (*apr*) and an enhancer of exotoxinA expression. Infect Immun 61: 1180–1184

Givskov M, De Nys R, Manefield M, Gram L, Maximilien R, Eberl L, Molin S, Steinberg PD, Kjelleberg S (1996) Eukaryotic interference with homoserine lactone-mediated prokaryotic signaling. J Bacteriol 178:6618–6622

Gram L, de Nys R, Maximilien R, Givskov SP, Kjelleberg S (1996) Inhibitory effects of secondary metabolites from the red alga *Delisea pulchra* on swarming motility of *Proteus mirabilis*. Appl Environ Microbiol 62(11):4284–4287

Harper MK, Bugni TS, Copp BR, James RD, Lindsay BS, Richardson AD, Schnabel PC, Tasdemir D, VanWagoner RM, Verbitski SM, Ireland CM (2001) Introduction to the chemical ecology of marine natural products. Mar Chem Ecol: 3–69

Hentzer M, Givskov M (2003) Pharmacological inhibition of quorum sensing for the treatment of chronic bacterial infections. J Clin Invest 112:1300–1307

Hentzer M, Riedel K, Rasmussen TB, Heydorn A, Andersen JB, Parsek MR, Rice SA, Eberl L, Molin S, Høiby N, Kjelleberg S, Michael G (2002) Inhibition of quorum sensing in *Pseudomonas aeruginosa* biofilm bacteria by a halogenated furanone compound. Microbiology 148(1):87–102

Hentzer M, Wu H, Andersen JB, Riedel K, Rasmussen TB, Bagge N, Kumar N, Schembri MA, Song Z, Kristofferson P, Manefield M, Costerton JW, Moiln S, Eberl L, Steinberg P, Kjelleberg S, Hoiby N, Givskov M (2003) Attenuation of *Pseudomonas aeruginosa* virulence by quorum sensing inhibitors. EMBO J 22:3803–3815

Husain FM, Ahmad I, Asif M, Tahseen Q (2013) Influence of clove oil on certain quorum sensing regulated functions and biofilm of *Pseudomonas aeruginosa* and *Aeromonas hydrophila*. J Biosci 38:1–10

Jha B, Kumari K, Westphal J, Hartmann A, Schmidtt-Koplin P (2013) Quorum sensing inhibition by *Asparagopsis taxiformis*, a marine macro alga: separation of the compounds that interrupts bacterial communication. Mar Drugs 11: 253–265

Kalia VC (2013) Quorum sensing inhibitors: an overview. Biotechnol Adv 31:224–245

Kanagasabhapathy M, Yamazaki G, Ishida A, Sasaki H, Nagata S (2009) Presence of quorum-sensing inhibitor-like compounds from bacteria isolated from brown alga *Colpomenia sinosa*. Lett Appl Microbiol 49: 573–579

Khan MSA, Zahin M, Hasan S, Husain FM, Ahmad I (2009) Inhibition of quorum sensing regulated bacterial functions by plant essential oils with special reference to clove oil. Lett Appl Microbiol 49:354–360

Kim JS, Kim YH, Seo YW, Park S (2007) Quorum sensing inhibitors from the red alga, *Ahnfeltiopsis flabelliformis*. Biotechnol Bioprocess Eng 12:308–311

Kjelleberg S, Steinberg P (2001) Defences against bacterial colonization of marine plants. In: Lindow SE, Hect-Poinar E, Elliot V (eds) Phyllosphere microbiology. APS Press, St. Paul, pp 157–172

Kwan JC, Teplitski M, Gunasekara SP, Paul VJ, Luesch H (2010) Isolation and biological evaluation of 8-epi-malyngamide C from the Floridian marine cyanobacterium *Lyngbya majuscula*. J Nat Prod 73:463–466

Kwan JC, Meickle T, Ladwa D, Teplitski M, Paul V, Luesch H (2011) Lyngbyoic acid, a "tagged" fatty acid from marine cyanobacterium, disrupts quorum sensing in *Pseudomonas aeruginosa*. Mol Biosyst 7:1205–1216

LaSarre B, Federle MJ (2013) Exploiting quorum sensing to confuse bacterial pathogens. Microbiol Mol Biol Rev 77:73–111

Majik MS, Naika D, Bhatb C, Tilveb S, Tilvia S, D'Souzaa L (2013) Synthesis of (*R*)-Norbgugaine and its potential as quorum sensing inhibitor against *Pseudomonas aeruginosa*. Bioorg Med Chem Lett 23:2353–2356

Manefield M, de Nys R, Kumar N, Read R, Givskov M, Steinberg P, Kjelleberg S (1999) Evidence that halogenated furanones from *Delisea pulchra* inhibit acylated homoserine lactone (AHL)-mediated gene expression by displacing the AHL signal from its receptor protein. Microbiology+ 145:283–291

Manefield M, Harris L, Rice SA, de Nys R, Kjelleberg S (2000) Inhibition of luminescence and virulence in the black tiger prawn (*Penaeus monodon*) pathogen *Vibrio harveyi* by intercellular signal antagonists. Appl Environ Microbiol 66(5):2079–2084

Manefield M, Rasmussen TB, Henzter M (2002) Halogenated furanones inhibit quorum sensing through accelerated LuxR turnover. Microbiology 148(4): 1119–1127

Marris E (2006) Marine natural products: drugs from the deep. Nature 443:904–905

Miller MB, Bassler BL (2001) Quorum sensing in bacteria. Annu Rev Microbiol 55:165–199

Morohoshi T, Ebata A, Nakazawa S, Kato N, Ikeda T (2005) N-Acylhomoserine lactone-producing or degrading bacteria isolated from the intestinal microbial flora of Ayu fish (*Plecoglossus altivelis*). Microbes Environ 20(4):264–268

Morohoshi T, Nakazawa S, Ebata A, Kato N, Ikeda T (2008) Identification and characterization of N-acylhomoserine lactone acylase from the fish intestinal *Shewanella* sp. strain MIB015. Biosci Biotechnol Biochem 72(7):1887–1893

Musthafa KS, Saroja V, Pandian SK, Ravi AV (2011) Antipathogenic potential of marine *Bacillus* sp. SS4on N-acyl-homoserine-lactone-mediated virulence factors production in *Pseudomonas aeruginosa* (PAO1). J Biosci 36:55–67

Musthafa KS, Pandian SK, Ravi AV (2012) Inhibition of quorum-sensing-dependent phenotypic expression in *Serratia marcescens* by marine sediment *Bacillus* spp. SS4. Ann Microbiol 62:443–447

Natrah FMI, Defoirdt T, Sorgeloos P, Bossier P (2011) Disruption of bacterial cell-to-cell communication by marine organisms and its relevance to aquaculture. Mar Biotechnol 13:109–126

Newman DJ, Cragg GM (2004a) Marine natural products and related compounds in clinical and advanced preclinical trials. J Nat Prod 67:1216–1238

Newman DJ, Cragg GM (2004b) Advanced preclinical and clinical trials of natural products and related compounds from marine sources. Curr Med Chem 11:1693–1713

Nielsen A, Mansson M, Bojer MS, Gram L, Larsen TO, Novick RP, Frees D, Frøkiær H, Ingmer H (2014) Solonamide B inhibits quorum sensing and reduces *Staphylococcus aureus* mediated killing of human neutrophils. PLoS One 9(1):e84992

Nithya C, Aravindraja C, Pandian SK (2010) *Bacillus pumilus* of Palk Bay origin inhibits quorum-sensing-mediated virulence factors in Gram-negative bacteria. Res Microbiol 161:293–304

Nylund GM, Cervin G, Hermansson M, Pavia H (2005) Chemical inhibition of bacterial colonization by the red algae *Bonnemaisonia hamifera*. Mar Ecol Prog Ser 302:27–36

Park SY, Kang HO, Jang HS, Lee JK, Koo BT, Yum DY (2005) Identification of extracellular N-acylhomoserine lactone acylase from a *Streptomyces* sp. and its application to quorum quenching. Appl Environ Microbiol 71:2632–2641

Paterson I, Anderson EA (2005) The renaissance of natural products as drug candidates. Science 310:451–453

Peters L, Konig GM, Wright AD, Pukall R, Stackebrandt E, Eberl L, Riedel K (2003) Secondary metabolites of *Flustra foliacea* and their influence on bacteria. Appl Environ Microbiol 69:3469–3475

Proksch P, Edrada RA, Ebel R (2002) Drugs from the seas-current status and microbiological implications. Appl Microbiol Biotechnol 59(2–3):125–134

Rasch M, Buch C, Austin B, Slierendrecht WJ, Ekmann KS, Larsen JL, Johansen C, Riedel K, Eberl L, Givskov M, Gram L (2004) An inhibitor of bacterial quorum sensing reduces mortalities caused by vibriosis in rainbow trout (*Oncorhynchus mykiss*, Walbaum). Syst Appl Microbiol 27:350–359

Rasmussen TB, Manefield M, Andersen JB, Eberl L, Anthoni U, Christopherse C, Steinberg P, Kjelleberg S, Givskov M (2000) How *Delisea pulchra* furanones affect quorum sensing and swarming motility in *Serratia liquefaciens* MG1. Microbiology+ 146:3237–3244

Rasmussen TB, Bjarnsholt BT, Skindersoe ME, Hentzer M, Kristoffersen P, Kote M, Nielsen J, Eberl L, Givskov M (2005) Screening for quorum-sensing inhibitors (QSI) by use of a novel genetic system, the QSI selector. J Bacteriol 187:1799–1814

Rasmussen TB, Givskov M (2006) Quorum sensing inhibitors: a bargain of effects. Microbiology 152: 895–904

Reacher MH, Shah A, Livermore DM, Wale MC, Graham C, Johnson AP, Heine H, Monnickendam MA, Barker KF, James D, George RC (2000) Bacteraemia and antibiotic resistance of its pathogens reported in England and Wales between 1990 and 1998: trend analysis. Br Med J 320:213–216

Ren D, Sims J, Wood TK (2001) Inhibition of biofilm formation and swarming of *Escherichia coli* by (5Z)-4-bromo-5-(bromomethylene)-3-butyl-2(5H)-furanone. Environ Microbiol 3:731–736

Skindersoe ME, Ettinger-Epstein P, Rasmussen TB, Bjarnsholt T, deNys R, Givskov M (2008) Quorum sensing antagonism from marine organisms. Mar Biotechnol 10:56–63

Teasdale ME, Liu J, Wallace J, Akhlaghi F, Rowley DC (2009) Secondary metabolites produced by the marine bacterium *Halobacillus salinus* that inhibit quorum sensing-controlled phenotypes in Gram-negative bacteria. Appl Environ Microbiol 75:567–572

Tenover FC, Hughes JM (1996) The challenges of emerging infectious diseases. Development and spread of multiply-resistant bacterial pathogens. JAMA 275:300–304

Teplitski M, Chen H, Rajamani S, Gao M, Merighi M, Sayre RT, Robinson JB, Rolfe BG, Bauer WD (2004) *Chlamydomonas reinhardtii* secretes compounds that mimic bacterial signals and interference with quorum sensing regulation in bacteria. Plant Physiol 134:137–146

Tinh NTN, Gunasekara RAYS, Boon N, Dierckens K, Sorgeloos P, Bossier P (2007a) N-Acylhomoserine lactone degrading microbial enrichment cultures isolated from *Penaeus vannamei* shrimp gut and their probiotic properties in *Brachionus plicatilis* cultures. FEMS Microbiol Ecol 62: 45–53

Tinh NTN, Lin ND, Wood TK, Dierckens K, Sorgeloos P, Bossier P (2007b) Interference with the quorum sensing systems in a *Vibrio harveyi* strain alters the growth rate of gnotobiotically cultured rotifer *Brachionus plicatilis*. J Appl Microbiol 103: 194–203

Toder DS, Gambello MJ, Iglewski BH (1991) *Pseudomonas aeruginosa* LasA: a second elastase under the transcriptional control of lasR. Mol Microbiol 5:2003–2010

Williams P (2007) Quorum sensing, communication and cross kingdom signaling in the bacterial world. Microbiology 153:3923–3928

Wu H, Song Z et al (2004) Synthetic furanones inhibit quorum-sensing and enhance bacterial clearance in *Pseudomonas aeruginosa* lung infection in mice. J Antimicrob Chemother 53(6):1054–1061

You J, Xue X, Cao L, Lu X, Wang J, Zhang L, Zhou S (2007) Inhibition of *Vibrio* biofilm formation by a marine actinomycete strain A66. Appl Microbiol Cell Physiol 76:1137–1144

Zahin M, Hasan S, Aqil F, Khan MSA, Husain FM, Ahmad I (2010) Screening of Indian medicinal plants for their anti-quorum sensing activity. Ind J Exp Biol 48:1219–1224

Zang T, Lee BWK, Cannon LM, Ritter KA, Dai S, Ren D, Wood TK, Zhou ZS (2009) A naturally occurring brominated furanone covalently modifies and inactivates LuxS. Bioorg Med Chem Lett 19: 6200–6204

Plant Quorum Sensing Inhibitors: Food, Medicinal Plants, and Others

Pragasam Viswanathan, Prasanth Rathinam, and Suneeva S.C.

Introduction

Plants living in an environment with high bacterial density were long suspected to have protective mechanisms against infections. Due to this reason, natural products were explored initially because of their broad therapeutic values in traditional medical practice. Since these plants can be consumed by humans, the active compounds that are having therapeutic activities should be safe toward human cells. However, as science progressed, toxicological studies were performed on these active compounds to avoid their toxicity. The interest to identify and understand the biological functions and the mode of action responsible for their therapeutic roles has been escalated. To date, biologically active constituents of natural products, especially plant-derived ones, have led to the discovery of new drugs used for the treatment of numerous diseases. However, the absence of an immune system in plants, unlike humans and animals against the invaders, demanded researchers to speculate other defense mechanisms to defeat the pathogens (Koh et al. 2013), and manipulation of quorum sensing (QS) systems is considered to be a mode of defense by the plants against pathogens.

Various studies showed that eukaryotes have evolved efficiently to maneuver bacterial QS systems and protect themselves from pathogen attack. Plant extracts have been found to act as quorum sensing inhibitors (QSI). The plant compounds usually target the bacterial QS system via three different ways (Fig. 1), by stopping the signaling molecules from being synthesized, by degrading the signaling molecules, or by targeting the signal receptor (Koh et al. 2013).

Koh et al. (2013), screened hundreds of plant extracts specifically for the QS signal synthase activity and couldn't find any, suggesting that it is rather rare for plants to possess anti-signal synthase activity. To interfere with signal reception, there can be competitive and noncompetitive molecules that can obstruct with the binding of the signal to its cognate receptor. It is logical to imagine that for competitive molecules to bind to receptor, these molecules must be structurally similar to the signal molecules. For noncompetitive binding to the signal receptor, conceivably, these molecules will bind to the site on the receptor other than the signal binding site. Plants can produce molecules that structurally mimic the signals, and such competitive binding is effective to block activation of signal-mediated QS. In addition to this, investigations have also proven that plants have the ability to degrade the signaling molecules produced by the bacteria and this will obstruct the bacteria virulence factors by disrupting their communication systems (Koh et al. 2013).

P. Viswanathan (✉) • P. Rathinam • S. S.C.
Renal Research Laboratory, Centre for Bio Medical Research and School of Bio Sciences and Technology, VIT University, Vellore 632 014, Tamil Nadu, India
e-mail: pragasam.v@vit.ac.in

V.C. Kalia (ed.), *Quorum Sensing vs Quorum Quenching: A Battle with No End in Sight*,
DOI 10.1007/978-81-322-1982-8_22, © Springer India 2015

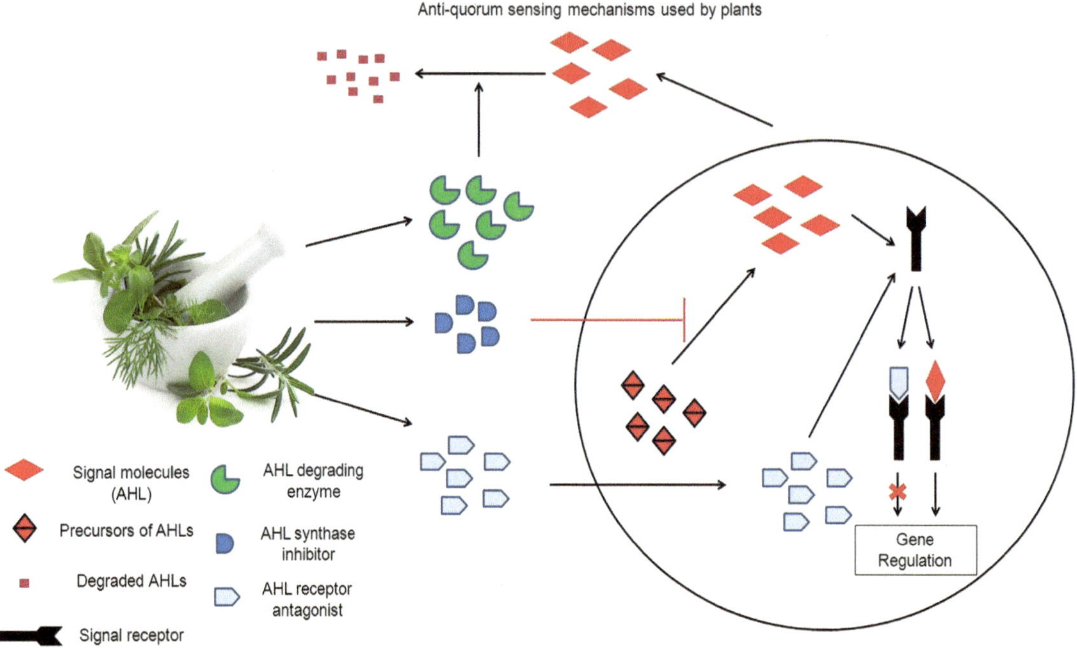

Fig. 1 Various mechanisms utilized by plants to inhibit quorum sensing

QSI Activity by Plants Used for Dietary Purpose

Medicago truncatula Gaertn. seedlings modulate AhyR, CviR, and LuxR reporter activities in different reporter strains like *Escherichia coli* JM109 [p (SB536)], *Chromobacterium violaceum* CV026, and *E. coli* JM109 [p (SB401)], respectively (Kalia 2013), and QS in general in *Pseudomonas aeruginosa* and *Sinorhizobium meliloti* by secreting AHL "signal-mimic" substances of unknown structure that specifically stimulate or inhibit AHL-regulated responses in bacteria (Kalia 2013). AHL-degrading abilities have been reported from a large number of legumes – *Medicago sativa* L. (alfalfa), clover, *Lotus corniculatus* L. seedlings, *Hordeum vulgare* L. (barley), *Pisum sativum* L. (pea), and *Pachyrhizus erosus* (L.) Urb. (yam beans) (Kalia 2013). Although the exact nature of enzyme is not elucidated, it is likely to be lactonase.

Seed exudate L-canavanine from alfalfa (*M. sativa* L.) affected the QS expression and exopolysaccharide (EPS) production by the nitrogen-fixing bacterium *Sinorhizobium meliloti* (Kalia 2013). The aqueous extracts of leaf and fruit of *Moringa oleifera* Lam. were found to inhibit violacein production, a QS-regulated behavior in *C. violaceum*. Compounds like chlorogenic acid, gallic acid, and quercetin that were present in appreciable quantities in the extract were expected to be responsible for the anti-QS property (Koh et al. 2013). Quorum sensing-mediated behaviors like violacein production by *C. violaceum*, biosurfactant production by *P. aeruginosa* PA01, and motility and EPS and biofilm production by *E. coli*, *Proteus mirabilis*, and *Serratia marcescens* were repressed by methanolic extract of *Capparis spinosa* Linn. dry fruits (Abraham et al. 2011).

Sweet basil (*Ocimum basilicum* L.) inhibited AHL activity-mediated violacein production in *C. violaceum* CV026 by 78 %. Rosamarinic acid, a multifunctional caffeic acid ester produced by the roots of *O. basilicum* L., reduced the biofilm formation by *P. aeruginosa* (Kalia 2013). Thyme (*Thymus* sp.) and kale (*Brassica oleracea* L.) decreased the pigment formation of *C. violaceum* by 60 % (Kalia 2013).

Exudates from *Arabidopsis* affected QS signals in *Agrobacterium tumefaciens* (Kalia 2013). Fatty acid amide hydrolase enzymes from *Arabidopsis* have been linked to AHL degradation through mutant studies. Mutants overexpressing this enzyme were more resistant to developmental changes by AHLs (Kalia 2013).

Herbal plants like *Camellia sinensis* (L.) (tea plant) produces catechins which affected the transfer of conjugative R plasmid in *E. coli* (Kalia 2013). Caffeine (1, 3, 7-trimethylxanthine), one of the few plant products with occurrence in beverages such as coffee and tea, could inhibit AHL production and swarming by *P. aeruginosa* PA01 (Norizan et al. 2013).

L-Canavanine from legume alfalfa (*M. sativa* L.) and iberin from horseradish (*Armoracia rusticana* G.Gaertn., B.Mey. & Scherb.) found to have anti-QS activity against *P. aeruginosa* PA01and *E. coli* QSIS1 system (Jakobsen et al. 2012). Aqueous extracts of edible plants and fruits such as *Ananas comosus* ((L.) Merr.), *Musa paradisiacal* L., *Manilkara zapota* (L.) P. Royen, and *O. sanctum* proved to be QSI against violacein production by *C. violaceum* and pyocyanin pigment, staphylolytic protease, elastase production, and biofilm formation abilities of *P. aeruginosa* PA01 (Musthafa et al. 2010). Sulforaphane and erucin, two natural isothiocyanates that are highly abundant in broccoli and other cruciferous vegetables, were found to strongly inhibit QS and virulence in *P. aeruginosa*. Mechanistic evaluations of these effects suggest that these isothiocyanates are antagonists of the transcriptional activator LasR (Ganin et al. 2013). The presence of AHL-mimic QS molecules in *Oryza sativa* L. (rice) and *Phaseolus vulgaris* L. (bean) plant samples showed the bean and rice seed extract contain molecules that lack the typical lactone ring of AHLs. Interestingly, these molecules specifically alter the QS-regulated biofilm formation of two plant-associated bacteria, *Sinorhizobium fredii* SMH12 and *Pantoea ananatis* AMG501 (Montaño et al. 2013). Growing onion bulbs found to inhibit the QS mechanism in *P. aeruginosa*, and the exogenous application of pantolactone and myristic acid, isolated from onion (*Allium ceppa* L.), significantly inhibited pyocyanin production, protease, and lipase and polygalacturonase activity but did not have any significant effects on bacterial growth (Abd-Alla and Bashandy 2012).

QSI Activity by Fruits

Among the many fruit extracts tested, raspberry (*Rubus idaeus* L.), blueberry (*Vaccinium angustifolium* Aiton), and grape (*Vitis* sp.) extracts inhibited AHL activity-mediated violacein production. The mechanism of inhibition appeared to be a combination of interfering with AHL activity and modulating the synthesis of AHLs (Kalia 2013). Naturally occurring furocoumarins from grapefruit showed >95 % inhibition of AI-1 and AI-2 activities based on the *Vibrio harveyi*-based autoinducer bioassay. Grapefruit juice and furocoumarins also inhibited biofilm formation by *E. coli* O157:H7, *Salmonella typhimurium*, and *P. aeruginosa*. Grapefruit juices showed 47–62 % inhibition against AI-1 and 16.8–27.5 % inhibition against AI-2, whereas purified furocoumarins dihydroxybergamottin and bergamottin caused AI inhibitions in the range of 94.6–97.7 %. Biofilm formation by *E. coli* was disrupted by grapefruit juice (furocoumarins). These lower inhibitory effects were assigned to low concentration of furocoumarins in grapefruit juice (Kalia 2013).

Sour orange seeds contain limonoids such as isolimonic acid, ichangin, and deacetyl nomilinic acid glucoside with abilities to interfere with cell–cell signaling and biofilm formation in *V. harveyi*. Furthermore, isolimonic acid and ichangin treatment resulted in induced expression of the response regulator gene *lux*O without affecting the *lux*R promoter activity. Limonoids were also effective against HAI- and AI-2-mediated bioluminescence. These furocoumarins and limonoids are unique secondary metabolites, characterized by a furan moiety with the synthetic furanones, which have QSI abilities (Vikram et al. 2010).

Aqueous extracts of edible fruits such as *A. comosus* (L.) Merr. (pineapple), *Musa paradisiacal* L. (banana), and *M. zapota* (L.) P.Royen. (sapodilla) proved to be QSI against violacein production by *C. violaceum* and pyocyanin pigment, staphylolytic protease, elastase production, and biofilm formation abilities of *P. aeruginosa* PA01 (Musthafa et al. 2010). The hexane extracts of *Kigelia africana* (Lam.) Benth. fruit showed QS inhibitory activity against *C. violaceum* and *A. tumefaciens* biosensor systems. Both LuxI and LuxR activities were affected by crude extracts suggesting that the phytochemicals target both QS signal and receptor (Chenia 2013).

QSI Activity by Plants Used as Spices

Curcuma longa L. (turmeric) produces curcumin, which inhibits the expression of virulence genes of *P. aeruginosa* PA01 in *A. thaliana/Caenorhabditis elegans* pathogenicity models (Kalia 2013). The antibiofilm potential of curcumin on uropathogens as well as its efficacy in disturbing the mature biofilms was examined and found to attenuate the QS-dependent factors, such as EPS production, alginate production, and swimming and swarming motility. It was documented that curcumin enhanced the susceptibility of a marker strain and uropathogens to conventional antibiotics (Packiavathy et al. 2014).

Cinnamaldehyde, an organic compound, given by *Cinnamomum verum* J. Presl (cinnamon) and its derivatives affect a wide range of QS-regulated activities such as biofilm formation in *P. aeruginosa* and AI-2-mediated QS in different *Vibrio* spp. by decreasing the DNA-binding ability of LuxR, resulting in several marked phenotypic changes, including reduced virulence and increased susceptibility to stress (Kalia 2013). The ethanolic extracts of *Laurus nobilis* L. (leaves, flowers, fruits, and bark) were found to possess anti-QS activities against *C. violaceum* (Kalia 2013).

Quorum sensing inhibition has been reported from *Allium sativum* L. (garlic), and it has been found that garlic extract has a preference for the genes belonging to the group (toxins, enzymes, and alginate), targeting 11 genes (22 % of the functional class genes of *P. aeruginosa*) like *LasA*, *LasB* (coding for elastase and protease), *rhlAB* (encoding rhamnolipid), *chiC* (encoding for chitinase), as well as *aprA*, *phzA1B*, *phzS*, *phzC2D2E2F2G2*, and *PA1L*. All these are involved in the virulence and pathogenesis of *P. aeruginosa* (Kalia 2013). QS inhibitory extract of garlic renders *P. aeruginosa* sensitive to tobramycin, respiratory burst, and phagocytosis by Polymorphonuclear Leucocytes, as well as leads to an improved outcome of pulmonary infections (Kalia 2013). Bioassay-guided fractionation of garlic extracts identified the primary QSI present in garlic to be ajoene, a sulfur-containing compound. Microarray studies of ajoene-treated *P. aeruginosa* cultures revealed a concentration-dependent attenuation of central QS-controlled virulence factors. Ajoene also demonstrated a clear synergistic effect with antibiotics on biofilm eradication and a cease in lytic necrosis by macrophages (Jakobsen et al. 2012).

Aqueous methanolic extract of vanilla beans (*Vanilla planifolia* Jacks. ex Andrews) demonstrated QSI activity in reporter strain *C. violaceum* CV026 (Kalia 2013). Essential oils from *Piper caucasanum* Bredemeyer, *P. brachypodon* Benth., and *P. bogotense* showed significant QSI property upon the production of violacein in *C. violaceum* CV026 (Olivero et al. 2011). Hexane and methanol extracts of clove (*Syzygium aromaticum* (L.) Merrill & Perry) inhibited the response of *C. violaceum* CV026 to exogenously supplied N-hexanoyl-homoserine lactone, in turn preventing violacein production. Chloroform and methanol extracts of clove significantly reduced bioluminescence production by *E. coli* [pSB1075] grown in the presence of N-(3-oxododecanoyl)-L-homoserine lactone, while the hexane extract inhibited QS-regulated phenotypes in *P. aeruginosa* PA01, including expression of lecA :: lux and pyocyanin production (Krishnan et al. 2012). Subinhibitory concentrations of the clove oil demonstrated significant reduction of Las- and rhl-regulated virulence factors such as LasB, total protease, chitinase and pyocyanin production, swimming motility, and EPS production in *P. aeruginosa* PA01. The biofilm-forming capability of *P. aeruginosa* and *A. hydrophila* WAF-38 was also reduced in a concentration-dependent manner (Husain et al. 2013). Eugenol, the major constituent of clove extract, was identified to have the anti-QS property and at subinhibitory concentrations inhibited the production of virulence factors, including violacein, elastase, pyocyanin, and biofilm formation. With the help of 2 *E. coli* biosensors, it has been found that the mechanism of eugenols' QSI property is related to the inhibition of the Las and PQS systems (Zhou et al. 2013).

QSI Activity by Medicinal Plants

QSI have also been reported from many other medicinal plants like *Prunus armeniaca* L., *Prunella vulgaris* L., *Nelumbo nucifera* Gaertn., *Punica granatum* L., *Areca catechu* L., and *Imperata cylindrical* (L.) P.Beauv. Among the acetone/water extracts of the medicinal plants, *A. catechu* L. seed extracts repressed QS-regulated characters in *P. aeruginosa* PA01 and *C. violaceum*. All the other plants, except *P. armeniaca* L., exhibited QSI property in *P. aeruginosa* PA01 and reduced violacein production in *C. violaceum*. *Panax notoginseng*, a Chinese medicinal plant and a dietary supplement, demonstrated its ability to counter the QS system by suppressing the protease activities and many other virulence factors from *P. aeruginosa* and violacein production in *C. violaceum*. It appears that the reduction of protease activity may be due to the decrease in the production of signal molecules, critical for the production of virulence factors (Song et al. 2010; Koh and Tham 2011).

O. sanctum L. proved to be QSI against violacein production by *C. violaceum* and pyocyanin pigment, staphylolytic protease, elastase production, and biofilm formation abilities of *P. aeruginosa* PA01 (Musthafa et al. 2010). The methanolic bark extract obtained from the mangrove plant *Rhizophora annamalayana* Kathiresan against *C. violaceum* exhibited a reduction in the QS-dependent violacein production and the QS-dependent bioluminescence production in the aquatic bacterial pathogen *V. harveyi*. The molecular docking analysis of the identified compounds like cyclononasiloxane octadecamethyl and cyclodecasiloxane eicosamethyl exhibited the best docking energy with the QS receptors of *C. violaceum* and *V. harveyi* with that of the natural ligand (Musthafa et al. 2013). The pigment production by *C. violaceum* strain was inhibited by the ethanolic fractions of medicinal plants like *Adhatoda vasica* L. (leaves), *Bauhinia purpurea* L. (leaves), *Lantana camara* L. (leaves), *Myoporum laetum* G. Forst. (leaves), *P. longum* L. (fruits), and *Taraxacum officinale* F.H. Wigg. (aerial parts) (Zaki et al. 2013).

Among the chloroform-soluble compounds extracted from 46 plant materials, extract from *Scorzonera sandrasica* Hartvig & Strid. proved effective in inhibiting violacein production in *C. violaceum* as well as the QS-regulated carbapenem antibiotic production in *Erwinia carotovora* (Kalia 2013). Hamameli tannin extracted from the bark of *Hamamelis virginiana* L. (witch hazel) like RNAIII-inhibiting peptide (RIP) did not affect the growth of *Staphylococcus* spp., but it did inhibit the QS regulator RNAIII and prevented biofilm formation and cell attachment in vitro. Further evidence was provided by implantation of grafts into the animal, which drastically decreased the bacterial load in comparison to the controls (Kalia 2013).

Flavan-3-ol catechin, one of the flavonoids from the bark and leaves of *Combretum albiflorum* Tul., had a significant negative effect on pyocyanin and elastase productions and biofilm formation, as well as on the expression of the QS-regulated genes like *lasR* and *rhlA*. Various biosensor-based assays indicated that catechin might interfere with the perception of the QS signal molecule by the respective receptor; thereby leading to a reduction of the production of QS factors (Vandeputte et al. 2010).

Pyrogallol extracted from medicinal plants such as *Phyllanthus emblica* L. and its analogues exhibit antagonism against AI-2 (Kalia 2013).

Botanical remedies from southern Italy are used to evaluate for their QSI properties by testing the biofilm inhibition by methicillin-resistant *S. aureus*. Among them, *Lonicera alpigena* L., *Castanea sativa* Mill., *Juglans regia* L., *Ballota nigra* L., *Rosmarinus officinalis* L., *Leopoldia comosa* (L.) Parl., *Malva sylvestris* L., *Cyclamen hederifolium* Aiton, *Rosa canina* L., and *Rubus ulmifolius* L. showed significant antibiofilm activity (Koh et al. 2013). Tannin-rich fraction from *Terminalia catappa* L. was found to have anti-QS activity against violacein pigment production in *C. violaceum* (Taganna et al. 2011; Jakobsen et al. 2012). The tannin-rich fraction also inhibited QS-regulated characters like LasA and biofilm formation in *P. aeruginosa* (Taganna et al. 2011).

Methanolic extract of *Terminalia chebula* Retz. fruits showed anti-QS activity using *A. tumefaciens* and downregulated the expression of LasI/R and rhlI/R genes with concomitant decrease in AHLs in *P. aeruginosa* PA01 causing attenuation of its virulence factors and enhanced sensitivity of its biofilm toward tobramycin (Sarabhai et al. 2013). The ethyl acetate fractions of *Syzygium cumini* L. and *Pimenta dioica* L. displayed significant anti-QS activity by inhibiting the pigment production by *C. violaceum* (Vasavi et al. 2013). *Lagerstroemia speciosa* (L.) Pers. fruit extract modulates QS-controlled virulence factors like LasA protease, LasB elastase, and pyoverdin production and biofilm formation in *P. aeruginosa* (Singh et al. 2012).

QSI Activity by Ornamental Plants

Extracts from different plants like *Tecoma capensis* (Thunb.) Lindl. (leaves, flowers), *R. officinalis* L. (leaves), and *Sonchus oleraceus* L. (aerial parts) were found to possess anti-QS activities against *C. violaceum* (Kalia 2013). Aqueous extracts of six southern Florida plants *Conocarpus erectus* L., *Chamaesyce hypericifolia* (L.) Millsp., *Callistemon viminalis* (Gaertn.) G.Don, *Bucida buceras* L., *Tetrazygia bicolor* (Mill.) Cogn., and *Quercus virginiana* Mill. displayed their QSI effects on two biomonitor strains, *C. violaceum* and *A. tumefaciens*. Each plant presented a distinctive action profile upon Las and Rhl QS genes and their respective signaling molecules, suggesting that different mechanisms are responsible for the efficacy of *P. aeruginosa* virulence factors and the QS systems (Kalia 2013). Essential oil and hydrosol fraction of *Satureja thymbra* L. and Polytoxinol™, a compound based on essential oil, were shown to be effective against biofilms formed by *Salmonella* sp., *Listeria* sp., *Pseudomonas* sp., *Staphylococcus* spp., and *Lactobacillus* spp. (Kalia 2013).

Essential oils from Colombian plants were assayed for anti-QS activity in bacteria sensor strains. Two major chemotypes were found for *Lippia alba* (Mill.) N.E.Br. ex Britton & P.Wilson, the limonene–carvone and the citral (geranial–neral). For other species, the main components included α-pinene (*Ocotea* sp.), β-pinene (*Swinglea glutinosa* Tabog), cineol (*E. cardamomum* (L.) Maton), α-zingiberene (*Zingiber officinale* Roscoe), and pulegone (*Minthostachys mollis* (Kunth) Griseb.). Several essential oils, in *L. alba*, in particular, presented promising inhibitory properties for the short-chain AHL quorum sensing (QS) system, in *E. coli* containing the biosensor plasmid pJBA132. Moderate activity as anti-QS using the same plasmid was also found for selected constituents of essential oils studied here, such as citral, carvone, and α-pinene, although solely at the highest tested concentration (250 μg/mL). Only citral presented some activity for the long-chain AHL QS system, in *Pseudomonas putida* containing the plasmid pRK-C12. In short, essential oils from Colombian flora have promising properties as QS modulators (Colorado et al. 2012).

Among the eight fractions of *Acacia nilotica* (L.) P.J.H.Hurter & Mabb (green pod), fraction which is rich in gallic acid, ellagic acid, epicatechin, rutin, etc., demonstrated QSI activity in *C. violaceum* (Koh et al. 2013). The ethanolic extract from *Scutellaria baicalensis* Georgi was found to inhibit violacein production, a

QS-regulated behavior in *C. violaceum* and *Pectobacterium carotovorum* (Song et al. 2012). A methanol-soluble extract of the bark of *Myristica cinnamomea* King was found to exhibit anti-QS activity, and subsequent bioassay-guided isolation led to the identification of the active compound malabaricone C (1). Malabaricone C (1) inhibited violacein production by *C. violaceum* when grown in the presence of a cognate signaling molecule, N-3-oxohexanoyl-homoserine lactone (Chong et al. 2011). Six sesquiterpene lactones of the goyazensolide and isogoyazensolide type isolated from the Argentine herb *Centratherum punctatum* Cassini inhibited the virulence factors of *P. aeruginosa* (Amaya et al. 2012).

Table 1 represents the abridged details about the anti-QS research in plants and related products.

QSI Activity by Various Phytocompounds

Phenolic plant secondary metabolites such as salicylic acid stimulate AHL-lactonase enzyme expression (Kalia 2013). Salicylic acid inhibits biofilm formation, motility, and N-acyl homoserine lactone production by *P. carotovorum* and *Pseudomonas syringae* (Lagonenko et al. 2013). The QSI action of salicylic acid might be due to the reduction in the signal molecules level produced by *Pseudomonas* sp. (Bandara et al. 2006).

A new biofilm inhibitor ursolic acid was identified from 13,000 samples of compounds purified from whole plants and separated parts such as fruits, leafs, roots, and stems (Kalia 2013). Ursolic acid added at the rate of 10 μg/ml decreased biofilm formation by 79 % in *E. coli* and 57–95 % in *V. harveyi* and *P. aeruginosa* PA01 depending up on the medium. Incidentally, in these experiments, ursolic acid was found to have no effect on QS as observed with *V. harveyi* AI-1 and AI-2 reporter systems (Kalia 2013). GABA (γ-aminobutyric acid) produced by plant acts as promoter for the degradation of AHL signal OHC8HSL by lactonase (AttM) of *A. tumefaciens*, attenuating the QS-dependent infection process (Kalia 2013; Zhang et al. 2002). However, under abiotic stress such as drought and salinity, accumulation of proline antagonizes the plant's GABA-defense mechanism (Kalia 2013).

Flavonoids have been the focus of research for their roles as antioxidant, anti-inflammatory, and anticancer agents. Keeping in view of these health benefits, flavonoids such as naringenin, kaempferol, quercetin, and apigenin were evaluated for their QSI activities. All these flavonoids inhibited HAI-1- or AI-2-mediated bioluminescence in *V. harveyi* BB886. Quercetin and naringenin were found to inhibit biofilm formation by *V. harveyi* BB120 and *E. coli* O157:H7. In addition, naringenin altered the expression of genes for type three secretory system in *V. harveyi* (Vikram et al. 2011).

Conclusion

It is concluded that anti-QS is as important to fight against infections, as it does not pose selection pressure as it will unlikely cause resistance problems. On the other hand, plants also prove to produce anti-QS compounds, effective in protecting crops from pathogenic bacteria. Other plants, which can provide effective support to protect human beings from infectious diseases, are traditionally known for their medicinal properties and can act as QSI. But it is important to establish the "modus operandi" of the different plant-derived QS antagonists against the pathogens in order to establish whether the antagonists are narrow or broad spectrum. Most antagonists from plant origin are reported to have narrow spectrum of activity which may be useful in specifically targeting a single type of pathogen in a polymicrobial environment such as infection sites. But, on the other hand, such a narrow action antagonist may have limited clinical value, and care must be taken to avoid the toxicity of these molecules to patients. Keeping these attributes, the plant-derived anti-QS antagonists may serve as the next-generation "magic bullets" against infectious diseases.

Table 1 List of the various plant compounds with anti-quorum sensing characteristics

Plant name	Part/active principle	Assay organisms	Test characters	Reference
QSI activity by dietary plants				
Medicago truncatula Gaertn.	Seedlings	*Escherichia coli* JM109 [p(SB536)], *Chromobacterium violaceum* CV026, *Escherichia coli* JM109 [p(SB401)]		Kalia (2013)
Medicago truncatula Gaertn.	Seed exudate (L-canavanine)	*Sinorhizobium meliloti*	Exopolysaccharide (EPS) production	Kalia (2013)
Lotus corniculatus L.	Seedlings	*C. violaceum* CV026, *Agrobacterium tumefaciens* NTL4	Violacein pigment, β-galactosidase activity	Kalia (2013)
Pisum sativum L. (pea)	Seedlings	*C. violaceum, Pseudomonas aeruginosa PA01*	Violacein pigment, swarming motility	Fatima et al. (2010)
Pachyrhizus erosus (L.) Urb. (yam beans)	Seeds		AHL aging and degradation test	Kalia (2013)
Hordeum vulgare L. (barley)	Seeds		AHL aging and degradation test	Kalia (2013)
Moringa oleifera Lam.	Leaf and fruit (aqueous extract)	*C. violaceum* 12472	Violacein pigment	Koh et al. (2013)
Capparis spinosa L.	Dry fruit (methanolic extract)	*C. violaceum, E. coli, Proteus mirabilis, Serratia marcescens, and P. aeruginosa PA01*	Violacein pigment, biosurfactant production	Abraham et al. (2011)
Ocimum basilicum L.	Root exudate (rosamarinic acid)	*C. violaceum* CV026, *P. aeruginosa* PA01	Violacein pigment, biofilm formation, swarming motility	Kalia (2013)
Arabidopsis thaliana (L.) Heynh.	Seedling exudates		Enzymatic degradation of AHLs	Kalia (2013)
Camellia sinensis (L.) Kuntze (tea)	Catechins (epigallocatechin gallate)	*E. coli* C600 (donor), *E. coli* RC85 (recipient)	Conjugative R plasmid transfer assay	Kaila (2013)
Brassica oleracea L. (broccoli)	Sulforaphane and erucin (isothiocyanates)	*P. aeruginosa* PA01	Virulence assays	Ganin et al. (2013)
Oryza sativa L. (rice)	AHL-mimic QS molecules	*Sinorhizobium fredii* SMH12, *Pantoea ananatis* AMG501	Biofilm formation	Montaño et al. (2013)
Phaseolus vulgaris L. (bean)	AHL-mimic QS molecules	*Sinorhizobium fredii* SMH12, *Pantoea ananatis* AMG501	Biofilm formation	Montaño et al. (2013)
Allium cepa L. (onion)	Pantolactone and myristic acid	*P. aeruginosa*	Virulence factors	Abd-Alla and Bashandy (2012)
QSI activity by fruits				
Rubus idaeus L. (raspberry)		*C. violaceum* CV026	Violacein production	Kalia (2013)

Vitis sp. (grape)	Furocoumarins	*E. coli* O157:H7, *Salmonella typhimurium, C. violaceum*	Biofilm formation, violacein production	Kalia (2013)
Vaccinium angustifolium Aiton (blueberry)		*Vibrio harveyi*	Autoinducer bioassay	Kalia (2013)
Sour orange (seeds)	Isolimonic acid, ichangin, deacetyl nomilinic acid glucoside	*E. coli* O157:H7, *Salmonella typhimurium*	Bioluminescence	Vikram et al. (2010)
Ananas comosus (L.) Merr. (pineapple), *Manilkara zapota* (L.) P.Royen (sapodilla), *Musa paradisiaca* L.(banana)	Fruit aqueous extract	*C. violaceum* CV026, *P. aeruginosa*	Violacein, Virulence factors	Musthafa et al. (2010)
Kigelia africana (Lam.) Benth.	Fruit hexane extract	*C. violaceum and A. tumefaciens biosensor systems*		Chenia (2013)
QSI activity by plants used as spices				
Curcuma longa L. (turmeric)	Curcumin	*P. aeruginosa* PA01	Virulence factor expression	Rudrappa and Bais (2008), Packiavathy et al. (2014)
Cinnamomum verum J. Presl (cinnamon)	Cinnamaldehyde	*P. aeruginosa*, *Vibrio* spp.	Biofilm formation, AI-2 mediated QS system	Brackman et al. (2008), Niu and Gilbert (2004), Niu et al. (2006)
Laurus nobilis L.	Ethanolic extract (leaves, flowers, fruits, and bark)	*C. violaceum*	Violacein	Al-Hussaini and Mahasneh (2009)
Allium sativum L. (garlic)	Ajoene	*P. aeruginosa*	QS-controlled virulence factors	Jakobsen et al. (2012)
Vanilla planifolia Jacks. ex Andrews (vanilla)	Aqueous methanolic extract of beans	*C. violaceum*	Violacein	Kalia (2013)
Piper caucasanum Bredemeyer, *P. bogotense, P. brachypodon* (Benth.)	Essential oils	C. violaceum CV026	Violacein	Olivero et al. (2011)
Syzygium aromaticum (L.) Merrill & Perry (clove)	Eugenol	*C. violaceum* CV026	Violacein	Husain et al. (2013), Krishnan et al. (2012), Zhou et al. (2013)
QSI activity by medicinal plants				
Prunus armeniaca L., *Prunella vulgaris* L.		*C. violaceum*	Violacein	Koh and Tham (2011)

(continued)

Table 1 (continued)

QSI activity by dietary plants				
Plant name	Part/active principle	Assay organisms	Test characters	Reference
Nelumbo nucifera Gaertn. *Imperata cylindrica* (L.) P.Beauv. *Punica granatum* L. *Areca catechu* L.	Acetone/water extracts	*P. aeruginosa* PA01 and *C. violaceum*	QS-controlled virulence factors and violacein pigment	Koh and Tham (2011)
Panax notoginseng		*P. aeruginosa* PA01 and *C. violaceum*	QS-controlled virulence factors, violacein pigment	Song et al. (2010), Koh and Tham (2011)
Ocimum sanctum L.		*P. aeruginosa PA01 and C. violaceum*	QS-controlled virulence factors, violacein pigment	Musthafa et al. (2010)
Rhizophora annamalayana Kathiresan (bark)	Cyclononasiloxane octadecamethyl, cyclodecasiloxane eicosamethyl	*C. violaceum* and *V. harveyi*	QS-dependent violacein, bioluminescence production	Musthafa et al. (2013)
Adhatoda vasica L. (leaves) *Bauhinia purpurea* L. (leaves) *Myoporum laetum* G. Forst. (leaves) *P. longum* L. (fruits) *Taraxacum officinale* F.H. Wigg. (aerial parts)	Ethanolic fraction	*C. violaceum*	Violacein production	Zaki et al. (2013)
Scorzonera sandrasica Hartvig & Strid.	Chloroform extract	*C. violaceum* and *Erwinia carotovora*	Violacein and carbapenem antibiotic production	Kalia (2013)
Hamamelis virginiana L. (witch hazel)	Hamameli tannin	*Staphylococcus* spp.	Biofilm formation and cell attachment	Kalia (2013)
Combretum albiflorum Tul. (bark and leaf)	Flavan-3-ol catechin	*P. aeruginosa* PA01	QS-controlled virulence factors	Vandeputte et al. (2010)
Phyllanthus emblica L.	Pyrogallol		Antagonism against AI-2	Kalia 2013
Terminalia catappa L.	Tannin-rich fraction	*P. aeruginosa* PA01 *and C. violaceum*	Violacein and carbapenem antibiotic production	Taganna et al. (2011), Jakobsen et al. (2012)
Terminalia chebula Retz. (fruits)	Methanolic extract	*A. tumefaciens* and *P. aeruginosa* PA01	QS-dependent characters	Sarabhai et al. (2013)
Syzygium cumini (L.) Skeels. *Pimenta dioica* (L.) Merr.	Ethyl acetate fractions	*C. violaceum*	Violacein production	Vasavi et al. (2013)
Lagerstroemia speciosa (L.) Pers. (fruit)		*P. aeruginosa* PA01	QS-controlled virulence factors	Singh et al. (2012)

Lonicera alpigena L. *Castanea sativa* Mill. *Juglans regia* L. *Ballota nigra* L. *Rosmarinus officinalis* L. *Leopoldia comosa* (L.) Parl. *Malva sylvestris* L. *Cyclamen hederifolium* Aiton *Rosa canina* L. *Rubus ulmifolius* L.		Methicillin-resistant *S. aureus*	Biofilm formation	Quave et al. (2008)
QSI activity by ornamental plants				
Tecoma capensis (Thunb.) Lindl. (leaves, flowers) *Sonchus oleraceus* L. (aerial parts) *R. officinalis* L. (leaves)		*C. violaceum*	Violacein production	Kalia (2013)
Conocarpus erectus L. *Chamaesyce hypericifolia* (L.) Millsp. *Quercus virginiana* Mill. *Tetrazygia bicolor* (Mill.) Cogn. *Bucida buceras* L. *Callistemon viminalis* (Gaertn.) G. Don	Aqueous extracts	*C. violaceum* and *A. tumefaciens*		Kalia (2013)
Satureja thymbra L.	Essential oil and hydrosol fraction	*P. aeruginosa* PA01	Biofilm formation	Kalia (2013)
Lippia alba (Mill.) N.E.Br. ex Britton & P. Wilson	Limonene–carvone and the citral	Biosensor strain	Anti-QS assay	Colorado et al. (2012)
E. cardamomum (L.) Maton	Cineol	Biosensor strain	Anti-QS assay	Colorado et al. (2012)
Zingiber officinale Roscoe	α-Zingiberene	Biosensor strain	Anti-QS assay	Colorado et al. (2012)
Minthostachys mollis (Kunth) Griseb.	Pulegone	Biosensor strain	Anti-QS assay	Colorado et al. (2012)
Acacia nilotica (L.) P.J.H.Hurter & Mabb (green pod)	Gallic acid, ellagic acid, epicatechin, rutin	*C. violaceum*	Violacein production	Koh et al. (2013)
Scutellaria baicalensis Georgi		*C. violaceum*, *Pectobacterium carotovorum*	QS-regulated behaviors	Song et al. (2012)
Myristica cinnamomea King (bark)	Malabaricone C (1)	*C. violaceum*	Violacein production	Chong et al. (2011)
Centratherum punctatum Cassini	Goyazensolide and isogoyazensolide	*P. aeruginosa*	QS-regulated behaviors	Amaya et al. (2012)

References

Abd-Alla MH, Bashandy SR (2012) Production of quorum sensing inhibitors in growing onion bulbs infected with *Pseudomonas aeruginosa* E (HQ324110). ISRN Microbiol 1:1–7

Abraham SV, Palani A, Ramaswamy BR, Shunmugiah KP, Arumugam VR (2011) Antiquorum sensing and antibiofilm potential of *Capparis spinosa*. Arch Med Res 42:658–668

Al-Hussaini R, Mahasneh AM (2009) Microbial growth and quorum sensing antagonist activities of herbal plant extracts. Molecules 14:3425–3435

Amaya S, Pereira JA, Borkosky SA, Valdez JC, Bardón A, Arena ME (2012) Inhibition of quorum sensing in *Pseudomonas aeruginosa* by sesquiterpene lactones. Phytomedicine 19:1173–1177

Bandara MB, Zhu H, Sankaridurg PR, Wilcox MD (2006) Salicylic acid reduces the production of several potential virulence factors of *Pseudomonas aeruginosa* associated with microbial keratitis. Invest Ophthalmol Vis Sci 47:4453–4460

Brackman G, Defoirdt T, Miyamoto C, Bossier P, Van Calenbergh S, Nelis H et al (2008) Cinnamaldehyde and cinnamaldehyde derivatives reduce virulence in Vibrio spp. by decreasing the DNA-binding activity of the quorum sensing response regulator LuxR. BMC Microbiol 8:149

Chenia HY (2013) Anti-quorum sensing potential of crude *Kigelia africana* fruit extracts. Sensors (Basel) 13:2802–2817

Chong YM, Yin WF, Ho CY, Mustafa MR, Hadi AH, Awang K (2011) Malabaricone C from *Myristica cinnamomea* exhibits anti-quorum sensing activity. J Nat Prod 74:2261–2264

Colorado J, Olivero-Verbel J, Stashenko EE, Wagner-Döbler I, Kunze B (2012) Anti-quorum sensing activity of essential oils from Colombian plants. Nat Prod Res 26:1075–1086

Fatima Q, Zahin M, Khan MSA, Ahmad I (2010) Modulation of quorum sensing controlled behavior of bacteria by growing seedling, seed and seedling extracts of leguminous plants. Ind J Microbiol 50:238–242

Ganin H, Rayo J, Amara N, Levy N, Krief P, Michael MM (2013) Sulforaphane and erucin, natural isothiocyanates from broccoli, inhibit bacterial quorum sensing. Med Chem Commun 4:175–179

Husain FM, Ahemad I, Asif M, Tahseen Q (2013) Influence of clove oil on certain quorum sensing regulated functions and biofilm of *Pseudomonas aeruginosa* and *Aeromonas hydrophila*. J Biosci 38:835–844

Jakobsen TH, Genninp M, Phipps RK, Shanmughan MS, Christensen LD, Alhede M et al (2012) Ajoene, a sulfur rich molecule from garlic, inhibits gene controlled by quorum sensing. Antimicrob Agents Chemother 56:2314–2325

Kalia VC (2013) Quorum sensing inhibitors: an overview. Biotechnol Adv 31(2):224–245

Koh KH, Tham FY (2011) Screening of traditional Chinese medicinal plants for quorum sensing inhibitors activity. J Microbiol Immunol Infect 44:144–148

Koh CL, Sam CK, Yin WF, Tan LY, Krishnan T, Chong YM et al (2013) Plant-derived natural products as sources of anti-quorum sensing compounds. Sensors (Basel) 13:6217–6228

Krishnan T, Yin WF, Chan KG (2012) Inhibition of quorum sensing-controlled virulence factor production in *Pseudomonas aeruginosa* PAO1 by Ayurveda spice clove (*Syzygium aromaticum*) bud extract. Sensors (Basel) 12:4016–4030

Lagonenko L, Lagonenko A, Evtushenkov A (2013) Impact of salicyclic acid on biofilm formation by plant pathogenic bacteria. J Biol Earth Sci 3:B176–B181

Montaño F, Jiménez-Guerrero I, Contreras Sánchez-Matamoros R, López-Baena FJ, Ollero FJ, Rodríguez-Carvajal MA et al (2013) Rice and bean AHL-mimic quorum-sensing signals specifically interfere with the capacity to form biofilms by plant-associated bacteria. Res Microbiol 164:749–760

Musthafa KS, Ravi AV, Annapoorani A, Packiavathy ISV, Pandian SK (2010) Evaluation of anti-quorum-sensing activity of edible plants and fruits through inhibition of the N-acyl-homoserine lactone system in *Chromobacterium violaceum* and *Pseudomonas aeruginosa*. Chemotherapy 56:333–339

Musthafa KS, Sahu SK, Ravi AV, Kathiresan K (2013) Anti-quorum sensing potential of the mangrove *Rhizophora annamalayana*. World J Microbiol Biotechnol 29:1851–1858

Niu C, Gilbert ES (2004) Colorimetric method for identifying plant essential oil components that affect biofilm formation and structure. Appl Environ Microbiol 70:6951–6956

Niu C, Afre S, Gilbert ES (2006) Subinhibitory concentrations of cinnamaldehyde interfere with quorum sensing. Lett Appl Microbiol 43:489–494

Norizan SN, Yin WF, Chan KG (2013) Caffeine as a potential quorum sensing inhibitor. Sensors (Basel) 13:5117–5129

Olivero VJT, Pájaro CNP, Stashenko E (2011) Anti-quorum sensing activity of essential oils isolated from different species of the genus *Piper*. Vitae 18: 77–82

Packiavathy IA, Priya S, Pandian SK, Ravi AV (2014) Inhibition of biofilm development of uropathogens by curcumin – an anti-quorum sensing agent from *Curcuma longa*. Food Chem 148:453–460

Quave CL, Plano LRW, Pantuso T, Bennett BC (2008) Effects of extracts from Italian medicinal plants on planktonic growth, biofilm formation and adherence of methicillin-resistant Staphylococcus aureus. J Ethnopharmacol 118(3):418–428

Rudrappa T, Bais HP (2008) Curcumin, a known phenolic from Curcuma longa, attenuates the virulence of *Pseudomonas aeruginosa* PAO1 in whole plant and animal pathogenicity models. J Agric Food Chem 56: 1955–1962

Sarabhai S, Sharma P, Capalash N (2013) Elagic acid derivatives from *Terminalia chebula* Retz. downregulate the expression of quorum sensing genes to attenuate *Pseudomonas aeruginosa* PA01 virulence. PLoS One 8:e53441

Singh BN, Singh HB, Singh A, Singh BR, Mishra A, Nautiyal CS (2012) *Lagerstroemia speciosa* fruit extract modulates quorum sensing controlled virulence factor production and biofilm formation in *Pseudomonas aeruginosa*. Microbiology 158:529–538

Song Z, Kong KF, Wu H, Maricic N, Ramalingam B, Priestap H et al (2010) *Panax ginseng* has anti-infective activity against opportunistic pathogen *Pseudomonas aeruginosa* by inhibiting quorum sensing, a bacterial communication processes critical for establishing infection. Phytomedicine 17:1040–1046

Song C, Ma H, Zhao Q, Song S, Jia Z (2012) Inhibition of quorum sensing activity by ethanol extract of *Scutellaria baicalensis* Georgi. J Pathol Microbiol S7:001

Taganna JC, Quanico JP, Perono RM, Amor EC, Rivera WL (2011) Tannin-rich fraction from *Terminalia catappa* inhibits quorum sensing (QS) in *Chromobacterium violaceum* and the QS-controlled biofilm maturation and LasA staphylolytic activity in *Pseudomonas aeruginosa*. J Ethnopharmacol 134:865–871

Vandeputte OM, Kiendrebeogo M, Rajaonson S, Diallo B, Mol A, Jaziri ME et al (2010) Identification of catechin as one of the flavonoids from *Combretum albiflorum* bark extract that reduces the production of quorum-sensing-controlled virulence factors in *Pseudomonas aeruginosa* PA01. Appl Environ Microbiol 71: 243–253

Vasavi HS, Arun AB, Rekha PD (2013) Inhibition of quorum sensing in *Chromobacterium violaceum* by *Syzygium cumini* L. and *Pimenta dioica* L. Asian Pac J Trop Biomed 3:954–959

Vikram A, Jayaprakasha GK, Jesudhasan PR, Pillai SD, Patil BS (2010) Suppression of bacterial cell-cell signaling, biofilm formation and type III secretion system by citrus flavonoids. J Appl Microbiol 109:515–527

Vikram A, Jesudhasan PR, Jayaprakasha GK, Pillai SD, Patil BS (2011) Citrus limonoid interfere with *Vibrio harveyi* cell-cell signaling and biofilm formation by modulating the response regulator LuxO. Microbiology 157:99–110

Zaki AA, Shaaban MI, Hashish NE, Amer MA, Lahloub MF (2013) Assessment of anti-quorum sensing activity for some ornamental and medicinal plants native to Egypt. Sci Pharm 81:251–258

Zhang HB, Wang LH, Zhang LH (2002) Genetic control of quorum-sensing signal turnover in *Agrobacterium tumefaciens*. Proc Natl Acad Sci USA 99:4638–4643

Zhou L, Zheng H, Tang Y, Yu W, Gong Q (2013) Eugenol inhibits quorum sensing at sub-inhibitory concentrations. Biotechnol Lett 35:631–637

Part V

Synthetic Quorum Sensing Inhibitors

Synthetic Quorum Sensing Inhibitors: Signal Analogues

Dimpy Kalia

Introduction

Quorum sensing (QS) is a bacterial regulatory mechanism that orchestrates communication between bacteria in response to their local population density. This coordinated behavior allows bacteria to regulate the expression of various phenotypes such as virulence, defense mechanisms, antibiotic resistance, and biofilm formation (Schuster et al. 2013, Ng and Bassler 2009; Guo et al. 2013; Geske et al. 2008b; Galloway et al. 2011). QS is orchestrated by a variety of small organic molecules called autoinducers which are broadly divided into three classes, namely, autoinducer-1 (AI-1), autoinducer-2 (AI-2), and autoinducing peptides (AIP). Whereas AI-1 compounds, also known as AHLs (*N*-acylated L-homoserine lactones), are the most common class of QS molecules in Gram-negative bacteria (Geske et al. 2008b), AIPs are the major autoinducers in Gram-positive bacteria (Sturme et al. 2002; Novick and Muir 1999). AI-2 on the other hand is utilized by both Gram-positive and Gram-negative bacteria and is therefore referred to as a "universal" signaling molecule (Lowery et al. 2008a; Waters and Bassler 2005). In addition to these compounds which are associated with a variety of bacterial strains, other small molecules have also been reported to play important roles in QS in certain strains. Prominent examples include PQS or 2-heptyl-3-hydroxy-4-quinolone in *Pseudomonas* (Dubern and Diggle 2008) and γ-butyrolactone and diffusible signal factors (DSF) such as CAI-1 and cis-decenoic acid in *P. aeruginosa* (Fig. 1) (Deng et al. 2011; Jimenez et al. 2012).

D. Kalia (✉)
DST-INSPIRE Faculty, Department of Chemistry, University of Pune, Pune 411007, Maharashtra, India
e-mail: dkalia@chem.unipune.ac.in; dimpychem@gmail.com

Quorum Sensing Inhibition

Several clinically relevant pathogens regulate virulence by QS-mediated pathways. For example, *P. aeruginosa*, an opportunistic pathogenic bacterium, uses QS to regulate biofilm formation and antibiotic tolerance (Jimenez et al. 2012; Davies et al. 1998; Bjarnsholt et al. 2005; Hentzer et al. 2003; Rasmussen et al. 2005). Considering the important roles of QS in virulence, perturbing bacterial QS pathways, also referred to as quorum quenching (QQ), has been proposed as a viable strategy to address the menacing problem of development of bacterial resistance toward antibiotics (Hentzer et al. 2003; Wu et al. 2004; Sintim et al. 2010). This idea is based on the notion that the use of antibiotics triggers evolution of resistant strains because they target essential bacterial cellular processes. Since QS is not essential for bacterial survival and is mainly utilized for controlling virulence, disruption of QS pathways may impose comparatively less evolutionary pressure on bacteria to develop resistant strains, thereby resulting in an

V.C. Kalia (ed.), *Quorum Sensing vs Quorum Quenching: A Battle with No End in Sight*, DOI 10.1007/978-81-322-1982-8_23, © Springer India 2015

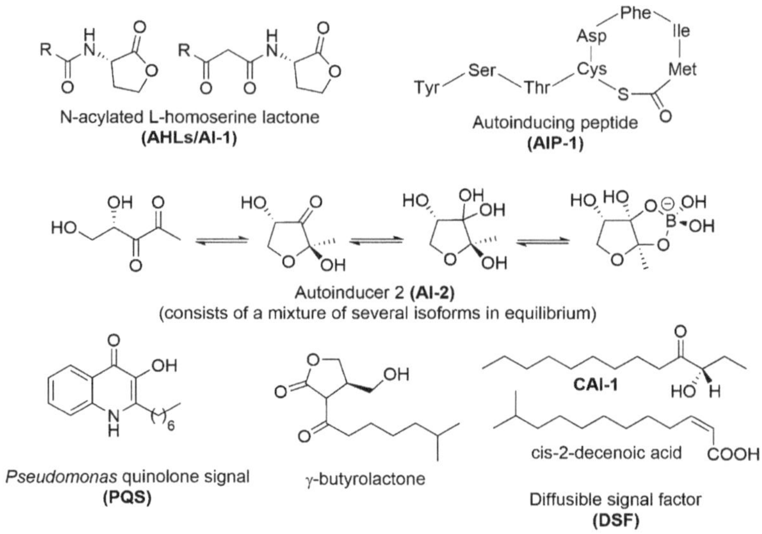

Fig. 1 Structures of selected autoinducers

efficacious antibacterial strategy (von Nussbaum et al. 2006; Clatworthy et al. 2007; Cegelski et al. 2008; Marra 2004; Lewis 2013). Two viable QQ approaches have been proposed—targeting the synthase enzymes of autoinducers and inhibiting the binding of autoinducers to their endogenous receptors. This section of the chapter discusses some of the highlights from the recent literature on the development of quorum quenchers based on both these approaches.

Targeting the Synthases of AI-1 (AHLs) and AI-2

One way to modulate QS pathways is to prevent the biosynthesis of autoinducers by inhibiting their synthase enzymes by using synthetic analogues of their biosynthetic intermediates. *S*-Adenosylmethionine (SAM) is a particularly attractive intermediate to focus on as it is a key precursor for the biosynthesis of both AI-1 (AHL) and AI-2 (Fig. 2).

AHL biosynthesis involves binding of SAM and acylated carrier protein (ACP) to AHL synthase (or LuxI-type protein) leading to acylation followed by lactonization to generate the AHL molecule (Fig. 2) (Rasmussen and Givskov 2006; Hentzer and Givskov 2003; Parsek et al. 1999; Watson et al. 2002). Various analogues of SAM have been reported that act as quorum sensing inhibitors (QSIs), presumably by interrupting this biosynthetic pathway of AHL molecules. Among them, *S*-adenosyl-homocysteine (SAH, **5**), sinefungin (**6**), and butyryl SAM are the most potent in vitro inhibitors reported for the *P. aeruginosa* AHL synthase, RhlI (Fig. 3) (Rasmussen and Givskov 2006; Hentzer and Givskov 2003).

Biosynthesis of AI-2 (Fig. 2), on the other hand, involves the enzymes 5′-methylthioadenosine nucleosidase (MTAN) and LuxS (Pereira et al. 2013; Parveen and Cornell 2011). The first step involves a methyl transferase-dependent demethylation reaction of SAM to give *S*-adenosylhomocysteine (SAH) which undergoes an MTAN (also known as Pfs nucleosidase)-catalyzed deadenylation reaction to give *S*-ribosylhomocysteine (SRH). Subsequently, the Zn metal-utilizing LuxS enzyme catalyzes the hydrolysis of SRH into 4,5-dihydroxy-2,3-pentanedione (DPD) (Zhu et al. 2013; Gopishetty et al. 2009; Alfaro et al. 2004; Shen et al. 2006). During the catalysis, the metal ion acts as a Lewis acid and catalyzes two consecutive aldose-ketose isomerization reactions to yield a 2-ketone intermediate (**2**) and ultimately a 3-ketone intermediate (**3**). A base in LuxS then abstracts a proton at the C4 position of **3** leading to the elimination of the enol intermediate (**4**) and homocysteinyl thiol (Hcy). The resultant

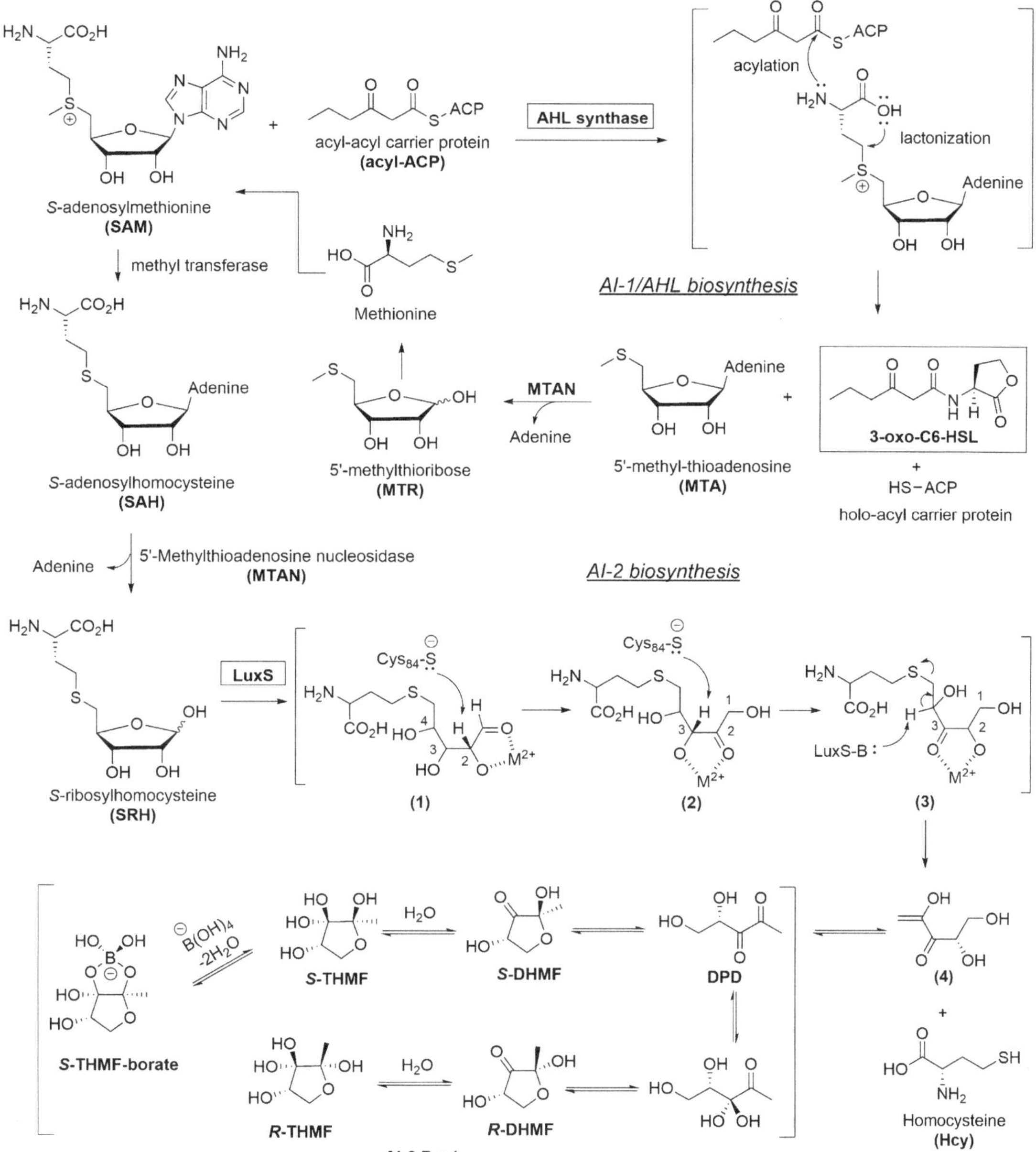

Fig. 2 Biosynthesis of AI-1 (AHL) and AI-2

Fig. 3 Selected SAM analogues as inhibitors of AHL synthase (Rh1I) in *P. aeruginosa*

S-adenosylhomocysteine (**SAH**)
(5)

sinefungin
(6)

Fig. 4 SRH analogues as LuxS inhibitors (Alfaro et al. 2004)

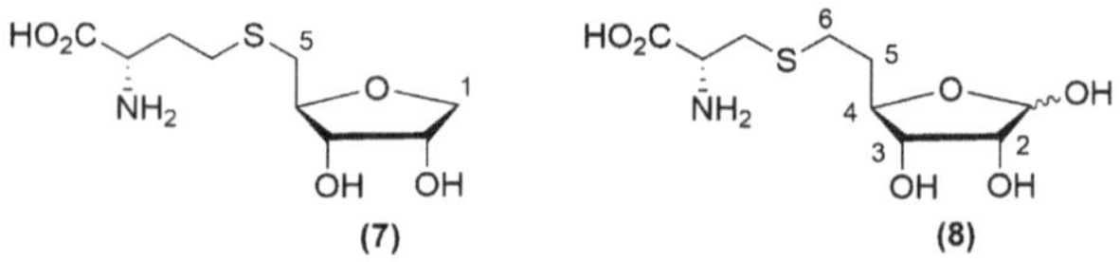

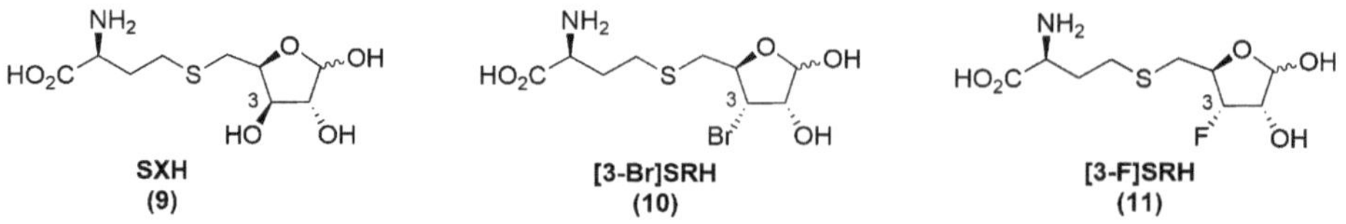

Fig. 5 Structures of LuxS inhibitors based on modifications at the ribose C-3 position (Wnuk et al. 2009)

Fig. 6 Hydroxamic acid analogues as LuxS inhibitors (Shen et al. 2006)

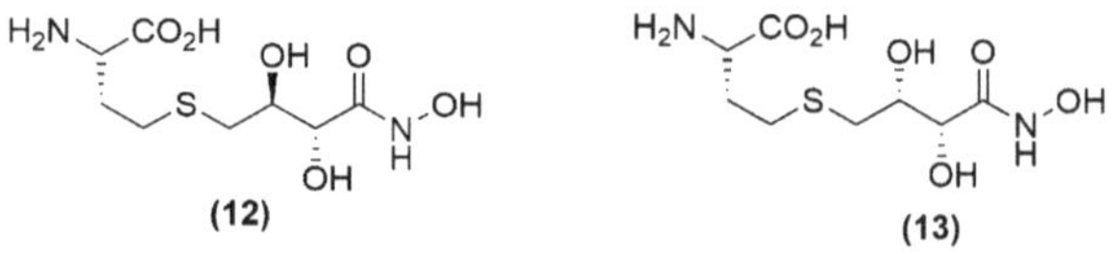

enol (**4**) spontaneously isomerizes to a highly reactive 1,2-diketone compound (DPD) which, in presence of water and borates, exists as a complex mixture of multiple species in a dynamic equilibrium, together comprising the AI-2 pool (Fig. 2) (Globisch et al. 2012; Worthington and Melander 2012).

Zhou and coworkers have reported the synthesis of various SRH analogues and evaluated them for their activity as AI-2 synthase (LuxS protein) inhibitors. Two compounds, **7** and **8,** were identified as potent LuxS inhibitors (Fig. 4) (Alfaro et al. 2004). In compound **7**, an ether group replaces the hemiacetal of SRH, presumably preventing the initial aldose-ketose isomerization reaction occurring during the LuxS catalyzed hydrolysis of SRH as discussed above (Fig. 2). In compound **8**, a C5-C6 carbon-carbon bond replaces the carbon-sulfur bond of SRH, whereas the rest of the molecule is identical to SRH, thereby precluding the cleavage of carbon-sulfur bond and hence inhibiting DPD synthesis.

Another interesting class of SRH analogues as AI-2 biosynthesis inhibitors was reported by Wnuk and coworkers who synthesized various SRH analogues modified at the ribose C3 position and evaluated them for their activity against the *Bacillus subtilis* LuxS (Wnuk et al. 2009). The goal was to prevent the enolization step that produces the 3-keto intermediate (**3**) (Fig. 2). The compounds with inverted stereochemistry **9** (SXH) or the compounds substituted with either bromine (e.g., compound **10** ([3-Br]SRH)) or fluorine (e.g., compound **11** ([3-F]SRH)) showed potent inhibition (Fig. 5).

Shen and coworkers have reported the synthesis of various hydroxamic acid derivatives as 2-keto intermediate (**4**) mimics involved in the hydrolysis of SRH. Two members of this library of compounds, **12** and **13,** were identified as potent inhibitors against LuxS, yielding K_I values of 0.72 μM and 0.37 μM, respectively (Fig. 6) (Shen et al. 2006).

MTAN is a crucial enzyme for both AI-1 and AI-2 biosyntheses (Fig. 2) and hence is an attractive target for developing quorum sensing inhibitors. Schramm and coworkers have reported transition state analogues that are femtomolar inhibitors of MTAN (Gutierrez et al. 2009; Thomas et al. 2012). These compounds are one of the most potent non-covalent enzyme inhibitors known and they also demonstrate potent quorum sensing inhibition. For example, late transition state analogues (R_2 = Me, Et, or *n*-Bu in Fig. 7) demonstrate nanomolar IC_{50} values for the inhibition of biosynthesis of both AI-1 and AI-2 (Gutierrez et al. 2009). Recently, early transition state analogues (R_1 = Me or 4-Cl-Ph in Fig. 7) have been shown to inhibit MTAN from *S. pneumoniae*, *E. coli*, and *V. cholerae* with picomolar K_I values (Thomas et al. 2012).

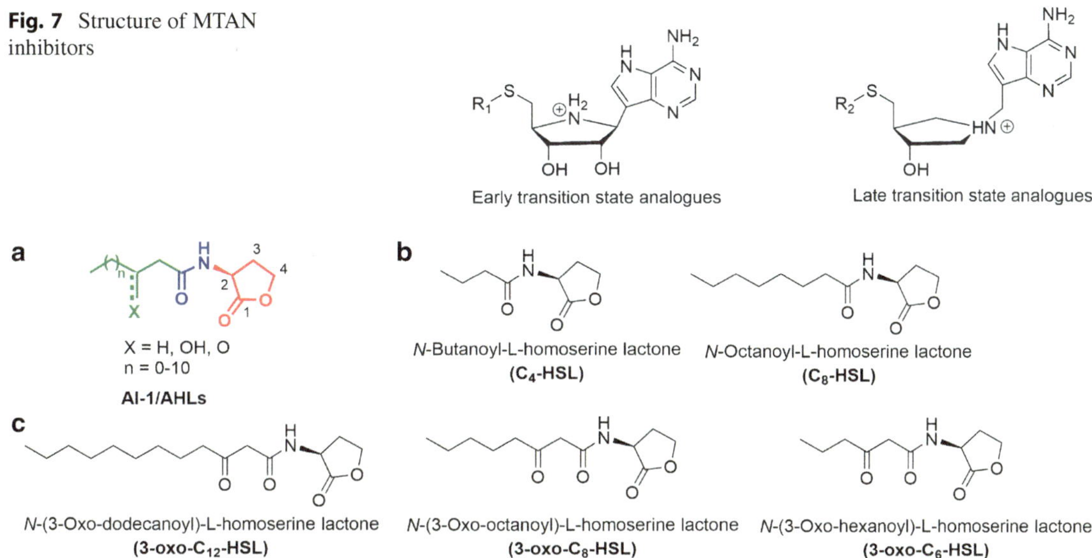

Fig. 7 Structure of MTAN inhibitors

Fig. 8 Structures of AI-1/AHLs (**a**) General structural features of AHL. (**b**) Selected C_n-HSLs discussed in the section. (**c**) Selected 3-oxo-C_n-HSLs discussed in this section

Targeting AHL Receptors with AHL Mimics

Another widely applied approach for modulating QS is to target the receptor proteins that bind to autoinducers by designing analogues of the autoinducers. Most of the natural AHLs reported share three conserved structural characteristics (Fig. 8a): (1) a homoserine lactone ring which is unsubstituted at the 3- and 4-positions (shown in red), (2) a central amide linkage (shown in blue), and (3) an acyl chain derived from fatty acid biosynthesis (shown in green). The AHLs discussed in this section are depicted in Fig. 8.

Acyl Chain Modifications

The first report of synthetic AHL analogues was published in 1986 by Eberhard and coworkers (Eberhard et al. 1986). This study focused on screening a set of ~20 synthetic AHL analogues for their agonist and antagonist activities against 3-oxo-C_6-HSL (the native autoinducer) in *Vibrio fischeri* for binding to the AHL receptor, the LuxR protein. This library was generated mainly by varying the acyl chain length and either incorporating or removing the 3-oxo group in the chain (Fig. 9). Length of the acyl chain was found to be a critical determinant of LuxR-modulation activity of these compounds. For agonistic activity, the optimal acyl chain length was six carbon atoms (the 3-oxo-C_8-HSL analogue). In contrast, C_9-HSL, an analogue without the oxo substituent, was a potent antagonist. Additionally, the 3-oxo group was found to be important for the agonistic activity. Indeed, moving the oxo group from C3 to C2 (e.g., compound **14**) completely abolished the agonistic activity. One limitation of this study is that all the compounds were tested in a bacterial reporter strain that expresses the AHL synthase (LuxI), and so the presence of native 3-oxo-C_6-HSL in these assays could influence the data interpretation.

To overcome this limitation of Eberhard's study (Eberhard et al. 1986), Greenberg and coworkers screened a set of AHL analogues very similar to those reported by Eberhard and coworkers against cognate 3-oxo-C_6-HSL for LuxR activation in an *E. coli* strain in which the *V. fischeri* LuxI synthase was inactivated (Fig. 10) (Schaefer et al. 1996). Similar trends in ligand activity were observed in this study. The 3-oxo-C_5-HSL analogue demonstrated potent activation of LuxR. Analogues with acyl chain

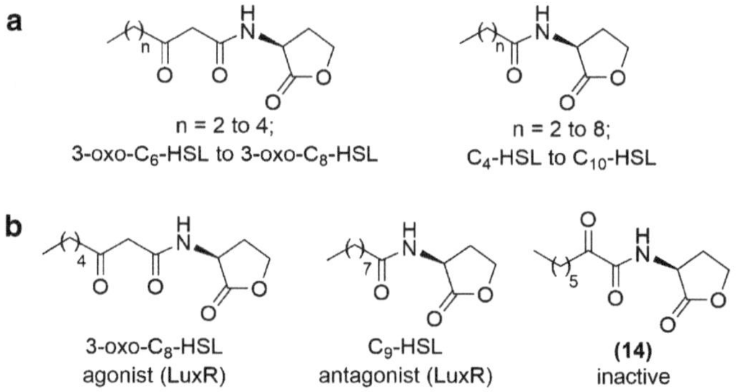

Fig. 9 AHL analogues screened against LuxR in *Vibrio fischeri* by Eberhard and coworkers (**a**) General structural features of AHLs tested. (**b**) Identified agonist, antagonist, and inactive analogues

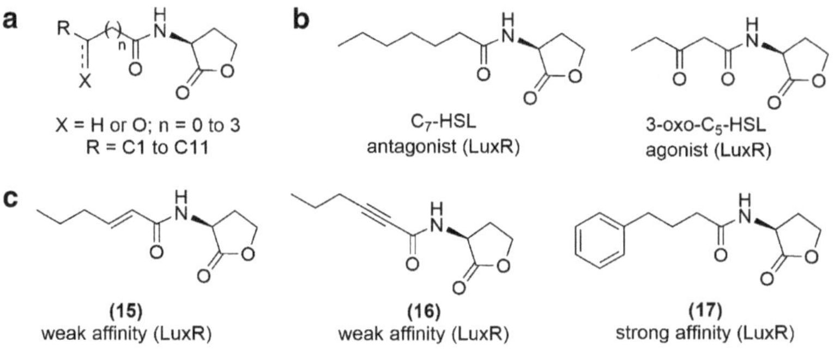

Fig. 10 AHL analogues evaluated against LuxR by Greenberg and coworkers (Schaefer et al. 1996). (**a**) General structural features of AHL analogues tested. (**b**) Identified antagonists and agonists. (**c**) Incorporation of unsaturated and aromatic groups in the acyl side chain

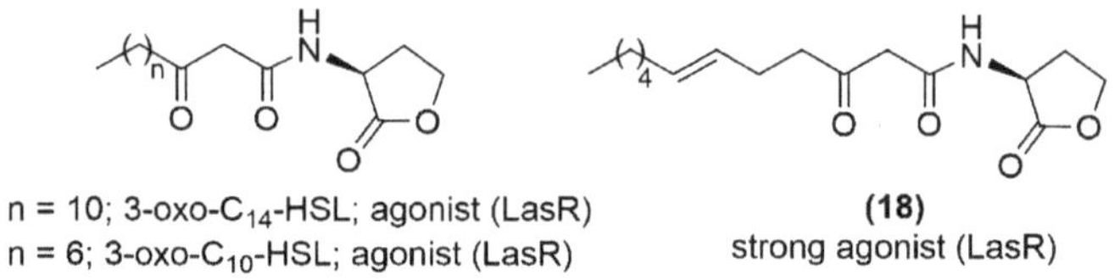

Fig. 11 Identified agonist against LasR by Passador and coworkers (Passador et al. 1996)

length greater than five carbons demonstrated potent inhibitory activity against LuxR. Among them, C_7-HSL was found to be the most efficient antagonist. A comparison of 3-oxo and corresponding 3-methylene-substituted analogues showed that 3-oxo derivatives were far superior agonists of LuxR. Furthermore, introduction of alkene or alkyne groups into the acyl chain next to the amide to yield compounds **15** and **16,** respectively, decreased the activity, suggesting that the flexibility of acyl chain is important for LuxR binding. Interestingly, the 3-oxo-C_4 derivative with a terminal phenyl group on the acyl chain (e.g., compound **17**) showed a strong affinity for LuxR, suggesting the tolerance of acyl chains for bulky substituents, which was further supported by several other studies (Reverchon et al. 2002; Geske et al. 2007; Geske et al. 2008a).

In another study, Passador and coworkers showed that the length of the acyl chain is the most critical parameter for the binding of structural analogues of the *P. aeruginosa* autoinducer, 3-oxo-C_{12}-HSL, to the AHL receptor, LasR (Passador et al. 1996). The results of this study revealed that both increasing (e.g., 3-oxo-C_{14}-HSL) or decreasing (e.g., 3-oxo-C_{10}-HSL) the acyl chain length by two carbon atoms as compared to the native AHL (3-oxo-C_{12}-HSL), while retaining the 3-oxo group, were well tolerated and gave the most potent agonists (Fig. 11). Interestingly, introduction of double bond at the C6-C7 position of the native AHL (e.g., compound **18**) resulted in an extremely potent agonist (EC_{50} 10 nM) and was comparable to the native 3-oxo-C_{12}-HSL (EC_{50} 7 nM). Increasing the chain length above 15 carbon atoms or reducing it below 6 carbon atoms resulted in a 300-fold decrease in activity. To test the antagonistic activity, all the analogues were analyzed for their ability to displace the

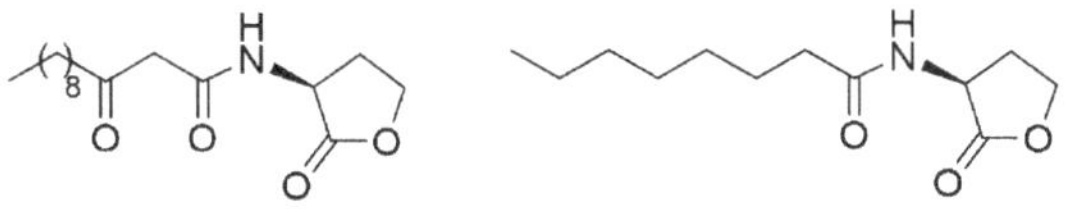

3-oxo-C_{12}-HSL
agonist (TraR)

C_8-HSL
antagonist (TraR)

Fig. 12 Identified agonist and antagonist against TraR by Zhu and coworkers (Zhu et al. 1998)

tritium-labeled bound native ligand from the LasR. The best agonist of the series, compound **18**, was a good competitor and excluded 80% of bound native AHL.

The significance of acyl chain length and the 3-oxo group was further seen in the study carried out by Zhu and coworkers (Zhu et al. 1998). In this work, the authors compared the ability of 32 acyl chain-modified AHL analogues to that of the native ligand (3-oxo-C_8-HSL) for activating TraR, an autoinducer-activated transcriptional factor in *Agrobacterium tumefaciens.* In general, AHL analogues with acyl chain length of 7 carbons or more containing the 3-oxo group were the most potent agonists. The most potent activator was 3-oxo-C_{12}-HSL (Fig. 12). Out of the 32 analogues tested, 28 did not show agonist activity but several were efficient antagonist. Interestingly, removing the 3-oxo group from the native AHL resulted in the most potent antagonist (C_8-HSL).

In 1999, Kline and coworkers explored the agonistic and antagonistic activities of constrained analogues of native AHL, 3-oxo-C_{12}-HSL, against LasR in the *P. aeruginosa* strain PAO-JP2 (Kline et al. 1999). The objective of this study was to probe the two possible tautomers (*E* and *Z*) of the β-ketoamide of native AHL (Fig. 13a). Both salicylamide (**19**) (*Z*-enol tautomer mimic) and the furan compound **20** (*E*-enol tautomer mimic) turned out to be inactive. On the other hand, *gem*-difluorinated analogues (e.g., compound **21**, Fig. 13b), in which keto-enol tautomerism is not possible, required a 10-fold higher concentration as compared to native AHL to display the same potency.

These studies revealed that an extended geometry of the acyl chain of 3-oxo-C_{12}-HSL analogues is essential to retain the agonistic activity.

Reverchon and coworkers synthesized C4-substituted analogues of 3-oxo-C_6-HSL (native AHL in *V. fischeri*) with cycloalkyl, branched, and aryl substituents (Fig. 14) (Reverchon et al. 2002). These analogues were screened for their agonistic activity against LuxR in recombinant *E. coli* pSB401, a bioluminescent reporter strain containing LuxR and the *lux box* from *V. fischeri* and the luminescence gene cluster from *Photorhabdus luminescens*. Apart from compounds having bulky substituents at the C4 position (e.g., **22**; R = naphthyl, *t*-butyl, or adamantyl), all other 4-alkyl-substituted analogues showed agonistic activity against LuxR. Cyclopentyl-substituted AHL (**23**) was found to be slightly more active than natural 3-oxo-C_6-HSL. Further, analogues substituted with phenyl group at the C4 position (**24**) displayed no agonistic activity and, instead, displayed potent inhibition of LuxR which was influenced by C4 substitution on the phenyl ring. For example, the IC_{50} value of the unsubstituted parent compound (**24**) was found to be 2 μM and suffered a 5-fold hit in IC_{50} when substituted with the 4-trifluoromethyl group (e.g., compound **25**, IC_{50} 10 μM). The authors suggested that this observed antagonist activity of aryl-substituted AHL analogues was due to specific interactions with the aromatic amino acids residue of the ligand receptor pocket. These interactions presumably prevent LuxR from adopting its active dimeric form. This hypothesis was further supported by an X-ray crystal structure of the TraR-3-oxo-C_8-HSL complex (Zhang et al. 2002).

Significant contributions have been made by Blackwell and coworkers on the design and the synthesis of acyl-chain-modified analogues of AHLs. In their initial efforts, the group developed a robust synthetic method for these compounds by utilizing solid phase and microwave-mediated synthetic technologies, resulting in the preparation of 11 native AHLs in addition to a small test library of nonnatural AHL analogues (Geske et al. 2005). Among the library, three AHL analogues (**26**, **27**, and **28**, Fig. 15a) showed significant antagonistic activity against receptor proteins, LasR (PAO-JP2 strain of *P. aeruginosa*) and TraR (WCF47 strain of

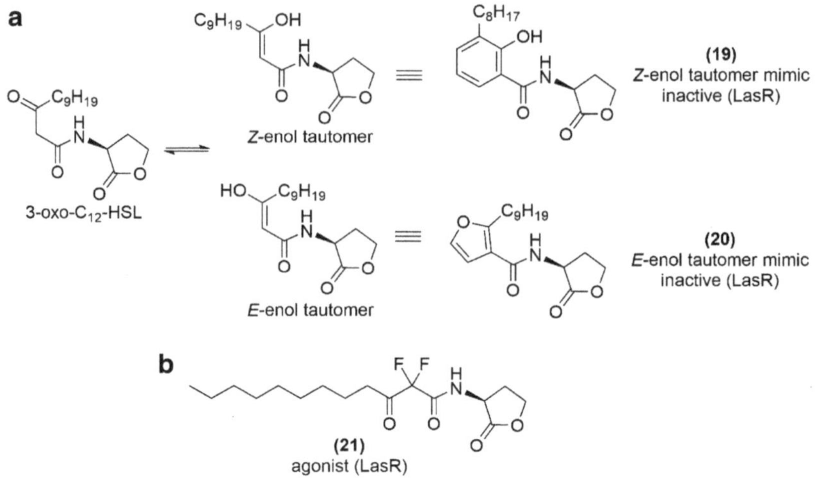

Fig. 13 Enol mimics and nonenolizable AHL analogues tested against LasR by Kline and coworkers (Kline et al. 1999). (**a**) *E*- and *Z*-enol tautomer mimics of native 3-oxo-C_{12}-HSL. (**b**) Nonenolizable *gem*-difluoro AHL analogue agonist

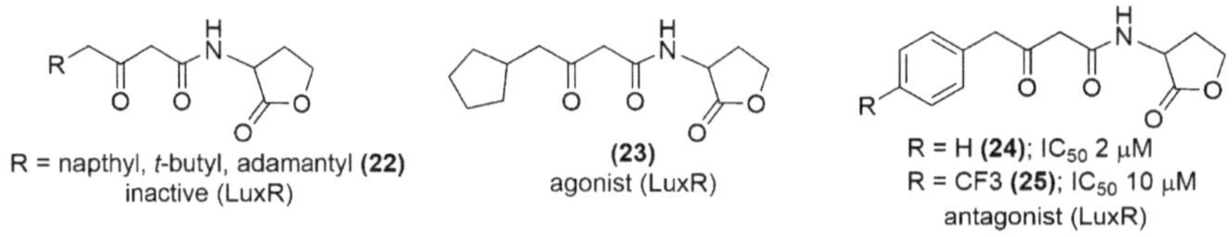

Fig. 14 Identified C4-substituted AHL analogues as inactive, agonist, and antagonist against LuxR by Ravenchon and coworkers (Reverchon et al. 2002)

A. tumefaciens). Encouraged by these results, the authors synthesized a larger library consisting of ~90 nonnative AHL analogues and evaluated their activity against QS regulator proteins, TraR (*A. tumefaciens*), LasR (*P. aeruginosa*), and LuxR (*V. fischeri*) (Geske et al. 2007). The main objective of this comprehensive study was to probe the key structural features required for small-molecule inhibition or activation of QS-modulating receptors in Gram-negative bacteria. A major highlight of this study was that a sub-library of 25 *N*-phenylacetyl-L-homoserine lactones (PHL) (particularly the 3- and 4-phenyl-substituted analogues) displayed a wide range of antagonistic and agonistic activities (Fig. 15b). The activity was found to greatly depend upon the nature of the substituents on the aromatic system. The most potent LuxR antagonists were obtained when the phenyl ring was substituted with the electron-withdrawing groups at the para position. For example, 4-I PHL (**29**) was the most potent antagonist against all the three receptors. Interestingly, while 3-bromo and 3-chloro PHLs displayed antagonistic activity against LasR, 3-NO_2-PHL (**30**) and 3-CN-PHL (**31**) proved to be potent agonists against LuxR. In fact, compound **30** was identified as the first nonnative superagonist (EC_{50} 0.35 μM), activating LuxR 10-fold more effectively than the native 3-oxo-C_6-HSL (EC_{50} 3 μM) in *V. fischeri*. Screening same libraries of synthetic nonnative AHL against the third LuxR-type protein in *P. aeruginosa*, QscR identified several potent agonists (e.g., compounds **32, 33**) and antagonists (e.g., compounds **34, 35**) (Fig. 15c) (Mattmann et al. 2008). Based on the lead compounds obtained during these studies, the Blackwell group then synthesized a more focused library of ~40 compounds of three basic structures, PHL, phenoxyacetyl-homoserine lactones (POHL) (Fig. 15d), and phenylpropionyl-homoserine lactones (PPHL) (Fig. 15e), with various substitutions on the aromatic ring (Geske et al. 2008a). 4-NO_2-POHL (**36**) and 4-CF_3-POHL (**37**) proved to be the potent antagonist against TraR. Against LasR, 3-I-PPHL (**38**) and 4-Cl-PPHL (**39**) were

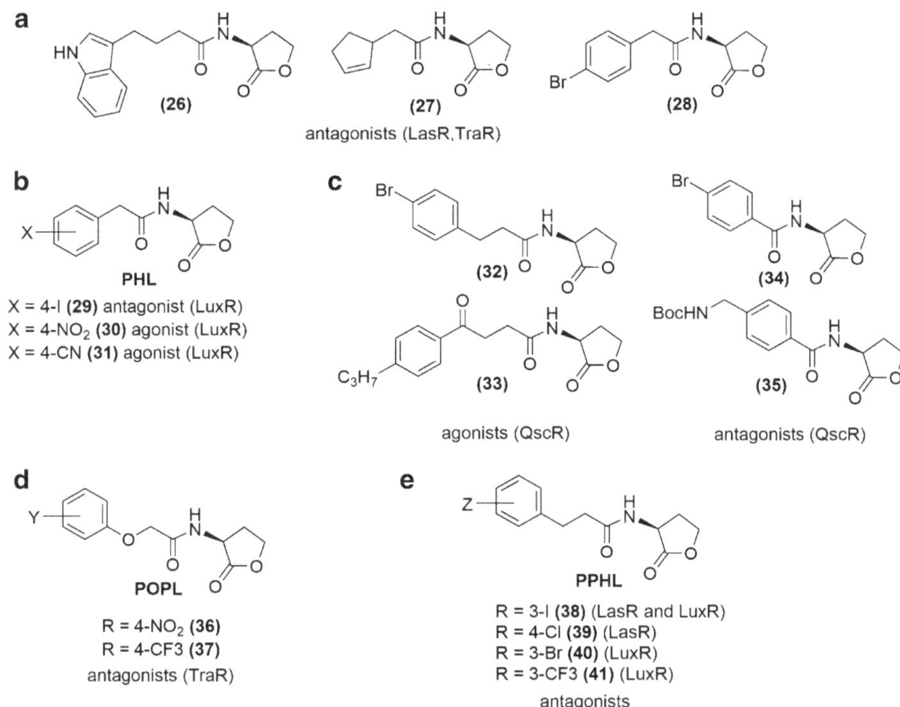

Fig. 15 AHL analogues screened by Blackwell and coworkers against various receptor proteins. (**a**) Identified antagonists against LasR and TraR. (**b**) Identified PHLs agonists and antagonists against LuxR. (**c**) Identified agonists and antagonists against QscR. (**d**) Identified POPLs as antagonists against TraR. (**e**) Identified PPHLs antagonists against LasR and LuxR

the potent inhibitors, whereas **38**, 3-Br-PPHL (**40**) and 3-CF3-PPHL (**41**) displayed potent antagonistic activity against LuxR.

Amide Function Modifications

One way to enhance the desired biological properties of a compound without making significant structural modification is to use bioisosteres (groups or substituents having similar physical or chemical properties) (Patani and LaVoie 1996). This approach has been employed to nonnative AHLs by replacing the amide function by its bioisosteres such as sulfonamides, ureas, and sulfonylureas.

Sulfonamides

In a 2003 patent (Givskov 2003), Giskov and coworkers have reported a variety of nonnative AHLs in which the amide bond has been substituted with sulfonamides. A year later, Douthean and coworkers reported their work on screening a series of racemic *N*-alkyl and phenylalkyl-sulfonyl HSL analogues for their agonistic or antagonistic activities against LuxR in *V. fischeri* (Fig. 16) (Castang et al. 2004). Interestingly, by replacing amide group with the sulfonyl group, none of the compounds showed agonistic activity, but instead were potent antagonists against LuxR. For *N*-alkyl-sulfonamide, the acyl chain length was found to be the critical parameter for the antagonistic activity. For example, the maximum inhibition was displayed by compound (**42**), containing five carbons in the acyl chain. Deviating (either increasing or decreasing) from a five carbon chain length resulted in a decrease in the activity. The antagonistic activity of phenylalkyl-sulfonamides varied with the length of the carbon spacer between the terminal phenyl group and the sulfonamide linkage. Compounds with one carbon spacer showed weak activity (e.g., compound **43**), whereas increasing the carbon spacer to two yielded the most potent antagonistic activity

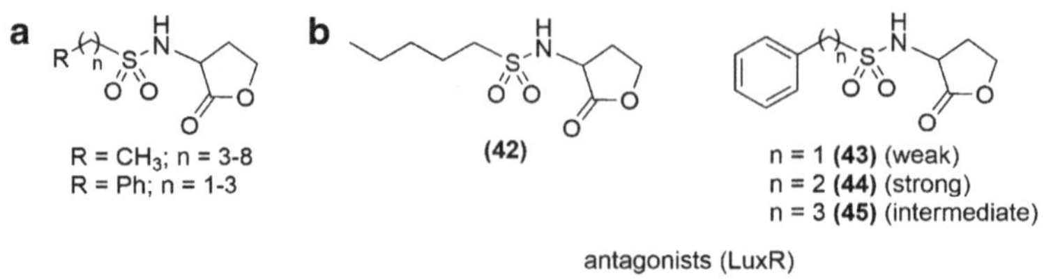

Fig. 16 AHL analogues tested against LuxR by Douthean and coworkers (Castang et al. 2004). (**a**) General structural features of AHL analogues tested. (**b**) Antagonists identified in the study

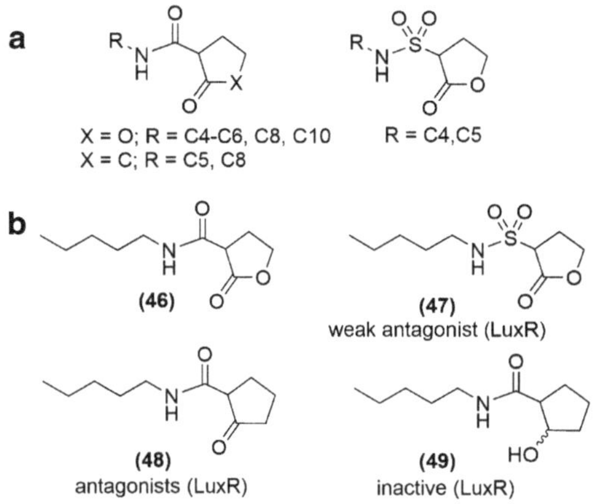

Fig. 17 Reverse amides and reverse sulfonamides derivatives of AHL tested against LuxR (Boukraa et al. 2011). (**a**) General structural features of AHL analogues tested. (**b**) Most potent antagonists and an inactive AHL analogue from the study

(compound **44**). Compound **45** which contains a three-carbon spacer yielded intermediate activity. Docking studies, using computational homology modeling of TraR X-ray structure, revealed the molecular basis of sulfonyl AHLs to bind to the 3-oxo-C_6-HSL binding domain of LuxR via hydrogen bonding interactions with the two amino acid residues, Y62 and S137 (Zhang et al. 2002).

Reverse Amides and Reverse Sulfonamides

Douthean and coworkers also synthesized AHL analogues in which the amide linkage was replaced with reverse amides and reverse sulfonamides and the lactone ring was replaced with the cyclopentanone and cyclopentanol (Fig. 17) (Boukraa et al. 2011). These compounds were evaluated for their agonistic and antagonistic activities with respect to 3-oxo-C_6-HSL in *V. fischeri* for binding to LuxR protein, using bioluminescence as the read out. Most of the compounds showed antagonistic activity against LuxR and none displayed agonistic activity up to a maximum concentration of 200 μM. Acyl chain length was found to be critical for the inhibition, and *N*-pentyl reverse amide **46** (IC_{50} 34 μM) was identified as the most potent inhibitor of the series, whereas the *N*-butyl derivative resulted in a ~3-fold reduction in activity, and compounds with chain lengths of 6, 8, or 10 carbon atoms were moderately less efficient antagonists. Replacing the reverse amide linkage in compound **46** with a reverse sulfonamide resulted in a 2-fold decrease in the inhibitory activity (e.g., compound **47**, IC_{50} 70 μM). Substituting lactone ring of the reverse amide **46** with the cyclopentanone (e.g., compound **48**) was well tolerated and showed the same range of activity as that of the lactone derivative. Replacement of the ketone functional group in **48** by an alcohol to yield cyclopentanol **49** resulted in complete loss of activity.

Ureas

Substituting the amide group by urea (Fig. 18) resulted in compounds that displayed only antagonistic activity and no agonistic activity for LuxR (Frezza et al. 2006). Among the phenyl-substituted ureas tested, the strongest inhibition was observed for compound **50**, in which the phenyl group and the urea group are separated by a three-carbon spacer. The antagonistic activity of *N*-alkyl ureas increases with the increasing chain length with pentyl urea (**51**) displaying the most potent inhibition. Furthermore, *N*-methylation of urea nitrogen (e.g., compound **52**) resulted in drastic decrease in the activity. These studies therefore revealed that replacement of the amide

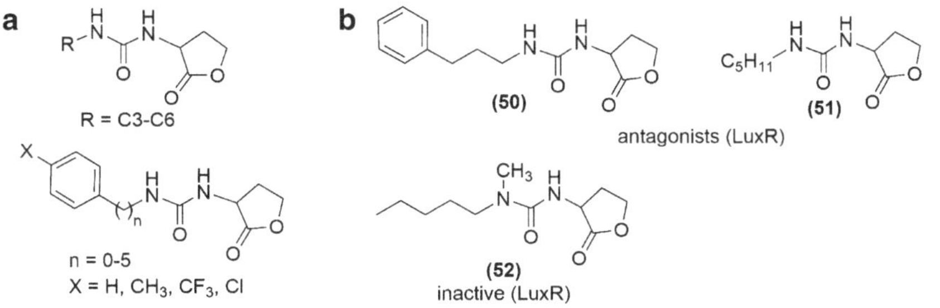

Fig. 18 Urea derivatives of AHL tested against LuxR (Frezza et al. 2006). (**a**) General structural features of AHL analogues tested. (**b**) Most potent antagonists and an inactive AHL analogue from the study

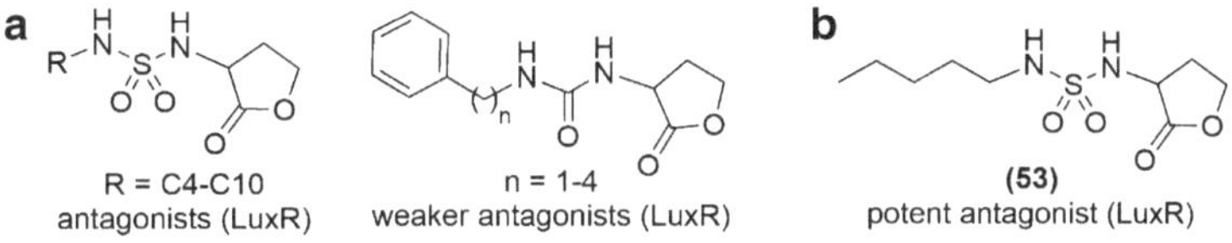

Fig. 19 Sulfonylurea derivatives of AHL tested against LuxR (Frezza et al. 2008). (**a**) Identified antagonists. (**b**) The most potent antagonist discovered in the study

Fig. 20 Thiolactones and lactams derivatives of AHL identified as agonist, antagonist, and inactive against LuxR and LasR

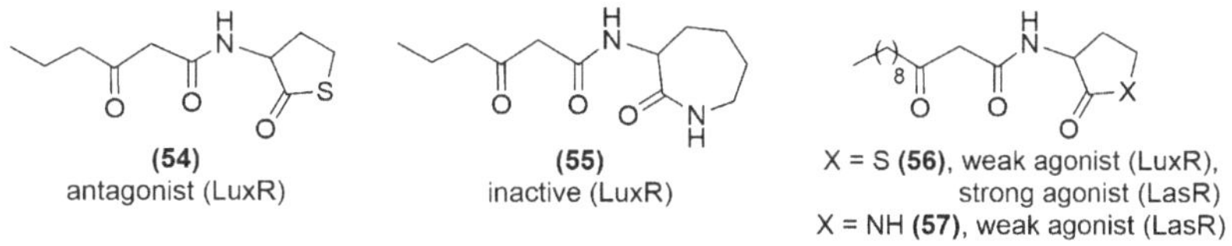

group of AHL with either sulfonyl amide or urea proved to be a viable strategy for designing antagonists against LuxR in *V. fischeri.*

Sulfonylureas

Douthean and coworkers widened the scope of designing inhibitors of LuxR by synthesizing nine *N*-alkyl- and phenylalkyl-sulfonylurea HSL analogues (Fig. 19) (Frezza et al. 2008). All these compounds showed antagonistic activity against LuxR, and the *N*-pentyl-sulfonylurea (compound **53**) proved to be the most potent inhibitor in the series (IC_{50} 2 μM).

Lactone Ring Modifications

To understand the importance of γ-lactone ring of the AHL, γ-lactone ring in the AHL analogues has been replaced with either carbocyclic or heterocyclic ring and was tested against their receptor protein. One major advantage of this modification is that such analogues would not be susceptible to degradation by the AHL-degrading enzymes.

Thiolactones and Lactams

In *V. fischeri*, replacing the lactone ring with the thiolactone ring in native 3-oxo-C_6-HSL (e.g., compound **54**) resulted in weak agonist, but strong antagonists against LuxR (Eberhard et al. 1986; Schaefer et al. 1996). However, switching from lactone ring to a 7-membered lactam ring (e.g., compound **55**) resulted in complete loss of antagonistic activity, suggesting that the lactone ring is important for binding to the LuxR receptor (Eberhard et al. 1986). Longer acyl chain (C_{12}) thiolactone (compound **56**) displayed weak agonistic activity against LuxR (Schaefer et al. 1996). In contrast, the same thiolactone (**56**) was as potent as native 3-oxo-C_{12}-HSL as an agonist for LasR in *P. aeruginosa* (Passador et al. 1996). Again, its lactam analogue (compound **57**) showed considerably less activity (Fig. 20).

A library of ~21 thiolactone derivatives of AHL was reported by Blackwell and coworkers (McInnis and Blackwell 2011). These nonnative

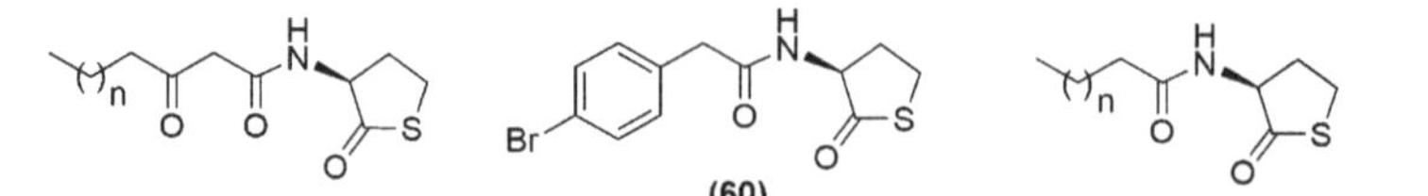

n = 3 (**58**), antagonist (LuxR), agonist (TraR)
n = 7 (**59**), agonist (LasR)
(**60**) antagonist (LasR)
n = 4 (**61**), antagonist (LasR)
n = 5 (**62**), antagonist (LasR)
n = 9 (**63**), agonist (LasR)

Fig. 21 Thiolactone derivatives of AHL identified as agonists and antagonists against LuxR, LasR, and TraR by Blackwell and coworkers (McInnis and Blackwell 2011)

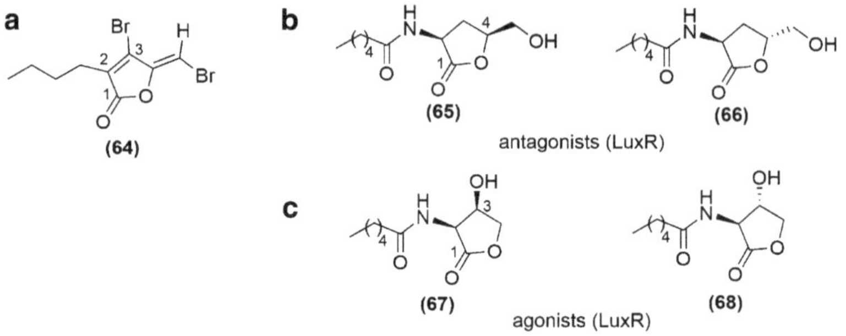

Fig. 22 3- and 4-Substituted γ-Lactone derivatives of AHL tested against LuxR by Neilsen and coworkers (Olsen et al. 2002). (**a**) Structure of halogenated furanone. (**b**) Identified antagonists. (**c**) Identified agonists

a (**64**)
b (**65**) (**66**) antagonists (LuxR)
c (**67**) (**68**) agonists (LuxR)

thiolactone AHL analogues were screened for their agonistic and antagonistic activities against LuxR-type QS receptors in *P. aeruginosa* (LasR), *V. fischeri* (LuxR), and *A. tumefaciens* (TraR) using bacterial reporter strains. Several new and potent QS modulators having multi-receptor activity were discovered in this study. Some of the identified potent agonists and antagonists (compounds **58-63**) are shown in Fig. 21.

3- and 4-Substituted γ-Lactone

Various halogenated furanoses (e.g., compound **64**, Fig. 22), isolated from the red marine alga *Delisea pulchra*, have been reported to inhibit several different QS systems (Givskov et al. 1996; Manefield et al. 1999). These compounds possessed structural homology with the 3- and 4-substituted AHL lactone rings (Fig. 22). Building on these observations, Nielsen and coworkers designed and synthesized AHL analogues that had various substitutions on the 3- and 4-positions of the lactone ring and evaluated their activity against LuxR in *V. fischeri* (Olsen et al. 2002). Compounds substituted at C4-position of the lactone ring (e.g., compounds **65** and **66**) were found to be weak activators of LuxR, but inhibitors at high concentration (50 μM). Compounds substituted at C3-position of the lactone (e.g., compounds **67** and **68**) and essentially having *3R* configuration (**67**) displayed potent agonistic activity, similar to that of the native ligand, 3-oxo-C_{12}-HSL.

Substituting Lactone Rings with Carbocycles and Heterocycles

Suga and coworkers designed and synthesized ~96 AHL mimics against LasR (3-oxo-C_{12}-HSL is its native ligand) and RhlR (C_4-HSL is its native ligand) in *P. aeruginosa* PAO-JP2 by replacing the AHL lactone ring with carbocycles and heterocycles, while keeping the acyl chain length unaltered with respect to the native ligands (Smith et al. 2003a; Smith et al. 2003b; Jog et al. 2006). These studies resulted in the discovery of several modulators of LasR and RhlR (Fig. 23) (Smith et al. 2003b). For example, the 3-oxo-C_{12} cyclopentanone (compound **69**) showed agonistic activity against LasR at potency levels close to the native ligand at high concentration (400 μM). However, compound **70**, the reduced variant of compound **69**, and notably the *S,S* derivative displayed agonistic activity comparable to native 3-oxo-C_{12}-HSL at 50 μM. Replacing the lactone ring with the cyclopentanone in C4-HSL (e.g., compound **71**) activated RhlR to the same level as native ligand at 100 μM concentrations. Switching from cyclopentanone (5-membered) to the cyclohexanone (6-membered) derivative, for

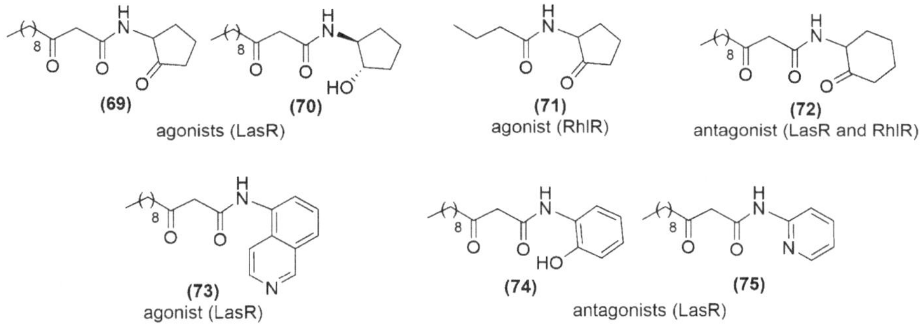

(69) (70) agonists (LasR) (71) agonist (RhlR) (72) antagonist (LasR and RhlR)

(73) agonist (LasR) (74) (75) antagonists (LasR)

Fig. 23 Carbocyclic and heterocyclic derivatives of AHL tested against LasR and RhlR tested by Suga and coworkers (Smith et al. 2003b)

example, compound **72**, resulted in an antagonist which inhibited LasR and RhlR by 35% and ~60% at 100 μM and 50 μM concentrations, respectively. Based on the initial findings, the authors then reported a more focused library of non-lactone AHL mimics against LasR (Jog et al. 2006). In this study, bicyclic analogues, for example, the isoquinoline derivative (**73**), showed strong agonistic activity against LasR, while monocyclic aromatic analogues, such as aniline derivatives, turned out to be antagonists. Out of eight identified inhibitors, five were aniline derivatives, and notably, anilines having a hydrogen bond acceptor at the ortho or meta position were identified as the most potent inhibitors of LasR (e.g., compounds **74** and **75**). Based on the results obtained from these studies (Jog et al. 2006; Smith et al. 2003b), the authors concluded that (1) AHL analogues with 5- and 6-membered ring containing hydrogen bond acceptors next to the amino group exhibited agonism against LasR and (2) AHL analogues containing aromatic ring in place of lactone displayed inhibition against LasR.

Kato and coworkers synthesized AHL analogues in which the lactone ring was replaced by the cyclopentane ring to ablate its hydrogen bonding capability and varied the alkyl chain from 3 to 12 carbon atoms (Fig. 24) (Ishida et al. 2007). All these compounds were tested against LasR in *P. aeruginosa*. These studies led to the discovery of C10-CPA (**76**) as one of the most potent antagonists against LasR. Furthermore, with cyclopentane analogue as the lead

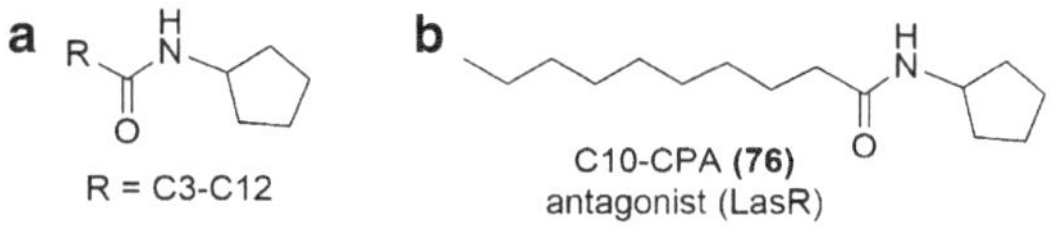

a R = C3-C12 **b** C10-CPA **(76)** antagonist (LasR)

Fig. 24 Cyclopentyl derivatives of AHL tested against LasR by Kato and coworkers (Ishida et al. 2007). (**a**) General structural features of AHLs testes. (**b**) The most potent antagonist identified in the study

compound, replacing the cyclopentane ring with other carbocycles such as cyclopropane, cyclobutane, cyclohexane, and cyclooctane resulted in a decrease in antagonistic activity. From these observations, the authors concluded that the both the acyl chain length and the ring size were important parameters to consider for designing inhibitors of LasR.

Targeting AI-2 Receptors with AI-2 Mimics

LsrB and LuxP are the two best characterized receptors for AI-2-mediated QS systems, which are usually employed for studying the nonnative AI-2-receptor interactions. LsrB regulates the β-galactosidase activity in *Salmonella typhimurium,* whereas LuxP plays an important role in the regulation of bioluminescence in *Vibrio harveyi*. In comparison to AHLs analogues, there is a relatively paucity of known synthetic nonnative small-molecule modulators of AI-2-regulated QS pathways.

As discussed earlier, DPD exhibits a complex equilibrium with the compounds comprising the

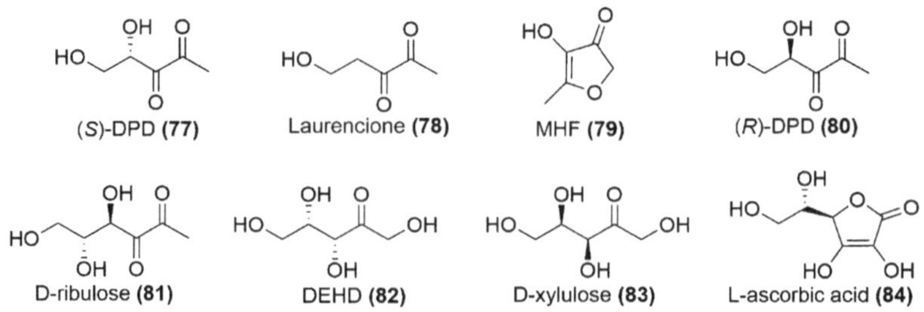

Fig. 25 DPD analogues tested for agonistic activity against LuxP by Janda and coworkers (Lowery et al. 2005)

(*S*)-DPD (**77**) Laurencione (**78**) MHF (**79**) (*R*)-DPD (**80**)
D-ribulose (**81**) DEHD (**82**) D-xylulose (**83**) L-ascorbic acid (**84**)

AI-2 pool, in the presence of water and borates (Fig. 2). Different bacteria recognize different DPD derivatives of this AI-2 pool, as their active signaling molecules. This interconversion of molecules within the AI-2 pool presumably allows bacteria to respond to both intra- and interspecies cell-to-cell communication. Therefore, synthesizing nonnative DPD analogues that will also display such complex equilibria has been a major focus of researchers engaged in developing small-molecule modulators of AI-2 signaling. Furthermore, despite its relative structural simplicity, the chemical synthesis of DPD and its analogues is nontrivial, as these compounds are stable only in dilute concentrations but dimerize to biologically inactive derivatives at higher concentrations (Smith et al. 2009; Meijler et al. 2004).

Janda and coworkers evaluated a series of natural and nonnatural DPD analogues (**77–84**) for their agonistic activity against the receptor, LuxP in the *V. harveyi* strain MM30, using bioluminescence induction as the read out (Fig. 25) (Lowery et al. 2005). The results obtained from this study showed that the natural *S* enantiomer of DPD (**77**) (EC_{50} 0.044 μM) was a 2000-fold more potent agonist than its nonnatural *L* enantiomer (**80**) (EC_{50} 84 μM). Furthermore, both the chelation of boron and the position of hydroxyl functions were found to be important for DPD analogue-LuxP binding.

A few years later, the same group reported a variety of C1-alkyl-substitued DPD analogues (**85–96**) (Fig. 26) (Lowery et al. 2008b). These analogues were evaluated for their QS-modulating activity by monitoring AI-2-regulated induction of β-galactosidase activity in *S. typhimurium* and bioluminescence in *V. harveyi*. In *S. typhimurium*, none of the compounds showed agonistic activity in the absence of DPD. Interestingly, however, in the presence of DPD (**77**), all the compounds displayed antagonistic activity, without affecting bacterial growth. Compounds **86** and **87** were found to be the most potent inhibitors in this study. In *V. harveyi*, in the absence of DPD, only **85** showed weak induction of bioluminescence (weak agonist). However, in the presence of DPD, a synergistic agonistic effect involving DPD-enhanced AI-2-induced bioluminescence was observed with the tested compounds. The authors hypothesized that instead of LuxP, some promiscuous receptors were responsible for the observed synergistic agonism displayed by DPD analogues in the presence of DPD. Similar results were obtained by Ganin and coworkers upon screening of C1-alkyl-DPD analogues by varying alkyl chain length from 1–7 carbons in *V. harveyi* (Ganin et al. 2009). Of note, two of the C1-alkyl-DPD analogues (compounds **91** and **92**) were identified as potent inhibitors for the production of virulence factor, pyocyanin in *P. aeruginosa*, a phenotype regulated by QS (Fig. 26). Synergistic agonism was also observed in DPD analogues synthesized by Smith and coworkers where the C1-alkyl groups were varied from linear (**85–87**) to branched (**93, 94**) to cyclic alkyl groups (**95, 96**), suggesting marked promiscuity of ligand binding (Smith et al. 2009). Compound **96** was identified as the most potent synergistic agonist in this study.

Future Perspectives

A plethora of synthetic analogues of autoinducers and those of their biosynthetic intermediates have been reported in the last 15 years, only

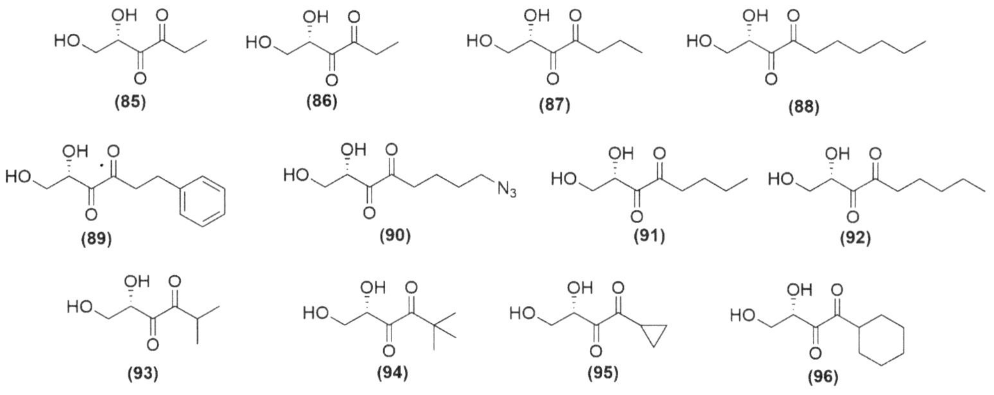

(85) (86) (87) (88) (89) (90) (91) (92) (93) (94) (95) (96)

Fig. 26 Structures of C1-alkylated DPD analogues

a few of which have been discussed here. Although these studies have resulted in the discovery of several potent modulators of QS, there is a paucity of mechanistic insight that underlies the biological effects of these compounds. Unraveling the molecular basis of binding and modulation of QS-orchestrating bacterial proteins by small molecules is critical for driving efforts in designing novel QS modulators. Recent breakthroughs in structural biology efforts focused at obtaining high-resolution crystal structures of QS-regulating bacterial proteins complexed with autoinducers or their cognate synthetic analogues are providing much needed molecular-level insights (Liu et al. 2005; Liu et al. 2013; Ha et al. 2013; Haapalainen et al. 2013; Ilangovan et al. 2013; Zou and Nair 2009). In addition to detailed information on the mode of binding of QS modulators to their target receptors, discovery of new QS pathways that are modulated by autoinducer analogues will be tremendously useful and will lead to the identification of new targets for designing antibacterial agents. A recent chemical biology-based study has utilized mass spectrometry to achieve this goal for the *S. typhimurium* QS antagonist, propyl-DPD (Lowery et al. 2013). Future studies focused on elucidating the molecular details of binding of quorum sensing modulators to their targets employing crystallography and other techniques, along with discovery of novel QS signaling mechanisms and pathways, are awaited with great anticipation. Such efforts are urgently required for developing novel QS inhibitors which can be ultimately utilized as antibacterials, thereby adding to our fast-depleting armamentarium against the scourge of bacterial diseases that threaten the very existence of mankind.

References

Alfaro JF, Zhang T, Wynn DP, Karschner EL, Zhou ZS (2004) Synthesis of LuxS inhibitors targeting bacterial cell-cell communication. Org Lett 6:3043–3046

Bjarnsholt T, Jensen PO, Burmolle M, Hentzer M, Haagensen JA, Hougen HP, Calum H, Madsen KG, Moser C, Molin S, Hoiby N, Givskov M (2005) *Pseudomonas aeruginosa* tolerance to tobramycin, hydrogen peroxide and polymorphonuclear leukocytes is quorum-sensing dependent. Microbiology 151:373–383

Boukraa M, Sabbah M, Soulere L, EL Efrit ML, Queneau Y, Doutheau A (2011) AHL-dependent quorum sensing inhibition: synthesis and biological evaluation of alpha-(N-alkyl-carboxamide)-gamma-butyrolactones and alpha-(N-alkyl-sulfonamide)-gamma-butyrolactones. Bioorg Med Chem Lett 21:6876–6879

Castang S, Chantegrel B, Deshayes C, Dolmazon R, Gouet P, Haser R, Reverchon S, Nasser W, Hugouvieux-Cotte-Pattat N, Doutheau A (2004) N-Sulfonyl homoserine lactones as antagonists of bacterial quorum sensing. Bioorg Med Chem Lett 14: 5145–5149

Cegelski L, Marshall GR, Eldridge GR, Hultgren SJ (2008) The biology and future prospects of antivirulence therapies. Nat Rev Microbiol 6:17–27

Clatworthy AE, Pierson E, Hung DT (2007) Targeting virulence: a new paradigm for antimicrobial therapy. Nat Chem Biol 3:541–548

Davies DG, Parsek MR, Pearson JP, Iglewski BH, Costerton JW, Greenberg EP (1998) The involvement of cell-to-cell signals in the development of a bacterial biofilm. Science 280:295–298

Deng Y, Wu J, Tao F, Zhang LH (2011) Listening to a new language: DSF-based quorum sensing in Gram-negative bacteria. Chem Rev 111:160–173

Dubern JF, Diggle SP (2008) Quorum sensing by 2-alkyl-4-quinolones in *Pseudomonas aeruginosa* and other bacterial species. Mol Biosyst 4:882–888

Eberhard A, Widrig CA, Mcbath P, Schineller JB (1986) Analogs of the autoinducer of bioluminescence in *Vibrio fischeri*. Arch Microbiol 146:35–40

Frezza M, Castang S, Estephane J, Soulere L, Deshayes C, Chantegrel B, Nasser W, Queneau Y, Reverchon S, Doutheau A (2006) Synthesis and biological evaluation of homoserine lactone derived ureas as antagonists of bacterial quorum sensing. Bioorg Med Chem 14: 4781–4791

Frezza M, Soulere L, Reverchon S, Guiliani N, Jerez C, Queneau Y, Doutheau A (2008) Synthetic homoserine lactone-derived sulfonylureas as inhibitors of *Vibrio fischeri* quorum sensing regulator. Bioorg Med Chem 16:3550–3556

Galloway WR, Hodgkinson JT, Bowden SD, Welch M, Spring DR (2011) Quorum sensing in Gram-negative bacteria: small-molecule modulation of AHL and AI-2 quorum sensing pathways. Chem Rev 111:28–67

Ganin H, Tang X, Meijler MM (2009) Inhibition of *Pseudomonas aeruginosa* quorum sensing by AI-2 analogs. Bioorg Med Chem Lett 19:3941–3944

Geske GD, Wezeman RJ, Siegel AP, Blackwell HE (2005) Small molecule inhibitors of bacterial quorum sensing and biofilm formation. J Am Chem Soc 127: 12762–12763

Geske GD, O'neill JC, Miller DM, Mattmann ME, Blackwell HE (2007) Modulation of bacterial quorum sensing with synthetic ligands: systematic evaluation of N-acylated homoserine lactones in multiple species and new insights into their mechanisms of action. J Am Chem Soc 129:13613–13625

Geske GD, Mattmann ME, Blackwell HE (2008a) Evaluation of a focused library of N-aryl L-homoserine lactones reveals a new set of potent quorum sensing modulators. Bioorg Med Chem Lett 18: 5978–5981

Geske GD, O'neill JC, Blackwell HE (2008b) Expanding dialogues: from natural autoinducers to non-natural analogues that modulate quorum sensing in Gram-negative bacteria. Chem Soc Rev 37:1432–1447

Givskov MNJ (2003) Compounds and methods for controlling bacterial virulence. *Patent No. WO2003106445 A1*

Givskov M, de Nys R, Manefield M, Gram L, Maximilien R, Eberl L, Molin S, Steinberg PD, Kjelleberg S (1996) Eukaryotic interference with homoserine lactone-mediated prokaryotic signalling. J Bacteriol 178:6618–6622

Globisch D, Lowery CA, Mccague KC, Janda KD (2012) Uncharacterized 4,5-dihydroxy-2,3-pentanedione (DPD) molecules revealed through NMR spectroscopy: implications for a greater signaling diversity in bacterial species. Angew Chem Int Ed Engl 51: 4204–4208

Gopishetty B, Zhu J, Rajan R, Sobczak AJ, Wnuk SF, Bell CE, Pei D (2009) Probing the catalytic mechanism of S-ribosylhomocysteinase (LuxS) with catalytic intermediates and substrate analogues. J Am Chem Soc 131:1243–1250

Guo M, Gamby S, Zheng Y, Sintim HO (2013) Small molecule inhibitors of AI-2 signaling in bacteria: state-of-the-art and future perspectives for anti-quorum sensing agents. Int J Mol Sci 14:17694–17728

Gutierrez JA, Crowder T, Rinaldo-Matthis A, Ho MC, Almo SC, Schramm VL (2009) Transition state analogs of 5′-methylthioadenosine nucleosidase disrupt quorum sensing. Nat Chem Biol 5:251–257

Ha JH, Eo Y, Grishaev A, Guo M, Smith JA, Sintim HO, Kim EH, Cheong HK, Bentley WE, Ryu KS (2013) Crystal structures of the LsrR proteins complexed with phospho-AI-2 and two signal-interrupting analogues reveal distinct mechanisms for ligand recognition. J Am Chem Soc 135:15526–15535

Haapalainen AM, Thomas K, Tyler PC, Evans GB, Almo SC, Schramm VL (2013) *Salmonella enterica* MTAN at 1.36 A resolution: a structure-based design of tailored transition state analogs. Structure 21: 963–974

Hentzer M, Givskov M (2003) Pharmacological inhibition of quorum sensing for the treatment of chronic bacterial infections. J Clin Invest 112: 1300–1307

Hentzer M, Wu H, Andersen JB, Riedel K, Rasmussen TB, Bagge N, Kumar N, Schembri MA, Song Z, Kristoffersen P, Manefield M, Costerton JW, Molin S, Eberl L, Steinberg P, Kjelleberg S, Hoiby N, Givskov M (2003) Attenuation of *Pseudomonas aeruginosa* virulence by quorum sensing inhibitors. EMBO J 22:3803–3815

Ilangovan A, Fletcher M, Rampioni G, Pustelny C, Rumbaugh K, Heeb S, Camara M, Truman A, Chhabra SR, Emsley J, Williams P (2013) Structural basis for native agonist and synthetic inhibitor recognition by the *Pseudomonas aeruginosa* quorum sensing regulator PqsR (MvfR). PLoS Pathog 9:e1003508

Ishida T, Ikeda T, Takiguchi N, Kuroda A, Ohtake H, Kato J (2007) Inhibition of quorum sensing in *Pseudomonas aeruginosa* by N-acyl cyclopentylamides. Appl Environ Microbiol 73:3183–3188

Jimenez PN, Koch G, Thompson JA, Xavier KB, Cool RH, Quax WJ (2012) The multiple signaling systems regulating virulence in *Pseudomonas aeruginosa*. Microbiol Mol Biol Rev 76:46–65

Jog GJ, Igarashi J, Suga H (2006) Stereoisomers of *P. aeruginosa* autoinducer analog to probe the regulator binding site. Chem Biol 13:123–128

Kline T, Bowman J, Iglewski BH, de Kievit T, Kakai Y, Passador L (1999) Novel synthetic analogs of the *Pseudomonas* autoinducer. Bioorg Med Chem Lett 9: 3447–3452

Lewis K (2013) Platforms for antibiotic discovery. Nat Rev Drug Discov 12:371–387

Liu D, Lepore BW, Petsko GA, Thomas PW, Stone EM, Fast W, Ringe D (2005) Three-dimensional structure of the quorum-quenching N-acyl homoserine lactone hydrolase from *Bacillus thuringiensis*. Proc Natl Acad Sci U S A 102:11882–11887

Liu CF, Liu D, Momb J, Thomas PW, Lajoie A, Petsko GA, Fast W, Ringe D (2013) A phenylalanine clamp controls substrate specificity in the quorum-quenching metallo-gamma-lactonase from *Bacillus thuringiensis*. Biochemistry 52:1603–1610

Lowery CA, Mckenzie KM, Qi L, Meijler MM, Janda KD (2005) Quorum sensing in Vibrio harveyi: probing the specificity of the LuxP binding site. Bioorg Med Chem Lett 15:2395–2398

Lowery CA, Dickerson TJ, Janda KD (2008a) Interspecies and interkingdom communication mediated by bacterial quorum sensing. Chem Soc Rev 37:1337–1346

Lowery CA, Park J, Kaufmann GF, Janda KD (2008b) An unexpected switch in the modulation of AI-2-based quorum sensing discovered through synthetic 4,5-dihydroxy-2,3-pentanedione analogues. J Am Chem Soc 130:9200–9201

Lowery CA, Matamouros S, Niessen S, Zhu J, Scolnick J, Lively JM, Cravatt BF, Miller SI, Kaufmann GF, Janda KD (2013) A chemical biology approach to interrogate quorum-sensing regulated behaviors at the molecular and cellular level. Chem Biol 20:903–911

Manefield M, de Nys R, Kumar N, Read R, Givskov M, Steinberg P, Kjelleberg S (1999) Evidence that halogenated furanones from *Delisea pulchra* inhibit acylated homoserine lactone (AHL)-mediated gene expression by displacing the AHL signal from its receptor protein. Microbiology 145(Pt 2):283–291

Marra A (2004) Can virulence factors be viable antibacterial targets? Expert Rev Anti Infect Ther 2:61–72

Mattmann ME, Geske GD, Worzalla GA, Chandler JR, Sappington KJ, Greenberg EP, Blackwell HE (2008) Synthetic ligands that activate and inhibit a quorum-sensing regulator in *Pseudomonas aeruginosa*. Bioorg Med Chem Lett 18:3072–3075

Mcinnis CE, Blackwell HE (2011) Thiolactone modulators of quorum sensing revealed through library design and screening. Bioorg Med Chem 19:4820–4828

Meijler MM, Hom LG, Kaufmann GF, Mckenzie KM, Sun C, Moss JA, Matsushita M, Janda KD (2004) Synthesis and biological validation of a ubiquitous quorum-sensing molecule. Angew Chem Int Ed Engl 43:2106–2108

Ng WL, Bassler BL (2009) Bacterial quorum-sensing network architectures. Annu Rev Genet 43:197–222

Novick RP, Muir TW (1999) Virulence gene regulation by peptides in *Staphylococci* and other Gram-positive bacteria. Curr Opin Microbiol 2:40–45

Olsen JA, Severinsen R, Rasmussen TB, Hentzer M, Givskov M, Nielsen J (2002) Synthesis of new 3- and 4-substituted analogues of acyl homoserine lactone quorum sensing autoinducers. Bioorg Med Chem Lett 12:325–328

Parsek MR, Val DL, Hanzelka BL, Cronan JE Jr, Greenberg EP (1999) Acyl homoserine-lactone quorum-sensing signal generation. Proc Natl Acad Sci U S A 96:4360–4365

Parveen N, Cornell KA (2011) Methylthioadenosine/S-adenosylhomocysteine nucleosidase, a critical enzyme for bacterial metabolism. Mol Microbiol 79:7–20

Passador L, Tucker KD, Guertin KR, Journet MP, Kende AS, Iglewski BH (1996) Functional analysis of the *Pseudomonas aeruginosa* autoinducer PAI. J Bacteriol 178:5995–6000

Patani GA, Lavoie EJ (1996) Bioisosterism: a rational approach in drug design. Chem Rev 96:3147–3176

Pereira CS, Thompson JA, Xavier KB (2013) AI-2-mediated signalling in bacteria. FEMS Microbiol Rev 37:156–181

Rasmussen TB, Givskov M (2006) Quorum-sensing inhibitors as anti-pathogenic drugs. Int J Med Microbiol 296:149–161

Rasmussen TB, Bjarnsholt T, Skindersoe ME, Hentzer M, Kristoffersen P, Kote M, Nielsen J, Eberl L, Givskov M (2005) Screening for quorum-sensing inhibitors (QSI) by use of a novel genetic system, the QSI selector. J Bacteriol 187:1799–1814

Reverchon S, Chantegrel B, Deshayes C, Doutheau A, Cotte-Pattat N (2002) New synthetic analogues of N-acyl homoserine lactones as agonists or antagonists of transcriptional regulators involved in bacterial quorum sensing. Bioorg Med Chem Lett 12:1153–1157

Schaefer AL, Hanzelka BL, Eberhard A, Greenberg EP (1996) Quorum sensing in Vibrio fischeri: probing autoinducer-LuxR interactions with autoinducer analogs. J Bacteriol 178:2897–2901

Schuster M, Sexton DJ, Diggle SP, Greenberg EP (2013) Acyl-homoserine lactone quorum sensing: from evolution to application. Annu Rev Microbiol 67:43–63

Shen G, Rajan R, Zhu J, Bell CE, Pei D (2006) Design and synthesis of substrate and intermediate analogue inhibitors of S-ribosylhomocysteinase. J Med Chem 49:3003–3011

Sintim HO, Smith JA, Wang J, Nakayama S, Yan L (2010) Paradigm shift in discovering next-generation anti-infective agents: targeting quorum sensing, c-di-GMP signaling and biofilm formation in bacteria with small molecules. Future Med Chem 2: 1005–1035

Smith KM, Bu Y, Suga H (2003a) Induction and inhibition of *Pseudomonas aeruginosa* quorum sensing by synthetic autoinducer analogs. Chem Biol 10:81–89

Smith KM, Bu Y, Suga H (2003b) Library screening for synthetic agonists and antagonists of a *Pseudomonas aeruginosa* autoinducer. Chem Biol 10:563–571

Smith JA, Wang J, Nguyen-Mau SM, Lee V, Sintim HO (2009) Biological screening of a diverse set of AI-2 analogues in *Vibrio harveyi* suggests that receptors which are involved in synergistic agonism of AI-2 and analogues are promiscuous. Chem Commun (Camb) 45:7033–7035

Sturme MH, Kleerebezem M, Nakayama J, Akkermans AD, Vaugha EE, de Vos WM (2002) Cell to cell communication by autoinducing peptides in gram-positive bacteria. Antonie Van Leeuwenhoek 81:233–243

Thomas K, Haapalainen AM, Burgos ES, Evans GB, Tyler PC, Gulab S, Guan R, Schramm VL (2012)

Femtomolar inhibitors bind to 5′-methylthioadenosine nucleosidases with favorable enthalpy and entropy. Biochemistry 51:7541–7550

von Nussbaum F, Brands M, Hinzen B, Weigand S, Habich D (2006) Antibacterial natural products in medicinal chemistry–exodus or revival? Angew Chem Int Ed Engl 45:5072–5129

Waters CM, Bassler BL (2005) Quorum sensing: cell-to-cell communication in bacteria. Annu Rev Cell Dev Biol 21:319–346

Watson WT, Minogue TD, Val DL, von Bodman SB, Churchill ME (2002) Structural basis and specificity of acyl-homoserine lactone signal production in bacterial quorum sensing. Mol Cell 9: 685–694

Wnuk SF, Robert J, Sobczak AJ, Meyers BP, Malladi VL, Zhu J, Gopishetty B, Pei D (2009) Inhibition of S-ribosylhomocysteinase (LuxS) by substrate analogues modified at the ribosyl C-3 position. Bioorg Med Chem 17:6699–6706

Worthington RJ, Melander C (2012) Deconvoluting interspecies bacterial communication. Angew Chem Int Ed Engl 51:6314–6315

Wu H, Song Z, Hentzer M, Andersen JB, Molin S, Givskov M, Hoiby N (2004) Synthetic furanones inhibit quorum-sensing and enhance bacterial clearance in *Pseudomonas aeruginosa* lung infection in mice. J Antimicrob Chemother 53:1054–1061

Zhang RG, Pappas KM, Brace JL, Miller PC, Oulmassov T, Molyneaux JM, Anderson JC, Bashkin JK, Winans SC, Joachimiak A (2002) Structure of a bacterial quorum-sensing transcription factor complexed with pheromone and DNA. Nature 417:971–974

Zhu J, Beaber JW, More MI, Fuqua C, Eberhard A, Winans SC (1998) Analogs of the autoinducer 3-oxooctanoyl-homoserine lactone strongly inhibit activity of the TraR protein of *Agrobacterium tumefaciens*. J Bacteriol 180:5398–5405

Zhu J, Hixon MS, Globisch D, Kaufmann GF, Janda KD (2013) Mechanistic insights into the LsrK kinase required for autoinducer-2 quorum sensing activation. J Am Chem Soc 135:7827–7830

Zou Y, Nair SK (2009) Molecular basis for the recognition of structurally distinct autoinducer mimics by the *Pseudomonas aeruginosa* LasR quorum-sensing signaling receptor. Chem Biol 16:961–970

Synthetic Quorum Sensing Inhibitors (QSIs) Blocking Receptor Signaling or Signal Molecule Biosynthesis in *Pseudomonas aeruginosa*

Christine K. Maurer, Cenbin Lu, Martin Empting, and Rolf W. Hartmann

Abbreviations

ACP	Acyl carrier protein
AHL	*N*-acyl homoserine lactone
AMP	Adenosine monophosphate
ATP	Adenosine triphosphate
C_4-HSL	*N*-butanoyl-l-homoserine lactone
CoA	Coenzyme A
HAQ	4-hydroxy-2-alkylquinoline
HHQ	2-heptyl-4-hydroxyquinoline
IC_{50}	Inhibitor concentration to achieve a half-maximal degree of inhibition
MTA	5-methylthioadenosine
NADH/NAD^+	Reduced/oxidized form of nicotinamide adenine dinucleotide
3-oxo-C_{12}-HSL	*N*-(3-oxo-dodecanoyl)-l-homoserine lactone
PP_i	Pyrophosphate
PQS	*Pseudomonas* quinolone signal
QS	Quorum sensing
QSI	Quorum sensing inhibitor
QZN	Quinazolinone
RNAP	RNA polymerase
SAM	l-*S*-adenosylmethionine

C.K. Maurer • C. Lu • M. Empting
Department Drug Design and Optimization, Helmholtz-Institute for Pharmaceutical Research Saarland (HIPS), Campus C2.3, 66123 Saarbrücken, Germany
e-mail: christine.maurer@helmholtz-hzi.de; cenbin.lu@helmholtz-hzi.de; martin.empting@helmholtz-hzi.de

R.W. Hartmann (✉)
Department Drug Design and Optimization, Helmholtz-Institute for Pharmaceutical Research Saarland (HIPS), Campus C2.3, 66123 Saarbrücken, Germany

Pharmaceutical and Medicinal Chemistry at the Saarland University, Campus C2.3, 66123 Saarbrücken, Germany
e-mail: rolf.hartmann@helmholtz-hzi.de

Introduction

Pseudomonas aeruginosa masters quorum sensing (QS) communication to coordinately regulate pathogenicity-associated group behaviors including the production of virulence factors and biofilm formation, which facilitate the invasion into the hosts, counteract host immune system, as well as promote the resistance/tolerance toward conventional antibiotics. Three main QS systems are employed by the pathogen, denoted as *las* (Gambello and Iglewski 1991; Passador et al. 1993), *rhl* (Ochsner et al. 1994; Ochsner and Reiser 1995), and *pqs* (Pesci et al. 1999). All the networks are hierarchically interconnected: *las* controls the other two systems; *pqs* positively regulates the *rhl* signaling, whereas *rhl* in turn puts a negative feedback upon *pqs* (Wilder et al. 2011; McGrath et al. 2004). Regarding the central role of QS for the infectious process, the interruption of these pathways by blocking the receptors or inhibiting the signal synthesis via small molecules is an attractive therapeutic strategy to attenuate the bacterial pathogenicity, thereby overcoming intractable *P. aeruginosa* infections (Rasmussen and Givskov 2006).

V.C. Kalia (ed.), *Quorum Sensing vs Quorum Quenching: A Battle with No End in Sight*, DOI 10.1007/978-81-322-1982-8_24, © Springer India 2015

Synthetic QSIs Blocking Receptor Signaling

The interference with the QS receptors via QSIs is a promising approach to efficiently interrupt the communication networks, thereby decreasing the QS-controlled pathogenicity (Kalia 2013; Rasmussen and Givskov 2006). Generally, such QSIs are derived from the natural ligands of the target receptors, structurally unrelated natural products, or small molecules/fragments. These QSIs could be either pure antagonists or weak agonists, all of which effectively compete with the natural agonists preventing a sufficient stimulation of the receptors.

N-Acyl l-Homoserine Lactone (AHL) Receptors as Targets

In Gram-negative bacteria, *N*-acyl l-homoserine lactones (AHLs) are the most commonly used signal molecules for QS. The innate receptors of these autoinducers belong to the LuxR type which act as transcriptional regulators upon activation by their native agonists. In *P. aeruginosa*, the so-called *las* and *rhl* systems have been identified to be the key AHL-based QS systems. The native ligands of the involved cytoplasmic receptors LasR and RhlR are *N*-(3-oxo-dodecanoyl)-l-homoserine lactone (3-oxo-C_{12}-HSL or OdDHL) and *N*-butanoyl-l-homoserine lactone (C_4-HSL or BHL), respectively (Fig. 1a).

Notably, a higher rank in the regulatory hierarchy is accounted to the *las* system. An additional LuxR-type receptor – the "orphan" receptor QscR also responding to 3-oxo-C_{12}-HSL – has been reported for this particular human pathogen (Chugani et al. 2001). However, the major part of scientific studies concentrates on the development of LasR antagonists to disrupt AHL signaling in *P. aeruginosa*. Desirable cellular effects of LasR- and/or RhlR-targeting QSIs would be the reduction of virulence factor production (e.g., elastase, hydrogen cyanide, pyocyanin, pyoverdine, rhamnolipids, or alkaline protease) as well as the attenuation of biofilm formation. Indeed, several groups have reported on the successful development of synthetic agents with promising *in cellulo* activities. The following section provides a detailed, yet not exhaustive, overview on the structural space covered by synthetic QSIs interfering with AHL signaling. In general, compounds addressing LasR and RhlR can be divided into two categories: (1) structural mimics of AHLs and (2) structurally unrelated substances.

AHL Mimics

Many nonnatural agonists and antagonists of LuxR-type receptors possess structural features very similar to the native ligands or are actually direct synthetic derivatives thereof (Fig. 1b, c).

Quite obviously, the AHL scaffold can be divided into two sections. The head group consists of the five-membered homoserine lactone moiety, while the tail region comprises a linear *N*-acyl residue of varying length. The amide-based linker between both segments facilitates modular approaches for straightforward synthesis and derivatization. Many reports in the literature dealing with synthetic AHL analogues focus on structural modifications in one of these two sections while leaving the other part of the molecule constant. Nevertheless, combinations of nonnatural head and tail modules have also been described.

Noteworthy, the native lactone pentacycles of 3-oxo-C_{12}-HSL and C_4-HSL are easily hydrolyzed under physiological conditions. The resulting linear product is QS inactive (McInnis and Blackwell 2011b). Thus, replacing this moiety by stable bioisosters is a worthwhile endeavor providing access to nonnatural QS modulators with an in vivo half-life superior to the natural ligands. The range of possible substituents is broad and includes penta- and hexacyclic thiolactones, homo- and heteroaromatic residues, as well as saturated ring systems (Fig. 1b). Interestingly, modifications within the head group of AHL-receptor agonists enable to achieve the desirable functional inversion to yield promising synthetic antagonists. Structural differences between agonists and antagonists can be very subtle. For example, Suga and coworkers

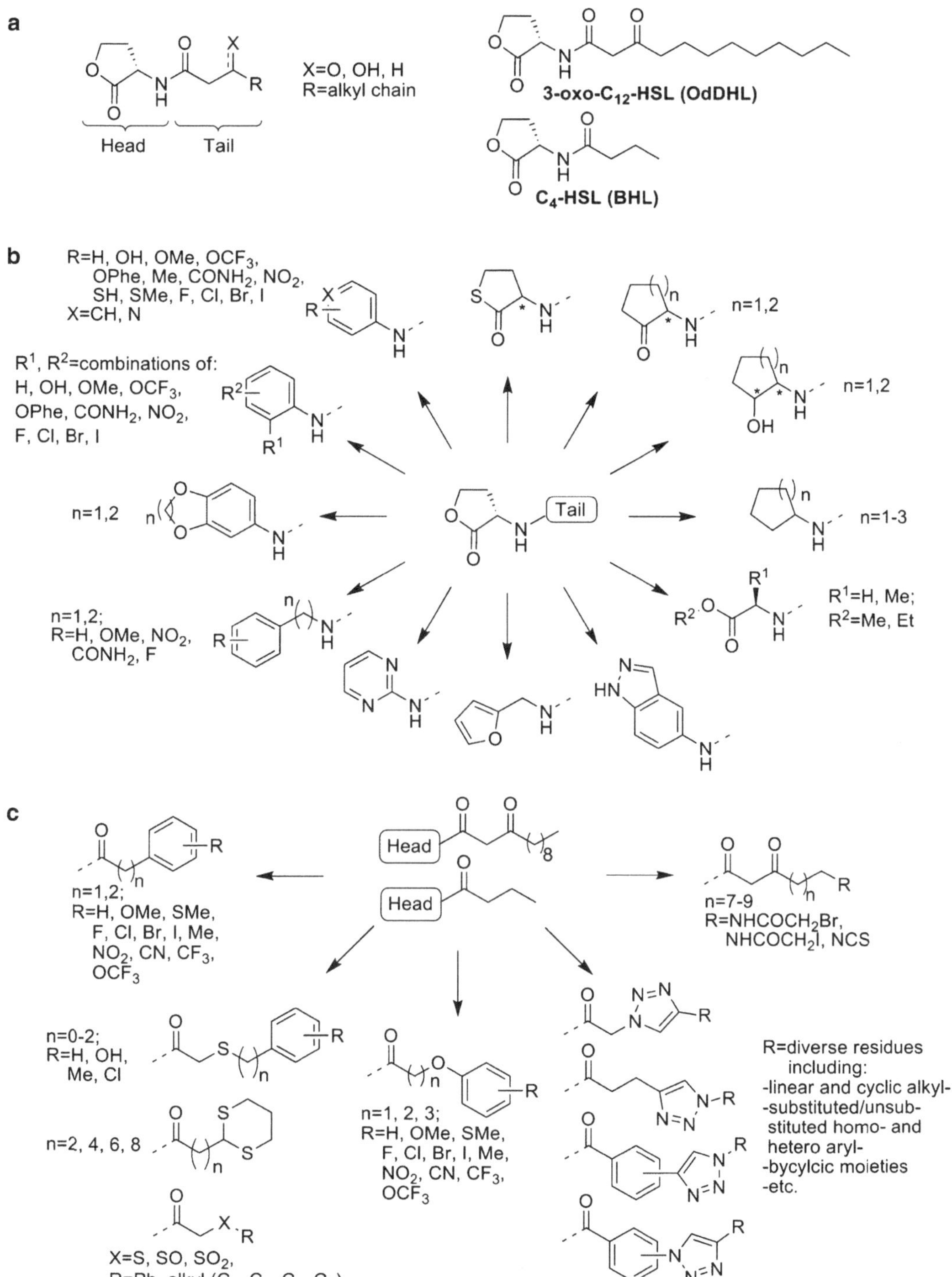

Fig. 1 (**a**) Structure of *N*-acyl l-homoserine lactone-based autoinducers in *P. aeruginosa*. (**b**) A selection of reported substituents as head group replacements found in the literature (McInnis and Blackwell 2011a, b; Jog et al. 2006; Ishida et al. 2007; Smith et al. 2003; Kim et al. 2009; Persson et al. 2005; Morkunas et al. 2012; Hodgkinson et al. 2012). (**c**) A selection of substituents as tail section replacements found in the literature (Persson et al. 2005; O'Loughlin et al. 2013; Geske et al. 2005, 2007, 2008; Stacy et al. 2013; Amara et al. 2009). Abbreviations: *3-oxo-C_{12}-HSL* *N*-(3-oxo-dodecanoyl)-l-homoserine lactone, *C_4-HSL* *N*-butanoyl-l-homoserine lactone

have shown that a compound possessing the natural 3-oxo-dodecanoyl tail and a nonnatural 2-aminocyclohexanoyl head group is a potent agonist and can be converted into an antagonist through head group replacement by a 2-aminocyclohexanone substituent (compound **1**) (Jog et al. 2006). A similar effect can be observed for the usage of a 2-aminophenol analogue (**2**) (Smith et al. 2003). The latter derivative has the advantage that it has no stereoisomers which abolishes the need for stereo control and/or racemate separation during synthesis/purification. Moreover, it has been reported that the incorporation of aromatic head groups usually yields an antagonistic functional profile (Hodgkinson et al. 2012). In summary, nonnatural head groups usually contain a cyclic motif which can have diverse electronic and/or chemical properties as well as varied substitution patterns providing control over agonist/antagonist functionality of the desired compound without compromising affinity to the respective LuxR-type receptor.

Reported variations in the tail region of synthetic AHL mimics are also quite numerous (Fig. 1c). Indeed, the difference between the native AHL signal molecules in *P. aeruginosa* lies in the length and chemistry of this section mediating receptor selectivity. The absence of the β-keto motif in the RhlR-selective autoinducer (C_4-HSL) inspired researchers to omit this structure also for the generation of LasR-addressing modulators. Interestingly, also short alkyl chains and even cyclic structures are accepted in this part of the molecule. Hence, substituted homo- and heteroaromatics with varying alkyl linker chains have been successfully incorporated into antagonists of LuxR-type AHL receptors in *P. aeruginosa* (Geske et al. 2008). Introduction of 1,2,3-triazole-based "click" linkers by Blackwell and coworkers allowed for additional synthetic modularity and combinatorial library generation (Stacy et al. 2013). However, antagonists possessing nonnatural tail groups that mimic the linear unbranched structure of the native signal molecules usually demonstrate higher potency than multi-cyclic or angled motifs (Geske et al. 2007). Incorporation of electrophilic reactive groups within such a linear tail section resulted in covalent QSI addressing LasR (e.g., compound **6**) (Amara et al. 2009).

Combining favorable nonnatural head and tail groups with each other may yield rather unexpected results. Spring and coworkers have shown that many of these "chimeric" compounds are of low potency or essentially inactive (Hodgkinson et al. 2012). Hence, conservation of the native 3-oxo-dodecanoyl chain in antagonists with nonnatural head groups can be mandatory for strong QS inhibition. A selection of promising AHL-mimicking LasR and RhlR antagonists and their effects on *P. aeruginosa* is given in Fig. 2.

Structurally Unrelated AHL-Interfering QSI

The application of experimental screening methodologies using compound libraries led to the identification of structurally diverse LasR/RhlR antagonists (Wu et al. 2004; Musthafa et al. 2012; Muh et al. 2006a, b). A selection of respective compounds together with biological activities is given in Fig. 3. One structure showing very promising effects in vitro is referred to as C-30 (**10**). This compound is a synthetic derivative of a marine natural compound found in *Delisea pulchra* and has been investigated for in vivo efficacy in murine infection models. **10** was capable of attenuating QS-mediated virulence in mice and improved bacterial clearance from infected animal lungs (Wu et al. 2004). However, the molecular mode of action by which C-30 (**10**) interacts with its target receptor (LasR) is not yet fully elucidated.

Finally, it has to be stated that means by which the potency of such LuxR-type antagonists can be determined are various and range from recombinant reporter gene assays, over the direct quantification of receptor-regulated downstream products, to the investigation of effects on biofilm formation. Each of these biological evaluation methodologies is highly dependent on experimental parameters like used cell culture media, effective concentration of organic cosolvents like dimethyl sulfoxide, or, importantly, chosen *P. aeruginosa* strain. Hence, a comparison of QSI interfering with AHL signaling developed by separate working groups is difficult and in many cases not practical. Additionally, not in all

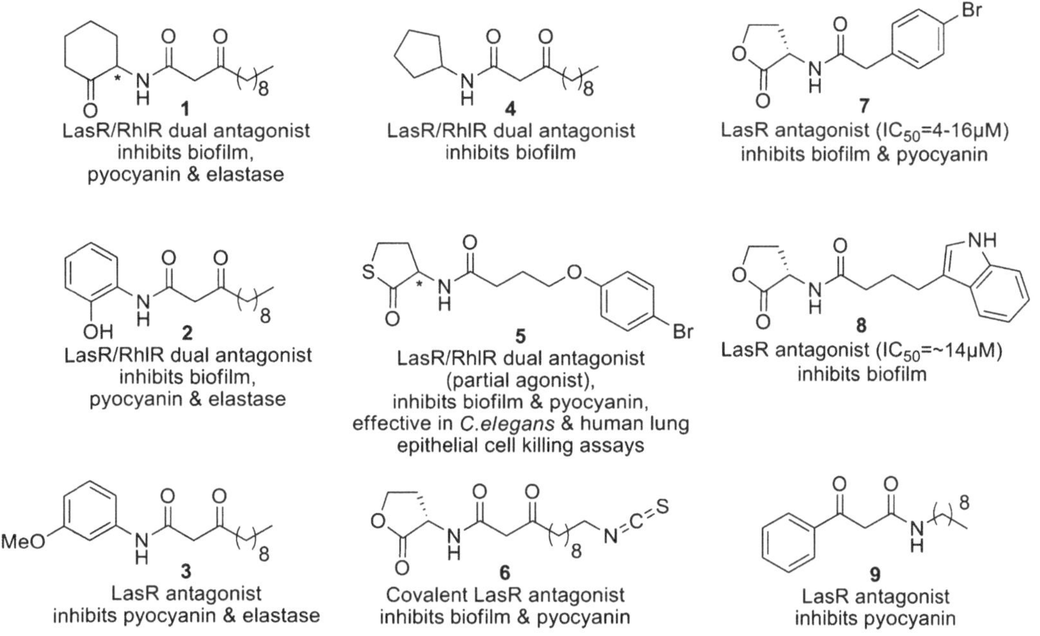

Fig. 2 A selection of LasR and/or RhlR antagonists with reported effects on *P. aeruginosa* cells (Jog et al. 2006; Smith et al. 2003; Hodgkinson et al. 2012; Ishida et al. 2007; O'Loughlin et al. 2013; Amara et al. 2009; Geske et al. 2005; Muh et al. 2006b)

10
LasR antagonist,
inhibits exoprotease &
chitinase activity,
pyoverdin, pyocyanin;
effective on biofilm and in
C.elegans killing assays
and murine lung infection
models

11
LasR postulated as target;
inhibits pyocyanin;
effective in *C.elegans*
killing assays

12
selective LasR antagonist

Fig. 3 A selection of LasR and/or RhlR antagonists with reported effects on *P. aeruginosa* cells (Hentzer et al. 2003; Wu et al. 2004; Musthafa et al. 2012; Muh et al. 2006a)

cases it has been analyzed whether a cellular effect (e.g., reduction of virulence factor production) was mediated via antagonism of LasR, RhlR, or both. Indeed, a very strong dual antagonizing agent might prove ineffective in the cellular context as the *las* and the *rhl* systems can act reciprocally on the production of key virulence factors. Thus, detailed biological studies including in vivo experiments will be necessary to guide further developments in the field of AHL-interfering QSI in *P. aeruginosa*.

4-Hydroxy-2-Alkylquinoline (HAQ) Receptor as Target

PqsR is the receptor of the *P. aeruginosa*-specific *pqs* QS circuit and functions as a critical regulator that fine-tunes a large set of pathogenicity-associated genes that encode for virulence factors, such as pyocyanin, elastase B, and hydrogen cyanide. *Pseudomonas* quinolone signal (PQS) and 2-heptyl-4-hydroxyquinoline (HHQ) – the two most predominant members of

Ligand-based approach

HHQ

PQS

PqsR natural ligands/agonists

Structural modification

13
44% reduction@15 μM
The first antagonist

Optimization

14
IC_{50} = 2 μM
The best antagonist

Structural modification

15
IC_{50} ~ 50 μM

Fragment-based approach

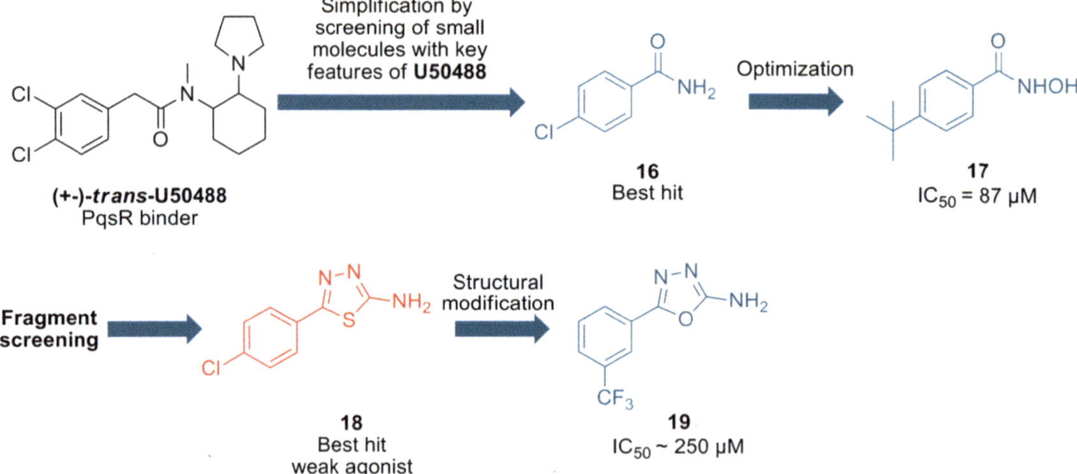

Fig. 4 Discovery of PqsR antagonists following ligand-based and fragment-based approaches and the biological effects on the production of virulence factor pyocyanin in *P. aeruginosa*. Abbreviations: *HHQ* 2-heptyl-4-hydroxyquinoline, *PQS Pseudomonas* quinolone signal, IC_{50} inhibitor concentration to achieve a half-maximal degree of inhibition

the 2-alkyl-4-hydroxyquinoline (HAQ) family – are the agonists of the receptor (Fig. 4) and serve as the signal molecules of the network. While QSIs interfering with receptors of *las* and *rhl* systems are being intensively investigated, only a few compounds targeting PqsR (PqsR antagonists) have been reported.

Following a ligand-based approach, two research groups have individually discovered potent PqsR antagonists. Hartmann and coworkers identified the first PqsR antagonists (e.g., compound **13**, Fig. 4) by means of introducing strong electron-withdrawing groups into the 6-position of HHQ (Lu et al. 2012). Interestingly, such in vitro highly active antagonists reveal opposite functionality (agonistic activity) in *P. aeruginosa* culture, which is attributed to an unexpected functional inversion mediated by a bacterial enzyme PqsH. Overcoming the problem via further structural optimization resulted in the most potent PqsR antagonist to date (inhibitor concentration to achieve a half-maximal degree of inhibition (IC_{50}) toward pyocyanin: 2 μM, compound **14**), which demonstrated anti-virulence efficacy in vivo, thereby providing the first proof of concept for the PqsR-targeting therapeutic strategy (Lu et al. 2014). Meanwhile, Williams and coworkers reported a series of novel PqsR antagonists based on a quinazolinone (QZN) core mimicking the quinolone scaffold of the natural ligands (compound **15**) (Ilangovan et al. 2013). The QZN compounds do not only strongly inhibit the *pqs* QS and production of pyocyanin but also attenuate the biofilm formation of *P. aeruginosa*. Importantly, this work provided the first co-crystal structures of the PqsR ligand-binding domain with either agonist or antagonist giving a deep insight into the ligand-receptor interactions (Ilangovan et al. 2013).

Application of fragment-based approaches is another promising way to discover potential PqsR antagonists. Based on the knowledge that the κ-opioid receptor agonist ($\pm$)-*trans*-U50488 stimulates the transcription of the PqsR-controlled operon, this compound was identified as a PqsR binder via biophysical methods (Klein et al. 2012). To simplify this compound into smaller molecules and modify it into potent antagonists, 106 fragments with key features derived from U50488 were screened and the best hit **16** was further optimized resulting in a hydroxamic acid **17** with both high activity and ligand efficiency. As expected, this antagonist significantly diminished the pyocyanin production. Similarly, attractive hits were discovered via fragment screening of a library collection composed of 720 small molecules (Zender et al. 2013). The most outstanding hit **18** having a 2-phenyl-1,3,4-thiadiazole core was subsequently transformed into potent PqsR antagonist **19**, which successfully suppressed *pqs* QS activity as well as production of virulence factor pyocyanin. Particularly, it is worth to note that **19** is more drug-like than other known PqsR antagonists regarding physicochemical properties.

Overall, PqsR has been attracting attention and the recent contributions highlight QSIs antagonizing PqsR as promising anti-virulence compounds combating *P. aeruginosa*.

Synthetic QSIs Blocking Signal Molecule Biosynthesis

An alternative approach for interference with QS is inhibition of the signal molecule synthesis. Although reports applying this strategy are far less numerous than those about signal reception inhibition, evidence indicates that inhibition of signal synthesis is feasible and effective both in vitro and in vivo (LaSarre and Federle 2013). Usually, signal molecules are generated via a cascade of enzymes, which can be targeted individually for inhibition of signal molecule production.

AHL Biosynthesis

In *P. aeruginosa*, AHL signals C_4-HSL and 3-oxo-C_{12}-HSL are produced by the LuxI-type AHL synthases LasI and RhlI, respectively (Raychaudhuri et al. 2005; Parsek et al. 1999; Gould et al. 2004a). Both enzymes use l-*S*-adenosylmethionine as substrate. The second substrates, butanoyl-acyl carrier protein (ACP) for RhlI and 3-oxo-dodecanoyl-ACP for LasI, are likely derived from fatty acid biosynthesis. Thereby, FabI, a nicotinamide adenine dinucleotide-dependent enoyl-ACP reductase, reduces *trans*-2-enoyl-ACPs to the corresponding acyl-ACPs. FabB, a β-ketoacyl-ACP synthase I, condenses malonyl-ACP with an acyl-ACP resulting in the respective 3-oxo-acyl-ACP (Fig. 5a) (Hoang and Schweizer 1999). So far, only a limited number of reports on AHL biosynthesis inhibition have been published.

Triclosan (**20,** Fig. 5b) was identified as potent inhibitor of FabI (IC_{50} of 0.2 μM) that could suppress the production of C_4-HSL in vitro. However, as *P. aeruginosa* was resistant to triclosan

a AHL biosynthesis

X FabI

S-ACP or C_7H_{15} S-ACP

NADH + H+ NAD+

S-ACP or C_7H_{15} S-ACP

X RhlI

SAM MTA + ACP

NH

C4-HSL

Malonyl-ACP

FabB

ACP + CO_2

C_7H_{15} S-ACP

LasI

SAM MTA + ACP

C_7H_{15} NH

C12-3-oxo-HSL

b QSIs interfering with AHL biosynthesis

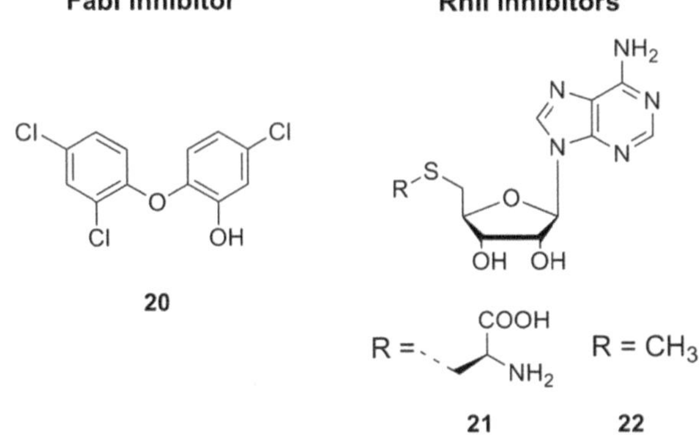

Fig. 5 Current model for *N*-acyl l-homoserine lactone biosynthesis in *P. aeruginosa* (**a**) and representative QSIs (**b**). Abbreviations: *AHL N*-acyl l-homoserine lactone, *ACP* acyl carrier protein, *NADH/NAD*+ reduced/oxidized form of nicotinamide adenine dinucleotide, *SAM* l-*S*-adenosylmethionine, *MTA* 5-methylthioadenosine, *C_4-HSL N*-butanoyl-l-homoserine lactone, *3-oxo-C_{12}-HSL N*-(3-oxo-dodecanoyl)-l-homoserine lactone

due to active efflux, the validity of FabI-targeting QS inhibition could not be demonstrated yet (Hoang and Schweizer 1999).

Interference with C_4-HSL production by inhibiting the synthase RhlI with substrate analogues, reaction intermediates, and final products was investigated by Greenberg and coworkers (Parsek et al. 1999). Inter alia, the substrate analogue l-*S*-adenosylcysteine (**21,** Fig. 5b) and the reaction product 5′-methylthioadenosine (**22,** Fig. 5b) strongly inhibited RhlI activity in vitro at micromolar concentrations (Parsek et al. 1999). However, these homologues have not been evaluated *in cellulo* as they are likely to affect the central pathways of amino acid and fatty acid metabolism (Scutera et al. 2014).

Although the crystal structure of LasI has been elucidated (Gould et al. 2004a, b), no inhibitor targeting the synthase in *P. aeruginosa* has been described so far. However, recent advances in inhibiting LuxI synthases in other bacterial species might pave the way for the development of such inhibitors in *P. aeruginosa* (Chung et al. 2011; Christensen et al. 2013).

HAQ Biosynthesis

The *pqs* QS system, which is unique to *P. aeruginosa*, makes use of HAQs as signal molecules, among which PQS and its precursor HHQ play a major role (Xiao et al. 2006). Their biosynthesis requires the enzymes PqsA-D and PqsH (Fig. 6a). Thereby, PqsA acts as a ligase catalyzing the formation of anthraniloyl coenzyme A (CoA) from anthranilate, adenosine triphosphate, and CoA (Coleman et al. 2008). PqsD, a β-ketoacyl-ACP synthase III (FabH)-type condensing enzyme, has been shown to catalyze the condensation reaction between anthraniloyl-CoA and β-ketodecanoic acid to give HHQ in vitro (Pistorius et al. 2011; Steinbach et al. 2013). However, recent studies revealed that in the cellular context of *P. aeruginosa*, PqsD more likely employs anthraniloyl-CoA and malonyl-CoA as substrates to form 3-(o-aminophenyl)-3-keto-propionic acid. This reactive intermediate is then condensated with octanoyl to yield HHQ by a PqsB/PqsC complex (Dulcey et al. 2013). As the exact mechanism of action of PqsB/PqsC still remains elusive, there have been no attempts so far to develop QSIs targeting PqsB/PqsC. Finally, HHQ is converted into PQS by the monooxygenase PqsH (Schertzer et al. 2010). However, PqsH has not been considered as suitable target as a *pqsH* mutant displayed wild-type virulence in mice (Xiao et al. 2006). Thus, most efforts have been put into the development of small molecule inhibitors blocking PqsA and PqsD.

The first reported inhibitor of PQS production in *P. aeruginosa* was methyl anthranilate (**23**, Fig. 6b), an analogue of the PqsA substrate anthranilate that was able to decrease the levels of PQS-dependent virulence factor elastase. However, concentrations in the millimolar range were necessary to see a pronounced inhibitory effect on PQS formation (Calfee et al. 2001). Although it could be excluded that methyl anthranilate was a substrate or inhibitor of PqsA (Coleman et al. 2008), the exact mechanism of action remained unknown.

Rahme and coworkers aimed at developing more potent substrate analogues based on the anthranilate structure (Lesic et al. 2007). Introduction of electron-withdrawing halogen

a HAQ biosynthesis

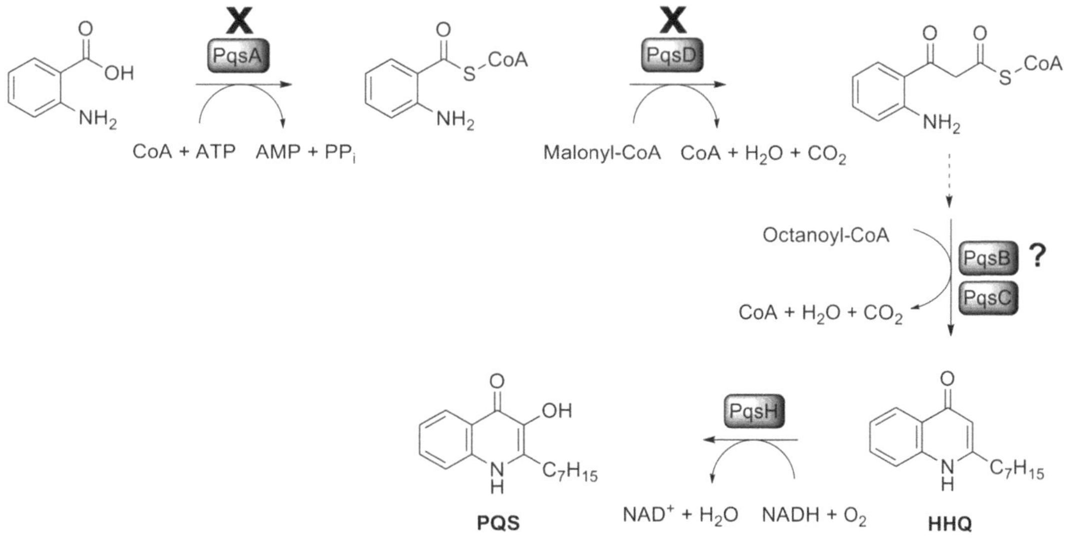

Fig. 6 Current model for 4-hydroxy-2-alkylquinoline biosynthesis in *P. aeruginosa* (**a**) and representative QSIs (**b**). Abbreviations: *HAQ* 4-hydroxy-2-alkylquinoline, *CoA* coenzyme A, *ATP* adenosine triphosphate, *AMP* adenosine monophosphate, *PP$_i$* pyrophosphate; II, *HHQ* 2-heptyl-4-hydroxyquinoline, *NADH/NAD$^+$* reduced/oxidized form of nicotinamide adenine dinucleotide, *PQS Pseudomonas* quinolone signal

b QSIs interfering with HAQ biosynthesis

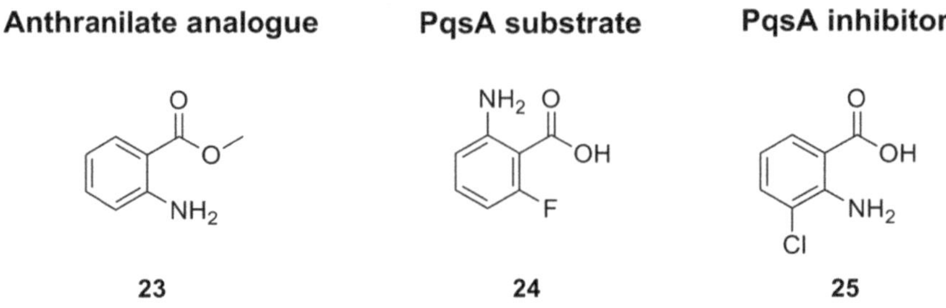

PqsD inhibitors

FabH inhibitor-based approach

$R^1 = H$, $R^2 = Br$ — **26**; $R_1 = OH$, $R_2 = H$ — **28**

$R^1 = H$, $R^2 =$ phenoxy — **27**; $R_1 = Cl$, $R_2 = H$ — **29**

Experimental screening

R = methoxy — **30**; R = - -HN(C=O)CH2Cl — **31**

Ligand-based strategy

32

Fig. 6 (continued)

atoms into the phenyl ring should restrict formation of an activated carbonyl. Indeed, these derivatives strongly inhibited HHQ and PQS formation (Lesic et al. 2007). For instance, 6-fluoroanthranilic acid (**24**, Fig. 6b) exhibited an IC_{50} of 109 μM regarding inhibition of PQS synthesis (Maurer et al. 2013). Excitingly, these compounds were shown to have therapeutic benefits in vivo, where they increased survival and limited systemic dissemination of *P. aeruginosa* in a thermal injury mouse model (Lesic et al. 2007). These effects were concluded to be likely due to PqsA inhibition as anthranilic acid accumulated in cultures grown in the presence of the compounds.

Pesci and coworkers purified PqsA for the first time and developed an in vitro assay for identifying substrates and inhibitors of PqsA (Coleman et al. 2008). For example, anthranilate derivatives bearing chloro- and fluoro-substituents in 4- to 6-position of the benzene ring, such as compound **24** (K_m of

11 μM), were found to be substrates of PqsA. In contrast, 3-chloroanthranilic (**25**, Fig. 6b) acid was identified as an inhibitor of PqsA (K_i of 12.9 μM). In general, the extent of inhibition of PQS synthesis achievable with both substrates and inhibitors did not correlate with respective K_m or K_i values. Finally, it remained to be elucidated whether the inhibition of PQS production by the substrates was due to competition with anthranilate or to the inhibition of downstream enzymes by formed CoA thioesters (Coleman et al. 2008).

In 2011, Müller and coworkers developed an in vitro enzyme assay with purified PqsD using anthraniloyl-CoA and β-ketodecanoic acid as substrates and detecting the product HHQ (Pistorius et al. 2011). As a starting point for the identification of PqsD inhibitors, known inhibitors of FabH, a structural and functional homologue of PqsD, were tested. Indeed, compounds **26** (IC_{50} of 65 μM) and **27** (IC_{50} of 35 μM) could be identified as the first inhibitors of PqsD (Fig. 6b). However, their activity was only moderate and they were not tested in cellular assays as they were expected to exhibit antibiotic activity (Pistorius et al. 2011).

Hartmann and coworkers initiated several rational design projects for the development of potent, selective, and non-bactericidal PqsD inhibitors. In the course of these studies, three classes of PqsD inhibitors have been identified following a design approach based on known inhibitors of FabH, an experimental screening, and a ligand-based strategy. For each class, a series of compounds was synthesized and evaluated for its inhibitory potency in an in vitro PqsD assay. Therefrom, structure-activity relationships were derived. Furthermore, molecular docking based on the crystal structure of PqsD (Bera et al. 2009), biochemical assays, and biophysical methods including surface plasmon resonance spectroscopy (Henn et al. 2012) were applied to characterize the compounds regarding binding site, binding mode, or molecular interactions with the target. Based on that knowledge, structural optimizations were performed that led to potent PqsD inhibitors with IC_{50} values in the single-digit micromolar to submicromolar range (Storz et al. 2012; Weidel et al. 2013; Sahner et al. 2013; Hinsberger et al. 2014; Storz et al. 2013).

The design approach based on known FabH inhibitors (Pistorius et al. 2011) resulted in two subclasses of compounds with 2-benzamidobenzoic acid core structure. The 3′-sulfonamide-substituted series was found to reversibly bind to the substrate access channel within PqsD. The most potent inhibitor **28** (Fig. 6b) exhibited an IC_{50} value of 1.2 μM (Weidel et al. 2013). The 3′-phenoxy/4′-phenyl-substituted 2-benzamidobenzoic acids, originally reported as inhibitors of bacterial RNA polymerase (RNAP), were systematically optimized regarding their activity and selectivity profile. The most promising compound **29** (Fig. 6b) strongly inhibited PqsD (IC_{50} of 6.2 μM) while not affecting RNAP (Hinsberger et al. 2014).

From the class of 5-aryl-ureidothiophene-2-carboxylic acids identified by experimental screening, compounds **30** (IC_{50} of 0.5 μM) and **31** (IC_{50} of 2 μM) turned out to be the most potent (Fig. 6b). As the latter binds covalently to the active site, it carries the potential of strong biological effects (Sahner et al. 2013).

The ligand-guided design strategy led to a potent class of PqsD inhibitors, the nitrophenylmethanols, which showed time-dependent inhibitory activity, tight-binding behavior, and active site binding (Storz et al. 2013). The most promising member of this series, compound **32** (Fig. 6b), exhibited high potency (IC_{50} of 3.2 μM) and high ligand efficiency (0.39). Applied at micromolar concentrations, this compound strongly inhibited HHQ and PQS synthesis as well as biofilm formation in *P. aeruginosa* without affecting growth. Thus, by the use of **32**, it was shown for the first time that inhibition of signal molecule synthesis is feasible with a PqsD inhibitor and that PqsD is a valid anti-biofilm target (Storz et al. 2012).

Conclusion

Numerous highly active blockers of QS receptors as well as inhibitors of QS signal synthases were identified using diverse design strategies during the last decade. Among the developed compounds, several showed anti-infective activities. However, studies providing detailed in vivo data are rare. Nevertheless, the recent scientific achievements emphasize the potential applicability of QSIs as a weapon to treat the recalcitrant infectious diseases caused by *P. aeruginosa*.

Opinion

Numerous scientific contributions demonstrate that inhibition of QS can be regarded as a promising strategy against *P. aeruginosa* infections, and the discovery of small molecules targeting QS is rapidly progressing. However, there are obstacles that hinder the successful translation of such QSIs into real anti-infective drugs. First, the methodologies used for biological evaluation of QSIs are various and the results are highly dependent on experimental conditions, especially on the chosen *P. aeruginosa* strain. This makes a comparison of QSIs developed by different working groups difficult or even impossible. Second, the molecular target of QSIs discovered by cellular screening approaches is often unknown, which hampers a directed optimization of their activity, selectivity, and pharmacokinetic profile. Third, QSIs derived from target-based drug discovery approaches often fail to exhibit cellular activity. Development of such QSIs overcoming the Gram-negative cell wall and escaping the widespread efflux pumps remains a very challenging task. Fourth, despite exciting in vitro and *in cellulo* activities achieved with QSIs, an in-depth evaluation of such inhibitors in advanced animal models or even clinical trials is still to be performed. Such studies are mainly hampered by lack of drug-like molecules. Many QSIs, especially those derived from natural products, suffer from chemical instability, exhibit toxic effects, or possess inappropriate pharmacokinetic properties. Thus, in our opinion, current QS research should focus on development of drug-like molecules applicable for in-depth in vivo studies providing the proof of concept for QS inhibition-based treatment of *P. aeruginosa* infections.

References

Amara N, Mashiach R, Amar D, Krief P, Spieser SAH, Bottomley MJ, Aharoni A, Meijler MM (2009) Covalent inhibition of bacterial quorum sensing. J Am Chem Soc 131(30):10610–10619. doi:10.1021/Ja903292v

Bera AK, Atanasova V, Robinson H, Eisenstein E, Coleman JP, Pesci EC, Parsons JF (2009) Structure of PqsD, a pseudomonas quinolone signal biosynthetic enzyme, in complex with anthranilate. Biochemistry 48(36):8644–8655. doi:10.1021/Bi9009055

Calfee MW, Coleman JP, Pesci EC (2001) Interference with *Pseudomonas* quinolone signal synthesis inhibits virulence factor expression by *Pseudomonas aeruginosa*. Proc Natl Acad Sci USA 98(20):11633–11637. doi:10.1073/pnas.201328498

Christensen QH, Grove TL, Booker SJ, Greenberg EP (2013) A high-throughput screen for quorum-sensing inhibitors that target acyl-homoserine lactone synthases. Proc Natl Acad Sci USA 110(34):13815–13820. doi:10.1073/pnas.1313098110

Chugani SA, Whiteley M, Lee KM, D'Argenio D, Manoil C, Greenberg EP (2001) QscR, a modulator of quorum-sensing signal synthesis and virulence in *Pseudomonas aeruginosa*. Proc Natl Acad Sci USA 98(5):2752–2757. doi:10.1073/pnas.051624298

Chung J, Goo E, Yu S, Choi O, Lee J, Kim J, Kim H et al (2011) Small-molecule inhibitor binding to an N-acyl-homoserine lactone synthase. Proc Natl Acad Sci USA 108(29):12089–12094. doi:10.1073/pnas.1103165108

Coleman JP, Hudson LL, McKnight SL, Farrow JM, Calfee MW, Lindsey CA, Pesci EC (2008) *Pseudomonas aeruginosa* PqsA is an anthranilate-coenzyme a ligase. J Bacteriol 190(4):1247–1255. doi:10.1128/Jb.01140-07

Dulcey CE, Dekimpe V, Fauvelle DA, Milot S, Groleau MC, Doucet N, Rahme LG, Lepine F, Deziel E (2013) The end of an old hypothesis: the pseudomonas signaling molecules 4-hydroxy-2-alkylquinolines derive from fatty acids, not 3-ketofatty acids. Chem Biol 20(12):1481–1491. doi:10.1016/j.chembiol.2013.09.021

Gambello MJ, Iglewski BH (1991) Cloning and characterization of the *Pseudomonas-aeruginosa* lasr gene, a transcriptional activator of elastase expression. J Bacteriol 173(9): 3000–3009

Geske GD, Wezeman RJ, Siegel AP, Blackwell HE (2005) Small molecule inhibitors of bacterial quorum sensing and biofilm formation. J Am Chem Soc 127(37):12762–12763. doi:10.1021/Ja0530321

Geske GD, O'Neill JC, Miller DM, Mattmann ME, Blackwell HE (2007) Modulation of bacterial quorum sensing with synthetic ligands: systematic evaluation of *N*-acylated homoserine lactones in multiple species and new insights into their mechanisms of action. J Am Chem Soc 129(44):13613–13625. doi:10.1021/Ja074135h

Geske GD, Mattmann ME, Blackwell HE (2008) Evaluation of a focused library of N-aryl L-homoserine lactones reveals a new set of potent quorum sensing modulators. Bioorg Med Chem Lett 18(22): 5978–5981. doi:10.1016/j.bmcl.2008.07.089

Gould TA, Schweizer HP, Churchill MEA (2004a) Structure of the *Pseudomonas aeruginosa* acyl-homoserinelactone synthase LasI. Mol Microbiol 53(4):1135–1146. doi:10.1111/j.1365-2958.2004.04211.x

Gould TA, Watson WT, Choi KH, Schweizer HP, Churchi MEA (2004b) Crystallization of *Pseudomonas aeruginosa* AHL synthase Lasl using beta-turn crystal engineering. Acta Crystallogr D Biol Crystallogr 60:518–520. doi:10.1107/S0907444903028300

Henn C, Boettcher S, Steinbach A, Hartmann RW (2012) Catalytic enzyme activity on a biosensor chip: combination of surface plasmon resonance and mass spectrometry. Anal Biochem 428(1):28–30. doi:10.1016/J.Ab.2012.05.024

Hentzer M, Wu H, Andersen JB, Riedel K, Rasmussen TB, Bagge N, Kumar N et al (2003) Attenuation of *Pseudomonas aeruginosa* virulence by quorum sensing inhibitors. Embo J 22(15):3803–3815. doi:10.1093/Emboj/Cdg366

Hinsberger S, de Jong JC, Groh M, Haupenthal J, Hartmann RW (2014) Benzamidobenzoic acids as potent PqsD inhibitors for the treatment of *Pseudomonas aeruginosa* infections. Eur J Med Chem 76:343–351. doi:10.1016/j.ejmech.2014.02.014 S0223-5234(14)00134-2 [pii]

Hoang TT, Schweizer HP (1999) Characterization of *Pseudomonas aeruginosa* Enoyl-Acyl carrier protein reductase (Fabl): a target for the antimicrobial triclosan and its role in acylated homoserine lactone synthesis. J Bacteriol 181(17): 5489–5497

Hodgkinson JT, Galloway WRJD, Wright M, Mati IK, Nicholson RL, Welch M, Spring DR (2012) Design, synthesis and biological evaluation of non-natural modulators of quorum sensing in *Pseudomonas aeruginosa*. Org Biomol Chem 10(30):6032–6044. doi:10.1039/C2ob25198a

Ilangovan A, Fletcher M, Rampioni G, Pustelny C, Rumbaugh K, Heeb S, Camara M et al (2013) Structural basis for native agonist and synthetic inhibitor recognition by the *Pseudomonas aeruginosa* quorum sensing regulator PqsR (MvfR). Plos Pathog 9(7). doi: ARTN e1003508 doi:10.1371/journal.ppat.1003508

Ishida T, Ikeda T, Takiguchi N, Kuroda A, Ohtake H, Kato J (2007) Inhibition of quorum sensing in *Pseudomonas aeruginosa* by N-acyl cyclopentylamides. Appl Environ Microbiol 73(10):3183–3188. doi:10.1128/Aem.02233-06

Jog GJ, Igarashi J, Suga H (2006) Stereoisomers of P-aeruginosa autoinducer analog to probe the regulator binding site. Chem Biol 13(2):123–128. doi:10.1016/j.chembiol.2005.12.013

Kalia VC (2013) Quorum sensing inhibitors: an overview. Biotechnol Adv 31(2):224–245. doi:10.1016/j.biotechadv.2012.10.004

Kim C, Kim J, Park HY, Lee JH, Park HJ, Kim CK, Yoon J (2009) Structural understanding of quorum-sensing inhibitors by molecular modeling study in *Pseudomonas aeruginosa*. Appl Microbiol Biotechnol 83(6):1095–1103. doi:10.1007/s00253-009-1954-3

Klein T, Henn C, de Jong JC, Zimmer C, Kirsch B, Maurer CK, Pistorius D, Muller R, Steinbach A, Hartmann RW (2012) Identification of small-molecule antagonists of the *Pseudomonas aeruginosa* transcriptional regulator PqsR: biophysically guided hit discovery and optimization. ACS Chem Biol 7(9):1496–1501. doi:10.1021/Cb300208g

LaSarre B, Federle MJ (2013) Exploiting quorum sensing to confuse bacterial pathogens. Microbiol Mol Biol Rev 77(1):73–111. doi:10.1128/Mmbr.00046-12

Lesic B, Lepine F, Deziel E, Zhang JW, Zhang QH, Padfield K, Castonguay MH et al (2007) Inhibitors of pathogen intercellular signals as selective anti-infective compounds. Plos Pathog 3(9):1229–1239. doi:ARTN e126. doi 10.1371/journal.ppat.0030126

Lu CB, Kirsch B, Zimmer C, de Jong JC, Henn C, Maurer CK, Musken M, Haussler S, Steinbach A, Hartmann RW (2012) Discovery of antagonists of PqsR, a key player in 2-alkyl-4-quinolone-dependent quorum sensing in *Pseudomonas aeruginosa*. Chem Biol 19(3):381–390. doi:10.1016/j.chembiol.2012.01.015

Lu CB, Maurer CK, Kirsch B, Steinbach A, Hartmann RW (2014) Overcoming the unexpected functional inversion of a PqsR antagonist in *Pseudomonas aeruginosa*: an in vivo potent antivirulence agent targeting pqs quorum sensing. Angew Chem Int Ed 53(4):1109–1112. doi:10.1002/anie.201307547

Maurer CK, Steinbach A, Hartmann RW (2013) Development and validation of a UHPLC-MS/MS procedure for quantification of the *Pseudomonas* Quinolone Signal in bacterial culture after acetylation for characterization of new quorum sensing inhibitors. J Pharm Biomed Anal 86:127–134. doi:10.1016/j.jpba.2013.07.047

McGrath S, Wade DS, Pesci EC (2004) Dueling quorum sensing systems in *Pseudomonas aeruginosa* control the production of the *Pseudomonas* quinolone signal (PQS). FEMS Microbiol Lett 230(1):27–34. doi:10.1016/S0378-1097(03)00849-8

McInnis CE, Blackwell HE (2011a) Design, synthesis, and biological evaluation of abiotic, non-lactone modulators of LuxR-type quorum sensing. Bioorg Med Chem 19(16):4812–4819. doi:10.1016/j.bmc.2011.06.072

McInnis CE, Blackwell HE (2011b) Thiolactone modulators of quorum sensing revealed through library design and screening. Bioorg Med Chem 19(16):4820–4828. doi:10.1016/j.bmc.2011.06.071

Morkunas B, Galloway WRJD, Wright M, Ibbeson BM, Hodgkinson JT, O'Connell KMG, Bartolucci N, Della Valle M, Welch M, Spring DR (2012) Inhibition of the production of the *Pseudomonas aeruginosa* virulence factor pyocyanin in wild-type cells by quorum sensing autoinducer-mimics. Org Biomol Chem 10(42):8452–8464. doi:10.1039/C2ob26501j

Muh U, Hare BJ, Duerkop BA, Schuster M, Hanzelka BL, Heim R, Olson ER, Greenberg EP (2006a) A structurally unrelated mimic of a *Pseudomonas aeruginosa* acyl-homoserine lactone quorum-sensing signal. Proc Natl Acad Sci USA 103(45):16948–16952. doi:10.1073/pnas.0608348103

Muh U, Schuster M, Heim R, Singh A, Olson ER, Greenberg EP (2006b) Novel *Pseudomonas aeruginosa* quorum-sensing inhibitors identified in an ultra-high-throughput screen. Antimicrob Agents Chemother 50(11):3674–3679. doi:10.1128/Aac.00665-06

Musthafa KS, Balamurugan K, Pandian SK, Ravi AV (2012) 2,5-Piperazinedione inhibits quorum sensing-dependent factor production in *Pseudomonas aeruginosa* PAO1. J Basic Microbiol 52(6):679–686. doi:10.1002/jobm.201100292

O'Loughlin CT, Miller LC, Siryaporn A, Drescher K, Semmelhack MF, Bassler BL (2013) A quorum-sensing inhibitor blocks *Pseudomonas aeruginosa* virulence and biofilm formation. Proc Natl Acad Sci U S A 110(44):17981–17986. doi:10.1073/pnas.1316981110

Ochsner UA, Reiser J (1995) Autoinducer-mediated regulation of rhamnolipid biosurfactant synthesis in *Pseudomonas-Aeruginosa*. Proc Natl Acad Sci USA 92(14):6424–6428. doi:10.1073/pnas.92.14.6424

Ochsner UA, Fiechter A, Reiser J (1994) Isolation, characterization, and expression in *Escherichia-Coli* of the *Pseudomonas-aeruginosa* Rhlab genes encoding a rhamnosyltransferase involved in rhamnolipid biosurfactant synthesis. J Biol Chem 269(31):19787–19795

Parsek MR, Val DL, Hanzelka BL, Cronan JE, Greenberg EP (1999) Acyl homoserine-lactone quorum-sensing signal generation. Proc Natl Acad Sci USA 96(8):4360–4365. doi:10.1073/pnas.96.8.4360

Passador L, Cook JM, Gambello MJ, Rust L, Iglewski BH (1993) Expression of *Pseudomonas-Aeruginosa* virulence genes requires cell-to-cell communication. Science 260(5111):1127–1130. doi:10.1126/science.8493556

Persson T, Hansen TH, Rasmussen TB, Skinderso ME, Givskov M, Nielsen J (2005) Rational design and synthesis of new quorum-sensing inhibitors derived from acylated homoserine lactones and natural products from garlic. Org Biomol Chem 3(2):253–262. doi:10.1039/B415761c

Pesci EC, Milbank JB, Pearson JP, McKnight S, Kende AS, Greenberg EP, Iglewski BH (1999) Quinolone signaling in the cell-to-cell communication system of *Pseudomonas aeruginosa*. Proc Natl Acad Sci U S A 96(20):11229–11234

Pistorius D, Ullrich A, Lucas S, Hartmann RW, Kazmaier U, Muller R (2011) Biosynthesis of 2-Alkyl-4(1H)-quinolones in *Pseudomonas aeruginosa*: potential for therapeutic interference with pathogenicity. Chembiochem 12(6):850–853. doi:10.1002/cbic.201100014

Rasmussen TB, Givskov M (2006) Quorum-sensing inhibitors as anti-pathogenic drugs. Int J Med Microbiol 296(2–3):149–161. doi:10.1016/j.ijmm.2006.02.005

Raychaudhuri A, Jerga A, Tipton PA (2005) Chemical mechanism and substrate specificity of RhlI, an acylhomoserine lactone synthase from *Pseudomonas aeruginosa*. Biochemistry 44(8):2974–2981. doi:10.1021/Bi048005m

Sahner JH, Brengel C, Storz MP, Groh M, Plaza A, Muller R, Hartmann RW (2013) Combining in silico and biophysical methods for the development of *Pseudomonas aeruginosa* quorum sensing inhibitors: an alternative approach for structure-based drug design. J Med Chem 56(21):8656–8664. doi:10.1021/Jm401102e

Schertzer JW, Brown SA, Whiteley M (2010) Oxygen levels rapidly modulate *Pseudomonas aeruginosa* social behaviours via substrate limitation of PqsH. Mol Microbiol 77(6):1527–1538. doi:10.1111/j.1365-2958.2010.07303.x

Scutera S, Zucca M, Savoia D (2014) Novel approaches for the design and discovery of quorum-sensing inhibitors. Expert Opin Drug Discov 9(4):353–366. doi:10.1517/17460441.2014.894974

Smith KM, Bu YG, Suga H (2003) Library screening for synthetic agonists and antagonists of a *Pseudomonas aeruginosa* autoinducer. Chem Biol 10(6):563–571. doi:10.1016/S1074-5521(03)00107-8

Stacy DM, Le Quement ST, Hansen CL, Clausen JW, Tolker-Nielsen T, Brummond JW, Givskov M, Nielsen TE, Blackwell HE (2013) Synthesis and biological evaluation of triazole-containing N-acyl homoserine lactones as quorum sensing modulators. Org Biomol Chem 11(6):938–954. doi:10.1039/C2ob27155a

Steinbach A, Maurer CK, Weidel E, Henn C, Brengel C, Hartmann RW, Negri M (2013) Molecular basis of HHQ biosynthesis: molecular dynamics simulations, enzyme kinetic and surface plasmon resonance studies. BMC Biophys 6. doi:Unsp 10. doi:10.1186/2046-1682-6-10

Storz MP, Maurer CK, Zimmer C, Wagner N, Brengel C, de Jong JC, Lucas S et al (2012) Validation of PqsD as an anti-biofilm target in *Pseudomonas aeruginosa* by development of small-molecule inhibitors. J Am Chem Soc 134(39):16143–16146. doi:10.1021/Ja3072397

Storz MP, Brengel C, Weidel E, Hoffmann M, Hollemeyer K, Steinbach A, Muller R, Empting M, Hartmann RW (2013) Biochemical and biophysical analysis of a Chiral PqsD inhibitor revealing tight-binding behavior and enantiomers with contrary thermodynamic signatures. ACS Chem Biol 8(12):2794–2801. doi:10.1021/Cb400530d

Weidel E, de Jong JC, Brengel C, Storz MP, Braunshausen A, Negri M, Plaza A, Steinbach A, Muller R, Hartmann RW (2013) Structure optimization of 2-benzamidobenzoic acids as PqsD inhibitors for *Pseudomonas aeruginosa* infections and elucidation of binding mode by SPR, STD NMR, and molecular docking. J Med Chem 56(15):6146–6155. doi:10.1021/Jm4006302

Wilder CN, Diggle SP, Schuster M (2011) Cooperation and cheating in *Pseudomonas aeruginosa*:

the roles of the las, rhl and pqs quorum-sensing systems. ISME J 5(8):1332–1343. doi:10.1038/ismej.2011.13

Wu H, Song Z, Hentzer M, Andersen JB, Molin S, Givskov M, Hoiby N (2004) Synthetic furanones inhibit quorum-sensing and enhance bacterial clearance in *Pseudomonas aeruginosa* lung infection in mice. J Antimicrob Chemother 53(6):1054–1061. doi:10.1093/Jac/Dkh223

Xiao GP, Deziel E, He JX, Lepine F, Lesic B, Castonguay MH, Milot S,Tampakaki AP, Stachel SE, Rahme LG (2006) MvfR, a key *Pseudomonas aeruginosa* pathogenicity LTTR-class regulatory protein, has dual ligands. Mol Microbiol 62(6):1689–1699. doi:10.1111/j.1365-2958.2006.05462.x

Zender M, Klein T, Henn C, Kirsch B, Maurer CK, Kail D, Ritter C, Dolezal O, Steinbach A, Hartmann RW (2013) Discovery and biophysical characterization of 2-amino-oxadiazoles as novel antagonists of PqsR, an important regulator of *Pseudomonas aeruginosa* virulence. J Med Chem 56(17):6761–6774. doi:10.1021/Jm400830r

Development of Quorum-Sensing Inhibitors Targeting the *fsr* System of *Enterococcus faecalis*

Ravindra Pal Singh and Jiro Nakayama

Introduction

Enterococcus spp. can cause illnesses such as bacteremia, endocarditis, urinary tract infections, posttreatment endophthalmitis, and endodontic infections (Murray 2000; Marothi et al. 2005). Medical treatment of these infectious diseases depends largely on bactericidal or bacteriostatic antibiotics. However, frequent use of such antibiotics has led to the development of drug-resistant bacterial strains, which are difficult to treat (Marothi et al. 2005; Murray 2000). As a result, the blockage of bacterial quorum-sensing (QS) systems has attracted attention owing to its potential to attenuate bacterial virulence without inducing bactericidal pressures that lead to drug resistance. This approach, called quorum quenching (QQ), can be undertaken partially or completely independently of antibiotic treatment. Among several QQ strategies, the use of QS inhibitors (QSIs), which are small molecules that have no adverse effects on bacteria, offers advantages in terms of drug delivery and decreased damage to commensal microbiota.

Enterococcus faecalis is the most prevalent species in the genus *Enterococcus*, especially at clinical sites. The expression of two pathogenicity-related proteases, namely, gelatinase (GelE) and serine protease (SprE), is regulated by a QS system encoded by the *fsr* gene cluster on the *E. faecalis* chromosome. Similar to the cognate QS systems frequently found in low-GC Gram-positive bacteria, the *fsr* QS system is mediated by a cyclic autoinducing peptide (AIP) called gelatinase biosynthesis-activating pheromone (GBAP). GBAP is an 11-residue cyclic peptide in which the hydroxyl group of the third serine forms a lactone bridge with the C-terminal carboxyl group of methionine (please refer to Chap. 14). The structure of GBAP is somewhat different from that of other cognate AIPs, which commonly have a thiolactone ring consisting of five amino acids instead of the lactone ring of nine amino acids found in GBAP. A nanomolar level of GBAP activates the FsrC-FsrA two-component regulatory system, which results in the expression of the *gelE-sprE* operon (Nakayama et al. 2001). QS-controlled protease activity is strongly associated with the pathogenicity of *E. faecalis* through promotion of biofilm formation, breakdown of complement systems, unlocking of collagen barriers, and modulation of cell surface proteins (Hirt et al. 2000; Chuang et al. 2009; Chuang-Smith et al. 2010).

To develop QSI compounds targeting the *fsr* system, we used two approaches: random screening from natural resources and rational drug design. In this chapter, the processes and outcomes of these approaches are summarized and discussed.

R.P. Singh • J. Nakayama (✉)
Laboratory of Microbial Technology, Department of Bioscience and Biotechnology, Faculty of Agriculture, Graduate School, Kyushu University, 6-10-1 Hakozaki, Higashi-ku, Fukuoka 812-8581, Japan
e-mail: nakayama@agr.kyushu-u.ac.jp

V.C. Kalia (ed.), *Quorum Sensing vs Quorum Quenching: A Battle with No End in Sight*,
DOI 10.1007/978-81-322-1982-8_25, © Springer India 2015

Screening of *fsr* Inhibitors from Natural Resources

We began QSI screening with a small number of *actinomycete* strains isolated from soil samples (Nakayama et al. 2007). In total, the culture extracts of 179 strains were subjected to screening for the *fsr* inhibitor. Among three samples showing *fsr* inhibitory activity, a culture filtrate of *Streptomyces* strain Y33-1 showed the most potent QSI activity, blocking the production of GBAP and gelatinase without growth-inhibitory effects. A QSI compound was isolated from the culture filtrate of Y33-1 and identified as a tricyclic peptide, siamycin I, which was originally discovered as an anti-HIV peptide (Nakayama et al. 2007). The half maximal inhibitory concentration (IC_{50}) of siamycin I for gelatinase production by *E. faecalis* is approximately 100 nM. At this concentration, siamycin I shows no cell growth-inhibitory activity, but it slightly inhibits the growth of *E. faecalis* at a concentration of 1 μM (80 % growth 5 h after inoculation) and completely inhibits growth at a concentration of 5 μM (no growth 12 h after inoculation). Siamycin I acts as noncompetitive inhibitor of GBAP and appears to disturb signal transduction of the *fsrC-fsrA* two-component regulatory system (Fig. 1). Siamycin I also inhibits biofilm formation of *E. faecalis* at sublethal doses (Nakayama et al. 2007). Another study has demonstrated that siamycin I blocks autophosphorylation of the histidine receptor kinase FsrC (Phillips-Jones et al. 2013). However, siamycin I also inhibits other ATP-dependent enzyme activities (Ma et al. 2011), which may explain the growth-inhibitory effect of siamycin I at the high dose of 200 μM.

Screening of fungal extracts yielded another attractive QSI. We assayed 153 butanol extracts from various fungal strains for *fsr* inhibition.

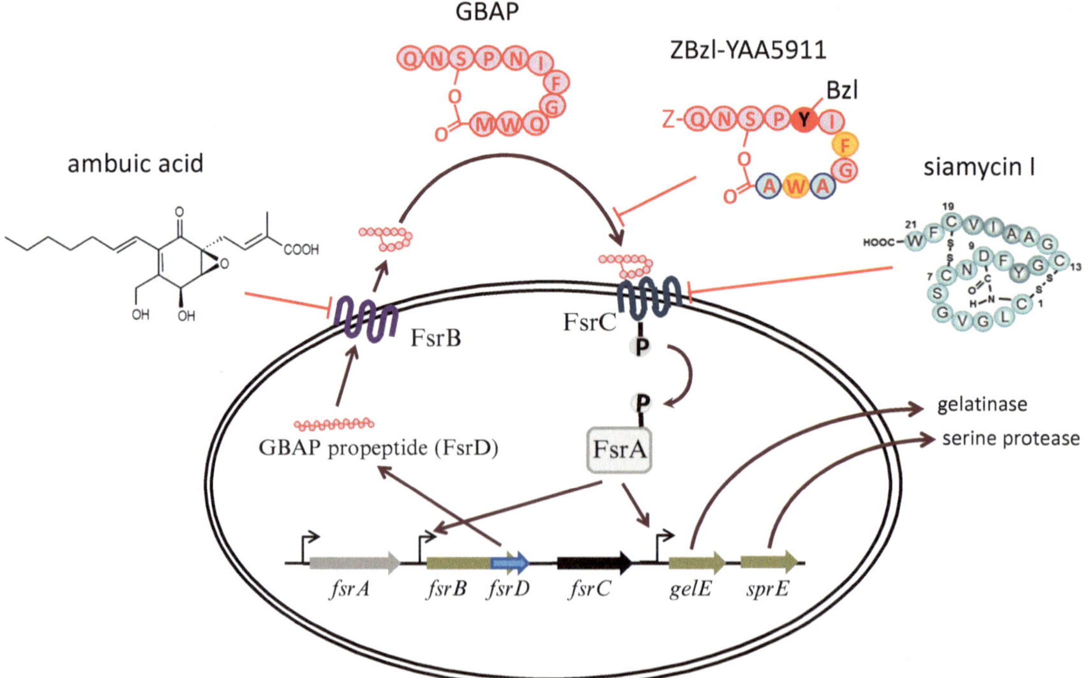

Fig. 1 Mode of action of developed inhibitors targeting the *fsr* quorum-sensing (QS) system of *Enterococcus faecalis*. Siamycin I, produced by *Streptomyces* strain Y33-1, inhibits phosphorylation of FsrC, and ambuic acid, produced by fungi, inhibits the activity of FsrB involved in biosynthesis of gelatinase biosynthesis-activating pheromone (GBAP). ZBzl-YAA5911, developed through peptide design, competitively inhibits the interaction of GBAP with FsrC. Bzl, benzyl

Ambuic acid, with an IC_{50} of approximately 10 μM, was identified from strain KAP-21 (Nakayama et al. 2009). Interestingly, ambuic acid targets GBAP biosynthesis rather than GBAP signal transduction. Specifically, ambuic acid inhibits the cysteine protease functions of FsrB, which are to process and cyclize FsrD, a GBAP propeptide (Fig. 1). FsrB belongs to the AgrB protein family that is widely conserved among low-GC Gram-positive bacteria, suggesting that cyclic AIP is produced via a similar molecular mechanism. Thus, the inhibitory effect of ambuic acid on the biosynthesis of cyclic AIPs was also examined in *Staphylococcus aureus* and *Listeria innocua*, whose genomes encode AgrB and its homologue, respectively. The results of liquid chromatography-mass spectroscopic analysis of culture filtrates revealed that ambuic acid inhibits the biosynthesis of AIPs in these Gram-positive bacteria. Furthermore, ambuic acid inhibits hemolysin production in *S. aureus*, which is one of the outputs controlled by the *agr* QS system (Nakayama et al. 2009).

The finding that ambuic acid has a broad anti-QS spectrum motivated us to perform large-scale screening of cyclic peptide-mediated QS in low-GC Gram-positive bacteria. For this purpose, a high-throughput screening (HTS) system was constructed by combing *agr* assay of the *S. aureus* with *fsr* systems of the *E. faecalis*; (Desouky et al. 2013; Shojima and Nakayama 2014). The screening system consisted of three steps. The first two steps used a dual-reporter strain for the *agr* system, and the last step was based on the conventional *E. faecalis* gelatinase assay. The reporter strain, *S. aureus* 8325-4, carried pSB2035 encoding the bacterial luciferase gene cluster and the green fluorescent protein (GFP) gene under the *agr*P3 promoter and generated luminescence and fluorescence, respectively, in response to AIP induction. In the first step, which was performed with a titer plate assay enabling HTS, we measured luciferase luminescence. To filter out pleiotropic inhibitors that indirectly affect luciferase activity, we performed a second screening with a GFP reporter which directly reflected the activity of the *agr*P3 promoter. The final step was to assay gelatinase production by *E. faecalis*, which is the output of the *fsr* QS system.

A total of 906 actinomycete culture extracts were subjected to the HTS, and 20 extracts were identified as positive in the first screening step. These 20 extracts were subjected to the second screening with GFP, and 16 extracts significantly inhibited GFP expression. After the final step of screening using the 16 extracts, 4 extracts named 608, 609, Y51, and Y67 were identified as potential QSIs against both the *agr* and the *fsr* systems. Some of these compounds were determined to be depsipeptides with molecular sizes approximately the same as those of staphylococcal AIP or GBAP. Determining the structure and mode of action will be a particularly interesting direction for further studies.

Apart from our studies, plant extracts have been assayed for anti-QS activity. *Salvadora persica*, typically used for teeth cleaning in Middle Eastern countries, is of interest for its antimicrobial activity against *Streptococcus mutans* and plaque control capability (Khalessi et al. 2004; Rezaei et al. 2011). Methanolic extracts (roots, leaves, bark, and shoots) of *S. persica* have been screened for QQ targeting *gelE* and *cylR1* expression in *E. faecalis* and *N*-acylhomoserine lactone-mediated QS in *Chromobacterium violaceum* (Rezaei et al. 2011). All extracts showed QQ activity against all targets without bactericidal activity. These results have attracted interest in the mining of QSIs from plant resources.

Development of GBAP Antagonists

S. aureus has four classes of AIPs that interfere with one another (Ji et al. 1997; Otto et al. 1999; Tal-Gan et al. 2013). The discovery of this cross inhibition has accelerated the rational design of AIP antagonists using the structure of natural antagonists as receptor-binding templates. On the contrary, AIP structure does not vary in *Enterococcus*. This structural consistency forced us to undertake de novo design of peptide antagonists. First, we gained knowledge of the structure-activity relationship of GBAP through an alanine scanning experiment. In contrast to those in staphylococcal AIPs, exocyclic amino acid residues are not critical for agonist activity for GBAP. Two aromatic residues, Phe-7 and Trp-10, are indispensable for receptor binding as well as agonist activity (Nishiguchi et al. 2009). Unexpectedly, no antagonistic peptide was obtained through single alanine substitution.

Therefore, we developed a novel reverse alanine scanning approach. Initially, we created a peptide called the receptor-binding scaffold (RBS), which has the minimum structure required for receptor binding. To create the RBS, we retained two aromatic acids, Phe-7 and Trp-10, which are essential for receptor binding, and replaced all other residues of the ring region with alanine (Nakayama et al. 2013). As expected, the synthetic RBS showed very weak but significant QSI activity (25 % inhibition at 100 μM). Thereafter, each alanine residue of RBS was reverted to the original GBAP amino acid. Concurrently, the QSI activity of each revertant was examined, and the peptide with the strongest QSI activity was selected for the next cycle of reverse alanine scanning. We obtained $[Ala^{5,9,11}]$-Z-GBAP (Z = benzyloxycarbonyl) after three cycles of screening. This compound had the strongest QSI activity (IC_{50} 8.7 μM), and none of the additional revertants created in the fourth round of scanning gained QSI activity, suggesting that $[Ala^{5,9,11}]$-Z-GBAP was the maximally reverted peptide (MRP). The structure of this MRP indicated that three residues, Asn^5, Gln^9, and Met^{11}, were critical in determining agonist/antagonist activity. Further experiments focusing on these three residues indicated that Asn^5 is the most critical for agonist/antagonist activity. Thus, Asn^5 was further substituted with other amino acids, including unnatural amino acids, to investigate high QSI activity. Consequently, ZBzl-YAA5911 ($[Tyr (Bzl)^5$, $Ala^{9,11}]$-Z-GBAP; Bzl = benzyl) was obtained as the most potent GBAP antagonist ($IC_{50} = 26$ nM and $K_d = 39$ nM, Fig. 1) (Nakayama et al. 2013).

It is found that enterococcal endophthalmitis is one of the most relentless sight-threatening adverse effects of cataract surgery (Scott et al. 2003; Chen et al. 2009). Studies have demonstrated that *fsr* QS-regulated expression of gelatinase and serine protease is significantly associated with the translocation of bacteria from the anterior chamber to the vitreous chamber, which is a key event in the progression of postoperative endophthalmitis and severe retinal damage (Mylonakis et al. 2002; Engelbert et al. 2004; Suzuki et al. 2008). Vancomycin and fluoroquinolones are commonly used to treat and prevent endophthalmitis, respectively. However, enterococci resistant to these antibiotics have appeared in clinical settings and the natural environment and are of significant concern to ophthalmologists. We challenged ZBzl-YAA5911 to block *E. faecalis* translocation in a rabbit endophthalmitis model, and the compound significantly inhibited the translocation without showing direct bactericidal activity (Nakayama et al. 2013). The destruction of retinal function was significantly rescued in treated animals. This result suggests that ZBzl-YAA5911 has potential as an anti-pathogenic agent. Combining drugs with ZBzl-YAA5911 to increase effectiveness and robustness could open the way for new chemotherapy for enterococcal endophthalmitis.

Conclusion

We used two approaches to identify QSIs targeting the enterococcal *fsr* system: random screening of natural compounds and drug design based on AIP structure. Small-scale screening of actinomycete culture supernatants revealed that siamycin I effectively inhibits *fsr* QS at sublethal concentrations. Siamycin I blocks autophosphorylation of the histidine receptor kinase FsrC. The compound also inhibits other ATP-dependent enzyme activities, which may explain the growth-inhibitory effect of siamycin I at high doses. Another study using fungal extracts showed that ambuic acid blocks QS not only in *E. faecalis* but also in *S. aureus* and *L. innocua*. Interestingly, ambuic acid inhibits the processing of AIP propeptides, which is necessary for the biosynthesis of mature AIP. The fact that the AgrB protein family, which is involved in the processing of AIP propeptides, is commonly encoded in the genomes of a number of low-GC Gram-positive bacteria suggests that thiolactone/lactone AIPs are produced via a common biosynthetic mechanism and that ambuic acid may be effective against QS in these bacteria. These findings related to ambuic acid led us to perform HTS to target cyclic peptide-mediated QS in low-GC Gram-positive bacteria.

In contrast to the varied AIPs of *S. aureus*, the AIPs of enterococci do not interfere with one another. To enable de novo design of GBAP antagonists, we used a novel approach of reverse alanine scanning based on an RBS. Using this approach, we developed an antagonist, ZBzl-YAA5911, with a potency ($IC_{50} = 26$ nM) higher than that of any of the QSIs thus far obtained in our studies. In vivo efficacy of ZBzl-YAA5911 was investigated using a rabbit endophthalmitis model. The compound inhibited *E. faecalis* translocation without direct bactericidal activity. To our knowledge, this example is the first of an anti-*fsr* QSI with in vivo anti-pathogenic efficacy.

Opinion

Throughout our studies of inhibitors targeting cyclic AIP-mediated QS in Gram-positive bacteria, both peptide design and natural compound screening approaches have been effective and characteristic. The peptide design approach has yielded a highly specific QSI, which shows no bactericidal effect or QSI activity against other species. On the contrary, natural compounds obtained have mostly pleiotropic effects, notably at high doses. It may appear that QSIs should be as specific as possible. However, in terms of anti-infectious chemotherapy, bactericidal effect is generally preferred. A strong attack against the QS system and a subsequent attack against pathogen growth would be the best combination. The availability of both specific synthetic QSIs and somewhat broad-spectrum QSIs can obtain this combination and should be central to future applications of these types of QSIs for chemotherapy.

Acknowledgments This research was supported in part by Grants-in-Aid for Scientific Research (B) No. 24380050 and Grant-in-Aid for JSPS Fellows No. 25-03389.

References

Chen KJ, Lai CC, Sun MH, Chen TL, Yang KJ, Kuo YH, Chao AN, Wu WC (2009) Postcataract endophthalmitis caused by *Enterococcus faecalis*. Ocul Immunol Inflamm 17:364–369

Chuang ON, Schlievert PM, Wells CL, Manias DA, Tripp TJ, Dunny GM (2009) Multiple functional domains of Enterococcus faecalis aggregation substance Asc10 contribute to endocarditis virulence. Infect Immun 77:539–548

Chuang-Smith O, Wells C, Henry-Stanley M, Dunny GM (2010) Acceleration of *Enterococcus faecalis* biofilm formation by aggregation substance expression in an *ex vivo* model of cardiac valve colonization. PLoS One 5:e15798

Desouky SE, Nishiguchi K, Zendo T, Igarashi Y, Williams P, Sonomoto K, Nakayama J (2013) High-throughput screening of inhibitors targeting Agr/Fsr quorum sensing in *Staphylococcus aureus* and *Enterococcus faecalis*. Biosci Biotechnol Biochem 77:923–927

Engelbert M, Mylonakis E, Ausubel FM, Calderwood SB, Gilmore MS (2004) Contribution of gelatinase, serine protease, and *fsr* to the pathogenesis of *Enterococcus faecalis* endophthalmitis. Infect Immun 72:3628–3633

Hirt H, Erlandsen SL, Dunny GM (2000) Heterologous inducible expression of *Enterococcus faecalis* pCF10 aggregation substance Asc10 in *Lactococcus lactis* and *Streptococcus gordonii* contributes to cell hydrophobicity and adhesion to fibrin. J Bacteriol 182:2299–2306

Ji G, Beavis R, Novick RP (1997) Bacterial interference caused by autoinducing peptide variants. Science 276:2027–2030

Khalessi AM, Pack AR, Thomson WM, Tompkins GR (2004) An *in vivo* study of the plaque control efficacy of Persica: a commercially available herbal mouthwash containing extracts of *Salvadora persica*. Int Dent J 54:279–283

Ma P, Nishiguchi K, Yuille HM, Davis LM, Nakayama J, Phillips-Jones MK (2011) Anti-HIV siamycin I directly inhibits autophosphorylation activity of the bacterial FsrC quorum sensor and other ATP-dependent enzyme activities. FEBS Lett 585:2660–2664

Marothi YA, Agnihotri H, Dubey D (2005) Enterococcal resistance- an overview. Indian J Med Microbiol 23:214–219

Murray BE (2000) Vancomycin-resistant enterococcal infections. N Engl J Med 342:710–721

Mylonakis E, Engelbert M, Qin X, Sifri CD, Murray BE, Ausubel FM, Gilmore MS, Calderwood SB (2002) The *Enterococcus faecalis fsrB* gene, a key component of the *fsr* quorum sensing system, is associated with virulence in the rabbit endophthalmitis model. Infect Immun 70:4678–4681

Nakayama J, Cao Y, Horii T, Sakuda S, Akkermans AD, de Vos WM, Nagasawa H (2001) Gelatinase biosynthesis activating pheromone: a peptide lactone that mediates a quorum sensing in *Enterococcus faecalis*. Mol Microbiol 41:145–154

Nakayama J, Tanaka E, Kariyama R, Nagata K, Nishiguchi K, Mitsuhata R, Uemura Y, Tanokura M, Kumon H, Sonomoto K (2007) Siamycin attenuates *fsr* quorum sensing mediated by a gelatinase biosynthesis-activating pheromone in *Enterococcus faecalis*. J Bacteriol 189:1358–1365

Nakayama J, Uemura Y, Nishiguchi K, Yoshimura N, Igarashi Y, Sonomoto K (2009) Ambuic acid inhibits the biosynthesis of cyclic peptide quormones in gram-positive bacteria. Antimicrob Agents Chemother 53:580–586

Nakayama J, Yokohata R, Sato M, Suzuki T, Matsufuji T, Nishiguchi K, Kawai T, Yamanaka Y, Nagata K, Tanokura M, Sonomoto K (2013) Development of a peptide antagonist against *fsr* quorum sensing of *Enterococcus faecalis*. ACS Chem Biol 8:804–811

Nishiguchi K, Nagata K, Tanokura M, Sonomoto K, Nakayama J (2009) Structure-activity relationship of gelatinase biosynthesis-activating pheromone of *Enterococcus faecalis*. J Bacteriol 191:641–650

Otto M, Submuth R, Vuong C, Jung G, Götz F (1999) Inhibition of virulence factor expression in *Staphylococcus aureus* by the *Staphylococcus epidermidis agr* pheromone and derivatives. FEBS Lett 450:257–262

Phillips-Jones MK, Patching SG, Edara S, Nakayama J, Hussain R, Siligardi G (2013) Interactions of the intact FsrC membrane histidine kinase with the tricyclic peptide inhibitor siamycin I revealed through synchrotron radiation circular dichroism. Phys Chem Chem Phys 15:444–447

Rezaei A, Oyong GG, Borja VB, Inoue M, Abe T, Tamamura R, Nagatsuka H, Setsu K, Buery RR (2011) Molecular screening of anti-quorum sensing capability of *Salvadora persica* on *Enterococcus faecalis*. J Hard Tissue Biol 20:115–124

Scott IU, Loo RH, Flynn HW Jr, Miller D (2003) Endophthalmitis caused by *Enterococcus faecalis*: antibiotic selection and treatment outcomes. Ophthalmology 110:1573–1577

Shojima A, Nakayama J (2014) Quorum sensing in gram positive bacteria: assay protocols for staphylococcal *agr* and enterococcal *fsr* systems. In: Donelli G (ed) Microbial biofilms: methods and protocols, vol 1147. Springer, Dordrecht, pp 33–41

Suzuki T, Wada T, Kozai S, Ike Y, Gilmore MS, Ohashi Y (2008) Contribution of secreted proteases to the pathogenesis of postoperative *Enterococcus faecalis* endophthalmitis. J Cataract Refract Surg 34:1776–1784

Tal-Gan Y, Stacy DM, Foegen MK, Koenig DW, Blackwell HE (2013) Highly potent inhibitors of quorum sensing in *Staphylococcus aureus* revealed through a systematic synthetic study of the group-III autoinducing peptide. J Am Chem Soc 135: 7869–7882

Part VI

Alternative Strategies as Quorum Sensing Inhibitors

An Alternative Strategy as Quorum-Sensing Inhibitor: Pheromone-Guided Antimicrobial Peptides

Yung-Hua Li and Xiao-Lin Tian

Abbreviations

AMP	Antimicrobial peptides
CSP	Competence-stimulating peptide
EDTA	Ethylenediaminetetraacetic acid
NMR	Nuclear magnetic resonance
PG-AMP	Pheromone-guided antimicrobial peptide
QSI	Quorum-sensing inhibitor

Introduction

Mankind fights microbial infections worldwide by using large quantities of antimicrobials (Geddes 2000). The success of these therapies in infectious diseases is largely based on the availability of antimicrobials that aim to kill target pathogens (Kohansk et al. 2010). Indeed, many bacterial infections can be treated effectively with antibiotics since Alexander Fleming discovered the first antibiotic penicillin in 1928 (Andersson and Hughes 2010). In the battles of fighting against microbial infections today, however, we have found ourselves facing a severe threat: many antibiotics have lost their effectiveness in treating infectious diseases due to the growing crisis of development of antibiotic-resistant microorganisms (Hogan and Kolter 2002; Hughes and Anderson 2012). Meanwhile, many conventionally available antimicrobials exhibit broad-spectral killing with regard to bacterial genera and species (Eckert et al. 2006a, b). Such indiscriminate killing of microbes can disrupt the ecological balance of the indigenous microflora in a natural ecosystem, resulting in various negative clinical consequences (Eckert 2011). The problems resulting from broad-spectrum antimicrobials combined with the emergency of antibiotic resistance highlight the urgent need for new antimicrobials that selectively target specific pathogens but less likely promote antibiotic resistance. Currently, there are at least two major research areas where researchers aim to discover such new antimicrobials as alternatives. In one field, antimicrobial peptides (AMPs) have been rigorously investigated as alternatives to small-molecule antibiotics, mainly because of their ability to kill antibiotic-resistant pathogens (Jenssen et al. 2006; Brogden 2005; Brogden and Brogden 2011). In another field that is extensively studied in recent years, scientists actively search for natural and synthetic compounds that act as quorum-sensing inhibitors (QSI) targeting bacterial cell-cell

Y.-H. Li (✉)
Department of Applied Oral Sciences, Dalhousie University, 5981 University Ave. Rm 5215, Halifax, NS B3H 1W2, Canada

Department of Microbiology and Immunology, Dalhousie University, Halifax, NS, Canada
e-mail: yung-hua.li@dal.ca

X.-L. Tian
Department of Applied Oral Sciences, Dalhousie University, 5981 University Ave. Rm 5215, Halifax, NS B3H 1W2, Canada

V.C. Kalia (ed.), *Quorum Sensing vs Quorum Quenching: A Battle with No End in Sight*, DOI 10.1007/978-81-322-1982-8_26, © Springer India 2015

signaling and its controlled pathogenic activities (LaSarre and Federie 2013). These compounds that quench quorum-sensing mechanisms are considered as promising alternatives to antibiotic-resistant microbes (Kalia and Purohit 2011). It is believed that QSIs target bacterial cell-cell signaling and coordinated activities required for infections, thereby, essentially disarming the bacteria and tipping the balance in favor of the host and allowing the immune system to clear the infectious pathogen (Li and Tian 2012). QSI therapies that specifically block bacterial quorum sensing can make the pathogens become "deaf," "mute," or "blind" rather than directly kill them. Therefore, QSI therapy may achieve the treatment but cause much less selective pressure to create resistant microbes (LaSarre and Federie 2013; Kalia 2013). More recently, a new class of antimicrobials, called pheromone-guided antimicrobial peptides (PG-AMP), has been developed as potential alternatives (Eckert et al. 2006a, b; Mai et al. 2011; Qiu et al. 2003, 2005). PG-AMPs are fusion peptides that consist of a targeting domain of a quorum-sensing signal pheromone from a specific pathogen and a killing domain of a known antimicrobial peptide. The targeting domain can guide such a fusion peptide to bind selectively to the target pathogen, leading to quorum-sensing interference and selective killing. Thereby, pheromone-guided AMPs have added an exciting opportunity to develop new antimicrobials that selectively target pathogens. However, pheromone-guided AMPs and their application as an alternative therapy are still in their infancy. This chapter briefly reviews the current advances in this field.

Rational Design of Pheromone-Guided Antimicrobial Peptides

To circumvent the problem that broad-spectral antimicrobials indiscriminately kill bacterial species, a few attempts have been made to achieve target-specific antimicrobial therapies by coupling antibiotics to species-specific monoclonal antibodies or creating fusion peptides that combine bactericidal domains with bacterial recognition domains (Qiu et al. 2003, 2005; Eckert et al. 2006a, b; Franzman et al. 2009; Mai et al. 2011). These narrow-spectrum antimicrobials can selectively target specific pathogens with little effect on the other members of the resident flora. Among these antimicrobials with "targeting activity," quorum-sensing pheromone-guided AMPs have shown the greater promise to improve the selectivity, effectiveness and applicability. It is well known that quorum-sensing signal pheromones can specifically bind to the same bacterial species that produce them and function at nano-molar concentrations even in a microbial community (Eckert 2011). The rationale of creating such antimicrobial agents is based on the addition of a targeting domain of a quorum-sensing signal pheromone from a target organism to the killing domain of an antimicrobial peptide. Both domains are fused via a small linker to generate a fusion AMP without detrimental change of their activities. Design of such target-specific AMPs requires the identification of two functionally independent molecules, a killing domain comprised of a known AMP and a targeting domain consisting of a species-specific, high-affinity binding molecule, such as quorum-sensing signal pheromones (Fig. 1). The major advantage of such a fusion AMP is that the targeting domain can guide the fusion molecule to bind selectively to the target organism, leading to quorum-sensing interference and selective killing. In recent years, such pheromone-guided antimicrobial peptides have been successfully developed against various bacterial pathogens, including MRSA *S. aureus*, vancomycin-resistant *E. faecalis*, and *Pseudomonas* spp., and cariogenic pathogen *S. mutans* (Qiu et al. 2003, 2005; Eckert et al. 2006a, b; Franzman et al. 2009; Mai et al. 2011).

Structure-Activity Analysis

Pheromone-guided AMPs are hybrid molecules that largely depend on chemical synthesis followed by experimental confirmation of their

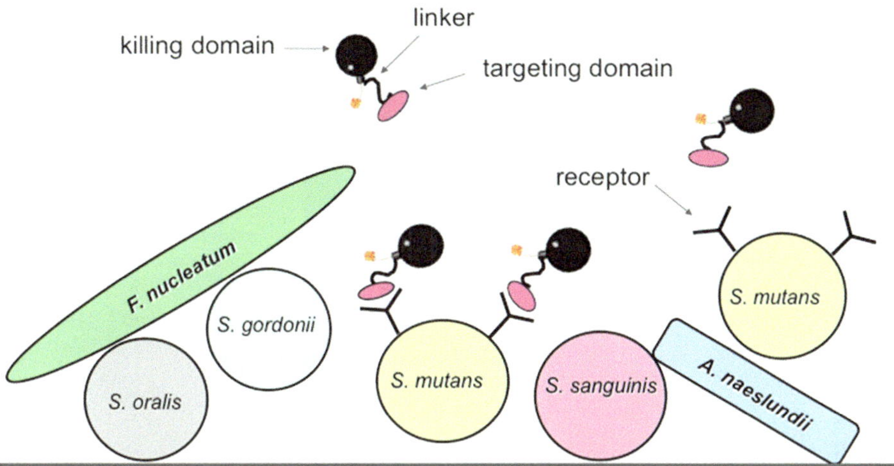

Fig. 1 A schematic diagram describes the rationale of a pheromone-guided antimicrobial peptide to achieve selective killing of a target bacterium, such as *S. mutans*, in a multispecies microbial community. The pheromone-guided AMP is a fusion peptide consisting of a targeting domain of a portion of quorum-sensing signal peptide pheromone specific to *S. mutans*, a killing domain of the portion from a known antimicrobial peptide, and a short linker enabling two functional domains to fuse together to form a fusion peptide. The targeting domain can guide the fusion peptide to bind to the target pathogen, *S. mutans*, thereby triggering the selective killing or quorum sensing interference

bactericidal activity against target pathogens. Therefore, design of pheromone-guided AMPs requires the identification of two functionally independent molecules, a killing domain comprising of a known AMP, and a targeting domain consisting of a species-specific binding domain of a quorum-sensing signal pheromone. However, previous knowledge of two functionally independent molecules may not be sufficient to predict targeted and killing activities of a newly constructed fusion AMP. In many cases, such fusion molecules may be lack of an expected functionality because of changes in primary sequences, charge, and hydrophobic features of a fusion molecule. In few cases, a linker sequence may also affect the targeted activity of a fusion molecule. Therefore, it is important to develop a rationale to design and synthesize pheromone-guided AMPs based on structure-activity analysis of two known functional molecules. Generally, a library containing a list of fusion molecules should be constructed and experimentally evaluated for their activities in terms of the selectivity and killing activity. Based on this concept, Mai et al. (2011) have constructed a library of fusion peptides consisting of a killing domain of the active portion of a marine-derived AMP (Patrzykat et al. 2003) and the targeting domain of a signaling peptide or CSP from *Streptococcus mutans*, a leading cariogenic pathogen (Syvitski et al. 2007). By screening the library, these researchers have identified several pheromone-guided AMPs that selectively target and kill *S. mutans* (Mai et al. 2011). The active fusion peptides consist of three parts: an 8-residue targeting domain of CSP, a GG linker, and a killing domain of a marine-derived AMP (Table 1). Structural analysis by NMR spectroscopy confirms that the active PG-AMP, such as IMB-2, form an amphipathic, α-helical character, which appears to favor the interaction of this peptide with the ComD receptor in the cell membrane of *S. mutans*, whereas the inactive peptide adopts a relatively unstructured, random coil form (Mai et al. 2011). In other studies, He et al. (2007; 2010) have developed a strategy to design and construct two libraries of fusion peptides that also target *S. mutans* by a tunable, building-block approach that use various combinations of antimicrobial, targeting, and linker regions. The rationale developed by this approach enables the researchers to

Table 1 Examples of pheromone-guided antimicrobial peptides (PG-AMP)

Name of peptide	Fusion peptide (targeting domain – linker – killing domain)	MIC	Target organism	Reference
PMC-EF	LVTLVFV (cCF10) – colicin Ia (175 residues)	4 μg/ml	Vancomycin-resistant *Enterococcus faecalis*	Qiu et al. (2005)
PMC-SA	YSTCDFIM (AgrD1) – colicin Ia (175 residues)	4 μg/ml	MSSA or MRSA *Staphylococcus aureus*	Qiu et al. (2005)
G10KHc	KNLRRIIRKGIHIIKKYG (G10) –GGSGGS– KKHRKHRKHRKH (KH)	2.8 μM	*Pseudomonas mendocina*	Eckert et al. (2006)
CSP-18[a]	SGSLSTFFRLFNRSFTQA	NA	*Streptococcus mutans*	Hossain and Biswas (2012)
CSP-21[a]	SGSLSTFFRLFNRSFTQALGK	NA	*S. mutans*	Syvitski et al. (2007)
C16G2	TFFRLFNRSFTQALGK (C16) –GGG– KNLRHRKGIHIIKKY	5.2 μM	*S. mutans*	Eckert et al. (2006)
M8-G2	TFFRLFNR -GGG– KNLRIRKGIHIIKKY	3.25 μM	*S. mutans*	Eckert et al. (2006)
IMB-1	TFFRLFNR –GG– GWGSF-FKKAAHVGKHVGKAALTHYL•NH_2	2.2 μM	*S. mutans*	Mai et al. (2011)
IMB-2	TFFRLFNR –GG– GWGSFFKKAAHVGKL•NH_2	2.8 μM	*S. mutans*	Mai et al. (2011)
IMB2-H/S14	TFFRLFNR –GG– GWG**H**FFKKAAHVGKL•NH_2	2.8 μM	*S. mutans*	Mai et al. (2011)

[a]CSP-18: competence-stimulating peptide from *S. mutans* after *C*-terminal cleavage by SepM; CSP-21: CSP before the *C*-terminal cleavage. Underlined is minimal targeting sequence from CSP. NH_2: *C*-terminal amidation. H/S14: a His-to-Ser14 substitution (bold). MIC: minimum inhibitory concentration. NA: not applicable

modify and improve the targeted killing activities of fusion peptides in the libraries. In both studies, the newly constructed peptides can be optimized through modification of their primary sequences, such as *C*-terminal amidation to reduce the end charge or residue substitutions to enhance their effectiveness and the stability in various physiological conditions, including salt concentrations, low pH, or proteolytic activity in body fluids (He et al. 2010; Li et al. 2010; Mai et al. 2011; Tian et al. 2009).

Targeted Killing Activity

All pheromone-guided AMPs newly constructed have to be experimentally tested and validated for their selective killing activities against target pathogens. The in vitro effectiveness and selectivity of a targeted peptide are usually determined by experimental tests under three culture conditions, mono-species culture, dual-species culture, or multispecies culture or even in biofilms. For example, a PG-AMP, IMB-2, shows strong bactericidal activity (MIC 2.5 μM) against *S. mutans* under a mono-culture condition. However, this may not be sufficient to determine the selectivity of this fusion peptide, so it is further tested in a dual-species culture condition or biofilm, which contains closely related species, such as *S. sanguinis* and *S. gordonii* found in the same ecosystem. This study confirms that IMB-2 selectively target *S. mutans* in the presence of these closely related species under tested culture conditions. The data suggest that the binding of the PG-AMPs to *S. mutans* cells is species-specific, since an enhanced killing activity against *S. mutans* cells is clearly observed in both monoculture and mixed culture. In addition, IMB-2 shows good killing activity against *S. mutans* cells grown in biofilms, although the higher concentration of the peptide is required to achieve desirable killing results. In addition to mono- and dual-species culture conditions, Eckert et al. (2006b) have also evaluated the effectiveness and selectivity of several PG-AMPs against *S. mutans* in multispecies biofilms. These researchers have confirmed that the fusion peptides, especially C16G2, are capable of eliminating *S. mutans* from multispecies biofilms without affecting closely related non-cariogenic oral streptococci. The work suggests that some of the PG-AMPs may have the potential to be developed into as "probiotic" antimicrobials.

In another study, a targeted peptide, G10KHc, exhibits a high degree of specificity for bacterium *Pseudomonas mendocina* and binds tightly to the surface of this species, resulting in rapid and selective killing of this species in a mixed culture condition (Eckert et al. 2006a). It has been found that the targeting domain (KH) and killing domain (G10) of G10KHc function separately: the targeting domain of G10KHc selectively delivers the peptide to the *P. mendocina* surface and the killing domain then rapidly disrupts the outer membrane, increasing its ability to access the cytoplasmic membrane and cause cell death. The data suggest that the specificity of G10KHc against *P. mendocina* results from an independent activity of the KH domain. The combination of these independent processes results in large overall enhancements in killing activity and selectivity against the target pathogen.

The Action Mechanisms

Theoretically, pheromone-guided AMPs should function to exhibit two types of activities, the targeting activity that guides the peptide to bind selectively to the target pathogen and bactericidal activity that triggers a rapid killing by a mechanism that primarily depends on the killing domain. However, the action mechanisms for targeted killing by pheromone-guided AMPs are not fully understood. Each newly constructed AMP requires experimental confirmation of its action mechanism. Currently, there are two types of the action mechanisms that may be involved in the activity of PG-AMPs.

Killing by Disrupting the Membrane

Kaplan et al. (2011) have described the action mechanism of a pheromone-guided AMP, C16G2, which contains a portion of the *S. mutans* CSP (CSP_{16}) and is designed to target *S. mutans*. By comparing C16G2 with a known AMP melittin B, these researchers confirm that C16G2 has a selective membrane-disrupting activity that results in loss of membrane potential ($\Delta\Psi$) of *S. mutans*, leading to cell death at a level of potency similar to that of melittin B. However, the exact mechanism by which C16G2 is selective for *S. mutans* remains unclear, since the ComD receptor that is supposed to be the binding site for C16G2 has no role in targeting of C16G2 to the surface of *S. mutans*. The evidence suggests that C16G2 may involve the membrane-binding or partition steps that enable C16G2 to retain on the *S. mutans* surface, independently of the ComD receptor.

However, Mai et al. (2011) have found that a PG-AMP IMB-2 consisting of eight residue targeting domain of CSP appears to enable the fusion peptide to bind to the ComD receptor of *S. mutans*, because deletion of *comD* results in significant reduction in the killing of *S. mutans* by IMB-2. In another word, the target domain of IMB-2 appears to play a true role in guiding the peptide to bind selectively to *S. mutans* cells and then triggers killing by the killing domain. This divergence was not clear until a recent discovery that an extracellular protease SepM on *S. mutans* cell surface is required for the processing of 21-residue peptide to generate an 18-residue CSP by cleavage of *C*-terminal three residues LGK (Hossain and Biswas 2012). Since the targeting domain of IMB-2 does not have the *C*-terminal three residues LGK, this implies that the *C*-terminal killing domain of IMB-2 would not be subject to the cleavage by the protease SepM (Table 1). In contrast, the fusion peptide C16G2 contains the intact *C*-terminal residues of CSP (CSP_{C16}), including LGK, suggesting that the *C*-terminal killing domain of C16G2 can be cleaved off by SepM, releasing two independent peptides, the targeting and the killing domains.

The discovery of the protease SepM explains the divergence in the action mechanisms between these two fusion peptides, C16G2 and IMB-2. Thus, the action mechanism of C16G2 observed by Kaplan et al. (2011) can be interpreted as the action mode of the *C*-terminal killing domain alone, which is actually consistent with their observation in the report. However, this does not exclude the possibility that fusion peptide IMB-2 may involve an additional unknown binding site on the surface of *S. mutans*, since it also kills about 20 % of the ComD deletion mutant cells. It is speculated that due to its amphipathic nature IMB-2 may interact mainly with the ComD receptor and, to a lesser extent, with the cell membrane of *S. mutans* before triggering the killing.

Quorum-Sensing Interference (QSI)

Pheromone-guided AMPs are chimeric molecules containing the binding domain of a quorum-sensing signal pheromone and the killing domain of a known antimicrobial peptide. The binding domain is expected to recognize and interact with the cognate receptor on a target organism. However, it remains an open question whether the binding of such a fusion AMP to the receptor affects quorum-sensing signaling transduction. Recent works by screening a library of PG-AMPs reveal that some fusion AMPs initiate rapid killing and these peptides are often selected for further investigation, as described in publications (Eckert et al. 2006b; Mai et al. 2011). However, the work also shows that some fusion peptides exhibit growth inhibition or slow "killing" of the target organism (Mai et al. 2011). Since the short-time killing is not so great, these fusion peptides are often not included for further study. To understand structure-activity relationships of newly constructed peptides, some of these molecules have been explored for their action mechanisms (unpublished data). Interestingly, some of these fusion peptides show growth inhibition or death and interference with quorum-sensing signaling, as detected by luciferase reporter activity assays of the promoters of quorum-sensing-controlled genes. The initial work suggests that these fusion peptides may exhibit dual activities, quorum-sensing interference (QSI), and targeted killing, although the mechanisms behind these observations remain to be determined. Therefore, it is reasonable to assume that fusion peptides that do not confer immediate killing may involve other mechanisms to affect the target organism and should be included for further investigation.

In recent year, quorum-sensing mechanisms have been recognized as a new target for development of quorum-sensing inhibitors as alternative antimicrobials (LaSarre and Federie 2013; Kalia and Purohit 2011; Kalia 2013). Any compound that interferes with interactions between quorum-sensing signal molecules and their cognate receptors might block bacterial cell-cell signaling and its controlled virulence. QSIs that specifically block or override bacterial quorum sensing can make the pathogens lose their capability to communicate for coordinated activities. It is believed that the QSI-interfered pathogens can be readily eliminated by the host defense mechanisms (LaSarre and Federie 2013). This approach potentially has several advantages, including exerting less selective pressure toward resistant microbes, preserving the normal microflora, and expanding the repertoire of bacterial targets. Clearly, further study is required to elucidate the action mechanisms of individual pheromone-guided AMPs.

The Stability in Physiological Conditions

It is known that many AMPs are sensitive to variations in salt concentrations, ionic strength, pH, and proteolytic activity in body fluids, such as in serum, saliva, or gastrointestinal fluid, resulting in reduction or loss of their bactericidal activities (Jenssen et al. 2006). Therefore, the ability to resist physiological salts, pH fluctuation, and proteolytic activity is crucial for pheromone-guided AMPs to retain their killing activities in the host. A pheromone-guided AMP IMB-2 that specifically target *S. mutans* has been assessed for their stability in physiological conditions in body fluids such as human saliva (Mai et al. 2011).

Human saliva is a mixed fluid that contains 99 % water and 1 % organic and inorganic molecules with an average pH 6.7 (Syeebny et al. 1992). Total electrolytes in saliva largely account for ionic strength or salt concentrations, which are normally lower than those in serum, but vary greatly with saliva flow and health status. In addition, whole saliva contains a number of components, such as antimicrobial compounds, proteins, glycoproteins, and bacterial products, which account for most of its proteolytic activity (Syeebny et al. 1992). It is reported that IMB-2 is capable of maintaining killing activity in physiological or even higher salt concentrations for about 15 min, which is sufficient for the peptide to exhibit nearly full killing activity, indicating the relative stability of this fusion peptide in human saliva (Mai et al. 2011). It is also found that a low dose of chelating agent EDTA can significantly improve the stability of IMB-2 in human saliva, suggesting that it should be an advantage by combining pheromone-guided AMP with a low concentration of EDTA. In addition to resistance to high salt and proteolytic activity, IMB-2 is also found to be relatively resistant to low pH. In particular, a modification of the peptide by a Ser-to-His substitution, named IMB2-H/S14, improves the killing activity of the fusion peptide by 10–20 % at lower pH levels. Thus, IMB-2H/S14 appears to be more desirable than IMB-2 for application under pH fluctuation conditions such as in the oral cavity.

Clinical Application and Efficacy

Currently, very few of studies have been carried out to investigate the efficacy and potential of pheromone-guided AMPs in vivo, although several studies have reported the killing activity of these fusion AMPs in laboratory conditions. There are considerable challenges in clinical application, because of some doubts about the ability to achieve high antimicrobial activity under physiological conditions, the susceptibility of peptides to proteolytic degradation, the lack of information about potential toxicities in vivo, and comparatively high costs associated with peptide production. Despite such challenges, a pilot study has been conducted in human volunteers to evaluate the clinical efficacy of a PG-AMP, C16G2, against *S. mutans* in a mouth rinse (Sullivan et al. 2011). These researchers have found that C16G2 is highly effective in stopping the growth of *S. mutans* in dental biofilms and reduces the formation of total dental biofilms during the course of the study. As a proof of concept, this study has achieved relatively good results of in vivo efficacy, the selectivity, the delivery method, and the stability in the human hosts.

Summary

Development of target-specific antimicrobial agents that selectively target specific pathogens may provide an opportunity to circumvent some clinical problems, such as ecological disruption and negative clinical consequences, due to the use of wide-spectrum antibiotics. Since bacterial quorum-sensing signal pheromones specifically bind to the same species at nano-molar concentrations, researchers have taken advantage to develop a new class of antimicrobials, called pheromone-guided antimicrobial peptides. This new class of AMPs consists of a targeting domain of a quorum-sensing signal pheromone from a pathogen and the killing domain of a known antimicrobial peptide, both of which are fused via a small linker to generate a pheromone-guided AMP without detrimental change of their activities. The major advantage of such an AMP is that the targeting domain can guide the fusion peptide to bind selectively to the target pathogen, leading to quorum-sensing interference and selective killing. Novel pheromone-guided AMPs, such as IMB-2 and C16G2, have been shown to be effective to kill their target organism in a mixed microbial community. These pheromone-guided AMPs have been demonstrated to retain relatively good killing activities under various physiological conditions and can be further modified to improve their stability or killing activity. Efforts have focused on increasing stability and retention, improving antimicrobial activity, reducing cytotoxicity, and incorporating AMPs in novel formulations. Because of their effectiveness, selectivity, relative stability, and minimal toxicity toward the host cells, this new class of target-specific AMPs has shown great potential for future development as alternatives in fighting against certain types of infectious diseases.

References

Andersson DI, Hughes D (2010) Antibiotic resistance and its cost: is it possible to reverse resistance? Nat Rev Microbiol 8:260–271

Brogden KA (2005) Antimicrobial peptides: pore formers or metabolic inhibitors in bacteria? Nat Rev Microbiol 3:238–250

Brogden NK, Brogden KA (2011) Will new generations of modified antimicrobial peptides improve their potential as pharmaceuticals? Int J Antimicrob Agents 38:217–225

Eckert R (2011) Road to clinical efficacy: challenges and novel strategies for antimicrobial peptide development. Future Microbiol 6:635–651

Eckert R, Qi F, Yarbrough DK, He J, Anderson MH, Shi WY (2006a) Adding selectivity to antimicrobial peptides: rational design of a multidomain peptide against *Pseudomonas* spp. Antimicrob Agents Chemother 50:1480–1488

Eckert R, He J, Yarbrough DK, Qi F, Anderson MH, Shi W (2006b) Targeted killing of *Streptococcus mutans* by a pheromone-guided "smart" antimicrobial peptide. Antimicrob Agents Chemother 50:3651–3657

Franzman MR, Burnell KK, Dehkordi-Vakil FH, Guthmiller JM, Dawson DV, Brogden KA (2009) Targeted antimicrobial activity of a specific IgG-SMAP28 conjugate against *Porphyromonas gingivalis* in a mixed culture. Int J Antimicrob Agents 33:14–20

Geddes A (2000) Infection in the twenty-first century: prediction and postulates. J Antimicrob Chemother 46:873–877

He J, Eckert R, Pharm T, Simanian MD, Hu C, Yarbrough DK, Qi F, Anderson MH, Shi W (2007) Novel synthetic antimicrobial peptides against *Streptococcus mutans*. Antimicrob Agents Chemother 51:1351–1358

He J, Yarbrough DK, Kreth J, Anderson MH, Shi W, Eckert R (2010) Systematic approach to optimizing specifically targeted antimicrobial peptides against *Streptococcus mutans*. Antimicrob Agents Chemother 54:2143–2151

Hogan D, Kolter R (2002) Why are bacteria refractory to antimicrobials? Curr Opin Microbiol 5:472–477

Hossain MS, Biswas I (2012) An extracellular protease, SepM, generates functional competence-stimulating peptide in *Streptococcus mutans* UA159. J Bacteriol 194:5886–5896

Hughes D, Anderson DI (2012) Selection of resistance at lethal and non-lethal antibiotic concentrations. Curr Opin Microbiol 15:555–560

Jenssen H, Hamill P, Hancock REW (2006) Peptide antimicrobial agents. Clin Microbiol Rev 19:491–511

Kalia VC (2013) Quorum sensing inhibitors: an overview. Biotechnol Adv 31:224–245

Kalia VC, Purohit HJ (2011) Quenching the quorum sensing system: potential antibacterial drug targets. Crit Rev Microbiol 37:121–140

Kaplan CW, Sim JH, Shah KR, Kolesnikova-Kaplan A, Shi WJ, Eckert R (2011) Selective membrane disruption: mode of action of C16G2, a specifically targeted antimicrobial peptide. Antimicrob Agents Chemother 55:3446–3452

Kohansk MA, Dwyer DJ, Collins JJ (2010) How antibiotics kill bacteria: from targets to network. Nat Rev Microbiol 8:423–435

LaSarre B, Federie MJ (2013) Exploiting quorum sensing to confuse bacterial pathogens. Microbiol Mol Biol Rev 77:73–111

Li YH, Tian XL (2012) Quorum sensing and bacterial social interactions in biofilms. Sensors 12:2519–2538

Li L, He J, Eckert R, Yarbrough D, Lux R, Anderson MH, Shi W (2010) Design and characterization of an acid-activated antimicrobial peptide. Chem Biol Drug Des 75:127–132

Mai J, Tian XL, Gallant JW, Merkley N, Biswas Z, Syvitski R, Douglas SE, Junqi Ling JQ, Li YH (2011) A novel target-specific, salt-resistant antimicrobial peptide against the cariogenic pathogen *Streptococcus mutans*. Antimicrob Agents Chemother 55:5205–5213

Patrzykat A, Gallant JW, Seo JK, Pytyck J, Douglas SE (2003) Novel antimicrobial peptides derived from flatfish genes. Antimicrob Agents Chemother 47:2464–2470

Qiu XQ, Wang H, Lu XF, Zhang J, Li SF, Cheng G, Wan L, Yang L, Zuo JY, Zhou YQ, Wang HY, Cheng X, Zhang SH, Ou ZR, Zhong ZC, Cheng JQ, Li YP, Wu GY (2003) An engineered multidomain bactericidal peptide as a model for targeted antibiotics against specific bacteria. Nat Biotechnol 21:1480–1485

Qiu XQ, Zhang J, Wang H, Wu GY (2005) A novel engineered peptide, a narrow-spectrum antibiotic, is effective against vancomycin-resistant *Enterococcus faecalis*. Antimicrob Agents Chemother 49:1184–1189

Sullivan R, Santarpia P, Lavender S, Gittins E, Liu Z, Anderson MH, He J, Shi W, Eckert R (2011) Clinical efficacy of a specifically targeted antimicrobial peptide mouth rinse: targeted elimination of Streptococcus mutans and prevention of demineralization. Caries Res 45:415–428

Syeebny LM, Banoczy J, Baum BJ, Edgar WM, Epstein JB, Fox PC, Larmas M (1992) Saliva: its role in health and disease. Int Dent J 42(Suppl 2):291–304

Syvitski RT, Tian XL, Sampara K, Salman A, Lee SF, Jakeman DL, Li YH (2007) Structure-activity analysis of quorum-sensing signaling peptides from *Streptococcus mutans*. J Bacteriol 189:1441–1450

Tian XL, Syvitski RT, Liu T, Livingstone N, Jakeman DL, Li YH (2009) A method for structure-activity analysis of quorum-sensing signaling peptides from naturally transformable streptococci. Biol Proced Online 11:207–226

Alternative Strategies to Target Quorum Sensing (QS): Combination of QS Inhibitors with Antibiotics and Nanotechnological Approaches

Divakara S.S.M. Uppu, Chandradhish Ghosh, and Jayanta Haldar

Introduction

Conventional antibiotics, targeting the essential cellular processes such as cell wall biosynthesis, protein synthesis, and DNA replication and repair, are limited by the ease at which bacteria develop resistance against them (Wright 2010). The increasing occurrence of resistant bacteria has become a major threat to public health worldwide (McKenna 2013). Consequently, much effort has been directed toward the identification of novel strategies such as targeting the bacterial cell membrane and attenuation of bacterial virulence to tackle bacterial resistance (Defoirdt et al. 2010). Quorum sensing (QS) is one such important phenomenon, which can be targeted to tackle the problems posed by multidrug-resistant bacteria. QS involves a mechanism of gene regulation in response to the presence or absence of small signal molecules (Miller and Bassler 2001). Thus, therapeutics targeting this nonessential pathway could prevent bacterial pathogenesis and in turn prevent the cause of a disease. Since QS is not essential for bacterial proliferation or survival, it is generally believed (although not proven yet) that pathogens are unlikely to develop resistance to such a strategy which poses no or little selective pressure to bacteria (Defoirdt et al. 2010). More importantly, this bacterial communication would be required for the transfer of genetic information (e.g., horizontal gene transfer) in the process of spreading bacterial resistance. Targeting such a communication would have a tremendous impact on stalling the spread of bacterial resistance. Extensive reviews on the development of quorum sensing inhibitors (QSIs) are available in the literature, which advocate the efficacy of this strategy (de Kievit 2009; Rasmussen and Givskov 2006; Kalia 2013; Tay and Yew 2013).

Biofilms, the most important outcome of QS in bacteria, often withstand the immune responses of the host and are markedly more resistant to various antimicrobial treatments. In most pathogens studied so far, blocking QS does not abolish biofilm formation, but does make the biofilm more susceptible to antimicrobials and immune reactions. So, targeting quorum sensing would help in the clearance of such stubborn biofilms (Rasmussen and Givskov 2006; de Kievit 2009; Kalia 2013; Tay and Yew 2013).

Antibiotics and Non-antibiotics as QSIs

Over the past few decades, the research related to the development of QSIs has gained increasing attention. However, none of the compounds made it yet to the phase II stage as the toxicity concerns limit their application in the clinical settings.

D.S.S.M. Uppu • C. Ghosh • J. Haldar (✉)
Chemical Biology and Medicinal Chemistry Laboratory, New Chemistry Unit, Jawaharlal Nehru Centre for Advanced Scientific Research, Jakkur, Bengaluru 560064, Karnataka, India
e-mail: jayanta@jncasr.ac.in

V.C. Kalia (ed.), *Quorum Sensing vs Quorum Quenching: A Battle with No End in Sight*,
DOI 10.1007/978-81-322-1982-8_27, © Springer India 2015

Table 1 Antibiotics and non-antibiotics as QSIs

Compound(s)	Inhibition of QS/biofilm formation	Organism	Reference
Antibiotics			
Kanamycin	Makes QS-deficient mutants more susceptible compared to wild-type strains	*P. aeruginosa*	Shih and Huang (2002)
Azithromycin Ciprofloxacin Ceftazidime	QS inhibition by reducing the production of virulence factors and altering the gene expression that regulates QS	*P. aeruginosa*	Skindersoe et al. (2008)
Azithromycin	Inhibits QS in vitro, reduces the expression of QS circuit and target genes in patient and favorable pharmacokinetics and metabolism	*P. aeruginosa*	Skindersoe et al. (2008) Delden et al. (2012)
Non-antibiotics			
Polyphenols	Quorum sensing ability has been reported to be a side effect of peroxide production of pyrogallol	*V. harveyi*	Huber et al. (2003) Ni et al. (2008) Defoirdt et al. (2013)
Niclosamide	Inhibits QS by inhibiting the production of virulence factors, biofilm formation, and effects motility. Increases the survival in *G. mellonella* insect model of infection	*P. aeruginosa*	Imperi et al. (2013)
4-Nitropyridine-*N*-oxide Halofuranones Penicillic acid Patulin Fimbrolides Nitric oxide (NO)	Blocks QS by inhibiting the production of various signal molecules and inhibits biofilm formation	*P. aeruginosa*	Rasmussen and Givskov (2006) Skindersoe et al. (2008) Straight and Kolter (2009) Romero et al. (2011) Kutty et al. (2013) Barraud et al. (2006)
2-Aminoimidazoles	Inhibit and disperse biofilms	*P. aeruginosa* *A. baumannii* *S. aureus*	Rogers et al. (2009) Rogers et al. (2010)

A promising way to overcome this problem is searching for QSIs among the thousands of drugs approved for clinical use in the treatment of different diseases. Antibiotics form the most important group of clinically approved drugs that would serve this purpose. Investigations have shown that biofilms of *P. aeruginosa* QS mutants have reduced tolerance to antibiotics compared with that of their wild-type parents, and inhibition of QS has been shown to promote the eradication of biofilms by antimicrobial treatments and make the biofilm more susceptible to phagocytosis by neutrophils (Rasmussen et al. 2005; Davies et al. 1998; Hentzer et al. 2003). It has been shown that biofilms formed by the QS-deficient mutants of *P. aeruginosa* were shown to be more susceptible to kanamycin compared to the wild-type strains (Table 1) (Shih and Huang 2002). Interestingly, antibiotics can act as QSIs at their subinhibitory concentrations and play a role in the expression of genes involved in key biological process including general stress response, exopolysaccharide production, virulence, QS, and biofilm formation (for more information, see Straight and Kolter (2009) and Romero et al. (2011)).

Screening of 12 antibiotics for QSI activities against *P. aeruginosa* resulted in the identification of three antibiotics, azithromycin (AZM), ciprofloxacin (CPR), and ceftazidime (CFT) (Table 1), that inhibited the production of the important virulence factors such as chitinase, protease, and elastase and the production of the heat-stable hemolysin rhamnolipid. Another important virulence component of *P. aeruginosa* is the T3S system, which is negatively regulated by QS in *P. aeruginosa*. The DNA microarray data showed that AZM increases the level of expression of T3S genes, in accordance with the QS-inhibiting properties of AZM. The data suggested that the underlying mechanism may be mediated by changes in membrane permeability, thereby influencing the flux of *N*-3-oxo-dodecanoyl-L-homoserine lactone (Skindersoe et al. 2008).

More importantly, macrolides, like azithromycin, at clinically achievable concentrations are neither bactericidal nor bacteriostatic for *P. aeruginosa* and are therefore unlikely to select resistant clones. However, at these concentrations, azithromycin inhibits QS in vitro (Skindersoe et al. 2008) and reduces the expression of QS circuit and target genes in patient and has a favorable profile in terms of pharmacokinetics and of metabolism (Table 1) (Delden et al. 2012).

Non-antibiotics such as polyphenols in general are known to have a broad range of biological properties including antimicrobial and antifouling properties. Several resistant interactions of fungi and bacteria involve toxic concentrations of polyphenolic compounds (Huber et al. 2003). Moreover, the boric complex, which is the active form of autoinducer AI-2 in *Vibrio harveyi*, interacts with the LuxP protein via two anionic OH groups. It was thus logical to explore polyphenols containing at least two ionizable OH groups as potential QSIs (Table 1) (Ni et al. 2008). In a study carried out with the *Vibrio harveyi* system, it was shown that polyphenols such as pyrogallol and its derivatives were quite active in inhibiting QS. Thus, the presence of the third OH group was important for activity. It was found that several other derivatives of pyrogallol were equally active (Ni et al. 2008). Polyphenols based on gallic acid moiety were also found to block AHL-mediated communication in bacteria. In one study, it was shown that common polyphenols produced by plants, such as epigallocatechin gallate, ellagic acid, and tannic acid, exhibited significant antagonistic effects (Huber et al. 2003).

Niclosamide (Fig. 1), an FDA-approved anthelmintic drug, has been shown to inhibit quorum sensing in *P. aeruginosa* by inhibiting the acyl homoserine lactone (AHL) QS signal molecules (Table 1). It suppressed the production of QS-regulated extracellular virulence factors such as elastase, pyocyanin, and rhamnolipids and effected the motility and biofilm formation of *P. aeruginosa*. In *G. mellonella* insect model of infection, niclosamide reduced the pathogenicity of *P. aeruginosa* (Imperi et al. 2013). Examples of other compounds that can block QS in *P. aeruginosa* by inhibiting the production of various signal molecules and/or inhibit biofilm formation are 4-nitropyridine-*N*-oxide, halofuranones, fimbrolides, 2-aminoimidazoles, and the fungal metabolites penicillic acid and patulin (Table 1) (Rasmussen and Givskov 2006).

Combination of Antibiotics and QSIs

Combination approach would be one of the most promising approaches to target the multidrug-resistant bacteria (Worthington and Melander 2013). The combination of antibiotics and QSIs (Table 2) can overcome the limitations of monotherapy as in: (1) the concentration of QSIs can be reduced, thus addressing its toxicity issues; (2) the concentration of antibiotic needed can also be brought down, thus decreasing its selective pressure on the bacteria; (3) using QSIs would yield low propensity to develop bacterial resistance; and (4) the combination of bactericidal antibiotics with QSIs would help in the effective bacterial clearance (Estrela and Abraham 2010).

Dissolving pathogenic biofilms exposes the bacteria to the immune system of the host. It may be possible that the immune reaction is sufficient to clear the pathogen, but it is more likely that some bacteria escape the immune response and reestablish the biofilms. To address this problem, several attempts have been made to kill the bacteria released from the biofilm (planktonic bacteria) by treatment with standard antibiotics.

Autoinducer (AI) analogues have been shown to inhibit QS responses in bacteria. It has been demonstrated that a AI-2 analogue, isobutyl-DPD (DPD- ((S)-4,5-dihydroxy-2,3-pentanedione)), and AI-1 analogue, phenyl-DPD (Fig. 1), act synergistically with gentamicin (Table 2) in the

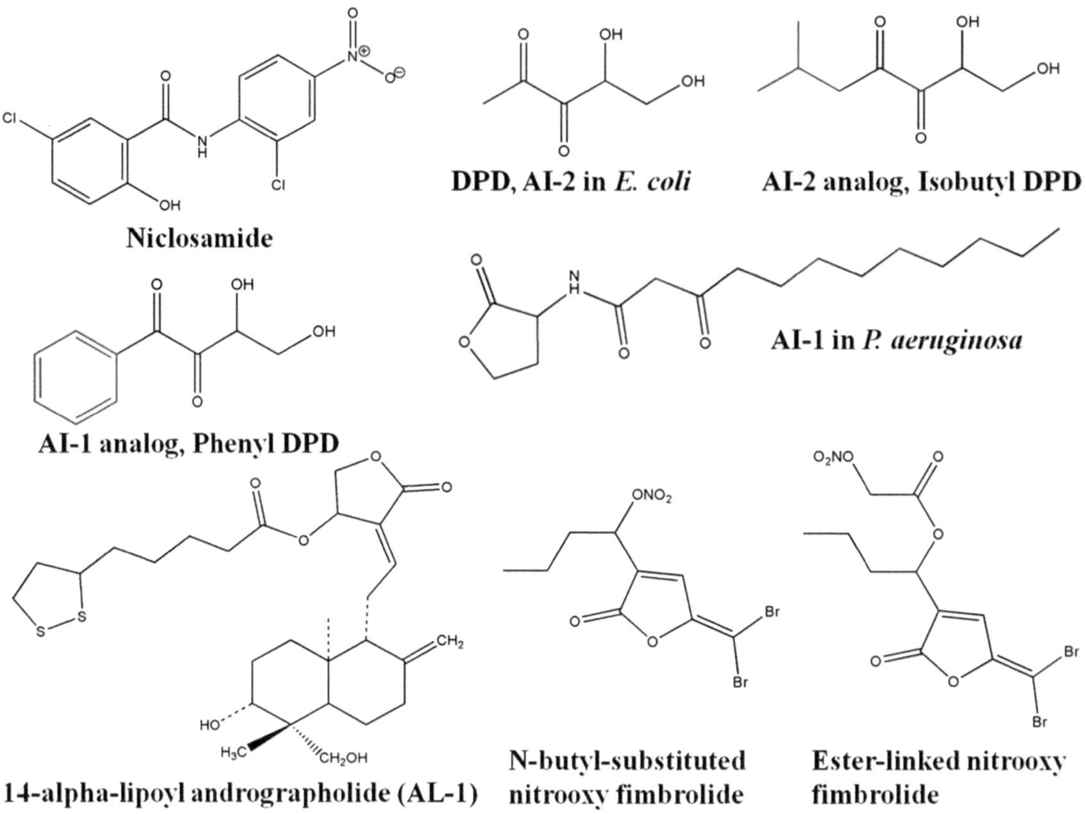

Niclosamide

DPD, AI-2 in *E. coli*

AI-2 analog, Isobutyl DPD

AI-1 analog, Phenyl DPD

AI-1 in *P. aeruginosa*

14-alpha-lipoyl andrographolide (AL-1)

N-butyl-substituted nitrooxy fimbrolide

Ester-linked nitrooxy fimbrolide

Fig. 1 Structures of various QSIs

clearance of preformed *E. coli* and *P. aeruginosa* biofilms, respectively (Fig. 2). Interestingly, the authors have used a microfluidic setting for real-time analysis of biofilms inhibition (Roy et al. 2013).

The effect of tobramycin (for *P. aeruginosa*, *B. cepacia* complex) and clindamycin or vancomycin (for *S. aureus*), alone or in combination with QSI, was studied on various in vitro and in vivo biofilm model systems (Table 2). The QSIs used here target the acylhomoserine lactone-based QS system present in *P. aeruginosa* and *B. cepacia* complex organisms (baicalin hydrate (BH), cinnamaldehyde) or the peptide-based system present in *S. aureus* (hamamelitannin (HM)). A significantly higher fraction of infected *Galleria mellonella* larvae and *Caenorhabditis elegans* survived infection, and the microbial load in the lungs of BALB/c mice was reduced following combined treatment, compared to treatment with an antibiotic alone. The data suggests that QSIs may increase the success of antibiotic treatment by increasing the susceptibility of bacterial biofilms and/or by increasing host survival following infection (Brackman et al. 2011).

14-Alpha-lipoyl andrographolide (AL-1) (Fig. 1) showed synergistic effects on antibiofilm and antivirulence factor activities when combined with azithromycin, ciprofloxacin, fosfomycin, streptomycin, and gentamicin. AL-1 also inhibited the production of the exopolysaccharide and pyocyanin components and (Zeng et al. 2011) altered the expression of genes that regulate the QS systems (Las and Rhl systems) (Table 2) (Ma et al. 2012).

Nitric oxide (NO), at low concentrations, causes *P. aeruginosa* biofilms to disperse, converting the bacteria in the planktonic phase to their normal susceptibility to antibiotics

Table 2 Combination strategy for multipronged QS inhibition

Compound(s)	Inhibition of QS/biofilm formation	Organism	Reference
Combination of antibiotics and QSIs			
Isobutyl-DPD + gentamicin	Acts synergistically with gentamicin in the clearance of preformed biofilms	*E. coli*	Roy et al. (2013)
Phenyl DPD + gentamicin	Acts synergistically with gentamicin in the clearance of preformed biofilms	*P. aeruginosa*	Roy et al. (2013)
Baicalin hydrate + tobramycin and hamamelitannin + vancomycin	Inhibits biofilm formation and decreases pathogenicity in vitro and in vivo (two invertebrate models and one mouse pulmonary infection model)	*P. aeruginosa* *B. cepacia* *S. aureus*	Brackman et al. (2011)
14-Alpha-lipoyl andrographolide (AL-1) + azithromycin, ciprofloxacin, fosfomycin, streptomycin, and gentamicin	AL-1 acts synergistically with antibiotics on antibiofilm and antivirulence factor activities by altering the QS-regulated genes	*P. aeruginosa*	Zeng et al. (2011) Ma et al. (2012)
Baicalein + ampicillin	Dissolves and eradicates biofilms	*P. aeruginosa*	Zeng et al. (2008)
Fimbrolide–NO hybrids	Antivirulent and antibiofilm agents	*P. aeruginosa*	Kutty et al. (2013)
2-Aminoimidazole/triazole conjugates + conventional antibiotics	Synergism in antibiofilm activity and dispersal of preestablished biofilms	*A. baumannii* *S. aureus*	Rogers et al. (2010)
Quorum quenching (QQ) enzymes + antibiotics			
Lactonase + ciprofloxacin and gentamicin	Lactonase increases the susceptibility of antibiotics to biofilms	*P. aeruginosa*	Kiran et al. (2011)
β-*N*-acetyl-glucos-aminidase dispersin B + cefamandole nafate	Dispersin B decreases biofilm formation and enhances the antibiotic activity against adherent cells on polyurethane surfaces	*S. epidermidis*	Donelli et al. (2007)

(Barraud et al. 2006). Fimbrolides are known for their potent QS inhibitory activities. Kutty et al. reported fimbrolide–NO hybrids, first dual-action antimicrobial agents based on the bacterial QS inhibition and NO signaling (Table 2). The fimbrolide–NO analogues like N-butyl-substituted nitrooxy compound (Fig. 1) and ester-linked nitrooxy compound (Fig. 1) possessed superior biofilm inhibition activity in *P. aeruginosa* compared to the corresponding non nitrooxy-containing fimbrolide and hydroxy fimbrolide, which were not effective in controlling biofilm formation. Compounds utilizing QS inhibition and NO release represent key strategies for controlling bacterial virulence without inhibiting bacterial growth (Kutty et al. 2013).

Another QSI, baicalein, could indeed dissolve *P. aeruginosa* biofilms and showed synergistic effects with ampicillin (Table 2), causing eradication of *P. aeruginosa* biofilms at 2 μg mL^{-1}, a concentration not effective to kill the bacteria without baicalein (Zeng et al. 2008). Melander and coworkers reported a dramatic increase in sensitivity toward conventional antibiotics, when biofilms of methicillin resistant *S. aureus* (MRSA) or *A. baumannii* were treated with biofilm-controlling 2-aminoimidazoles (Table 1) (Rogers et al. 2009) in combination with antibiotics (Table 2) (Rogers et al. 2010).

Nanotechnological Approaches Toward Quorum Sensing

Nanotechnology provides an exciting approach toward quorum sensing as antimicrobial therapeutics. The main idea behind this approach is to encapsulate a smart protein factory which would be able to produce quorum sensing molecules as a stimuli response to the presence of bacteria in the environment. Essentially, this factory should be able to target specific bacteria and coax a quorum

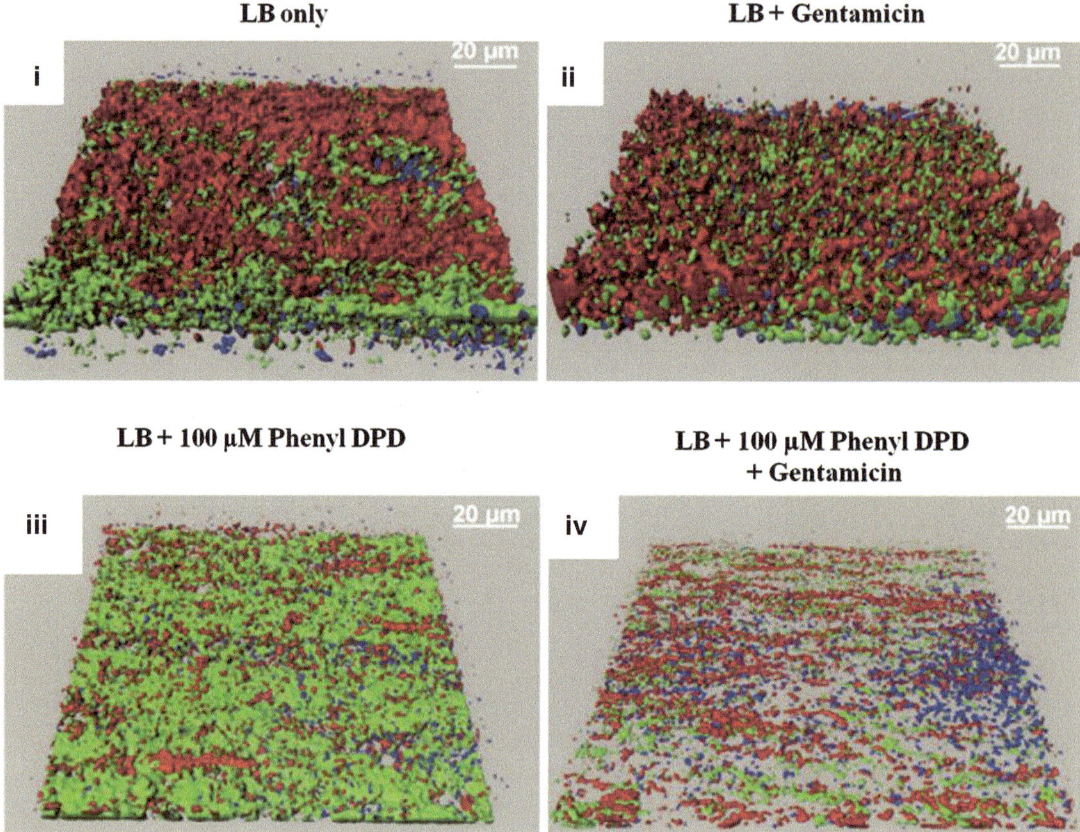

Fig. 2 Efficacy of combination approach in antibiofilm activity. Synergistic action of phenyl DPD and gentamicin in the clearance of *P. aeruginosa* biofilms (Adapted from Roy et al. 2013)

sensing response from that species which will interfere with the sustenance of other harmful bacteria.

Over the years, Bentley and coworkers have developed several nanoparticle systems to elicit interspecies communication in bacteria. Initially, they made magnetic nanoparticles functionalized with enzymes Pfs and LuxS (responsible for the synthesis of AI-2, universal bacterial signal) individually for localized manufacture and delivery on the bacterial surface (Fernandes et al. 2007). Then, they constructed a plasmid capable of expressing the AI-2 synthetic module as a single protein complex of Pfs and LuxS (Fernandes and Bentley 2009), which was subsequently conjugated with antibody to make nanofactories that target a specific bacteria and elicit QS from them (Fernandes et al. 2010). In the latest development, they have made chitosan–alginate capsules containing nanofactories that allow small molecules (substrates) to enter the capsules and also allow easy diffusion of the AI-2 molecules after their synthesis. The ultimate goal of this approach is to create artificial cells that communicate with the human microbial flora to perform the desired functions and to selectively harness useful bacteria over the harmful ones.

Conclusions and Future Perspectives

Antibiotics that have long been known to exert antibacterial effects are being discovered as QSIs or antibiofilm agents. Non-antibiotics such as natural polyphenols and the compounds from the screening of the already approved drugs represent a promising strategy for the development of novel QSIs. Despite the fact that QSIs have a potential to be alternatives for conventional antibiotics, their use is largely limited due to their toxicity. Combination approaches of antibiotics and QSIs would overcome many shortfalls of the application of QS inhibition alone. This approach is based on the observation that most QSIs enhance the sensitivity of pathogens to antibiotics, even if the QSIs could not achieve complete dissolution of the biofilms. In animal models, this combination strategy has already been successfully validated and could overcome the disadvantages of the monotherapy. An emerging field, nanotechnology has a lot to offer in the field of quorum sensing.

One of the most intriguing problems is that many QSIs can prevent the formation of biofilms but are not effective in dissolving existing ones and are therefore of little use for the treatment of already existing biofilm infections. More importantly, high specificity of autoinducers combined with a complex network of often redundant regulation circuits and an increased virulence of pathogens released from biofilms are real challenges in the development of QSIs for medical applications. Biofilm dispersing agents like NO, D-amino acids, and fatty acids have been found to be effective due to their interactions with different targets of the biofilms. The combination of such biofilm dispersing agents and QSIs can be effective in dissolving the existing biofilms as well as inhibiting the quorum sensing. The addition of antibiotics to such combination systems would complete the cocktail to effectively eradicate the preformed biofilms and prevent their formation through inhibition of quorum sensing. Such a combination is attractive and holds high expectations, but one should also bear in mind that we are dealing here with drugs influencing each other and complicating pharmacokinetics.

Nanotechnological approaches toward quorum sensing definitely point toward future success stories in the field. Although it is still in its nascent form, proper implementation of the concepts laid down in such approaches will allow targeting specific bacteria in a group of many. Thus, the activity of beneficial bacteria can be switched on while that of harmful bacteria is switched off.

References

Barraud N, Hassett DJ, Hwang SH, Rice SA, Kjelleberg S, Webb JS (2006) Involvement of nitric oxide in biofilm dispersal of *Pseudomonas aeruginosa*. J Bacteriol 188(21):7344–7353

Brackman G, Cos P, Maes L, Nelis HJ, Coenye T (2011) Quorum sensing inhibitors increase the susceptibility of bacterial biofilms to antibiotics in vitro and in vivo. Antimicrob Agents Chemother 55(6):2655–2661

Davies DG, Parsek MR, Pearson JP, Iglewski BH, Costerton JW, Greenberg EP (1998) The involvement of cell-to-cell signals in the development of a bacterial biofilm. Science 280(5361):295–298

de Kievit TR (2009) Quorum sensing in *Pseudomonas aeruginosa* biofilms. Environ Microbiol 11(2): 279–288

Defoirdt T, Boon N, Bossier P (2010) Can bacteria evolve resistance to quorum sensing disruption? PLoS Pathog 6(7):e1000989

Defoirdt T, Pande GSJ, Baruah K, Bossier P (2013) The apparent quorum-sensing inhibitory activity of Pyrogallol is a side effect of peroxide production. Antimicrob Agents Chemother 57(6):2870

Delden C, Köhler T, Brunner-Ferber F, François B, Carlet J, Pechère J-C (2012) Azithromycin to prevent Pseudomonas aeruginosa ventilator-associated pneumonia by inhibition of quorum sensing: a randomized controlled trial. Intensive Care Med 38(7): 1118–1125

Donelli G, Francolini I, Romoli D, Guaglianone E, Piozzi A, Ragunath C, Kaplan JB (2007) Synergistic activity of dispersin B and cefamandole nafate in inhibition of staphylococcal biofilm growth on polyurethanes. Antimicrob Agents Chemother 51(8):2733–2740

Estrela AB, Abraham W-R (2010) Combining biofilm-controlling compounds and antibiotics as a promising new way to control biofilm infections. Pharmaceuticals 3(5):1374–1393

Fernandes R, Bentley WE (2009) AI-2 biosynthesis module in a magnetic nanofactory alters bacterial response via localized synthesis and delivery. Biotechnol Bioeng 102(2):390–399

Fernandes R, Tsao CY, Hashimoto Y, Wang L, Wood TK, Payne GF, Bentley WE (2007) Magnetic nanofactories: localized synthesis and delivery of quorum-sensing signaling molecule autoinducer-2 to bacterial cell surfaces. Metab Eng 9(2):228–239

Fernandes R, Roy V, Wu HC, Bentley WE (2010) Engineered biological nanofactories trigger quorum sensing response in targeted bacteria. Nat Nanotechnol 5(3):213–217

Hentzer M, Wu H, Andersen JB, Riedel K, Rasmussen TB, Bagge N, Kumar N, Schembri MA, Song Z, Kristoffersen P, Manefield M, Costerton JW, Molin S, Eberl L, Steinberg P, Kjelleberg S, Hoiby N, Givskov M (2003) Attenuation of *Pseudomonas aeruginosa* virulence by quorum sensing inhibitors. Embo J 22(15):3803–3815

Huber B, Eberl L, Feucht W, Polster J (2003) Influence of polyphenols on bacterial biofilm formation and quorum-sensing. Z Naturforsch C 58(11–12): 879–884

Imperi F, Massai F, Ramachandran Pillai C, Longo F, Zennaro E, Rampioni G, Visca P, Leoni L (2013) New life for an old drug: the anthelmintic drug niclosamide inhibits *Pseudomonas aeruginosa* quorum sensing. Antimicrob Agents Chemother 57(2):996–1005

Kalia VC (2013) Quorum sensing inhibitors: an overview. Biotechnol Adv 31(2):224–245

Kiran S, Sharma P, Harjai K, Capalash N (2011) Enzymatic quorum quenching increases antibiotic susceptibility of multidrug resistant *Pseudomonas aeruginosa*. Iran J Microbiol 3(1):1–12

Kutty SK, Barraud N, Pham A, Iskander G, Rice SA, Black DS, Kumar N (2013) Design, synthesis, and evaluation of fimbrolide-nitric oxide donor hybrids as antimicrobial agents. J Med Chem 56(23):9517–9529

Ma L, Liu X, Liang H, Che Y, Chen C, Dai H, Yu K, Liu M, Yang CH, Song F, Wang Y, Zhang L (2012) Effects of 14-alpha-lipoyl andrographolide on quorum sensing in *Pseudomonas aeruginosa*. Antimicrob Agents Chemother 56(12):6088–6094

McKenna M (2013) Antibiotic resistance: the last resort. Nature 499(7459):394–396

Miller MB, Bassler BL (2001) Quorum sensing in bacteria. Annu Rev Microbiol 55:165–199

Ni N, Choudhary G, Li M, Wang B (2008) Pyrogallol and its analogs can antagonize bacterial quorum sensing in Vibrio harveyi. Bioorg Med Chem Lett 18(5): 1567–1572

Rasmussen TB, Givskov M (2006) Quorum-sensing inhibitors as anti-pathogenic drugs. Int J Med Microbiol 296(2–3):149–161

Rasmussen TB, Bjarnsholt T, Skindersoe ME, Hentzer M, Kristoffersen P, Kote M, Nielsen J, Eberl L, Givskov M (2005) Screening for quorum-sensing inhibitors (QSI) by use of a novel genetic system, the QSI selector. J Bacteriol 187(5):1799–1814

Rogers SA, Huigens RW 3rd, Melander C (2009) A 2-aminobenzimidazole that inhibits and disperses gram-positive biofilms through a zinc-dependent mechanism. J Am Chem Soc 131(29):9868–9869

Rogers SA, Huigens RW 3rd, Cavanagh J, Melander C (2010) Synergistic effects between conventional antibiotics and 2-aminoimidazole-derived antibiofilm agents. Antimicrob Agents Chemother 54(5):2112–2118

Romero D, Traxler MF, Lopez D, Kolter R (2011) Antibiotics as signal molecules. Chem Rev 111(9): 5492–5505

Roy V, Meyer MT, Smith JA, Gamby S, Sintim HO, Ghodssi R, Bentley WE (2013) AI-2 analogs and antibiotics: a synergistic approach to reduce bacterial biofilms. Appl Microbiol Biotechnol 97(6): 2627–2638

Shih PC, Huang CT (2002) Effects of quorum-sensing deficiency on *Pseudomonas aeruginosa* biofilm formation and antibiotic resistance. J Antimicrob Chemother 49(2):309–314

Skindersoe ME, Alhede M, Phipps R, Yang L, Jensen PO, Rasmussen TB, Bjarnsholt T, Tolker-Nielsen T, Hoiby N, Givskov M (2008) Effects of antibiotics on quorum sensing in *Pseudomonas aeruginosa*. Antimicrob Agents Chemother 52(10): 3648–3663

Straight PD, Kolter R (2009) Interspecies chemical communication in bacterial development. Annu Rev Microbiol 63:99–118

Tay SB, Yew WS (2013) Development of quorum-based anti-virulence therapeutics targeting Gram-negative bacterial pathogens. Int J Mol Sci 14(8): 16570–16599

Worthington RJ, Melander C (2013) Combination approaches to combat multidrug-resistant bacteria. Trends Biotechnol 31(3):177–184

Wright GD (2010) Q&A: antibiotic resistance: where does it come from and what can we do about it? BMC Biol 8:123

Zeng Z, Qian L, Cao L, Tan H, Huang Y, Xue X, Shen Y, Zhou S (2008) Virtual screening for novel quorum sensing inhibitors to eradicate biofilm formation of *Pseudomonas aeruginosa*. Appl Microbiol Biotechnol 79(1):119–126

Zeng X, Liu X, Bian J, Pei G, Dai H, Polyak SW, Song F, Ma L, Wang Y, Zhang L (2011) Synergistic effect of 14-alpha-lipoyl andrographolide and various antibiotics on the formation of biofilms and production of exopolysaccharide and pyocyanin by *Pseudomonas aeruginosa*. Antimicrob Agents Chemother 55(6):3015–3017

Heterologous Expression of Quorum Sensing Inhibitory Genes in Diverse Organisms

Prasun Kumar, Shikha Koul, Sanjay K.S. Patel, Jung-Kul Lee, and Vipin C. Kalia

Abbreviations

AHL	Acyl homoserine lactone
C4-HSL	N-butanoyl-L-HSL
C6-HSL	N-hexanoyl HSL
C8-HSL	N-octanoyl HSL
C10-HSL	N-decanoyl HSL
C12-HSL	N-dodecanoyl HSL
C14-HSL	N-tetradecanoyl-HSL
3OC6-HSL	3-oxo-N-hexanoyl-HSL
3OC8-HSL	3-oxo-N-octanoyl-HSL
3OC10-HSL	3-oxo-N-decanoyl-HSL
3OC12-HSL	3-oxo-N-dodecanoyl- HSL
HSL	Homoserine lactone
QS	Quorum sensing
QSS	Quorum sensing systems
QQ	Quorum quenching

P. Kumar
Microbial Biotechnology and Genomics, CSIR-Institute of Genomics and Integrative Biology, Mall Road, Delhi 110007, India

S. Koul • V.C. Kalia (✉)
Microbial Biotechnology and Genomics, CSIR-Institute of Genomics and Integrative Biology, Mall Road, Delhi 110007, India

Academy of Scientific and Innovative Research (AcSIR), 2, Rafi Marg, Anusandhan Bhawan, New Delhi 110001, India
e-mail: vckalia@igib.res.in; vc_kalia@yahoo.co.in

S.K.S. Patel • J.-K. Lee
Department of Chemical Engineering, A1414, Konkuk University, 1 Hwayang-Dong, Gwangjin-Gu, Seoul 143-701, South Korea

Introduction

The discovery of antibiotics was a wonderful solution to provide relief to human beings from infectious diseases. However, indiscriminate usage of antibiotics turned out to be counterproductive. It was observed that patients were not getting cured in spite of the systematic use of antibiotics. In fact, microbes had developed resistance to antibiotics. This perturbation has been in operation even with antibiotics subsequently developed during the next 6–7 decades (D'Costa et al. 2006). Pharmaceutical companies are no longer interested in investing money into this business (Spellberg et al. 2004; Courvalin 2008). It obliged scientists to look for alternative drugs and new drug targets. It was realised that more than 80 % of the infectious diseases are caused by microbial pathogens, through specialised structures – biofilms. It enables bacteria to survive the lethal effect of drugs, as they "become" up to 1,000 times more resistant to antibiotics (Kalia 2013; Gui et al. 2014; Kalia et al. 2014a, b). These biofilms are developed by bacteria in a population density-dependent process called quorum sensing (QS) (Dong and Zhang 2005). Most Gram-negative bacteria operate through a QS system termed as LuxR/I-type, where acylated homoserine lactones (AHLs) acts as signals. QS signals consist of the lactone ring with varying acyl chains (Yang et al. 2012; Shang et al. 2014). QS regulates the expression of virulence factors, antibiotic production, nitrogen fixation, sporulation,

V.C. Kalia (ed.), *Quorum Sensing vs Quorum Quenching: A Battle with No End in Sight*,
DOI 10.1007/978-81-322-1982-8_28, © Springer India 2015

conjugation, swarming, etc. (Borlee et al. 2008; Kalia and Purohit 2011; Kalia 2013; Wang et al. 2013; Zhang et al. 2013; Kalia et al. 2014a, b). These properties allow such bacteria to dominate the community structure. It is thus no surprise that the competing organisms have also developed mechanisms to interfere with the QSS and degrade these signals – a phenomenon termed as quorum quenching (QQ) (Kalia and Purohit 2011; Annapoorani et al. 2012; Bakkiyaraj et al. 2013; Kalia 2013; Agarwala et al. 2014).

Signal Degradation

QS signals – AHLs – can be degraded through: (1) chemical, (2) metabolic and (3) enzymatic routes. In the chemical degradation route, alkaline pH leads to the opening of the lactone ring which thus inactivates the AHL signals produced by plant pathogen – *Erwinia* species (Byers et al. 2002; Yates et al. 2002). This process can be reversed at acidic pH, where the cyclisation of the lactone ring leads to the reformation of an active signal molecule. In addition to chemical inactivation, bacteria, such as *Variovorax paradoxus* and *Pseudomonas aeruginosa* PAI-A can metabolise AHLs to use them as an energy source (Leadbetter and Greenberg 2000; Huang et al. 2003). In addition to these mechanisms, QS signals can be degraded enzymatically through AHL-lactonase, AHL-acylase, oxidoreductases and lactonase-like enzymes (paraoxonases) (Tables 1 and 2), which hydrolyze either the lactone ring or the amide bond of the AHL (Dong et al. 2000; Lee et al. 2002; Lin et al. 2003).

Diversity of Organisms Possessing AHL-Lactonase

The distribution of AHL-lactonase has been reported among diverse taxa: *Actinobacteria*, *Bacteroidetes*, and *Firmicutes*. It is interesting to learn that members of *Acidobacteria*, *Planctomycetes*, *Sphingobacteria*, and *Spirochaetales* possess only AHL-lactonase. However, most of bacteria belonging to these taxa do not show the presence of AHL-acylase (Kalia et al. 2011). AHL-lactonase belonging to the superfamily – metallohydrolase – has been reported to be produced by *Bacillus*, *Arthrobacter*, *Acidobacteria*, *Agrobacterium*, *Klebsiella*, *Streptomyces*, *Rhodococcus*, *Pseudomonas*, *Comamonas*, *Shewanella*, etc. (Kalia et al. 2011; Chen et al. 2013). The activity of the enzyme, AHL-lactonase, is influenced by a wide range of metal ions and chelating reagents. The variation in enhancing the activity has been observed with certain metal ions such as Mg^{2+} and Zn^{2+} at high concentration of 10 mM. However, this activity was observed to decline dramatically at lower metal ion concentration of 1 mM (Chen et al. 2010). In other cases, the enhancement in enzyme activity was similar due to the presence of metal ions such as Na^{+}, K^{+}, Ca^{2+}, Fe^{3+} and Mn^{2+} in the range of 1–10 mM. In contrast, quite a few ions at concentrations ranging between 0.2 and 2 mM did not have any effect on the activity of this enzyme: Ca^{2+}, Mn^{2+}, Co^{2+}, Ni^{2+} and Cd^{2+} (Wang et al. 2004). There was complete inhibition in AHL-lactonase activity in the presence of heavy metals (1–10 mM): Cu^{2+}, Cr^{3+}, Hg^{2+} and Ag^{+} (Chen et al. 2010). Sodium do-decyl sulphate inhibited the enzyme activity whereas reagents such as ethylenediaminetetraacetic acid helped to enhance AHL-lactonase activity at 10 mM (Chen et al. 2010). In previous studies, reagents like 2,2′-bipyridine and *o*-phenanthroline were found to have no impact on this enzyme (Wang et al. 2004). It is important to know the determinants that regulate the broad substrate specificity of AHL-lactonases. In case of *Bacillus thuringiensis* AiiA, F107 residue was found to have an important role in the selective interaction preferably for longer acyl-chain substrates (Liu et al. 2013). Identification of such residues may help in the development of highly selective QSIs as potential therapeutics.

Hydrolysis of the lactone ring inactivates the QS signal and thus influences the QS-mediated characteristics (Dong et al. 2000, 2002, 2004; Lee et al. 2002; Park et al. 2003; Ulrich 2004; Dong and Zhang 2005; Liu et al. 2005; Thomas et al. 2005; Bai et al. 2008; Riaz et al. 2008;

Table 1 Effect of heterologous expression of acyl homoserine lactone (AHL)-lactonase in diverse organisms on their quorum sensing systems

Source organism and gene	Host	QS signals and characteristic affected	References
Bacillus amyloliquefaciens PEBA20, *aii*A	*Escherichia coli* BL21(DE3)	Interferes with QS-mediated functions in *P. carotovorum* subsp. *carotovorum*	Yin et al. (2010)
Bacillus anthracis (Ames), *aii*A	*Burkholderia thailandensis*	In activated C6HSL, C8HSL and C10HSL	Ulrich (2004)
*Bacillus cereus, aii*A			
Bacillus subtilis subsp. *endophyticus* BS1, *aii*A	*E. coli* BL21(DE3) pLysS	Attenuated the soft rot symptoms caused by *Erwinia carotovora* var. carotovora	Pan et al. (2008)
Bacillus thuringiensis subsp. *morrisoni* and subsp. *kyushuensis, aii*A	*E. coli* BL21(DE3)	3OC6HSL, C6HSL and C8HSL Attenuates the pathogenicity caused by *E. carotovora*	Lee et al. (2002)
Bacillus sp. A24, *aii*A	*Pseudomonas aeruginosa* PAO1	3OC12HSL; reduces virulence gene expression and swarming motility in *P. aeruginosa* PAO1	Reimmann et al. (2002)
	Pseudomonas fluorescens P3	C4HSL and C6HSL; reduction in protease production and diseases caused by plant pathogens – *A. tumefaciens* and *Erwinia carotovora*	Molina et al. (2003)
	Serratia plymuthica HRO-C48	Abolished AHL production; reduced pyrrolnitrin and chitinase production	Müller et al. (2009)
	Lysobacter enzymogenes OH11A	Strong reduction of *Pectobacterium carotovorum* virulence on Chinese cabbage	Qian et al. (2010)
	S. plymuthica strain G3	Modification of the adhesion and biofilm forming abilities	Liu et al. (2011)
Bacillus sp. 240B1, *aii*A	*E. carotovora* SCG1 (E7-R3)	Reduced release of AI signals; decrease extracellular level of pectolytic enzymes and polygalacturonase; attenuation in pathogenicity on wide range of host plants	Dong et al. (2000)
	Transgenic tobacco and potato plants	Resistant to infection by the plant pathogen – *E. carotovora*	Dong et al. (2001)
Bacillus sp., *aii*A	*B. thuringiensis* BMB171	Delayed sporulation	Zhou et al. (2006)
Bacillus sp. BC6, *aiiA*	*E. coli* BL21(DE3)pLysS	Biofilm inhibition of *V. cholerae*	Augustine et al. (2010)
Bacillus sp. B546, *aii*A	*Pichia pastoris*	Resistance against *Aeromonas hydrophila*	Chen et al. (2010)
Agrobacterium tumefaciens; *att*M	*Azospirillum lipoferum* strain B518	Abolished pectinase activity, increased siderophore synthesis and reduced indole acetic acid production	Boyer et al. (2008)
Ochrobactrum sp. T63, *aid*H	*E. coli* BL21(DE3)	Interferes with QS-mediated functions in *P. fluorescens* 2P24 and *P. carotovorum*	Mei et al. (2010)
Pseudoalteromonas byunsanensis 1A01261, *qsd*H	*E. coli* BL21(DE3)	Attenuates the pathogenicity caused by *E. carotovora*	Huang et al. (2012)

(continued)

Table 1 (continued)

Source organism and gene	Host	QS signals and characteristic affected	References
Rhodococcus erythropolis W2	*E. coli* DH5α	C6-14 HSL with or without substitution at C3. Quench violacein synthesis in reporter strain	Uroz et al. (2008)
	P. aeruginosa 1855-344	Reduction in symptom severity of *P. carotovorum* PCC797	
	Agrobacterium tumefaciens 15955	Prevented accumulation of HSLs and conjugal transfer of Ti plasmid	
Chryseobacterium sp. StRB126, *aid*C	*E. coli* DH5α	C6HSL and C10HSL	Wang et al. (2012)
Metagenome-derived clones; *bpi*B01, *bpi*B04, *bpi*B07	*P. aeruginosa* PAO1	Inhibit motility and biofilm formation in *P. aeruginosa*	Schipper et al. (2009)

Uroz et al. 2009; Chan et al. 2010; Han et al. 2010; Deng et al. 2011; Yin et al. 2012). Among the different organisms known to produce lactonase, *Bacillus* has a broad substrate specificity with a preference for signals with (s)-configuration (Fuqua et al. 2001; Thomas et al. 2005). *Bacillus* spp. could degrade AHL signal HAI-1 of *V. harveyi* (Dong et al. 2002; Bai et al. 2008). Many homologues of this AHL-lactonase have been identified (Ulrich 2004) (Table 1). In a recent effort to look for diversity of AHL-lactonases, screening of 800 different bacteria revealed 42 strains of different *Bacillus* spp. to have *aiiA* gene (Huma et al. 2011). Other *Bacillus* species reported to express AHL-lactonase activity belonged to *B. amyloliquefaciens*, *B. subtilis*, *B. mycoides* and *B. microestinctum* (Dong et al. 2002; Pan et al. 2008; Han et al. 2010; Yin et al. 2010). Comparative genomics (*in silico*) study has also proved helpful in providing insights into genetic variability of genes for AHL-lactonase: *B. weihenstephanensis* KBAB4, *B. licheniformis* ATCC 14580, *Geobacillus* sp. WCH70, *Dorea longicatena* DSM 13814, *Pelotomaculum thermopropionicum* SI, *Moorella thermoacetica* ATCC 39073, *Clostridium beijerinckii* NCIMB 8052, *Lysinibacillus sphaericus* C3-41, *Bacillus* sp. B14905, *Staphylococcus saprophyticus* subsp. *saprophyticus* ATCC 15305, *Bacillus cereus* W, *Bacillus subtilis*, *B. thuringiensis*, *B. thuringiensis* serovar *israelensis* ATCC 35646, *B. thuringiensis* str. Al Hakam, *Clostridium kluyveri* DSM 555, *Staphylococcus aureus* RF122, *S. aureus* subsp. *aureus* MSSA476, *S. aureus* Mu50, *S. aureus* MW2, *Thermosinus carboxydivorans* Nor1, *Caldicellulosiruptor saccharolyticus* DSM 8903 and *Clostridium scindens* ATCC 35704 (Huma et al. 2011; Kalia et al. 2011).

Eukaryotic organisms are also known to exhibit activities similar to AHL-lactonases. *Phialocephala*, *Ascomycetes* and *Meliniomyces* are fungi which have lactonase activities for QS signals C6-HSL and 3OC6-HSL (Uroz and Heinon 2008). Human epithelial cells show activities for enzymes paraoxonases – PON1, PON2 and PON3 (Ng et al. 2005; Dong et al. 2007). These enzymes inactivate QS signals 3OC12-HSL produced by *P. aeruginosa* infecting human respiratory system (Chun et al. 2004; Hastings 2004). Human paraoxonases (PON2) hydrolyze and inactivate QS signals – 3OC6-HSL of pathogenic bacteria. PON lactonases are specific to six-member ring lactones than 5-member ring analogs (Draganov et al. 2005).

Diversity of Organisms Possessing AHL-acylase

The enzyme AHL-acylase, present in Gram-positive and Gram-negative bacteria, acts by cleaving the side chain of the signal molecule – AHL. The activity of AHL-acylase depends on acyl-chain lengths (Table 1). AHL-acylase reported from *Ralstonia eutropha* although quite specific, however has preference for long-chain AHLs more than 8 carbons (Lin et al. 2003).

Table 2 Effect of heterologous expression of acyl homoserine lactone (AHL)-acylase in diverse organisms on their quorum sensing systems

Source organism and gene	Host	QS signals and characteristic affected	References
Anabaena sp. PCC7120, *aii*C	*Escherichia coli*	Specificity for broad acyl-chain length HSLs	Romero et al. (2008)
Brucella melitensis, *aii*D	*B. melitensis*	Clumping phenotype and produce exopolysaccharide	Godefroid et al. (2010)
Ochrobactrum A44, *aii*O	*E. coli* BL21(DE3)	Inactivates C4 to C14 HSLs with or without 3-oxo or 3-hydroxy substituents	Czajkowski et al. (2011)
Pseudomonas aeruginosa PAO1, *qui*P (PA1032)	*E. coli* *P. aeruginosa*	Long-chain AHLs Decreased accumulation of the 3OC12HSL	Huang et al. (2006)
P. aeruginosa PAO1, *pvdQ* (PA2385)	*P. aeruginosa* *E. coli* DH10B	Dismantle the biofilm formation C7HSL, C8HSL, 3OC10HSL and 3OC12HSL	Sio et al. (2006)
P. aeruginosa, *pvdQ* (PA2385)	*P. aeruginosa* PAO1	Less virulent in *Caenorhabditis elegans* infection model; modulate its own pathogenicity	Papaioannou et al. (2009)
P. aeruginosa, *pa0305*	*P. aeruginosa* *E. coli*	Reduction in 3OC12HSL accumulation and the expression of virulence factors	Wahjudi et al. (2011)
Pseudomonas syringae, *hac*A, *hac*B and *psyr*3871	*E. coli*	AHLs with differing substrate specificities	Shepherd and Lindow (2009)
P. aeruginosa, PAI-A		Degrade long-chain AHLs but not short chain AHLs	Huang et al. (2003)
P. aeruginosa, *pvdQ*	*E. coli*	Rapid inactivation of long-chain AHLs	Huang et al. (2003)
	P. aeruginosa PAO1	Did not accumulate 3OC12HSL	
Ralstonia sp. XJ12B, *aiiD*	*E. coli*	Inactivated 3OC8HSL, 3OC10HSL and 3OC12HSL	Lin et al. (2003)
	P. aeruginosa PAO1	Influenced AHL accumulation, extracellular secretion and swarming motility, attenuated elastase and pyocyanin production, paralysed nematodes	
Ralstonia solanacearum GMI1000, *aac*	*E. coli* DH10B *Chromobacterium violaceum* CV026	C6HSL, C7HSL, C8HSL Inhibited violacein and chitinase activity	Chen et al. (2009)
Shewanella sp. MIB015, *aac*	*E. coli* *Vibrio anguillarum*	Long-chain AHLs Reduced AHL production and biofilm formation	Morohoshi et al. (2008)
*Streptomyces sp.*M664, *ahlM* gene	*Streptomyces lividans*	Long acyl chains; degrades penicillin G; decreased the production of virulence factors, including elastase, total protease and LasA	Park et al. (2005)
Metagenome-derived clone, *bpi*B09	*P. aeruginosa* PAO1	Reduced pyocyanin production, decreased motility and poor biofilm formation	Bijtenhoorn et al. (2011)

AHL-acylase from *Streptomyces* sp. M664 can be exploited for degrading AHLs with less than 8 carbons (Park et al. 2005). *P. aeruginosa* PAO1 can also degrade long-chain AHLs (Lamont and Martin 2003; Zhang and Dong 2004; Huang et al. 2006; Sio et al. 2006). The diversity of organisms showing AHL-acylase is reflected by their presence in *Ralstonia solanacearum* GMI1000 (Chen et al. 2009) and *Shewanella* sp. (Morohoshi et al. 2005). *Comamonas* can degrade acyl-chain lengths between 4 and 16 carbons with varying substitutions (Uroz et al. 2003, 2007). Homology to the acylase was seen in two diverse organisms such as nitrogen-fixing cyanobacterium *Anabaena* (*Nostoc*) sp. PCC7120 and QuiP of *P. aeruginosa* PAO1 (Romero et al. 2008). In spite of a wide diversity of taxa having organisms possessing AHL-lactonase or AHL-acylase, however, cyanobacterial members have been reported to possess only AHL-acylase (Kalia et al. 2011).

Organisms Possessing Multiple AHL-Degrading Enzymes

Rhodococcus erythropolis is unique with a wide range of QQ abilities (Table 1). *R. erythropolis* W2 is one of those strains which possess activities for AHL-lactonase, AHL-acylase and oxidoreductase (Uroz et al. 2005; Park et al. 2006). In silico studies have also reported the presence of organisms with multiple AHL hydrolytic enzymes: (1) *Deinococcus radiodurans*, (2) *Hyphomonas neptunium* and (3) *Photorhabdus luminescens* (Kalia et al. 2011).

Expression of Prokaryotic Genes for AHL-Lactonase in Different Hosts

Attempts to enhance the activity of AHL-lactonase have been made by expressing the genes in different organisms (Table 1). Most of the studies have been targeted to express *aii*A of the *Bacillus* species (Kumar et al. 2013; Tinh et al. 2013). These heterologous expressions have proved effective in manipulating a wide range of QS-mediated characteristics in different organisms.

Expression of *Bacillus aiiA*

In *Escherichia*

Gene *aiiA* from *Bacillus subtilis* subsp. *endophyticus* BS1 expressed in *Escherichia coli* BL21 (DE3) pLysS proved effective in attenuating the soft rot symptoms caused by plant pathogen *Erwinia carotovora* (Pan et al. 2008). Using the same host, the expression of genes *aii*A from (1) *B. thuringiensis* and (2) *Bacillus amyloliquefaciens* interfered with the pathogenicity caused by *E. carotovora* by inhibiting the activities of the QS signals – 3OC6-HSL, C6-HSL and C8-HSL (Lee et al. 2002; Yin et al. 2010), whereas (3) *Bacillus* sp. BC6 could inhibit biofilm formation abilities of *Vibrio cholerae* (Augustine et al. 2010). The expression of *aii*A from *B. thuringiensis* subsp. *morrisoni* was weak in comparison to that of *B. thuringiensis* subsp. *kyushuensis*, as observed by their AHL degradation capacity. However, when *E. coli* BL21 (DE3) was used as a host, these differences were no longer evident with different QS signals: 3OC6-HSL, C6-HSL and C8-HSL (Lee et al. 2002).

In *Pseudomonas*

Expression of *aiiA* gene from *Bacillus* sp. A24 has been tested in a wide range of host organisms. There was a direct inhibitory impact on QS-mediated functions of *P. aeruginosa* PAO1, which include accumulation of QS signals (3OC12-HSL to approximately 0.10 μM), and expression of properties like swarming, motility and secretion of virulence factors (Reimmann et al. 2002). Similarly, soft rot disease of potatoes and eggplants caused by *Pectobacterium carotovorum* and crown gall disease of tomatoes caused by *Agrobacterium tumefaciens* could be prevented by expressing *aiiA* in *Pseudomonas fluorescens* (Dong et al. 2000; Molina et al. 2003). Vascular wilt of tomato plants caused by *Fusarium oxysporum* could be controlled when co-inoculated with *Pseudomonas chlororaphis*. However, this advantage was lost in the presence of AHL-lactonase producing *Bacillus* sp. A24 (Molina et al. 2003).

In *Burkholderia*

AHL-lactonase gene *aiiA* from *Bacillus anthracis* and *B. cereus* was effective in degrading AHL signal molecules when expressed in *Burkholderia thailandensis*. AHL signal molecules, such as C6- to C10-HSL, were significantly reduced from 2.4 to 300 pmol to undetectable levels. Subsequently, this gene was shown to be instrumental in retarding the growth rate such that generation time of *B. thailandensis* increased from 48 min in the wild-type to 243 min in the genetically modified strain. In addition, this genetically engineered strain also affected the swarming and twitching motility of this pathogen. It has been envisaged that this genetic modification can prove effective in developing vaccine against Gram-negative pathogenic bacteria (Ulrich 2004).

In *Erwinia*

E. carotovora SCG1 (E7-R3) expressing *aiiA* gene of *Bacillus* sp. 240 B1 could effectively inhibit the release of AI signals into the milieu. The activities of extracellular pectolytic enzymes such as pectate lyase, pectin lyase and polygalacturonase were 3–10-fold lower than the wild type. This genetic change in *E. carotovora* was pivotal for reducing its ability to cause disease on eggplant, potato, celery and leafy vegetables such as cabbage, Chinese cabbage, carrot and cauliflower (Dong et al. 2000).

In *Serratia and Lysobacter*

Endophytic bacteria, such as *Serratia plymuthica*, are closely associated with plant rhizosphere and phyllosphere. It has been proposed to be a potential biocontrol agent against fungal diseases. The influence of *Bacillus* A24 *aiiA* gene on QSS of *S. plymuthica* G3 was recorded in terms of the modified adhesion and biofilm-forming abilities. QS signals were completely degraded, which diminished its antifungal activity but augmented its indole acetic acid biosynthesis (Liu et al. 2011). Similar impact on AHL signal concentration and consequent depletion in production of chitinase and pyrrolnitrin were evident in *S. plymuthica* HRO-C48 expressing *aiiA* gene (Müller et al. 2009). *aiiA* gene of *Bacillus* has been expressed in *Lysobacter enzymogenes*. Here it affected pathogenicity caused by *E. carotovora* on Chinese cabbage (Qian et al. 2010).

In Eukaryotes

Aeromonas spp. are known to cause infections in fishes through biofilm formation and efforts are being made to control it (Chu et al. 2013; Mahanty et al. 2013). A unique approach of overproducing AHL-lactonase using eukaryotic high yielding expression system led to many advantages. Expression of *aii*A in *Pichia pastoris* made the fish less susceptible to infection by the bacterial pathogen *Aeromonas hydrophila* (Chen et al. 2010; Chu et al. 2013). The secreted lactonase was found to be stable and active on a wide pH of 6.5–8.9, having thermal stability at 70 °C, and most importantly was protease resistant. This gave an advantage of having a QSI that can work strongly and efficiently on field trials, without the need to clone genes using biocontrol agents (Chen et al. 2010). *Bacillus* spp. with AHL-lactonase could protect *Macrobrachium rosenbergii*, a giant freshwater prawn, from infection caused by *Vibrio harveyi* infection (Nhan et al. 2010). Tobacco and potato plants modified by the introduction of *aii*A gene were resistant to *E. carotovora* infection (Dong et al. 2001).

In *Bacillus* spp.

In general, *aiiA* from *Bacillus* spp. shows higher expression in a heterologous host such as *E. coli.* However, certain mutants of the *B. thuringiensis* BMB171 were found to perform better, but the mutant strains were slow in sporulation process. This property was linked to modifications in its membrane channel. This scenario can be envisaged to allow sustainable commercial production of AHL-lactonase (Zhou et al. 2006).

Expression of Homologues of *aiiA* in Different Hosts

In Gram-positive bacterium *R. erythropolis* W2 a unique AHL-lactonase encoded by gene *qsd*A belonging to phosphotriesterases was found. Like other classes of lactonases, it also has broad substrate specificity for acyl-chain length upto C14. All the clones (*E. coli*) harbouring *qsd*A alleles efficiently inactivated AHL signals. *P. fluorescens* 1855–344 expressing *qsd*A$_{W2}$ also conferred resistance to infection against *P. carotovorum* PCC797. It completely abolished the QS-mediated Ti-plasmid conjugal transfer ability of *A. tumefaciens* expressing *qsd*A (Uroz et al. 2008). Similarly, another lactonase variant *aid*H was found in Gram-negative bacterium *Ochrobactrum* sp. T63. The encoded protein is a metal-dependent (Mn^{2+}) hydrolase belonging to α/β hydrolase family. Cloning of *aid*H into *P. fluorescens* 2P24 and *P. carotovorum* Z3-3 curtailed biofilm formation and abolished pathogenicity in the tested plants (Mei et al. 2010). The absence of AHL-type substrate specificity makes AidH a unique AHL-lactonase to be used against QS-mediated harmful phenotypes. Similarly, expression of *qsd*H gene from *Pseudoalteromonas byunsanensis* in *E. coli* BL21 (DE3) was effective in inhibiting the QS-regulated functions and pathogenicity caused by *E. carotovora* (Huang et al. 2012). Heterologous expression of *att*M (a homologue of *aii*A) from *A. tumefaciens* was able to reduce the QS-mediated pectinase enzyme activity of *Azospirillum lipoferum*. It also influenced the siderophore and indole acetic acid production (Boyer et al. 2008). An *attM* paralogous gene (*aiiB*) was also found in *A. tumefaciens*, having potent activity against AHLs that can be exploited (Carlier et al. 2003).

Recent developments in metagenomic approaches to look for AHL-lactonase from uncultured bacteria are expected to add to the limited diversity of this enzyme (Williamson et al. 2005). Metagenome of a bacterial community from soil is another genetic resource of great potential to be exploited. Screening of various metagenomic clones allowed the identification of genes such as *qlc*A, *bpi*B01 and *bpi*B04, which had very low or no similarities to any known AHL-lactonases. Expression of *bpi*B01 in *P. aeruginosa* PAO1 inhibited its biofilm-forming ability (Riaz et al. 2008; Schipper et al. 2009).

Hetreologous Expression of Eukaryotic Lactonase

Expression of genes for three mouse paraoxonases (PONs) in mammalian cells led to the degradation of AHL in a manner which was quite similar to that of lactonases (Yang et al. 2005). Expression of human paraoxonase in *Drosophila melanogaster* allowed the organism to survive the onslaught of the lethal action of *P. aeruginosa* infection (Stoltz et al. 2008).

Heterologous Expression of AHL-Acylase Gene from *Ralstonia*

Cloning and expression of gene *aiiD* from *Ralstonia* sp. into *E. coli* was very effective in abolishing the 3 QS signals: 3OC8-HSL, 3OC10-HSL and 3OC12-HSL, while its overexpression within *Brucella melitensis* affected the QS-mediated process and clumping phenotypes (Godefroid et al. 2010) (Table 2). This inactivation was quite rapid as it was observed within 3 h of incubation. Wild-type *P. aeruginosa* PAO1 possessing plasmid pUCM9-PAO1 showed accumulation of QS signal in normal concentrations. In contrast, introduction of pUCaiiD encoding AHL-acylase gene into *P. aeruginosa* PAO1 affected a wide range of QS regulated expressions. The pathogenic bacterium could not accumulate AHL signals: C4-HSL and 3OC12-HSL. The genetically engineered *P. aeruginosa* was not able to produce elastase and pyocyanin in normal quantities, and its ability to swarm was significantly restricted. The pathogen could no longer paralyse the nematode, *Caenorhabditis elegans*, which had more than 80 % survival rate (Lin et al. 2003). Subsequently, an aculeacin A acylase was reported from *Ralstonia solanacearum* GMI1000. The *aac* gene had 83 % identity with acylase gene *aiiD*, but no significant resemblance with lactonase gene *aiiA* sequence (Chen et al. 2009). This enzyme is active against AHLs having side chains more than 6 carbons. Expression of *aac* gene in *C. violaceum* CV026 could inhibit AHL-mediated activities such as production of violacein and chitinase enzyme (Chen et al. 2009).

Over Expression of AHL-Acylase Gene from *Pseudomonas*

P. aeruginosa genes coding for AHL-acylase have the potential to be exploited for developing novel therapies against the infections caused by this pathogen. Over expression of gene PA2385 also known as *qsc112* and *pvdQ* (Whiteley et al. 1999; Lamont and Martin 2003) has differential behaviour against AHL signals.

It could not hydrolyse QS signals: C10-HSL, C12-HSL and C14-HSL (Sio et al. 2006). This gene was first cloned into the *P. aeruginosa-E. coli* shuttle vector pME6032 and electroporated in *P. aeruginosa* PAO1. Introduction of gene PA2385 absolutely extripated 3OC12-HSL, but there was no impact on the level of C4-HSL. Over expression of PA2385 abolished activities of enzymes, LasB, and also minimised the expression of *lecA* gene. It, however, did not affect production of pyoverdin (Sio et al. 2006).

A novel AHL-acylase coded by *aiiC* (*all3924*) was reported from filamentous nitrogen fixing cyanobacterium, *Anabaena* sp. PCC7120. It showed homology to *quiP* of *P. aeruginosa* PAO1. This gene was also cloned into shuttle vector pME6032 and expressed in *E. coli*. AiiC has broad specificity with respect to the acyl-chain length, but did not degrade short chain AHLs (Romero et al. 2008). Although *P. aeruginosa* PAO1 has been established to possess *pvdQ* and *qui*P, however, two more genes, pa1893 and pa0305 belonging to the Ntn hydrolase superfamily, have been predicted to encode for penicillin acylase (Wahjudi et al. 2011). Biosensor assays showed their ability to degrade C6 to C14-HSLs. Pa0305 was found to be 56 times more efficient in degrading 3OC12-HSL than C8-HSL. The enzyme was effective in killing *C. elegans* as well (Table 2).

Gene *ahlM* from *Streptomyces* sp. M664 was able to express and degrade AHLs in *Streptomyces lividans* (Park et al. 2005). AhlM enzyme had higher deacylation activity towards long acyl-chain AHLs. It could also degrade penicillin by the same mechanism. It had high activity, even at low concentration of 2 μg/ml. It could significantly affect the production level of virulence factors such as elastase, protease and LasA of *P. aeruginosa* (Park et al. 2005).

Horizontal Gene Transfer

Screening for AHL-degrading proteins has revealed the presence of *aiiA* gene in a large number of organisms. These organisms belong to taxonomically diverse genera. AHL-lactonases from different sources may share homology among themselves. Such genetic similarity can be explained on the basis of horizontal gene transfer (HGT). The occurrence of HGT among bacterial kingdom and its impact of genetic and phenotypic changes has been well reported (Lal et al. 2008). The unique AHL-lactonase of *R. erythropolis* W2 seems to be a case of HGT (Uroz et al. 2008).

In silico studies have also highlighted the occurrence of HGT involving AHL-degrading enzymes (Kalia et al. 2011). Genetic variability in AHL-lactonase coding gene *aii*A was reported on the basis of their phylogenetic relationships and the restriction digestion patterns (Huma et al. 2011; Kalia et al. 2011). Taxonomically diverse organisms were found to show discrepancies in their phylogenetic relationships for *aii*A. Genes for AHL-lactonase from *Firmicutes*, *Moorella thermoacetica* ATCC 39073, and β-*Proteobacteria*, *Burkholderia graminis* C4D1M, were very close to each other with a bootstrap value, BV of 1000. Similar correlations with high BVs were also recorded among: (i) *Actinobacteria* such as *Mycobacterium* sp. MCS and *Rubrobacter xylanophilus* DSM9941 on one hand and members of α-*Proteobacteria* (*Granulibacter bethesdensis* CGDNIH1, *Acidiphilium cryptum* JF-5 and *A. tumefaciens*) on the other and (ii) also between *D. radiodurans* R1 (*Deinococcus-Thermus*) and *Xylella fastidiosa* (γ-*Proteobacteria*).

Taxonomic discrepancies in the phylogenetic tree of the gene for AHL-acylase were reported between different groups: (1) *D. radiodurans* R1 and *Ralstonia* spp. and (2) δ-*Proteobacteria*, *Plesiocystis pacifica* SIR-1, and *Acidobacteria*, *Solibacter usitatus*. This emphasises that in nature, organisms live in symbiotic relationships and share these genes through HGTs for evolving better survival strategies (Kalia et al. 2011).

Conclusion

Heterologous expression of QS signal degrading enzymes can be exploited for producing them on a commercial scale. The immediate application is to use them as potential drugs against biofilm-forming bacteria. Their usage in a range of biotechnological applications has been discussed in a subsequent chapter.

Acknowledgement The authors wish to thank the Director of CSIR-Institute of Genomics and Integrative Biology (IGIB), CSIR-INDEPTH (BSC0111), Government of India for providing the necessary funds and facilities. Authors are also thankful to Academy of Scientific and Innovative Research (AcSIR), New Delhi. PK is thankful to CSIR for granting Senior Research Fellowship.

References

Agarwala M, Choudhury B, Yadav RNS (2014) Comparative study of antibiofilm activity of copper oxide and iron oxide nanoparticles against multidrug resistant biofilm forming uropathogens. Indian J Microbiol. doi:10.1007/s12088-014-0462-z

Annapoorani A, Jabbar AKKA, Musthafa SKS, Pandian SK, Ravi AV (2012) Inhibition of quorum sensing mediated virulence factors production in urinary pathogen *Serratia marcescens* PS1 by marine sponges. Indian J Microbiol 52:160–166. doi:10.1007/s12088-012-0272-0

Augustine N, Kumar P, Thomas S (2010) Inhibition of *Vibrio cholerae* biofilm by AiiA enzyme produced from *Bacillus* spp. Arch Microbiol 192:1019–1022. doi:10.1007/s00203-010-0633-1

Bai F, Han Y, Chen J, Zhang X-H (2008) Disruption of quorum sensing in *Vibrio harveyi* by the AiiA protein of *Bacillus thuringiensis*. Aquaculture 274:36–40. doi:10.1016/j.aquaculture.2007.11.024

Bakkiyaraj D, Sivasankar C, Pandian SK (2013) Anti-pathogenic potential of coral associated bacteria isolated from Gulf of Mannar against *Pseudomonas aeruginosa*. Indian J Microbiol 53:111–113. doi:10.1007/s12088-012-0342-3

Bijtenhoorn P, Mayerhofer H, Müller-Dieckmann J, Utpatel C, Schipper C, Hornung C, Szesny M, Grond S, Thürmer A, Brzuszkiewicz E, Daniel R, Dierking K, Schulenburg H, Streit WR (2011) A novel metagenomic short-chain dehydrogenase/reductase attenuates *Pseudomonas aeruginosa* biofilm formation and virulence on *Caenorhabditis elegans*. PLoS One 6:e26278. doi:10.1371/journal.pone.0026278

Borlee BR, Geske GD, Robinson CJ, Blackwell HE, Handelsman J (2008) Quorum-sensing signals in the microbial community of the cabbage white butterfly larval midgut. ISME J 2:1101–1111. doi:10.1038/ismej.2008.70

Boyer M, Bally R, Perrotto S, Chaintreuil C, Wisniewski-Dyé F (2008) A quorum-quenching approach to identify quorum-sensing-regulated functions in *Azospirillum lipoferum*. Res Microbiol 159:699–708. doi:10.1016/j.resmic.2008.08.003

Byers JT, Lucas C, Salmond GP, Welch M (2002) Nonenzymatic turnover of an *Erwinia carotovora* quorum–sensing signaling molecule. J Bacteriol 184:1163–1171. doi:10.1128/jb.184.4.1163-1171.2002

Carlier A, Uroz S, Smadja B, Fray R, Latour X, Dessaux Y, Faure D (2003) The Ti plasmid of *Agrobacterium tumefaciens* harbors an *attM*-paralogous gene, *aiiB*, also encoding N-acyl homoserine lactonase activity. Appl Environ Microbiol 69:4989–4993. doi:10.1128/AEM.69.8.4989-4993.2003

Chan KG, Wong CS, Yin WF, Sam CK, Koh CL (2010) Rapid degradation of N-3-oxo-acylhomoserine lactones by a *Bacillus cereus* isolate from Malaysian rainforest soil. Antonie Van Leeuwenhoek 98:299–305. doi:10.1007/s10482-010-9438-0

Chen C-N, Chen C-J, Liao C-T, Lee C-Y (2009) A probable aculeacin A acylase from the *Ralstonia solanacearum* GMI1000 is N-acyl-homoserine lactone acylase with quorum-quenching activity. BMC Microbiol 9:89. doi:10.1186/1471-2180-9-89

Chen R, Zhou Z, Cao Y, Yao B (2010) High yield expression of an AHL-lactonase from *Bacillus* sp. B546 in *Pichia pastoris* and its application to reduce *Aeromonas hydrophila* mortality in aquaculture. Microb Cell Fact 9:39

Chen F, Gao Y, Chen X, Yu Z, Li X (2013) Quorum quenching enzymes and their application in degrading signal molecules to block quorum sensing-dependent infection. Int J Mol Sci 14:17477–17500. doi:10.3390/ijms140917477

Chu W, Liu Y, Jiang Y, Zhu W, Zhuang X (2013) Production of N-acyl homoserine lactones and virulence factors of waterborne *Aeromonas hydrophila*. Indian J Microbiol 53:264–268. doi:10.1007/s12088-013-0381-4

Chun CK, Ozer EA, Welsh MJ, Zabner J, Greenberg EP (2004) Inactivation of a *Pseudomonas aeruginosa* quorum-sensing signal by human airway epithelia. Proc Natl Acad Sci USA 101:3587–3590. doi:10.1073/pnas.0308750101

Courvalin P (2008) Predictable and unpredictable evolution of antibiotic resistance. J Intern Med 264:4–16. doi:10.1111/j.1365-2796.2008.01940.x

Czajkowski R, Krzyżanowska D, Karczewska J, Atkinson S, Przysowa J, Lojkowska E, Williams P, Jafra S (2011) Inactivation of AHLs by *Ochrobactrum* sp. A44 depends on the activity of a novel class of AHL acylase. Environ Microbiol Rep 3:59–68. doi:10.1111/j.1758-2229.2010.00188.x

D'Costa VM, McGrann KM, Hughes DW, Wright GD (2006) Sampling the antibiotic resistome. Science 311:374–377. doi:10.1126/science.1120800

Deng Y, Zhu Y, Wang P, Zhu L, Zheng J, Li R, Ruan L, Peng D, Sun M (2011) Complete genome sequence of *Bacillus subtilis* BSn5, an endophytic bacterium of *Amorphophallus konjac* with antimicrobial activity for the plant pathogen *Erwinia carotovora* subsp. carotovora. J Bacteriol 193:2070–2071. doi:10.1128/JB.00129-11

Dong YH, Zhang LH (2005) Quorum sensing and quorum quenching enzymes. J Microbiol 43: 101–109

Dong YH, Xu JL, Li XZ, Zhang LH (2000) AiiA, an enzyme that inactivates the acylhomoserine lactone quorum-sensing signal and attenuates the virulence of *Erwinia carotovora*. Proc Natl Acad Sci USA 97:3526–3531. doi:10.1073/pnas.97.7.3526

Dong YH, Wang LH, Xu JL, Zhang HB, Zhang XF, Zhang LH (2001) Quenching quorum-sensing dependent bacterial infection by an N-acyl homoserine lactonase. Nature 411:813–817. doi:10.1038/35081101

Dong YH, Gusti AR, Zhang Q, Xu JL, Zhang LH (2002) Identification of quorum-quenching N-acyl homoserine lactonases from *Bacillus* species. Appl Environ Microbiol 68:1754–1759. doi:10.1128/AEM.68.4.1754-1759.2002

Dong YH, Zhang XF, Xu JL, Zhang LH (2004) Insecticidal *Bacillus thuringiensis* silences *Erwinia carotovora* virulence by a new form of microbial antagonism, signal interference. Appl Environ Microbiol 70:954–960. doi:10.1128/AEM.70.2.954-960.2004

Dong YH, Wang LH, Zhang LH (2007) Quorum-quenching microbial infections: mechanisms and implications. Phil Trans R Soc Lond B Biol Sci 362:1201–1211. doi:10.1098/rstb.2007.2045

Draganov DI, Teiber JF, Speelman A, Osawa Y, Sunahara R, LaDu BN (2005) Human paraoxonases (PON1, PON2 and PON3) are lactonases with overlapping and distinct substrate specificities. J Lipid Res 46:1239–1247. doi:10.1194/jlr.M400511-JLR200

Fuqua C, Parsek MR, Greenberg EP (2001) Regulation of gene expression by cell-to-cell communication: acyl-homoserine lactone quorum sensing. Annu Rev Genet 35:439–468. doi:10.1146/annurev.genet.35.102401.090913

Godefroid M, Svensson MV, Cambier P, Uzureau S, Mirabella A, De Bolle X, Van Cutsem P, Widmalm G, Letesson JJ (2010) *Brucella melitensis* 16M produces a mannan and other extracellular matrix components typical of a biofilm. FEMS Immunol Med Microbiol 59:364–377. doi:10.1111/j.1574-695X.2010.00689.x

Gui Z, Wang H, Ding T, Zhu W, Zhuang X, Chu W (2014) Azithromycin reduces the production of α-hemolysin and biofilm formation in *Staphylococcus aureus*. Indian J Microbiol 54:114–117. doi:10.1007/s12088-013-0438-4

Han Y, Chen F, Li N, Zhu B, Li X (2010) *Bacillus marcorestinctum* sp. nov., a novel soil acylhomoserine lactone quorum-sensing signal quenching bacterium. Int J Mol Sci 11:507–520. doi:10.3390/ijms11020507

Hastings JW (2004) Bacterial quorum–sensing signals are inactivated by mammalian cells. Proc Natl Acad Sci U S A 101:3993–3994. doi:10.1073/pnas.0400874101

Huang JJ, Han J-I, Zhang L-H, Leadbetter JR (2003) Utilization of acylhomoserine lactone quorum signals for growth by a soil pseudomonad and *Pseudomonas aeruginosa* PAO1. Appl Environ Microbiol 69:5941–5949. doi:10.1128/AEM.69.10.5941-5949.2003

Huang JJ, Petersen A, Whiteley M, Leadbetter JR (2006) Identification of QuiP, the product of gene PA1032, as the second acyl-homoserine lactone acylase of *Pseudomonas aeruginosa* PAO1. Appl Environ Microbiol 72:1190–1197. doi:10.1128/AEM.72.2.1190-1197.2006

Huang W, Lin Y, Yi S, Liu P, Shen J, Shao Z, Liu Z (2012) QsdH, a novel AHL lactonase in the RND-type inner membrane of marine *Pseudoalteromonas byunsanensis* strain 1A01261. PLoS One 7:e46587. doi:10.1371/journal.pone.0046587.t001

Huma N, Shankar P, Kushwah J, Bhushan A, Joshi J, Mukherjee T, Raju SC, Purohit HJ, Kalia VC (2011) Diversity and polymorphism in AHL-lactonase gene (*aiiA*) of *Bacillus*. J Microbiol Biotechnol 21:1001–1011. doi:10.4014/jmb.1105.05056

Kalia VC (2013) Quorum sensing inhibitors: an overview. Biotechnol Adv 31:224–245. doi:10.1016/j.biotechadv.2012.10.004

Kalia VC, Purohit HJ (2011) Quenching the quorum sensing system: potential antibacterial drug targets. Crit Rev Microbiol 37:121–140. doi:10.3109/1040841X.2010.532479

Kalia VC, Raju SC, Purohit HJ (2011) Genomic analysis reveals versatile organisms for quorum quenching enzymes: acyl-homoserine lactone-acylase and -lactonase. Open Microbiol J 5:1–13. doi:10.2174/1874285801105010000

Kalia VC, Kumar P, Pandian SK, Sharma P (2014a) Biofouling control by quorum quenching. In: Kim SK (ed) Handbook of marine biotechnology. Springer

Kalia VC, Wood TK, Kumar P (2014b) Evolution of resistance to quorum-sensing inhibitors. Microb Ecol 68:13–23. doi:10.1007/s00248-013-0316-y

Kumar P, Patel SKS, Lee JK, Kalia VC (2013) Extending the limits of *Bacillus* for novel biotechnological applications. Biotechnol Adv 31:1543–1561. doi:10.1016/j.biotechadv.2013.08.007

Lal S, Cheema S, Kalia VC (2008) Phylogeny vs genome reshuffling: horizontal gene transfer. Indian J Microbiol 48:228–242. doi:10.1007/s12088-008-0034-1

Lamont IL, Martin LW (2003) Identification and characterization of novel pyoverdine synthesis genes in *Pseudomonas aeruginosa*. Microbiology 149:833–842. doi:10.1099/mic.0.26085-0

Leadbetter JR, Greenberg EP (2000) Metabolism of acylhomoserine lactone quorum–sensing signals by *Variovorax paradoxus*. J Bacteriol 182:6921–6926. doi:10.1128/JB.182.24.6921-6926.2000

Lee SJ, Park SY, Lee JJ, Yum DY, Koo BT, Lee JK (2002) Genes encoding the N–acyl homoserine lactone–degrading enzyme are widespread in many subspecies of *Bacillus thuringiensis*. Appl Environ Microbiol 68:3919–3924. doi:10.1128/AEM.68.8.3919-3924.2002

Lin YH, Xu JL, Hu J, Wang LH, Ong SL, Leadbetter JR, Zhang LH (2003) Acyl–homoserine lactone acylase from *Ralstonia* strain XJ12B represents a novel and potent class of quorum–quenching enzymes.

Mol Microbiol 47:849–860. doi:10.1046/j.1365-2958.2003.03351.x

Liu D, Lepore BW, Petsko GA, Thomas PW, Stone EM, Fast W, Ringe D (2005) Three-dimensional structure of the quorum quenching N-acyl homoserine lactone hydrolase from *Bacillus thuringiensis*. Proc Natl Acad Sci USA 102:11882–11887. doi:10.1073/pnas.0505255102

Liu X, Jia J, Popat R, Ortori CA, Li J, Diggle SP, Gao K, Cámara M (2011) Characterization of two quorum sensing systems in the endophytic *Serratia plymuthica* strain G3: differential control of motility and biofilm formation according to life style. BMC Microbiol 11:26. doi:10.1186/1471-2180-11-26

Liu CF, Liu D, Momb J, Thomas PW, Lajoie A, Petsko GA, Fast W, Ringe D (2013) A distal phenylalanine clamp in a hydrophobic channel controls the substrate specificity in the quorum-quenching metallo-γ-lactonase (AiiA) from *Bacillus thuringiensis*. Biochemistry 52:1603–1610. doi:10.1021/bi400050j

Mahanty A, Mishra S, Bosu R, Maurya UK, Netam SP, Sarkar B (2013) Phytoextracts-synthesized silver nanoparticles inhibit bacterial fish pathogen *Aeromonas hydrophila*. Indian J Microbiol 53:438–446. doi:10.1007/s12088-013-0409-9

Mei GY, Yan XX, Turak A, Luo ZQ, Zhang LQ (2010) AidH, an alpha/beta-hydrolase fold family member from an *Ochrobactrum* sp. strain, is a novel N-acylhomoserine lactonase. Appl Environ Microbiol 76:4933–4942. doi:10.1128/AEM.00477-10

Molina L, Constantinescu F, Michel L, Reimmann C, Duffy B, Défago G (2003) Degradation of pathogen quorum-sensing molecules by soil bacteria: a preventive and curative biological control mechanism. FEMS Microbiol Ecol 45:71–81. doi:10.1016/S0168-6496(03)00125-9

Morohoshi T, Ebata A, Nakazawa S, Kato N, Ikeda T (2005) N-acyl homoserine lactone-producing or -degrading bacteria isolated from the intestinal microbial flora of Ayu fish (*Plecoglossus altivelis*). Microb Environ 20:264–268. doi:10.1264/jsme2.20.264

Morohoshi T, Nakazawa S, Ebata A, Kato N, Ikeda T (2008) Identification and characterization of N-acylhomoserine lactone-acylase from the fish intestinal *Shewanella* sp. strain MIB015. Biosci Biotechnol Biochem 72:1887–1893. doi:10.1271/bbb.80139

Müller H, Westendorf C, Leitner E, Chernin L, Riedel K, Schmidt S, Eberl L, Berg G (2009) Quorum-sensing effects in the antagonistic rhizosphere bacterium *Serratia plymuthica* HRO-C48. FEMS Microbiol Ecol 67:468–478. doi:10.1111/j.1574-6941.2008.00635.x

Ng CJ, Shih DM, Hama SY, Villa N, Navab M, Reddy ST (2005) The paraoxonase gene family and atherosclerosis. Free Radical Biol Med 38:153–163. doi:10.1016/j.freeradbiomed.2004.09.035

Nhan DT, Cam DTV, Wille M, Defoirdt T, Bossier P, Sorgeloos P (2010) Quorum quenching bacteria protect *Macrobrachium rosenbergii* larvae from *Vibrio harveyi* infection. J Appl Microbiol 109:1007–1016. doi:10.1111/j.1365-2672.2010.04728.x

Pan J, Huang T, Yao F, Huang Z, Powell CA, Qiu S, Guan X (2008) Expression and characterization of *aiiA* gene from *Bacillus subtilis* BS-1. Microbiol Res 163:711–716. doi:10.1016/j.micres.2007.12.002

Papaioannou E, Wahjudi M, Nadal-Jimenez P, Koch G, Setroikromo R, Quax WJ (2009) Quorum-quenching acylase reduces the virulence of *Pseudomonas aeruginosa* in a *Caenorhabditis elegans* infection model. Antimicrob Agents Chemother 53:4891–4897. doi:10.1128/AAC.00380-09

Park SY, Lee SJ, Oh TK, Oh JW, Koo BT, Yum DY, Lee JK (2003) AhlD, an N-acylhomoserine lactonase in *Arthrobacter* sp., predicted homologues in other bacteria. Microbiology 149:1541–1550. doi:10.1099/mic.0.26269-0

Park SY, Kang HO, Jang HS, Lee JK, Koo BT, Yum DY (2005) Identification of extracellular N-acylhomoserine lactone acylase from a *Streptomyces* sp. and its application to quorum quenching. Appl Environ Microbiol 71:2632–2641. doi:10.1128/AEM.71.5.2632-2641.2005

Park SY, Hwang BJ, Shin MH, Kim JA, Kim HK, Lee JK (2006) N-acyl homoserine lactone producing *Rhodococcus* sp. with different AHL-degrading activities. FEMS Microbiol Lett 261:102–108. doi:10.1111/j.1574-6968.2006.00336.x

Qian GL, Fan JQ, Chen DF, Kang YJ, Han B, Hu BS, Liu FQ (2010) Reducing *Pectobacterium* virulence by expression of an *N*-acyl homoserine lactonase gene P_{lpp}-*aiiA* in *Lysobacter enzymogenes* strain OH11. Biol Control 52:17–23. doi:10.1016/j.biocontrol.2009.05.007

Reimmann C, Ginet N, Michel L, Keel C, Michaux P, Krishnapillai V, Zala M, Heurlier K, Triandafillu K, Harms H, Défago G, Haas D (2002) Genetically programmed autoinducer destruction reduces virulence gene expression and swarming motility in *Pseudomonas aeruginosa* PAO1. Microbiology 148:923–932

Riaz K, Elmerich C, Moreira D, Raffoux A, Dessaux Y, Faure D (2008) A metagenomic analysis of soil bacteria extends the diversity of quorum-quenching lactonases. Environ Microbiol 10:560–570. doi:10.1111/j.1462-2920.2007.01475.x

Romero M, Diggle SP, Heeb S, Cámara C, Otero A (2008) Quorum quenching activity in *Anabaena* sp. PCC7120: identification of AiiC, a novel AHL-acylase. FEMS Microbiol Lett 280:73–80. doi:10.1111/j.1574-6968.2007.01046.x

Schipper C, Hornung C, Bijtenhoorn P, Quitschau M, Grond S, Streit WR (2009) Metagenome-derived clones encoding two novel lactonase family proteins involved in biofilm inhibition in *Pseudomonas aeruginosa*. Appl Environ Microbiol 75:224–233. doi:10.1128/AEM.01389-08

Shang Z, Wang H, Zhou S, Chu W (2014) Characterization of N-Acyl-homoserine lactones (AHLs)-deficient clinical isolates of *Pseudomonas aeruginosa*. Indian J Microbiol 54:158–162. doi:10.1007/s12088-014-0449-9

Shepherd RW, Lindow SE (2009) Two dissimilar N-acyl-homoserine lactone acylases of *Pseudomonas syringae* influence colony and biofilm morphology. Appl Environ Microbiol 75:45–53. doi:10.1128/AEM.01723-08

Sio CF, Otten LG, Cool RH, Diggle SP, Braun PG, Bos R, Daykin M, Cámara M, Williams P, Quax WJ (2006) Quorum Quenching by an N-Acyl-Homoserine lactone acylase from *Pseudomonas aeruginosa* PAO1. Infect Immun 74:1673–1682. doi:10.1128/IAI.74.3.1673-1682.2006

Spellberg B, Powers JH, Brass EP, Miller LG, Edwards JE Jr (2004) Trends in antimicrobial drug development: implications for the future. Clin Infect Dis 38:1279–1286. doi:10.1086/420937

Stoltz DA, Ozer EA, Taft PJ, Barry M, Liu L, Kiss PJ, Moninger TO, Parsek MR, Zabner J (2008) *Drosophila* are protected from *Pseudomonas aeruginosa* lethality by transgenic expression of paraoxonase-1. J Clin Invest 118:3123–3131. doi:10.1172/JCI35147

Thomas PW, Stone EM, Costello AL, Tierney DL, Fast W (2005) The quorum-quenching lactonase from *Bacillus thuringiensis* is a metalloprotein. Biochemistry 44:7559–7569. doi:10.1021/bi050050m

Tinh NTN, Dung NV, Trung CT (2013) In vitro characterization of a recombinant AHL-Lactonase from *Bacillus cereus* isolated from a striped catfish (*Pangasianodon hypophthalmus*) pond. Indian J Microbiol 53:485–487. doi:10.1007/s12088-013-0415-y

Ulrich RL (2004) Quorum quenching: enzymatic disruption of N-acylhomoserine lactone-mediated bacterial communication in *Burkholderia thailandensis*. Appl Environ Microbiol 70:6173–6180. doi:10.1128/AEM.70.10.6173-6180.2004

Uroz S, Heinon J (2008) Degradation of N-acyl homoserine lactone quorum sensing signal molecules by forest root associated fungi. FEMS Microbiol Ecol 65:271–278. doi:10.1111/j.1574-6941.2008.00477.x

Uroz S, DAngelo–Picard C, Carlier A, Elasri M, Sicot C, Petit A, Oger P, Faure D, Dessaux Y (2003) Novel bacteria degrading N-acylhomoserine lactones and their use as quenchers of quorum-sensing-regulated functions of plant-pathogenic bacteria. Microbiology 149:1981–1989. doi:10.1099/mic.0.26375-0

Uroz S, Chhabra SR, Cámara M, Wiliams P, Oger PM, Dessaux Y (2005) N-Acylhomoserine lactone quorum–sensing molecules are modified and degraded by *Rhodococcus erythropolis* W2 by both amidolytic and novel oxidoreductase activities. Microbiology 151:3313–3322. doi:10.1099/mic.0.27961-0

Uroz S, Oger P, Chhabra SR, Cámara M, Williams P, Dessaux Y (2007) N-acyl-homoserine lactones are degraded via an amidolytic activity in *Comamonas* sp. strain D1. Arch Microbiol 187:249–256. doi:10.1007/s00203-006-0186-5

Uroz S, Oger PM, Chapelle E, Adeline MT, Faure D, Dessaux Y (2008) A *Rhodococcus qsdA*-encoded enzyme defines a novel class of large-spectrum quorum-quenching lactonases. Appl Environ Microbiol 74:1357–1366. doi:10.1128/AEM.02014-07

Uroz S, Dessaux Y, Oger P (2009) Quorum sensing and quorum quenching: the Yin and Yang of bacterial communication. ChemBiochem 10:205–216. doi:10.1002/cbic.200800521

Wahjudi M, Papaioannou E, Hendrawati O, van Assen AH, van Merkerk R, Cool RH, Poelarends GJ, Quax WJ (2011) PA0305 of *Pseudomonas aeruginosa* is a quorum quenching acylhomoserine lactone acylase belonging to the Ntn hydrolase superfamily. Microbiology 157:2042–2055. doi:10.1099/mic.0.043935-0

Wang LH, Weng LX, Dong YH, Zhang LH (2004) Specificity and enzyme kinetics of the quorum-quenching N-acyl homoserine lactone lactonase (AHL-lactonase). J Biol Chem 279:13645–13651. doi:10.1074/jbc.M311194200

Wang WZ, Morohoshi T, Someya N, Ikeda T (2012) AidC, a novel N-acylhomoserine lactonase from the potato root-associated *Cytophaga-Flavobacteria-Bacteroides* (CFB) group bacterium *Chryseobacterium* sp. strain StRB126. Appl Environ Microbiol 78:7985–7992. doi:10.1128/AEM.02188-12

Wang H, Tu F, Gui Z, Lu X, Chu W (2013) Antibiotic resistance profiles and quorum sensing-dependent virulence factors in clinical isolates of *Pseudomonas aeruginosa*. Indian J Microbiol 53:163–167. doi:10.1007/s12088-013-0370-7

Whiteley M, Lee KM, Greenberg EP (1999) Identification of genes controlled by quorum sensing in *Pseudomonas aeruginosa*. Proc Natl Acad Sci USA 96:13904–13909. doi:10.1073/pnas.96.24.13904

Williamson LL, Borlee BR, Schloss PD, Guan C, Allen HK, Handelsman J (2005) Intracellular screen to identify metagenomic clones that induce or inhibit a quorum-sensing biosensor. Appl Environ Microbiol 71:6335–6344. doi:10.1128/AEM.71.10.6335-6344.2005

Yang F, Wang LH, Wang J, Dong YH, Hu JY, Zhang LH (2005) Quorum quenching enzyme activity is widely conserved in the sera of mammalian species. FEBS Lett 579:3713–3717. doi:10.1016/j.febslet.2005.05.060

Yang Q, Han Y, Tinh NTN, Hien NT, Bossier P (2012) Detection of quorum sensing signal molecules in *Edwardsiella ictaluri* Ei-151. Indian J Microbiol 52:581–586. doi:10.1007/s12088-012-0312-9

Yates EA, Philipp B, Buckley C, Atkinson S, Chhabra SR, Sockett RE, Goldner M, Dessaux Y, Cámara M, Smith H, Williams P (2002) N–Acylhomoserine lactones undergo lactonolysis in a pH–, temperature–, acyl chain length–dependent manner during growth of *Yersinia pseudotuberculosis* and *Pseudomonas aeruginosa*. Infect Immun 70:5635–5646. doi:10.1128/IAI.70.10.5635-5646.2002

Yin XT, Xu L, Fan SS, Xu LN, Li DC, Liu ZY (2010) Isolation and characterization of an AHL lactonase gene from *Bacillus amyloliquefaciens*. World J Microbiol Biotechnol 26:1361–1367. doi:10.1007/s11274-010-0308-8

Yin WF, Tung HJ, Sam CK, Koh CL, Chan KG (2012) Quorum quenching *Bacillus sonorensis* isolated from soya sauce fermentation brine. Sensors 12:4065–4073. doi:10.3390/s120404065

Zhang L-H, Dong Y-H (2004) Quorum sensing and signal interference: diverse implications. Mol Microbiol 53:1563–1571. doi:10.1111/j.1365-2958.2004.04234.x

Zhang Y, Zhang Y, Yang Y, Wang L, Weng L (2013) Identification of a *Pseudomonas* sp. that inhibits RHL system of quorum sensing. Indian J Microbiol 53:28–35. doi:10.1007/s12088-012-0340-5

Zhou Y, Ye WX, Zhou Y, Zhu CG, Sun M, Yu ZN (2006) Ethanol tolerance, yield of melanin, swarming motility and growth are correlated with the expression levels of *aii*A gene in *Bacillus thuringiensis*. Enzyme Microb Technol 38:967–974. doi:10.1016/j.enzmictec.2005.08.037

Part VII

Biotechnological Applications of Quorum Sensing Inhibitors

Potential Applications of Quorum Sensing Inhibitors in Diverse Fields

Vipin C. Kalia and Prasun Kumar

Abbreviations

AHL	Acyl homoserine lactone
HSL	Homoserine lactone
3OC6-HSL	3-oxo-N-hexanoyl-HSL
3OC12-HSL	3-oxo-N-dodecanoyl- HSL
QQ	Quorum quenching
QS	Quorum sensing
QSI	Quorum sensing inhibitors
QSS	Quorum sensing systems

Introduction

All organisms are susceptible to attack by other organisms, and it may even spell death for the recipient. However, each organism is also bestowed with an inherent ability to protect itself by developing self-defence mechanisms. Human beings have a strong immune system but are quite susceptible to infections by fungi, bacteria and viruses. Microbial infections have been a constant worry for health departments (Kalia 2013, 2014). The economy of a nation is dramatically affected by the health of its residents. The discovery of antibiotics was a great boon to mankind. However, microbes have been developing resistance against antibiotics. So much so that during the last seven to eight decades, there has been a need to find new antibiotics. Now the scenario is quite depressing as almost all antibiotics are proving ineffective. The evolution of multidrug resistance among pathogenic microbes has taken a new dimension (Davies and Davies 2010). Bacteria expressing their virulent behaviour through the phenomenon of quorum sensing (QS) develop a biofilm. Bacteria inside the biofilm are up to 1000 times more resistant to antibiotics compared to their planktonic counterparts (Kalia et al. 2014a, b). Research efforts during the last four decades have brought hope by providing alternatives and supplements to antibiotics. Quorum sensing inhibitors (QSIs) are seen as novel drugs especially against infectious bacteria. Although, the search for QSIs was intended for helping human beings to fight against diseases, the applications can be extended to other fields as well: agriculture, aquaculture, water treatment, fisheries, etc. A few examples of applications of QSIs have been described below (Table 1).

V.C. Kalia (✉) • P. Kumar
Microbial Biotechnology and Genomics, CSIR-Institute of Genomics and Integrative Biology, Mall Road, Delhi 110007, India
e-mail: vckalia@igib.res.in; vc_kalia@yahoo.co.in

Agriculture

Associations between plant and bacteria are well known and prove helpful to each other. Bacteria living as epiphytes – *Pseudomonas, Pantoea* and *Erwinia* – seem to help the plant by manipulating the QS behaviour of plant pathogens. Premature induction of QS can allow the host to activate its defence mechanisms. Epiphytic bacteria having an inherent genetic make-up to produce 10-fold

V.C. Kalia (ed.), *Quorum Sensing vs Quorum Quenching: A Battle with No End in Sight*,
DOI 10.1007/978-81-322-1982-8_29, © Springer India 2015

Table 1 Biotechnological applications of quorum sensing inhibitors from diverse sources

Source of quorum quenchers	Biotechnological applications	Reference
Agriculture		
Erwinia, Pantoea and *Pseudomonas* (epiphytic bacteria)	Reduction in pathogenicity of *Pseudomonas syringae* on *Nicotiana* (tobacco) plant	Quinones et al. (2004), Dulla and Lindow (2009)
Microbacterium testaceum (epiphyte on potato leaf)	Effective against soft rot disease-causing pathogen *Pectobacterium carotovorum*	Wang et al. (2010)
Recombinant *Erwinia carotovora*	Reduced infection on plants: cauliflower, Chinese cabbage and *Nicotiana* (tobacco)	Dong et al. (2000)
Erwinia carotovora subsp. *atroseptica* (AHL-lactonase genes: *attM, aiiB*)	Reduce maceration in potato	Carlier et al. (2003)
Furocoumarins from grapefruit juice	Inhibitor of AI-1 and AI-2 activities in *S. typhimurium* and *P. aeruginosa*, may act as food preservatives	Girennavar et al. (2008)
Water treatment		
Secondary metabolites from *Delisea, Asparagopsis* and *Bonnemaisonia*	Antifouling agent against bacteria	Nylund et al. (2005, 2008, 2010), de Nys et al. (2006), Paul et al. (2006a, b)
Pyranone (kojic acid) from *Aspergillus*	Inhibit the growth of bacteria and diatoms involved in community formation	Dobretsov et al. (2011)
Biomembranes immobilized with porcine kidney acylase 1	Reduction in biofouling during water treatment	Swift et al. (1997), Paul et al. (2009), Yeon et al. (2009a, b), Choudhary and Schmidt-Dannert (2010)
Filtration membrane coated with porcine kidney acylase I	Antibiofouling, inhibits biofilm formation and production of exopolysaccharides	Yeon et al. (2009a), Kim et al. (2011)
Fisheries and aquaculture		
Gut microbes from shrimp – marine organism (*Penaeus vannamei*)	Quench the QS signal molecule HAI-I of *Vibrio harveyi*. Enhanced growth of rotifers and survival of turbot larvae	Tinh et al. (2007a, 2008)
Gut microbes from European sea bass, *Dicentrarchus labrax* L, and Asian sea bass, *Lates calcarifer*	Increase in the survival rate of larvae of *Macrobrachium rosenbergii*	Cam et al. (2009), Nhan et al. (2010)
Gut microbes (*Shewanella* sp.) from ayu fish, *Plecoglossus altivelis*	Inhibited biofilm formation by fish pathogen – *Vibrio anguillarum*	Morohoshi et al. (2005, 2008)
Natural and synthetic QS inhibitors such as brominated furanones	Protect brine shrimp, *Artemia franciscana*; rotifer, *Brachionus plicatilis*; and rainbow trout from pathogenic *Vibrio* spp.	Defoirdt et al. (2006), Tinh et al. (2007b), Rasch et al. (2004)
Expression of *aii*A in *Pichia pastoris*	Increase survival against the fish pathogen *Aeromonas hydrophila*	Chen et al. (2010)
Medical		
Mammalian serum paraoxonases	Degrades broad range of AHLs	Yang et al. (2005)
Garlic	QS inhibitory activity in animal models and human trials	Smyth et al. (2010)
Transgenic *Drosophila* – expressing human paraoxonase 1	Reduced pathogenicity of *Pseudomonas aeruginosa* and *Serratia marcescens*	Stoltz et al. (2008)
Homologues of QS signals (AHL)	Anticancer	Oliver et al. (2009)
Recombinant *Escherichia coli*	As vehicles in cancer therapy	Anderson et al. (2006)
Recombinant *Burkholderia thailandensis* (*aiiA* from *Bacillus*)	Significant reduction in bacterial activities: swarming and twitching motility. Prevents the β-haemolysis of sheep erythrocytes	Ulrich (2004)
Synthetic AIP-II	Mouse become resistant to *S. aureus*	Mayville et al. (1999)
Furanone	Inhibits pathogenicity of *P. aeruginosa*	Hentzer et al. (2003)
Candida albicans (DSF)	Inhibits dimorphism linked to fungal virulence	Wang et al. (2004)

higher quantities of QS signal (acyl homoserine lactones [AHLs]) – 3OC6-HSL – caused *Pseudomonas syringae* to prematurely induce its quorum sensing system (QSS). It allows tobacco plants to become resistant to pathogenic attack of *P. syringae* (Quinones et al. 2004; Dulla and Lindow 2009) (Table 1). Potato leaf surface harboured by *Microbacterium testaceum* can protect the potato plant from soft rot disease caused by plant pathogen *Pectobacterium carotovorum*. It is primarily because of the AHL-degrading capacity of the bacterium – hence reduced QS-regulated expression of virulence genes (Wang et al. 2010). Another approach prevalent among agricultural scientists is the production of transgenic plants. Genetic modification of plants with bacterial enzymes known to inhibit QSS of pathogens can be effective tools for crop protection. In fact, transgenic plants of tobacco and potato expressing AHL-lactonase enzyme were helpful in reducing the impact of infection caused by *Erwinia carotovora* on tobacco, on potato and also on cauliflower and Chinese cabbage (Dong et al. 2000, 2001; Carlier et al. 2003). Nitrogen-fixing plants like *Medicago truncatula* and *Pisum sativum* are known to exude chemical compounds which act as QS mimics in response to bacterial infections. These compounds provide protection to plants against pathogens by manipulating their QS-regulated expression of genes responsible for their virulence behaviour (Teplitski et al. 2000; Cirou et al. 2012; LaSarre and Federle 2013). For agricultural managers, these genetic tools can help to control bacterial infection and, consequently, help them to achieve higher crop yield.

Water Treatment

Biofilm formation is generally linked to the infectious bacteria (Vejborg and Klemm 2008). In fact, more than 80 % of the infectious diseases are caused by biofilm-forming bacteria (Donlan 2002; Martinez and Fries 2010; Kalia et al. 2014a, b). Industrial houses involved in reclamation and desalination of brackish and sea water face a very strange scenario (Cooley et al. 2006). They employ membrane bioreactor to treat water (Stamper et al. 2003). These membrane filters get choked by bacterial growth and biofilm formation (Azizi et al. 2013). This exerts unnecessary pressure on the membrane filters, a phenomenon called as biofouling. This adverse effect on membrane permeability obviously adds to treatment costs (Paul et al. 2009; Kalia et al. 2014a, b). Among a large number of bacteria, the most prevalent are *Aeromonas hydrophila* and *Pseudomonas putida* (Swift et al. 1997; Chen et al. 2010; Zhang et al. 2013). Mechanisms to distort the biofilm have been through coating of QSIs or immobilizing enzymes. Others have employed genetically engineered marine organisms, which release QSI to act as antifouling agents. This non-toxic antifouling technology can be used effectively in a wide variety of areas (Qian et al. 2007; Choudhary and Schmidt-Dannert 2010). The use of acylase I (porcine kidney) was effective in reducing biofilm formed by *A. hydrophila* by a magnitude of 20–24 % (Paul et al. 2009) (Table 1). Although this antibiofouling technique appears promising, it is limited by the short life of the enzyme and difficulty in recovering the enzyme for re-usage. Immobilization of quorum quenching (QQ) enzyme (acylase I) on membrane was shown to control biofouling (Yeon et al. 2009a). Other innovative steps such as immobilization of the enzymes on magnetic particles also improve the efficiency of water treatment process (Yeon et al. 2009b; Kim et al. 2011).

Adhesion and biofilm formation by *Pseudoalteromonas* sp. D41 could be abolished by hydrolases, which act upon extracellular polysaccharide (EPS) and QS molecules. Similarly, enzymes – mutanase, dextranase, Savinase and Alcalase – were instrumental in reducing bacterial adhesion and inhibiting dental biofilms formed by *Streptococcus mutans* and *Streptococcus sobrinus* (Leroy et al. 2008). *Rhodococcus* sp. BH4, producing QQ enzymes, was encapsulated in alginate beads (cell entrapping beads [CEBs]). These CEBs colliding with biofilm on membrane surface were effective in removing biofilms in a more efficient manner than the physical phenomenon of friction

(Kim et al. 2013). Secondary metabolites from *Asparagopsis*, *Bonnemaisonia* and *Delisea* have been shown to act as antifouling agents (Nylund et al. 2005, 2008, 2010; de Nys et al. 2006; Paul et al. 2006a, b; Lane et al. 2009). It has been envisaged that these biological approaches are likely to tackle biofouling problems (Malaeb et al. 2013).

Biotransformations

The exopolysaccharide succinoglycan is produced by bacteria. It has biochemical properties similar to xanthan, which has applications in oil wells (Clarke-Sturman et al. 1989). Symbiotic nitrogen-fixing *Sinorhizobium meliloti* in association with *Medicago* plants produce exopolymers. Under QS regulatory mechanism, the bacterium produces important exopolysaccharides, succinoglycan and EPS II, which have great biotechnological applications (Marketon et al. 2003). The most widely used bacterial host for expression of a range of genes has been *Escherichia coli*. With its potential to be used on industrial scale, the pharmaceutical industry has used it for producing biochemicals and proteins of therapeutic value (Huang et al. 2012). However, the associated risk in biochemical transformation carried out by single cultures is its getting contaminated. It obviously demands strict monitoring. Genetically engineering cyclic AMP into *E. coli* allowed uninterrupted production of indole, which plays a dual role. Firstly, it supports the growth of the host and, secondly, inhibits QS mediated virulence factors including pyocyanin production of contaminating bacteria such as *Pseudomonas aeruginosa* (Chu et al. 2012; Lee et al. 2012). The role of indole in regulating the process of biofilm formation especially in *E. coli* has been shown by manipulating specific proteins. SdiA is a transcription regulator which is influenced by indole and AHLs (van Houdt et al. 2006; Yao et al. 2006; Jayaraman and Wood 2008; Wood 2009). QS signal *N*-butyryl-*DL*-homoserine lactone at low concentration of 10 μM decreased biofilm formation of *E. coli* (25 %) (Lee et al. 2007). SdiA variants formed 5-fold lower biofilms in comparison to the wild-type cells, while some variants had enhanced biofilm-forming capacity. It has also led to the construction of the first synthetic signalling circuit, which can manipulate biofilm formation. Indole also acts as an interspecies signal between *E. coli* and pseudomonads (Lee et al. 2007).

Most biochemical reactions need very stringent control and demand efforts. It has been realized that synthetic biology can offer better solutions and easy manipulation on QS-based bacterial biofilms (Hooshangi and Bentley 2008). This self-regulatory system can be easily automated and will need little supervision, if any. For biochemical transformations on large scale, especially in biorefineries, biofilms play a significant role. They provide the platform for biocatalysis, which need to be reused for achieving high economy. *P. aeruginosa* QS signal 3OC12-HSL plays an important role in the formation of biofilm. Dispersal of the biofilm can be achieved with the aid of two proteins. Engineering the global regulator (Hha13D6) and cyclic diguanylate binding (BdcAE50Q) proved instrumental in dispersing the cells forming the biofilm. These cells had a retarded growth rate. Incidentally, these slow growers could enhance their QS signal molecule (3OC12-HSL)-producing capacity by a magnitude of 14-fold compared to their planktonic counterparts. The system of removing the old biofilm paved way for a fresh round of biofilm formation. This strategy can be effectively exploited for biotechnological applications: biofouling, bioremediation, biocorrosion, biosensors, etc. (Hong et al. 2012a, b).

Fisheries and Aquaculture

Fisheries and aquaculture departments are primarily viewed as source of food especially along coastal areas and water bodies (Smith et al. 2010). A worry which looms large on the managers of these industries is the diseases caused by different pathogenic bacteria (Defoirdt et al. 2011; Tinh et al. 2013). QSIs such as floridoside, betonicine and isethionic acid produced by a macro-alga – *Ahnfeltiopsis flabelliformis* – can be effective in aquacultures. Bromo-tryptamine-based alkaloids have been shown to be produced by Bryozoa, *Flustra foliacea* isolated from North Sea, which can inhibit QS-regulated gene expression (Peters et al. 2003). Other equally important sources of QSI such as secomanoalide and manoalide are sponges – *Luffariella variabilis* – and corals (Ettinger-Epstein et al. 2008; Bakkiyaraj et al. 2013). Waterborne pathogens such as *A. hydrophila* are responsible for infections among fishes (Chen et al. 2010; Chu et al. 2013). Rotifer *Brachionus plicatilis* used as feed for fish larvae get infected by bacteria belonging to *Aeromonas*, *Pseudomonas* and *Vibrio* (Skjerma and Vadstein 1993). An easy option seems to be the use of antibiotics to protect fishes and other organisms from bacterial attack. It has been well established that persistent and indiscriminate use of antibiotics results in their acquiring multiple drug resistance traits (Davies and Davies 2010). Since these bacteria cause disease primarily through gene regulated by QS, it can be envisaged that disruption of this system can be a better alternative (Defoirdt et al. 2004).

Members of the red algal family Bonnemaisoniaceae produce secondary metabolites having strong antifouling activities and retard the growth of pathogenic bacteria (Nylund et al. 2005, 2008, 2010; de Nys et al. 2006; Paul et al. 2006a, b). Brominated furanones have been found to be helpful in higher survival and growth of rotifers in the presence of *Vibrio harveyi* strains (Tinh et al. 2007a, b). A similar protective effect at reasonably low concentrations of this QSI was reported against: (1) *Vibrio* spp. infecting gnotobiotic brine shrimp *Artemia franciscana* (Defoirdt et al. 2006); (2) *V. anguillarum* infecting rainbow trout, *Oncorhynchus mykiss* (Rasch et al. 2004); and (3) *V. harveyi* (Tinh et al. 2007b). Gut microbes (*Shewanella* sp.) isolated from ayu fish (*Plecoglossus altivelis*) could inhibit biofilm formation by *Vibrio anguillarum* (Morohoshi et al. 2005, 2008). QS mediated virulence genes of *V. harveyi* were also found to be suppressed by *Halobacillus salinus* strain isolated from seagrass. The bacterial strain produces two phenethylamide metabolites, which act as QSIs (Teasdale et al. 2009). Microbes from the guts of marine organisms Asian sea bass, *Lates calcarifer*, and European sea bass, *Dicentrarchus labrax* L., used as inoculants in prawn rearing water, were effective in enhancing the survival rate of larvae of *Macrobrachium rosenbergii* (Cam et al. 2009).

Bacillus spp. being a very good source of AHL-lactonase are among the most deserving candidates, which can be exploited on an industrial scale (Kumar et al. 2013). The use of AHL degraders such as *Bacillus* spp. was instrumental in increasing the survival rate of larvae of: (1) *Scophthalmus maximus* (turbot) (Tinh et al. 2008), (2) *Sparus aurata* (gilthead sea beam) (Avella et al. 2010) and (3) *M. rosenbergii* (the giant freshwater prawn) (Nhan et al. 2010). QQ enzyme – acylase from fish pathogen *Tenacibaculum maritimum* NCIMB2154T belonging to Bacteroidetes – provides more opportunities for biotechnological application of the enzyme-based QSIs (Romero et al. 2010). The use of QSIs is likely to improve the economy of the food industry by increasing the shelf life of the food. Well-preserved food and reduced chances of the evolution of MDRs are good for health and fishery departments as well (Rasch et al. 2004; Cam et al. 2009; Nhan et al. 2010; Craigen et al. 2011; Jamuna Bai and Rai 2011; Annapoorani et al. 2012). Since marine animals such as big fishes and whales do not allow microbes to stick to their bodies, we can learn to develop surfaces for avoiding biofouling (Natrah et al. 2011; Hong et al. 2012a, b).

Medical

In Humans

Microorganisms are very closely associated with human beings: (1) harbouring the skin surface, (2) present in the gastrointestinal system and (3) infecting different parts of the body (Grice and Segre 2011). We take notice of the microbes which cause a number of diseases. Bacteria operating through QSS are responsible for metabolic disturbances. The growth of *E. coli* and *P. aeruginosa* is regulated by different bioactive molecules: adenosine, indole, QS signal (3OC12-HSL) and noradrenaline. Adenosine regulates secretion of electrolytes, downregulates inflammation and disrupts epithelial cells (Ye and Rajendran 2009). All these activities allow pathogens like *Staphylococcus aureus* and *Bacillus anthracis* to escape phagocytosis (Thammavongsa et al. 2009; Tonello and Zornetta 2012). Adenosine is also instrumental in inhibiting the expression of pathogenicity of *P. aeruginosa* on *Caenorhabditis elegans*. It operates by repressing the acquisition of iron and in turn influences bacterial growth. It also regulates the precursor of signal molecule PQS and thus acts as a QSI (Sheng et al. 2012).

QS signal (3OC12-HSL) from *P. aeruginosa* can inhibit proliferation of human breast cancer cell lines and induce apoptosis (Li et al. 2004). Its potential as anticancer therapy can be better exploited through the use of synthetic AHL homologues, which may not allow virulence gene expression in *P. aeruginosa* especially in immunocompromised patients (Oliver et al. 2009). Genetically engineered bacteria with QSS targeting only cancerous cells are a better option (Choudhary and Schmidt-Dannert 2010). Recombinant *E. coli* with invasion gene of *Yersinia pestis* regulated by QS signals of *Vibrio fischeri* was useful in invading human cancer-derived cell lines (Anderson et al. 2006). Similarly, a recombinant strain of *Burkholderia thailandensis* expressing AHL-lactonase gene of *Bacillus* was shown to be affecting the QS mediated motility activities of the bacterium (Ulrich 2004). Garlic extracts have been demonstrated as QSIs against *P. aeruginosa* infections by inhibiting biofilms produced by them (Rasmussen et al. 2005; Jakobsen et al. 2012). In fact, QSI treatment made the biofilm more susceptible to antibiotics such as azithromycin and tobramycin and phagocytosis by neutrophils (Bjarnsholt et al. 2005; Gui et al. 2014). The use of diffusible signal factor and its structural analogues can be used to enhance the efficiency of conventional antibiotic for treating infectious diseases (Deng et al. 2014). The first clinical trial on human patients suffering from cystic fibrosis was based on this QSI (Smyth et al. 2010). *P. aeruginosa* biofilms can be treated using synthetic furanone and consequently reduce the production of enzymes elastase and chitinase responsible for manifesting the disease (Hentzer et al. 2002, 2003). Thus, genetically engineered bacteria can be used for large-scale production of QSIs in a self-regulatory manner (Basu et al. 2004; Brenner et al. 2007; Chan et al. 2010; Choudhary and Schmidt-Dannert 2010; Danino et al. 2010).

In clinical settings, *P. aeruginosa* resistant to antibiotics and even those deficient in AHL production are also effective in causing infections (Wang et al. 2013; Shang et al. 2014). In clinics, Dispersin B (hexosaminidase) produced by *Actinobacilli* and *Aggregatibacter actinomycetemcomitans* can prevent biofilm formation by *Staphylococcus epidermidis* on medical implants, plastics and polyurethane coatings. Synergistic effect of Dispersin B along with antiseptics such as triclosan or chlorhexidine was seen by their antibiofilm and antimicrobial activities against *S. aureus*, *S. epidermidis* and *E. coli* (Arciola et al. 2011; Boles and Horswill 2011). Recombinant phages producing Dispersin B can be applied to hydrolyze bacterial adhesion molecules and other biofilm-forming molecules (Lu and Collins 2007; Darouiche et al. 2009; Arciola et al. 2011). Hence, surgical implants and catheters coated with QSIs can be a viable strategy for medical applications (Choudhary and Schmidt-Dannert 2010).

In Non-humans

Transgenic *Drosophila* harbouring human paraoxonases 1 (PON1) were quite resistant to infections caused by *P. aeruginosa* and *Serratia marcescens* (Table 1). PON1 acts by degrading AHL signals of the pathogenic bacteria, which thus prevents pathogenesis. These QQ mechanisms can be exploited as therapeutic strategy for infections and inflammations (Stoltz et al. 2008). Besides the enzymatic QSIs well appreciated for its broad QS targets, synthetic molecules were also employed. Usage of synthetic molecules such as AIP-II, furanones and other diffusible signal factors led to host organisms such as mice becoming resistant against *S. aureus*, *P. aeruginosa* and *Candida albicans*, respectively (Mayville et al. 1999; Hentzer et al. 2003; Wang et al. 2004).

Shipping Industry

The growth of sessile organisms on ships is a serious cause of concern for the shipping industry. Biofouling caused by marine organisms affects the efficiency of ships and propellers. This leads to increase in the weight and consequently higher consumption of fuels. QQ producing marine organisms can be effective in developing antifouling technologies and thus tackling these problems (Qian et al. 2007). Application of QQ biomolecules on the surfaces of ship hulls, pipes, tanks, etc. can prove effective in reducing biofouling (Choudhary and Schmidt-Dannert 2010). Incorporation of kojic acid produced by *Aspergillus* into a non-toxic paint can be applied on the base of the ships for controlling bacteria and diatoms (Dobretsov et al. 2011).

Conclusion

Antibiotics like penicillin have been proving to be quite ineffective in recent years. Now, a novel strategy has been proposed by combining the drug with a polymer, which is expected to make these drugs effective against MDRs (Zhang et al. 2014). The broad range of applications of QSIs also demands effort to improve their efficiency. Nanoparticles have been found to be effective against biofilm-forming MDRs (Mahanty et al. 2013; Agarwala et al. 2014). Even gamma-radiations seem to help in enhancing the antimicrobial properties of these bacteria (Pathak et al. 2013). So far the use of QSIs has played an effective role in agriculture, aquaculture, fisheries and shipping industries. However, their use as therapeutics against bacteria causing infections in human being is still limited. A word of caution has also been given by the recent discoveries that bacteria are likely to become resistant to QSI (Kalia et al. 2014a, b).

Acknowledgements The authors wish to thank the director of CSIR-Institute of Genomics and Integrative Biology (IGIB), CSIR-INDEPTH (BSC0111), Government of India, for providing the necessary funds and facilities. PK is thankful to CSIR for granting Senior Research Fellowship.

References

Agarwala M, Choudhury B, Yadav RNS (2014) Comparative study of antibiofilm activity of copper oxide and iron oxide nanoparticles against multidrug resistant biofilm forming uropathogens. Indian J Microbiol. doi:10.1007/s12088-014-0462-z

Anderson JC, Clarke EJ, Arkin AP, Voigt CA (2006) Environmentally controlled invasion of cancer cells by engineered bacteria. J Mol Biol 355:619–627. doi:10.1016/j.jmb.2005.10.076

Annapoorani A, Jabbar AKKA, Musthafa SKS, Pandian SK, Ravi AV (2012) Inhibition of quorum sensing mediated virulence factors production in urinary pathogen *Serratia marcescens* PS1 by marine sponges. Indian J Microbiol 52:160–166. doi:10.1007/s12088-012-0272-0

Arciola CR, Montanaro L, Costerton JW (2011) New trends in diagnosis and control strategies for implant infections. Int J Artif Organs 34:727–736. doi:10.5301/IJAO.2011.8784

Avella MA, Gioacchini G, Decamp O, Makridis P, Bracciatelli C, Carnevali O (2010) Application of multispecies of *Bacillus* in sea bream larviculture. Aquaculture305:12. doi:10.1016/j.aquaculture.2010.03.029

Azizi S, Valipour A, Sithebe T (2013) Evaluation of different wastewater treatment processes and development of a modified attached growth bioreactor as a decentralized approach for small communities. The Scientific World Journal 2013. Article ID 156870, 8 pgs. doi:10.1155/2013/156870

Bakkiyaraj D, Sivasankar C, Pandian SK (2013) Anti-pathogenic potential of coral associated bacteria isolated from Gulf of Mannar against *Pseudomonas aeruginosa*. Indian J Microbiol 53:111–113. doi:10.1007/s12088-012-0342-3

Basu S, Mehreja R, Thiberge S, Chen MT, Weiss R (2004) Spatiotemporal control of gene expression with pulse-generating networks. Proc Natl Acad Sci U S A 101:6355–6360. doi:10.1073/pnas.0307571101

Bjarnsholt T, Jensen PØ, Rasmussen TB, Christophersen L, Calum H, Hentzer M, Hougen H-P, Rygaard J, Moser C, Eberl L, Høiby N, Givskov M (2005) Garlic blocks quorum sensing and promotes rapid clearing of pulmonary *Pseudomonas aeruginosa* infections. Microbiology 151:3873–3880. doi:10.1099/mic.0.27955-0

Boles BR, Horswill AR (2011) Staphylococcal biofilm disassembly. Trends Microbiol 19:449–455. doi:10.1016/j.tim.2011.06.004

Brenner K, Karig DK, Weiss R, Arnold FH (2007) Engineered bidirectional communication mediates a consensus in a microbial biofilm consortium. Proc Natl Acad Sci USA 104:17300–17304. doi:10.1073/pnas.0704256104

Cam DTV, Nhan DT, Ceuppens S, Hao NV, Dierckens K, Wille M, Sorgeloos P, Bossier P (2009) Effect of N-acyl homoserine lactone-degrading enrichment cultures on *Macrobrachium rosenbergiii* larviculture. Aquaculture 294:5–13

Carlier A, Uroz S, Smadja B, Fray R, Latour X, Dessaux Y, Faure D (2003) The Ti plasmid of *Agrobacterium tumefaciens* harbors an *attM*-paralogous gene, *aiiB*, also encoding N-acyl homoserine lactonase activity. Appl Environ Microbiol 69:4989–4993. doi:10.1128/AEM.69.8.4989-4993.2003

Chan K-G, Yin W-F, Sam C-K, Koh C-L (2010) A novel medium for the isolation of N-acylhomoserine lactone-degrading bacteria. J Ind Microbiol Biotechnol 36:247–251. doi:10.1007/s10295-008-0491-x

Chen R, Zhou Z, Cao Y, Yao B (2010) High yield expression of an AHL-lactonase from *Bacillus* sp. B546 in *Pichia pastoris* and its application to reduce *Aeromonas hydrophila* mortality in aquaculture. Microb Cell Fact 9:39

Choudhary S, Schmidt-Dannert C (2010) Applications of quorum sensing in biotechnology. Appl Microbiol Biotechnol 86:1267–1279. doi:10.1007/s00253-010-2521-7

Chu W, Zere TR, Weber MM, Wood TK, Whiteley M, Hidalgo-Romano B, Valenzuela EJ, McLean RJ (2012) Indole production promotes *Escherichia coli* mixed-culture growth with *Pseudomonas aeruginosa* by inhibiting quorum signaling. Appl Environ Microbiol 78:411–419. doi:10.1128/AEM.06396-11

Chu W, Liu Y, Jiang Y, Zhu W, Zhuang X (2013) Production of N-acyl homoserine lactones and virulence factors of waterborne *Aeromonas hydrophila*. Indian J Microbiol 53:264–268. doi:10.1007/s12088-013-0381-4

Cirou A, Mondy S, An S, Charrier A, Sarrazin A, Thoison O, DuBow M, Faure D (2012) Efficient biostimulation of native and introduced quorum-quenching *Rhodococcus erythropolis* populations is revealed by a combination of analytical chemistry, microbiology, and pyrosequencing. Appl Environ Microbiol 78:481–492. doi:10.1128/AEM.06159-11

Clarke-Sturman AJ, Den Ottelander D, Sturla PL (1989) Succinoglycan: a new biopolymer for the oil field. Oil-Field Chem, Prepr; (United States), 33 396:157–168. doi:10.1021/bk-1989-0396.ch008

Cooley H, Gleick PH, Wolff G (2006) Desalination, with a grain of salt – A california perspective. Pacific Institute for Studies in Development, Environment, and Security. Oakland, CA

Craigen B, Dashiff A, Kadouri DE (2011) The use of commercially available alpha-amylase compounds to inhibit and remove *Staphylococcus aureus* biofilms. Open Microbiol J 5:21–31. doi:10.2174/1874285801105010021

Danino T, Mondrag6n-Palomino O, Tsimring L, Hasty J (2010) A synchronized quorum of genetic clocks. Nature 463:326–330. doi:10.1038/nature08753

Darouiche RO, Mansouri MD, Gawande PV, Madhyastha S (2009) Antimicrobial and antibiofilm efficacy of triclosan and DispersinB® combination. J Antimicrob Chemother 64:88–93. doi:10.1093/jac/dkp158

Davies J, Davies D (2010) Origins and evolution of antibiotic resistance. Microbiol Mol Biol Rev 74:417–433. doi:10.1128/MMBR.00016-10

de Nys R, Givskov M, Kumar N, Kjelleberg S, Steinberg PD (2006) Furanones. Prog Mol Subcell Biol 42:55–86. doi:10.1007/3-540-30016-3_2

Defoirdt T, Boon N, Bossier P, Verstraete W (2004) Disruption of bacterial quorum sensing: an unexplored strategy to fight infections in aquaculture. Aquaculture 240:69–88

Defoirdt T, Crab R, Wood TK, Sorgeloos P, Verstraete W, Boossier P (2006) Quorum sensing-disrupting brominated furanones protect the gnotobiotic brine shrimp *Artemia franciscana* from pathogenic *Vibrio harveyi*, *Vibrio campbellii* and *Vibrio parahaemolyticus* isolates. Appl Environ Microbiol 72:6419–6423. doi:10.1128/AEM.00753-06

Defoirdt T, Sorgeloos P, Bossier P (2011) Alternatives to antibiotics for the control of bacterial disease in aquaculture. Curr Opin Microbiol 14:251–258. doi:10.1016/j.mib.2011.03.004

Deng Y, Lim A, Lee J, Chen S, An S, Dong YH, Zhang LH (2014) Diffusible signal factor (DSF) quorum sensing signal and structurally related molecules enhance the antimicrobial efficacy of antibiotics against some bacterial pathogens. BMC Microbiol 14:51. doi:10.1186/1471-2180-14-51

Dobretsov S, Teplitski M, Bayer M, Gunasekra S, Proksch P, Paul VJ (2011) Inhibition of marine biofouling by bacterial quorum sensing inhibitors. Biofouling 27:893–905. doi:10.1080/08927014.2011.609616

Dong YH, Xu JL, Li XZ, Zhang LH (2000) AiiA, an enzyme that inactivates the acylhomoserine lactone quorum-sensing signal and attenuates the virulence of *Erwinia carotovora*. Proc Natl Acad Sci U S A 97:3526–3531. doi:10.1073/pnas.97.7.3526

Dong YH, Wang LH, Xu JL, Zhang HB, Zhang XF, Zhang LH (2001) Quenching quorum-sensing dependent bacterial infection by an N-acyl homoserine lactonase. Nature 411:813–817

Donlan RM (2002) Biofilms: microbial life on surfaces. Emerg Infect Dis 8:881–890

Dulla GFJ, Lindow SE (2009) Acyl-homoserine lactone-mediated cross talk among epiphytic bacteria modulates behaviour of *Pseudomonas* on leaves. ISME J 3:825–834. doi:10.1038/ismej.2009.30

Ettinger-Epstein P, Tapiolas DM, Motti CA, Wright AD, Battershill CN, de Nys R (2008) Production of manoalide and its analogues by the sponge *Luffariella variabilis* is hardwired. Marine Biotechnol 10:64–74. doi:10.1007/s10126-007-9037-x

Girennavar B, Cepeda ML, Soni KA, Vikram A, Jesudhasan P, Jayaprakasha GK, Pillai SD, Patil BS (2008) Grapefruit juice and its furocoumarins inhibits autoinducer signaling and biofilm formation in bacteria. Int J Food Microbiol 125:204–208. doi:10.1016/j.ijfoodmicro.2008.03.028

Grice EA, Segre JA (2011) The skin microbe. Nat Rev Microbiol 9:244–253. doi:10.1038/nrmicro2537

Gui Z, Wang H, Ding T, Zhu W, Zhuang X, Chu W (2014) Azithromycin reduces the production of α-hemolysin and biofilm formation in *Staphylococcus aureus*. Indian J Microbiol 54:114–117. doi:10.1007/s12088-013-0438-4

Hentzer M, Riedel K, Rasmussen TB, Heydorn A, Andersen JB, Parsek MR, Rice SA, Eberl L, Molin S, Høiby N, Kjelleberg S, Givskov M (2002) Inhibition of quorum sensing in *Pseudomonas aeruginosa* biofilm bacteria by a halogenated furanone compound. Microbiology 148:87–102

Hentzer M, Wu H, Andersen JB, Riedel K, Rasmussen TB, Bagge N, Kumar N, Schembri MA, Song Z, Kristoffersen P, Manefield M, Costerton JW, Molin S, Eberl L, Steinberg P, Kjelleberg S, Høiby N, Givskov M (2003) Attenuation of *Pseudomonas aeruginosa* virulence by quorum sensing inhibitors. EMBO J 22:3803–3815. doi:10.1093/emboj/cdg366

Hong KW, Koh CL, Sam CK, Yin WF, Chan KG (2012a) Quorum quenching revisited – from signal decays to signalling confusion. Sensors 12:4661–4696. doi:10.3390/s120404661

Hong SH, Hegde M, Kim J, Wang J, Jayaram A, Wood TK (2012b) Synthetic quorum-sensing circuit to control consortial biofilm formation and dispersal in a microfluidic device. Nat Commun 3:613. doi:10.1038/ncomms1616

Hooshangi S, Bentley WE (2008) From unicellular properties to multicellular behavior: bacteria quorum sensing circuitry and applications. Curr Opin Biotechnol 19:550–555. doi:10.1016/j.copbio.2008.10.007

Houdt R, Aertsen A, Moons P, Vanoirbeek K, Michiels CW (2006) N-acyl-l-homoserine lactone signal interception by *Escherichia coli*. FEMS Microbiol Lett 256:83–89. doi:10.1111/j.1574-6968.2006.00103.x

Huang CJ, Lin H, Yang X (2012) Industrial production of recombinant therapeutics in *Escherichia coli* and its recent advancements. J Ind Microbiol Biotechnol 39:383–399. doi:10.1007/s10295-011-1082-9

Jakobsen TH, van Gennip M, Phipps RK, Shanmugham MS, Christensen LD, Alhede M, Skindersoe ME, Rasmussen TB, Friedrich K, Uthe F, Jensen PØ, Moser C, Nielsen KF, Eberl L, Larsen TO, Tanner D, Høiby N, Bjarnsholt T, Givskov M (2012) Ajoene, a sulfur-rich molecule from garlic, inhibits genes controlled by quorum sensing. Antimicrob Agents Chemother 56:2314–2325. doi:10.1128/AAC.05919-11

Jamuna Bai A, Rai VR (2011) Bacterial quorum sensing and food industry. Compr Rev Food Sci Food Saf 10:183–194. doi:10.1111/j.1541-4337.2011.00150.x

Jayaraman A, Wood TK (2008) Bacterial quorum sensing: signals, circuits, and implications for biofilms and disease. Annu Rev Biomed Eng 10:145–167. doi:10.1146/annurev.bioeng.10.061807.160536

Kalia VC (2013) Quorum sensing inhibitors: an overview. Biotechnol Adv 31:224–245. doi:10.1016/j.biotechadv.2012.10.004

Kalia VC (2014) Microbes, antimicrobials and resistance: the battle goes on. Indian J Microbiol 54:1–2. doi:10.1007/s12088-013-0443-7

Kalia VC, Kumar P, Pandian SK, Sharma P (2014a) Biofouling control by quorum quenching. In: Kim SK (ed) Handbook of marine biotechnology. Springer

Kalia VC, Wood TK, Kumar P (2014b) Evolution of resistance to quorum-sensing inhibitors. Microb Ecol 68(1):13–23. doi:10.1007/s00248-013-0316-y

Kim J-H, Choi D-C, Yeon K-M, Kim S-R, Lee C-H (2011) Enzyme-immobilized nanofiltration membrane to mitigate biofouling based on quorum quenching. Environ Sci Technol 45:1601–1607. doi:10.1021/es103483j

Kim SR, Oh HS, Jo SJ, Yeon KM, Lee CH, Lim DJ, Lee CH, Lee JK (2013) Biofouling control with bead-entrapped quorum quenching bacteria in membrane bioreactors: physical and biological effects. Environ Sci Technol 47:836–842. doi:10.1021/es303995s

Kumar P, Patel SKS, Lee JK, Kalia VC (2013) Extending the limits of *Bacillus* for novel biotechnological applications. Biotechnol Adv 31:1543–1561. doi:10.1016/j.biotechadv.2013.08.007

Lane AL, Nyadong L, Galhena AS, Shearer TL, Stout EP, Parry RM, Kwasnika M, Wang MD, Hay ME, Fernandez FM, Kubanek J (2009) Desorption electrospray ionization mass spectrometry reveals surface-mediated antifungal chemical defense of a tropical seaweed. Proc Natl Acad Sci 106:7314–7319. doi:10.1073/pnas.0812020106

LaSarre B, Federle MJ (2013) Exploiting quorum sensing to confuse bacterial pathogens. Microbiol Mol Biol Rev 77:73–111. doi:10.1128/MMBR.00046-12

Lee J, Jayaraman A, Wood TK (2007) Indole is an inter-species biofilm signal mediated by SdiA. BMC Microbiol 7:42. doi:10.1186/1471-2180-7-42

Lee J, Attila C, Cirillo SL, Cirillo JD, Wood TK (2012) Indole and 7-hydroxyindole diminish *Pseudomonas aeruginosa* virulence. Microb Biotechnol 2:75–90. doi:10.1111/j.1751-7915.2008.00061.x

Leroy C, Delbarre C, Ghillebaert F, Compere C, Combes D (2008) Effects of commercial enzymes on the adhesion of a marine biofilm-forming bacterium. Biofouling 24:11–22. doi:10.1080/08927010701784912

Li L, Hooi D, Chhabra SR, Pritchard D, Shaw PE (2004) Bacterial N-acylhomoserine lactone-induced apoptosis in breast carcinoma cells correlated with down-modulation of STAT3. Oncogene 23:4894–4902. doi:10.1038/sj.onc.1207612

Lu TK, Collins JJ (2007) Dispersing biofilms with engineered enzymatic bacteriophage. Proc Natl Acad Sci 104:11197–11202. doi:10.1073/pnas.0704624104

Mahanty A, Mishra S, Bosu R, Maurya UK, Netam SP, Sarkar B (2013) Phytoextracts-synthesized silver nanoparticles inhibit bacterial fish pathogen

Aeromonas hydrophila. Indian J Microbiol 53:438–446. doi:10.1007/s12088-013-0409-9

Malaeb L, Le-Clech P, Vrouwenvelder JS, Ayoub GM, Saikaly PE (2013) Do biological-based strategies hold promise to biofouling control in MBRs? Water Res 47:5447–5463. doi:10.1016/j.watres.2013.06.033

Marketon MM, Glenn SA, Eberhard A, González JE (2003) Quorum sensing controls exopolysaccharide production in *Sinorhizobium meliloti*. J Bacteriol 185:325–331. doi:10.1128/JB.185.1.325-331.2003

Martinez LR, Fries BC (2010) Fungal biofilms: relevance in the setting of human disease. Curr Fungal Infect Rep 4:266–275. doi:10.1007/s12281-010-0035-5

Mayville P, Ji G, Beavis R, Yang H, Goger M, Novick RP, Muir TW (1999) Structure–activity analysis of synthetic autoinducing thiolactone peptides from *Staphylococcus aureus* responsible for virulence. Proc Natl Acad Sci USA 96:1218–1223. doi:10.1073/pnas.96.4.1218

Morohoshi T, Ebata A, Nakazawa S, Kato N, Ikeda T (2005) N-Acylhomoserine lactone-producing or -degrading bacteria isolated from the intestinal microbial flora of Ayu fish (*Plecoglossus altivelis*). Microbes Environ 20:264–268

Morohoshi T, Nakazawa S, Ebata A, Kato N, Ikeda T (2008) Identification and characterization of N-acylhomoserine lactone-acylase from the fish intestinal *Shewanella* sp. strain MIB015. Biosci Biotechnol Biochem 72:1887–1893. doi:10.1271/bbb.80139

Natrah FMI, Defoirdt T, Sorgeloos P, Bossier P (2011) Disruption of bacterial cell-to-cell communication by marine organisms and its relevance to aquaculture. Marine Biotechnol 13:109–126. doi:10.1007/s10126-010-9346-3

Nhan DT, Cam DTV, Wille M, Defoirdt T, Bossier P, Sorgeloos P (2010) Quroum quenching bacteria protect *Macrobrachium rosenbergii* larvae from *Vibrio harveyi* infection. J Appl Microbiol 109:1007–1016. doi:10.1111/j.1365-2672.2010.04728.x

Nylund GM, Cervin G, Hermansson M, Pavia H (2005) Chemical inhibition of bacterial colonization by the red alga *Bonnemaisonia hamifera*. Mar Ecol Prog Ser 302:27–36

Nylund GM, Cervin G, Persson F, Hermansson M, Steinberg PD, Pavia H (2008) Seaweed defence against bacteria: a poly-brominated 2-heptanone from the red alga *Bonnemaisonia hamifera* inhibits bacterial colonisation. Mar Ecol Prog Ser 369:39–50. doi:10.3354/meps07577

Nylund GM, Persson F, Lindegarth M, Cervin G, Hermansson M, Pavia H (2010) The red alga *Bonnemaisonia asparagoides* regulates epiphytic bacterial abundance and community composition by chemical defence. FEMS Microbiol Ecol 71:84–93. doi:10.1111/j.1574-6941.2009.00791.x

Oliver CM, Schaefer AL, Greenberg EP, Sufrin JR (2009) Microwave synthesis and evaluation of phenacylhomoserine lactones as anticancer compounds that minimally activate quorum sensing pathways in *Pseudomonas aeruginosa*. J Med Chem 52:1569–1575. doi:10.1021/jm8015377

Pathak R, Kumar R, Gautam HK (2013) Cross-species induction and enhancement of antimicrobial properties in response to gamma irradiation in *Exiguobacterium* sp. HKG 126. Indian J Microbiol 53:130–136. doi:10.1007/s12088-013-0369-0

Paul NA, Cole L, de Nys R, Steinberg PD (2006a) Ultra structure of the gland cells of the red alga *Asparagopsis armata* (Bonnemaisoniaceae). J Phycol 42:637–645. doi:10.1111/j.1529-8817.2006.00226.x

Paul NA, de Nys R, Steinberg PD (2006b) Chemical defence against bacteria in red alga *Asparagopsis armata* linking structure with function. Mar Ecol Prog Ser 306:87–101. doi:10.3354/meps306087

Paul D, Kim YS, Ponnusamy K, Kweon JH (2009) Application of quorum quenching to inhibit biofilm fermentation. Environ Eng Sci 26:1319–1324. doi:10.1089/ees.2008.0392

Peters L, König GM, Wright AD, Pukall R, Stackebrandt E, Eberl L, Riedel K (2003) Secondary metabolites of *Flustra foliacea* and their influence on bacteria. Appl Environ Microbiol 69:3469–3475. doi:10.1128/AEM.69.6.3469-3475.2003

Qian PY, Lau SCK, Dahms HU, Dobretsov S, Harder T (2007) Marine biofilms as mediators of colonization by marine macroorganisms: implications for antifouling and aquaculture. Marine Biotechnol 9:399–410. doi:10.1007/s10126-007-9001-9

Quinones B, Pujol CJ, Lindow SE (2004) Regulation of AHL production and its contribution to epiphytic fitness in *Pseudomonas syringae*. Mol Plant Microbe Interact 17:521–531. doi:10.1094/MPMI.2004.17.5.521

Rasch M, Buch C, Austin B, Slierendrecht WJ, Ekmann KS, Larsen JL, Johansen C, Riedel K, Eberl L, Givskov M, Gram L (2004) An inhibitor of bacterial quorum sensing reduces mortalities caused by Vibriosis in Rainbow trout (*Oncorhynchus mykiss*, Walbaum). Syst Appl Microbiol 27:350–359. doi:10.1078/0723-2020-00268

Rasmussen TB, Bjarnsholt T, Skindersoe ME, Hentzer M, Kristoffersen P, Köte M, Nielsen J, Eberl L, Givskov M (2005) Screening for quorum-sensing inhibitors (QSI) by use of a novel genetic system, the QSI selector. J Bacteriol 187:1799–1814. doi:10.1128/JB.187.5.1799-1814.2005

Romero M, Avendaño-Herrera R, Magariños B, Cámara M, Otero A (2010) Acylhomoserine lactone production and degradation by the fish pathogen *Tenacibaculum maritimum*, a member of the Cytophaga-Flavobacterium-Bacteroides (CFB) group. FEMS Microbiol Lett 304:131–139. doi:10.1111/j.1574-6968.2009.01889.x

Shang Z, Wang H, Zhou S, Chu W (2014) Characterization of N-Acyl-homoserine lactones (AHLs)-deficient clinical isolates of *Pseudomonas aeruginosa*. Indian J Microbiol 54:158–162. doi:10.1007/s12088-014-0449-9

Sheng L, Pu M, Hegde M, Zhang Y, Jayaraman A, Wood TK (2012) Interkingdom adenosine signal reduces *Pseudomonas aeruginosa* pathogenicity. Microb Biotechnol 2:560–572. doi:10.1111/j.1751-7915.2012.00338.x

Skjerma J, Vadstein O (1993) Characterization of the bacterial flora of mass cultivated *Brachionus plicatilis*. Hydrobiologia 255:185–191

Smith MD, Roheim CA, Crowder LB, Halpern BS, Turnipseed M, Anderson JL, Asche F, Bourillón L, Guttormsen AG, Kahn A, Liguori LA, McNevin A, O'Connor M, Squires D, Tyedemers P, Brownstein C, Carden K, Klinger DH, Sagarin R, Selkoe KA (2010) Sustainability and global seafood. Science 327:784–786. doi:10.1126/science.1185345

Smyth AR, Cifelli PM, Ortori CA, Righetti K, Lewis S, Erskine P, Holland ED, Givskov M, Williams P, Cámara M, Barrett DA, Knox A (2010) Garlic as an inhibitor of *Pseudomonas aeruginosa* quorum sensing in cystic fibrosis – a pilot randomized controlled trial. Pediatr Pulmonol 45:356–362. doi:10.1002/ppul.21193

Stamper DM, Walch M, Jacobs RN (2003) Bacterial population changes in a membrane bioreactor for graywater treatment monitored by denaturing gradient gel electrophoretic analysis of 16S rRNA gene fragments. Appl Environ Microbiol 69:852–860. doi:10.1128/AEM.69.2.852-860.2003

Stoltz DA, Ozer EA, Taft PJ, Barry M, Liu L, Kiss PJ, Moninger TO, Parsek MR, Zabner J (2008) *Drosophila* are protected from *Pseudomonas aeruginosa* lethality by transgenic expression of paraoxonase-1. J Clin Invest 118:3123–3131. doi:10.1172/JCI35147

Swift S, Karlyshev AV, Fish L, Durant EL, Winson MK, Chhabra SR, Williams P, Macintyre S, Stewart GS (1997) Quorum sensing in *Aeromonas hydrophila* and *Aeromonas salmonicida*: identification of the LuxRI homologs AhyRI and AsaRI and their cognate N-acylhomoserine lactone signal molecules. J Bacteriol 179:5271–5281

Teasdale ME, Liu J, Wallace J, Akhlaghi F, Rowley DC (2009) Secondary metabolites produced by marine bacterium *Halobacillus salinus* that inhibit quorum sensing controlled phenotypes in gram-negative bacteria. Appl Environ Microbiol 75:567–572. doi:10.1128/AEM.00632-08

Teplitski M, Robinson JB, Bauer WD (2000) Plants secrete substances that mimic bacterial N-acyl homoserine lactone signal activities and affect population density-dependent behaviors in associated bacteria. Mol Plant Microbe Interact 13:637–648. doi:10.1094/MPMI.2000.13.6.637

Thammavongsa V, Kern JW, Missiakas DM, Schneewind O (2009) *Staphylococcus aureus* synthesizes adenosine to escape host immune responses. J Exp Med 206:2417–2427. doi:10.1084/jem.20090097

Tinh NT, Asanka Gunasekara RA, Boon N, Dierckens K, Sorgeloos P, Bossier P (2007a) N-acyl homoserine lactone-degrading microbial enrichment cultures isolated from *Penaeus vannamei* shrimp gut and their probiotic properties in *Brachionus plicatilis* cultures. FEMS Microbiol Ecol 62:45–53. doi:10.1111/j.1574-6941.2007.00378.x

Tinh NTN, Lin ND, Wood TK, Dierckens K, Sorgeloos P, Bossier P (2007b) Interference with the quorum sensing systems in a *Vibrio harveyi* strain alters the growth rate of gnotobiotically cultured rotifer *Brachionus plicatilis*. J Appl Microbiol 103:194–203. doi:10.1111/j.1365-2672.2006.03217.x

Tinh NTN, Yen VHN, Dierckens K, Sorgeloos P, Bossier P (2008) An acyl homoserine lactone-degrading microbial community improves the survival of first-feeding turbot larvae (*Scophthalmus maximus* L.). Aquaculture 285:56–62. doi:10.1016/j.aquaculture.2008.08.018

Tinh NTN, Dung NV, Trung CT (2013) In vitro characterization of a recombinant AHL-Lactonase from *Bacillus cereus* isolated from a striped catfish (*Pangasianodon hypophthalmus*) pond. Indian J Microbiol 53:485–487. doi:10.1007/s12088-013-0415-y

Tonello F, Zornetta I (2012) *Bacillus anthracis* factors for phagosomal escape. Toxins 4:536–553. doi:10.3390/toxins4070536

Ulrich RL (2004) Quorum quenching: enzymatic disruption of N-acylhomoserine lactone-mediated bacterial communication in *Burkholderia thailandensis*. Appl Environ Microbiol 70:6173–6180. doi:10.1128/AEM.70.10.6173-6180.2004

Vejborg RM, Klemm P (2008) Blocking of bacterial biofilm formation by a fish protein coating. Appl Environ Microbiol 74:3551–3558. doi:10.1128/AEM.00279-08

Wang LH, Weng LX, Dong YH, Zhang LH (2004) Specificity and enzyme kinetics of the quorum quenching N-acyl homoserine lactone lactonase (AHL lactonase). J Biol Chem 14:13645–13651. doi:10.1074/jbc.M311194200

Wang WZ, Morohoshi T, Ikenoya M, Somaya N, Ikeda T (2010) AiiM, a novel class of N-acylhomoserine lactonase from the leaf-associated bacterium *Microbacterium testaceum*. Appl Environ Microbiol 76:2524–2530. doi:10.1128/AEM.02738-09

Wang H, Tu F, Gui Z, Lu X, Chu W (2013) Antibiotic resistance profiles and quorum sensing-dependent virulence factors in clinical isolates of *Pseudomonas aeruginosa*. Indian J Microbiol 53:163–167. doi:10.1007/s12088-013-0370-7

Wood TK (2009) Insights on *Escherichia coli* biofilm formation and inhibition from whole-transcriptome profiling. Environ Microbiol 11:1–15. doi:10.1111/j.1462-2920.2008.01768.x

Yang F, Wang LH, Wang J, Dong YH, Hu JY, Zhang LH (2005) Quorum quenching enzyme activity is widely conserved in the sera of mammalian species. FEBS Lett 579:3713–3717. doi:10.1016/j.febslet.2005.05.060

Yao Y, Martinez-Yamout MA, Dickerson TJ, Brogan AP, Wright PE, Dyson HJ (2006) Structure of the *Escherichia coli* quorum sensing protein SdiA: activation of the folding switch by acyl homoserine lactones. J Mol Biol 355:262–273. doi:10.1016/j.jmb.2005.10.041

Ye JH, Rajendran VM (2009) Adenosine: an immune modulator of inflammatory bowel diseases. World J Gastroenterol 15:4491–4498. doi:10.3748/wjg.15.4491

Yeon K-M, Cheong W-S, Oh H-S, Lee W-N, Hwang B-K, Lee C-H, Beyenal H, Lewandowski Z (2009a) Quorum sensing: a new biofouling control paradigm in a membrane bioreactor for advanced waste water treatment. Environ Sci Technol 43:380–385. doi:10.1021/es8019275

Yeon K-M, Lee C-H, Kim J (2009b) Magnetic enzyme carrier for effective biofouling control in the membrane bioreactor based on enzymatic quorum quenching. Environ Sci Technol 43:7403–7409. doi:10.1021/es901323k

Zhang Y, Zhang Y, Yang Y, Wang L, Weng L (2013) Identification of a *Pseudomonas* sp. that inhibits RHL system of quorum sensing. Indian J Microbiol 53:28–35. doi:10.1007/s12088-012-0340-5

Zhang J, Chen YP, Miller KP, Ganewatta MS, Bam M, Yan L, Nagarkatti M, Decho AW, Tang C (2014) Antimicrobial metallopolymers and their bioconjugates with conventional antibiotics against multidrug-resistant bacteria. J Am Chem Soc 136:4873–4876. doi:10.1021/ja5011338

Biotechnological Applications of Quorum-Sensing Inhibitors in Aquacultures

Faseela Hamza, Ameeta Ravi Kumar, and Smita Zinjarde

Introduction

For more than 4.3 billion people, fish are a source of animal protein. The fishery and aquaculture sectors are significant as they increase nutritional security, generate economic growth, and alleviate poverty. In the year 2011, the exports associated with fish and fishery products exceeded US$125 billion (FAO 2012). In the developing world, capture fisheries are the main source of nutrition security (Hall et al. 2013). Diminishing resources, energy crisis, and resultant high cost of fishing have led to an increased realization of the potential of aquaculture as a viable and cost-effective alternative to capture fisheries (Ayyappan and Jena 2001; Pillai and Katiha 2004: Campbell and Pauly 2013). Aquaculture is thus being considered as a vital complement to global capture fisheries. The United Nations Food and Agriculture Organization (FAO) defines aquaculture as the farming of individually or corporately owned aquatic organisms such as fish, mollusks, crustaceans, and aquatic plants with interventions in the rearing process so as to enhance production. This is one of the fastest growing sectors in the world (Subasinghe et al. 1998). According to FAO (global aquaculture production database), aquaculture production has increased at an average rate of around 8 % in the past three decades, and this growth is higher than any other major animal food production sector. By the year 2021, world fisheries production is projected to be about 172 million tonnes.

There are some problems associated with the increased intensification and commercialization of aquatic procedures. For example, disease outbreaks are a major cause of serious economic losses. According to a World Bank Report, global losses due to shrimp diseases alone amount to US$ 3,000 million (Lundin 1996). The aquaculture industry has been seriously affected by a variety of viral, bacterial, fungal, parasitic, and other undiagnosed and emerging pathogens. Diseases are thus primary constraints during the cultivation of aquatic species (Subasinghe et al. 2001). Some of the reasons for the increased incidence of diseases are (1) the release of aquatic animals raised in hatcheries into the marine and coastal areas, (2) the use of pathogen containing stocks, (3) unanticipated negative interactions between cultured and wild populations, (4) lack of effective biosecurity measures, (5) unawareness regarding emerging diseases, and (6) climate change (Olivier 2002; Bartley et al. 2006; Bell et al. 2006).

Role of Bacterial Pathogens and Their Biofilms in Aquaculture

Bacterial pathogens are a major cause of diseases in fish and aquaculture (Gonzalez et al. 2004; Bondad-Reantaso et al. 2005). Such diseases

F. Hamza • A.R. Kumar • S. Zinjarde (✉)
Institute of Bioinformatics and Biotechnology, University of Pune, Pune 411007, Maharashtra, India
e-mail: smita@unipune.ac.in

V.C. Kalia (ed.), *Quorum Sensing vs Quorum Quenching: A Battle with No End in Sight*,
DOI 10.1007/978-81-322-1982-8_30, © Springer India 2015

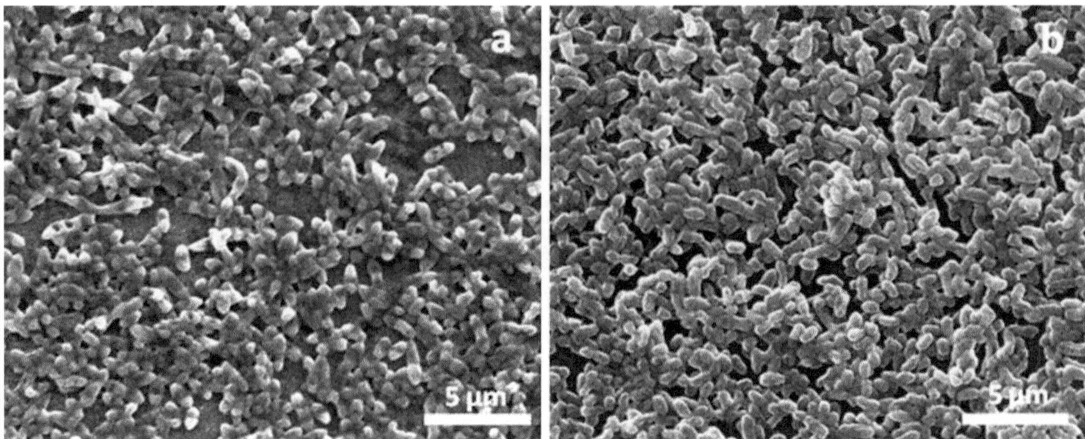

Fig. 1 Representative scanning electron micrographs of the major aquatic pathogens *Vibrio harveyi* (**a**) and *Pseudomonas aeruginosa* (**b**) forming biofilms on glass surfaces

cause growth retardation and mass mortalities. Bacteria belonging to the genus *Vibrio* are important pathogens. They are widely distributed in freshwater, estuarine, and marine environments (Otta et al. 1999, 2001; Austin and Zhang 2006). A few of these are human pathogens (e.g., *V. cholerae, V. parahaemolyticus,* and *V. vulnificus*) and some are pathogenic to aquatic animals (*V. harveyi, V. spendidus, V. penaecida, V. anguillarum, V. parahaemolyticus,* and *V. vulnificus*). Although most *Vibrio* species are regarded as opportunistic pathogens, some of them (e.g., *V. harveyi*) are primary pathogens and can result in mass mortality. *V. harveyi* are luminous bacteria that are found in coastal and marine waters. They are generally associated with the surface and gut of organisms found in such environments (Moriarty 1998; Zhang and Austin 2000; Muroga 2001). *Pseudomonas* and *Aeromonas* are other pathogenic bacteria. They cause black spot diseases and bacterial necrosis in juveniles of the freshwater prawns *Macrobrachium rosenbergii* (Lombardi and Labao 1991; Jayasree et al. 1999). These two organisms along with *Bacillus* species have also been isolated from hatcheries wherein they exist as biofilms (Dass et al. 2007). Representative scanning electron micrographs of the major aquatic pathogens *V. harveyi* and *P. aeruginosa* forming biofilms on glass surfaces are depicted in Fig. 1.

The conventional approaches of using antibiotics and sanitizers have a limited success in preventing aquatic diseases (King et al. 2008). Their indiscriminate and frequent use has led to the development of resistant strains. For example, bacterial strains isolated from aquaculture environments have displayed the presence of transferable antibiotic resistant genes (Bruun et al. 2000; Jayaprakash et al. 2006; Agersø et al. 2007; Akinbowale et al. 2007). The importance of biofilms for survival, virulence, and stress resistance is well-documented (Karunasagar et al. 1996; Wai et al. 1998; Watnick and Kolter 1999; Watnick et al. 2001; Wang et al. 2003; Zhu and MeKalanos 2003; Faruque et al. 2006) and their significance in fish processing, aquaculture industry is being increasingly recognized (Joseph et al. 2001).

Bacteria are known to settle on the surface of submerged material, pipes, and tanks of hatcheries; form biofilms; and act as reservoirs (Karunasagar et al. 1996). In general, they are difficult to eliminate as they resist antibiotics and sanitizer treatments (Costerton et al. 1987, 1999; Thompson et al. 2002). On account of the aforementioned factors, there is need to explore alternative means of controlling biofilms. In this regard, novel quorum-sensing inhibitors may be a lucrative option. Such compounds negatively modulate bacterial cell-to-cell communication via by small signal molecules and decrease virulence and biofilm formation in aquaculture-related bacteria.

Quorum-Sensing Inhibition: An Anti-infective Strategy in Aquaculture

Most of the bacteria coordinate gene expression using quorum sensing (QS), a process of cell-to-cell communication. It is a mechanism of gene regulation in which bacteria coordinate the expression of certain genes in response to the presence/absence of small signal molecules. Quorum sensing in general involves the production of extracellular signaling molecules (autoinducers), their detection by bacterial populations, and the elicitation of appropriate responses. This mechanism was first discovered in *V. fischeri* (Nealson et al. 1970). QS systems are known to control diverse functions such as bioluminescence, conjugation, biofilm formation, antibiotic production, swarming, nodulation, sporulation, and expression of virulence factors such as toxins, siderophores, lytic enzymes, and adhesion molecules (Dunny and Leonard 1997; de Kievit and Iglewski 2000; Jayaraman and Wood 2008). Virulence, pathogenicity, and biofilm formation can be controlled by employing QS inhibitors. This strategy is particularly important in recent years when several pathogenic bacteria are developing resistance toward antibiotics. Aquatic pathogens such as *V. harveyi* and *P. aeruginosa* display quorum-sensing systems as summarized in the following sections.

Quorum Sensing in *V. harveyi*

Three QS systems have been demonstrated in this marine bacterium. System 1 consists of the LuxM-dependent autoinducer HAI-1 and the HAI-1sensor LuxN. System 2 comprises the LuxS-dependent autoinducer AI-2 and the AI-2 detector LuxPQ. System 3 includes the CqSA-dependent autoinducer CAI-1 and the sensor CqsS. These three communication systems involving HAI-1, AI-2, and CAI-1 work in a coordinated manner and regulate a variety of genes through the Lux network (Henke and Bassler 2004).

P. aeruginosa Quorum-Sensing Systems

There are main two acyl homoserine lactone (AHL)-mediated QS systems in *P. aeruginosa*, namely, *las* and *rhl*. The former system is responsible for the synthesis of the autoinducer and the production of LasI synthase via the *las*I gene. The *las*R gene codes for the transcriptional regulator and LasR synthase. LasA protease precursor and elastase LasB are gene products of *las*A and *las*B, respectively. The *rhl* system includes the *rhl*I gene that is responsible for the production of the autoinducer synthesis protein and Rh1I synthase. The *rhl*R gene encodes for a transcriptional regulator and Rh1R synthase. In addition, *rhl*AB, *rhl*C, and *rhl*G code for rhamnosyl transferase 1, rhamnosyl transferase 2, and β-ketoacyl reductase, respectively. The *pqs* system found in this bacterium acts as a link between the *las* and *rhl* systems (Dusane et al. 2010). Interplay between these systems regulates biofilm formation and synthesis of metabolites such as toxins, rhamnolipids, and pigments.

QS Inhibitors Effective Against Aquaculture Pathogens

There are certain molecules referred to as quorum-sensing inhibitors (QSIs) that disrupt cell-to-cell communication. As stated earlier, the major aquatic pathogens are *V. harveyi* and *P. aeruginosa*. There are reports on the effectiveness of a variety of QSIs in controlling biofilm formation, virulence, and pathogenesis in these bacteria. The QSIs effective against these aquatic pathogens are of either of biological or chemical origin as discussed in the following sections.

QSIs of Biological Origin

There are reports on the production of QSIs from a wide variety of biological forms such as bacteria, algae, tunicates, corals, and higher plants. Bacteria are able to interfere with QS-related phenomena in two ways (1) by degrading AHL molecules and (2) by producing QSIs.

A number of bacteria can utilize AHL molecules as carbon and nitrogen sources. Such bacteria have been isolated and used for controlling aquaculture infections. When mixtures of AHL were added in rearing water, a sharp reduction in the turbot larvae survival was observed (Tinh et al. 2008). This effect could be counteracted by introducing a microbial community that degraded the AHL molecules. In some later studies, similar observations were made with the prawn *M. rosenbergii* (Cam et al. 2009a; Nhan et al. 2010). Two enrichment cultures (EC) degrading AHL were obtained. EC5(D) was isolated from the European sea bass *Dicentrarchus labrax* and consisted of *Bacillus circulans*, *Bacillus* sp., and *Vibrio* sp. EC5(L) was obtained from the Asian sea bass *Lates calcarifer* and included members of the family Enterobacteriaceae. Both these ECs used AHL as carbon and nitrogen sources and were equally effective in enhancing prawn larvae survival under experimental conditions when 10^6 colony-forming units ml^{-1} were added on a daily basis. In another report, Cam et al. (2009b) obtained an enrichment culture capable of degrading homoserine lactone and accumulating poly-β-hydroxybutyrate (PHB) from the European sea bass. This EC was used as a biocontrol agent against *Vibrio* infections in *Artemia*. Such studies on the application of natural populations in combating infections could provide an alternative for sustainable aquaculture practices.

Some bacteria produce QSIs that can interfere with and inhibit QS signals in other bacteria. Kanagasabhapathy et al. (2009) screened several epibiotic bacteria associated with the brown alga *Colpomenia sinuosa* for their ability to produce QSIs against AHL-based QS systems. Out of the 96 bacteria screened, 12 % strains belonging to the Bacillaceae, Pseudomonadaceae, Pseudoalteromonadaceae, and Vibrionaceae families produced QSIs. In another study, Clark et al. (2008) isolated tumonoic acid F (Fig. 2a) from the cyanobacterium, *Blennothrix cantharidosmum*. This compound inhibited bioluminescence in *V. harveyi* by acting as a QSI.

There are also reports on the synthesis of QSIs from micro- and macroalgae. For example, Natrah et al. (2011) have investigated the effect of some microalgae strains (commonly used in aquaculture) on AHL-regulated QS in *V. harveyi* and *C. violaceum* (CV026). In particular, extracts of *Chlorella saccharophila* CCAP211/48 inhibited violacein production in CV026 and interfered with bioluminescence in *V. harveyi* without affecting cell densities. *Chlamydomonas reinhardtii* is another microalga that inhibited AHL-mediated luminescence. The substances inhibiting QS were considerably higher in phototrophically cultured algae than in cultures grown on acetate (Teplitki et al. 2004).

Delisea pulchra a red macroalga of marine origin is reported to produce halogenated furanones that act as antagonists for AHL-mediated QS. This halogenated furanone (Fig. 2b) has structural similarity with AHLs and most probably binds LuxR-type proteins. Intramuscular administration of such halogenated furanones in the shrimp *P. monodon* reduced mortality by 50 % (Manefield et al. 2000). In later report, AHL antagonists were discovered from the red macroalga *Ahnfeltiopsis flabelliformes*. Three compounds, floridoside, betonicine, and isethionic acid (Fig. 2c–e), were purified via bioactivity-guided fractionation (Kim et al. 2007). In a later study, these authors (Liu et al. 2008) also evaluated the effectiveness of commercially available isethionic acid and chemically synthesized floridoside and betonicine individually and in combinations. It was found that a mixture of floridoside and isethionic acid inhibited QS in a dose-dependent manner. Skindersoe et al. (2008) screened 284 extracts from the Great Barrier Reef for QSIs. It was observed that 23 % of the marine organisms including corals, algae, and sponges inhibited bacterial AHL-mediated QS and 56 % of these were specifically active against

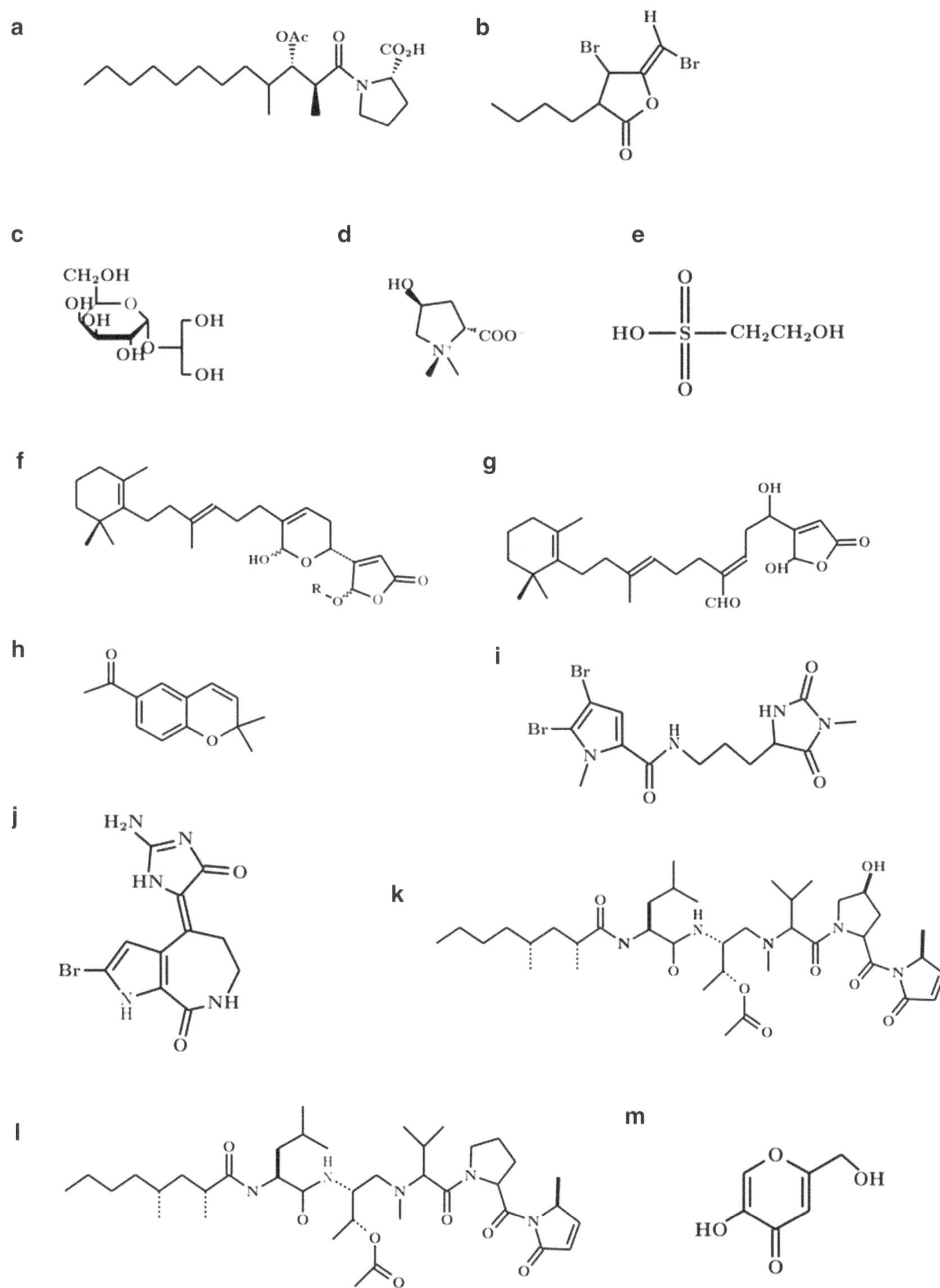

Fig. 2 Summary of the quorum-sensing inhibitors of biological origin effective against selected aquatic pathogens (**a**) tumonoic acid F, (**b**) furanone derived from *Delisea pulchra*, (**c**) floridoside, (**d**) betonicine, (**e**) isethionic acid, (**f**) manoalide (R = H) or manoalide monoacetate R = Ac, (**g**) secomanoalide, (**h**) demethoxy encecalin, (**i**) midpacamide, (**j**) hymenialdisin, (**k**) microcolin A, (**l**) microcolin B, (**m**) kojic acid

a b c

d e

f g

h i j

k l

m n o

p q

r s

P. aeruginosa. During this study, three related C25 sesterterpene metabolites (manoalide, manoalide monoacetate, and secomanolaide) were purified from the sponge *Luffariella variabilis* (Fig. 2f, g). In another investigation, Dobretsov et al. (2011) screened natural products from marine organisms (sponges, algae, fungi, tunicates, and cyanobacteria) and terrestrial plants for QSI activity. Compounds such as demethoxy encecalin, midpacamide, hymenialdisin, microcolins (A, B), and kojic acid (Fig. 2h–m, respectively) displayed QSI activity.

Extracts from higher plants such as pea, rice, soybean, tomato, and barrel clover were found to inhibit AHL-mediated quorum sensing. Several AHL-mimicking substances were present in these extracts (Teplitki et al. 2000). The essential oil from *Syzygium aromaticum* (clove) is also reported to significantly decrease the *las*- and *rhl*-regulated virulence factors (LasB elastase, chitinase, pyocyanin, protease production) and biofilm formation in *P. aeruginosa* (Husain et al. 2013). Thus, it is evident that a vast variety of molecules derived from different biological forms act as QSIs.

Chemically Synthesized QS Inhibitors

There are reports on the synthesis and screening of chemical compounds for QSI activity. In general, furanones, chlorolactones, thiophenones, cinnamaldehyde and its derivatives, nucleoside analogues, pyrogallol, phenothiazine, boronic acid, and S-phenyl-L-cysteine derivatives display such activities (Fig. 3).

Furanones are heterocyclic lactones that share structural similarity with AHL and display QS inhibitory activity. A natural furanone [(5Z)-4-bromo-5-(bromomethylene)-3-butyl-2(5H)-furanone] (Fig. 3a) and a brominated furanone [(5Z)-4-bromo- 5-(bromomethylene) -2 (5H)-furanone] (Fig. 3b) inhibiting AI-2-based quorum sensing in *V. harveyi, V. campbellii,* and *V. parahaemolyticus* has been reported by Defoirdt et al. (2006). During the in vivo challenge tests involving the pathogenic isolates in brine shrimp (*Artemia franciscana*), it was found that the furanones offered some protection from pathogens. Particularly *V. harveyi* and *V. parahaemolyticus* strains were completely inhibited. The brominated furanone inhibited *V. harveyi* and *V. campbellii* when used in a range between 5 and 20 mgl^{-1}. The authors have further elucidated the mode of action of these halogenated furanones in quorum-sensing inhibition (Defoirdt et al. 2007). In the *V. harveyi* model system, these compounds decreased the DNA-binding activity of the quorum-sensing master regulator LuxR. Some other analogues of furanones have also been synthesized and assessed for their QS inhibitory activities. Kim et al. (2008) synthesized a series of analogs of (5-oxo-2,5-dihydrofuran-3-yl) methyl alkanoates and checked them for inhibiting QS signaling and biofilm formation in *P. aeruginosa.* The methyl substituent (Fig. 3c) was found to be most effective. In addition, some substituted [1^1-bromo, 1^1-acetoxy 3-alkyl-5-methylene - 2(5H)-furanones] have also been evaluated for QS-regulated bioluminescence inhibition in *V. harveyi.* Although the introduction of an acetoxy group did not enhance the activity, with bromide group substitution, the action was improved (Steenackers et al. 2010).

Fig. 3 Summary of different chemical compounds displaying QSI activity (**a**) (5Z)-4-bromo-5-(bromomethylene)-3-butyl-2(5H)-furanone, (**b**) (5Z)-4-bromo-5-(bromomethylene)-2 (5H)-furanone, (**c**) derivative of (5-oxo-2,5-dihydrofuran-3-yl)methyl alkanoate, (**d**) chlorolactone, (**e**) chloro-thiolactone, (**f**) meta-chloro-thiolactone (mCTL), (**g**) meta-bromo-thiolactone, (**h**) TF101 [(z)-5-(bromomethylene)thiophen-2(5H)-one], (**i**) TF310 [(z)-4-((5-(bromomethylene)-2-oxo-2,5-dihydrothiophen-3-yl)methoxy)-4-oxobutanoic acid], (**j**) LMC 21 (*p*-methoxy phenyl propionamide derivative of adenosine), (**k**) cinnamaldehyde, (**l**) cinnamaldehyde derivatives R1 = 3,4-Cl (3,4-dichloro-cinnamaldehyde), R1 = 2,3,4,5,6-F (2,3,4,5,6-pentafluoro-cinnamaldehyde) and R1 = 3-CF_3,4Cl (4-chloro-3-trifluoromethyl-cinnamaldehyde), (**m**) pyrogallol (1,2,3-trihydroxybenzene), (**n**) phenothiazine, (**o**) boronic acid, (**p**) S-phenyl-L-cysteine sulfoxide, (**q**) diphenyl disulfide, (**r**) PD12, (**s**) V-06-018

The effectiveness of chlorolactone (CL) as a QSI was first demonstrated by Swem et al. (2009) during preliminary studies related to CviR-dependent QS virulence pathway. In a more recent study, O'Loughlina et al. (2013) analyzed the QSI activities of CL (Fig. 3d) and its halogenated derivatives chloro-thiolactone (Fig. 3e), meta-chloro-thiolactone (Fig. 3f), and meta-bromo-thiolactone (Fig. 3g) against *P. aeruginosa* (PA14) with respect to the production of the virulence factor pyocyanin and biofilm formation. Among the different CL derivatives, meta-bromo-thiolactone was found to be most effective in inhibiting the aforementioned phenomena.

Thiophenones are another class of compounds that are reported to display QS inhibitory action. Two thiophenones (1) TF101 (z)-5-(bromomethylene)thiophen-2(5H)-one (Fig. 3h) and (2) TF310 (z)-4-((5-(bromomethylene)-2-oxo-2,5-dihydrothiophen-3-yl)methoxy)-4-oxobutanoic acid (Fig. 3i) were tested (Defoirdt et al. 2012). These inhibitors could modulate the virulence of *V. harveyi* toward brine shrimp in a negative manner. The quorum-sensing blocking ability was monitored by measuring the bioluminescence of wild-type strain (BB120) and quorum-sensing mutants (JAF553, JAF 483, and BNL 258). When these organisms were grown to high cell densities (to activate quorum-sensing-regulated bioluminescence) and treated with 2.5 μM thiophenones, bioluminescence was blocked although there was no effect on the growth of the microorganisms. Fluorescence anisotropy assays showed that these compounds disrupted quorum sensing by decreasing the ability of quorum-sensing regulator LuxR to bind its target promoter DNA in vitro. In vivo, with the brine shrimp larvae, TF101 was found to be highly toxic. However, TF 310 completely protected them from *V. harveyi* infection when 2.5 μM was used.

Some nucleoside analogues have also been effective QSIs. For example, Brackman et al. (2009) screened a small panel of nucleoside analogues for their ability to inhibit AI-2-based quorum-sensing systems. An adenosine derivative with a *p*-methoxy phenyl propionamide moiety at C-3[1] (LMC-21) was found to be effective (Fig. 3j). The mode of inhibition by this compound was investigated by measuring bioluminescence in a series of *V. harveyi* AI-2 QS mutants. It was observed that the derivative blocked AI-2-based QS without interfering with bacterial growth. The compound interfered with the signal transduction pathway at the level of LuxPQ. This nucleoside analogue reduced *Vibrio* species starvation response and affected biofilm formation in *V. anguillarum, V. vulnificus,* and *V. cholera*. It was also found to reduce pigment and protease production in *V. anguillarum*. This was effective in preventing vibriosis in *A. franciscana* nauplii larvae during in vivo studies.

Cinnamaldehyde (Fig. 3k) and its analogs are also shown be effective in inhibiting QS-regulated virulence in *Vibrio* species (Brackman et al. 2008). Later the authors (Brackman et al. 2011) screened several cinnamaldehyde analogs (Fig. 3l) for their inhibitory effect on QS. In particular, halogenated compounds such as 3,4-dichloro-cinnamaldehyde, 2,3,4,5,6-pentafluoro-cinnamaldehyde, and 4-chloro-3-trifluoromethyl-cinnamaldehyde were found to be more effective than the unsubstituted cinnamaldehyde. 3,4-dichloro-cinnamaldehyde reduced QS-regulated bioluminescence in *V. harveyi* BB170 by 99 % without interfering with the bacterial growth. It was concluded that cinnamaldehyde and its analogs disrupted AI-2 by decreasing the DNA-binding ability of LuxR to its promoter DNA. Incubation of LuxR with this DNA fragment in the absence of QS inhibitors resulted in a significant increase in anisotropy. When LuxR was incubated with this DNA fragment in the presence of cinnamaldehyde and its analogs, binding to DNA was strongly inhibited thereby indicating the role of these compounds in inhibiting AI-2-mediated QS (by decreasing the DNA-binding ability of LuxR). Cinnamaldehyde analogs also affected starvation response, pigment synthesis, protease production, and biofilm formation in *Vibrio* species and exhibited low cytotoxicity.

Pyrogallol (1,2,3-trihydroxybenzene) a polyphenolic compound (Fig. 3m) that has been reported to be an antimicrobial and antiquorum-sensing agent (Ni et al. 2008). Defoirdt et al. (2013) reported the impact of pyrogallol on the virulence of *V. harveyi* toward brine shrimp larvae (*A. franciscana*) and giant river prawn (*M. rosenbergii*). It was demonstrated that the quorum-sensing disrupting ability is a side effect of peroxide producing activity of pyrogallol rather than true QS inhibition. Addition of pyrogallol (at a concentration of 10 mgl^{-1}) protected both the types of larvae from *V. harveyi* and this compound was also relatively less toxic.

Phenothiazine (Fig. 3n) and its analogs were screened for QS inhibition using *V. harveyi* as a model system (Ni et al. 2009a). The authors checked quorum-sensing-controlled bioluminescence in the *V. harveyi* strain BB886 (mutated with respect to the AI-1 pathway lacking the LuxP receptor required for AI-2 response) and the mutant strain MM32 (mutated with regard to the AI-2 pathway lacking the LuxS enzyme needed for producing DPD and the LuxN receptor need to respond to AI-1). It was found that phenothiazine inhibited both AI-1- and AI-2-mediated quorum sensing without affecting bacterial growth. Some closely related analogs did not show any inhibitory activity which meant that tolerance for structural modification on the aryl ring was low. Oxidizing sulfur to sulfone moiety and replacing phenothiazine nitrogen by an oxygen atom also resulted in loss of activity. Phenothiazine analog with an ionizable side-chain group (amino) showed modest activity; however, it affected bacterial growth.

Ni et al. (2009b) also screened several boronic acid (Fig. 3o) compounds for the ability to inhibit QS in the *V. harveyi* model system. The authors tested bioluminescence production in this bacterium and used BB886 strain for AI-1 pathway and MM32 strain for AI-2 pathway-mediated QS. Five compounds showed inhibitory effect in the single digit μM range for MM32 strain, and most of the tested compounds showed some inhibitory effect with IC_{50} values in the range of 0.7–100 μM without effecting cell density. Most of the boronic acid showed moderate inhibition of BB886 strain although the IC_{50} values were generally two- to fourfold higher than against MM32.

Cady et al. (2012) synthesized a library of *S*-substituted cysteine sulfoxides and their corresponding disulfide derivatives and evaluated them as QSIs. Two compounds S-phenyl-L-cysteine sulfoxide (Fig. 3p) and its breakdown product diphenyl disulfide (Fig. 3q) were found to be effective in inhibiting biofilm formation and QS without reducing the planktonic growth. The inhibition was monitored in *E. coli*-based reporter strains expressing *P. aeruginosa lasR rhlR* response protein and an endogenous *P. aeruginosa* reporter from *las I/las R* QS system. While S-phenyl-L-cysteine sulfoxide inhibited biofilm formation in both the systems, diphenyl sulfide showed only limited inhibition in the *P. aeruginosa* reporter system.

Müh et al. (2006) screened a library of approximate 200,000 compounds for inhibition of LasR-dependent genes expression. Among these, two compounds that were structurally related to the native Las-dependent signal molecule *N*-3-oxododecanoyl homoserine lactone (3OC12-HSL) were identified as potential lead molecules. PD12 a tetrazole compound with a 12-carbon alkyl tail (Fig. 3r) and V-06-018 containing a phenyl ring with a 12-carbon alkyl tail (Fig. 3s) showed QS inhibition. Both the compounds were found to affect LasR-dependent genes responsible for the production of elastase and pyocyanin. From the foregoing discussion, it is evident that a wide variety of molecules of biological and chemical origin are effective QSIs.

Conclusions

The aquaculture sector has been witnessing speedy growth in the past few decades. It is emerging as the major animal food production sector and is projected to increase rapidly in the years to come. This expanding sector is also challenged with certain serious problems such as disease outbreaks and the emergence of bacterial strains resisting antibiotics and sanitizers. The significance of quorum sensing in formation of biofilms that are better adapted for survival and in the production of virulence factors is well known. An understanding on the molecular basis of quorum sensing has opened up new lines of investigations and a search for new classes of molecules that can control aquatic pathogens has been initiated. Molecules interfering with quorum-sensing systems are capable of providing alternative strategies for controlling diseases. Most of the studies on the effect of QSIs on aquatic pathogens are restricted to the laboratory, although on a few occasions, their effectiveness has also been proven under field conditions. QSIs have been obtained from bacteria, algae, tunicates, corals, and higher plants. An interesting fact is that some natural bacterial consortia also bring about degradation of quorum-sensing molecules. In addition, some algae commonly encountered in aquaculture setups are also reported to inherently produce QSIs. By seeding bacterial consortia (capable of degrading signaling molecules) or algae (producing such inhibitors), self-regulated systems limiting the growth of aquatic pathogens may be developed. The variety of marine forms implicated in producing such inhibitors would be of significance in marine aquacultures. In addition, the varied chemical entities including furanones, chlorolactones, thiophenones, cinnamaldehyde and its derivatives, nucleoside analogues, pyrogallol, phenothiazine, boronic acid, and S-phenyl-L-cysteine provide alternative modes of controlling the pathogenesis of aquatic bacteria. To conclude, the use of QSIs in controlling aquatic pathogens may in the future provide an effective and sustainable alternative to the use of commonly used antibiotics or sanitizers.

Acknowledgments All authors thank University Grants Commission for funding under UPE Phase II. FH thanks DST, India, for financial support under WOS-B scheme.

References

Agersø Y, Bruun MS, Dalsgaard I, Larsen JL (2007) Tetracycline resistance gene *tet*(E) is frequently occurring and present on large horizontally transferable plasmids in *Aeromonas* spp. from fish farms. Aquaculture 266:47–52

Akinbowale OL, Peng H, Barton MD (2007) Diversity of tetracycline resistance genes in bacteria from aquaculture sources in Australia. J Appl Microbiol 103:2016–2025

Austin B, Zhang XH (2006) *Vibrio harveyi*: a significant pathogen of marine vertebrates and invertebrates. Lett Appl Microbiol 43:119–124

Ayyappan S, Jena JN (2001) Sustainable freshwater aquaculture in India. In: Pandian TJ (ed) Sustainable Indian fisheries. National Academy of Agricultural Sciences, New Delhi, pp 83–133

Bartley DM, Bondad-Reantaso MG, Subasinghe RP (2006) A risk analysis framework for aquatic animal health management in marine stock enhancement programmes. Fish Res 80:28–36

Bell JD, Bartley DM, Lorenzen K, Loneragan NR (2006) Restocking and stock enhancement of coastal fisheries: potential, problems and progress. Fish Res 80:1–8

Bondad-Reantaso MG, Subasinghe RP, Arthur JR, Ogawa K, Chinabut S, Adlard R, Tan Z, Shariff M (2005) Disease and health management in Asian aquaculture. Vet Parasitol 132:249–272

Brackman G, Defoirdt T, Miyamoto C, Bossier P, Van Calenbergh S, Nelis H, Coenye T (2008) Cinnamaldehyde and cinnamaldehyde analogs reduce virulence in *Vibrio* spp. by decreasing the DNA-binding activity of the quorum sensing response regulator LuxR. BMC Microbiol 8:149

Brackman G, Celen S, Baruah K, Bossier P, Van Calenbergh S, Nelis HJ, Coenye T (2009) AI-2 quorum sensing inhibitors affect the starvation response and reduce virulence in several *Vibrio* species, most likely by interfering with LuxPQ. Microbiology 155:4114–4122

Brackman G, Celen S, Hillaert U, Calenbergh SV, Cos P, Maes L, Nelis HJ, Coenye T (2011) Structure–activity relationship of cinnamaldehyde analogs as inhibitors of AI-2 based quorum sensing and their effect on virulence of *Vibrio* spp. PLoS One 6:1–10

Bruun MS, Schmidt AS, Madsen I, Dalsgaard I (2000) Antimicrobial resistance patterns in Danish isolates of *Flavobacterium psychrophilum*. Aquaculture 187:201–212

Cady NC, McKean KA, Behnke J (2012) Inhibition of biofilm formation, quorum sensing and infection in *Pseudomonas aeruginosa* by natural products-inspired organosulfur compounds. PLoS One 7:1–12

Cam DTV, Nhan DT, Ceuppens S, Hao NV, Dierckens K, Wille M, Sorgeloos P, Bossier P (2009a) Effect of *N*-acyl homoserine lactone-degrading enrichment cultures on *Macrobrachium rosenbergii* larviculture. Aquaculture 294:5–13

Cam DTV, Hao NV, Dierckens K, Defoirdt T, Boon N, Sorgeloos P, Bossier P (2009b) Novel approach of using homoserine lactone degrading and poly-β-hydroxybutyrate accumulating bacteria to protect *Artemia* from the pathogenic effects of *Vibrio harveyi*. Aquaculture 291:23–30

Campbell B, Pauly D (2013) Mariculture: a global analysis of production trends since 1950. Mar Policy 39:94–100

Clark BR, Engene N, Teasdale ME, Rowley DC, Matainaho T, Valeriote FA, Gerwick WH (2008) Natural products chemistry and taxonomy of the marine Cyanobacterium *Blennothrix cantharidosmum*. J Nat Prod 71:1530–1537

Costerton JW, Cheng KJ, Geesey GG, Ladd T, Nickel JC, Dasgupta M, Marrie TJ (1987) Bacterial biofilms in nature and disease. Annu Rev Microbiol 41:435–464

Costerton JW, Stewart PS, Greenberg EP (1999) Bacterial biofilms: a common cause of persistent infections. Science 284:1318–1322

Dass BKM, Venugopal MN, Karunasagar I, Karunasagar I (2007) Bacteria associated with biofilms in a *Macrobrachium rosenbergii* hatchery. Asian Fish Sci 20:299–307

De Kievit TR, Iglewski BH (2000) Bacterial quorum sensing in pathogenic relationships. Infect Immun 68:4839–4849

Defoirdt T, Crab R, Wood TK, Sorgeloos P, Verstraete W, Bossier P (2006) Quorum sensing-disrupting brominated furanones protect the gnotobiotic brine shrimp *Artemia franciscana* from Pathogenic *Vibrio harveyi*, *Vibrio campbellii*, and *Vibrio parahaemolyticus* isolates. Appl Environ Microbiol 72:6419–6423

Defoirdt T, Miyamoto CM, Wood TK, Meighen EA, Sorgeloos P, Verstraete W, Bossier P (2007) The natural furanone (5Z)-4-bromo-5-(bromomethylene)-3-butyl-2(5H)-furanone disrupts quorum sensing–regulated gene expression in *Vibrio harveyi* by decreasing the DNA-binding activity of the transcriptional regulator protein luxR. Environ Microbiol 9:2486–2495

Defoirdt T, Benneche T, Brackman G, Coenye T, Sorgeloos P, Scheie AA (2012) A quorum sensing-disrupting brominated thiophenone with a promising therapeutic potential to treat luminescent Vibriosis. PLoS One 7:1–7

Defoirdt T, Pande GSJ, Baruah K, Bossier P (2013) The apparent quorum sensing inhibitory activity of pyrogallol is a side effect of peroxide production. Antimicrob Agents Chemother. doi:10.1128/AAC.00401-13

Dobretsov S, Teplitski M, Bayer M, Gunasekera S, Proksch P, Paul VJ (2011) Inhibition of marine biofouling by bacterial quorum sensing inhibitors. Biofouling 27:893–905

Dunny GM, Leonard BAB (1997) Cell-cell communication in gram-positive bacteria. Annu Rev Microbiol 51:527–564

Dusane DH, Zinjarde SS, Venugopalan VP, McLean RJC, Weber MM, Rahman PKSM (2010) Quorum sensing: implications on rhamnolipid biosurfactant production. Biotechnol Genet Eng Rev 27:159–184

FAO (2012) The state of world fisheries and aquaculture 2012. U.N. Food and Agricultural Organization, Rome

Faruque SM, Biswas K, Udden SM, Ahmad QS, Sack DA, Nair GB, Mekalanos JJ (2006) Transmissibility of cholera: *in vivo*-formed biofilms and their relationship to infectivity and persistence in the environment. Proc Natl Acad Sci USA 103:6350–6355

Gonzalez SF, Krug MJ, Nielsen ME, Santos Y, Call DR (2004) Simultaneous detection of marine fish pathogens by using multiplex PCR and a DNA microarray. J Clin Microbiol 42:1414–1419

Hall SJ, Hilborn R, Andrew NL, Allison EH (2013) Innovations in capture fisheries are an imperative for nutrition security in the developing world. Proc Natl Acad Sci USA 110:8393–8398

Henke JM, Bassler BL (2004) Three parallel quorum–sensing systems regulate gene expression in *Vibrio harveyi*. J Bacteriol 186:6902–6914

Husain FM, Ahmad I, Asif M, Tahseeni Q (2013) Influence of clove oil on certain quorum-sensing-regulated functions and biofilm of *Pseudomonas aeruginosa* and *Aeromonas hydrophila*. J Biosci 38:835–844

Jayaprakash NS, Kumar R, Philip VJ, Bright Singh IS (2006) *Vibrio*s associated with *Macrobrachium rosenbergii* (De Man, 1879) larvae from three hatcheries on the Indian southwest coast. Aquac Res 37:351–358

Jayaraman A, Wood TK (2008) Bacterial quorum sensing: Signals, circuits, and implications for *biofilms and disease*. Annu Rev Biomed Eng 10:145–167

Jayasree LP, Janakiram PR, Madhavi R (1999) Shell disease in freshwater prawn *Macrobrachium rosenbergii* (de Man) aetiology, pathogenicity and antibiotic sensitivity. J Aquac Trop 14:289–298

Joseph B, Otta SK, Karunasagar I, Karunasagar I (2001) Biofilm formation by *Salmonella* spp. on food contact surfaces and their sensitivity to sanitizers. Int J Food Microbiol 64:367–372

Kanagasabhapathy M, Yamazaki G, Ishida A, Sasaki H, Nagata S (2009) Presence of quorum-sensing inhibitor-like compounds from bacteria isolated from the brown alga *Colpomenia sinuosa*. Lett Appl Microbiol 49:573–579

Karunasagar I, Otta SK, Karunasagar I (1996) Biofilm formation by *Vibrio harveyi* on surfaces. Aquaculture 140:241–245

Kim J, Kim Y, Seo Y, Park S (2007) Quorum sensing inhibitors from the red alga, *Ahnfeltiopsis flabelliformis*. Biotechnol Bioprocess Eng 12:308–311

Kim C, Kim J, Park HY, Park HJ, Lee JH, Kim CK, Yoon J (2008) Furanone derivatives as quorum–sensing antagonists of *Pseudomonas aeruginosa*. Appl Microbiol Biotechnol 80:37–47

King RK, Flick GJ, Smith SA, Pierson MD, Boardman GD, Coale CW (2008) Response of bacterial biofilms in recirculating aquaculture systems to various sanitizers. J Appl Aquac 20:79–92

Liu HB, Koh KP, Kim JS, Seo Y, Park S (2008) The effects of betonicine, floridoside, and isethionic acid from the red alga *Ahnfeltiopsis flabelliformis* on quorum-sensing activity. Bioprocess Biosyst Eng 13:458–463

Lombardi JV, Labao V (1991) Diseases and conditioning factors of mortality in larval culture of prawns of the genus *Macrobrachium*. In: Proceedings of the 3rd Brazilian symposium on shrimp culture, Joao Pes–sao, Paraiba, Brazil, pp 401–408

Lundin CG (1996) Global attempts to address shrimp disease. Marine/Environmental paper no. 4 Land, Water and Natural Habitats Division, Environment Department, The World Bank, p 45

Manefield M, Harris L, Rice SA, de Nys R, Kjelleberg S (2000) Inhibition of luminescence and virulence in the black tiger prawn (*Penaeus monodon*) pathogen *Vibrio harveyi* by intercellular signal antagonists. Appl Environ Microbiol 66:2079–2084

Moriarty DJW (1998) Control of luminous *Vibrio* species in penaeid aquaculture ponds. Aquaculture 164:351–358

Müh U, Schuster M, Heim R, Singh A, Olson ER, Greenberg EP (2006) Novel *Pseudomonas aeruginosa* quorum-sensing inhibitors identified in an ultra-high-throughput screen. Antimicrob Agents Chemother 50:3674–3679

Muroga K (2001) Viral and bacterial diseases of marine fish and shellfish in Japanese hatcheries. Aquaculture 202:23–44

Natrah FMI, Kenmegne MM, WiyotoW SP, Bossier P, Defoirdt T (2011) Effect of micro-algae commonly used in aquaculture on acyl homoserine lactone quorum sensing. Aquaculture 317:53–57

Nealson KH, Platt T, Hastings W (1970) Cellular control of the synthesis and activity of the bacterial biolumionescent system. J Bacteriol 104:313–322

Nhan DT, Cam DTV, Wille M, Defoirdt T, Bossier P, Sorgeloos P (2010) Quorum quenching bacteria protect *Macrobrachium rosenbergii* larvae from *Vibrio harveyi* infection. J Appl Microbiol 109:1007–1016

Ni N, Choudhary G, Li M, Wang B (2008) Pyrogallol and its analogs can antagonize bacterial quorum sensing in *Vibrio harveyi*. Bioorg Med Chem Lett 18:1567–1572

Ni V, Choudhary G, Li M, Wang B (2009a) A new phenothiazine structural scaffold as inhibitors of bacterial quorum sensing in *Vibrio harveyi*. Biochem Biophys Res Commun 382:153–156

Ni N, Choudhary G, Peng H, Li M, Chou HT, Lu CD, Gilbert ES, Wang B (2009b) Inhibition of quorum sensing in *Vibrio harveyi* by boronic acids. Chem Biol Drug Des 74:51–56

O'Loughlina CT, Miller LC, Siryaporn A, Drescher K, Semmelhack MF, Bassler BL (2013) A quorum-sensing inhibitor blocks *Pseudomonas aeruginosa* virulence and biofilm formation. Proc Natl Acad Sci USA 110:17981–17986

Olivier G (2002) Disease interactions between wild and cultured fish-perspectives from the American Northeast (Atlantic Provinces). Bull Eur Assoc Fish Pathol 22:103–108

Otta SK, Karunasagar I, Karunasagar I (1999) Bacterial flora associated with shrimp culture ponds growing *Penaeus monodon* in India. J Aquac Trop 14:309–318

Otta SK, Karunasagar I, Karunasagar I (2001) Bacterial flora associated with shrimp culture ponds, *Penaeus monodon* Fabricius hatcheries in India. J Appl Ichtyol 17:59–63

Pillai NGK, Katiha PK (2004) Evolution of fisheries and aquaculture in India. Niseema Printers and Publishers, Kochi

Skindersoe ME, Epstein PE, Rasmussen TB (2008) Quorum-sensing antagonism from marine organisms. Mar Biotechnol 10:56–63

Steenackers HP, Levin J, Janssens JC, De Weerdt A, Balzarini J, Vanderleyden J, De Vos DE, De Keersmaecker SC (2010) Structure-activity relationship of brominated 3-alkyl-5-methylene-2(5H)-furanones and alkylmaleic anhydrides as inhibitors of *Salmonella* biofilm formation and quorum sensing regulated bioluminescence in *Vibrio harveyi*. Bioorg Med Chem 18:5224–5233

Subasinghe R, Bartly DM, Mcgladdery S, Barg U (1998) Sustainable shrimp culture development: biotechnological issues and challenges. In: Flegel TW (ed) Advances in shrimp biotechnology. National Centre for Genetic Engineering and Biotechnology, Bangkok, pp 13–18

Subasinghe RP, Bonad-Reantaso MG, McGladdery ME (2001) Aquaculture development, health and wealth. In: Subasinghe RP, Bueno P, Phillips MJ, Hough C, McGladdery SE, Arthur JR (eds), Aquaculture in the third millennium. Technical proceedings of the conference on aquaculture in the third millennium, Bangkok, Thailand, February 2000. Network of Aquaculture Centers in Asia and the Pacific, Bangkok, pp 167–192

Swem LR, Swem DL, O'Loughlina CT, Gatmaitan R, Zhao B, Ulrich SM, Bassler BL (2009) A quorum–sensing antagonist targets both membrane–bound and cytoplasmic receptors and controls bacterial pathogenicity. Mol Cell 35:143–153

Teplitki M, Robinson JB, Bauer WD (2000) Plants secrete substances that mimic bacterial *N*-acyl homoserine lactone signal activities and affect population density-dependent behaviors in associated bacteria. Mol Plant Microbe Interact 13:637–648

Teplitki M, Chen H, Rajamani S, Gao M, Merighi M, Sayre RT, Robinson JB, Rolfe BG, Bauer WD (2004) *Chlamydomonas reinhardtii* secretes compound that mimic bacterial signals and interfere with quorum sensing regulation in bacteria. Plant Physiol 134:137–146

Thompson FL, Abreu PC, Wasielesky W (2002) Importance of biofilm for water quality and nourishment in intensive shrimp culture. Aquaculture 203:263–278

Tinh NTN, Yen VHN, Dierckens K, Sorgeloos P, Bossier P (2008) An acyl homoserine lactone-degrading microbial community improves the survival of first feeding turbot larvae (*Scophthalmus maximus* L.). Aquaculture 285:56–62

Wai SN, Mizunoe Y, Takade A, Kawabata SI, Yoshida SI (1998) *Vibrio cholerae* O1 strain TSI-4 produces the exopolysaccharide materials that determine colony morphology, stress resistance, and biofilm formation. Appl Environ Microbiol 64:3648–3655

Wang SY, Lauritz J, Jass J, Milton DL (2003) Role for the major outer-membrane protein from *Vibrio anguillarum* in bile resistance and biofilm formation. Microbiology 149:1061–1071

Watnick PI, Kolter R (1999) Steps in the development of a *Vibrio cholerae* El Tor biofilm. Mol Microbiol 34:586–595

Watnick PI, Lauriano CM, Klose KE, Croal L, Kolter R (2001) The absence of a flagellum leads to altered colony morphology, biofilm development and virulence in *Vibrio cholerae* O139. Mol Microbiol 39: 223–235

Zhang XH, Austin B (2000) Pathogenicity of *Vibrio harveyi* to salmonids. J Fish Dis 23: 93–102

Zhu J, Mekalanos JJ (2003) Quorum sensing-dependent biofilms enhance colonization in *Vibrio cholerae*. Dev Cell 5:647–656

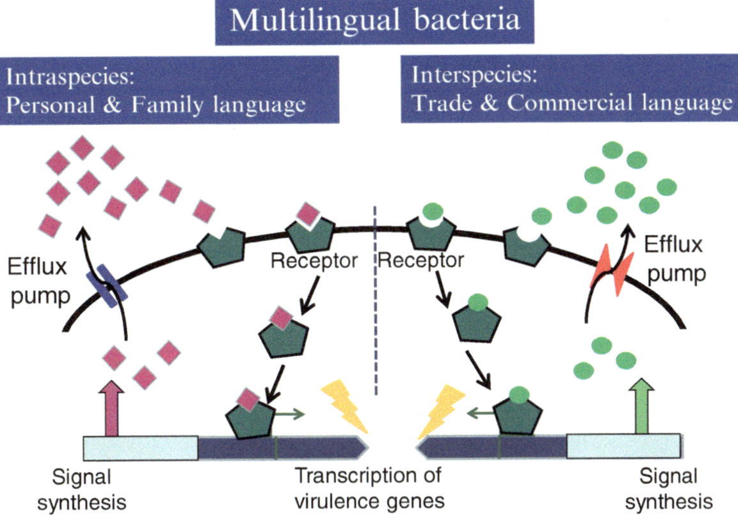

Fig. 1 Diversity of microbial quorum-sensing systems

Signal-Based Modified QS Behaviour

QS-regulated genes of *P. aeruginosa* primarily those responsible for biofilm formation and partial repression of Type III secretion system (TTSS) are observed in patients of cystic fibrosis and keratitis. It has been realized that AHLs may repress TTSS, while another type of signal molecule may modify its expression (Mikkelsen et al. 2009; Njoroge and Sperandio 2009). The QS-mediated functions like swarming motility of *Serratia liquefaciens* could be inhibited by brominated furanone; however, the negative effect of QSI was annulled by just supplying N-butanoyl-HSL (C4HSL) at higher concentration (Givskov et al. 1996). A similar reversal of the negative impact of Furanone C-30 on QS-operated PluxI-gfp (ASV) fusion in a mouse model was achieved by increasing the dose of 3-hydroxy-N-hexanoyl-HSL (Hentzer et al. 2003). Although C-30 affected a wide array of genes, however, it did not influence *lasI/R* and *rhlI/R* gene clusters, which are under the control of 3-oxo-N-dodecanoyl-HSL (3OC12HSL) (Hentzer et al. 2003). Ahl-acylase gene PA2385 of *P. aeruginosa* PAO1 shows the differential behaviour against AHLs. Functional analysis revealed that this gene could completely abolish QS signal 3OC12-HSL but did not affect the levels of C4-HSL. Consequently, a certain QS-mediated virulent gene expression such as that of *lecA* was reduced, whereas the production of pyoverdin remained unaffected. It has been proposed that this acylase may enable *P. aeruginosa* PAO1 to modulate the abilities of its QSS (Sio et al. 2006). Thus, multiple QSS and signals may allow bacteria to evade QSIs and give an impression of their ability to have gained resistance (Hentzer et al. 2003; Schuster et al. 2013). It also amounts to stating that for efficient control of virulent behaviour of *P. aeruginosa*, it may be imperative to block both the *las*- and *rhl*-mediated QSS. Similar mechanisms to evade the effect of QSI seem to be operative in *Escherichia coli,* where transcription of virulence genes is under the control of the membrane-bound sensor kinase (QseC). The kinase can however be activated by any one of the three optional signal molecules: (1) aromatic autoinducer (AI-3) and (2) hormones – epinephrine and norepinephrine (Walters and Sperandio 2006).

Genetic Mutations

Genetic changes in the bacterial genome through mutations and/or acquisition of genes via horizontal gene transfer is a well-accepted phenomena (Kalia et al. 2007; Lal et al. 2008). Mutations in the different components of QSS can be envisaged to allow bacteria to elucidate a modified behaviour. Mutation studies carried out on QSS of *Agrobacterium tumefaciens*, *P. aeruginosa* and *Rhizobium etli* revealed that inducer genes can be activated by any one of the multiple signal molecules (Rosemeyer et al. 1998; Galloway et al. 2011; Jimenez et al. 2012). Reverse mutations in the defective *lasR* gene of *P. aeruginosa* allowed bacteria to carry the infection process even in the absence of *las* system. Mutations in the QSS of *Vibrio cholerae* strains may lead to a constitutive behaviour (Joelsson et al. 2006). *lasI* mutant of *P. aeruginosa* PAO1-JP1 could form biofilms, in spite of its having lost its ability to produce 3OC12HSL (Purevdorj et al. 2002). *P. aeruginosa* mutant PAO-R1, defective in LasR-PAI-1, expressed upregulation in QSS – RhlR-PAI-2 (Sandoz et al. 2007). QscR regulator of *P. aeruginosa* inhibits QS gene expression by acts of forming inactive heterodimers with LasR and/or RhlR, at low cell densities (Ledgham et al. 2003). A mutation in this gene can be expected to allow the bacteria to lead a constitutive expression of virulence genes. Similarly, a mutation in the *rsaL* gene, which encodes for the global regulator of QSS in *P. aeruginosa*, will lead to an uninterrupted supply of QS signal 3OC12HSL (Rampioni et al. 2007). A scenario may thus arise where destruction of signal molecules by QSI may be nullified by such mutants (Mattmann and Blackwell 2010).

Efflux Pumps

Bacteria manifest resistance to antibiotics, toxins, heavy metals and biocides by exploiting their efflux pumps (Köhler et al. 2001; Poole 2005). Antibiotics such as azithromycin, ceftazidime and ciprofloxacin act by altering membrane permeability, which can also be instrumental in affecting the efflux of QS signal – 3OC12HSL (Skindersoe et al. 2009). It may either inhibit QS or, under certain conditions, for example, at subinhibitory concentrations, may even promote biofilm formation (Köhler et al. 2001; Hoffman et al. 2005). The efflux of tryptophan, the precursor molecule leading to the production of *Pseudomonas* quinolone signal (PQS) led to reduced transcription of *rhlI* and finally the functioning of QSS. Bacteria under such environmental stresses can evade the QSIs by retarding QS signal synthesis (Köhler et al. 2001; Aendekerk et al. 2002). This expression was attributed to mutations in the genes, *mexR* and *nalC*, that allowed *P. aeruginosa* to withstand pressure caused by QSI – brominated furanone C-30. Such mutants could infect *Caenorhabditis elegans* in the presence of C-30 (Maeda et al. 2012). In fact, the first evidence of evolution of bacterial resistance to QSIs came from this study, where addition of C-30 impaired the growth of *P. aeruginosa* on adenosine as the sole carbon source. Under these rigorous conditions, *P. aeruginosa* can grow only by operating special QS, responsible for producing only private goods (Maeda et al. 2012). In these experiments, cells resistant to the QSI were observed to evolve within four sequential dilutions after transposon mutagenesis. The genetic changes were recorded to be in the repressors of an efflux pump, which led to higher efflux of QSI. These mutations were reported among clinical isolates from patients with cystic fibrosis (Maeda et al. 2012). A previous study based on QS mimics had anticipated the occurrence of this phenomenon (Mellbye and Schuster 2011). Support for the evolution of resistance to QSI has been also provided by clinical strains resistant to C-30 (García-Contreras et al. 2013).

The Social Cheats

Bacteria are perceived to hold out as individuals with no social life. Nevertheless, the discovery of QS has made us understand that bacteria also possess a social life, where they act in a co-operative fashion. It makes goods for all members of the community to take an advantage. Like all social systems, QS is also exploited by certain individuals, which can be marked as free loaders or cheaters. These individuals can be a drain on the energy spent on producing enzymes and other compounds, which may adversely affect QS (West et al. 2006; Diggle et al. 2007; Sandoz et al. 2007; Buckling and Brockhurst 2008). QS cheats can be either (1) signal negative- the *lasI* mutants, which do not produce QS signals; or (2) signal blind- the *lasR* mutants, which can produce QS signals but do not use them for transcribing genes responsible for pathogenesis (Rumbaugh et al. 2009). Their roles were deciphered using two different carbon sources: (1) bovine serum albumin (BSA) and (2) adenosine. For metabolizing BSA, bacteria need to produce and release proteases, which can be used by all, i.e. public goods. In contrast, adenosine can be degraded only in the periplasmic space and is accessible only to the producer cell, i.e. private goods (Mellbye and Schuster 2011). Coculturing the wild types and the mutants in either of the C sources led to differential growth (Mellbye and Schuster 2011). Wild-type strains of *P. aeruginosa* could not grow well in the presence of *lasR* mutants, since iron-scavenging siderophores were not available for public benefit (Griffin et al. 2004). These evidences support the fact that bacteria can become resistant to QSIs by operating their QS for either public or personal benefit. Bacterial resilience allows them to check the growth of cheaters – *lasR* mutants – by undergoing compensatory mutation. In this scenario, enhanced production QS signals enabled bacteria to produce pathogenicity factors (Sandoz et al. 2007; Köhler et al. 2010).

The Battle Continues

Antibiotics launched an attack on microbial growth. Still, it did not take them long to blunt the onslaught. They built up resistance to whatever human beings could envisage and discover. It seems nature has already exposed the bacteria to all that we can imagine. Bacteria have the necessary arsenal to take on all the battles (Kalia 2014). We should not forget that up to 10 % of the bacterial genome is always geared up for expressing its virulent behaviour. QSIs are proving quite promising as novel drugs. The target should be to develop non-competitive or uncompetitive inhibitors, which may exert relatively less pressure on QSS and should not affect efflux pumps. The need is to be innovative and judicious (Allen et al. 2014).

Acknowledgements The authors wish to thank the Director of CSIR-Institute of Genomics and Integrative Biology (IGIB), CSIR-INDEPTH (BSC0111), and the Government of India for providing the necessary funds and facilities. PK is thankful to CSIR for granting Senior Research Fellowship.

References

Aendekerk S, Ghysels B, Cornelis P, Baysse C (2002) Characterization of a new efflux pump, MexGHI-OpmD, from *Pseudomonas aeruginosa* that confers resistance to vanadium. Microbiology 148: 2371–2381

Allen RC, Popat R, Diggle SP, Brown SP (2014) Targeting virulence: can we make evolution- proof drugs? Nat Rev Microbiol 12:300–308. doi:10.1038/nrmicro3232

Antunes L, Caetano M, Ferreira RBR, Buckner MMC, Finlay BB (2010) Quorum sensing in bacterial virulence. Microbiology 156:2271–2282. doi:10.1099/mic.0.038794-0

Bakkiyaraj D, Sivasankar C, Pandian SK (2013) Anti-pathogenic potential of coral associated bacteria isolated from Gulf of Mannar against *Pseudomonas aeruginosa*. Indian J Microbiol 53:111–113. doi:10.1007/s12088-012-0342-3

Buckling A, Brockhurst MA (2008) Kin selection and the evolution of virulence. Heredity 100:484–488. doi:10.1038/sj.hdy.6801093

Case RJ, Labbate M, Kjelleberg S (2008) AHL-driven quorum-sensing circuits: their frequency and function among the Proteobacteria. ISME J 2:345–349. doi:10.1038/ismej.2008.13

Chu W, Liu Y, Jiang Y, Zhu W, Zhuang X (2013) Production of *N*-acyl homoserine lactones and virulence factors of waterborne *Aeromonas hydrophila*. Indian J Microbiol 53:264–268. doi:10.1007/s12088-013-0381-4

Davies J, Davies D (2010) Origins and evolution of antibiotic resistance. Microbiol Mol Biol Rev 74:417–433. doi:10.1128/MMBR.00016-10

Davies J, Spiegelman GB, Yim G (2006) The world of subinhibitory antibiotic concentrations. Curr Opin Microbiol 9:445–453. doi:10.1016/j.mib.2006.08.006

Decho AW, Norman RS, Visscher PT (2010) Quorum sensing in natural environments: emerging views from microbial mats. Trends Microbiol 18:73–80. doi:10.1016/j.tim.2009.12.008

Defoirdt T, Boon N, Sorgeloos P, Verstraete W, Bossier P (2008) Quorum sensing and quorum quenching in *Vibrio harveyi* lessons learned from *in vivo* work. ISME J 2:19–26. doi:10.1038/ismej.2007.92

Defoirdt T, Boon N, Bossier P (2010) Can bacteria evolve resistance to quorum sensing disruption? PLoS Pathog 6:e1000989. doi:10.1371/journal.ppat.1000989

Defoirdt T, Brackman G, Coenye T (2013) Quorum sensing inhibitors: how strong is the evidence? Trends Microbiol 21:619–624. doi:10.1016/j.tim.2013.09.006

Dembitsky VM, Al Quntar AAA, Srebnik M (2011) Natural and synthetic small boron-containing molecules as potential inhibitors of bacterial and fungal quorum sensing. Chem Rev 111:209–237. doi:10.1021/cr100093b

Diggle SP, Griffin AS, Campbell GS, West SA (2007) Cooperation and conflict in quorum-sensing bacterial populations. Nature 450:411–414. doi:10.1038/nature06279

Galloway WRJD, Hodgkinson JT, Bowden SD, Welch M, Spring DR (2011) Quorum sensing in Gram-negative bacteria: small-molecule modulation of AHL and AI-2 quorum sensing pathways. Chem Rev 111:28–67. doi:10.1021/cr100109t

García-Contreras R, Martinez-Vazquez M, Velazquez Guadarrama N, Villegas Paneda AG, Hashimoto T, Maeda T, Quezada H, Wood TK (2013) Resistance to the quorum-quenching compounds brominated furanone C-30 and 5-fluorouracil in *Pseudomonas aeruginosa* clinical isolates. Pathog Dis 68:8–11. doi:10.1111/2049-632X.12039

Geske GD, O'Neill JC, Blackwell HE (2008) Expanding dialogues: from natural autoinducers to non-natural analogues that modulate quorum sensing in Gram-negative bacteria. Chem Soc Rev 37:1432–1447. doi:10.1039/b703021p

Givskov M, de Nys R, Manefield M, Gram L, Maximilien R, Eberl L, Molin S, Steinberg PD, Kjelleberg S (1996) Eukaryotic interference with homoserine lactone-mediated prokaryotic signalling. J Bacteriol 178:6618–6622

Griffin AS, West SA, Buckling A (2004) Cooperation and competition in pathogenic bacteria. Nature 430:1024–1027. doi:10.1038/nature02744

Hentzer M, Wu H, Andersen JB, Riedel K, Rasmussen TB, Bagge N, Kumar N, Schembri MA, Song Z, Kristoffersen P, Manefield M, Costerton JW, Molin S, Eberl L, Steinberg P, Kjelleberg S, Høiby N, Givskov M (2003) Attenuation of *Pseudomonas aeruginosa* virulence by quorum sensing inhibitors. EMBO J 22:3803–3815. doi:10.1093/emboj/cdg366

Hoffman LR, D'Argenio DA, MacCoss MJ, Zhang Z, Jones RA, Miller SI (2005) Aminoglycoside antibiotics induce bacterial biofilm formation. Nature 436:1171–1175. doi:10.1038/nature03912

Jimenez PN, Koch G, Thompson JA, Xavier KB, Cool RH, Quax WJ (2012) The multiple signaling systems regulating virulence in *Pseudomonas aeruginosa*. Microbiol Mol Biol Rev 76:46–65. doi:10.1128/MMBR.05007-11

Joelsson A, Liu Z, Zhu J (2006) Genetic and phenotypic diversity of quorum–sensing systems in clinical and environmental isolates of *Vibrio cholerae*. Infect Immun 74:1141–1147. doi:10.1128/IAI.74.2.1141-1147.2006

Kalia VC (2013) Quorum sensing inhibitors: an overview. Biotechnol Adv 31:224–245. doi:10.1016/j.biotechadv.2012.10.004

Kalia VC (2014) Microbes, antimicrobials and resistance: the battle goes on. Indian J Microbiol 54:1–2. doi:10.1007/s12088-013-0443-7

Kalia VC, Purohit HJ (2011) Quenching the quorum sensing system: potential antibacterial drug targets. Crit Rev Microbiol 37:121–140. doi:10.3109/1040841X.2010.532479

Kalia VC, Rani A, Lal S, Cheema S, Raut CP (2007) Combing databases reveals potential antibiotic producers. Expert Opin Drug Discov 2:211–224. doi:10.1517/17460441.2.2.211

Kalia VC, Wood TK, Kumar P (2014) Evolution of resistance to quorum-sensing inhibitors. Microb Ecol 68:13–23. doi:10.1007/s00248-013-0316-y

Köhler T, van Delden C, Curty LK, Hamzehpour MM, Pechere JC (2001) Overexpression of the MexEF-OprN multidrug efflux system affects cell-to-cell signaling in *Pseudomonas aeruginosa*. J Bacteriol 183:5213–5222. doi:10.1128/JB.183.18.5213-5222.2001

Köhler T, Perron GG, Buckling A, van Delden C (2010) Quorum sensing inhibitors selects for virulence and cooperation in *Pseudomonas aeruginosa*. PLoS Pathog 6:e1000883. doi:10.1371/journal.ppat.1000883

Kumar P, Patel SKS, Lee JK, Kalia VC (2013) Extending the limits of *Bacillus* for novel biotechnological applications. Biotechnol Adv 31:1543–1561. doi:10.1016/j.biotechadv.2013.08.007

Lal S, Cheema S, Kalia VC (2008) Phylogeny vs genome reshuffling: horizontal gene transfer. Indian J Microbiol 48:228–242. doi:10.1007/s12088-008-0034-1

Ledgham F, Ventre I, Soscia C, Foglino M, Strugis JN, Lazdunski A (2003) Interaction of the quorum sensing regulator QscR: interaction with itself and the other regulators of *Pseudomonas aeruginosa* LasR and RhlR. Mol Microbiol 48:199–210. doi:10.1046/j.1365-2958.2003.03423.x

Maeda T, García-Contreras R, Pu M, Sheng L, Garcia LR, Tomás M, Wood TK (2012) Quorum quenching quandary: resistance to antivirulence compounds. ISME J 6:493–501. doi:10.1038/ismej.2011.122

Mattmann ME, Blackwell HE (2010) Small molecules that modulate quorum sensing and control virulence in *Pseudomonas aeruginosa*. J Org Chem 75:6737–6746. doi:10.1021/j0101237e

McDougald D, Rice SA, Kjelleberg S (2007) Bacterial quorum sensing and interference by naturally occurring biomimics. Anal Bioanal Chem 387:445–453. doi:10.1007/s00216-006-0761-2

Mellbye B, Schuster M (2011) The sociomicrobiology of antivirulence drug resistance: a proof of concept. mBio 2:e00131–11. doi:10.1128/mBio.00131-11

Mikkelsen H, Bond NJ, Skindersoe ME, Givskov M, Lilley KS, Welch M (2009) Biofilms and type III secretion are not mutually exclusive in *Pseudomonas aeruginosa*. Microbiology 155:687–698. doi:10.1099/mic.0.025551-0

Nadell CD, Xavier JB, Levin SA, Foster KR (2008) The evolution of quorum sensing in bacterial biofilms. PLoS Biol 16:e14. doi:10.1371/journal.pbio.0060014

Nazzaro F, Fratianni F, Coppola R (2013) Quorum sensing and phytochemicals. Int J Mol Sci 14:12607–12619. doi:10.3390/ijms140612607

Ng WL, Bassler BL (2009) Bacterial quorum-sensing network architectures. Annu Rev Genet 43:197–222. doi:10.1146/annurev-genet-102108-134304

Njoroge J, Sperandio V (2009) Jamming bacterial communication: new approaches for the treatment of infectious diseases. EMBO Mol Med 1:201–210

Otto M (2004) Quorum-sensing control in Staphylococci – a target for antimicrobial drug therapy? FEMS Microbiol Lett 241:135–141. doi:10.1016/j.femsle.2004.11.016

Poole K (2005) Efflux-mediated antimicrobial resistance. J Antimicrob Chemother 56:20–51. doi:10.1093/jac/dki171

Purevdorj B, Costerton JW, Stoodley P (2002) Influence of hydrodynamics and cell signaling on the structure and behavior of *Pseudomonas aeruginosa* biofilms. Appl Environ Microbiol 68:4457–4464. doi:10.1128/AEM.68.9.4457-4464.2002

Rampioni G, Schuster M, Greenberg EP, Bertani I, Grasso M, Venturi V, Zennaro E, Leoni L (2007) RsaL provides quorum sensing homeostasis and functions as a global regulator of gene expression in *Pseudomonas aeruginosa*. Mol Microbiol 66:1557–1565. doi:10.1111/j.1365-2958.2007.06029x

Rasmussen TB, Givskov M (2006) Quorum-sensing inhibitors: a bargain of effects. Microbiology 152:895–904. doi:10.1016/j.fitote.2007.03.009

Rosemeyer V, Michiels J, Verreth C, Vanderleyden J (1998) *lux*I- and *lux*R-homologous genes of *Rhizobium etli* CNPAF512 contribute to synthesis of autoinducer molecules and nodulation of *Phaseolus vulgaris*. J Bacteriol 180: 815–821

Rumbaugh KP, Diggle SP, Watters CM, Gillespie AR, Griffin AS, West SA (2009) Quorum sensing and the social evolution of bacterial virulence. Curr Biol 19:341–345. doi:10.1016/j.cub.2009.01.050

Sandoz KM, Mitzimberg SM, Schuster M (2007) Social cheating in *Pseudomonas aeruginosa* quorum sensing. Proc Natl Acad Sci U S A 104:15876–15881. doi:10.1073/pnas.0705653104

Schuster M, Sexton DJ, Diggle SP, Greenberg EP (2013) Acyl-homoserine lactone quorum sensing: from evolution to application. Annu Rev Microbiol 67:43–63. doi:10.1146/annurev-micro-092412-155635

Scutera S, Zucca M, Savoia D (2014) Novel approaches for the design and discovery of quorum-sensing inhibitors. Expert Opin Drug Discov 9:353–366. doi:10.1517/17460441.2014.894974

Sio CF, Otten LG, Cool RH, Diggle SP, Braun PG, Bos R, Daykin M, Cámara M, Williams P, Quax WJ (2006) Quorum quenching by an *N*-Acyl-homoserine lactone acylase from *Pseudomonas aeruginosa* PAO1. Infect Immun 74:1673–1682

Skindersoe ME, Zeuthen LH, Brix S, Fink LN, Lazenby J, Whittall C, Williams P, Diggle SP, Froekiaer H, Cooley M, Givskov M (2009) *Pseudomonas aeruginosa* quorum-sensing signal molecules interfere with dendritic cell-induced T-cell proliferation. FEMS Immunol Med Microbiol 55:335–345. doi:10.1111/j.1574-695X.2008.00533.x

Teplitski M, Robinson JB, Bauer WD (2000) Plants secrete substances that mimic bacterial *N*-acyl homoserine lactone signal activities and affect population density-dependent behaviors in associated bacteria. Mol Plant Microbe Interact 13:637–648. doi:10.1094/MPMI.2000.13.6.637

Tinh NTN, Dung NV, Trung CT, Thuy VT (2013) *In vitro* characterization of a recombinant AHL-Lactonase from *Bacillus cereus* isolated from a striped catfish (*Pangasianodon hypophthalmus*) Pond. Indian J Microbiol 53:485–487. doi:10.1007/s12088-013-0415-y

Walters M, Sperandio V (2006) Quorum sensing in *Escherichia coli* and *Salmonella*. Int J Med Microbiol 296:125–131. doi:10.1073/pnas.96.4.1639

West SA, Griffin AS, Gardner A, Diggle SP (2006) Social evolution theory for microorganisms. Nat Rev Microbiol 4:597–607. doi:10.1038/nrmicro1461